AF683532

Molecular Chaperones *in the* Life Cycle *of* Proteins

Molecular Chaperones *in the* Life Cycle *of* Proteins

Structure, Function, and Mode of Action

edited by

Anthony L. Fink
University of California, Santa Cruz
Santa Cruz, California

Yuji Goto
Osaka University
Osaka, Japan

MARCEL DEKKER, INC. NEW YORK • BASEL • HONG KONG

Fink, Anthony L.
Molecular chaperones in the life cycle of proteins: structure, function, and mode of action/Anthony L. Fink, Yuji Goto
p. cm.
Includes index.
ISBN 0-8247-0100-3 (hardcover)
1. Molecular chaperones—Physiological effect. I. Goto, Yuji. II. Title.
QP552.M64F56 1997
572'.645—dc21

97-34429
CIP

The publisher offers discounts on this book when ordered in bulk quantities. For more information, write to Special Sales/Professional Marketing at the address below.

This book is printed on acid-free paper.

Copyright © 1998 by MARCEL DEKKER, INC. All Rights Reserved.

Neither this book nor any part may be reproduced or transmitted in any form or by any means, electronic or mechanical, including photocopying, microfilming, and recording, or by any information storage and retrieval system, without permission in writing from the publisher.

MARCEL DEKKER, INC.
270 Madison Avenue, New York, New York 10016
http://www.dekker.com

Current printing (last digit):
10 9 8 7 6 5 4 3 2 1

PRINTED IN THE UNITED STATES OF AMERICA

Preface

Proteins come and go; in the cell, their lifetimes and steady-state concentrations are determined by the relative rates of protein synthesis and degradation. The life of a protein may be considered to be the sum of events and processes beginning with its synthesis on the ribosome and ending (normally) with its final degradation by proteases into its constituent amino acids. Throughout its lifetime, but especially at the beginning and end, the protein is assisted by chaperones, various other proteins that facilitate the desired course of events for the target protein. These molecular chaperones may be highly specific in that they interact with only one target protein, or a very limited number of target proteins. One such example is PapD, which is involved in the assembly of pili. They also may be very promiscuous and interact with most of the proteins in a cell, such as Hsp60 and Hsp70, which play an important role in the folding of nascent polypeptide chains. In this book, we have taken the unusual approach of considering the molecular chaperones from the perspective of their role in guiding proteins through their life cycle, "from the cradle to the grave."

The heat shock protein and molecular chaperone field is an extremely active one. It encompasses studies ranging from basic biophysical properties of the chaperone proteins themselves, through their enzyme-like functions and reaction cycles, their genetic regulation, their key role in combating stress at the cellular level, and their role in a wide variety of diseases. In the context of the life cycle of proteins, the emphasis of this book is on the former three areas. The major classes of general chaperones (as opposed to dedicated or specific chaperones) are Hsp100, Hsp90, Hsp70, Hsp60, Hsp40, and the small heat shock proteins. Recent investigations have shown that not only do the major classes of chaperones often function with protein cofactors, but interactions between members of the Hsp40, Hsp60, Hsp70, and Hsp90 families may be frequent.

The basic paradigm of molecular chaperones is that they bind unfolded, or relatively unfolded, polypeptides (but not native or native-like conformations) to form relatively stable complexes. In some cases, these are dissociated by the binding and hydrolysis of adenosine triphosphate (ATP). The more well-defined *constitutive* roles of heat shock proteins are facilitating (chaperoning) the folding and assembly of newly synthesized polypeptides, and facilitating the translocation of proteins across membranes, such as those of the mitochondria. The induced roles of heat shock proteins involve their ability to help the cell cope with stress, and are not emphasized in this volume. Although we now know a great deal about the major chaperone families, as indicated in this treatise, we know much less about the many cofactors that are also suspected of being involved. Although many chaperones are heat shock proteins or stress proteins, many are not, and are implicated only in various aspects of the *normal* functioning of the cell.

Although a few monographs have been published on molecular chaperones, these have emphasized their role as stress proteins, rather than facilitators of protein folding and degradation. Sufficient recent progress has been made in our understanding of the role of the molecular chaperones in various aspects of protein synthesis and degradation, and the biophysical, structural, and functional properties of the chaperones themselves, to justify the present volume. Putting this material in the context of the life cycle of proteins provides an interesting and unique perspective. A detailed understanding of how proteins fold to their native structures, both in vitro and in vivo, and how they are degraded, remain major challenges of cellular and molecular biology. In one sense this volume is a status report on recent progress in these areas.

Why do we need chaperones? After all, a basic paradigm of in vitro protein folding has been the seminal work of Anfinsen with small globular proteins, who demonstrated 40 years ago that formation of the native protein from the unfolded state is a spontaneous process determined by the global free energy minimum. The results indicated that the native state of small globular proteins is determined by their amino acid sequence. It is important to note, however, that successful in vitro refolding has been limited to relatively small globular proteins, and at relatively low protein concentrations. This reflects the fact that large proteins are more difficult to fold in vitro. For example, immunoglobulin G (IgG, M_r ~180,000 Da) cannot be refolded in vitro once it is unfolded and its disulfide bonds reduced, whereas 100% refolding occurs for the *isolated immunoglobulin domains*. Thus, although folding experiments of small globular proteins in vitro are essential to understand the basic features of protein folding, there appear to be additional factors that are essential for the successful folding of many proteins in vitro, as well as in vivo. When one considers the environment within a cell, it becomes clear that in vitro folding experiments at low protein concentrations are very poor models for what happens in the cell, where a newly synthesized protein is in a crowded environment with little or no "free" water, very high concentrations of other proteins and metabolites, and typically membranes, cytoskeletal elements, and other cellular components. Thus the need becomes evident for chaperones to prevent aggregation during the folding of nascent chains, to prevent nonproductive interactions with other cell components, to direct the assembly of larger proteins and multiprotein complexes, and to prevent premature folding prior to membrane translocation.

Substantial progress has been made recently in the elucidation of the structures of a number of molecular chaperones using X-ray diffraction and nuclear magnetic resonance (NMR). The availability of molecular structures greatly facilitates determination of the functional mechanisms of these proteins, for example, the proposed mechanism for the concerted action of GroEL and GroES, coupled with the hydrolysis of ATP. The basic knowledge accumulated from studies of protein behavior in vitro, although still far from complete, is essential to improve our understanding of the role of molecular chaperones in vivo. It is apparent that studies of protein folding in vitro can contribute significantly to our knowledge of corresponding processes in vivo.

The development of the book has been broken down into a short initial treatment of the fundamentals of protein folding in vitro in the absence of molecular chaperones, followed by the main part of the book that focuses on the interaction of proteins with chaperones during various stages of their lives, especially their synthesis and degradation. It is convenient to consider the life of a protein being divided into three parts: birth, maturation, and death. We begin in Chapter 3, with interactions of chaperones with the nascent chain. This is followed by a detailed discussion of the structural and functional properties of the Hsp40, Hsp60, Hsp70, and Hsp100 families, in particular, during the folding and maturation of the protein to its native state. The assembly of oligomeric proteins, membrane translocation of proteins, and specialized chaperoning functions, such as pro-peptides, and proline and disulfide isomerization are then considered. Finally, the role of chaperones in the death of the protein, by either proteolytic degradation or irreversible insolubilization, is described.

The book begins with a summary of the current picture of in vitro protein folding, where substantial progress has been made recently, especially due to advances in rapid reaction techniques and NMR. In addition, site-directed mutagenesis has enabled analysis of the role of individual residues in the folding reaction. Current controversies regarding the role of partially folded intermediates in protein folding are also discussed in Chapter 1. Several observations have indicated that although small globular proteins may fold rapidly without significant populations of partially-folded intermediates, this is not true for most proteins both in vivo and in vitro. For example, the in vivo formation of inclusion bodies represents a combination of protein folding and chaperone problems, because the origin of the inclusion bodies is likely due to insufficient availability of chaperones to prevent the aggregation of partially folded intermediates (Chapter 7). Oligomeric proteins provide additional problems for the protein folding process both in vivo and in vitro, because the process involves both association as well as folding of each subunit (Chapter 2). Because the monomeric intermediates have exposed hydrophobic surfaces that are shielded from solvent upon formation of the oligomer, folding of oligomeric proteins is often a competition between correct association resulting in the native structure and incorrect association resulting in misfolded aggregates.

A key question regarding in vivo protein folding concerns the involvement of factors that facilitate the folding of the nascent chain. The bulk of the monograph is taken up with the role of these chaperones in the folding and proper assembly of new proteins. Substantial evidence now supports the presence of large molecular weight complexes associated with the nascent chain as it emerges from the ribosome (Chapter 3). Hsp70 has been shown to be a key component of such complexes, along with other chaperones. The structure and properties of the Hsp70 family are considered in Chapter 4. These strongly conserved proteins consist of two domains. Crystallographic structures are now available for both domains, but not the intact protein, and provide insight into the mechanism of ATPase activity and peptide binding specificity, which is considered in more detail in Chapter 6. The main role of Hsp70 in facilitating protein folding appears to be to sequester unfolded and partially folded newly synthesized proteins, thereby preventing their aggregation (Chapter 5). The main details of the Hsp70 reaction cycle are now known, although there is still considerable uncertainty about many of the finer points, especially the contributions of the cofactors (DnaJ and GrpE in *Escherichia coli*). DnaJ and other members of the Hsp40 family appear to be chaperones in their own right, binding to at least some unfolded proteins and nascent chains, as well as critical cofactors in the Hsp70 reaction cycle (Chapter 11). In contrast to Hsp70 and Hsp40, Hsp47 is an example of a chaperone with a more specific function, namely the folding and processing of procollagen in the endoplasmic reticulum (ER) (Chapter 10). Both Hsp40 and Hsp90 have been found in complexes with Hsp70.

Although there are a number of similarities between Hsp90 and Hsp70, the former has several identified specific interactions—for example, with cytoskeleton elements, steroid hormone receptors, and protein kinases (Chapter 8). Hsp104, a member of the Hsp100/Clp family, has a number of very intriguing properties (Chapter 9), including, in contrast to most of the other chaperones, the ability to dissociate protein aggregates. The Hsp100/Clp family of chaperones carry out a diverse set of functions, including proteolysis. One of the most fascinating aspects of Hsp104 is its role in triggering a prion-like disorder in yeast, involving the extrachromosomal elements [PSI^+] and [URE3].

GroEL, a member of the Hsp60 family, is probably the most studied of all molecular chaperones (Chapter 12); in combination with its co-chaperonin GroES, it has been shown to facilitate protein folding, especially by preventing aggregation. The availability of a high resolution crystallographic structure, in conjunction with mutagenesis studies, has helped in the elucidation of the details of the reaction cycle. However, there are still many points of controversy, reflecting the complexity of the mechanism of this large chaperone. Points of similarity and disparity between members of the Hsp60 family are evident by comparison of GroEL with a thermophilic chaperonin (Chapter 13).

Two other cofactors that play critical roles in the folding of many proteins are: protein disulfide isomerase (PDI) (Chapter 14) and peptidyl prolyl isomerase (PPI) (Chapter 15), which are found in the ER. Both these chaperones function to enhance the rate at which native proteins are produced during peptide biosynthesis. PDI has a number of interesting properties, including the ability to act both as a chaperone in the sense of prevention of aggregation and as an antichaperone, by stimulating aggregation, depending on the relative concentrations and experimental conditions. Under in vitro (and presumably in vivo) conditions, proline cis-trans isomerization may become rate-limiting in the folding of proteins, especially at lower temperatures; in many cases the presence of PPI and related proteins (e.g., immunophilins) will enhance the rate of folding. Recently trigger factor has been shown to be a chaperone, and may be a key PPI in vivo.

There are many large complex protein machines in cells; in some of these cases specific molecular chaperones have been shown to be involved in their assembly (Chapter 16). Some of the best-studied systems are bacteriophage capsids and bacterial pili and flagella. Many critical cells are specialized for protein export, and thus have substantial secretory pathways. The ER is a key compartment in such cells, and contains many chaperones that are essential for the production of functional proteins for export (Chapter 17). These chaperones range from general ones of the Hsp70 and Hsp90 families (BiP and Grp94) to more specific chaperones such as calnexin, which is involved in the biosynthesis of glycoproteins. Mitochondrial import also involves a rich array of molecular chaperones (Chapter 18), many of which have been identified. For such a complicated system we know a good deal about the molecular details of the mechanism of protein import into the mitochondria; major players include members of the Hsp70 and Hsp40 (DnaJ) families, as well as a number of cofactors binding to the mitochondrial presequence. Several models have been proposed for the mechanism whereby proteins are actually translocated across the mitochondrial membranes (which involves the mitochondrial Hsp70).

The final chapter concerning the protein folding aspect of the life of a protein concerns proteins that use intramolecular chaperones in the form of pro-peptide sequences (Chapter 19). The majority of known members of this class are various protease families, the best studied being α-lytic protease and subtilisin.

If one considers the life of a protein being divided into three parts—birth, maturation, and death—then the latter usually comes in one of two forms, degradation or irreversible aggregation. More and more evidence is accumulating for a role of molecular chaperones in the degradation of proteins. Chaperones have been implicated in two pathways of degradation, the lysosomal pathway of proteolysis (Chapter 20) and the proteasome ("the death machine") (Chapter 21). As mentioned previously, a number of chaperones are in fact proteases, e.g. the Clp family, and La and Pim1, homologous ATP-dependent proteases from bacteria and yeast mitochondria. Hsp70 has been observed in the lysosome and is implicated in the uptake of proteins by lysosomes. The recent determination of the crystal structure of the proteasome has facilitated understanding its mode of action in the degradation of ubiquitin-labeled proteins. Several observations suggest that Hsp40 and Hsp70 may be involved in the degradation of proteins via this pathway.

The small heat shock proteins are abundant and ubiquitous; however, they differ from other chaperones in that there is limited evidence that they release their cargo, once they have bound an unfolded protein (Chapter 22). Structurally the α-crystallins belong to this family; their major role is to sequester denatured proteins in the lens of the eye and to prevent their aggregation (which would cause a cataract). The small heat shock proteins also are unusual in their structure, which involves large oligomeric complexes and on which there is no current consensus. The book concludes with a discussion of the irreversible aggregation of proteins in the form of amyloid fibrils and other pathological forms (Chapter 23). More than 40 diseases are known in which abnormal protein deposition occurs. Most of these are lethal, and many cause neurodegeneration (e.g., prion diseases, Alzheimer's disease). Chaperones are involved in two ways in many such disorders. First, it seems likely that the absence of sufficient chaperones may enhance the accumulation of aggregates, due to the transient presence of partially folded intermediates with a propensity to aggregate, which would be expected to be substrates for many of the heat shock (or stress) proteins (chaperones). Second, accessory proteins are found in many of these abnormal deposits, suggesting at least a possible chaperone-like involvement (although true chaperones, at least in the sense of the normal definition, would be expected to be absent from the final complex).

Thus we see that molecular chaperones of one sort or another are intimately involved with most proteins throughout much of their lives, and especially the critical beginning and ending. This book is the first to present the story of molecular chaperones from this perspective. It is meant as a resource for scientists interested in protein folding, both in vivo and in vitro, as well as those interested in the molecular chaperones from both a biophysical and physiological point of view. Unlike previous monographs on chaperones, recent advances in our understanding of the molecular details of the structure and function of the molecular chaperones have allowed us to place more emphasis on their biophysical properties and the molecular mechanisms in this book.

Anthony L. Fink
Yuji Goto

Contents

Contributors

Joseph W. Brewer, Ph.D. Department of Tumor Cell Biology, St. Jude Children's Research Hospital, Memphis, Tennessee

Alexander Buchberger, Ph.D. Zentrum für Molekulare Biologie, Universität Heidelberg, Heidelberg, Germany

Johannes Buchner, Ph.D. Institut für Biophysik und Physikalische Biochemie, Universität Regensburg, Regensburg, Germany

Bernd Bukau, Ph.D. Institut für Biochemie und Molekularbiologie, Universität Freiburg, Freiburg, Germany

Ana Maria Cuervo, M.D., Ph.D. Department of Physiology, Tufts University School of Medicine, Boston, Massachusetts

J. Fred Dice, Ph.D. Department of Physiology, Tufts University School of Medicine, Boston, Massachusetts

Daryl K. Eggers Department of Cellular and Molecular Pharmacology, University of California, San Francisco, San Francisco, California

Monika Ehrnsperger, Ph.D. Institut für Biophysik und Physikalische Biochemie, Universität Regensburg, Regensburg, Germany

Toshiya Endo, Ph.D. Department of Chemistry, Faculty of Science, Nagoya University, Nagoya, Japan

Anthony L. Fink, Ph.D. Department of Chemistry and Biochemistry, University of California, Santa Cruz, Santa Cruz, California

Matthias Gaestel, Ph.D. Department of Cell Biology, Max-Delbrück-Centrum für Molekulare Medizin, Berlin, Germany

Hiram F. Gilbert, Ph.D. Verna and Marrs McLean Department of Biochemistry, Baylor College of Medicine, Houston, Texas

John R. Glover, Ph.D. Department of Molecular Genetics and Cell Biology, Howard Hughes Medical Institute, The University of Chicago, Chicago, Illinois

Yuji Goto, Ph.D. Department of Biology, Graduate School of Science, Osaka University, Osaka, Japan

Alexander Gragerov, Ph.D. ABL-Basic Research Program, NCI-Frederick Cancer Research and Development Center, Frederick, Maryland

Jeung-Hoi Ha, Ph.D. Department of Structural Biology, Stanford University School of Medicine, Stanford, California

***Yoshihisa Hagihara, Ph.D.** Department of Biology, Graduate School of Science, Osaka University, Osaka, Japan

William J. Hansen, Ph.D. Department of Medicine, University of California, San Francisco, San Francisco, California

Sandra A. Hayes, Ph.D. Department of Physiology, Tufts University School of Medicine, Boston, Massachusetts

Linda M. Hendershot, Ph.D. Department of Tumor Cell Biology, St. Jude Children's Research Hospital, Memphis, Tennessee

Lawrence E. Hightower, Ph.D. Department of Molecular and Cell Biology, The University of Connecticut, Storrs, Connecticut

Nobuko Hosokawa, M.D. Department of Cell Biology, Chest Disease Research Institute, Kyoto University, Kyoto, Japan

Paul M. Horowitz, Ph.D. Department of Biochemistry, The University of Texas Health Science Center, San Antonio, Texas

Masayori Inouye, Ph.D. Department of Biochemistry, University of Medicine and Dentistry of New Jersey, Robert Wood Johnson Medical School, Piscataway, New Jersey

†Noriyuki Ishii, Ph.D. Department of Biophysical Chemstry, National Institute of Bioscience and Human Technology, Tsukuba, Japan

*Current affiliation: Kim Laboratory, Whitehead Institute for Biomedical Research, Massachusetts Institute of Technology, Cambridge, Massachusetts

†Current affiliation: Skirball Institute of Biomolecular Medicine, New York University Medical Center, New York, New York

Rainer Jaenicke, Dr. phil. nat. Institut für Biophysik und Physikalische Biochemie, Universität Regensburg, Regensburg, Germany

Eric R. Johnson, Ph.D. Department of Structural Biology, Stanford University School of Medicine, Stanford, California

Hiroyuki Kawahara, Ph.D. Department of Cancer Therapeutics, The Tokyo Metropolitan Institute of Medical Science, Tokyo, Japan

***Thomas Laufen, Dipl. Biol.** Zentrum for Molekulare Biologie, Universität Heidelberg, Heidelberg, Germany

Sau-Mei Leung, Ph.D. Department of Molecular and Cell Biology, The University of Connecticut, Storrs, Connecticut

Susan L. Lindquist, Ph.D. Department of Molecular Genetics and Cell Biology, Howard Hughes Medical Institute, The University of Chicago, Chicago, Illinois

Jianjun Liu, M.Sc. Department of Biochemistry, University of Medicine and Dentistry of New Jersey, Robert Wood Johnson Medical School, Piscataway, New Jersey

David B. McKay, Ph.D. Department of Structural Biology, Stanford University School of Medicine, Stanford, California

Andrew D. Miller, B.Sc., Ph.D. Department of Chemistry, Imperial College of Science, Technology and Medicine, London, United Kingdom

Hiroshi Nagata, M.D., Ph.D. Department of Medicine, University of California San Francisco, San Francisco, California

Kazuhiro Nagata, Ph.D. Department of Cell Biology, Chest Disease Research Institute, Kyoto University, Kyoto, Japan

Mamoru Satoh, Ph.D. Department of Cell Biology, Chest Disease Research Institute, Kyoto University, Kyoto, Japan

Eric C. Schirmer, Ph.D. Department of Molecular Genetics and Cell Biology, Howard Hughes Medical Institute, The University of Chicago, Chicago, Illinois

Franz Xaver Schmid, Dr. rer. nat. Laboratorium für Biochemie, Universität Bayreuth, Bayreuth, Germany

Robert Seckler, Dr. rer. nat. Department of Biophysics and Physical Biochemistry, University of Regensburg, Regensburg, Germany

---

*Current affiliation: Institut für Biochemie und Molekularbiologie, Universität Freiburg, Freiburg, Germany

Ujwal Shinde, Ph.D. Department of Biochemistry, University of Medicine and Dentistry of New Jersey, Robert Wood Johnson Medical School, Piscataway, New Jersey

Mike A. Singer Department of Pathology, The University of Chicago, Chicago, Illinois

Marcelo C. Sousa, Ph.D. Department of Structural Biology, Stanford University School of Medicine, Stanford, California

Hideki Taguchi, Ph.D. Research Laboratory of Resources Utilization, Tokyo Institute of Technology, Yokohama, Japan

Shigeki Takeda, Ph.D. Department of Structural Biology, Stanford University School of Medicine, Stanford, California

Keiji Tanaka, Ph.D. Department of Cancer Therapeutics, The Tokyo Metropolitan Institute of Medical Science, Tokyo, Japan

Kenneth W. Walker, Ph.D. Verna and Marrs McLean Department of Biochemistry, Baylor College of Medicine, Houston, Texas

William J. Welch, Ph.D. Department of Medicine, University of California San Francisco, San Francisco, California

***Ronald Wetzel, Ph.D.** Chemistry Department, Temple University, Philadelphia, Pennsylvania

Sigurd M. Wilbanks, Ph.D. Department of Structural Biology, Stanford University School of Medicine, Stanford, California

Ichiro Yahara, Ph.D. Department of Cell Biology, The Tokyo Metropolitan Institute of Medical Science, Tokyo, Japan

Mitsuko Yamakoshi, M.S. Research Laboratory of Resources Utilitzation, Tokyo Institute of Technology, Yokohama, Japan

Masasuke Yoshida, Ph.D. Research Laboratory of Resources Utilization, Tokyo Institute of Technology, Yokohama, Japan

Ulrich Zuber, Ph.D. Institute for Microbiology and Molecular Biology, University of Greifswald, Greifswald, Germany

*Current affiliation: Graduate School of Medicine, University of Tennessee Medical Center at Knoxville, Knoxville, Tennessee

1

Folding of Monomeric Proteins In Vitro: Physicochemical and Biological Roles of the Intermediate States

***Yoshihisa Hagihara and Yuji Goto**

Osaka University, Osaka, Japan

I. INTRODUCTION

Folding is a process in which a conformational ensemble of extended polypeptide chains acquires a unique folded conformation with biological activity [1,2]. The classic work of Anfinsen indicated that it is a thermodynamically reversible process in which all the information required for folding under physiological conditions is coded in its amino acid sequence [3]. How, then, do proteins fold into its unique native structure? There have been numerous studies attempting to address this problem. Since Anfinsen's work, many of the studies used the reversible unfolding–refolding reactions of small globular proteins in test tube (i.e., in vitro), for which the complicated environment of cellular machineries is missing. The folding in vivo must be affected, at least to some extent, by the fact that protein is synthesized from the N-terminal to the C-terminal. It is generally difficult to refold the larger proteins in vitro without the help of molecular chaperones. Thus, we know that the refolding of proteins in test tubes may not necessarily reproduce the folding in vivo. Nevertheless, the unfolding–refolding reactions in vitro are crucial to improve our understanding of the mechanism of

*Current affiliation: Massachusetts Institute of Technology, Cambridge, Massachusetts

protein folding, because in this markedly simplified system, we can close up many of the principles that govern the protein folding.

A variety of experimental methods have supported the significant progress in the study of protein folding in the last decade [4,5]. Whereas the previous approaches largely depended on the classical biophysical and biochemical methods, the first breakthrough was introduced by the stopped-flow circular dichroism (CD) [6] and then by high-resolution two-dimensional nuclear magnetic resonance (NMR) coupled with hydrogen–deuterium (H–D) exchange [7–9]. Protein-engineering methods with various mutant proteins have been used to monitor systematically the role of individual amino acids in the folding process [10]. Solution X-ray scattering [11] and mass spectroscopy [12] are joining in the field as promising techniques complementary to other methods. The theoretical studies are contributing to the progress in this field much more than before [2,5,13]. Indeed, a new view of protein folding [13–15], which envisages the protein folding through a funnel-like energy landscape, would not have emerged without such studies.

In this chapter, we overview the unfolding–refolding reactions of monomeric proteins in vitro. Because there are so many excellent reviews on protein folding [1,5,13,16–18], we do not intend to give another comprehensive review of this area. Instead, we try to focus our attention on the roles of the intermediate conformational states, because such states may be involved in the major functions of molecular chaperones. Starting with the discussion on the "physiochemical roles" of the intermediate states, we direct our discussion to their "biological roles." Because it is difficult to cite all the related topics, we will often use our own studies to close up the topics.

II. FOLDING PATHWAY

We first discuss how protein-folding studies came out through the efforts trying to characterize the "folding pathway." As often cited, Levinthal [19] suggested that, whereas the native state of a protein is a thermodynamically stable state, finding the native state by random search would be impossible. [see also 13]. This Levinthal's paradox implicated the presence of a unique folding pathway leading to the native state. On the other hand, one of the most important characteristics of protein folding is its high cooperativity, as approximated by a two-state transition between the native and denatured states (Fig. 1A) [20]. In spite of the inherent complexity of the folding reaction, this simple model is adequate for explaining a large number of experimental data. This means that the intermediate states that might be useful for characterizing the folding pathway are rarely observed. Therefore, a variety of efforts trying to

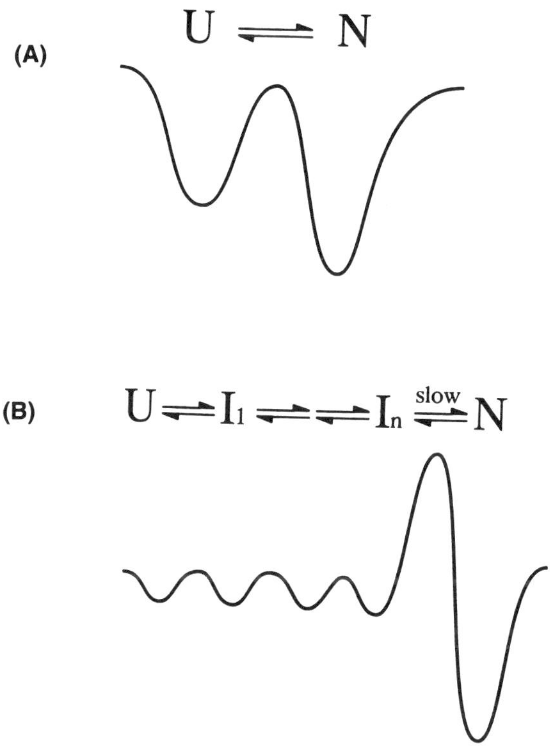

Figure 1 Various models of protein folding and their free energy profiles: (A) A two-state mechanism; (B) a sequential mechanism with the rate-limiting step located at the later stage (e.g., folding model with the molten globule intermediate); (C) a multiple pathway mechanism with various rate-limiting steps; (D) a folding mechanism with an off-pathway intermediate; (E) an idealized funnel model; (F) a rugged energy landscape with a unique pathway.

detect otherwise unstable intermediates started. There are three major approaches in this context.

1. Since Anfinsen's work, the reduction–oxidation of the disulfide bond has been recognized as a unique chemical reaction directly coupled to the protein-folding transition. The intermediates of disulfide bond formation may tell us about the pathway of protein folding.
2. Although the equilibrium transition can be approximated by a two-state mechanism, kinetic intermediates may accumulate. The kinetic approach should also be useful for characterizing the transition state of folding (i.e., the activated state separating the native and denatured states).

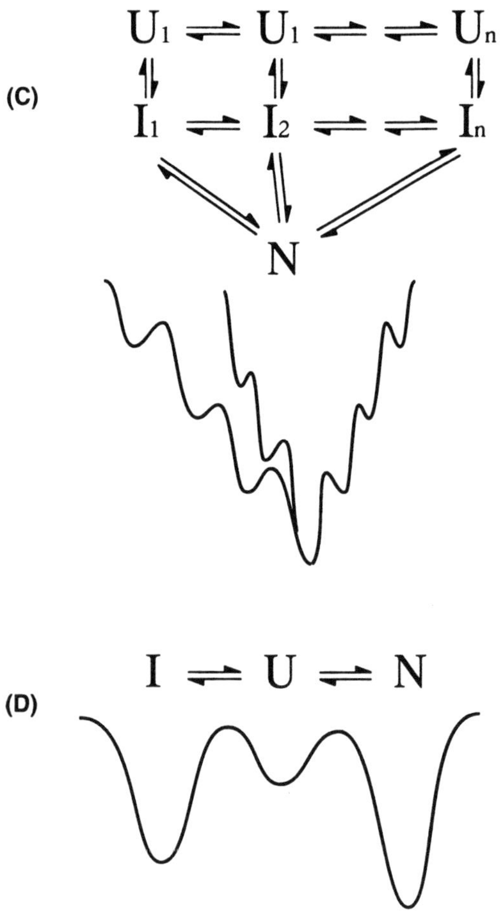

Figure 1 Continued

3. The equilibrium intermediate conformation detected under the mildly denaturing conditions may reproduce the real intermediate located on the direct folding pathway.

A. Disulfide Bond Formation

1. Pathway

The earlier studies of the analysis of the disulfide bond formation in proteins, such as ribonuclease A (RNase; 124 amino acids, 4 disulfide bonds) suggested that complete elucidation of the pathway of disulfide bond formation in proteins

(E)

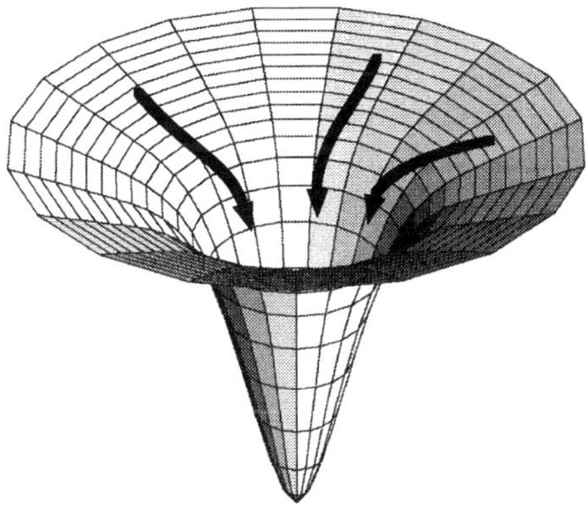

(F)

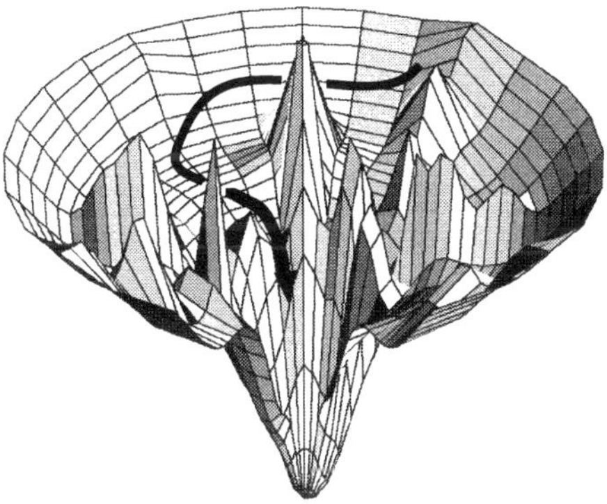

Figure 1 Continued

with multiple disulfide bonds is beyond the analytical resolution because of the complexities of the intermediate species [21]. However, Creighton and coworkers studied the oxidation–reduction of bovine pancreatic trypsin inhibitor (BPTI; 57 amino acids, 3 disulfide bonds) catalyzed by the thiol–disulfide exchange reaction and identified the intermediate states of disulfide bond formation. From extensive analysis of the equilibria and kinetics of the reactions, they reached the following conclusions [22,23]: First, the formation of disulfide bonds in BPTI does not occur randomly. Instead, there is a unique pathway leading to the native disulfide bonds. Second, obligatory intermediates with nonnative disulfide bonds exist before formation of the native species. Third, the rate-limiting step of the reaction is located close to the native state, suggesting that the transition state is a distorted form of the native state. Their conclusions emphasized an importance of unique folding pathway (see Fig. 1B).

2. *Multiple Pathways*

On the other hand, Scheraga and co-workers [24,25] studied disulfide bond formation in RNase A, also catalyzed by thiol–disulfide-exchange reaction. Their scheme of oxidation–reduction mechanism emerged with RNase A was different from that proposed for BPTI. For RNase A, because of the presence of four disulfide bonds, the population of intermediates was much more complicated than that of BPTI. When they analyzed the intermediates, they could not find a unique pathway. In contrast, they proposed the presence of multiple pathways that are classified into two types [24] (Fig. 1C). One is the "growth-type" folding and the other is the "rearrangement-type" folding. In the growth-type folding, the rate-limiting step is specific to the respective pathways (i.e., the formation of a critical disulfide bond, the nucleation site, or other). Once such reaction occurred, folding proceeds smoothly and rapidly to the native state. In rearrangement-type folding, the intermediates accumulate because of kinetic trapping and the rate-limiting step is the rearrangement (or disordering) of the structure preventing further folding.

They suggested that the mechanism is common to the folding reactions without the oxidation–reduction of disulfide bonds, where there are many possibilities of kinetic trapping [24]. As an example, they considered that the fast-folding of RNase A with the correct prolyl isomer represents growth-type folding, whereas the slow-folding owing to the wrong prolyl isomer is rearrangement-type folding. It is noted that their models can accommodate the multiple pathways, kinetic trapping, and nucleation–condensation model which are popular today.

The idea that we can study the mechanism of protein folding by studying the disulfide bond formation is based on the assumption that disulfide bond formation is closely coupled to the conformational folding. However, it has become evident that this assumption is not always true because the reactivity of the thiol group depends on various factors. Among them, the exposure of the thiol group is a critical factor determining the rate of disulfide bond formation. The first clear demonstration of this fact came from the disulfide bond formation in the isolated constant domain of the immunoglobulin light chain (C_L fragment) [26,27].

3. *Trapping by Protein Folding*

The domains of immunoglobulin have one intrachain disulfide bond buried in the hydrophobic core of the protein molecule (Fig. 2A). It was indicated with the C_L fragment (110 amino acids) that the reduction of the disulfide bond does not change the apparent conformation of this protein, although the stability is decreased substantially [26]. Therefore, the two thiols of the reduced C_L fragment are buried in the interior of the protein molecule. Disulfide bond formation in the reduced C_L fragment by the thiol–disulfide-exchange reaction with gluthathione was very slow because of the burial of the thiol groups (Fig. 2B). To

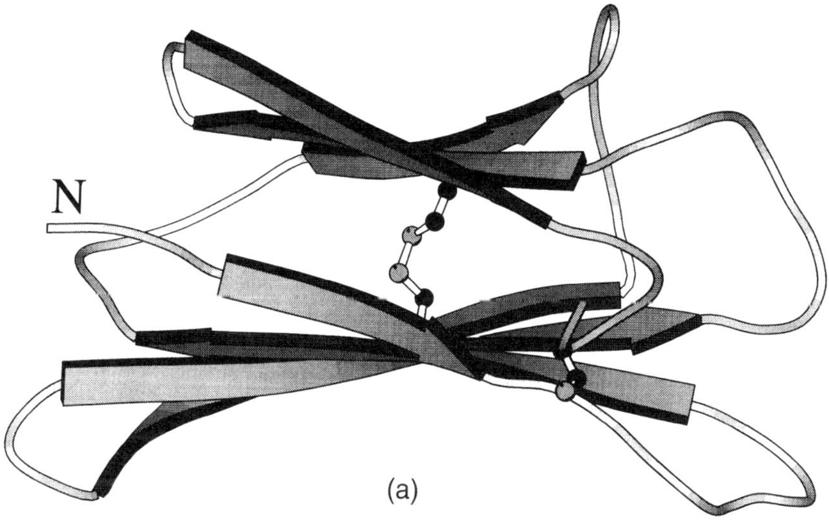

Figure 2 Structure of the C_L fragment (a) and the mechanism of the disulfide bond formation in the reduced C_L fragment (b). Ribbon illustrations of protein structures shown in Figs. 2A, 3A, 5, 6, and 7 were created using Molscript [108]. The coordinates of the C_L fragment (3FAB) were taken from the Brookhaven databank.

form the disulfide bond, substantial unfolding of the reduced C_L fragment by which thiol groups are exposed to the solvent, was necessary to react with the reagents. Thus, the folded reduced C_L fragment is a trapped state (or off-pathway intermediate) relative to the disulfide bond formation, even though the protein folding is completed.

Later, a series of studies of the disulfide bond formation in BPTI was carried out by Kim and co-workers [28–30]. Many of the intermediates with one or two of the native disulfide bonds were shown to assume a native-like conformation in which the thiol groups are buried and, consequently, cannot directly form the remaining native disulfide bonds. They indicated that the pathway and energetics of disulfide bond formation in BPTI are determined, to a great extent, by the reactivity of thiol groups in the intermediate states. This situation is the same as that for the C_L fragment.

If disulfide bond formation is exactly related to the proximity of two thiols and, hence, the extent of protein conformational folding, we can consider the folding mechanism directly by probing the pathway of the disulfide bond formation. Unfortunately, the examples discussed in the foregoing suggest that, probably, the two are frequently independent. The pathway of disulfide bond formation

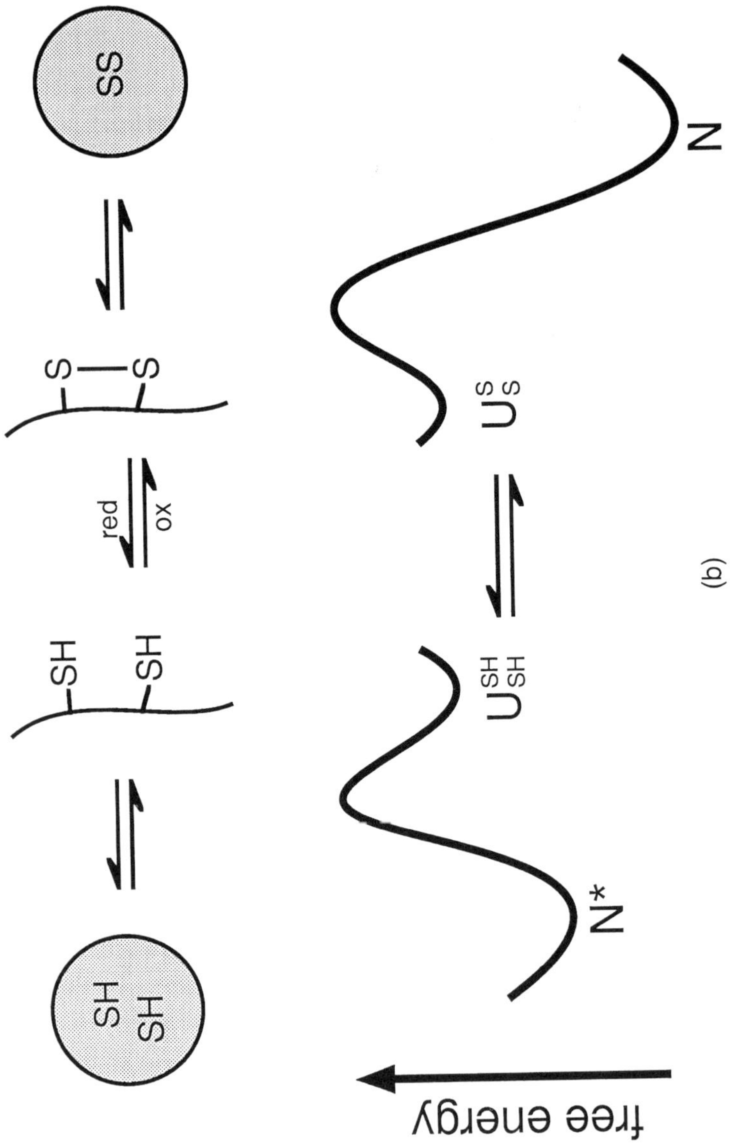

Figure 2 Continued

is not equal to the pathway of protein folding (i.e., folding and disulfide bond formation are often independent).

An important implication is that, because the stability of the folded state is determined by a subtle balance of various factors, it may be easy to expect that the folded (or intermediate) state lacking some of the components of the native state (other than the disulfide bond) accumulates under particular conditions. According to the proline hypothesis, the unfolded species with nonnative isomers of X-Pro peptide bonds cannot refold to the native state because the nonnative peptide bonds interfere with the folding [31]. However, the exact role of the nonnative X-Pro peptide bonds depends on the magnitude of folding forces. Under strongly native conditions, it is expected that protein folding proceeds regardless of Pro cis–trans-isomerism, resulting in the formation of the intermediate state. Indeed, with RNase T1, Kiefhaber et al. [32] demonstrated that the wrong Pro-39 isomer is trapped in the intermediate state by the specific interaction including Trp-59, indicating that specific tertiary interactions in folding intermediates do not necessarily facilitate folding. Kinetic trapping of ovalbumin intermediate with the wrong disulfide bond is another case (see Section IV.C). Once such a state is stabilized, the protein has to climb up the high-energy barrier to complete the folding. Whereas such state can be called a dead-end or off-pathway intermediate relative to a specific structural property (i.e., disulfide bond formation), it might be indistinguishable from the true native state for other properties. The distinction between the off-pathway and on-pathway intermediates is one of the most important topics today [33]. However, the results discussed here suggest that, occasionally when proteins assume alternative folds, such a distinction may be misleading because both can be considered as the products of real folding transition.

B. Equilibrium Intermediates

Strong evidence for the existence of equilibrium intermediate came from the studies of protein unfolding using denaturants such as guanidine hydrochloride (Gdn-HCl) or acid, in particular with α-lactalbumin (123 amino acids, 4 disulfide bonds) [6,34] and cytochrome *c* (104 amino acids, covalently bound heme) [35]. These studies proposed a novel conformational state named the *molten globule* (MG) [34,35]. Similar states have since been observed for many other proteins [6,17]. According to the original definition, it is a compact denatured state with a significant amount of the native-like secondary structure, but a substantially disordered tertiary structure. The rate-limiting step of protein folding was proposed to be located between the MG state and the native state (Fig. 1B). In other words, the packing of the side chains is the barrier of protein folding. The finding [6] that the conformational state similar to the MG state accumulates during the kinetic refolding of several proteins triggered the subsequent extensive studies.

1. Structure

However, detailed characterization of the so-called equilibrium MG states indicated that their exact conformation is different from the definition, and depends very much on protein species [18,36,37]. NMR characterization of the MG state of several proteins in conjunction with the H–D exchange of amide protons indicated that uniform melting of the tertiary structure and total persistence of the secondary structure as suggested by the definition is not true (Fig. 3A). Instead, whereas some regions of the native structure are disordered substantially, some

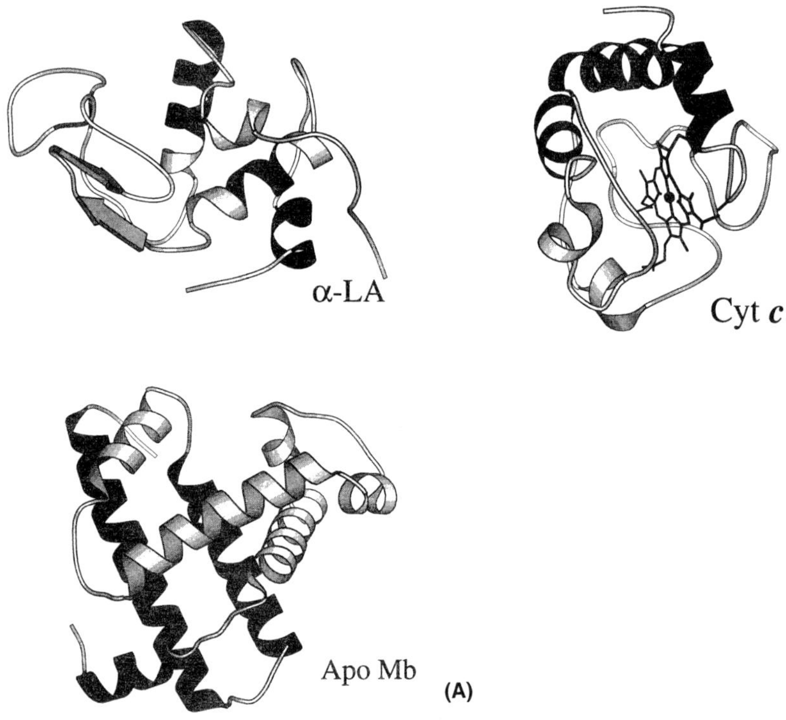

Figure 3 Schematic illustration of the MG states of α-lactalbumin, cytochrome *c*, and apomyoglobulin elucidated by (A) NMR and (B) solution X-ray scattering. In panel A, the native structures are shown and the darkly shaded regions are proposed to be folded in a native-like conformation in the MG states; in panel B, the size of the native state is shown for comparison. The coordinates of bovine α-lactalbumin (1ALC) and sperm whale apomyoglobin (1MBO) were taken from the Brookhaven databank. Those of horse cytochrome *c* were provided by G.D. Brayer.

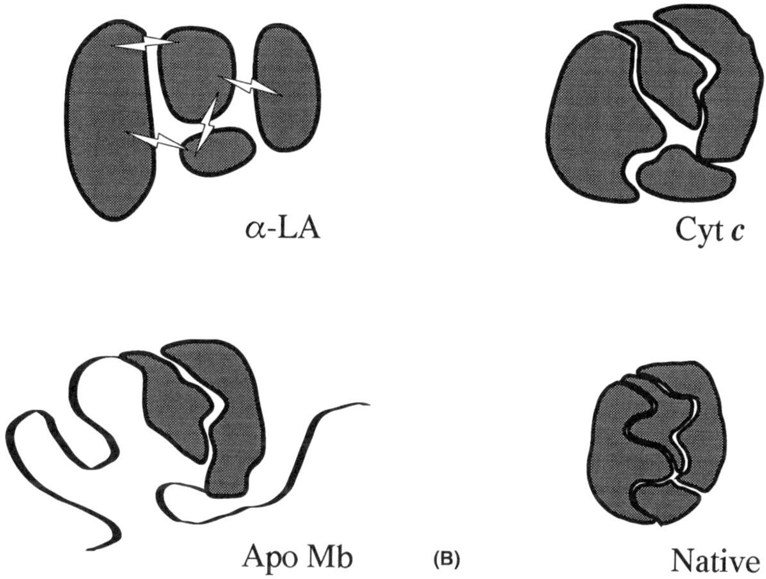

Figure 3 Continued

remain relatively intact in terms of both the secondary and tertiary structures. In cytochrome *c,* whereas the hydrophobic core including the helices and heme are remained, other loop regions are largely unfolded [38]. α-Lactalbumin consists of α-helix and β-sheet domains. In the MG state, whereas the α-domain is relatively intact, the β-domain is largely disordered [39]. For apomyoglobin (153 amino acids) which consists of A–H helices, only A, G, and H helices are formed in the MG state accumulated at pH 3 [40]. In contrast, the acid-denatured state of the four helix bundle protein interleukin-4 (IL-4) retains a highly ordered hydrophobic core in which most, but not all, of the secondary structure is preserved, whereas extensive disorder exists only in regions of the polypeptide chain linking the helices [41]. The term *highly ordered molten globule* was used to distinguish it from the more "disordered" molten globule.

Protein folding is a process in which an extended polypeptide chain acquires the maximal packing through the formation of secondary and tertiary structures. Therefore, the size and shape are the important properties characterizing the intermediate state and hence the folding pathway [2]. Solution X-ray scattering may be the best technique available today for characterizing these properties of the intermediate states. Kataoka and co-workers studied with this technique a variety of the denatured states (Fig. 3B) [11,42,43]. Whereas the MG states of both cytochrome *c* and α-lactalbumin are expanded compared with those of the native state, the expansion in R_g relative to the native state was about 30% for cytochrome *c* and only 10% for α-lactalbumin. The relatively small expansion in R_g of α-lactalbumin is due to the presence of four disulfide bonds. In the MG state of apomyoglobin, the conformational image with a core and flaring tail was elucidated. It is intriguing that whereas other methods suggested the bipartite shape of the MG state of α-lactalbumin [9,44], X-ray scattering provided a compact globule. The discrepancy may be explained by the presence of four disulfide bonds by which the unfolded β-domain collapses to the structured α-domain. The results emphasize the importance of using various methods for characterizing the intermediate conformational states.

2. *Stability*

One of the major arguments related to the MG state has been the mechanism of conformational stability [36]. The major force stabilizing the native state is hydrophobic interactions, including van der Waals interactions, although the contribution of polar groups is also important [2,45]. The role of the hydrophobic interactions in stabilizing the MG state was controversial. However, it is now evident that hydrophobic interaction is the major factor stabilizing the MG state and that its magnitude is much less than that of the native state and varies largely depending on the protein species [46–48]. The acidic MG state of cytochrome *c* is a good example illustrating how the conformational stability of the intermediate state is determined.

Whereas cytochrome *c* is largely unfolded at pH 2 in the absence of salt, the addition of salts or acids stabilizes the MG state (Fig. 4). A series of work carried out by Goto and Fink demonstrated the mechanism responsible for the salt-induced stabilization of the MG state [49–51]. They clarified the importance of anion binding by comparing the effects of various salts and acids. The mechanism of the salt-induced stabilization of the MG state is summarized as follows: Many acid-denatured proteins are unfolded substantially at pH close to 2 in the absence of salt because of the electrostatic repulsion between positively charged groups. The anions either from acids or salts bind directly to the protein positive groups and shield the charge repulsion. Consequently, the intrinsic forces favoring the MG state (i.e., hydrophobic interaction) manifests themselves.

The cooperative thermal unfolding of the MG state of cytochrome *c* has been reported by several groups [52,53], and the contribution of hydrophobic interactions in stabilizing this state was estimated to be 20–40% of that of the native state. Hamada et al. [47] studied the salt-induced formation of the

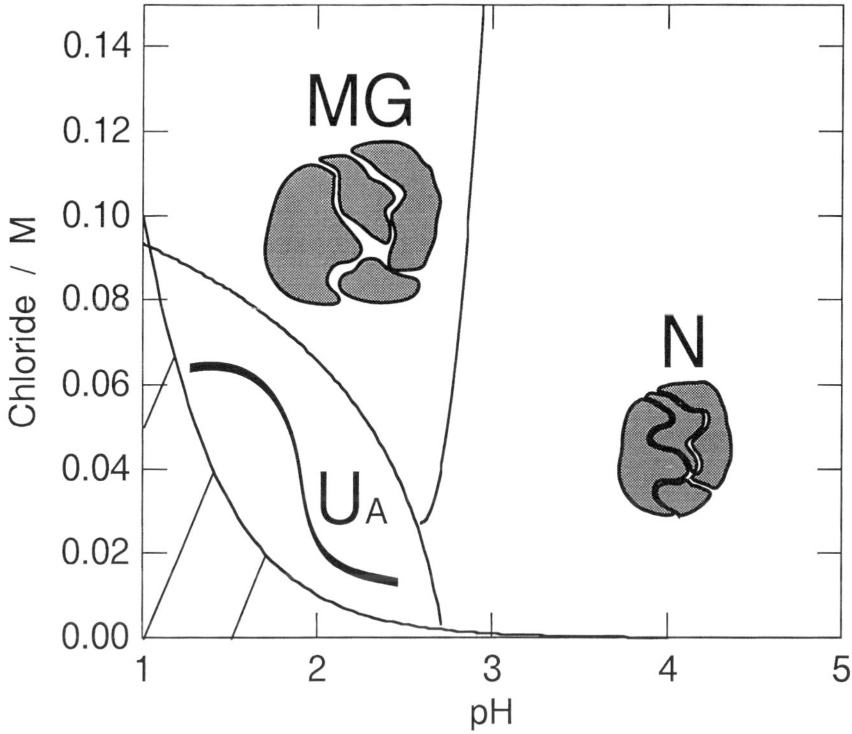

Figure 4 Phase diagram of the chloride- and pH-dependent conformational states of cytochrome *c*. The hatched area is prohibited owing to the increase in the minimal ionic strength with decrease in pH. The structural images are taken from Fig. 3B.

MG state at pH 2 by isothermal titration calorimetry (ITC). Whereas differential scanning calorimetry (DSC) measures the heat of unfolding of the MG state, ITC measures the heat of folding to the MG state. The results obtained with ITC agreed well with those from DSC, confirming the contribution of hydrophobic interaction.

If the charge repulsion critically destabilizes the MG state, the removal of the positive charges should result in the stabilization in MG state even in the absence of salt. A series of cytochrome *c* species with various extents of acetylation of Lys amino groups was prepared and their conformations were examined [54,55]. Indeed, even in the absence of salt at pH 2, the MG state was cooperatively stabilized with the progress of acetylation (i.e., with the decrease in protein net charge) and the contribution of charge repulsion in determining the equilibrium between the MG and the unfolded states was estimated. The results indicated that the unfavorable electrostatic repulsive force is proportional to the square root of the protein net charge. Although the charge repulsion is also an important factor of the stability of the native state, that of the native state is much more complicated than that approximated by the net charge.

These results suggest that the conformational stability of the intermediate state is simpler than that of the native state. Although the exact mechanism of the thermal unfolding of MG states, such as those of α-lactalbumin or apomyoglobin, is controversial [48,56,57], there is no doubt that hydrophobic interaction is the major force stabilizing such marginal states. On the other hand, the importance of specific interactions in the MG states was also shown in several cases [58,59]. The difference in the mechanism of stability between the native and MG states resides in the relative contributions of specific and nonspecific interactions: whereas the specific interactions are critically important in the native state, their contributions are less in the MG state. This situation may be not surprising if we consider that the intermediate state is not tightly packed in many parts of the molecule.

One important implication from the study of the MG state under the artificial conditions is that the MG state may represent the denatured state under the physiological conditions [54]. Suppose that the denatured cytochrome *c* is in an equilibrium between the MG and fully unfolded states. The results obtained with the acetylated cytochrome *c* at pH 2 indicate that the equilibrium depends on the net charge repulsion. At pH 7, the net charge of the denatured cytochrome *c* is +7 according to the amino acid composition. Under such conditions, the MG state is much more stable than the fully unfolded state. Therefore, most of the denatured cytochrome *c* under physiological conditions exist in the MG state and, consequently, may play important biological roles. H–D-exchange study of the native cytochrome *c* indicates that the helices present in the acidic MG state are strongly protected from H/D exchange, consistent with this prediction [60].

3. *Molten Globule Versus Subdomain*

Kim and Baldwin [61] considered two types of folding intermediate. One is the framework intermediate, with an ordered secondary structure, but disordered tertiary structure. The other is the subdomain intermediate in which some of the native-like highly organized subdomains are formed [see also 21,56]. The MG model is consistent with the framework intermediate although the framework model does not specify the compactness. As discussed earlier, recent studies emphasize that there are various types of intermediate states. The real picture may be one that is intermediate between the two extremes, and significant variation may exist among the protein species. In addition, it is important to consider that the intermediate state might be an ensemble of various conformational states. Although a bipartite model of the MG state of α-lactalbumin is popular, this does not mean that the α-domain in the MG state is totally native-like. On the contrary, NMR and X-ray scattering strongly indicate that substantial disordering and conformational change occur in the structured α-domain [43,62]. The bipartite model has been argued for the folding intermediate of staphylococcal nuclease which consists of α- and β-domains [56]. Both the calorimetric and kinetic studies indicate that the intermediate of staphylococcal nuclease contains a structured β-sheet domain whereas the α-helical domain is unfolded. Nevertheless, solution X-ray scattering suggests substantial disordering in the β-sheet domain in the intermediate state [63].

We still do not know the exact picture, whereas we know that both extremes are incorrect. Even for the intermediates with a bipartite structure, substantial disordering in the structured core regions is indicated, suggesting that the image of the original MG definition might still be valid for such subregions. Therefore, we consider that we can continue to use the term MG state if we could recognize that the structure and stability are still under study and that they are different from the original definition.

C. Kinetic Intermediates

Since the pioneering work of Kuwajima and co-workers in the mid-1980s [6], there have been enthusiastic attempts to characterize the kinetic intermediates of small globular proteins. Their experiments with CD stopped-flow clearly demonstrated the accumulation of the MG-like intermediate in the first 10 ms after refolding started. The burst-phase intermediates have substantial secondary structure but disordered tertiary interactions, although the amount of secondary structure in the intermediate depends on the protein species [64]. It is important to see that, for the proteins such as hen lysozyme, that do not have the equilibrium intermediate, the MG-like kinetic intermediate has been elucidated [6]. The results suggested that the rate-limiting state of protein-folding pathway is indeed located after the MG state.

Soon afterward, the pulsed H–D exchange, coupled with two-dimensional NMR, was successfully applied to characterize the secondary structure present in the kinetic intermediate state [7,8,65–68]. The results obtained with several small globular proteins are basically consistent with the MG hypothesis: the kinetic intermediates have particular native-like secondary structures, but disordered tertiary interactions. At the same time, these results further emphasized that the original image of the MG is misleading and the exact structure depends substantially on protein species (see Fig. 3).

D. Transition State

Kinetic studies have been extended in various directions, one of which is the more detailed characterization of the folding pathway, which will be discussed later. Another is the characterization of the transition state.

The pathway hypothesis of protein folding implies the presence of a transition state (see Fig. 1). The transition state has been considered to be an activated barrier separating the native and denatured states. Previously, the transition state was explored from the dependence of the unfolding–refolding rates on temperature or denaturant concentration. The mutational analysis was introduced to analyze the transition state [10,69,70]. In particular, extensive studies with barnase and its mutants by Fersht and co-workers [10,69] characterized the structure of the transition state as well as the intermediate state. Several secondary structural elements and a part of the hydrophobic cores are formed in the intermediate state. These structural elements are further stabilized in the transition state. Surface loops and tertiary interactions are formed only after the transition state. This scheme can be considered basically to be consistent with the MG hypothesis.

Chymotrypsin inhibitor 2 (CI2) is a small protein (64 amino acids) that lacks disulfide bonds and folds rapidly without any intermediates. Fersht and co-workers [70,71], using many mutants, characterized the transition state of folding of CI2. Their results showed that the folding of CI2 is a highly concerted reaction and the secondary and tertiary structures are formed concomitantly and are consolidated throughout the folding reaction. They suggested that the transition state is an expanded state of the native state and that the folding is controlled by the formation of a specific nucleus (nucleation–condensation model). The small structural nucleus is weakly structured by the adjacent interaction and the rest of the protein then condenses around the nucleus, which leads to more stabilization of the nucleus. Alanine-16 is proposed to be a key residue of the folding nucleus. The nucleation–condensation model differs from the classical nucleation–growth mechanism in which a nucleus is formed and then a structure propagates from it.

Although the nucleation–condensation model is evidently contrary to the sequential model with an MG intermediate, it is noted that the model also assumes a unique folding pathway. Whereas the MG hypothesis assumes that the rate-limiting step is the final packing of side chains (see Fig. 1B), the nucleation–

condensation model locates the transition state to the formation of a specific nucleus or microdomain occurring relatively early in the folding process. The roles of the intermediate or transition states are discussed in analogy with the enzyme reaction, for which accumulation of the intermediate (i.e., the enzyme–substrate complex) is unfavorable for the rapid conversion of the substrate to product [72].

III. FOLDING LANDSCAPE

A. Kinetic Trap

Denaturant-unfolded cytochrome *c* at neutral pH refolds into the native state through several kinetic phases with a time scale of seconds, typical of many small proteins [65]. The main fraction of this process proceeds with formation of an intermediate state mimicking the MG state. However, refolding from the unfolded state at a lower pH is very rapid [73] being completed in milliseconds. The refolding is more rapid if one chooses the salt-induced acidic MG state as the initial state. This finding suggested that the kinetic intermediate at neutral pH interferes with the rapid formation of the native state. The investigation concluded that the misorganization after the early condensation of the polypeptide chain results in an accumulation of kinetically trapped intermediate states that are the off-pathway of fast folding. The kinetic trap is due to the incorrect ligation of histisine residues with the heme which is ligated by Met-80 in the native state. Elöve et al [74] also reported the same results, although they considered the trapped state as an obligatory on-pathway intermediate.

Folding mechanism of lysozyme (129 amino acids, 4 disulfide bonds) has been extensively studied and well characterized by various methods, revealing the presence of multiple folding pathways and accumulation of the MG-like intermediate on the major pathway [9,75]. Kiefhaber [76] investigated the folding of lysozyme using interrupted-refolding experiments, which is a variation of the double-jump stopped-flow technique and useful in detecting only the formation of the native state. He observed the two kinetic phases at low concentration of denaturant. At 0.6 M Gdn-HCl, about 14% of all molecules formed the native structure ($\tau = 50$ ms) through a "fast-track" pathway. The rest of molecules reached the native state on a slower pathway ($\tau = 420$ ms). He showed that the slow folding was due to the kinetic trapping under strongly native conditions.

Another interesting report is the rapid folding of ubiquitin. Ubiquitin consists of 76 amino acids and refolds very rapidly, within 10 ms. Khorasanizadeh et al. [77] studied the Gdn-HCl dependencies of refolding kinetics of a pseudo-wild type and three mutants with altered hydrophobic cores. Similar to lysozyme, in low concentrations of denaturant, kinetics were approximated by double exponential curve and the plot of slow-folding rate constant against Gdn-HCl concentration showed a rollover. The rollover mechanism can be explained by both the on-pathway and off-pathway models. The authors concluded that the sequential on-pathway model provides the simplest and physically most satisfactory interpretation of their data.

With cytochrome *c*, it is clear that the kinetic intermediate with incorrect heme ligation accumulates at neutral pH as a kinetically trapped state. [see 78 for more recent results]. The example of lysozyme shows that the kinetically observed major intermediate is a trapped state and there is a more rapid and direct pathway to the native conformation. For ubiquitin, the experimentally observed transient intermediate state can be considered an obligatory intermediate state on the folding pathway. As discussed here, contrary to the pathway hypothesis, there is no clear case that the intermediate accelerates protein folding. In fact, several small globular proteins, including CI2 [70,71] the cold-shock protein CspB [79], and Arc repressor [80], refold rapidly (sub-milliseconds to milliseconds) without the intermediate. The results strongly suggest that the intrinsic folding rate of small globular proteins is much faster than previously imagined [5,81,82].

B. Pathway Versus Funnel

One of the previous central questions was why protein folds so rapidly. To answer this question, several models, as well as the MG hypothesis, have been proposed [reviewed in 61]. Among these, two models that were neglected previously have become important today. One is the nucleation–limited mechanism as described above. The other is the biased random-search model.

In the biased random-search model, compaction of the polypeptide chain is considered to be important. Even the interactions are nonspecific, once the chain became compact, self-avoided conformational space is drastically reduced, inducing the effective formation of the native structure [83]. In other words, rapid, nonspecific collapse permits the random search of all the possible conformational space, although the transition state is unclear. This idea has been elaborated to the "smooth funnel" folding mechanism [2,5,13,15,84]. The smooth funnel model suggests that there is no specific transition state or intermediate state (see Fig. 1E). In this model, the random-coil-like polypeptide chain is dropped into the lowest-energy conformation (i.e., the native state) through nonspecific pathways, suggesting the specific interactions between molecules are important but do not determine the folding pathway. The ideal smooth funnel model is considered as an infinite multiple pathway folding mechanism.

One of the important differences of the smooth funnel model from others is their standpoints on the transition state: namely, whether the transition state is specific or not [13]. In the smooth funnel model, no specific transition states are needed in protein folding, implying that the conformation of transition state cannot be determined. In the nucleation–condensation and other pathway models, the conformation of transition states should be specific and can be characterized.

As shown above, there are many types of folding patterns, even in small proteins. Folding from the unfolded state to the native state includes various events such as hydrophobic collapse, formation of secondary structures and tight packing of side chains. We have to also take into account the rearrangement-type folding. Observed kinetic intermediates should correspond to the low-energy state along the structural coordinate. If there are many energy valleys, the entire energy landscape (i.e., the deepness and position of the valleys, raggedness of the funnel, and so forth) determines which intermediates can be observed. Also, the landscape should depend on the amino acid sequence and the environmental conditions.

The advantage of the funnel model (or folding energy landscape) is that it can accommodate a variety of events occurring during protein-folding process. The assumption of a unique folding pathway is an extreme case of the folding funnel in which only a single channel is allowed because of many barriers (see Fig. 1F). Basically, the funnel concept is similar to the folding with multiple pathways (see Fig. 1C). However, the folding rate is not necessarily determined by the specific energetic barrier although such might be involved. In the discussion of the intermediate state with the pathway concept, we often try to address whether the intermediate is on-pathway or off-pathway (see Fig. 1D). The funnel model suggests that such discussion may be misleading: although the intermediate might not be on the direct pathway to the native state, it might be very close in its structure to the native state, and it will never go back to the original unfolded state, as implicated in the term *off-pathway.*

Intuitively, small globular proteins can fold by a variety of routes as shown for lysozyme [9,75] and cytochrome *c* [73,74] (see Fig. 1E). With an increase in size of the protein molecule, the barrier gets higher in many routes. Because the larger globular proteins consist of subdomains, the association of subdomains may introduce additional complexity. So, with the increase in size, the pathway will become more general and important (see Fig. 1F).

IV. FOLDING OF RELATIVELY LARGE MONOMERIC PROTEINS

A. Apomyoglobin

The removal of the heme from myoglobin results in the formation of apomyoglobin in which most of the helices are still intact, although the stability is decreased significantly. The MG state of apomyoglobin, which is accumulated at pH 4, or pH 2 in the presence of high salt, contains A, G, and H helices [40] (see Fig. 3). Jennings and Wright [68] demonstrated, by the refolding experiment with pulsed H–D exchange, that these helices are formed rapidly within the dead time (5 ms) during the refolding experiment. Subsequently, B-helix is formed and then entire apomyoglobin structure is stabilized (Fig. 5). Sodium trichloroacetate stabilizes the additional B-helix in the equilibrium MG state [42,85]. This intermediate was shown to be compact but largely solvated to water judging from solution X-ray scattering and thermodynamic analysis [42,46,86]. Equilibrium and kinetic results are consistent each other indicating that apomyoglobin structure is formed sequentially. Recently, using a new temperature-jump technique, very rapid process of the refolding of apomyoglobin was followed [87]. The results suggest that the assembly of the A, G, and H helices is complete in a few microseconds.

The folding of apomyoglobin is considered to proceed sequentially through a series of intermediates, each with progressively higher stability and structure content (see Fig. 5). As described in the previous section, the folding pathway of apomyoglobin is not inconsistent with the folding funnel concept. Many pathways are possible. Because of the raggedness and barriers preventing most of the pathways, a unique order of structural formation emerges. The pathway may be a consequence avoiding such barriers (see Fig. 1F).

B. β-Lactoglobulin

In any popular model of protein folding, the folding has been considered to proceed sequentially with hierarchical formation of the native structure (i.e., the intermediate and transition states have a part of the structures present in the native state). Indeed, as shown above, much recent experimental data support this idea. On the other hand, if we consider the fact that the conformational stability of the intermediate state may be different from and simpler than that of the native state [46,55], it is possible that the structure and interactions present in the intermediate state is different from those of the native state. An interesting example is β-lactoglobulin (Fig. 6).

β-Lactoglobulin is a predominantly β-sheet protein (162 amino acids, 2 disulfide bonds and 1 free thiol). However, this protein has been predicted to have a surprisingly high α-helical preference [88]. In accordance with the prediction, β-lactoglobulin exhibits extremely high helical propensity in trifluoroethanol (TFE), an alcohol strongly stabilizing the α-helical structure [89,90]. The helical propensity in TFE is determined by the local helical propensity. β-Lactoglobulin may be a case that the structure of the global free energy minimum (i.e., β-sheet) determined by both the local and nonlocal interactions is very different from its local preference (i.e., α-helix). Then, what is the mechanism of the native β-sheet formation starting from the unfolded state?

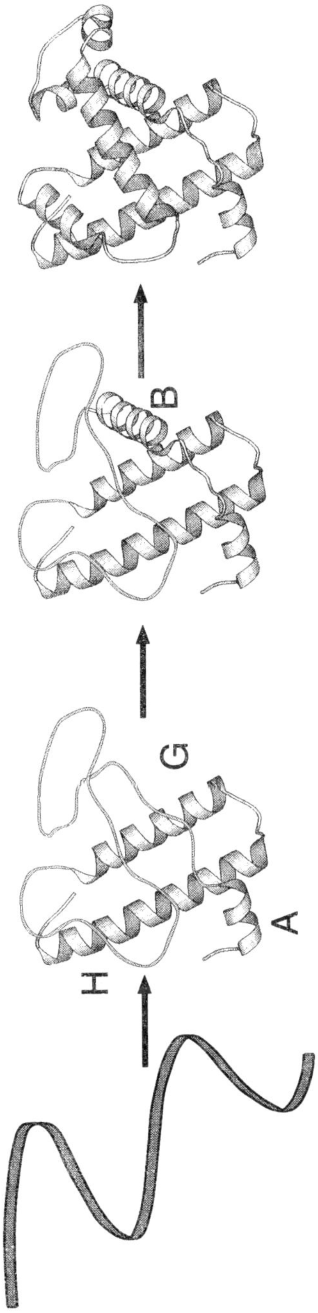

Figure 5 Schematic representation of the folding mechanism of apomyoglobin where the helices are formed sequentially [68].

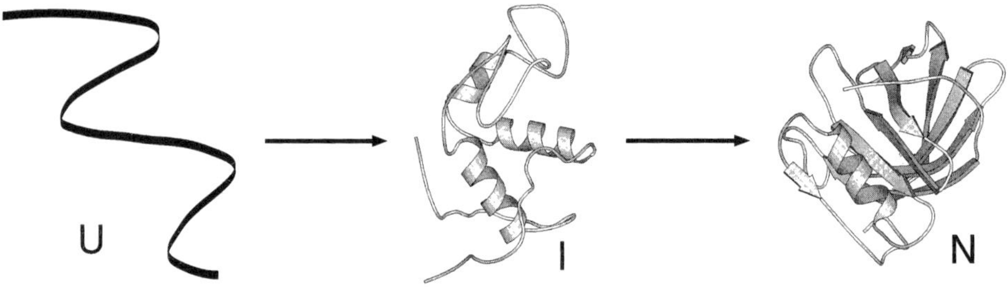

Figure 6 Schematic representation of the folding mechanism of β-lactoglobulin where the partly α-helical intermediate accumulates before the formation of the native predominantly β-structure [92]. The coordinates of bovine β-lactoglobulin were provided by L. Sawyer.

Kuwajima et al. [91] with the CD stopped-flow method first indicated that β-lactoglobulin shows an unusual refolding kinetics: during the dead time (18 ms) of stopped-flow mixing, the far-UV CD intensity exceeded that of the native structure. This has been called the "overshoot phenomenon." Hamada et al. [92] measured the wavelength dependence of the refolding kinetics. The plot of ellipticity values of overshoot, which were extrapolated to time zero, against wavelength, clearly showed the spectrum characteristic of α-helical structure. This suggests that an intermediate with a larger amount of α-helix than the native state accumulates at the early stage of folding (see Fig. 6). From the analysis of the spectra, the helical content of this "burst-phase" intermediate is 38%. They also measured the refolding kinetics in 10% TFE, in which the protein still assumes the native conformation. Under this condition, the α-helical content of the burst-phase intermediate was increased to 52%. The increase in the helical content in TFE is the strong evidence supporting the accumulation of helical intermediate.

The α-helical intermediate before the formation of the native state is obviously contradictory to the generally accepted hierarchical folding models. What is the role of the helical intermediate for the formation of β-sheet structure? Is it a trapped state which prevents the rapid formation of β-sheet structure? Or is it an off-pathway intermediate? The answers to these questions are still unknown. It is important to consider the β-lactoglobulin folding keeping the folding landscape in mind (see Fig. 1F). It is possible that the formation of the complicated native β-structure is guided by a partly helical intermediate in which the protein molecule becomes compact. Probably, the helical intermediate is an obligatory intermediate, even though it does not accelerate the entire folding process.

C. Ovalbumin

Ovalbumin, a major protein of chicken egg white, is a glycophosphoprotein with molecular weight of 45,000 (385 amino acids, 1 disulfide bond, 4 thiols). The structure of ovalbumin is considered to be a model of active inhibitory serpin, which will be described below [93] (Fig. 7). Tani et al. [94] studied the refolding of ovalbumin from the heat-denatured state at 80°C at pH 7.5. The slow cooling below 50°C allowed the heat-denatured ovalbumin to refold to its native state suggesting that heat-induced denaturation of ovalbumin is a thermodynamically reversible process.

The situation is different in the refolding experiment by rapid cooling. The rapid cooling to near 0°C resulted in the accumulation of metastable intermediate. This intermediate state is less stable than the native state against trypsin digestion. Intriguingly, this metastable state can be refolded to the native state by reheating to 50°C, at which the native state is still stable. The intermediate state is stable for several weeks at the low temperature. All of these data suggest that the metastable state formed by the rapid-cooling is the kinetically trapped intermediate. It was confirmed that the trapped state is monomeric. From measurements of CD, intrinsic tryptophan fluorescence and 8-anilino-1-naphthalenesulfonate (ANS) binding, the authors proposed that this metastable intermediate is the MG-like state.

Ovalbumin has a single disulfide bond between Cys-73 and Cys-120 and four free cysteine thiols (see Fig. 7). Tani et al. [94] examined the positions of disulfide bond in the native, trapped intermediate and heat-denatured states. The heat-denatured state at pH 7.5 is an ensemble of the isomers with the random distribution of a disulfide bond induced by intramolecular thiol–disulfide exchange. The slowly refolded species has a correct disulfide bond, indicating that the exchange reaction occurred during the refolding. On the other hand, the kinetically trapped state has a mixture of the disulfide isomers. Therefore, the nonnative disulfide bond is the origin of the kinetic trap. Tani et al. [94] concluded that nonnative disulfide bonds introduced the barrier in the folding pathway, and that the intramolecular thiol–disulfide-exchange reaction is coupled to the release of the barrier.

In the refolding reaction from the heat denatured state of ovalbumin, the refolding conditions change the folding kinetics (i.e., the energy landscape). Under the marginally stable conditions, the correct disulfide bond formation is coupled with the protein folding to the native state. Under the strongly native conditions, the protein can fold with incorrect disulfide isomers, but results in the MG-like intermediate in which the thiol–disulfide-exchange reaction is trapped. This situation is similar to the trapping of proline cis–trans-isomerization by protein folding under the strongly native conditions observed for RNase T1 [32].

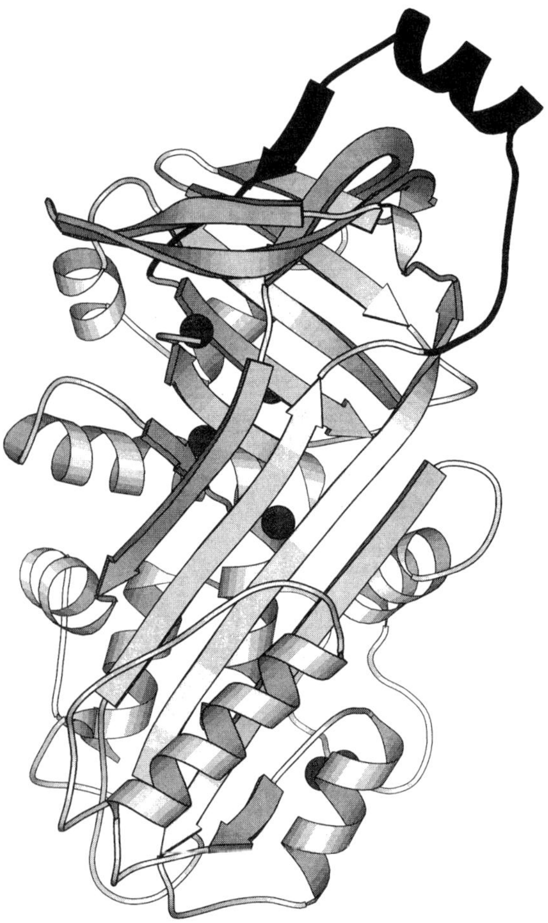

Figure 7 Structure of ovalbumin. The positions of Cys residues (11, 30, 73, 120, 367, and 382) are indicated by the closed circles. The disulfide bond between Cys 73 and Cys 120 is located at the bottom of the molecule. Ovalbumin is modeled as an active form of PAI-1. In the conformational transition from the active form to the latent form of PAI-1, the reactive helical segment of PAI-1, indicated by the black shade, is inserted as a central β-strand to the largest four-stranded β-sheet [98]. The coordinates of ovalbumin (1OVA) were taken from the Brookhaven databank.

V. BIOLOGICAL ROLES

Examples have been known that proteins can assume alternative conformation [95,96]. The domain swapping can be considered to be a case of alternative conformation [97]. From a viewpoint of protein folding kinetics, there are cases that such alternative conformation can be regarded as a consequence of kinetic trapping which is of particular importance for large proteins.

A. Plasminogen Activator Inhibitor-1

Active inhibitory serpins inhibit serine proteases. These inhibitors interact with proteases with a 1:1 stoichiometric manner, then form a covalent bond between the hydroxyl group of the reactive-site serine of the protease and the carboxyl group of the P1 residue at the reactive center of the serpin. The cleavage of the reactive-site loop induces a large conformational change of inhibitory serpin. In the cleaved inhibitor, the NH_2-terminal region of the reactive-site helical loop is known to be inserted into the β-sheets, forming a new antiparallel β-strand [98] (see Fig. 7).

Plasminogen activator inhibitor I (PAI-I) (379 amino acids) has a unique feature as compared to other serpins that such a large conformational change occurs without the cleavage of the peptide bond. The inhibitor is synthesized in its active form by endothelial cells in culture. After release into culture medium the active PAI-1 spontaneously converts to the inactive latent form [99]. The active form can be regenerated from the latent form by denaturant, heat, and sodium dodecyl sulfate (SDS) [100]. This indicates that the conversion between the active and latent forms is reversible, suggesting the active form of PAI-1 is a kinetically trapped state.

Berkenpas et al. [101] constructed the random-mutated recombinant PAI-1 expression library in bacteriophage and screened for increased functional stability. Most stable mutant exhibited the functional half-lives exceeding that of wild-type by up to 72-folds. The authors indicated that the increased functional half-lives is, at least partly due to structural stabilization of the active (trapped) form, decreasing the thermodynamic drive toward the more stable latent structure.

PAI-1 is interesting because the kinetically trapped state is functionally important and most stable state has no activity. It is also interesting that the helical strand is formed first in the active (intermediate) state and ultimately the β-strand is stabilized. This situation is similar to the folding of β-lactoglobulin suggesting that the rapid formation of the helical structure and subsequent stabilization of β-structure is common to the folding of several proteins. It is also noted that the α-to-β transition of prion protein has been proposed to be critical for the spread of prion disease [102].

B. Antitrypsin

α_1-Antitrypsin is also a member of serpin and its major role is to protect lung elastin fibers from proteolytic attack by neutrophil elastase. Some genetic variants of α_1-antitrypsin including Z type variation (Glu-342 $\rightarrow$ Lys) cause aggregation of protein [103–105]. This variant has been studied extensively because of its high frequency, the severity of its secretory defect, extensive intracellular accumulation, and association of diseases. Under mild denaturing condition at 37°C, wild-type protein also tends to form aggregation. In the aggregation form of both the wild and Z-type proteins, the reactive center loop of one molecule inserts in the large β-sheet of another molecule, which reaction is called loop-sheet polymerization [104] and is resembling the α-to-β conformational transition of PAI-1. In the native condition at 37°C, although wild-type antitrypsin is not polymerized, Z-type protein forms aggregate, explaining the accumulation of Z-type antitrypsin in the liver [103]. It should be noted that the Z-type protein can fold into the inhibitory active state at 30°C. Also, once folded at 30°C, the native Z-type protein is substantially stable even at 37°C [105]. These results suggest that a high tendency to aggregate of Z-type protein comes from the accumulation of kinetically trapped intermediate state.

VI. CONCLUSION

We still have no clear perspective of the general roles of intermediate state of protein folding. Although the accumulation of various intermediate states were detected, there has been no clear evidence that the intermediate promotes the folding reaction. On the contrary, the data are accumulating that the escape from its accumulation accelerates the refolding rates. To demonstrate that the kinetic intermediate always slows down the folding reaction, it is necessary to know the intrinsic folding rate, which is an important topic of the physicochemical study of protein folding [5].

Irrespective of its kinetic role, the accumulation of the intermediate states is obligatory for the larger proteins and the pathway becomes important because of the complicated protein-folding landscape. In some cases, the nature takes advantage of the intermediate state or trapped state, resulting in the functional diversity. The case of PAI-1 is an example for which the activity control is introduced by kinetic trapping. In other cases, the accumulation of intermediates that often exhibit high aggregation tendency is not useful for life and, similar to antitrypsin, may be responsible for the formation of amyloid fibrils [96].

One of the important roles of the molecular chaperones is to prevent the aggregation of the intermediate state by interacting with the exposed hydrophobic surface of the substrate proteins [106]. More than preventing the aggregation, the recent landscape perspective of protein folding suggests that the molecular chaperones rescue the trapped intermediates by multiple rounds of binding and release to give a chance for folding to restart from a more unfolded conformation [13,107]. Thus, the molecular chaperones may modulate the folding energy landscape so that even a protein which cannot find out its native state by itself can fold smoothly to the native state.

ACKNOWLEDGMENTS

We thank M. Kataoka, D. Hamada, and M. Hoshino for discussion, and Hue Sun Chan and H. Tani for sending preprints. This work was supported by the Ministry of Education, Science, Culture, and Sports of Japan.

REFERENCES

1. R. H. Pain, *Mechanism of protein folding,* Oxford University Press, Oxford. (1994).
2. K. Dill, S. Bromberg, K. Yue, K. M. Fiebig, D. P. Yee, P. D. Thomas, and H. S. Chan, Principles of protein folding—a perspective from simple exact model, *Protein Sci. 4*:561 (1995).
3. C. B. Anfinsen, Principles that govern the folding of protein chains, *Science 181*:223 (1973).
4. K. W. Plaxco and C. M. Dobson, Time-resolved biophysical methods in the study of protein folding, *Curr. Opin. Struct. Biol. 6*:630 (1996).
5. W. A. Eaton, P. Thompson, C.-K. Chan, S. J. Hagen, and J. Hofrichter, Fast events in protein folding, *Structure 4*:1133 (1996).
6. K. Kuwajima, The molten globule state as a clue for understanding the folding and cooperativity of globular-protein structure, *Proteins Struct. Funct. Genet. 6*:87 (1989).
7. R. L. Baldwin, Pulsed H/D-exchange studies of folding intermediates, *Curr. Opin. Struct. Biol. 3*:84 (1993).
8. S. W. Englander and L. Mayne, Protein folding studied using hydrogen-exchange labeling and two-dimensional NMR, *Annu. Rev. Biophys. Biomol. Struct. 21*:243 (1992).
9. S. E. Radford and C. M. Dobson, Insights into protein folding using physical techniques: Studies of lysozyme and α-lactalbumin, *Philos. Trans. R. Soc. Lond. Biol. 348*:17 (1995).
10. A. R. Fersht, Characterizing transition states in protein folding: An essential step in the puzzle, *Curr. Opin. Struct. Biol. 5*:79 (1995).
11. M. Kataoka and Y. Goto, X-ray solution scattering studies of protein folding, *Folding & Design 1*:R107 (1996).

12. A. Miranker, C. V. Robinson, S. E. Radford, and C. M. Dobson, Investigation of protein folding by mass spectrometry, *FASEB J. 10*:93 (1996).
13. K. A. Dill and H. S. Chan, From Levinthal to pathways to funnels: The "new view" of protein folding kinetics, *Nature Struct. Biol. 4*:10 (1997).
14. R. L. Baldwin, The nature of protein folding pathways: The classical versus the new view, *J. Biomol. NMR 5*:103 (1995).
15. P. G. Wolynes, J. N. Onuchic, and D. Thirumalai, Navigating the folding routes, *Science 267*:1619 (1995).
16. P. S. Kim and R. L. Baldwin, Intermediates in the folding reactions of small proteins, *Annu. Rev. Biochem. 59*:631 (1990).
17. O. B. Ptitsyn, Molten globule and protein folding, *Adv. Protein Chem. 47*:83 (1995).
18. A. L. Fink, Compact intermediate states in protein folding, *Annu. Rev. Biophys. Biomol. Struct. 24*:495 (1995).
19. C. Levinthal, Are there pathways for protein folding? *J. Chim. Phys. 65*:44 (1968).
20. P. L. Privalov, Stability of proteins: Small globular proteins, *Adv. Protein Chem. 33*:167 (1979).
21. C. B. Anfinsen and H. A. Scheraga, Experimental and theoretical aspects of protein folding, *Adv. Protein Chem. 29*:205 (1975).
22. D. P. Goldenberg and T. E. Creighton, Energetics of protein structure and folding, *Biopolymers 24*:167 (1985).
23. T. E. Creighton, Folding pathways determined using disulfide bonds, *Protein folding* (ed. T. E. Creighton), W. H. Freeman, New York, 1992, p. 801.
24. Y. Konishi, T. Ooi, and H. A. Scheraga, Regeneration of RNaseA from the reduced protein: Models of regeneration pathway, *Proc. Natl. Acad. Sci. USA 79*:5734 (1982).
25. H. A. Scheraga, Y. Konishi, D. M. Rothwarf, and P. W. Mui, Toward an understanding of the folding of ribonuclease A, *Proc. Natl. Acad. Sci. USA 84*:5740 (1987).
26. Y. Goto and K. Hamaguchi, The role of the intrachain disulfide bond in the conformation and stability of the constant fragment of the immunoglobulin light chain, *J. Biochem. 86*:1433 (1979).
27. Y. Goto and K. Hamaguchi, Formation of the intrachain disulphide bond in the constant fragment of the immunoglobulin light chain, *J. Mol. Biol. 146*:321 (1981).
28. J. S. Weissman and P. S. Kim, Reexamination of the folding of BPTI: Predominance of native intermediates, *Science 253*:1386 (1991).
29. P. J. Staley and P. S. Kim, Complete folding of bovine pancreatic trypsin inhibitor with only a single disulfide bond, *Proc. Natl. Acad. Sci. USA 89*:1519 (1992).
30. M. Dadlez and P. S. Kim, A third native one-disulphide intermediate in the folding of bovine pancreatic trypsin inhibitor, *Nature Struct. Biol. 2*:674 (1995).
31. J. F. Brandts, H. R. Halvorson, and M. Brennan, Consideration of the possibility that the slow step in protein denaturation reactions is due to cis-trans isomerism of proline residues, *Biochemistry 14*:4953 (1975).
32. T. Kiefhaber, H. P. Grunert, U. Hahn, and F. X. Schmid, Folding of RNase T1 is

decelerated by a specific tertiary contact in a folding intermediate, *Proteins 12*:171 (1992).

33. R. L. Baldwin, On-pathway versus off-pathway folding intermediates, *Folding & Design 1*:R1 (1996).
34. O. B. Ptitsyn, Protein folding: Hypotheses and experiments, *J. Prot. Chem. 6*:273 (1987).
35. M. Ohgushi and A. Wada, 'Molten-globule state': A compact form of globular proteins with mobile side chains, *FEBS Letters 164*:21 (1983).
36. R. L. Baldwin, Molten globules: Specific or nonspecific folding intermediates? *Chemtract Biochem. Mol. Biol. 2*:379 (1991).
37. C. M. Dobson, Unfolded proteins, compact states and molten globules, *Curr. Opin. Struct. Biol. 2*:6 (1992).
38. M. F. Jeng, S. W. Englander, G. A. Elöve, A. J. Wand, and H. Roder, Structural description of acid-denatured cytochrome *c* by hydrogen exchange and 2D NMR, *Biochemistry 29*:10433 (1990).
39. C. L. Chyan, C. Wormald, C. M. Dobson, P. A. Evans, and J. Baum, Structure and stability of the molten globule state of guinea-pig α-lactalbumin: A hydrogen exchange study, *Biochemistry 32*:5681 (1993).
40. F. M. Hughson, P. E. Wright, and R. L. Baldwin, Structural characterization of a partly folded apomyoglobin intermediate, *Science 249*:1544 (1990).
41. C. Redfield, R. A. G. Smith, and C. M. Dobson, Structural characterization of a highly-ordered 'molten globule' at low pH, *Nature Struct. Biol. 1*:23 (1994).
42. M. Kataoka, I. Nishii, T. Fujisawa, T. Ueki, F. Tokunaga, and Y. Goto, Structural characterization of molten globule and native states of apomyoglobin by solution X-ray scattering, *J. Mol. Biol. 249*:215 (1995).
43. M. Kataoka, K. Kuwajima, F. Tokunaga, and Y. Goto, Structural characterization of the molten globule of α-lactalbumin by solution X ray scattering, *Protein Sci. 6*:422 (1997).
44. L. C. Wu, Z. Y. Peng, and P. S. Kim, Bipartite structure of the α-lactalbumin molten globule, *Nat Struct Biol 2*:281 (1995).
45. G. I. Makhatadze and P. L. Privalov, Energetics of protein structure, *Adv. Protein Chem. 47*:307 (1995).
46. I. Nishii, M. Kataoka, F. Tokunaga, and Y. Goto, Cold-denaturation of the molten globule states of apomyoglobin and a profile for protein folding, *Biochemistry 33*:4903 (1994).
47. D. Hamada, S. Kidokoro, H. Fukada, K. Takahashi, and Y. Goto, Salt-induced formation of the molten globule state of cytochrome c studied by isothermal titration calorimetry, *Proc. Natl. Acad. Sci. USA 91*:10325 (1994).
48. Y. V. Griko and P. L. Privalov, Thermodynamic puzzle of apomyoglobin unfolding, *J. Mol. Biol. 235*:1318 (1994).
49. Y. Goto and A. L. Fink, Phase diagram for acidic conformational states of apomyoglobin, *J. Mol. Biol. 214*:803 (1990).

50. Y. Goto, L. J. Calciano, and A. L. Fink, Acid-induced folding of proteins, *Proc. Natl. Acad. Sci. USA 87*:573 (1990).
51. Y. Goto, N. Takahashi, and A. L. Fink, Mechanism of acid induced folding of proteins, *Biochemistry 29*:3480 (1990).
52. S. Potekhin and W. Pfeil, Microcarolimetric studies of conformational transition of ferricytochrome c in acidic solution, *Biophys. Chem. 34*:55 (1989).
53. Y. Kuroda, S. Kidokoro, and A. Wada, Thermodynamic characterization of cytochrome c at low pH: Observation of the molten globule state and the cold denaturation process, *J. Mol. Biol. 223*:1139 (1992).
54. Y. Goto and S. Nishikiori, Role of electrostatic repulsion in the acidic molten globule of cytochrome c, *J. Mol. Biol. 222*:679 (1991).
55. Y. Hagihara, Y. Tan, and Y. Goto, Comparison of the conformational stability of the molten globule and native states of horse cytochrome c. Effects of acetylation, heat, urea and guanidine-hydrochloride, *J Mol Biol 237*:336 (1994).
56. P. L. Privalov, Intermediate states in protein folding, *J. Mol. Biol. 258*:707 (1996).
57. K. Kuwajima, The molten globule state of α-lactalbumin, *FASEB J. 10*:102 (1996).
58. J. L. Marmorino and G. J. Pielak, A native tertiary interaction stabilizes the A state of cytochrome *c*, *Biochemistry 34*:3140 (1995).
59. D. Hamada, Y. Kuroda, M. Kataoka, S. Aimoto, T. Yoshimura, and Y. Goto, Role of heme axial ligands in the conformational stability of the native and molten globule states of horse cytochrome *c*, *J. Mol. Biol. 256*:172 (1996).
60. Y. Bai, T. R. Sosnick, L. Mayne, and S. W. Englander, Protein folding intermediates: Native-state hydrogen exchange, *Science 269*:192 (1995).
61. P. S. Kim and R. L. Baldwin, Specific intermediates in the folding reactions of small proteins and the mechanism of protein folding, *Annu. Rev. Biochem. 51*:459 (1982).
62. J. Balbach, V. Forge, N. A. J. van Nuland, S. L. Winder, P. J. Hore, and C. M. Dobson, Following protein folding in real time using NMR spectroscopy, *Nature Struct. Biol. 2*:865 (1995).
63. J. M. Flanagan, M. Kataoka, D. Shortle, and D. M. Engelman, Truncated staphylococcal nuclease is compact but disordered, *Proc. Natl. Acad. Sci. USA 89*:748 (1992).
64. T. Sugawara, K. Kuwajima, and S. Sugai, Folding of staphylococcal nuclease A studied by equilibrium and kinetic circular dichroism spectra, *Biochemistry 30*:2698 (1991).
65. H. Roder, G. A. Elöve, and S. W. Englander, Structural characterization of folding intermediates in cytochrome *c* by H-exchange labelling and proton NMR, *Nature 335*:700 (1988).
66. J. B. Udgaonkar and R. L. Baldwin, NMR evidence for an early framework intermediate on the folding pathway of ribonuclease A, *Nature 335*:694 (1988).
67. A. Miranker, S. E. Radford, M. Karplus, and C. M. Dobson, Demonstration by NMR of folding domains in lysozyme, *Nature 349*:633 (1991).
68. P. A. Jennings and P. E. Wright, Formation of a molten globule intermediate early in the kinetic folding pathway of apomyoglobin, *Science 262*:892 (1993).

69. A. Matouschek, J. T. J. Kellis, L. Serrano, M. Bycroft, and A. R. Fersht, Transient folding intermediates characterized by protein engineering, *Nature 346*:440 (1990).
70. J. L. Neira, B. Davis, A. G. Ladurner, A. M. Buckle, G. de Prat Gay, and A. R. Fersht, Towards the complete structural characterization of a protein folding pathway: The structures of the denatured, transition and native states for the association/folding of two complementary fragments of cleaved chymotrypsin inhibitor 2. Direct evidence for a nucleation-condensation mechanism, *Folding & Design 1*:189 (1996).
71. L. S. Itzhaki, D. E. Otzen, and A. R. Fersht, The structure of the transition state for folding of chymotrypsin inhibitor 2 analysed by protein engineering methods: evidence for a nucleation-condensation mechanism for protein folding, *J. Mol. Biol. 254*:260 (1995).
72. A. R. Fersht, Optimization of rate of protein folding: the nucleation condensation mechanism and its implications, *Proc. Natl. Acad. Sci. USA 92*:10869 (1995).
73. T. R. Sosnick, L. Mayne, R. Hiller, and S. W. Englander, The barriers in protein folding, *Nature Struct. Biol. 1*:149 (1994).
74. G. A. Elöve, A. K. Bhuyan, and H. Roder, Kinetic mechanism of cytochrome c folding: involvement of the heme and its ligands, *Biochemistry 33*:6925 (1994).
75. S. E. Radford, C. M. Dobson, and P. A. Evans, The folding of hen lysozyme involves partially structured intermediates and multiple pathways, *Nature 335*:700 (1992).
76. T. Kiefhaber, Kinetic traps in lysozyme folding, *Proc. Natl. Acad. Sci. USA 92*:9029 (1995).
77. S. Khorasanizadeh, I. D. Peters, and H. Roder, Evidence for a three-state model of protein folding from kinetic analysis of ubiquitin variants with altered core residues, *Nature Struct. Biol. 3*:193 (1996).
78. S.-R. Yeh, S. Takahashi, B. Fan, and D. L. Rousseau, Ligand exchange during cytochrome c folding, *Nature Struct. Biol. 4*:51 (1997).
79. T. Schindler, M. Herrler, M. A. Marahiel, and F. X. Schmid, Extremely rapid protein folding in the absence of intermediates, *Nature Struct. Biol. 2*:663 (1995).
80. C. R. Robinson and R. T. Sauer, Equilibrium stability and sub-millisecond refolding of a designed single-chain Arc repressor, *Biochemistry 35*:13878 (1996).
81. R. L. Baldwin, Why is protein folding so fast?, *Proc. Natl. Acad. Sci. USA 93*:2627 (1996).
82. A. D. Miranker and C. M. Dobson, Collapse and cooperativity in protein folding, *Curr. Opin. Struct. Biol. 6*:31 (1996).
83. A. Sali, E. Shakhnovich, and M. Karplus, How does a protein fold?, *Nature 369*:248 (1994).
84. J. D. Bryngelson, J. N. Onuchis, N. D. Socci, and P. G. Wolynes, Funnels, pathways, and the energy landscape of protein folding: a synthesis, *Proteins: Struct. Funct. Genet. 21*:167 (1995).
85. S. N. Loh, M. S. Kay, and R. L. Baldwin, Structure and stability of a second molten globule intermediate in the apomyoglobin folding pathway, *Proc. Natl. Acad. Sci. USA 92*:5446 (1995).

86. I. Nishii, M. Kataoka, and Y. Goto, Thermodynamic stability of the molten globule states of apomyoglobin, *J. Mol. Biol. 250*:223 (1995).
87. R. M. Ballew, J. Sabelko, and M. Graubele, Direct observation of fast protein folding: the initial collapse of apomyoglobin, *Proc. Natl. Acad. Sci. USA 93*:5759 (1996).
88. K. Nishikawa and T. Noguchi, Predicting protein secondary structure based on amino acid sequence, *Methods Enzymol. 202*:31 (1991).
89. K. Shiraki, K. Nishikawa, and Y. Goto, Trifluoroethanol-induced stabilization of the α-helical structure of β-lactoglobulin: Implication for non-hierarchical protein folding, *J. Mol. Biol. 245*:180 (1995).
90. D. Hamada, Y. Kuroda, T. Tanaka, and Y. Goto, High helical propensity of the peptide fragments derived from β-lactoglobulin, a predominantly β-sheet protein, *J. Mol. Biol. 254*:737 (1995).
91. K. Kuwajima, H. Yamaya, S. Miwa, S. Sugai, and T. Nagamura, Rapid formation of secondary structure framework in protein folding studied by stopped flow circular dichroism, *FEBS Lett. 221*:115 (1987).
92. D. Hamada, S. Segawa, and Y. Goto, Non-native α-helical intermediate in the refolding of β-lactoglobulin, a predominantly β-sheet protein, *Nature Struct. Biol. 3*:868 (1996).
93. P. E. Stein, A. G. W. Leslie, J. T. Finch, W. G. Turnell, P. J. McLaughlin, and R. W. Carrell, Crystal structure of ovalbumin as a model for the reactive centre of serpins, *Nature 347*:99 (1990).
94. F. Tani, N. Shirai, T. Onishi, F. Venelle, K. Yasumoto, and E. Doi, Finding the right fold by alternative pathway for ovalbumin heat-denatured at the monomeric state, *Protein Sci.*: in press (1997).
95. C. M. Dobson, Finding the right fold, *Nature Struct. Biol. 2*:513 (1995).
96. J. W. Kelly, Alternative conformations of amyloidogenic proteins govern their behavior, *Curr. Opin. Struct. Biol. 6*:11 (1996).
97. M. J. Bennett, M. P. Schlunegger, and D. Eisenberg, 3D domain swapping: A mechanism for oligomer assembly, *Protein Sci. 4*:2455 (1995).
98. J. Mottonen, A. Strand, J. Symersky, R. M. Sweet, D. Danley, K. F. Geoghegan, R. D. Gerard, and E. J. Goldsmith, Structural basis of latency in plasminogen activator inhibitor-1, *Nature 355*:270 (1992).
99. T. Kooistra, E. D. Sprengers, and V. W. M. van Hinsbergh, Rapid inactivation of the plasminogen-activator inhibitor upon secretion from cultured human endothelial cells, *Biochem. J. 239*:497 (1986).
100. K. Katagiri, K. Okada, H. Hattori, and M. Yano, Bovine endothelial cell plasminogen activator inhibitor. Purification and heat activation, *Eur. J. Biochem. 176*:81 (1988).
101. M. B. Berkenpas, D. A. Lawrence, and D. Ginsburg, Molecular evolution of plasminogen activator inhibitor-1 functional stability, *EMBO J. 14*:2969 (1995).
102. Z. Huang, S. B. Prusiner, and F. E. Cohen, Scrapie prions: a three-dimensional model of an infectious fragment, *Folding & Design 1*:13 (1996).

103. D. A. Lomas, D. L. Evans, J. T. Finch, and R. W. Carrell, The mechanism of Z α_1-antitrypsin accumulation in the liver, *Nature 357*:605 (1992).
104. D. A. Lomas, D. L. Evans, S. R. Stone, W.-S. W. Chang, and R. W. Carrell, Effect of the Z mutation on the physical and inhibitory properties of α1-antitrypsin, *Biochemistry 32*:500 (1993).
105. M.-H. Yu, K. N. Lee, and J. Kim, The Z-type variation of human α_1-antitrypsin causes a protein folding disease, *Nature Struct. Biol. 2*:363 (1995).
106. M. Hoshino, Y. Kawata, and Y. Goto, Interaction of GroEL with conformational states of horse cytochrome *c*, *J. Mol. Biol. 262*:575 (1996).
107. M. J. Todd, G. H. Lorimer, and D. Thirumalai, Chaperonin-facilitated protein folding: Optimization of rate and yield by an iterative annealing mechanism, *Proc. Natl. Acad. Sci. USA 93*:4030 (1996).
108. P. J. Kraulis, MOLSCRIPT: A program to produce both detailed and schematic plots of protein structures, *J. Appl. Crystallog. 24*:946 (1991).

2

Oligomeric Proteins

Rainer Jaenicke
Universität Regensburg, Regensburg, Germany

I. INTRODUCTION

Proteins as the main components of biological systems guide us from low molecular weight compounds, such as proteohormones, to the microscopic and macroscopic world of flagella, bacterial surface layers, capsids of viruses, giant multienzyme complexes, and muscle. The fundamental principle behind this wide range of sizes and functions is the hierarchical order of protein structure (Fig. 1). As has been established by the early observation that proteins fold autonomously, the information for their defined spatial organization must reside in their individual amino acid sequence, together with the specific solvent environment. Given a sequence, next-neighbor interactions along the polypeptide chain lead first to regular structural elements of secondary and supersecondary structure. They merge subsequently into domains as independent folding units, which at the next higher level coalesce to form the compact tertiary structure. In quaternary assemblies, *structured monomers* with native-like tertiary structure, drop to their final state of lowest potential energy by sequential association and shuffling processes. The given structural levels describe in a qualitative way the sequence of events in protein folding. At the same time, they illustrate the contributions of the various types of local through-chain and through-space interactions to the free energy of stabilization ($\Delta G_{N \rightarrow U}$) of the native protein.

In dealing with the in vitro folding of oligomeric proteins, the significant difference from monomeric proteins is not only the superposition of folding and association, but also that most oligomeric proteins are large, commonly consisting of more than one structural domain. This means that folding of oligomers, in general, involves specific recognition at the level of domains and subunits. Both

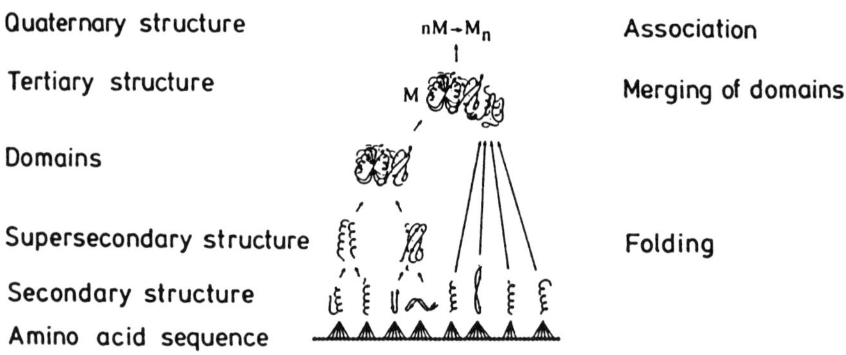

Figure 1 Hierarchy of protein structure and protein folding.

require high specificity. Their correlation has been the subject of extensive research that cannot be covered in detail in this chapter. Instead, representative examples will be selected to illustrate the important issues (for further information, see [1–3]).

II. DEFINITIONS

Because of the previously mentioned relation between the docking of multidomain and multisubunit proteins, in the following discussion we shall include proteins consisting of more than one entity, also certain examples in which the "compact regions" are linked covalently. Such *domain proteins* may give rise to the same kinds of side reactions in terms of wrong interactions that have also been observed for oligomeric proteins. In addition, they allow the evolution of oligomeric proteins from monomeric precursors to be understood in terms of *domain swapping* [4,5].

There are at least six definitions of *domains* [1,2,6]: (a) Domains are stable, compact globular *substructures of a polypeptide chain that fold independently*; they are stabilized by noncovalent interactions [7–10]. (b) Cooperative folding–unfolding of the previously mentioned substructures allows domains to be defined as *thermodynamic units,* separable, for example, by differential scanning calorimetry, or by hydrodynamic and spectroscopic measurements [11–13]. (c) Domains are *topological entities visible at the atomic level,* which exhibit more pronounced interactions within the structural unit than with other parts of the polypeptide chain [6]. (d) Sequence homologies at the level of DNA or amino acid sequences may allow domains to be defined a *genetic units* based on the consensus role that one-dimensional sequence homology or identity commonly translates into three-dimensional relations [14,15]. (e) Closely related to the latter definition is the *evolutionary* concept that correlates domains with *exons* [16,17]. (f) Domains represent substructures with specific *functional properties,* leading to the concept that proteins are composed of exchangeable building blocks with partial functions, such as binding of substrates, coenzymes, effectors, and such [18].

Evidently, the given definitions overlap to a large extent. In the present context, in general, the term *domain* refers to units that allow the polypeptide chain to fold sequentially "by parts," thereby enhancing the process of protein self-organization from an astronomic to the biological time range [19,20]. Essentially, folding-by-parts must be considered a most significant acquisition of evolution, because of the enhancement of the folding rate by synchronous folding at multiple sites along the nascent polypeptide chain. There are several additional advantages: (a) Domain folding is an efficient way of excluding wrong intermolecular interactions in large protein molecules; (b) it protects the nascent polypeptide chain from proteolysis; and (c) it may be considered a simple mechanism to proceed from monomeric to multimeric proteins by "domain swapping" (see Section III).

In contrast to the term "domain," the *oligomeric state* (quaternary structure) is semantically well-defined as the native state of a protein composed of more than one polypeptide chain, associated through noncovalent interactions. The chains can be identical or nonidentical (homo- or hetero-oligomers), and may vary in size and state of association in a wide range, to the extent that certain proteins represent polymer *distributions*; rather than monodisperse systems (e.g., glutamate dehydrogenase, and acetyl-CoA carboxylase). The unifying principle is the geometry of association and, in enzymes, the correlation of assembly structure and function. The term *association–dissociation* will be used to describe the chemical reaction underlying quaternary structure formation and disintegration; *assembly* refers to the more general phenomena involved in the generation of biological superstructures; *aggregation* stands for "wrong association" as a side reaction of association, leading to loss of native protein in inclusion bodies, and so on [21]. In considering refolding, reassociation, and reactivation in vitro as the typical approach to mimic protein self-organization in the cell, *reconstitution* will be used to describe the sequential folding–association reaction of oligomeric systems.

When we consider the evolution of oligomeric or multimeric proteins, several advantages can be easily visualized: economy of the genome, elimination of misfolded polypeptide chains as pieces in the assembly puzzle, stability, and osmotic effects, to name a few. However, it seems much more difficult to understand how oligomers may have evolved. How could random mutations alter the surface of monomeric proteins such that the entropy loss accompanying dimerization or multimer formation is paid by the gain in enthalpy caused by the formation of weak intersubunit bonds? In asking this question, the tacit assumption that monomers preceded oligomers in evolution is generally accepted.

III. DOMAIN SWAPPING

The problem of how oligomeric proteins have evolved from monomeric precursor proteins is challenging because our present view that quaternary structure formation is governed by hydrophobic recognition sites in the subunit interface leaves unanswered the question of how the oligomers could have been assembled in the first place. As has been pointed out [5], the gradual accumulation of random mutations that are required to stabilize a dimer of a spherical single-domain "sticky billard ball molecule," and to select for its improved stability cannot be accomplished within biologically feasible time. At this point, the evolution of domains offers a key to our understanding. By starting from a monomeric domain protein, "domain swapping" allows the transition from monomers to assembly structures of any size simply by switching from interdomain to intersubunit interactions. Here, the interface is a priori available (perhaps by the duplication of an ancestral minigene), such that natural selection can now improve the stability by additional dimer stabilizing mutations (Fig. 2).

Both the low probability to gradually acquire dimer stability by random mutagenesis, and the concept of domain swapping may be illustrated using crystallins from the eye lens as examples [22]. In the case of αA-crystallin, the extreme conservation during evolution (1% over 30 million years) suggests a highly specific structure with extreme requirements in terms of intrinsic stability and protein interactions [23]. This translates into an extremely slow accumulation of random mutations, rendering significant dimer stabilization most unlikely. βγ-Crystallins form a superfamily of homologous proteins that is possibly related to bacterial spore coat proteins and dormant proteins of primitive eukarya. Their structural data suggest two gene duplication steps from an ancestral protein folded to a Greek key (see Fig. 2B). β-Crystallin forms dimers or higher polymers, whereas γ-crystallins are exclusively monomeric; both show highly homologous two-domain structures. Comparing the two topologies, it is most tempting to interpret the β-dimers as the product of domain swapping and association of a γ-ancestor [24]. An important detail supporting the swapping hypothesis is that two-thirds of the residues in the interfaces between the NH_2- and COOH-terminal domains in γB-crystallin and between the NH_2- and COOH-terminal domains of the two subunits in β_{B2}-crystallin are identical or highly conserved [25–27].

Evidently, domain swapping is not restricted to dimerization: linear and cyclic higher oligomers can arise both in two or three dimensions [4]. However, there are data that do not support the domain-swapping hypothesis: First, three-dimensional domain swapping cannot explain hetero-oligomers, because the subunits do not share common interdomain interfaces and, second, there are oligomers in which domains share active sites at subunit interfaces. Obviously, there are stable monomeric proteins in which the domain-swapping ability is revealed only under particular circumstances. On the other hand, there are

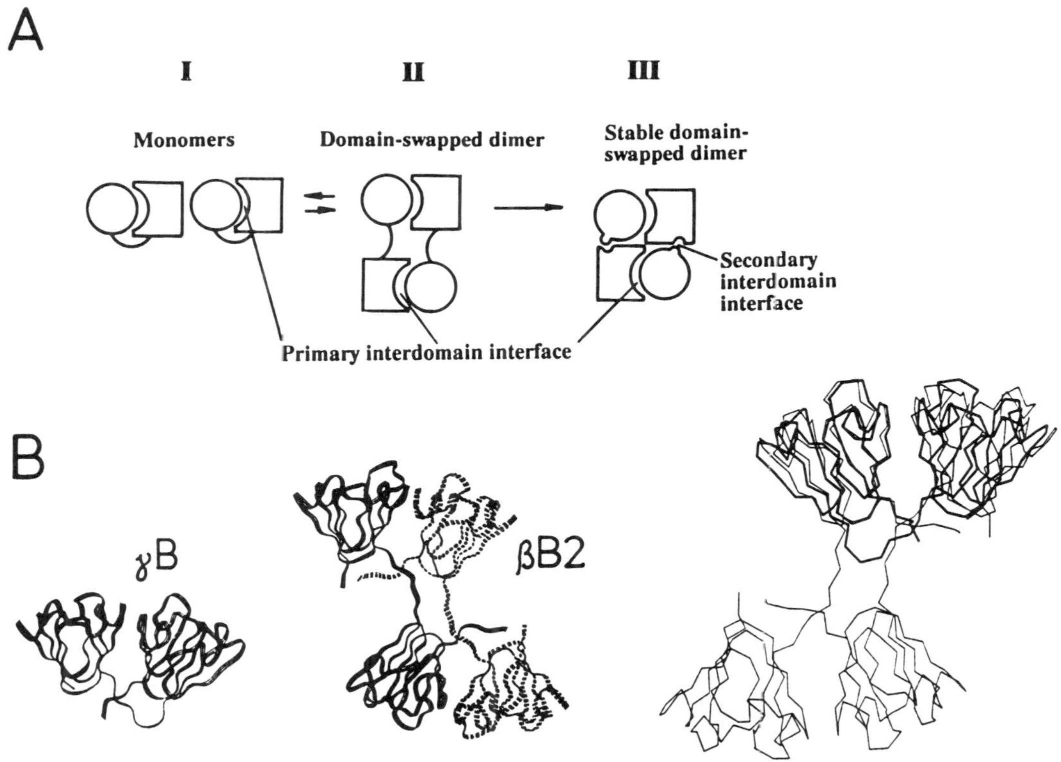

Figure 2 Domain swapping hypothesis: (A) The monomer (I) with its primary interdomain interface may participate in interchain interactions, generating domain-swapped dimers (II). Stabilizing mutations in the linker region allow stable dimers (III) to be formed. (B) βγ-Crystallins as potential products of domain swapping; the ribbon diagrams (left) are superimposed in the backbone structure (right) to indicate the structural homology. (From Ref. 4.)

oligomeric proteins in which swapping is an established structural feature. Finally, there are oligomeric proteins in which the domain-swapped form is not the exclusive conformation; instead, states with alternative structures coexist. It is obvious that unregulated domain swapping can lead to aggregation, as well as abnormal precipitates (e.g., in inclusion bodies and amyloids) [5].

The in vitro association pathways in the process of domain-swapped oligomers are unknown, even for dimerization as the simplest example. Only in systems, such as diphtheria toxin, in which monomers are temporarily subjected to conditions favoring metastable domain-swapped dimers, the large activation barrier between this species and the closed monomer allows the transition to be analyzed [4]. If association of folded domains is slow relative to domain folding, or if proteins encounter transient denaturing conditions (e.g., during translocation or compartmentation), it may be visualized that monomeric proteins might undergo three-dimensional domain swapping in vivo. As an example, the cell adhesion protein CD2 undergoes domain swapping when it is expressed as a fusion protein with glutathion *S*-transferase (GST) [28]. The tethering of two CD2 molecules to a GST dimer during folding permits domain swapping, perhaps mimicking interactions between folding intermediates at high local concentration in the cell [5]. That is, three-dimensional domain swapping may occur in the folding pathways of proteins if intermediates accumulate that have folded but unassociated domains.

IV. METHODS TO PROBE FOLDING AND ASSOCIATION

It is obvious that the in vitro folding and association of oligomeric proteins must involve multiple intermediate states that may be populated at different levels along the folding pathway, according to

$$U \rightarrow I_i \rightarrow N \tag{1}$$

with I_i as intermediates, and U and N as unfolded and native states. For multisubunit proteins, U and I_i have to be replaced by nU and nI_i, and N by N_n. Here, intermediate states along the pathway of folding and association must include *structured monomers* that are characterized by the capacity to recognize each other to form the multimeric native state.

By starting from fully denatured polypeptide chains, and switching to quasiphysiological conditions, the formation of structured monomers and their assembly may be quantitatively analyzed in terms of sequential folding–association events [1]. In this context, it is important that available methods reflect "renativation" at different levels: fluorescence emission monitors changes in the local environment of aromatic residues; ANS binding, alterations in accessible hydrophobic surface area; far- and near-UV circular dichroism, the recovery of native secondary and tertiary structure; chemical cross-linking, the pathway of subunit association.

Ligand binding, biological function (reactivation), and allosteric regulation allow the native state to be monitored in an unambiguous way. Thus, the latter probes are the most stringent criteria if experiments are intended to explore whether the native structure is regained after a complete unfolding–refolding cycle. For enzymes, measurements related to catalytic activity are not only a better index of minor conformational adjustments or deviations than the usual physical parameters, but they are also well-suited for the low protein concentrations that are required to avoid side reactions, such as kinetic competition between folding and aggregation. Regulatory properties, such as cooperative binding of ligands or allosteric behavior, are good criteria for the unperturbed native conformation and quaternary structure, because they require that multiple functional sites be formed and associated, and they interact through subtle conformational changes of the entire structure. Here, the structural requirements are most demanding, such that the initial native protein and the product of reconstitution may be compared unequivocally. Crystal structure determination might be considered the final proof for authenticity. However, in contrast with *specific* activity or ellipticity, one has to keep in mind that crystallization may be a fractionation step that does not necessarily represent the bulk of the reconstituted protein in solution.

The outcome of denaturation–renaturation studies on a wide variety of proteins has shown that in vitro protein folding after preceding denaturation leads to recovery of the native state for at least a major fraction of the molecules. This holds for monomeric, oligomeric, and multimeric systems. The latter two commonly require careful optimization of the renaturation conditions to accomplish high yields. However, there seems to be no size limit for the protein, nor limits of complexity that would absolutely require accessory proteins [3,29]. As a consequence, denaturation–renaturation has become a routine in the downstream processing of proteins of therapeutic importance. Within the cell, where crowding of cytosolic components and high local concentrations of proteins may cause kinetic competition of folding–association, on one hand, and misfolding–misassembly, on the other, the nascent polypeptide chain may require folding catalysts or chaperones to reach the native state.

To follow the kinetics of folding and association, monitoring biological (catalytic) activity is the only approach that can be easily normalized without perturbations by kinetic partitioning or chemical modification. Such characteristics as spectral and conformational properties are affected by the incompleteness of the renaturation reaction. This holds also for ligand binding and immunological methods (including enzyme-linked immunosorption assay [ELISA] techniques) which are prone to artifacts caused by induced conformational effects. With oligomeric proteins, reactivation kinetics must be supplemented by a direct kinetic analysis of the reassociation process using chemical cross-linking, hybridization, and high-performance liquid chromatography (HPLC). In this connection, glutardialdehyde has been most successfully applied, because it reacts fast and complete as a heterogeneous polymer, thereby providing a whole spectrum of distances between the reactive groups [30–32]. There are proteins for which quantitative cross-linking has not been accomplished, probably because of the unfavorable distribution or absence of reactive lysine residues in the surface of the refolding structured monomers; if fixation does not go to completion, reassociation may still be determined, provided that the data are corrected accordingly (e.g., on the basis of the particle distribution obtained for the native oligomer) [33,34].

V. THERMODYNAMICS

In discussing the folding and association of oligomeric proteins, thermodynamics need to be considered in connection with (a) the *stability* of proteins in solution at the various levels of the structural hierarchy, (b) the *forces* involved in stabilization, and (c) the *specificity* of quaternary structure formation.

A. Stability

In spite of intense efforts during almost 40 years [11,12,35–40], the stability of proteins is still not fully understood. One of the reasons is that empirical potential functions are insufficient to make unequivocal predictions about the effective energy surface of unknown proteins. Additional problems come from the in vitro or in vivo issue, on the one hand, and the questionable two-state situation in the process of folding and association, on the other.

1. The In Vitro or In Vivo Issue and Extremophilic Adaptation

The question of how in vitro studies relate to the situation in the cell refers, first, to potential effects of the cellular environment (i.e., solvent components, specific ligands, crowding, or kinetic partitioning with the help of chaperones), and second, to the problem of whether and to what extent the translation machinery is involved in the formation and stabilization of the native conformation of the nascent polypeptide chain.

There is still no way to design proteins with specific stability characteristics. Therefore, it seems useful to try to elucidate the long-term experiment that has been going in nature, where adaptation to extremes of physical conditions during 3.5 billion years has led to proteins with highly specified stabilities toward temperature, pressure, pH, and low water activity. In comparing mesophiles and extremophiles, the range of viability may be broadened or shifted by minute mutational alterations using the canonical amino acid repertoire [47]. When focusing on homologous proteins from organisms taken from different extreme environments, adaptation to temperatures close to the boiling point of water, or extremes of pH and salt concentrations tends to maintain "corresponding states" relative to overall structure, flexibility, and ligand affinity. Basically, evolution needs to compromise in optimizing multifunctionality; for example, enzymes have to *fold* to stable three-dimensional structures; these have to be sufficiently flexible to *perform catalysis*; finally, they undergo *degradation* for reasons of regeneration or as nutrient in periods of starvation [48]. As taken from systematic studies of common proteins and proteins from a wide variety of extremophiles, only a few weak interactions are required to shift the limits of stability [49]. Thus, an amazingly high degree of conservation is observed when comparing the topology of the proteins isolated (e.g., from mesophiles, psychrophiles, thermophiles, and even hyperthermophiles). As a consequence, no unambiguous general strategies of adaptation at the sequence level of proteins have yet been discovered [47,50]. When considering the quaternary structure, one mechanism of stabilization seems to be an increased state of association: several enzymes from hyperthermophiles, which are commonly monomeric, form either oligomers or fusion proteins with anomalous assembly structures [50].

Another issue in considering protein folding and association in the cell refers to cotranslational or posttranslational events connected with codon frequencies and secondary structure at the mRNA and protein level. There are significant differences in the codon usage, especially in linker regions between secondary structural elements or domains, that lead to nonuniform polypeptide synthesis [51,52]. Whether they are important in protein folding and association remains to be shown.

2. Two-State or Multistate?

Even in the most simple case of a single-chain, one-domain protein, such as ribonuclease (RNase), the concentration of intermediate states in the thermal unfolding transition is about 5% [41]. Thus, the thermodynamic description in terms of only two macroscopic states, the unfolded and the completely folded one, holds to only a first approximation. In the unfolded state U, it is evident that U stands for a large number of energetically similar substates; considering the phenomena of static and dynamic disorder in X-ray and nuclear magnetic resonance (NMR) analyses, the native state also contains substates, separated from each other by well-defined activation barriers [42]. The cooperativity of conformational transitions suggests that the native conformation requires all its attractive interactions to maintain its stability, or vice versa, all essentially stabilizing elements responsible for the native conformation seem to occur simultaneously late on the pathway of folding and association. The controversial question whether they do so along a single pathway, or whether multiple pathways are involved in structure formation, can be ignored in the present context (for reviews see [43–46]).

3. *Intrinsic Stability*

Proteins have to cope with extremes of temperature, pressure, and solvent components. Correspondingly, the metabolic network has to compensate for shifts in activation energies and activation volumes, on the one hand, and dissociation equilibria caused by electrostriction or "iceberg" formation, on the other.

Relative to the intrinsic stability of proteins, the quantity to normalize experimental data is the free energy of stabilization, $\Delta G_{N \rightarrow U}$. Its determination requires reversible two-state conditions, (i.e., thermodynamic equilibrium). The best criteria to distinguish between two-state and multistate transitions are parallel and reversible denaturation–renaturation profiles for different structural or functional characteristics, and their comparisons with calorimetric data. In monitoring the equilibrium constant K in a two-state equilibrium

$$N \overset{K}{\rightleftharpoons} U \tag{2}$$

for example, by catalytic activity, fluorescence emission, circular dichroism, or specific binding of ligands, K values are obtained for which temperature dependence allows the van't Hoff enthalpy $\Delta H_{\text{van't Hoff}}$ to be calculated. Correlating this quantity to the calorimetrically determined ΔH_{cal}, gives a quantitative measure for two-state or multistate behavior of a given system under denaturing conditions. If the ratio $\Delta H_{\text{cal}}/\Delta H_{\text{van't Hoff}}$ is close to unity, the two-state model can be applied, allowing the free energy of stabilization ΔG_{stab} to be determined. For domain proteins and oligomeric proteins, large deviations of the $\Delta H_{\text{cal}}/\Delta H_{\text{van't Hoff}}$ ratio from unity have been observed, proving that independent cooperative folding of separate parts of a protein molecule may occur.

The two-state order–disorder transition is accompanied by a large positive change in the specific heat ΔC_p owing to the exposure of hydrophobic residues in the process of unfolding [37]. On the other hand, the molar enthalpy of protein denaturation and subunit dissociation may either be positive or negative. Because of the large ΔC_p, it increases markedly with temperature. If $\Delta H_{N \rightarrow U}$ is positive, the conformation entropy ΔS_{conf} must be the driving force in the overall denaturation process. Basically, $\Delta S_{N \rightarrow U}$ is expected to be positive and to increase, as far as alterations in chain conformation on unfolding are concerned; on the other hand, the solvation term ΔS_{hydr}, which is attributable to the exposure of hydrophobic residues and the disruption of interior ion pairs, will be negative owing to the ordering effects of charged groups and hydrophobic residues on water molecules in their surroundings. Similar to $\Delta H_{N \rightarrow U}$, the conformational entropy also increases with temperature. If we consider folding and association instead of denaturation, $\Delta H_{N \rightarrow U}$ and $\Delta S_{N \rightarrow U}$ change signs: now, endothermic association (accompanied by a decrease in ΔS_{conf} of the polypeptide chain) is driven by water release from nonpolar residues and charged groups involved in intra- and intermolecular hydrophobic and ion-pair interactions [1]. Experimental proof for this mechanism goes back to Kauzmann [35] and Lauffer [53].

From the previous discussion of the temperature dependence of ΔH and ΔS, the temperature dependence of the Gibbs free energy follows in a straightforward manner; it yields maximum curves that suggest that there is a low-temperature analogue of heat denaturation. This *cold denaturation* has been established for both monomeric and oligomeric proteins [54,55]. At pH values corresponding to optimum stability, the temperature maxima of the free energy of stabilization are close to physiological temperatures.

The $\Delta G_{N \rightarrow U}$ is a direct measure of the stability of the native macroscopic state under the condition that the free energy is calculated for the entire cooperative system. In domain proteins, as well as in oligomeric and multimeric proteins, this cannot be generally assumed [13,56]. Strikingly, numerical data for $\Delta G_{N \rightarrow U}$ cluster in a narrow range close to 50 kJ mol^{-1} (12 kcal mol^{-1}), independent of both the size of the protein and mode of denaturation [57]. Calculating the stability increment per residue, the result ($\approx$ 0.5 kJ mol^{-1}) immediately tells us that the overall stability must involve cooperativity, because the thermal energy (kT) exceeds the foregoing increment by almost one order of magnitude.

Cooperativity refers to all levels of the hierarchy of protein structure: secondary structure formation, tertiary collapse to form the hydrophobic core, and subunit assembly. Figure 3 shows examples for the mutual stabilization of domains and subunits. For γB-crystallin the complete molecule and the separate recombinant domains are compared. For porcine lactate dehydrogenase, the native tetramer is stabilized by the NH_2-terminal decapeptide; after removing this, dimers are formed that show decreased stability and cooperativity of unfolding; the monomer is accessible only as a folding intermediate with no catalytic activity and only marginal stability [58,59]. There are numerous examples in which quaternary structure formation leads to significant stabilization. If we take viruses, chromatin, or ferritin as examples, the conjugates show low turnover, whereas the constituent subunits have only average stability and high flexibility, allowing their multiple functions: processing, translocation, morphopoiesis, turnover, and so on.

B. Forces

There is still considerable uncertainty about the forces stabilizing globular proteins in solution. The players in the game, hydrogen bonds (H-bonds), ion pairs, hydrophobic interactions, and van der Waals forces, have been classified [35,60]; however, their relative roles have yet to be settled [37,47,56,61–63]. Results from calorimetric analysis, and experiments modeling the hydrophobic effect by the transfer of nonpolar substances to water, yield distinct differences

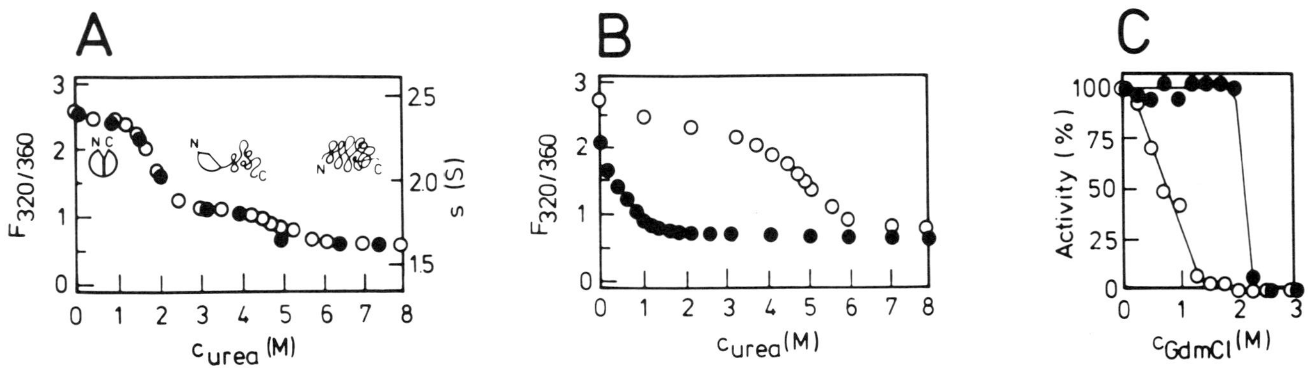

Figure 3 Mutual stabilization of domains and subunits: γβ-crystallin and lactate dehydrogenase. (A) Urea-dependent equilibrium transitions of γβ-crystallin at pH 2, 20°C, monitored by sedimentation analysis (●) and fluorescence (○). (B) Fluorescence profiles for the transitions of the isolated NH_2-terminal (○) and COOH-terminal (●) domains [27]. (C) Guanidine-dependent deactivation of native (tetrameric) LDH (●) and the "proteolytic dimer" lacking the NH_2-terminal decapeptide (○); 0.1 M phosphate pH 7.6, 1.5 M ammonium sulfate, 20°C [59].

because the protein interior resembles a crystal-like solid phase, rather than a nonpolar liquid [64]. Apart from that, there is a significant enthalpic contribution to hydrophobic interactions (attributable to van der Waals forces) in addition to the increase in entropy caused by water release [38]. If one considers the average energy contributions of the foregoing weak interactions involved in protein stabilization (Table 1), on balance, $\Delta G_{N \rightarrow U}$ is based on the equivalent of only a few hydrogen bonds, ion pairs, or hydrophobic interactions. Inspection of known three-dimensional structures of globular proteins clearly indicates that a large number of interactions are involved in the formation of the native secondary structure, as well as the packing and docking of both the hydrophobic inner core and the subunit interfaces [63]. Evidently, the free energy of stabilization represents a marginal difference between large numbers of attractive and repulsive forces. Making use of the mutational analysis of the contributions of individual residues or types of interactions to protein structure and stability, quantitative assignments have been successfully attempted for bacteriophage T4 lysozyme [40], staphylococcal nuclease [65], the *C*1 and *Cro* repressors of bacteriophage λ [66,67], cytochrome *c* [68], bovine pancreatic trypsin inhibitor (BPTI) [69], and various other systems [70].

A number of safe conclusions may be drawn from these studies: (a) In case of multiple mutants, if there is no interaction between the sites of the replacements in the native protein, the free-energy changes are additive. (b) $\Delta G_{N \rightarrow U}$ is equivalent to the energy required to break a maximum of about 1% of the total number of hydrogen bonds in the native state. (c) In the unfolded state, a 10-kDa protein exposes about 440 polar sites, half of which are involved in internal hydrogen bonds in the native state. As a consequence, even a marginal difference in hydrogen bond strength between water–water and water–protein will be magnified to a significant energy change. (d) Water release from polar and nonpolar sites will lead to an increase in entropy, which is known to be the driving force in endothermic folding and assembly processes [35,53]. (e) Despite the tendency of the folding polypeptide to accomplish maximum packing density and minimum hydrophobic surface area, only about 70% of the theoretically available

Table 1 Intermolecular Interactions in Biopolymers[a]

Hydrogen bond (dipol interaction)	≈ 20 kJ mol^{-1}
Hydrophobic interaction (–CH_2-transfer)	10–20
Ion pairs (coulombic interaction)	20–50
van der Waals	1–10

[a]Note: The given values do not represent net contributions to ΔG_{stab}, because interactions are also possible with solvent in the unfolded state.

hydrophobic interactions are realized in forming the inner core of a globular protein. (f) As charged groups are commonly exposed to the aqueous solvent, coulombic interactions in the interior of a globular protein cannot be of major importance in protein stabilization. On the average, only 1 ion pair in 150 amino acid residues is buried in the interior core; thus, only surface and intersubunit ion pairs or clusters of ion pairs are involved in stabilization. Considering the solvation of the separate charges, entropic effects (owing to proton and water release) may contribute more to stability than coulombic charge interactions. (g) When internal polar sites are incapable of forming hydrogen bonds, structured water molecules in cavities may bridge gaps [47].

C. Specificity

The aforementioned weak interactions determine the hierarchy of protein structure at all levels: secondary structure, domains, and subunits. Their intramolecular and intermolecular coupling is highly specific owing to (a) amphipathic surfaces of helices, (b) van der Waals contacts and ion pairing between domains, (c) domain swapping, and (d) subunit complementarity at the quaternary level. The specificity of recognition may be illustrated using complementation of ribonuclease S and a few homologous NAD-dependent dehydrogenases and bacterial luciferase as examples.

1. Fragments

Ribonuclease S (RNaseS) is native RNaseA with its NH_2-terminal 20-residue "S-peptide" nicked by subtilisin. The S-protein together with the S-peptide retain full enzymatic activity. Removal of the S-peptide from the inactive 21–124 fragment, and reconstitution of the "scrambled" S-protein with structurally altered S-peptides have been extensively used to characterize the essential properties responsible for the pairing of both constituents [71,72]. Complete restoration of catalytic function of the fragments clearly proved that the information in the S-peptide is required for the stabilization of the S-protein; the S-peptide alone retains its conformation only in the presence of its parent protein. A covalent connection linking the two parts is not required. Thus, monitoring secondary structure formation in the 1–13 helix of the S-peptide, on the one hand, and reactivation of complete RNaseS, on the other, allows the mutual stabilization of the fragments to be applied to model the correlation of folding and association in oligomeric proteins [73].

As in RNaseA [74], the refolding kinetics of RNaseS are governed by fast- and slow-folding species. In the fast one, recombination of the two fragments occurs before folding of S-protein is complete; in the slow one, proline isomerization occurs as a rate-limiting late event. Binding of the structured S-peptide to an early intermediate of the S-protein precedes complete folding; it stabilizes the

S-protein moiety at a stage when its native-like conformation has not yet been formed [75]. The following simplified scheme describes the kinetic analysis:

$$\mathrm{U} \rightleftharpoons \mathrm{I} \underset{p}{\overset{p}{\rightleftharpoons}} p\mathrm{I} \rightleftharpoons p\mathrm{N} \tag{3}$$

The folding–combination scheme is a limiting case for the general mechanism underlying the structure formation of oligomeric proteins. Evidently, an alternative would be the complementation at an early stage of folding that would eliminate concentration-dependent steps, and lead to mere first-order shuffling (see Section VI.A).

2. Domains

Octopine dehydrogenase (ODH) is a single-chain homologue of lactate dehydrogenase in molluscs. Both the gene sequence and the three-dimensional structure of the enzyme are still unknown. However, the monophasic deactivation transition and the biphasic unfolding transitions at elevated guanidinium chloride concentration, clearly indicate independent-folding modules, expected for two topologically well-separated coenzyme and substrate-binding domains (Fig. 4). Partial unfolding in the first transition leads to an inactive intermediate, whereas in the second step, complete unfolding takes place (see Fig. 4A). Refolding yields only 70%, even under optimum conditions, suggesting kinetic partitioning between multiple conformational states (N*, N′) according to

$$\begin{array}{ccc} \mathrm{N} & = & \mathrm{N}^* \\ & \searrow \; \nearrow & \\ & \mathrm{U} & \\ & \downarrow & \\ & \mathrm{N}' & \end{array} \tag{4}$$

Repetitive unfolding–refolding (with n cycles) yields 0.7^n of the initial activity (see Fig. 4B). N* and N′ differ in their spectral properties: the 70% fraction is native, whereas the 30% fraction consists of incorrectly folded monomers with nonnative fluorescence characteristics, but native-like far-UV circular dichroism [76,77]. Refolding of the enzyme under essentially irreversible conditions obeys a simple sequential scheme

$$\mathrm{U} \xrightarrow{\text{fast}} \mathrm{I} \xrightarrow{k} \mathrm{N}^* \tag{5}$$

As shown in Figure 4C, the recovery of native helicity and fluorescence are too fast to be followed by manual mixing, even at low temperature and high viscosity. Although the fast U → I transition represents the major part of structure formation, it yields inactive enzyme. Activity appears only with a half-time of 20 min (20°C). It is slowed down drastically by solvent viscosity, suggesting that the

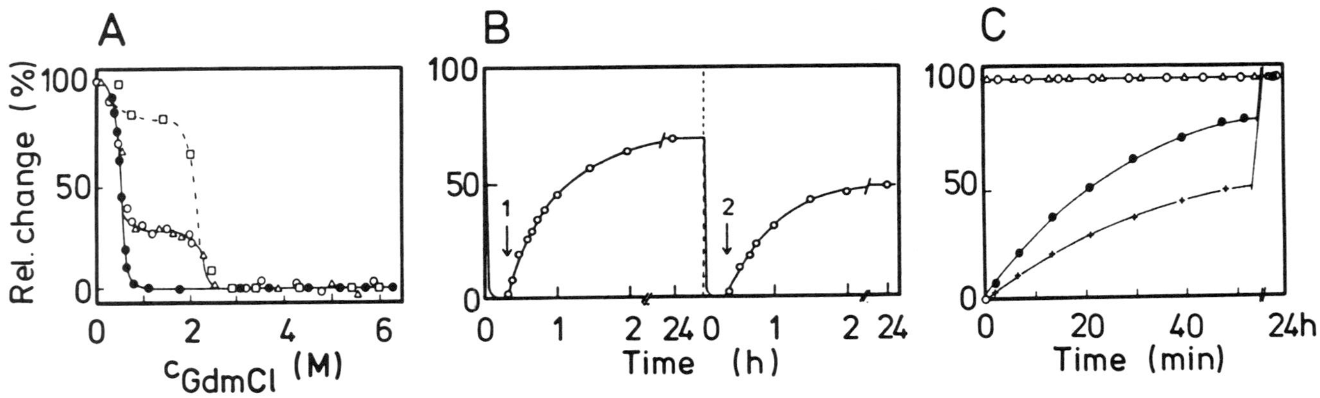

Figure 4 Denaturation–renaturation of octopine dehydrogenase (ODH) from *Pecten jacobaeus*: (A) Guanidine (GdmCl)-dependent deactivation (●) and denaturation monitored by relative fluorescence (△), maximum fluorescence (□), and ellipticity θ_{222nm} (○) [76]; (B) reactivation after repetitive denaturation in 6 M GdmCl, 20°C [29,76]; (C) renaturation after 24 h denaturation in 6 M GdmCl, 20°C, monitored by fluorescence (△), θ_{222nm} (○), activity under standard conditions (●) and in the presence of 25% (v/v) glycerol (+); there is a linear relation of the reactivation rate and viscosity up to 8 cP [77].

rate-limiting process involves domain *pairing,* rather than domain *folding.* With this result in mind, it is tempting to assume that the foregoing kinetic partitioning refers to differences in domain recognition in terms of correct (N*) and incorrect (N′) domain pairing. The strong temperature dependence of the rate-limiting step ($\Delta G^{\ddagger} \approx 100$ kJ mol^{-1}), as well as results from limited proteolysis experiments at the intermediate level, prove that slow-folding steps participate in the reaction. Whether proline isomerization within the connecting peptide is involved in the partitioning reaction remains to be shown.

3. Subunits

Lactate dehydrogenase (LDH) has been a paradigm for the recognition of topological units at the level of domains and subunits. The extreme stability of its native quaternary structure (which does not undergo dissociation at concentrations far below the nanomolar range), has become clear from Rossmann's pioneering crystallographic studies [78]. As it turns out, the enzyme shows absolute complementarity of its domains and subunits, with close packing of the modules and no space for water molecules in its interior. Figure 5 summarizes the state of association of LDH and its relatives in the dehydrogenase family, as well as the subunit contacts responsible for the high association constant. Considering the conditions that shift the $M_4 \rightleftharpoons 4M$ equilibrium toward the monomeric state, it becomes clear that hydrophobic interactions and ion pairs must be involved [79–81].

To study the specificity of domain and subunit recognition, limited proteolysis and joint reconstitution of mixtures of related dehydrogenases have been applied. By removing the NH_2-terminal decapeptide of the porcine skeletal muscle enzyme, inactive "proteolytic dimers" are obtained that are partially nicked to 14- and 21-kDa domain fragments. They gain catalytic activity in the presence of structure-making ions and therefore, may be studied for their structure–function relation [58,82]. Subjected to unfolding–refolding, the fragments recognize each other forming "nicked monomers," if the two complementary units are reconstituted in one batch; obviously, there is a mutual chaperone effect, as reconstitution of the separate domains and subsequent mixing of the products are unsuccessful [59].

The question of whether domain recognition can be extended to homologues from different organisms has been investigated for phosphoglycerate kinase, for which chimeras with domains from the human and yeast enzymes fold to the native enzyme, exhibiting full catalytic activity.

Specific recognition at the subunit level was investigated, making use of the topologically related dehydrogenases, with the question in mind of whether chimeric end products or kinetic intermediates might be detectable. In contrast with homologous lactate dehydrogenases (which are known for their combinatorial exchange of subunits in tissue-specific isoenzyme patterns), attempts to

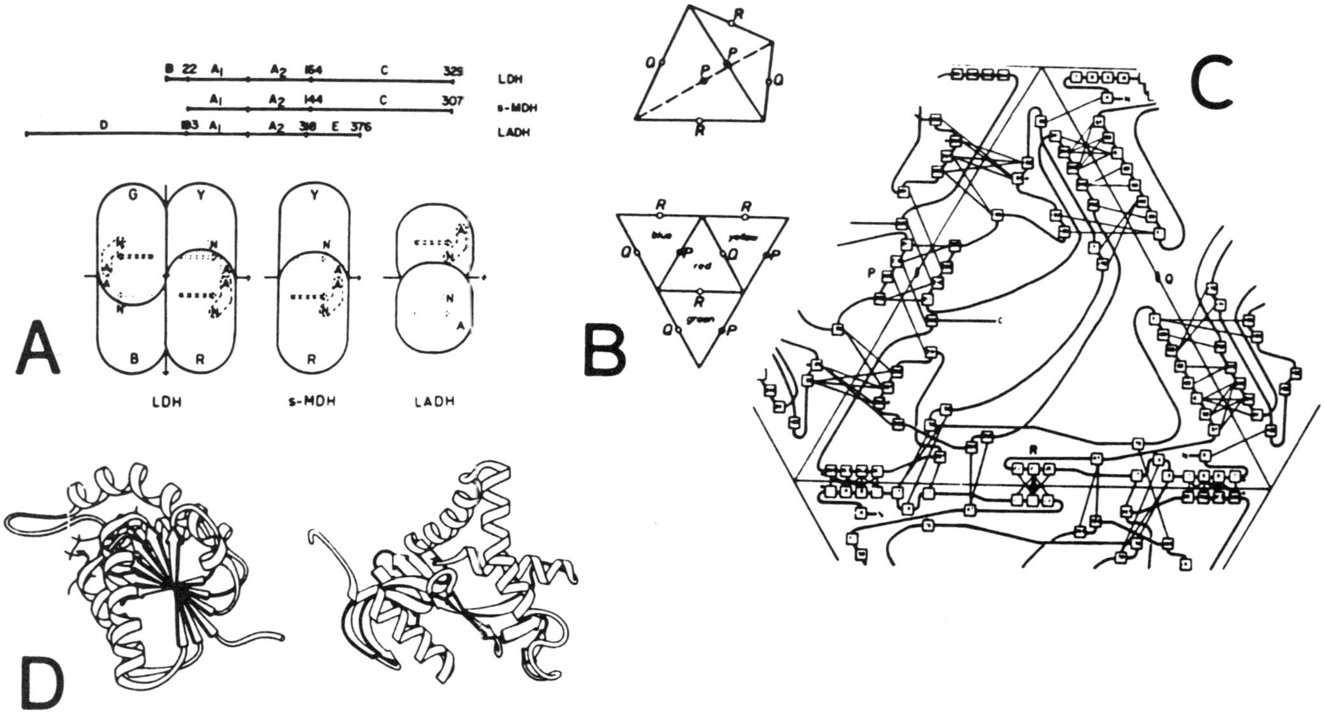

Figure 5 Schematic representation of the three-dimensional structure of lactate dehydrogenase: (A; Upper part) topological comparison of various dehydrogenases. Numbers refer to residue numbers in linker regions between the NAD-binding domains (A_1, A_2) and the catalytic domains (C,D), B to the "NH_2-terminal arm," stabilizing LDH as dimer of dimers. (Lower part) Association patterns of subunits in LDH, s-MDH, and LADH (cf. [78]). (B) Spatial and planar representation of the 222 symmetry of tetrameric LDH with three twofold axes *P, Q, R*. Each triangle stands for a subunit, and each edge for a subunit interface. (C) Main-chain interactions between $^{\alpha}C$ atoms of the "red subunit" and neighboring subunits. (D) Tertiary structure of the NAD^+- and substrate-binding domains [1].

produce hybrids of tetrameric porcine LDH and dimeric mitochondrial or cytosolic malate dehydrogenases (m-MDH and s-MDH) failed [84]. When using dimeric LDH from *Limulus polyphemus* and m-MDH under strictly synchronized reactivation conditions, again, neither hybrid intermediates nor chimeric end products were detectable [85]. That this finding cannot be generalized is obvious, considering the above-mentioned, tissue-specific isoenzyme patterns of LDH, or multifunctional enzymes in which certain gene products have been found as subunits in different complexes (e.g., protein disulfide isomerase in prolyl hydroxylase). In the isoenzymes of LDH, the structural homology is extremely high, such that complementary subunit interfaces can easily be rationalized. For isoenzymes in which different compartments are involved, both target sequences and different-folding mechanisms may contribute to specificity. For example, in mitochondrial and cytosolic MDHs, the second-order subunit assembly of m-MDH is preceded by slow folding, whereas for s-MDH first-order reconstitution indicates diffusion-controlled association (see later discussion). It might be attributed to these mechanistic differences that so far all attempts to form hydrid MDHs have been unsuccessful.

VI. KINETICS OF FOLDING AND ASSOCIATION

In going from monomeric multidomain proteins to multisubunit proteins (i.e., from the molecular level to cellular substructures and finally to the microscopic and macroscopic level), most of the findings discussed for domain folding and domain pairing can be transferred to subunits. Actually, in oligomeric proteins, subunit assembly corresponds to domain pairing, with the sole difference that we are now dealing with unlinked polypeptide chains, in which the "local concentration" can be varied at will. The steps on the assembly pathway are as before: folding of subdomains and domains to form *structured monomers,* and subsequent association to the native quaternary structure. The intermolecular interactions are the same. It is obvious that, after the structured monomers have joined, the collision complex may (and most probably will) undergo intramolecular rearrangements to reach the final state of maximum packing density. Thus, folding and association may be followed by further transconformation steps; however, frequently, the kinetics of reconstitution can be fitted with a simple unibimolecular folding–association scheme, according to

$$n\mathcal{M} \overset{k_1}{\rightarrow} n\mathrm{M} \overset{k_2}{\rightarrow} \mathrm{M}_n \tag{6}$$

with one first-order rate constant k_1 for the rate-limiting folding reaction, and one second-order rate constant k_2 for subsequent association, $\mathcal{M}$, M, and M_n standing for the unfolded and structured monomer and the oligomer, respectively. At this point, it becomes clear why reconstitution experiments have been a powerful tool in elucidating the mechanism of protein self-assembly: in contrast with dentura-tion (where both unfolding and dissociation follow first-order kinetics), renaturation often allows one to distinguish between folding and association, making use of the concentration dependence of the collision frequency. Thus, the kinetic analysis of the properties of the refolding polypeptide chains may provide a means to monitor the recovery of the secondary, tertiary, and quaternary structure, as well as the occurrence of catalytic efficiency along the folding–association pathway, thereby enabling the structure–function relation in the process of structure formation to be investigated. Only rarely have structured monomers been obtained as stable entities; for example, at high dilution, low temperature, and elevated hydrostatic pressure, occasionally dissociation yields subunits that still exhibit native-like properties. As kinetic intermediates during reconstitution, collapsed or structured monomers, or both, have been frequently identified and analyzed by spectral techniques; hence, at least some global characteristics have become available (for examples, see [86–90]).

Which of the consecutive steps in Eq. (6) determines the overall kinetics depends on the protein. If association is rate-limiting, this will hold only up to the concentration limit, at which the folding of structured monomers becomes rate-limiting. If association is diffusion-controlled, reconstitution will simply follow first-order kinetics; the same holds if association is followed by slow-reshuffling steps. Figure 6 illustrates the limiting cases.

In the following, representative examples will be selected to illustrate the range of variability relative to the mechanisms of reconstitution. All belong to the family of NAD-dependent dehydrogenases; hence, they are structurally and functionally related, differing mainly in their state of association. For a detailed discussion of other assembly systems, from dimers to multimers, see [1,2].

A. Monomer: Octopine Dehydrogenase

Octopine dehydrogenase has been discussed in connection with domain recognition and may serve as a monomeric reference system. As expected for a single-chain protein, its reactivation obeys simple first-order kinetics, with fast formation of the backbone structure preceding the rate-determining acquisition of the native three-dimensional structure [76,77] (see Figs. 4 and 6A).

B. Dimers: Malate Dehydrogenase and Bacterial Luciferase

The MDHs are stable dimers with association constants higher than 10^9 L mol^{-1}. According to the sequential mechanism

$$2\mathcal{M} \xrightarrow{\text{fast}} 2\text{M}' \xrightarrow{k_1} 2\text{M} \xrightarrow{k_2} \text{M}_2^* \xrightarrow{k_1'} \text{M}_2 \tag{7}$$

their reconstitution should consist of one association step ($2\text{M} \rightarrow \text{M}_2^*$), apart from

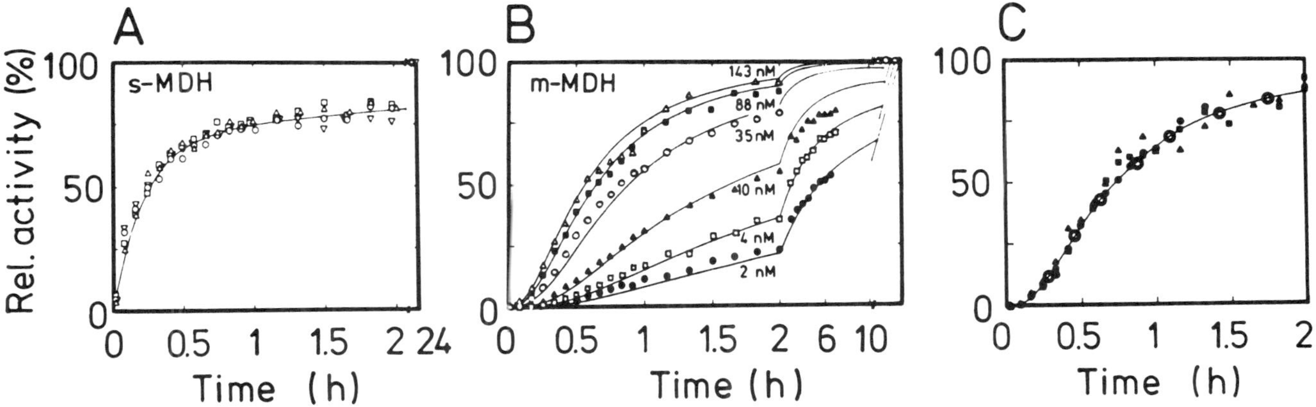

Figure 6 Reconstitution kinetics of cytoplasmic and mitochondrial porcine malate dehydrogenase: (A) s-MDH: Denaturation in 6 M GdmCl; renaturation in 0.2 M phosphate buffer pH 7.6 at 1.3 (○), 3.7 (△), 6.4(□), and 12.8 (▽) μg ml^{-1}. (B) m-MDH: Denaturation at pH 2.3 or in 6 M GdmCl; renaturation as in A, at 0.07 (●), 0.14 (□), 0.35 (△), 1.2 (○), 3.1 (■), and 5.0 (△) μg ml^{-1}. Curves calculated according to Eq. (6) (cf. Table 2). (C) m-MDH: Reactivation at 2.1 μg ml^{-1} after denaturation in 1 M gly/H_3PO_4 pH 2.3 (●), 6 M GdmCl (■), 8 M urea (▲), and in the presence of 10 mM NAD^+ (≥ 95% saturation) (*o*). Curve calculated according to Eq. (6) [92].

two possible folding reactions at the monomer or dimer level ($M' \rightarrow M$ and $M_2^* \rightarrow M_2$). Porcine cytosolic and mitochondrial MDH (s-MDH and m-MDH) are examples for both mechanisms: s-MDH follows a unimolecular multiple pathway mechanism with both structured monomers and nonnative dimers as intermediates [91], quite different from m-MDH, which is a prototype of sequential unibimolecular kinetics [92] (see Fig. 6). In the presence of structure-making salts, the folding rate is enhanced so that the sigmoidicity of the reactivation profiles vanishes. On the other hand, different modes of denaturation do not affect the kinetics, indicating that the collapsed state (M′) is a common structured intermediate on the pathway of reconstitution. In both isoenzymes, the folding kinetics are multiphasic and inaccessible to a detailed analysis in terms of specific residues or structural elements involved in the single kinetic steps.

Another model system that has been studied in great detail relative to its folding pathway, both in vitro and in vivo, is bacterial luciferase. The enzyme from *Vibrio harveyi* is a heterodimer composed of homologous polypeptide chains, which contains a single active center on the α-subunit, whereas the β-subunit is responsible for the high quantum yield of bioluminescence [93]. Urea denaturation of the enzyme, and subsequent refolding on dilution, suggested the existence of multiple intermediates leading to the native state by an inactive heterodimeric species [corresponding to M_2* in Eq. (7); 94,95]. Investigating the folding of the separate α- and β-chains showed that the lag in formation of the active αβ-dimer is caused by slow-folding steps of the individual subunits; these differ in their folding rates. Because separate folding of α and β leads to a decrease in reactivation (i.e., αβ-dimer formation), one may assume that the faster-folding β-chain provides a "matrix" for the slower-folding α-chain, this way trapping α and shifting the equilibrium toward the native state. Thus, it appears that in bacterial luciferase, kinetic competition of folding and association constitutes a "trap" on the folding pathway, which guides the individual subunits to the global energy minimum of the active dimer [96,97].

C. Tetramers: Lactate Dehydrogenase

Folding studies on tetrameric porcine LDH isoenzymes from heart and muscle have been focusing mainly on three questions: Is dimerization of the monomer or dimer rate-limiting in the assembly? Do intermediates of association accumulate during reconstitution? Are the intermediates catalytically active? The following kinetic scheme summarizes the essential results:

$$\underset{\text{inactive}}{4\mathcal{M}} \xrightarrow{\text{fast}} \underset{\text{inactive}}{4M'} \xrightarrow{k_1} \underset{\text{inactive}}{4M} \xrightarrow{\text{fast}} \underset{\text{inactive}}{2M_2} \xrightarrow{k_2} \underset{\text{active}}{M_4} \tag{8}$$

As shown by cross-linking experiments, dimer formation ($2M \rightarrow M_2$) is predominant in the early phase of reconstitution. The corresponding decrease in monomers follows first-order kinetics, with a rate constant in agreement with the simulation of the time- and concentration-dependent reactivation profiles. Thus, both the monomer and the dimeric intermediate are catalytically inactive; the association of the structured monomer (M) to the dimeric intermediate (M_2) is fast, and dimerization of the dimer ($2M_2 \rightarrow M_4$) parallels reactivation [33,34].

D. Hexamer: UDP-Glucose Dehydrogenase

Uridine diphosphate glucose dehydrogenase (UDP-GlcDH), an NADP-dependent 306-kDa homohexamer, is a trimer of dimers that form in a sequential uni-bimolecular reaction according to

$$6\mathcal{M} \xrightarrow{\text{fast}} 6M' \xrightarrow{k_1} 6M \xrightarrow{k_2} 3M_2' \xrightarrow{\text{fast}} \rightarrow M_6 \tag{9}$$

As taken from cross-linking experiments, the decrease in monomer concentration during reconstitution deviates markedly from normal exponential behavior; hence, slow-folding must be involved as a rate-determining step before dimerization takes place. The fast collapse in forming the native-like secondary structure precedes the transconformation at the monomer level. As in the other example, the monomer is inactive; the hexamer is the only active species [98].

E. Multimer: Pyruvate Dehydrogenase

Pyruvate dehydrogenase (PDH) from *Bacillus stearothermophilus,* a giant multienzyme complex of 10^7-Da molecular mass (consisting of four different polypeptide chains) can be partially reconstituted after complete denaturation and dissociation [99]. Applying stoichiometric amounts, the joint reactivation obeys first-order kinetics; for the rate limiting step, the pyruvate decarboxylase moiety (E1α,E1β) seems to be essential: neither the fast folding of the lipoamide dehydrogenase component (E3) nor the addition of the lipoate acyltransferase core (E2) enhances the reactivation reaction. Quaternary structure formation needs structured monomers to allow specific subunit assembly. As no significant concentration dependence of the reactivation process is detectable, the overall reaction must contain isomerization reactions at the monomer and multimer level; the association reaction itself seems to be fast compared with the final shuffling reaction.

Summarizing the given examples, it turns out that the sequential uni-bimolecular mechanism allows one to quantify the kinetics of reconstitution in a satisfactory manner. Usually, the assumption that the collapsed and structured monomer, or other intermediates on the folding pathway, do not exhibit catalytic activity is supported by the fit of the kinetic mechanism with the experimental data. As taken from the kinetic constants (Table 2), the various systems differ in their in vitro folding–association properties. The differences are smoothed in the table, because different temperatures were used for the different enzymes to

Table 2 Reconstitution Kinetics of Oligomeric Enzymes According to the Sequential Unibimolecular Folding–Association Mechanism[a]

$$nM' \xrightarrow{k_1} nM \xrightarrow{k_2} M_n$$

Enzyme	n	Denaturant	Temp. (°C)	$10^3 \cdot k_1$ (s^{-1})	$10^{-4} \cdot k_2$ ($M^{-1} \cdot s^{-1}$)	Comments
LADH	2	6 M G, pH 7.6	25	7.0	0.16	cf. [114]
YADH	4	6 M G, pH 7.6	25	Fast	0.04	cf. [114]
Aldolase	4	pH 2.3	0	1.6	0.18	Monomer; 50% active
Y-GAPDH	4	6 M G, pH 2.3	15	11.0	4.5	
		0°C, 5 mM ATP	25	9.0	7.0	Monomer binds NAD^+
LDH-H_4	4	pH 2.3	20	1.5	0.5	
		6 M G, pH 2.3	20	1.5	0.5	
		1.2 kbar pH 7	20	1.5	0.35	cf. [113]
LDH-M_4	4	pH 2.3	20	Fast	2.3	
		6 M G, pH 6.0	20	0.8	3.0	
		1.4 kbar pH 7	20	Fast	0.5	cf. [113]
s-MDH	2	6 M G, pH 7.6	20	1.3 0.07	> 100	
m-MDH	2	pH 2.3	20	0.65	3.0	cf. [92]
		6 M G, pH 7.6	20	0.65	3.0	cf. [92]
PDH	$(ABC_2)_{60}$	pH 2.3	20–50	0.02	Fast	cf. [94]
TIM	2	7 M G, pH 7.5	0	19	30	$^c TIM^{<10 nm}$
TSase	β_2	5 M G, pH 2.3	12	≈ 0.6	Fast	[1,86,87]
		2 kbar pH 7	10	0.65	> 40	cf. [88]
UDP-GlcDH	6	4 M G, pH 7.8	20	0.45	1.6	cf. [93]

[a]Abbreviations: LADH, YADH, horse liver and yeast alcohol dehydrogenase; Y-GAPDH, yeast glyceraldehyde-3-phosphate dehydrogenase; LDH-H_4, LDH-M_4, porcine heart and muscle lactate dehydrogenase; s-MDH, m-MDH, porcine cytosolic and mitochondrial malate dehydrogenase; PDH, pyruvate dehydrogenase multienzyme complex from *Bacillus stearothermophilus*; TIM, rabbit muscle triose phosphate isomerase; TSase, tryptophan synthase from *E. coli*; UDP-GlcDH, UDP-glucose dehydrogenase from *E. coli*; G, guanidinium chloride (molar concentration); k_1, k_2, first- and second-order rate constants.

optimize the kinetic analysis. The activation energies are always high; thus, the rates differ over several orders of magnitude. Whether this also holds true under conditions in vivo remains to be shown.

VII. ASSOCIATION VERSUS AGGREGATION

If we consider the concentration dependence of m-MDH in Figure 6B, it is obvious that, even for the unibimolecular process, at sufficiently high protein concentration, folding dominates the overall reconstitution kinetics. However, the presentation of the data in terms of *relative reactivation* may be deceptive, because at high concentration, not only a change in the kinetic mechanism, but also a decrease in the yield of active protein may be observed. This is caused by the formation of aggregates (A) competing kinetically with the foregoing sequential reaction [see Eq. (6)], according to

$$\begin{array}{ccccccccc} n\mathcal{M} & \rightarrow & nM' & \rightarrow & nM'' & \overset{k_1}{\rightarrow} & nM & \overset{k_2}{\rightarrow} & M_n \\ \downarrow & & \downarrow & & \downarrow & & & & \\ A & & A & & A & & & & \end{array} \tag{10}$$

The side reaction is at least second-order, so that, at high protein concentration, it will be favored owing to "kinetic partitioning" (Fig. 7). The molecular basis of kinetic trapping is twofold: (a) During the first steps along the folding path, the refolding protein exposes hydrophobic residues that are hidden in the interior of

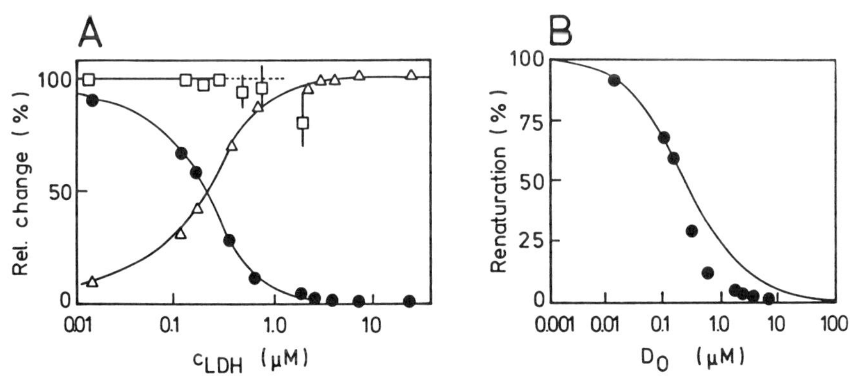

Figure 7 Kinetic competition of reconstitution and aggregation: Porcine muscle lactate dehydrogenase after denaturation in 0.1 M H_3PO_4, pH 2.0, and renaturation in phosphate buffer, pH 7.0, 10°C. (A) Partitioning between reactivation (●) and aggregation (△). The far-UV circular dichroism θ_{222nm} (□) reflects the helicity in the transition range [108]. (B) Concentration dependence of the yield of reactivation (data from A), calculated based on Eq. (11) (From Ref. 100.)

the molecule at a later stage; (b) the folding polypeptide chain does not differentiate between intra- and intermolecular interactions. If these explanations are correct, and if the side reaction is indeed a *kinetic* phenomenon, it should be possible to fit the profile of the concentration-dependent aggregation using the simple partitioning equation

$$\mathrm{N} \overset{k_1}{\leftarrow} \mathrm{U} \overset{k_2}{\rightarrow} \mathrm{A} \tag{11}$$

with k_1 and k_2 as rate constants of the rate-determining folding reaction and diffusion-controlled aggregation, respectively. As shown in Figure 7B, the calculated curve actually fits the experimental data. Another argument that supports the view that the native protein is in a lower-energy minimum than the aggregates, and that aggregation and inclusion body formation are determined by kinetic competition, comes from the possibility to shuffle aggregates back to the native state by destabilization in the presence of moderate denaturant concentrations. What the physical explanation is for the latter optimization, especially for the effect of arginine, is still enigmatic [100,101]. The same holds for the mechanisms underlying the "helper function" of accessory proteins in the cell. That kinetic partitioning occurs in vivo has long been known [102–105]. In prokaryotes, it leads to *inclusion body formation,* in eukaryotes to granules in the endoplasmic reticulum (ER). In connection with protein targeting in cells of higher organisms, it seems that oligomers are generally rapidly and selectively transported, whereas unassembled subunits and misassembled proteins (aggregates) are retained. Those not passing the "quality control" in the ER include proteins with deletions, insertions, point mutations, and defective glycosylation, altogether typical alterations also responsible for aggregation in vitro [106].

In dealing with aggregation as the most significant side reaction along the sequential folding–association pathway, three questions need to be answered: (a) What is the committed step in aggregate formation? (b) When is the structured monomer committed to form the native protein? and (c) What is the structure of aggregates and their constituent polypeptide chains? For the first two questions, commitment to aggregation is a fast reaction preceding the commitment to renaturation; this parallels the overall kinetics of reactivation, to the effect that the fast collapse still allows aggregation. After a certain intermediate state has been reached, slow shuffling leads exclusively to the native state [107]. For the structure of aggregates, electron microscopy and circular dichroism have shown that wrong subunit interactions give rise to irregular networks with a broad distribution of highly structured particles at least ten times the size of the native protein. They resemble the native protein in its spectral properties, as far as turbidity allows a quantitative analysis (see Fig. 7A) [108]. Inclusion bodies commonly represent surprisingly homogeneous precipitates, often with less than a dozen components (including RNA) (for their structural characteristics, see [105,109,110]).

VIII. EXTRINSIC FACTORS INVOLVED IN PROTEIN FOLDING

Proteins form their three-dimensional structure based solely on the translational relation between gene sequence and protein sequence. In the foregoing discussion, most folding and association experiments were performed far from physiological conditions, to accomplish high yields of reconstitution and sufficiently slow kinetics: for example, protein concentrations and temperature were anomalously low, specific ligands or accessory components were lacking, and no attempts were made to mimic crowding. As has been shown by systematic variations of the relevant physical parameters in folding experiments, these factors usually seem to affect the kinetics, rather than the nature, of the final product. This holds not only for the cytosolic solvent conditions, but also for accessory proteins such as folding catalysts and molecular chaperones.

A. Solvent Conditions

The physical parameters of the cytosol in common (mesophilic) cells vary only in a narrow range, compatible with the conditions used in in vitro "renativation" experiments. As shown by the successful expression of active recombinant proteins from extremophiles in mesophilic hosts even extreme deviations from the norm do not have drastic effects [50,81,111]. This holds, at least for temperature and pH, for the latter because, even in extreme acidophiles and alkalophiles, the cytosolic pH is close to neutrality [47]. Little is known about other solvent parameters. For halophilic proteins, salt is required not only as stabilizing agent, but also for folding and assembly [112–114]. With respect to hydrostatic pressure, no data on proteins from barophiles are yet available; evidently, the adaptive effort to cope with abyssal conditions had to focus mainly on the low in situ temperatures; effects of compressibility on the assembly of proteins seem to be of minor importance [115].

Given the high cellular concentrations of biopolymers, the cytosol is highly viscous. In stimulating the frictional characteristics in vitro by adding glycerol or carbohydrates, the interpretation is rendered difficult because polyols not only increase the viscosity, but also stabilize proteins. Thus, the reduced folding rate in viscous solvents (see Fig. 4C) may be explained either by tightening the structure or by frictional effects on domains or subunits according to Stokes–Einstein's equation. That different proteins differ in their response seems to indicate that the hydrodynamic effect is predominant [2,77].

B. Conjugation and Ligand Binding

Glycoproteins are worth mentioning in connection with the effect of polyols on protein folding. To be effective, a sufficiently high degree of glycosylation is

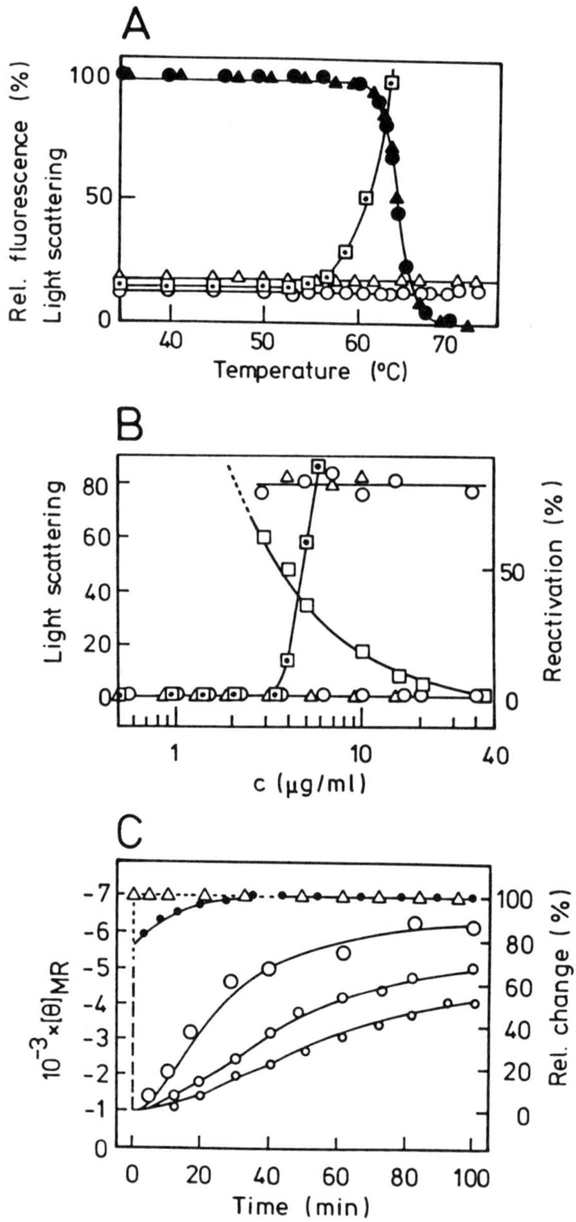

Figure 8 Effect of *O*-glycosylation on the stability and folding of invertase. Denaturation–renaturation of internal, core-glycosylated, and external invertase from *Saccharomyces cerevisiae,* with 0, 34, and 65% carbohydrate content. (A) Thermally induced unfolding

required. In RNaseB (the 6% mannosylated form of RNaseA), the two forms of the enzyme do not differ either in stability or in the kinetics and mechanism of folding [116]. In the invertase from yeast, the "internal," "core-glycosylated," and "external" mutant forms of the enzyme (with 0, 34, and 65% carbohydrate) differ significantly in stability, quaternary structure, and the tendency to aggregate. The glycomoiety as a stabilizing and solubilizing component inhibits aggregation (e.g., by heat). However, again, there is no effect on the kinetics and mechanism of folding (Fig. 8). In the nonglycosylated, internal form of the enzyme, the lack of the covalently bound carbohydrate can be substituted by high concentrations of polyols; on the other hand, glycosylation mimics chaperon action [89,90,117].

Ligands, such as cofactors, substrates, or specific ions, may stabilize the native structure of proteins. A physiologically relevant example is the stabilization of lactate dehydrogenase against high-pressure dissociation in the presence of phosphate ions [118]. Because certain enzymes require metal ions or other ligands for the attainment of their functional state, the question arises whether and at which stage ligands may affect their folding and association. In principle, effectors may enhance the kinetics of folding and association either by lowering the activation energy of rate-limiting steps, or by shifting conformational equilibria toward the native state. For complex ligands, such as NAD, the effect on folding refers mainly to the *yield* of reconstitution; the *rate* is affected to a smaller extent because coenzyme binding (e.g., to MDH), requires practically the final product of the folding process (see Fig. 6C). The binding of Zn^{+2} to horse liver alcohol dehydrogenase (LADH) may serve as an example illustrating that ligands may enter the folding–association reaction at various levels. Because Zn^{2+} ions in LADH participate in catalysis and stabilization of the dimeric structure, reconstitution in the absence of Zn^{2+} yields neither recovery of enzyme activity nor proper assembly. Instead, partial refolding (easily detectable as a first-order recovery of native fluorescence), takes place. Competing aggregation is quenched immediately after Zn^{2+} is added; at the same time, catalytic activity is recovered in a second-order reaction, which indicates that subunit assembly is required for

---

of the internal (⊡), core-glycosylated (○, ●), and external enzyme (△, ▲) at 10 μg ml^{-1} in 50 mM acetate, pH 5.0. Open symbols: light scattering at 500 nm (arbitrary units); closed symbols: relative fluorescence at 335 nm (λ_{exc} = 280 nm) (%). (B) Effect of protein concentration on the reactivation for internal (□) and glycosylated enzyme (upper plateau at ~ 80%) and aggregation at 20°C, symbols as in A. (C) Reconstitution kinetics of core-glycosylated invertase at 20°C, followed by θ_{222nm} (△), fluorescence (•), and enzymatic activity at 8.5 (○), 17 (○), and 68 (○) nM subunit concentration. Solid lines calculated based on Eq. (6) with $k_1 = 0.16 s^{-1}$ and $k_2 = 1.2 \times 10^4\ M^{-1}s^{-1}$. (From Refs. 89, 90.)

catalytic activity. Folding and association may be analyzed independently by adding Zn^{2+} after completion of the slow-folding steps [119].

IX. OPEN QUESTIONS

In recent years, the elucidation of the self-organization of single-chain proteins has reached a level that allows elementary steps in the millisecond range and below to be explained in terms of well-defined local events. In contrast, the analysis of oligomeric proteins is still in its infancy, mainly because of the size and complexity of the systems, and because of the superposition of conformational effects at all levels of the hierarchy of protein structure. The energetic reasons for the difficulties in predicting protein structure have been mentioned; they still leave us with the notorious *give me the answer, and I shall predict it.* Similarly, insight into the self-organization in vivo are still rudimentary. What has been resolved is that the resulting three-dimensional structure of proteins is the same in vitro and in the cell. However, the question of whether the folding–association of the nascent polypeptide chain leaving the ribosome proceeds by the same pathway as the one observed in vitro, is still unresolved. Another problem that is still awaiting a solution is the question how the rates of translation and folding are tuned within the cell. Because molecular chaperones enter the game during translation, one would suspect that in vitro folding of large multi-chain proteins is preferably a biotechnological and not a biological problem. Again, given the different families of chaperones and their concentrations in different organisms and stress conditions [120,121], no definite answer is available. In considerations dealing with the concentrations of accessory proteins and their substrates, interacting groups along the polypeptide chain may reach extremely high "local concentrations" so that even 55 M water becomes negligible [122]. As a consequence, studies devised to simulate local structures in protein folding have to consider this unusual situation, apart from problems of crowding and compartmentation.

In the evolution of oligomeric proteins, several examples seem to confirm the switch from intra- to intermolecular interactions making use of complementary domain surfaces [4,5]. In expanding this attractive idea, the origin of heterologous systems remains open. Finally, defining the folding problem for a given oligomeric protein, such as an allosteric enzyme or a multienzyme complex, as the sequence of events, starting from the formation of local next-neighbor interactions at the primary structural level, and ending with the final assembly of a 100- to 1000-kDa particle, it becomes clear how far we have to go to give the final answer. It might be consoling to remember the analogy of the self-organization of biomolecules to simple physical processes, such as the crystallization of ice: The problem has been around since Johannes Kepler published his treatise *Strena seu de nive sexangula* in 1611; it still awaits its exact solution.

REFERENCES

1. Jaenicke, R., Folding and association of proteins, *Prog. Biophys. Mol. Biol. 49*:117 (1987).
2. Garel, J.-R., Folding of large proteins: Multidomain and multisubunit proteins, *Protein Folding* (T. E. Creighton, ed.), W. H. Freeman, New York, 1992, p. 405.
3. Jaenicke, R., Protein folding and association: In vitro studies for self-organization and targeting in the cell, *Curr. Top. Cell. Regul. 34*:209 (1996).
4. Bennett, M. J., Choe, S., and Eisenberg, D., Domain swapping, *Proc. Natl. Acad. Sci. USA 91*:3127 (1995).
5. Bennett, M. J., Schlunegger, M. P., and Eisenberg, D., 3D domain swapping: A mechanism for oligomer assembly, *Protein Sci. 4*:2455 (1995).
6. Janin, J. and Wodak, S. J., Structural domains in proteins and their role in the dynamics of protein function, *Prog. Biophys. Mol. Biol. 42*:21 (1983).
7. Goldberg, M. E., Tertiary structure of *E. coli* β-D-galactosidase, *J. Mol. Biol. 46*:441 (1969).
8. Porter, R. R., Structural studies of immunoglobulins, *Science 180*:713 (1973).
9. Wetlaufer, D. B., Nucleation, rapid folding and globular intrachain regions in proteins, *Proc. Natl. Acad. Sci. USA 70*:697 (1973).
10. Wetlaufer, D. B., Folding of protein fragments, *Adv. Protein Chem. 34*:61 (1981).
11. Tanford, C., Protein denaturation, *Adv. Protein Chem. 23*:121 (1968).
12. Tanford, C., Protein denaturation, *Adv. Protein Chem. 24*:1 (1970).
13. Privalov, P. L., Stability of proteins, *Adv. Protein Chem. 33*:167 (1979).
14. Bajaj, M. and Blundell, T. L., Evolution and the tertiary structure of proteins, *Annu. Rev. Biophys. Bioeng. 13*:453 (1984).
15. Chothia, C. and Lesk, A., The relation between the divergence of sequence and structure in proteins, *EMBO J. 5*:823 (1986).
16. Gilbert, W., Genes-in-pieces revisited, *Science 228*:823 (1985).
17. Stoltzfus, A., Spencer, D. F., Zucker, M., Logsdon, J. M., Jr., and Doolittle, W. F., Testing the exon theory of genes: The evidence from protein structure, *Science 265*:202 (1994).
18. Rossmann M. G. and Argos, P., Protein folding, *Annu. Rev. Biochem. 50*:497 (1981).
19. Levinthal, C., Are there pathways for protein folding? *J. Chem. Phys. 65*:44 (1968).
20. Wetlaufer, D. B. and Ristow, S., Acquisition of three-dimensional structure of proteins, *Annu. Rev. Biochem. 42*:135 (1973).
21. Jaenicke, R., Folding and association vs misfolding and aggregation of proteins, *Philos. Trans. R. Soc. (Lond.) 348*:97 (1995).
22. Jaenicke, R., Eye-lens proteins: Structure, superstructure, stability and genetics, *Naturwissenschaften 81*:423 (1994).
23. Groenen, P. J. T. A., Merck, K. B., de Jong, W. W., and Bloemendal, H., Structure and modifications of the junior chaperone α-crystallin, *Eur. J. Biochem. 225*:1 (1994).
24. Wistow, G. J. and Piatigorsky, J., Lens crystallins: The evolution and expression of a highly specialized tissue, *Annu. Rev. Biochem. 57*:479 (1988).
25. Bax, B., Lapatto, R., Nalini, V., Driessen, H., Lindley, P. F., Mahadevan, M.,

Blundell, T. L., and Slingsby, C., X-ray analysis of βB2-crystallin and evolution of oligomeric lens proteins, *Nature 347*:776 (1990).

26. Trinkl, S., Glockshuber, R., and Jaenicke, R., Dimerization of βB2-crystallin: The role of the linker peptide and the N- and C-terminal extensions, *Protein Sci. 3*:1392 (1994).
27. Mayr, E.-M., Domänen and Domänen-Wechselwirkungen in γ-Kristallinen, Dissertation, University of Regensburg (1995).
28. Murray, A. J., Lewis, S. J., Barcley, A. N., and Brady, R. L., One sequence, two folds: A metastable structure of CD-2, *Proc. Natl. Acad. Sci. USA 92*:7337 (1995).
29. Jaenicke, R., Is there a code for protein folding? *Protein Structure and Protein Engineering* (R. Huber and E. L. Winnacker, eds.), Springer Verlag, Berlin, 1989, p. 16.
30. Hermann, R., Rudolph, R., and Jaenicke, R., Kinetics of in vitro reconstitution of oligomeric enzymes by cross-linking, *Nature 277*:243 (1979).
31. Jaenicke, R. and Rudolph, R., Folding and association of oligomeric proteins, *Methods Enzymol. 131*:218 (1986).
32. Rudolph, R., Böhm, G., Lilie, H., and Jaenicke, R., Folding proteins, *Protein Structure: A Practical Approach* (T. E., Creighton, ed.) IRL Press, Oxford, (in press).
33. Hermann, R., Jaenicke, R., and Rudolph, R., Analysis of the reconstitution of oligomeric enzymes by cross-linking with glutaraldehyde: Kinetics of reassociation of LDH, *Biochemistry 20*:5195 (1981).
34. Hermann, R., Rudolph, R., Jaenicke, R., Price, N. C., and Scobbie, A., The reconstitution of denatured phosphoglycerate mutase, *J. Biol. Chem. 258*:11014 (1983).
35. Kauzmann, W., Some factors in the interpretation of protein denaturation, *Adv. Protein Chem. 14*:1 (1959).
36. Privalov, P. L., Stability of proteins which do not present a single cooperative unit, *Adv. Protein Chem. 35*:1 (1982).
37. Privalov, P. L. and Gill, S. J., Stability of protein structure and hydrophobic interaction, *Adv. Protein Chem. 39*:191 (1988).
38. Makhatadze, G. I. and Privalov, P. L., Energetics of protein structure, *Adv. Protein Chem. 47*:308 (1995).
39. Alber, T., Stabilization energies of protein conformation, *Prediction of Protein Structure and the Principles of Protein Conformation* (G. D. Fasman, ed.), Plenum Press, New York, 1989, p. 161.
40. Matthews, B. W., Studies on protein stability with T4 lysozyme, *Adv. Protein Chem. 46*:249 (1995).
41. Freire, E. and Biltonen, R. L., Statistical mechanical deconvolution of thermal transitions in macromolecules, *Biopolymers 17*:468 (1978).
42. Frauenfelder, H., Sligar, S. G., and Wolynes, P. G., The energy landscapes and motions of proteins, *Science 254*:1598 (1991).
43. Baldwin, R. L., The nature of protein folding pathways: The classical versus the new view, *J. Biomol. NMR 5*:103 (1995).
44. Creighton, T. E., Darby, N. J., and Kemmink, J., The roles of partly folded intermediates in protein folding, *FASEB J. 10*:110 (1996).

45. Dobson, C. and Fersht, A. R., eds. Protein folding, *Philos. Trans. R. Soc. Lond. 348*:1 (1995).
46. Sali, A., Shakhnovich, E. I., and Kaplan, M., Kinetics of protein folding: A lattice model study of the requirements for folding to the native state, *J. Mol. Biol. 235*:1614 (1994).
47. Jaenicke, R., Protein stability and molecular adaptation to extreme conditions, *Eur. J. Biochem. 202*:715 (1991).
48. Wetlaufer, D. B., Practical consequences of protein folding mechanisms, *Protein Folding* (R. Jaenicke, ed.), Elsevier/North-Holland, Amsterdam, 1980, p. 323.
49. Zuber, H., Temperature adaptation of LDH: Structural, functional and genetic aspects, *Biophys. Chem. 29*:171 (1988).
50. Jaenicke, R., Schurig, H., Beaucamp, N., and Ostendorp, R., Structure and stability of "hyperstable proteins": Glycolytic enzymes from the hyperthermophilic bacterium *Thermotoga maritima*, *Adv. Protein Chem. 48*:181 (1996).
51. Komar, A. A. and Jaenicke, R., Kinetics of translation of γ-crystallin and its circularly permuted variant in an in vitro cell-free system: Possible relations to codon distribution and protein folding, *FEBS Lett. 376*:195 (1995).
52. Thanaraj, T. A. and Argors, P., Ribosome-mediated translational pause and domain organization, *Protein Sci. 5*:1594 (1996).
53. Lauffer, M. A., *Entropy-Driven Processes in Biology*, Springer Verlag, Berlin, 1975.
54. Jaenicke, R., Protein structure and function at low temperature, *Philos. Trans. R. Soc. Lond. B 316*:535 (1990).
55. Franks, F., Protein destabilization at low temperatures, *Adv. Protein Chem. 46*:105 (1995).
56. Jaenicke, R., Protein folding: Local structures, domains, subunits and assemblies, *Biochemistry 30*:3147 (1991).
57. Pfeil, W., Unfolding of proteins, *Thermodynamic Data for Biochemistry and Biotechnology* (H.-J. Hinz, ed.), Springer Verlag, Berlin, 1986, p. 349.
58. Girg, R., Jaenicke, R., and Rudolph, R., Dimers of porcine skeletal muscle LDH produced by limited proteolysis during reassociation are enzymatically active in the presence of stabilizing salt, *Biochem. Int. 7*:433 (1983).
59. Opitz, U., Rudolph, R., Jaenicke, R., Ericsson, L., and Neurath, H., Proteolytic dimers of porcine muscle LDH: Characterization, folding and reconstitution of the truncated and nicked polypeptide chain, *Biochemistry 26*:1399 (1987).
60. Bernal, J. D., Structure arrangements of macromolecules, *Discuss. Faraday Soc. 25*:7 (1958).
61. Burley, S. K. and Petsko, G. A., Weakly polar interactions in proteins, *Adv. Protein Chem. 39*:125 (1988).
62. Dill, K. A., Dominant forces in protein folding, *Biochemistry 29*:7133 (1990).
63. Pace, C. N., Heinemann, U., Hahn, U., and Saenger, W., Ribonuclease T1: Structure, function, and stability, *Angew. Chem. Int. Ed. Engl. 30*:343 (1991).
64. Hvidt, A., Interactions of water with nonpolar solutes, *Annu. Rev. Biophys. Bioeng. 12*:1 (1983).
65. Shortle, D., Staphylococcal nuclease: A showcase of m-value effects, *Adv. Protein Chem. 46*:217 (1995).
66. Hecht, M. H., Nelson, H. C. M., and Sauer, R. T., Mutations in λ-repressor's

N-terminal domain: Implications for protein stability and DNA binding, *Proc. Natl. Acad. Sci. USA 80*:2676 (1983).
67. Pakula, A. A. and Sauer, R. T., Genetic analysis of protein stability and function, *Annu. Rev. Genet. 23*:289 (1986).
68. Hampsey, D. M., Das, G., and Sherman, F., Yeast iso-1-Cyt *c*: Genetic analysis of structural requirements, *FEBS Lett. 231*:275 (1988).
69. Coplen, L. J., Frieden, R. W., and Goldenberg, D. P., A genetic screen to identify variants of BPTI with altered folding energetics, *Proteins Struct. Funct. Genet. 7*:16 (1990).
70. Goldenberg, D. P., Mutational analysis of protein folding and stability, *Protein Folding* (T. E., Creighton, ed.) W. H. Freeman, New York, 1992, p. 353.
71. Kato, I. and Anfinsen, C. B., On the stabilization of RNase S-protein by RNase S-peptide, *J. Biol. Chem. 244*:1004 (1969).
72. Finn, F. M. and Hofmann, K., The synthesis of peptides by solution methods with emphasis on peptide hormones, *The Proteins*, Vol. 2, 3rd ed. (H. Neurath, R. L. Hill, and C.-L. Boeder, eds.), Academic Press, New York, 1976, p. 106.
73. Labhardt, A. M. and Baldwin, R. L., Recombination of S-peptide with S-protein during folding of RNase S, *J. Mol. Biol. 135*:231 (1979).
74. Garel, J.-R. and Baldwin, R. L., Both the fast and slow refolding reactions of RNaseA yield native enzyme, *Proc. Natl. Acad. Sci. USA 70*:3347 (1973).
75. Labhardt, A. M., Measurement of the refolding combination reaction between S-peptide and S-protein, *Biochemistry 22*:321 (1983).
76. Zettlmeissl, G., Teschner, W., Rudolph, R., Jaenicke, R., and Gäde, G., Isolation, physicochemical properties and folding of ODH from *Pecten jacobaeus*, *Eur. J. Biochem. 143*:401 (1984).
77. Teschner, W., Rudolph, R., and Garel, J.-R., Intermediates on the folding pathway of ODH from *Pecten jacobaeus*, *Biochemistry 26*:2791 (1987).
78. Abad-Zapatero, C., Griffith, J. P., Sussman, J. L., and Rossmann, M. G., Refined crystal structure of dog-fish M_4 apo-LDH, *J. Mol. Biol. 198*:445 (1987).
79. Jaenicke, R. and Koberstein, R., High pressure dissociation of LDH, *FEBS Lett. 17*:351 (1971).
80. Müller, K., Lüdemann, H.-D., and Jaenicke, R., Thermodynamics and mechanism of high pressure deactivation and dissociation of porcine LDH, *Biophys. Chem. 16*:1 (1982).
81. Ostendorp, R., Auerbach, G., and Jaenicke, R., Extremely thermostable L(+) LDH from *Thermotoga maritima*, *Protein Sci. 5*:862 (1996).
82. Girg, R., Rudolph, R., and Jaenicke, R., The dimeric intermediate on the pathway of reconstitution of LDH is enzymatically active, *FEBS Lett. 163*:132 (1983).
83. Mas, M. T., Chen, C. Y., Hitzeman, R. A., and Riggs, A. B., Active human–yeast chimeric PGKs engineered by domain interchange, *Science 233*:788 (1986).
84. Jaenicke, R., Rudolph, R., and Heider, I., Specificity in the subunit assembly of oligomeric enzymes: Synchronous reconstitution of mammalian LDH and MDH, *Biochem. Int. 2*:23 (1981).
85. Gerl, M., Rudolph, R., and Jaenicke, R., Mechanism and specificity of reconstitution of dimeric LDH from *Limulus polyphemus*, *Biol. Chem. Hoppe-Seyler 366*:447 (1985).

86. Lane, A. N. and Kirschner, K., The quaternary structure of TSase from *E. coli*, *Eur. J. Biochem. 129*:675 (1983).
87. Chaffotte, A. F., Cadieux, C., Guillou, Y., and Goldberg, M. E., A possible initial folding intermediate: The C-terminal proteolytic domain of TSase β-chains folds in less than 4 ms into a condensed state with non native-like secondary structure, *Biochemistry 31*:4303 (1992).
88. Seifert, T., Bartholmes, P., and Jaenicke, R., Influence of PLP on the reversible high pressure denaturation of the isolated β_2-dimer of the TSase bienyzme complex from *E. coli*, *Biochemistry 24*:339 (1985).
89. Kern, G., Schülke, N., Schmid, F. X., and Jaenicke, R., Quaternary structure and stability of internal, external and core-glycosylated invertase from yeast, *Protein Sci. 1*:120 (1992).
90. Kern, G., Kern, D., Jaenicke, R., and Seckler, R., Kinetics of folding and association of differently glycosylated variants of invertase from *S. cerevisiae*, *Protein Sci. 2*:1862 (1993).
91. Rudolph, R., Fuchs, I., and Jaenicke, R., Reassociation of dimeric c-MDH is determined by slow and very slow folding reactions, *Biochemistry 25*:1662 (1986).
92. Jaenicke, R., Rudolph, R., and Heider, I., Quaternary structure, subunit activity, and in vitro association of porcine m-MDH, *Biochemistry 18*:1217 (1979).
93. Baldwin, T. O. and Ziegler, M. M., Bacterial luciferase, *Chemistry and Biochemistry of Flavoenzymes*, Vol. 3 (F. Müller, ed.) CRC Press, Boca Raton, 1992, p. 467.
94. Ziegler, M. M., Goldberg, M. E., Chaffotte, A. F., and Baldwin, T. O., Refolding of luciferase subunits from urea and assembly of the active heterodimer, *J. Biol. Chem. 268*:10760 (1993).
95. Clark, A. C., Sinclair, J. F., and Baldwin, T. O., Folding of bacterial luciferase involves a non-native heterodimeric intermediate in equilibrium with the native enzyme and the unfolded subunits, *J. Biol. Chem. 268*:10773 (1993).
96. Baldwin, T. O., Ziegler, M. M., Chaffotte, A. F., and Goldberg, M. E., Contribution of folding steps involving the individual subunits of bacterial luciferase to the assembly of the heterodimeric enzyme, *J. Biol. Chem. 268*:10766 (1993).
97. Sinclair, J. F., Ziegler, M. M., and Baldwin, T. O., Kinetic partitioning during protein folding yields multiple native states, *Struct. Biol. 1*:320 (1994).
98. Jaenicke, R., Rudolph, R., and Feingold, D. S., The dissociation and in vitro reconstitution of bovine liver UDP-GlcDH, *Biochemistry 25*:7283 (1986).
99. Jaenicke R. and Perham, R. N., Reconstitution of the PDH multienzyme complex from *B. stear.*, *Biochemistry 21*:3378 (1982).
100. Rudolph, R., Renaturation of recombinant, disulfide-bonded proteins from "inclusion bodies," *Modern Methods in Protein and Nucleic Acid Research* (H. Tschesche, ed.), de Gruyter Verlag, Berlin, 1990, p. 149.
101. Buchner, J., Brinkmann, U., and Pastan, I., Renaturation of a single-chain immunotoxin facilitated by chaperones and protein disulfide isomerase, *Biotechnology 10*:682 (1992).
102. Schachtele, C. F. and Rogers, P., Mechanism of canavanine death in *E. coli*, *J. Mol. Biol. 34*:843 (1968).
103. Cheng, Y.-S. E., Increased cell buoyant densities of protein overproducing *E. coli* cells, *Biochem. Biophys. Res. Commun. 111*:104 (1983).

104. Haase-Pettingell, C. A. and King, J., Formation of aggregates from a thermolabile in vivo folding intermediate in P22 tailspike maturation, *J. Biol. Chem. 263*:4977 (1988).
105. Mitraki, A. and King, J., Protein folding intermediates and inclusion body formation, *Biotechnology 7*:690 (1989).
106. Hurtley, A. M. and Helenius, A., Protein oligomerization in the ER, *Annu. Rev. Cell. Biol. 5*:277 (1989).
107. Goldberg, M. E., Rudolph, R., and Jaenicke, R., A kinetic study of the competition between renaturation and aggregation during the refolding of denatured-reduced egg-white lysozyme, *Biochemistry 30*:2790 (1991).
108. Zettlmeissl, G., Rudolph, R., and Jaenicke, R., Reconstitution of LDH: Non-covalent aggregation vs reactivation, *Biochemistry 18*:5567 (1979).
109. Bowden, G. A., Paredes, A. M., and Georgiou, G., Structure and morphology of protein inclusion bodies in *E. coli*, *Biotechnology 9*:725 (1991).
110. Valax, P. and Georgiou, G., Characterization and refolding of β-lactamase inclusion bodies in *E. coli*, *Biocatalyst Design for Stability and Specificity* (M. E. Himmel and G. Georgiou, eds.), *ACS Symp. Ser. 516*:126 (1993).
111. Jaenicke, R., GAPDH from *Thermotoga maritima*: Strategies of protein stabilization, *FEMS Microbiol. Rev. 18*:215 (1996).
112. Hecht, K. and Jaenicke, R., MDH from the extreme halophilic archaebacterium *Hal. marismortui*: Reconstitution of the enzyme after denaturation and dissociation in various denaturants, *Biochemistry 28*:4979 (1989).
113. Cendrin, F., Chroboczek, J., Zacchai, G., Eisenberg, H., and Mevarech, M., Cloning, sequencing, and expression in *E. coli* of the gene for MDH from *Hal. marismortui*, *Biochemistry 32*:4308 (1993).
114. Dym, O., Maverech, M., and Sussman, J. L., Structural features that stabilize halophilic MDH from an archaebacterium, *Science 267*:1344 (1995).
115. Gross, M. and Jaenicke, R., Proteins under pressure: The influence of high hydrostatic pressure on structure, function and assembly of proteins and protein complexes, *Eur. J. Biochem. 221*:617 (1994).
116. Krebs, H., Schmid, F. X., and Jaenicke, R., Native-like folding intermediates of homologous RNases, *Biochemistry 24*:3846 (1985).
117. Kern, G., Schmidt, M., Buchner, J., and Jaenicke, R., Glycosylation inhibits the interaction of invertase with the chaperone GroEL, *FEBS Lett. 305*:203 (1992).
118. Schade, B., Lüdemann, H.-D., Rudolph, R., and Jaenicke, R., Reversible high pressure dissociation of LDH from porcine muscle, *Biochemistry 19*:1121 (1980).
119. Rudolph, R., Gerschitz, J., and Jaenicke, R., Effect of Zn^{2+} on the refolding and reactivation of L-ADH, *Eur. J. Biochem. 87*:601 (1978).
120. Lorimer, G. H., A quantitative assessment of the role of the chaperonin proteins in protein folding in vivo, *FASEB J. 10*:5 (1996).
121. Trent, J. D., Kagawa, H. K., Olle, E. W., and Zaluzec, N. J., Chaperonin filaments: The archaeal cytoskeleton, *Proc. Natl. Acad. Sci. USA* (in press).
122. Page, M. I. and Jencks, W. P., Entropic contribution to rate acceleration in enzymic and intramolecular reactions and the chelate effect, *Proc. Natl. Acad. Sci. USA 68*:1678 (1971).

3

Chaperone Interactions with Nascent and Newly Synthesized Proteins

William J. Welch, Daryl K. Eggers, William J. Hansen, and Hiroshi Nagata
University of California, San Francisco, San Francisco, California

I. INTRODUCTION

New insights into the mechanisms by which nascent polypeptides acquire their properly folded state have been realized over the course of the last few years. Although the primary amino acid sequence of a polypeptide provides all of the necessary information for the folding process, we now know that protein folding in vivo oftentimes requires additional components. Many of these components, now being referred to as *molecular chaperones*, participate in the folding event by reducing the possibility of the nascent polypeptide going "off-pathway" (i.e., misfolding) during the early stages of its synthesis and folding. Herein we review our current understanding of some of the earliest events in protein maturation. We begin by discussing some of the potential problems a nascent chain might encounter during the early stages of its synthesis and folding. Next we review what is currently known about cellular components that interact with *nascent polypeptides* (i.e., polypeptides still attached to the ribosome) and *newly synthesized polypeptides* (i.e., those just released from the ribosome), both in eukaryotes and prokaryotes. We discuss how different nascent chain-binding components might influence the pathway by which a newly synthesized protein adopts its final folded state. Finally, we review briefly why many molecular chaperones also are members of the heat-shock or stress protein family. Portions of this review article have been published in an article entitled "Early events in the synthesis and maturation of polypeptides." In: *Molecular Biology of Chaperones*, (B. Bukau, ed., Harwood Academic Publishers, Amsterdam).

II. IN VITRO VERSUS IN VIVO PROTEIN FOLDING

Much of our understanding about the process of protein folding has resulted from the use of Anfinsen-like protein-folding experiments [Anfinsen, 1973]. Here a purified protein is first denatured by its exposure to a protein chaotrope. Following subsequent dilution of the chaotrope, the protein undergoes spontaneous refolding back into its native conformation. Both the rate and the efficiency of the refolding reaction can vary significantly depending on the size (i.e., number of amino acids) and composition of the polypeptide being studied. Nevertheless, from these types of experiments some of the rate-limiting steps in the overall pathway of protein folding have been realized. Extrapolating the results of in vitro-folding experiments to understand protein folding inside the cell, however, has limitations. First, the folding of a newly synthesized protein in vivo has obvious restrictions because polypeptide synthesis or translocation into organelles occurs as a vectorial process. In an Anfinsen folding reaction, all of the information for folding (i.e., the entire amino acid sequence) is already present at the time the protein-folding event is initiated. In the cell, however, the synthesis of a polypeptide occurs as a sequence of vectorial steps; that is, synthesis proceeds from the amino(NH_2)- to the carboxy(COOH)-terminus of the polypeptide. Similarly, the translocation of nascent or newly synthesized proteins from their site of synthesis in the cytosol in the lumen of organelles (e.g., endoplasmic reticulum [ER] or mitochondria) is thought to occur in a linear-like fashion, with the polypeptide being in an extended-type conformation during the actual translocation event. If we consider that individual protein domains usually comprise at least 100 or more amino acids [Berman, 1994], one wonders what the nascent chain looks like when there are not enough amino acids polymerized (or translocated into the organelle) to permit a productive folding reaction. At the early stages of its synthesis, is the nascent chain undergoing a continuous series of folding transitions; or, alternatively, might it be stabilized by an interaction with one or more molecular chaperones?

Another problem not addressed by the Anfinsen model of protein folding concerns the issue of molecular crowding inside the cell. In vitro-folding reactions typically employ a single polypeptide species under relatively dilute conditions. After removal of the chaotrope, the protein is free to initiate folding without the nuances of encountering a different macromolecule. Even here, however, intermolecular interactions with different intermediates in the folding pathway could result in misfolding or aggregation. In the cell, where many different species of polypeptides are being synthesized (or being translocated into an organelle) at any one time, the possibility of a particular nascent chain inadvertently interacting with another nascent polypeptide, or with any other macromolecule in its local vicinity, seems quite high. Such interactions could conceivably lead to errors in the folding pathway and the generation of protein aggregates.

Finally, we do not know whether the kinetics of protein folding observed in vitro are at all relevant to the situation in vivo. Obviously, this is a difficult question to test. The rate of protein folding certainly can be faster than the rate of protein synthesis. It seems plausible, therefore, that the folding of a nascent chain can occur co-translationally. For example, in the multidomain proteins, folding of the most NH_2-terminal domain might commence once a suitable number of amino acids (e.g., 100–200 amino acids) have been polymerized and cleared out of the immediate vicinity of the ribosomal subunits. In the single-domain proteins, one suspects that the final folding of the polypeptide will likely occur only after release of the polypeptide from the ribosome. For both single- and multidomain proteins, however, the very early events of their biogenesis will likely be similar; both will require stabilization of their NH_2-terminus until enough amino acids have been polymerized to initiate a productive-folding reaction (e.g., see Figs. 2 and 3).

Despite these potential problems, the fidelity of protein-folding reactions, either in the cell or using in vitro translation systems, appears to be quite high. As we discuss later, the high efficiency of protein folding in vivo is probably due to a group of macromolecules, the primary task of which is to monitor the earliest stages of protein maturation. Some of these components appear dedicated to interacting with only nascent chains that are targeted to specific organelles, or are transported out of the cell. These dedicated factors recognize their targets by virtue of particular amino acid sequences present within the polypeptide chain. Examples of dedicated factors include the signal-recognition particle, a ribonucleoprotein complex involved in directing polysomes translating secretory or integral membrane proteins to the membrane of the endoplasmic reticulum; the related prokaryotic signal-recognition particle and the SecB machinery, both of which are involved in protein export in bacteria; different mitochondrial-targeting factors (e.g., mitochondrial import factor) that participate in the import of proteins from the cytosol into the mitochondria; and finally, a large family of proteins involved in the translocation of proteins into the nucleus. We do not discuss these types of "dedicated" nascent chain or newly synthesized binding factors here, but rather refer the reader to several other excellent articles that review these types of components in detail [Mihara and Omura, 1996; Lill et al., 1996; Pugsley, 1993; Rapoport et al., 1996; Walter and Johnson, 1994]. Instead, we will limit our discussion to those components that interact with either nascent or newly synthesized proteins in a more generic-like fashion, presumably by their ability to recognize "nonnative" or thermodynamically unstable structures. Members of this family, most of which are now referred to as molecular chaperones, likely function in all aspects of protein maturation, be it folding, transport across membranes, or higher-ordered assembly (e.g., see Fig. 2).

III. NASCENT CHAIN BINDING COMPONENTS IN THE EUKARYOTIC CYTOPLASM

A. The Hsp70 Family

The first components suggested to interact in a general fashion with nascent polypeptides were the 70-kDa heat-shock or stress proteins. Before describing this pathway, it is important to point out that there exists multiple and highly related forms of Hsp70 that are present within all cellular compartments. Unfortunately, the nomenclature used to describe this family of proteins is somewhat confusing and, therefore, merits a brief discussion [also see reviews by Craig, 1990; Welch, 1991; Georgopoulos and Welch, 1993]. In animal cells the predominant form of cytosolic–nuclear Hsp70, which is expressed under normal growth conditions, is referred to as the *h*eat *s*hock *c*ognate (Hsc70, and also referred to as Hsp73). The levels of Hsc70 are relatively high, it being one of the more abundant proteins in the cytosol of most animal cells. The second major form of the cytosolic–nuclear chaperone is Hsp70 (also referred to as Hsp72). For the most part, Hsp70 is expressed in cells only in response to heat-shock treatment, or some other type of metabolic stress (note, however, there are several exceptions to this rule). Both Hsc 70 and Hsp 70 exhibit a high degree of sequence homology and appear to interact with one another in some cells [Brown et al., 1993]. It is still unclear why eukaryotic cells (in contrast with prokaryotes) produce a novel form of Hsp70 in response to heat shock and other metabolic stresses (therefore, in the following discussion, we will use Hsp70 and Hsc70 interchangeably). As is discussed elsewhere in this volume, other Hsp70 family members include the 78-kDa *g*lucose *r*egulated *p*rotein (Grp78), now more commonly referred to as BiP and present within the lumen of the ER. Within the matrix of either mitochondria or chloroplasts is Grp75, or what we often refer to as mito-BiP. Finally, other Hsp70 homologues, still not fully characterized, continue to be identified.

All members of the Hsp70 family possess a highly conserved nucleotide-binding domain at their NH_2-terminus, the structure of which has been solved by X-ray crystallography [Flaherty et al., 1990]. At the COOH-terminus, where the different family members show the most divergence in their amino acid sequence, is the substrate-binding site. The different members of the Hsp70 family interact with small peptides or with proteins that appear nonnative [Flynn et al., 1989; Palleros et al., 1991; Sadis and Hightower, 1992]. The ability of Hsp70 family members to discriminate between a properly folded versus a misfolded polypeptide is the key property by which these proteins function as molecular chaperones.

Exactly how they carry out this discrimination at the molecular level is still under intense study. However, insights into the substrate-binding domain of the bacterial DnaK protein have been realized from the recent crystallization of the COOH-terminus of that protein. Here peptides, in a relatively extended conformation, appear to bind within a "channel" present near the COOH-terminus of the DnaK protein [Zhu et al., 1996].

B. Interactions of Hsp70 with Nascent Polypeptides In Vivo

The impetus for examining the potential interaction of cytosolic Hsc70 with nascent polypeptides was based on earlier studies showing that Grp78/BiP interacted with at least some newly synthesized proteins as they were being translocated into the lumen of the endoplasmic reticulum. For example, BiP binds to the monomeric forms of IgG heavy (H) and light (L) chains and orchestrates their assembly into the mature H_2L_2 structure [Haas and Wabl, 1983; Munro and Pelham, 1986]. Subsequently, BiP was observed to interact with newly synthesized influenza hemagglutinin monomers and facilitate their assembly into the mature trimer [Copeland et al., 1986; Gething et al., 1986]. In each case, BiP was believed to stabilize or maintain the translocating proteins in a conformation amenable for their subsequent folding or higher-ordered assembly. Prompted by these observations, our laboratory examined whether the cytosolic forms of Hsp70, such as BiP, might interact with nascent or newly synthesized proteins within the eukaryotic cytosol [Beckmann et al., 1990]. To address this issue, animal cells were subjected to metabolic pulse–chase radiolabeling, and the possible interaction of cytosolic Hsc70 with the radiolabeled nascent or newly synthesized proteins was examined by immunoprecipitation. For these studies the cells were solubilized under conditions chosen to minimize the destruction of relevant protein–protein interactions. Moreover, knowing that BiP interacted with its protein substrates in an ATP-dependent manner, we included the ATP-hydrolyzing enzyme apyrase in the buffers used to solubilize the cells. Immunoprecipitation of Hsc70 from the cells pulse-labeled with [^{35}S]methionine resulted in the capture of Hsc70 along with a collection of newly synthesized or nascent polypeptides (Fig. 1B). Analysis of the material by sodium dodecyl sulfate–polyacrylamide gel electrophoresis (SDS–PAGE) revealed a "smear" of radioactive material migrating from the top to the bottom of the gel, indicative of a heterogeneous mixture of nascent polypeptides. In contrast, immunoprecipitation of Hsc70 from the cells pulse-labeled and then chased in the absence of the radiolabel resulted in the coprecipitation of only a few radiolabeled proteins (see Fig. 1A).

The capture of nascent and newly synthesized proteins in a complex with the Hsc70 chaperone was dependent on the rapid depletion of ATP levels. Simply adding excess ATP to the cell lysate resulted in the release of Hsc70 from its

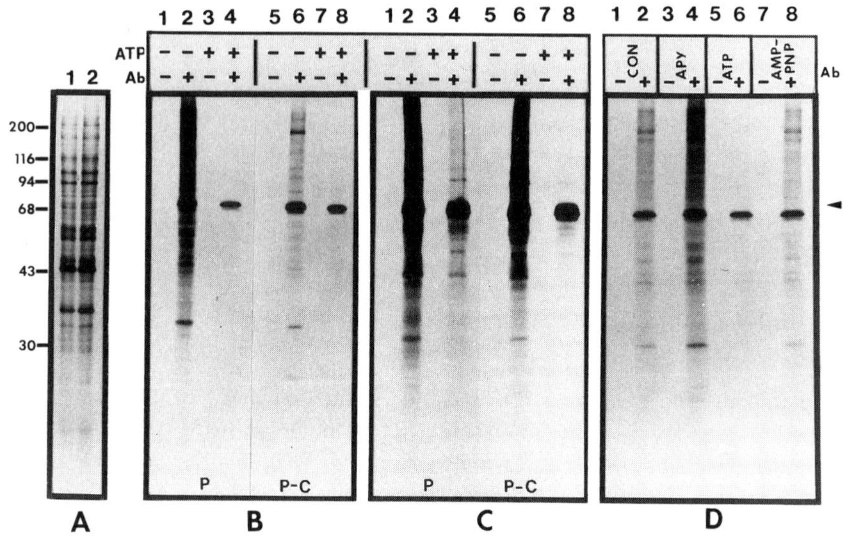

Figure 1 Examining the interaction of Hsp70 with nascent and newly synthesized proteins: (A and B) Cytosolic Hsp70 interacts transiently with nascent and newly synthesized proteins. HeLa cells were pulse-labeled with [^{35}S]methionine for 20 min. One plate of the cells was harvested immediately (i.e., pulsed). From another plate of pulse-labeled cells, the media was removed, and the cells extensively washed with and further incubated in culture medium (not containing any radiolabel) for 90 min before being harvested (i.e., pulsed–chased). The cell lysates were divided in half: to one half of the lysate was added ATP (final 10 mM, ATP^{+}) ; to the other half ATP levels were depleted (ATP^{-}) by addition of the enzyme apyrase. From the pulsed and pulsed–chased cell lysates, immunoprecipitation reactions employing antibodies against Hsp70 were performed and the resultant immunoprecipitates analyzed by SDS–PAGE. (A) the starting cell lysates of the cells pulsed (lane 1) or pulsed–chased (lane 2); (B) the Hsp70 immunoprecipitates. Note that many nascent and newly synthesized proteins coprecipitated with the Hsp70 chaperone following the pulse labeling, providing that ATP levels were first depleted (ATP^{-}). On the subsequent chase period significantly fewer radiolabeled proteins were now observed to coprecipitate with Hsp70. (C) Proteins synthesized in the presence of an amino acid analogue remain bound to the Hsp70 chaperone. In an experiment almost identical with that shown in B, HeLa cells first were incubated in the presence of an amino acid analogue of proline (L-azetidine-2-carboxylic acid, AZC) and then subjected to the pulse, and pulse–chase labeling with [^{35}S]methionine. Lysates were prepared as described in the foregoing and analyzed by immunoprecipitation using the Hsp70 antibody. Note that the newly synthesized proteins made in the presence of the amino acid analogue again interacted with the Hsp70 chaperone after the pulse-labeling period. After the subsequent chase period, however, the newly synthesized and analogue-containing proteins failed to be released from their Hsp70 chaperone. (D) The capture of Hsp70 with nascent or newly synthesized proteins requires ATP depletion. HeLa cells were pulse-labeled as described in B. The lysate was then divided into four equal portions and either treated with nothing (con); the ATP depleting enzyme apyrase (APY); 10 mM ATP (ATP); or 10 mM of the non-hydrolyzable ATP analogue (AMP–PNP). After 30 min of incubation on ice, immunoprecipitations using the Hsp70 antibody were performed.

nascent or newly synthesized protein targets (see Fig. 1B and D). Note that even the addition of nonhydrolyzable ATP analogues was sufficient to elicit the release of the nascent chain substrates from the Hsc70 chaperone. Presumably, binding (but not hydrolysis) of ATP induced a conformational change within Hsc70 that resulted in substrate release. Nascent polypeptides that were released prematurely from the ribosome, by treatment of cells with the antibiotic puromycin, could be captured in a complex with the Hsc70 chaperone. Finally, from highly purified preparations of polysomes, radiolabeled nascent chains again were observed to coprecipitate with Hsc70. Given all of these observations we suggested that Hsc70 likely interacted with most nascent polypeptides while they were being synthesized on the ribosome [Beckmann et al., 1990] (Fig. 2). Through such an interaction with the Hsc70 chaperone, the nascent chain would be maintained in a "protected conformation," perhaps being restricted from interacting with other components in its vicinity, which might lead to its misfolding. Only after a sufficient amount of information was present (i.e., a sufficient number of amino acids synthesized and free of the ribosome channel), would the nascent polypeptide be able to initiate the folding process. Folding would commence after the ATP-dependent release of the Hsc70 chaperone. Importantly, the interaction of Hsc70 with the nascent chain is likely a very dynamic process. We suspect that the chaperone is constantly binding to and then releasing from its target (Fig. 3). Otherwise the nascent chain would never have a chance to initiate the folding process. For other newly synthesized proteins, for example, those that were to be assembled into higher-ordered structures or perhaps be transferred into an organelle, such as the mitochondria, interaction with the Hsp70 chaperone might still occur even after release of the protein from the ribosome. Here, Hsp70 might act to stabilize such proteins in an "intermediate conformation" until completion of their final assembly, or until the initiation of their translocation into the organelle [Chirico et al., 1988; Deshaies et al., 1988].

Work over the past 6 years has supported and extended these original observations. Craig and co-workers reported that a portion of yeast cytosolic Hsc70 fractionated with translating ribosomes [Nelson et al., 1992]. Prior treatment of the cells with puromycin (to elicit release of the nascent chains from the ribosome) resulted in a significant loss of Hsc70 from the polysomes, consistent with the idea that Hsc70 was present because of its interaction with the nascent chains. Yeast strains depleted of cytosolic Hsp70 grow slowly, are hypersensitive to drugs that cause errors in protein synthesis, and contain relatively low levels of active polysomes. A suppressor, which appeared to restore at least some wild-type growth properties to the cells depleted of Hsc70, was isolated. Interestingly, the suppressor gene encoded a homologue of a protein synthesis factor involved in translational elongation. These observations, in sum, provided further support for the idea that cytosolic Hsp70 plays a role in the early stages of polypeptide maturation.

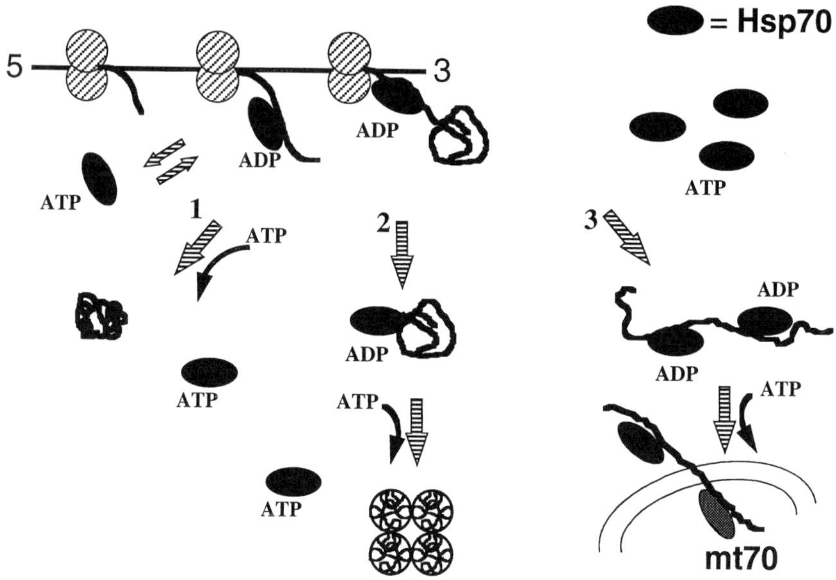

Figure 2 Summarization of the interactions of Hsp70 with nascent and newly synthesized proteins: As a nascent chain emerges from the ribosome it does so in a relatively unfolded conformation. As a consequence, the nonnative nascent polypeptide is recognized by and becomes bound to the Hsp70 chaperone. Such an interaction acts to stabilize the nascent polypeptide, perhaps by preventing either its premature folding or its inadvertent interactions with other macromolecules that might lead to its misfolding. We suspect that under normal cellular conditions (high levels of ATP) the interaction of Hsp70 with the nascent chain is very dynamic (i.e., the chaperone undergoes repeated cycles of binding and release to the nascent chain substrate). Once enough amino acids have been synthesized to provide for the productive folding of the first domain, folding commences on the next cycle of Hsp70 release (1). For some proteins whose final destiny is oligomeric assembly, Hsp70 might be needed to stabilize the monomer until the assembly process is initiated (2). For some proteins that are to be transferred into an organelle, such as the mitochondria, the newly synthesized protein is maintained in a relatively unfolded or "translocation-competent" conformation by an interaction with the Hsp70 chaperone (3).

C. Examining Nascent Chain Biogenesis by In Vitro Approaches

The pathways of protein folding now are being examined primarily using in vitro translation systems. The identification of proteins that interact with nascent polypeptides, as well as the physical properties of the nascent polypeptide itself, are being explored with different model proteins. Our own studies have shown that nascent chloramphenicol acetyl transferase (CAT) chains of fewer than 200

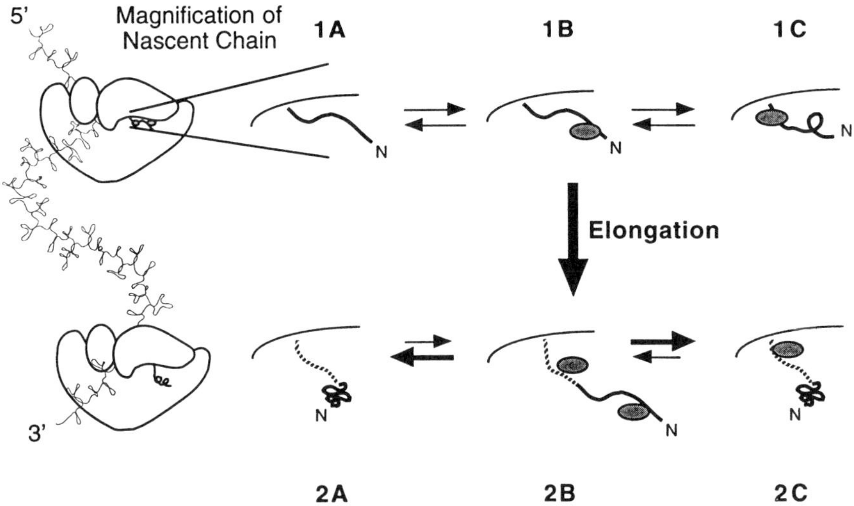

Figure 3 Cotranslational folding is dependent on dynamic interactions with the Hsp70 chaperone. As elongation of the nascent chain proceeds, we suspect that the polypeptide undergoes many changes in its structure and interaction with the Hsp70 chaperone. Depending on the relative kinetics of protein synthesis, protein-folding, and the binding of molecular chaperones, the nascent chain could exist in more than one conformation (1A–1C). Early during translation, a sufficient number of amino acids may not be present to allow the formation of any stable tertiary structures (1A). Consequently, at this stage in its synthesis the nascent chain would be recognized by and become bound to a molecular chaperone such as Hsp70 (1B and 1C). Note however, that the interaction of the Hsp70 chaperone is dynamic, with the chaperone continually binding and releasing from its substrate to allow for productive folding. As elongation proceeds and more folding information exists in the ribosome, the nascent chain may begin to collapse into stable intermediates in which both secondary and tertiary structures become apparent (2A and 2C). Although folding of the NH_2-terminal domain has been completed, new amino acids emerge from the ribosome and represent the next target for interaction with the Hsp70 chaperone (indicated by the dashed line). Thus, even late in translation (and depending on the complexity of the protein) the nascent polypeptide may contain already-folded domains, as well as domains still requiring continued interactions with the Hsp70 chaperone.

amino acids exhibit a very large native molecular mass when released from the ribosome [Hansen et al., 1994]. By chemical cross-linking we found that a significant portion of the nascent CAT chains were bound to the Hsc70 chaperone. With limited proteolysis, the nascent chains were relatively protected from added protease, provided that ATP levels were first depleted. For example, on adding excess ATP, and thereby, eliciting release of the Hsp70 chaperone from the nascent chains, the nascent CAT molecules were now digested completely by the added protease. Frydman et al. [1994] examined the early events surrounding the synthesis and folding of firefly luciferase. Either an mRNA encoding full length luciferase, or truncated luciferase mRNAs lacking a stop codon, were translated in rabbit reticulocyte lysate and analyzed by various means, including limited proteolysis, coprecipitation with antibodies to specific molecular chaperones, and determination of the specific activity of the newly synthesized luciferase enzyme. In addition to Hsc70, they reported two other cytosolic chaperones that interacted with the nascent luciferase polypeptide. One of these Hsp40, represents one of many eukaryotic homologues of the bacterial DnaJ protein, a known cofactor for the DnaK (i.e., bacterial Hsp70) chaperone [for a review of eukaryotic homologues of the bacterial DnaJ protein, see Caplan et al., 1993]. The second was the TCP-1-containing cytosolic chaperonin complex (CCT). Cytosolic chaperonin is weakly related to both the bacterial GroEL complex and the mitochondrial–chloroplast Hsp60 chaperonin [for a review of CCT, see Kubota et al., 1994].

To examine the relevance of these different chaperones interacting with nascent luciferase, depletion experiments were performed. No adverse effects on the overall rate, or extent of translation was observed following the depletion of the individual chaperones. Rather, removal of any one of the three different chaperone components (e.g., Hsp70, Hsp40, or CCT) resulted in a proportional reduction in the specific activity of the newly synthesized luciferase enzyme, presumably owing to the newly synthesized protein failing to fold properly. Readdition of Hsp70 to the depleted lysate was effective in restoring the efficient folding of the enzyme, provided that it was added before the start of translation. Luciferase folding could not be rescued when Hsp70 was added posttranslationally, consistent with the idea that it interacts cotranslationally with the nascent luciferase molecules. In contrast, addition of purified CCT posttranslationally was partially effective in restoring the activity of the newly synthesized enzyme, indicative that the chaperonin could act on the enzyme late in the folding process. From these and other results, the investigators suggested a model in which the three different chaperones cooperated in sequential fashion to facilitate the folding of nascent luciferase [Frydman et al., 1994]. They suggest that, as the nascent luciferase chain emerges from the ribosome, with as few as 40 amino acids exposed, it is recognized by and subsequently becomes bound to Hsp40. Targeting of the substrate by Hsp40 leads to the subsequent recruitment of Hsp70 and the formation of a ternary complex in which both Hsp40 and Hsp70 are now bound to and presumably stabilizing the nascent chain. Only after approximately 250–300 amino acids have been synthesized would the nascent luciferase chain now contain enough information to allow the folding of its first domain. At this stage in its synthesis, the nascent luciferase chain would be transferred over to the CCT chaperonin. In their model, subsequent folding of the nascent luciferase would then occur within the confines of the chaperonin particle.

Additional studies over the last 2 years, many originating from the Hartl

Laboratory, have resulted in some modifications of this model [Frydman and Hartl, 1996; Hartl, 1996]. For example, another component has been suggested to be involved in the nascent chain interaction, an approximate 45-kDa protein termed Hip (for Hsc70-*i*nteracting *p*rotein) [Höhfeld et al., 1995]. Sequencing of Hip revealed its homology to the so-called tetratricopeptide repeat (TPR) family of proteins, all of which contain multiple degenerate repeats of a 34-amino acid motif. One such protein is the Sti1 protein, a stress-inducible protein of unknown function in *Saccharomyces cerevisiae* [Nicolet and Craig, 1989]. In animals, a Sti1 homologue, termed p48, interacts with steroid hormone receptors in collaboration with both Hsp70 and Hsp90 [Prapapanich et al., 1996]. Owing to its reported ability to interact with Hsp70, it was suggested that the tetrameric Hip protein might function to stabilize Hsc70 in its ADP-bound state, thereby favoring Hsc70's interaction with a nascent chain [Höhfeld et al., 1995].

Our own work (both published and unpublished) continues to support the idea that nascent chains, in general, enter into high molecular weight complexes during the early stages of their biosynthesis. When mRNAs (either full length or truncated) encoding β-tubulin (an α/β-heterodimer), firefly luciferase (a monomer), chloramphenicol acetyl transferase (CAT; a homo-trimer) and β-galactosidase (a homotetramer) are translated in rabbit reticulocyte lysate, Hsc70 represents the most prominent nascent chain-binding component. In some experiments for which we reduce the levels of either Hsc70 or Hsp40 from the rabbit reticulocyte lysate, we observe a corresponding reduction in the overall extent of protein synthesis activities [W. J. Hansen, unpublished observations]. These latter observations appear in concert with those of Matts and colleagues [1992], who showed that the relative extent of protein synthesis activities in rabbit reticulocyte lysates was directly correlated with the levels of Hsc70. Simply lowering Hsc70 levels by the addition of an unfolded protein (to provide a target for Hsc70 binding), resulted in a reduction of protein synthesis activity [Matts et al., 1993].

A useful reagent we have developed for our studies is an antibody specific for the antibiotic puromycin. Following release of the nascent chains by puromycin, we can rapidly isolate the puromycin-containing nascent chains with the puromycin antibody, and then examine both their physical properties as well as any associated proteins. Results from these types of in vivo experiments continue to support the idea that it is the Hsc70 chaperone that represents the most prominent nascent chain-binding protein in the eukaryotic cytosol. We have not yet had much success in capturing nascent chains bound to either Hsp40 or Hip. Although we do observe small amounts of the cytosolic CCT chaperonin bound to the puromycin-released chains, the levels (relative to the total amount of chains) indicate to us that a general or obligative cotranslational interaction of CCT with nascent chains seems unlikely. Consistent with these observations are our findings that, in isolated polysomes, the levels of CCT (as well as both Hsp40 and p48/Hip) are extremely low, be it relative to the mass of either the nascent chains or Hsc70 [W. J. Hansen et al., in preparation; H. Nagata et al., in preparation; D. K. Eggers et al., in preparation].

The idea that the folding of at least some polypeptides can occur cotranslationally is supported by some recent observations using rather clever methods. First, the observation that radiolabeled hemin cosediments with polysomes carrying nascent α- and β-globin is consistent with the idea of globin chains undergoing cotranslational folding as they bind to their prosthetic group [Komar et al., 1993]. In other studies, modified mRNAs were prepared in which additional amino acid-coding information (but no information for a stop codon) was engineered into the 3′ ends of mRNAs encoding either luciferase or rhodanese. The idea was to generate ribosomal-bound forms of the proteins with these COOH-terminal extensions, and to determine whether the newly synthesized proteins were enzymatically active. In both cases, COOH-terminal extensions of approximately 25 amino acids resulted in the ribosome-bound forms of the proteins being enzymatically active [Kudlicki et al., 1995; Makeyev et al., 1996]. Presumably, the COOH-terminal extensions allowed the entirety of the luciferase or rhodanese molecules to escape the "ribosomal tunnel" and initiate folding, even while still bound to the ribosome. Finally, the concept of cotranslational folding takes on a whole different level of complexity when one considers the results of recent studies examining the folding of a particular viral protein. For example, the reovirus cell attachment protein is known to exist as a homotrimer, with its trimerization being driven both at the NH_2- and COOH-termini. Using in vitro translation approaches, Gilmore et al., [1996] have presented data showing that trimerization of the reovirus attachment protein occurred as the protein was being synthesized on the ribosome. Presumably, the nascent chains initiated their interaction with one another cotranslationally (within the same polysome?) once a sufficient number of amino acids had been synthesized and extruded from the ribosome.

IV. COMPONENTS INTERACTING WITH NASCENT CHAINS OR NEWLY SYNTHESIZED PROTEINS IN PROKARYOTES

Support for a role of one or more members of the bacterial chaperone systems in protein folding and maturation follows from studies showing that *Escherichia coli* mutants, deficient in the expression of either the DnaK–DnaJ–GrpE chaperone machine, or the GroEL–ES chaperonin, fail to support the growth of bacteriophages and exhibit high levels of protein aggregates [reviewed by Georgopoulos et al., 1990; Georgopoulos and Welch, 1993]. Overexpression of different members of these chaperone machines helped reverse such defects and restore normal protein solubility [Van Dyk et al., 1989]. By using a slightly different approach, Horwich and colleagues [1993] created a bacterial cell line expressing a temperature-sensitive form of GroEL. At the nonpermissive temperature at which GroEL was no longer functional, the cells grew very poorly, exhibited significant reductions in their ability to support phage biogenesis, and showed a reduction in overall protein synthesis activities. Moreover, at the nonpermissive temperature, many newly synthesized proteins apparently failed to reach their native state. Instead, a large fraction of the newly synthesized proteins produced in the absence of functional GroEL partitioned into the detergent-insoluble fraction, likely because of their misfolding and subsequent aggregation.

In vitro translation, along with the use of either photoactivated or chemical-based cross-linking methods, has been used to identify bacterial proteins interacting with nascent or newly synthesized proteins. For example, using an initiator methionine containing a photoreactive cross-linker, Bochkareva and associates [1988,1992] reported the interaction of newly synthesized lactamase (in its precursor form) and chloramphenicol acetyl transferase (CAT) with the GroEL chaperonin. Hendrick et al. [1993], also employing cross-linking methods, reported that the bacterial DnaJ protein, when added to a rabbit reticulocyte lysate, could be cross-linked to both nascent luciferase and CAT. Here, addition of high levels of DnaJ to the lysate appeared to arrest the folding of the newly synthesized proteins, unless the bacterial cofactors of DnaJ, the DnaK and GrpE proteins, also were added. Results of more recent cross-linking experiments have led two groups to conclude that the most prominent nascent chain-binding proteins in *E. coli* are the signal-recognition particle (SRP) and trigger factor (TF). Valent et al. [1995] found that most presecretory proteins made in vitro could be cross-linked to SRP. These investigators also reported that TF could be cross-linked to most any nascent chain being synthesized in the in vitro system, whether the proteins contained a signal sequence or not. The interaction of TF with nascent polypeptides was no longer observed, however, when the nascent chains were released prematurely from the ribosomes by puromycin. Hesterkamp et al. [1996] also concluded that nascent polypeptides could be efficiently cross-linked to TF in a bacterial extract. These investigators noted that TF exhibits homology to and can function as a peptidyl prolyl-*cis/trans*-isomerase. Studies by Stoller et al. [1995] have shown that at least some of TF is associated with the bacterial 50S ribosomal subunit.

It is still not clear why these different groups have arrived at somewhat different conclusions using cross-linking methods to identify nascent chain-binding proteins in bacteria. To reiterate the results, GroEL, DnaJ, SRP, and TF, all were reported to cross-link with a nascent polypeptides (sometimes with the same polypeptide substrate). In the discussion of their work Valent et al. [1995], prompted by the results of Hendrick et al., mentioned that they looked for, but could not find, evidence for an interaction of DnaJ with nascent polypeptides. Finally, in light of the observations in eukaryotes during which Hsc70 appears to bind nascent polypeptides, it is somewhat surprising that none of the aforementioned studies in bacteria reported the cross-linking of DnaK with nascent or newly synthesized proteins. Yet, as is described later, other investigators have noted the presence of DnaK within isolated polysomes. Our ability to capture Hsp70 bound to nascent chains in eukaryotic cells was dependent on the rapid depletion of ATP levels immediately after lysing the cells. Hence, it would be of interest to see whether DnaK could be cross-linked to nascent chains under conditions for which ATP levels are depleted immediately after the translation reaction.

In yet another approach to identify components that might interact with nascent or newly synthesized proteins, Langer et al. [1992] diluted a chaotrope-denatured substrate, rhodanese, into buffers containing different members of the bacterial chaperone system. These investigators concluded that DnaK, DnaJ, and GroEL cooperated together in the refolding of rhodanese. Results from order-of-addition experiments led them to conclude that DnaK was the first to bind to the unfolded protein, with DnaJ participating shortly thereafter. It was noted that for certain substrates DnaJ might interact first, followed by DnaK. Final folding of rhodanese would involve the transfer of the protein from the DnaK machinery (i.e., DnaK–DnaJ–GrpE) over to the GroEL–ES chaperonin machinery.

Although the GroEL–ES chaperonin machinery clearly has been implicated as being important for the folding of newly synthesized proteins, all of the available evidence indicate that its interactions with folding substrates occurs posttranslationally. Gaitanaris et al. [1994], analyzing the composition of polysomes prepared from *E. coli*, noted the presence of both DnaK and DnaJ, but not the GroEL chaperonin. Although some newly synthesized proteins were found in a complex with GroEL, it was suggested that such interactions occurred only following release of the newly synthesized proteins from the ribosome. Vysokanov [1995] has reported an intimate role for both DnaK and DnaJ in the early stages of CAT folding. Here DnaK and DnaJ cofractionated with polysomes and, similar to the situation in eukaryotes, both were lost from the peak polysome fraction on prior treatment of the lysate with puromycin. Interestingly, in those lysates depleted of both DnaK and DnaJ, newly synthesized CAT (released from the ribosome) was found either as an inactive monomer, or bound to the GroEL chaperonin. By using rhodanese as the test protein, Reid and Flynn [1996] similarly have concluded that the newly synthesized protein interacts only posttranslationally with the GroEL chaperonin. Kudlicki et al. [1994] also have examined the protein-folding pathway of rhodanese synthesized in vitro. These investigators were surprised to find that as much as 60% of the newly synthesized full-length rhodanese remained bound to the ribosome, apparently in the form of a peptidyl-tRNA. Whereas this form of the protein appeared enzymatically inactive, addition of the DnaK–DnaJ–GrpE and the GroEL–ES chaperone machineries resulted in the release of the ribosome-bound rhodanese and its acquisition of biological activity. Curiously, addition of either DnaK or GroES promoted the release of the ribosome-bound protein in a form that was enzymatically inactive. Given these and other observations, Kudlicki et al. suggested that full-length rhodanese accumulates on the ribosome as peptidyl-tRNA, and that chaperone-mediated interactions are necessary both for translational termination and the release of the properly folded protein from the ribosome.

V. SUMMARIZING THE RESULTS OBTAINED FROM EUKARYOTIC AND PROKARYOTIC STUDIES

It is clear from the discussion presented here that no consensus on the components that interact with nascent polypeptides, either in eukaryotic or prokaryotic systems, has been reached. Nor do we have any firm conclusions for the physical properties of a nascent polypeptide. Finally, whether nascent chains, in general, undergo folding either co- or posttranslationally (or both) still awaits further study. We are not too surprised by this lack of consensus considering the complexity of the problem, the different techniques and systems being employed to attack the problem, and that these types of studies are relatively new. Indeed, when reading textbooks published within the last 10–20 years, little mention is even made of the status of the nascent chain itself, or the early events involving the folding of a newly synthesized protein inside the cell.

Although there may not be a consensus on what components interact with the majority of nascent polypeptides, there does seem to be some general agreement that a nascent chain might require component(s) to chaperone the early stages of its synthesis. Specifically, it seems plausible that the nascent chain, lacking a suitable number of amino acids to initiate productive folding, would need some type of escort as it emerges from the ribosome. Such an escort would help ensure that the nascent chain does not initiate an interaction with another macromolecule in its local vicinity that could interfere with its eventual folding. Currently the best candidates for such an escort(s) in eukaryotes (prokaryotes) are Hsp70 (DnaK); Hsp40 (DnaJ); CCT (GroEL); and TF (trigger factor).

Whether all nascent polypeptides and or newly synthesized proteins require an escort or molecular chaperone for their efficient synthesis and folding is still open to debate. Indeed, a similar question relating to the participation of members of chaperonin family for protein folding in both prokaryotes and eukaryotes has recently been discussed by workers in the field. For example, Lorimer [1996] has published calculations from which he concludes that the levels of the GroEL–ES chaperonin machinery could accommodate no more than 5% of the newly synthesized proteins being made at any one time in bacteria maintained under normal growth conditions. Rospert et al. [1996] demonstrated that of four different proteins imported into the mitochondria, only one apparently required the participation of the Hsp60 chaperonin for its efficient folding. For the eukaryotic cytosolic chaperonin CCT, our own studies indicate that the levels of this molecular chaperone are rather low compared, for example, with the Hsp70 chaperone [W. J. Hansen and D. K. Eggers, unpublished observations]. Our own suspicion is that CCT has a limited role in protein-folding events, likely being restricted to a rather small number of polypeptide substrates, the most notable being actin and tubulin. Even here, however, the CCT chaperonin by itself may not be sufficient. Recent studies from Cowan and colleagues have shown that the TCP-1 containing chaperonin, along with four other factors, are required for the efficient folding of β-tubulin (Tian et al., 1996).

VI. WHY ARE MANY MOLECULAR CHAPERONES ALSO MEMBERS OF THE HEAT-SHOCK PROTEIN FAMILY?

Long before their structures and functions were elucidated, most of the molecular chaperones were originally identified as members of the heat-shock protein family [see reviews by Craig, 1985; Lindquist, 1986; Welch, 1991, 1993; Morimoto et al., 1990]. Simply subjecting cells to temperatures slightly above those that are optimal for growth resulted in the cells rapidly changing their pattern of gene expression. The net result being the high level expression and accumulation of a limited number of proteins, the heat-shock proteins. Subsequent studies showed that a large number of different metabolic insults, including exposure to various metals, inhibitors of energy metabolism, amino acid analogues, and perturbants of protein synthesis (e.g., puromycin), all resulted in the high-level expression of the heat-shock proteins (hence, their more recent designation as stress proteins). A substantial amount of work, both in tissue culture systems as well as in animals, has shown that the stress proteins, collectively, are important for the ability of the cells to survive different metabolic insults. Examples of heat-shock or stress proteins that have now been shown to function as molecular chaperones include the DnaK (DnaK, DnaJ, GrpE) and the GroEL–ES chaperone machines in bacteria; in eukaryotes, the Hsp70 family along with its cohorts of DnaJ homologues (present in the cytosol, nucleus, ER, and mitochondria); the Hsp60 chaperonins (present in the mitochondria and chloroplast); and Hsp104 (so far only identified in yeast). Other eukaryotic Hsps, including Hsp90, Hsp28, and Hsp47, all have been reported to exhibit some chaperone-like properties, perhaps acting in a substrate-specific fashion and apparently not requiring ATP [reviewed by Jakob and Buchner, 1994]. Interestingly, the different subunits that constitute the CCT cytosolic chaperonin themselves are not induced in cells subjected to either heat-shock or other agents or treatments that result in an increase in the other stress proteins [our own unpublished observations, and see Kubota et al., 1994].

Exactly how do the different heat-shock proteins function to protect cells against different metabolic traumas? Now knowing that many of the heat-shock proteins act as molecular chaperones, the answer to this question appears straightforward. Most of the different stressors that result in the activation of the stress response fall under the category of protein denaturants. Consequently, it was suggested back in the early 1980s (and later proved) that whenever cells were placed under conditions where abnormally folded proteins accumulated, a stress response would be initiated [Hightower, 1980; Ananthan et al., 1986]. Presumably, the heat-shock proteins themselves might somehow aid in the repair or replacement of those cellular proteins adversely affected by the particular environmental insult [Hightower, 1980; Pelham, 1986]. Thus, even before the concept of molecular chaperones had been presented, investigators already had begun to suspect that the heat-shock proteins likely played some role in protein-folding events inside the cell.

All of the available evidence now supports the idea that cells do, in fact, increase their levels of the heat-shock proteins in response to any significant changes in protein metabolism. As we have described, under normal growth conditions, the constitutively expressed heat-shock proteins, acting as molecular chaperones, participate in various aspects of protein synthesis, protein trafficking, and higher-ordered protein assembly. Consequently, whenever there is a demand on the cell for increased protein biogenesis, the expression of those heat-shock proteins that function as molecular chaperones also increases. The cellular demand for the heat-shock proteins increases even more when the cell is confronted with heat-shock treatments. First, some mature proteins (i.e., those that already have achieved their final and properly folded state) begin to undergo partial denaturation. As a result, these proteins now likely display localized protein domains that appear nonnative and, therefore, that represent targets for chaperone interaction. Whether such proteins are restored to their native and biologically active conformation through the help of the chaperones, or eventually are targeted for degradation, has not been examined in any great detail. Likely both scenarios are operative. Depending on the extent of denaturation some proteins might be "rescueable," whereas others may simply be degraded, probably by the ATP-dependent proteosome pathway. For example, in *E. coli* at least two proteins, RNA polymerase and the bacteriophage lambda repressor protein, have been "reactivated or resurrected" back into their native conformation after heat shock, presumably by the actions of the DnaK chaperone machinery [Gaitanaris et al., 1990; Skowra et al., 1990]. Yet another labile target to the effects of heat shock and other metabolic stresses are nascent or newly synthesized proteins. Under conditions of heat shock, many nascent and newly synthesized proteins tend to fractionate within the detergent-insoluble fraction, indicative perhaps of their denaturation or aggregation [Beckmann et al., 1992; Eggers et al., unpublished observations]. Perhaps because of these types of problems, animal cells oftentimes shut down all protein synthesis activities following a relatively severe heat-shock treatment. Only after several hours back at their normal temperature will the cells resume protein synthesis activities.

Once resumed, the cells respond by the high-level expression of the heat-shock proteins.

Most investigators within the field believe that the activation of the heat-shock response is tightly coupled to the absolute levels of the heat-shock proteins [reviewed by Craig and Gross, 1991]. For example, under normal growth conditions, the constitutively expressed Hsp70 chaperone exists in equilibrium between two states: its "substrate-bound" state; and its "free" or unoccupied form. Whenever the balance is shifted such that the amounts of free Hsp70 are reduced to some critical level, the cell responds by the activation of the stress response. Thus, in heat shock treatment, mature proteins that begin to denature now become targets for Hsp70 interaction. The net result is a reduction in the available levels of the Hsp70 chaperone and a corresponding activation of the stress response to increase and, thereby, restore, the levels of the Hsp70 chaperone (as well as the other Hsps that function as chaperones). In cells treated with amino acid analogues, many newly synthesized proteins incorporating the analogue appear unable to adopt a thermodynamically stable structure. As a consequence some of the newly synthesized and analogue-containing proteins continue to appear nonnative and, therefore, remain as a target for Hsp70 interactions (see Fig. 1C) [Beckmann et al., 1992]. Over time and as more of these analogue-containing proteins accumulate, the free levels of Hsp70 are reduced and the cell therefore responds by the activation of the stress response. (Unfortunately, in the presence of the amino acid analogues the newly synthesized stress proteins themselves also incorporate the analogue and, therefore, are rendered nonfunctional. On removal of the analogue, functional forms of the Hsps eventually are produced; and the cell then gradually returns to its normal patterns of protein synthesis). Thus, in some respects Hsp70 might be considered as a type of thermometer; as its "free" levels decline, the stress response is activated.

VII. SUMMARY

Results from experiments performed both in vitro and in vivo have begun to provide us with new insights into the mechanisms by which proteins achieve their biologically active conformation. Although it is still accepted that the primary sequence of a polypeptide contains all of the necessary information for its folding, we now know that protein folding inside the cell involves the participation of accessory components, now commonly referred to as molecular chaperones. In general, the different molecular chaperones appear to influence the folding pathway by reducing, or preventing, interactions of the nascent polypeptide with other components, which then might lead to its misfolding or aggregation. Such interactions with the different molecular chaperones oftentimes occur as the protein is still being synthesized, or shortly following the release of the newly synthesized protein from the ribosome. The search for proteins that interact with nascent polypeptides is still relatively new and, therefore, no real consensus has yet been reached concerning all of the cellular components that interact with proteins during the early stages of their synthesis, folding, and higher-ordered assembly. At least in animal cells, all of the available data implicate one or more members of the family of heat-shock proteins as being major participants in the early stages of protein biogenesis. Considering the complexity of protein folding inside the cell, however, we suspect that additional components that feature in the synthesis and folding of proteins await identification.

REFERENCES

Ananthan, J., Goldberg, A. L., and Voellmy, R. (1986), Abnormal proteins serve as eukaryotic stress signals and trigger the activation of the heat shock genes, *Science 232*:252–254.

Anfinsen, C. B. (1973), Principles that govern the folding of protein chains, *Science 181*:223–230.

Beckmann, R. P., Lovett, M., and Welch, W. J. (1992), Examining the function and regulation of Hsp70 in cells subjected to metabolic stress, *J. Cell Biol. 117*:1137–1150.

Beckmann, R. P., Mizzen, L. A., and Welch, W. J. (1990), Interaction of Hsp70 with newly synthesized proteins: Implications for protein folding and assembly, *Science 248*:850–854.

Berman, A. L., Kolker, E., and Trifonov, E. N. (1994), Underlying order in protein sequence organization, *Proc. Natl. Acad. Sci. USA 91*:4044–4047.

Bochkareva, E. S. and Girshovich, A. S. (1992), A newly synthesized protein interacts with GroES on the surface of chaperonin GroEL, *J. Biol. Chem. 267*:25672–25675.

Bochkareva, E. S., Lissin, N. M., and Girshovich, A. S. (1988), Transient association of newly synthesized unfolded proteins with the heat-shock GroEL protein, *Nature 336*:254–257.

Brown, C. R., Martin, R. L., Hansen, W. J., Beckmann, R. P., and Welch, W. J. (1993), The constitutive and stress inducible forms of Hsp70 exhibit functional similarities and interact with one another in an ATP-dependent fashion, *J. Cell Biol. 120*:1101–1112.

Caplan, A. J., Cyr, D. M., and Douglas, M. G. (1993), Eukaryotic homologues of *Escherichia coli* DnaJ: A diverse protein family that functions with Hsp70 stress proteins, *Mol. Biol. Cell 4*:555–563.

Chirico, W., Waters, G., and Blobel, G. (1988), 70K heat shock related proteins stimulate protein translocation into microsomes, *Nature 332*:805–810.

Copeland, C. S., Doms, R. W., Bolzau, E. M., Webster, R. G., and Helenius, A. (1986), Assembly of influenza hemagglutinin trimers and its role in intracellular transport, *J. Cell Biol. 103*:1179–1191.

Craig, E. A. (1985), The heat shock response, *CRC Crit. Rev. Biochem. 18*:239–280.

Craig, E. A. (1990), Regulation and function of Hsp70 multigene family of *Saccharomyces cerevisiae*, *Stress Proteins in Biology and Medicine* (R. Morimoto, A. Tissieres, and C. Georgopoulos, eds.), Cold Spring Harbor Press, Cold Spring Harbor, NY, pp. 301–321.

Craig, E. A. and Gross, C. A. (1991), Is Hsp70 the cellular thermometer? *Trends Biochem. Sci. 16*:135–139.

Deshaies, R. J., Koch, B. D., Werner-Washburne, M., Craig, E. A., and Schekman, R. (1988), A subfamily of stress proteins facilitates translocation of secretory and mitochondrial precursor polypeptides, *Nature 332*:800–805.

Flaherty, K. M., DeLuca, C., and McKay, D. B. (1990), Three-dimensional structure of the ATPase fragment of a 70 kDa heat-shock cognate protein, *Nature 346*:623–628.

Flynn, G. C., Chappell, G. T., and Rothman, J. E. (1989), Peptide binding and release by proteins implicated as catalysts of protein assembly, *Science 245*:385–390.

Frydman, J. and Hartl, F. U. (1996), Principles of chaperone-assisted protein folding: Differences between in vitro and in vivo mechanisms, *Science 272*:1497–1501.

Frydman, J., Nimmesgern, E., Ohtsuka, K., and Hartl, F. U. (1994), Folding nascent polypeptide chains in a high molecular mass assembly with molecular chaperones, *Nature 370*:111–117.

Gaitanaris, G. A., Papavassiliou, A. G., Rubock, P., Silverstein, S. J., and Gottesman, M. E. (1990), Renaturation of denatured lambda repressor requiries heat shock proteins, *Cell 61*:1013–1020.

Gaitanaris, G. A., Vysokanov, A., Hung, S. C., Gottesman, M. E., and Gragerov, A. (1994), Successive action of *E. coli* chaperones in vivo, *Mol. Microbiol. 14*:861–869.

Georgopoulos, C., Ang, D., Liberek, K., and Zylicz, M. (1990), Properties of the *Escherichia coli* heat shock proteins and their role in bacteriophage lambda growth, *Stress Proteins in Biology and Medicine* (R. Morimoto, A. Tissieres, and C. Georgopoulos, eds.), Cold Spring Harbor Press, Cold Spring Harbor, NY, pp. 191–222.

Georgopoulos, C. and Welch, W. (1993), Role of the major heat shock proteins as molecular chaperones. *Annu. Rev. Cell Biol. 9*:601–634.

Gething, M. J., McCammon, K., and Sambrook, J. (1986), Expression of wild type and mutant forms of influenza hemagglutinin: The role of folding in intracellular transport, *Cell 46*:939–950.

Gilmore, R., Coffey, M. C., Leone, G., McLure, K., and Lee, P. W. (1996), Co-translational trimerization of the reovirus cell attachment protein, *EMBO J. 15*:2651–2658.

Haas, I. G. and Wabl, M. (1983), Immunoglobulin heavy chain binding protein, *Nature 306*:387–389.

Hansen, W. J., Lingappa, V. R., and Welch, W. J. (1994), Complex environment of nascent polypeptide chains, *J. Biol. Chem. 269*:26610–26613.

Hartl, F. U. (1996), Molecular chaperones in cellular protein folding, *Nature 381*:571–580.

Hendrick, J. P., Langer, T., Davis, T. A., Hartl, F. U., and Wiedmann, M. (1993), Control of folding and membrane translocation by binding of the chaperone DnaJ to nascent polypeptides, *Proc. Natl. Acad. Sci. USA 90*:10216–10220.

Hesterkamp, T., Hauser, S., Lutcke, H., and Bukau, B. (1996), *E. coli* trigger factor is a prolyl isomerase that associates with nascent polypeptide chains, *Proc. Natl. Acad. Sci. USA 93*:4437–4441.

Hightower, L. E. (1980), Cultured cells exposed to amino acid analogues or puromycin rapidly synthesize several polypeptides, *J. Cell Physiol. 102*:407–424.

Höhfeld, J., Minami, Y., and Hartl, F. U. (1995), Hip, a novel cochaperone involved in the eukaryotic Hsc70/Hsp40 reaction cycle, *Cell 83*:589–598.

Horwich, A. L., Low, K. B., Fenton, W. A., Hirshfield, I. N., and Furtak, K. (1993), Folding in vivo of bacterial cytoplasmic proteins: Role of GroEL, *Cell 74*:909–917.

Komar, A. A., Kommer, A., Krasheninnikov, I. A., and Spirin, A. S. (1993), Cotranslational heme binding to nascent globin chains, *FEBS Lett. 326*:261–263.

Jakob, U. and Buchner, J. (1994), Assisting spontaneity: The role of Hsp90 and small Hsps as molecular chaperones, *Trends Biochem. Sci. 19*:205–211.

Kubota, H., Hynes, G., Carne, A., Ashworth, A., and Willison, K. (1994), Identification of six *Tcp-1*-related genes encoding divergent subunits of the TCP-1-containing chaperonin, *Curr. Biol. 4*:89–99.

Kudlicki, W., Chirgwin, J., Kramer, G., and Hardesty, B. (1995), Folding of an enzyme into an active conformation while bound as peptidyl-tRNA to the ribosome, *Biochemistry 34*:14284–14287.

Kudlicki, W., Odom, O. W., Kramer, G., and Hardesty, B. (1994), Chaperone-dependent folding and activation of ribosome-bound nascent rhodanese, *J. Mol. Biol. 244*:319–331.

Langer, T., Lu, C., Echols, H., Flanagan, J., Hayer, M. K., and Hartl, F. U. (1992), Successive action of DnaK, DnaJ and GroEL along the pathway of chaperon-mediated protein folding, *Nature 356*:683–689.

Lill, R., Hargang, F., and Neupert, W. (1996), Biogenesis of mitochondrial proteins, *Curr. Opin. Cell Biol. 8*:505–512.

Lindquist, S. (1986), The heat shock response, *Annu. Rev. Biochem. 55*:1151–1191.

Lorimer, G. H. (1996), A quantitative assessment of the role of the chaperonin proteins in protein folding in vivo, *FASEB J. 10*:1–5.

Makeyev, E. V., Kolb, V. A., and Spirin, A. S. (1996), Enzymatic activity of the ribosome-bound nascent polypeptide, *FEBS Lett 378*:166–170.

Matts, R. L. and Hurst, R. (1992), The relationship between protein synthesis and heat shock proteins levels in rabbit reticulocyte lysates, *J. Biol. Chem. 267*:18168–18174.

Matts, R. L., Hurst, R., and Xu, Z. (1993), Denatured proteins inhibit translation in hemin-supplemented rabbit reticulocyte lysate by inducing the activation of the heme-regulated eIF-2a kinase, *Biochemistry 32*:7323–7328.

Mihara, K. and Omura, T. (1996), Cytoplasmic chaperones in precursor targeting to mitochondria: The role of MSF and HSP70, *Trends Cell Biol. 6*:103–108.

Morimoto, R. I., Tissieres, A., and Georgopoulos, C. (1990), *Stress Proteins in Biology and Medicine*, Cold Spring Harbor Laboratories, Cold Spring Harbor, NY.

Munro, S. and Pelham, H. R. B. (1986), An Hsp70-like protein in the ER: Identity with the 78 kD glucose-regulated protein and immunoglobulin heavy chain binding protein, *Cell 46*:294–300.

Nelson, R. J., Ziegelhoffer, T., Nicolet, C., Werner-Washburne, M., and Craig, E. A. (1992),

The translation machinery and 70 kD heat shock protein cooperate in protein synthesis, *Cell 71*:97–105.

Nicolet, C. M. and Craig, E. A. (1989), Isolation and characterization of STI1, a stress inducible gene from *Saccharomyces cerevisiae*, *Mol. Cell. Biol. 9*:3638–3648.

Palleros, D. R., Welch, W. J., and Fink, A. L. (1991), Interaction of Hsp70 with unfolded proteins: Effects of temperature and nucleotides on the kinetics of binding. *Proc. Natl. Acad. Sci. USA 88*:5719–5723.

Pelham, H. R. B. (1986), Speculations on the functions of the major heat shock and glucose regulated proteins, *Cell 46*:959–961.

Prapapanich, V., Chen, S., Nair, S. C., Rimerman, R. A., and Smith, D. F. (1996), Molecular cloning of human p48, a transient component of progesterone receptor complexes and an Hsp-70 binding protein, *Mol. Endocrinol. 10*:420–431.

Pugsley, A. (1993), The complete general secretory pathway in gram-negative bacteria, *Microbiol. Rev. 57*:50–108.

Rapoport, T. A., Rolls, M. M., and Jungnickel, B. (1996), Approaching the mechanism of protein transport across the ER membrane, *Curr. Opin. Cell Biol. 8*:499–504.

Reid, B. G. and Flynn, G. C. (1996), GroEL binds to and unfolds rhodanese posttranslationally, *J. Biol. Chem. 271*:7212–7217.

Rospert, S., Looser, R., Dubaquie, Y., Matouschek, A., Glick, B., and Schatz, G. (1996), Hsp60-independent protein folding in the matrix of yeast mitochondria, *EMBO J 15*:764–774.

Sadis, S. and Hightower, L. E. (1992), Unfolded proteins stimulate molecular chaperone Hsc70 ATPase by accelerating ADP/ATP exchange, *Biochemistry 31*:9406–9412.

Skowra, D., Georgopoulos, C., and Zylicz, M. (1990), The *E. coli dnaK* gene product, the Hsp70 homolog, can reactivate heat-inactivated RNA polymerase in an ATP hydrolysis-dependent manner, *Cell 62*:939–944.

Stoller, G., Rucknagel, K., Nierhaus, K. H., Schmid, F. X., Fischer, G., and Rahfeld, J. U. (1995), A ribosome-associated peptidyl-prolyl isomerase identified as the trigger factor, *EMBO J 14*:4939–4948.

Tian, G., Huang, Y., Rommelaere, H., Vandekerckhove, J., Ampe, C., and Cowan, N. J. (1996), Pathway leading to correctly folded β-tubulin, *Cell 86*:287–296.

Valent, Q. A., Kendall, D. A., High, S., Kusters, R., Oudega, B., and Luirink, J. (1995), Early events in preprotein recognition in *E. coli*: Interaction of SRP and trigger factor with nascent polypeptides, *EMBO J 14*:5494–5505.

Van Dyk, T. K., Gatenby, A. A., and LaRossa, R. A. (1989), Demonstration by genetic suppression of interaction of GroE products with many proteins, *Nature 342*:451–453.

Vysokanov, A. V. (1995), Synthesis of chloramphenicol acetyltransferase in a coupled transcription–translation in vitro system lacking the chaperones DnaK and DnaJ, *FEBS Lett. 375*:211–214.

Walter, P. and Johnson, A. E. (1994), Signal sequence recognition and protein targeting to the endoplasmic reticulum, *Annu. Rev. Cell Biol. 10*:87–120.

Welch, W. J. (1990), The mammalian stress response: Cell physiology and biochemistry of stress proteins, *Stress Proteins in Biology and Medicine* (R. Morimoto, A.

Tissieres, and C. Georgopoulos, eds.), Cold Spring Harbor Press, Cold Spring Harbor, NY, pp. 223–278.

Welch, W. J. (1991), The role of heat shock proteins as molecular chaperones, *Curr. Opin. Cell Biol. 3*:1033–1038.

Welch, W. J. (1993), How cells respond to stress, *Sci. Am. 268*:56–64.

Zhu, X., Zhao, X., Burkholder, W. F., Gragerov, A., Ogata, C. M., Gottesman, M. E., and Hendrickson, W. A. (1996), Structural analysis of substrate binding by the molecular chaperone DnaK, *Science 272*:1606–1614.

4

Structure and Properties of the 70-Kilodalton Heat-Shock Proteins

Jeung-Hoi Ha, Eric R. Johnson,
David B. McKay, Marcelo C. Sousa,
Shigeki Takeda, and Sigurd M. Wilbanks
Stanford University School of Medicine, Stanford, California

I. INTRODUCTION

The 70-kDa heat–shock-related proteins, or "Hsp70s"—taken to include the endoplasmic reticulum-resident BiPs, cytoplasmic and nuclear Hsp70s, prokaryotic DnaKs, and others—compose a family of molecular chaperones that are essential for cell viability. Their biochemical and biological functions are discussed in depth in other chapters of this volume; hence, they will not be elaborated extensively here. In brief, the Hsp70s cyclically bind and release "unstructured" segments of polypeptides; the binding presumably suppresses aggregation or misfolding of the peptide segments, whereas their release gives them the opportunity to coalesce into a correctly folded tertiary structure. The kinetics of the binding and release cycle is modulated by ATP–ADP; in the presence of ATP, peptide release is relatively rapid, whereas in the presence of ADP, Hsp70–peptide complexes are stable and long-lived. In vivo, unstructured segments of polypeptide may present themselves in many different contexts, ranging from protein folding (e.g., nascent polypeptides emerging from a ribosome during translation [1]; polypeptides translocating across a membrane into an intracellular organelle [2]; to protein denaturation (e.g., heat-denatured proteins arising from thermal stress of cells [3].

For simplicity, it is convenient to delineate the in vitro activities of the Hsp70 proteins as (a) an ATPase activity; (b) a peptide-binding activity; (c) a mechanism of coupling binding and release of peptides to the enzymatic ATPase cycle; and (d) mechanisms of modulating the Hsp70 activities through self-association, posttranslational modification, and interaction with accessory proteins. The focus of this chapter is the structure and properties of Hsp70 proteins as they relate to biochemical mechanism. The reaction cycle and peptide binding specificity of Hsp70s are discussed in other chapters of this volume; hence, they are not discussed in detail here.

II. STRUCTURE

Approximately 100-amino acid sequences (derived from DNA sequences) are currently on deposit in databases for Hsp70 proteins [Swiss-prot release 34.0, 4/96]. The mature proteins are typically 630- to 650-amino acid residues in length. Conservation of primary structure within the family is quite high, with the most distantly separated members showing 50% identity in amino acid sequence. The ATPase activity is localized in the NH_2-terminal of approximately 385 residues, whereas the peptide-binding activity resides in approximately 160 residues following the ATPase domain (Fig. 1). Different subfamilies of the Hsp70s have different sequence characteristics in the COOH-terminal region. The last 10–20 residues of DnaKs are predominantly charged and primarily acidic. BiP proteins have a glycine, proline-rich region, followed by a primarily acidic region, terminating in an endoplasmic reticulum (ER) retention signal, DEL. The eukaryotic Hsp70s and Hsc70s have a glycine, proline-rich segment of variable length, often with an approximate (GGMP) repeat, terminating with a signature EEVD which has been shown to be important for interaction with denatured polypeptides [4].

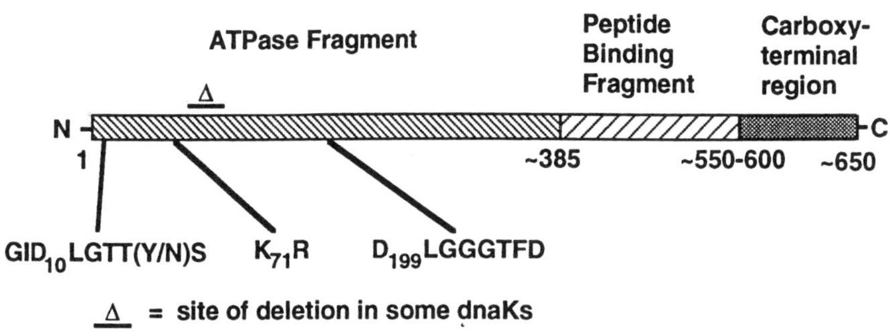

Figure 1 Schematic delineation of functional domains within the primary sequence of Hsc70 proteins. Conserved amino acid sequence motifs that interact with the nucleotide phosphates are shown, with sequence numbering corresponding to the bovine Hsc70 protein.

Although the three-dimensional structure of a full-length Hsp70 molecule has not been solved, the X-ray crystallographic structures of the isolated ATPase domain of bovine Hsc70 (Fig. 2) [5] and the peptide-binding domain of *Escherichia coli* DnaK (Fig. 3) [6] have been determined. The ATPase domain has two lobes with a deep cleft between them; nucleotide is bound at the base of the cleft and is almost entirely inaccessible to solvent. The tertiary structure of the ATPase domain has a striking similarity to that of actin [7]. Although the overall sequence similarity between actin and Hsc70 is relatively low, many of the residues that interact with MgADP and MgATP are identical [8].

Contrary to an early proposal that the structure of the peptide-binding domain of Hsp70s would be similar to that of the peptide-binding portion of major histocompatibility complex (MHC) proteins [9,10], and in concordance with the secondary structure assignment determined by nuclear magnetic resonance (NMR) [11], the X-ray crystallographic structure of the DnaK peptide-binding domain (residues 389–607) complexed with a seven-residue peptide (Asn-Arg-Leu-Leu-Leu-Thr-Gly), has revealed a novel tertiary fold that is unrelated to the MHC structure [6] (see Fig. 3). The peptide-binding moiety is composed of an eight-stranded antiparallel β-structure that forms a crevice with a single, deep pocket that accommodates the central leucine of the bound peptide. Although the topology diagram of the secondary structure depicts a pair of four-stranded sheets, only one of these segments (strands 3 and 6–8; see Fig. 3) forms a continuous sheet. In the other segment, the first two strands (strands 1 and 2; see Fig. 3) are over the top of the peptide, whereas the other two strands (strands 4 and 5; see Fig. 3) run under it. Although biochemical studies suggested that a minimum of seven residues was required for optimal peptide affinity [12], only the central five residues of the peptide make substantial interaction with the peptide binding domain in the crystallographic structure. One residue appears to be "buried" in a hydrophobic-binding pocket of the protein; it is the central leucine of the peptide, which is sandwiched between the two pairs of β-strands, the side chain of which protrudes downward into the crevice between the strands in the view shown in Figure 3. A long α-helix lies across the crevice, but does not actually contact the peptide; it has been suggested that this helix acts as a "latch" to stabilize peptide complexes and to control the kinetics of peptide binding and release.

The functional requirement of Hsp70 proteins to recognize a diversity of unstructured polypeptides and to modulate their peptide-binding and release rates fits well with the structure of the DnaK peptide-binding domain. The hydrophobic-binding pocket for the peptide is consistent with the preference for aliphatic hydrophobic amino acids in peptides bound by Hsp70s [12]. The long helix that lies over the top of the peptide-binding crevice suggests a mechanism whereby peptide binding and release would be relatively slow under circumstances in which the helix was latched over the crevice, but relatively rapid if the helix swung outward to allow unhindered access.

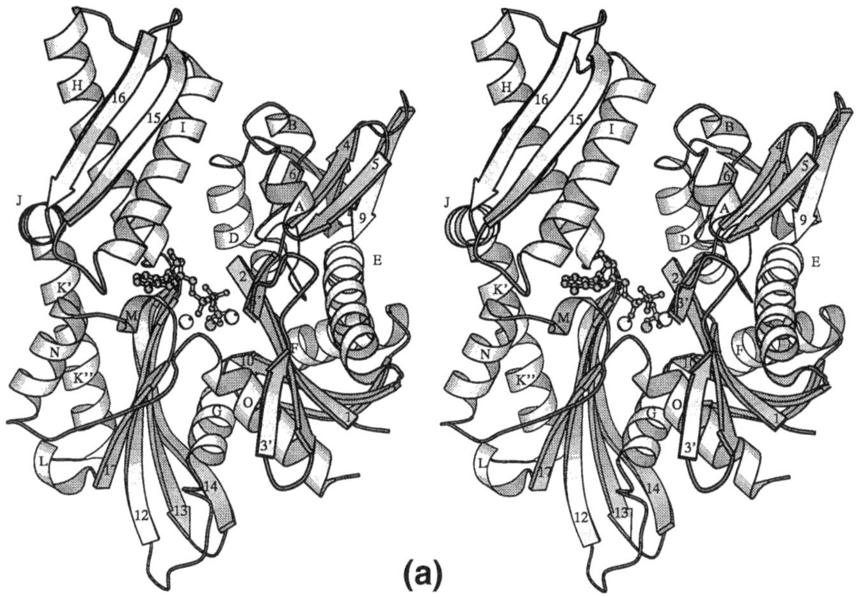

Figure 2 Structure of the ATPase domain. (a) Schematic drawing, in stereo, of the crystallographic structure of bovine Hsc70 ATPase domain. ADP and P_i are shown as ball-and-stick models; Mg^{2+} is shown as a small, shaded sphere; K^+ ions are larger, unshaded spheres. Drawing made with Molscript [57]. (b) Folding topology: Secondary structure assignment determined with the program DSSP; numbering referenced to bovine Hsc70 primary sequence.

III. ATPase MECHANISM

A pathway for chemical hydrolysis of ATP has been suggested (Fig. 4), based on both the effects of mutagenesis of residues in the nucleotide-binding cleft (Table 1) and the comparison of crystallographic structures of the Hsc70 ATPase fragment that have ATP or slowly hydrolyzable analogues bound (AMPPNP bound to wild-type protein; ATP bound to several different mutant proteins) to those that have ADP bound [13]. In the suggested scenario, MgATP would initially bind in a conformation that was not optimally positioned for hydrolysis, possibly with an H_2O molecule intercalated between the Mg^{2+} ion and the γ-phosphate of the nucleotide, as observed crystallographically. Then, the γ-phosphate would rearrange to form a "tight" β,γ-bidentate complex with the Mg^{2+} ion. This is proposed to align the terminal phosphate of the nucleotide for in-line attack by an H_2O molecule or OH^- ion, ultimately yielding ADP and P_i as products. Model-building of a hypothetical transition state of the hydrolysis reaction suggested an experimentally observed H_2O molecule that is hydrogen-

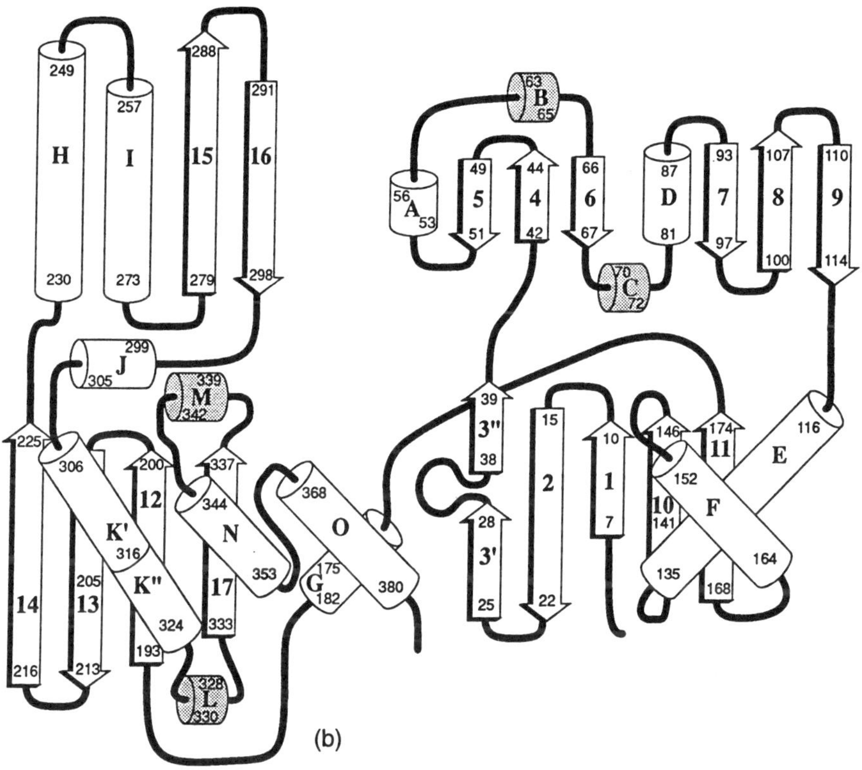

Figure 2 Continued

bonded to the ε-amino group of the side chain of Lys-71 in the Hsc70 ATPase fragment as a candidate catalytic H_2O molecule. The Lys-71 could then act either to accept a proton from the H_2O molecule or to stabilize an OH^- ion. Subsequent studies have shown that mutagenesis of Lys-71 completely abolishes ATPase activity, lending plausibility to the proposed scenario [14].

The steady-state ATPase turnover rate (Table 2) and nucleotide affinity (Table 3) have been reported for several Hsp70-related proteins, as well as for the NH_2-terminal ATPase fragment of bovine and rat Hsc70. The general consensus that emerges from these studies is (a) the steady-state ATPase turnover rate is relatively slow; reported k_{cat} values range from 0.02 min^{-1} to 1.0 min^{-1}, with the faster rates generally measured under conditions of peptide stimulation of the ATPase activity; K_m values for MgATP range about 0.1–1 μM; (b) within the enzymatic cycle, ATP hydrolysis is also relatively slow and, under many conditions, it appears to be the rate-limiting step of the cycle [15,16] (although, by contrast, P_i release is rate-limiting for the isolated ATPase domain of recombinant

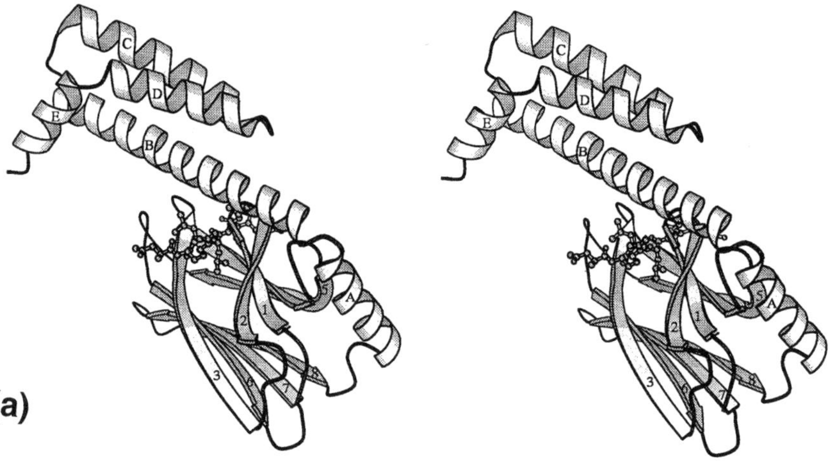

(a)

Figure 3 Structure of the peptide-binding domain: (a) Schematic drawing of the crystallographic structure of *E. coli* DnaK peptide-binding domain. Peptide is shown as a ball-and-stick model. (b) Folding topology: Secondary structure assignment determined with the program DSSP; numbering referenced to *E. coli* DnaK primary sequence.

bovine Hsc70 [17]); (c) affinities of both MgATP and MgADP for Hsp70 proteins is submicromolar, with reported values ranging between 0.01 and 1 μM; in absence of divalent ion, the affinities are significantly weaker, about 1–10 μM; (d) following ATP hydrolysis, product release is ordered, with P_i preceding ADP [17]. The ATPase activity requires Mg^{2+}; the $K_{0.5}$ for the Mg^{2+}-dependence of the activity is 1–10 μM [18]. The ATPase activity is also dependent on monovalent ion with a $K_{0.5}$ of approximately 20–50 mM; K^+ (ionic radius = 1.33 Å) induces maximal activity, whereas significantly smaller (e.g., Na^+ or Li^+ with ionic radii 0.97 Å and 0.68 Å, respectively) or larger (e.g., Cs^+, ionic radius = 1.67 Å) monovalent ions give substantially lower ATPase activity [19].

The requirement of the triphosphate moiety of ATP forming a specific β,γ-bidentate complex with Mg^{2+} or similar divalent ion for hydrolysis, rationalizes the requirement for Mg^{2+} in the ATPase reaction. Additionally, crystallographic studies have identified two specific monovalent ion-binding sites in the nucleotide-binding cleft of the ATPase fragment; ions at these sites influence the geometry and electrostatic interactions of the protein–metal–nucletide complex [20]. Substitution of one monovalent ion with another of significantly different size (e.g., Na^+ for K^+) can be expected to perturb the geometry and electrostatic interactions of the catalytic transition state, thereby affecting the ATP hydrolysis reaction rate, leading to the effects that have been observed [19].

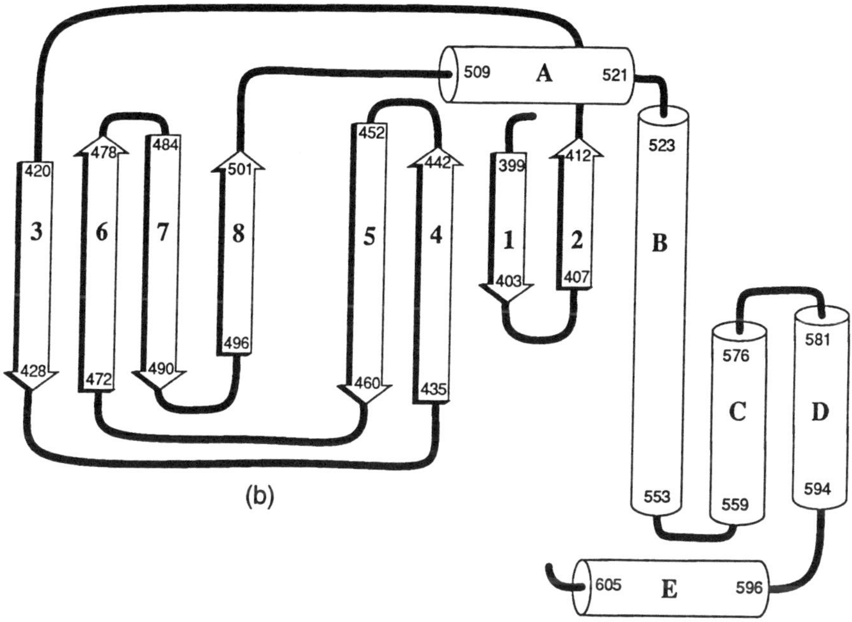

(b)

Figure 3 Continued

IV. COUPLING OF ATP-BINDING TO THE PEPTIDE-BINDING ACTIVITY

Several lines of evidence (proteolytic digestion patterns of *E. coli* DnaK [21–23] and hamster BiP [24]; changes in intrinsic tryptophan fluorescence of DnaK [25,26] and bovine Hsc70 [27]; changes in infrared spectra of DnaK [28]; solution small-angle X-ray scattering of DnaK [29] and bovine Hsc70 [30]) indicate that Hsp70 proteins access at least two significantly different conformations in a nucleotide-dependent manner. Similarly, 60-kDa fragments of both DnaK (isolated in a genetic screen) [22] and bovine Hsc70 (constructed by rational design) [30] lacking about 100 COOH-terminal amino acids undergo a nucleotide-dependent conformational change. The two predominant conformations appear to correspond to (a) a state that forms long-lived, stable Hsp70–peptide complexes, often referred to as the "peptide-binding" state, and (b) a state in which Hsp70–peptide complexes are relatively transient, or labile, often called the "peptide-release" state. It should be emphasized that the ATPase and peptide binding cycles are not stoichiometrically coupled in the sense of one ATP molecule inducing the release of one peptide; rather, the nucleotide can be thought to regulate the kinetics of peptide binding and release [31]. It should also be emphasized that describing the Hsp70 proteins as "two-state" systems is a first-order approxima-

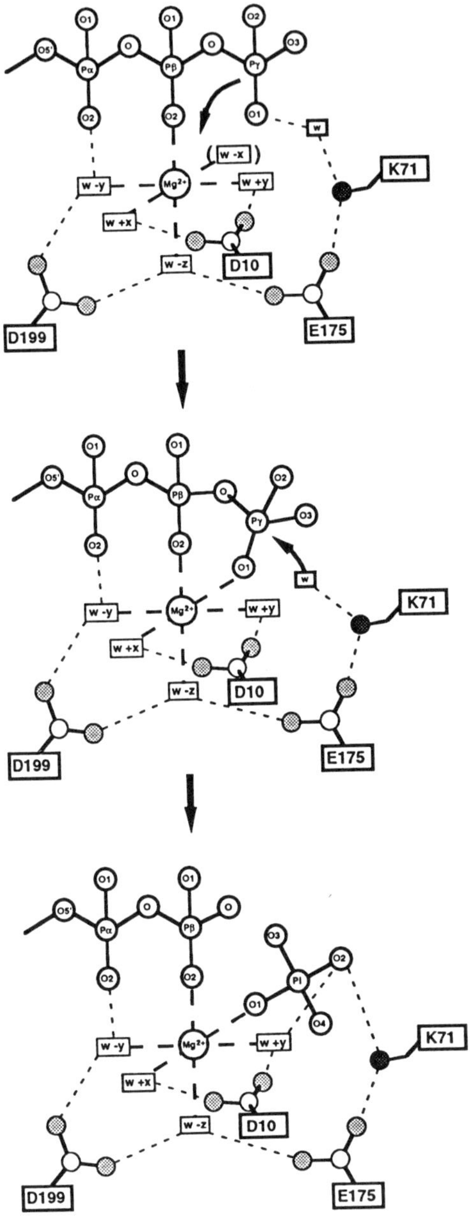

Figure 4 Suggested pathway of ATP hydrolysis by Hsp70 proteins: Only a few amino acid side chains are included for clarity. See text for discussion of details.

Table 1 Mutants in the Isolated ATPase Domain of Hsc70 Derived from Data in the Literature

Protein	Mutation	*E. coli* DnaK equivalent residue	k_{cat} (min^{-1})	$\frac{k_{cat}(mutant)}{k_{cat}(w.t.)}$	K_m(μM)	$\frac{k_m(mutant)}{k_m(w.t.)}$	Ref.
Bovine	wt		0.8 ± 0.1		0.7 ± 0.1		58
Rat[a]	wt		0.28				59
Bovine	wt		0.53 ± 0.07		0.49 ± 0.26		18
Bovine	wt (25°C)		0.14 ± 0.012		0.37 ± 0.09		17
Rat[a]	D10N	D8	0	0			59
Bovine	D10N	D8	0.021 ± 0.008	0.04	10.1 ± 4.3	20	18
Bovine	D10S	D8	0.009 ± 0.003	0.02	5.9 ± 2.3	12	18
Bovine	K71A	K70	< 0.003	< 0.01			14
Bovine	K71E	K70	< 0.003	< 0.01			14
Bovine	K71M	K70	< 0.003	< 0.01			14
Bovine	E175Q	E171	0.066 ± 0.022	0.12	46.5 ± 17.6	95	18
Bovine	E175S	E171	0.014 ± 0.010	0.03	16.2 ± 3.1	33	18
Bovine	D199N	D194	0.011 ± 0.002	0.02	0.73 ± 0.49	1.5	18
Bovine	D199S	D194	0.008 ± 0.11	0.02	0.45 ± 0.19	0.82	18
Bovine	T204E	T199	1.9 ± 0.3	2.4	93 ± 14	133	58
Bovine	T204V	T199	1.9 ± 0.2	2.4	89 ± 12	127	58
Bovine	D206N	D201	0.064 ± 0.012	0.12	0.27 ± 0.25	0.55	18
Bovine	D206S	D201	0.076 ± 0.030	0.14	0.27 ± 0.25	0.55	18

wt, wild-type.

Assays done in 40 mM HEPES, 75 mM KCl, 4.5 mM Mg $(CH_3COO)_2$, pH 7.0; or, (a) 50 mM MES, 5 mM $MgSO_4$, 10 mM $CaCl_2$, 1 mM DTT, pH 6.5. Assay temperature was 37°C except where noted. Bovine constructs include residues 1–386 (= Δ387–650); rat constructs include residues 1–384 (= Δ385–646). The amino acid sequences for bovine and rat Hsc70 are identical in the ATPase domain.

Table 2 Kinetic Parameters of Steady-State ATPase Activity of Wild-Type Hsp70s Derived from Data in the Literature

No.	Protein	Ligand (conc.)	k_{cat} (min^{-1})	K_m (μM)	Assay condition	Ref.
1	Bovine Hsc70 natural	0.5 mg mL^{-1} triton-washed clathrin cages	1	0.7	20 mM HEPES 25 mM KCl 10 mM $(NH_4)_2SO_4$ 2 mM $Mg(OAc)_2$ 0.8 mM DTT pH 7.0, 37°C	60
2	Bovine Hsc70 natural	Vesicular stomatitis virus glyco-protein peptides	≈ 0.2		20 mM HEPES 20 mM NaCl 2 mM $MgCl_2$ 1% sodium cholate 8 μM ATP pH 7.0, 37°C	61
3	Bovine Hsc70 natural	Clathrin cages and various peptides de-rived from clathrin	1	nd	40 mM HEPES 75 mM KCl 4.5 mM $Mg(OAc)_2$ 550 μM ATP pH 7.0, 37°C	62
4	Bovine Hsc70 natural	Apo cyt *c*: 0 μg mL^{-1} 400 μg mL^{-1}	 0.075 0.19	 1.37 1.44	20 mM HEPES-KOH 25 mM KCl 10 mM $(NH_4)_2SO_4$ 2 mM $Mg(OAc)_2$ 0.1 mM EDTA 1 mM DTT pH 7.0, 37°C	63
5	Bovine Hsc70 natural	Peptide C: No peptide 1.2 mM	 0.028 0.29		Reportedly same as for 3 above	64
6	Bovine Hsc70 natural		0.037 ± 0.002	nd	20 mM imidazole 25 mM KCl 10 mM $(NH_4)_2SO_4$ 2 mM $Mg(OAc)_2$ 1 mM DTT 0.1 mM ATP pH 7.0, 25°C	65
7	Bovine Hsc70 natural	No peptide: 1 mM of cyt *c* peptide peptide C S peptide KFERQ	≈ 0.02 ≈ 0.12 ≈ 0.08 ≈ 0.04 ≈ 0.03		20 mM imidazole 25 mM KCl 10 mM $(NH_4)_2SO_4$ 2 mM $Mg(OAc)_2$ 1 mM DTT 0.2 mM ATP pH 7.0, 25°C	66

Table 2 Continued

No.	Protein	Ligand (conc.)	k_{cat} (min^{-1})	K_m (μM)	Assay condition	Ref.
8	Rat Hsc70 re-	S-peptide:			40 mM HEPES	64
	combinant	no peptide	0.16		75 mM KCl	
		1.2 mM	0.38		4.5 mM $Mg(OAc)_2$	
					pH 7.0, 37°C	
9	Canine BiP		0.34	0.11	50 mM MES	67
	natural				25 mM NaCl	
					1 mM $MgCl_2$	
					1 mg mL^{-1} BSA	
					pH 6.0, 37°C	
10	Bovine BiP	No peptide	0.02		20 mM HEPES	61
	natural	Various peptides	0.08–0.16		20 mM NaCl	
					2 mM $MgCl_2$	
					1% sodium cholate	
					8 μM ATP	
					pH 7.0, 37°C	
11	Bovine BiP	Peptide A:			20 mM HEPES	46
	natural	no peptide	0.027		20 mM KCl	
		1 mM	0.15		10 mM $(NH_4)_2SO_4$	
					2 mM $Mg(OAc)_2$	
					0.5 mM DTT	
					0.25 mM PMSF	
					pH 7.0, 37°C	
12	Murine BiP re-	Peptide A:			20 mM HEPES	46
	combinant	no peptide	0.1	0.4	20 mM KCl	
		1 mM	0.15		10 mM $(NH_4)_2SO_4$	
					2 mM $Mg(OAc)_2$	
					0.5 mM DTT	
					0.25 mM PMSF	
					pH 7.0, 37°C	
13	Hamster BiP		0.4 ± 0.1	1.5 ± 0.2	20 mM HEPES	24
	recombinant				25 mM KCl	
					2 mM $MgCl_2$	
					0.1 mM EDTA	
					0.5 mM DTT	
					pH 7.0, 37°C	
14	*E. coli* DnaK		0.15		100 mM HEPES	35
	over-				10% (v/v) glycerol	
	expressed				35 mM KCl	
					5 mM $MgCl_2$	
					5 mM β-mercapto-	
					ethanol	
					70 μM ATP	
					pH 8.1, 37°C	

Table 2 Continued

No.	Protein	Ligand (conc.)	k_{cat} (min^{-1})	K_m (μM)	Assay condition	Ref.
15	*E. coli* DnaK over-expressed		1	200	50 mM Tris-HCl 2 mM $MgCl_2$ 0.2 mg mL^{-1} BSA 5 mM β-mercapto-ethanol 5% (v/v) glycerol pH 8.0, 30°C	68
16	*E. coli* DnaK		0.23	20	30 mM HEPES 40 mM KCl 50 mM NaCl 7 mM $Mg(OAc)_2$ 2 mM DTT 0.29 mg mL^{-1} BSA pH 7.6, 30°C	69
17	*E. coli* DnaK over-expressed	a-pp peptide: no peptide 5 μM	 0.13 0.13		25 mM HEPES 100 mM KCl 5 μM DnaK pH 7.0, 30°C	31
18	*E. coli* DnaK over-expressed	Peptide C: no peptide 500 μM	 0.035 0.24	3.9 ± 0.7	25 mM HEPES-KOH 50 mM KCl 11 mM $Mg(OAc)_2$ pH 7.6, 30°C	70
19	*E. coli* DnaK over-expressed	Peptide C: no peptide 200 μM	 0.045 0.41		25 mM HEPES-KOH 50 mM KCl 10 mM $Mg(OAc)_2$ 200 μM ATP pH 7.6, 30°C	16
20	*E. coli* DnaK over-expressed	No peptide 0.5 mM peptide A 0.5 mM peptide C	0.03 0.12 0.21		50 mM HEPES-KOH 50 mM KCl 10 mM $Mg(OAc)_2$ 0.2 mM ATP pH 7.6, 30°C	23
21	*S. cerevisiae* Ssa1P Natural		2.5 mM KCl 0.028 ± 0.003 15 mM KCl 0.046 ± 0.003 200 mM KCl 0.045 ± 0.007	 4.2 ± 1.0 0.68 ± 0.006 0.19 ± 0.07	50 mM Tris-HCl various [KCl] 2 mM $MgCl_2$ 5 mM β-mercapto-ethanol pH 7.5, 33°C	71

Table 2 Continued

No.	Protein	Ligand (conc.)	k_{cat} (min^{-1})	K_m (μM)	Assay condition	Ref.
22	*S. cerevisiae*	Peptide A7:			50 mM Tris-HCl	71
	Ssa1P	no peptide	5 mM KCl	0.01	various [KCl]	
			50 mM KCl	0.04		
	Natural	200 μM	5 mM KCl	0.01	2 mM $MgCl_2$	
			50 mM KCl	0.1	5 mM β-mercapto-ethanol	
					10 mM ATP	
					pH 7.5, 33°C	

Sequence of peptide A, KRQIYTDLEMNRLGK; a-pp peptide, CALLQSRLLLSAPRAAATARA; peptide C, KLIGVLSSLFRPK; peptide A7, RRLIEDAETAARG; S peptide, KETAAAKFERQHMDSSTSAA; cytochrome *c* peptide, IFAGIKKKAERADLIAYLKQATAK.

tion that will likely be supplanted at some time in the future by a more complex scheme, as may arise when Hsp70 proteins interact with accessory proteins [32].

It has been established at the phenomenological level that ATP binding (rather than ATP hydrolysis) induces a conformational change in Hsp70s [33]. Nonhydrolyzable ATP analogues fail to induce the change; furthermore, the transition is discriminatory relative to monovalent ions: Hsp70 proteins will undergo the conformational change to the ATP-induced form in the presence of K^+ ions, but not in the presence of Na^+ [33]. Mutagenesis studies have identified residues in the nucleotide-binding cleft, as seen in the isolated ATPase fragment, the alteration of which abolishes the ability of Hsp70 proteins to undergo an ATP-induced conformational change (Table 4; Fig. 5). In hamster BiP, mutagenesis of Gly-226 or Gly-227 (corresponding to Gly-201 and Gly-202, respectively, in bovine Hsc70) to aspartic acid results in a protein that has a substantially lower ATP hydrolysis rate and that does not undergo an ATP-induced transition, as monitored by proteolytic digestion patterns [24,34]. These results can be rationalized by suggesting that introduction of a large side chain at a position normally occupied by glycine results in a clash between the side chain and ATP when it binds. Mutagenesis of Thr-37 in BiP (corresponding to Thr-13 of Hsc70) to glycine also yields a protein that will not undergo the normal conformational change, suggesting that the side chain (possibly the hydroxyl) of this residue is essential for the coupling mechanism [24,34]. In contrast, some residues have been mutated without causing a loss of coupling; these include Thr-199 [35] and E-171 [36] in DnaK (corresponding to T-204 and E-175 in Fig. 6).

Combining these data with the observations showing that the coupling mechanism can be defeated by relatively subtle changes in the nucleotide

Table 3 Binding Affinities of Adenine Nucleotides for Hsp70 Proteins

Protein	Nucleotide	K_d (μM)	Assay conditions	Method/comments	Ref.
Bovine Hsc70 natural	ATP ADP	0.7 1.35	40 mM HEPES 75 mM KCl 4.5 mM $Mg(OAc)_2$ 0.8 mM DTT, 5% BSA pH 7.0, 37°C	Nucleotide binding proceeded for 5′; then, free nucleotide removed by charcoal	50
Bovine Hsc70 natural	ATP ADP	9.5 ± 3.9 1.6 ± 0.5	20 mM Na phosphate 200 mM KCl pH 6.5, 4°C	Equilibrium dialysis	49
Bovine Hsc70 natural	ATP ADP AMPPNP dATP	0.012 0.018 2.8 23	20 mM imidazole, 25 mM KCl 10 mM $(NH_4)_2SO_4$ 2 mM $Mg(OAc)_2$ 1 mM DTT pH 7.0, 25°C	Equilibrium dialysis: AMPPNP and dATP measured directly; ATP and ADP measured by competition with AMPPNP	65
Bovine Hsc70 E543K recombinant	ATP ADP	0.042 ± 0.007 0.11 ± 0.02	40 mM HEPES 75 mM KCl 4.5 mM $Mg(OAc)_2$ pH 7.0, 25°C	Equilibrium binding	17
Bovine Hsc70 E543K recombinant	ATP ADP	Two-step binding: $K_1 = 1.5$ μM $K_2 = 0.03$ 0.020 ± 0.002	40 mM HEPES 75 mM KCl 4.5 mM $Mg(OAc)_2$ pH 7.0, 25°C	Kinetics of change in tryptophan fluorescence when nucleotides bind	27
Bovine Hsc70 E543K recombinant	ATP	No peptide 0.011 ± 0.005 With peptide 0.022 ± 0.001	20 mM HEPES 150 mM KCl 3 mM $Mg(OAc)_2$ pH 7.0, 25°C	Kinetics of nucleotide binding	72

	ADP	No peptide 0.048 ± 0.005 With peptide 0.099 ± 0.014	(peptide = fluorescein-FYQLALT)		
Bovine Hsc70 44-kDa ATPase fragment	ADP	0.094 ± 0.012	40 mM HEPES 75 mM KCl 4.5 mM $Mg(OAc)_2$ pH 7.0, 25°C	Equilibrium binding and kinetics of nucleotide binding	17
Rat Hsc70 recombinant	ATP/ADP	0.2–0.3	25 mM Tris-HCl 120 mM NaCl 5 mM $MgCl_2$ pH 7.0, room temp.	Binding of nucleotide to protein immobilized on Sepharose	64
Rat Hsc70 recombinant WT, D10N	ATP/ADP	0.3	40 mM HEPES 75 mM KCl 5 mM $Mg(OAc)_2$ pH 7.0, 4°C	Binding of nucleotide to protein immobilized on Sepharose	59
E. coli DnaK	ATP	0.5	20 mM HEPES 100 mM NaCl 100 mM KCl ~ 10 nM Mg^{2+}, controlled with 20 mM EDTA 0.1 mg mL^{-1} BSA 5% glycerol, 5 mM DTT pH 7.6, 4°C	Equilibrium dialysis	73
Hamster BiP recombinant wt	ATP ADP	0.2 ± 0.1 0.29 ± 0.12	20 mM HEPES 25 mM KCl, 2 mM $MgCl_2$ 0.1 mM EDTA 0.5 mM DTT pH 7.0, 4°C	Equilibrium dialysis	24
T229G	ATP ADP	0.11 ± 0.1 0.15 ± 0.11			
T37G	ATP ADP	0.12 ± 0.11 0.16 ± 0.12			
G227, 226D	ATP,ADP	Not detectable			

Table 4 ATPase Activity of Missense Mutants of Hsp70 Proteins[a]

Protein	Mutation (origin)	Bov. hsc70 equivalent	k_{cat} (min^{-1})	$\frac{k_{cat}(\text{mutant})}{k_{cat}(\text{wt})}$	K_m(μM)	$\frac{K_m(\text{mutant})}{K_m(\text{wt})}$	Phenotype/ in vitro effects	Ref.
Rat Hsc70	D10N (sd)	D10	"Extremely low"	"Extremely low"	0.3 (K_d)	0.3 (K_d)	Reduced ability to autophosphorylate	59
E. coli DnaK	Δ28–33 (sd)		0.32 (V_{max})	2.2				74
Hamster BiP	T37G (sd)	T13	0.077 ± 0.038	0.19	0.12 ± 0.11 (K_d)	0.6	ATP-induced conformational change undetectable by limited proteolysis; binds peptide C, not released + ATP; ATPase not stimulated by peptide C	24
Hamster BiP	T37G (sd)	T13	Not reported	0.022 ± 0.049			No ATP-dependent release of IgG	75
Hamster BiP	T37G (sd)	T13	Not reported	0.025 ± 0.010			No autokinase activity	40
Hamster BiP	T37S (sd)	T13	Not reported	0.094 ± 0.029			3.8% w.t. autokinase activity	40
Hamster BiP	Δ175–201 (sd)	Y159–E175	Not reported	0.002 ± 0.005				75
Hamster BiP	E201G (sd)	E175	Not reported	0.026 ± 0.020			No ATP-dependent release of IgG	75
Hamster BiP	E201G (sd)	E175	Not reported	0.025 ± 0.008			65% w.t. autokinase activity	40
E. coli DnaK	E171K (sel)	E175	0.4	13	30 and 175 (two K_m observed)	30 and 175	Dominant, does not support λ replication, ≤ 1% activity in mini F′ replication; viable up to 34°C;	23

							ATPase activity 44-fold lower in absence of acetate; ATPase activity insensitive to peptides	
E. coli DnaK	E171A (sd)	E175	0.14	4.0	26 ± 8	6.7	Peptide C binding does not stimulate ATPase; ATP does not release peptide	70
E. coli DnaK	E171L (sd)	E175	0.09	2.6	48 ± 10	12	Peptide C binding does not stimulate ATPase; ATP does not release peptide	70
E. coli DnaK	E171K (sd)	E175	1.0	28	70 ± 7	18	Peptide C binding does not stimulate ATPase; ATP does not release peptide	70
E. coli DnaK	A174T (sel)	A178	0.014	0.46	1.5 and 23 (two K_m observed)	1.5 and 23	Recessive {does not support λ replication} ≤ 1% activity in mini F′ replication viable up to 40°C; ATPase stimulated by peptide	23
Hamster BiP	G226D (sd)	G201	0.13 ± 0.01	0.31	97 ± 14	65	ATP-induced conformational change undetectable by limited proteolysis; binds peptide C, not released when ATP added	24

Table 4 Continued

Protein	Mutation (origin)	Bov. hsc70 equivalent	k_{cat} (min^{-1})	$\frac{k_{cat}(\text{mutant})}{k_{cat}(\text{wt})}$	K_m(μM)	$\frac{K_m(\text{mutant})}{K_m(\text{wt})}$	Phenotype/ in vitro effects	Ref.
Hamster BiP	G227D (sd)	G202	0.11 ± 0.01	0.28	93 ± 17	62	ATP-induced conformational change undetectable by limited proteolysis; binds peptide C, not released when ATP added	24
Hamster BiP	T229G (sd)	T204	Not reported	0.037 ± 0.004			No ATP-dependent release for IgG	75
Hamster BiP	T229G (sd)	T204	None detected		0.11 ± 0.10 (K_d reported)	0.55	ATP-induced conformational change detectable by limited proteolysis; binds peptide C, and releases in an ATP-dependent fashion	24
Hamster BiP	T229G (sd)	T204	Not reported	0.061 ± 0.001			No autokinase activity	40
Hamster BiP	T229D (sd)	T204	Not reported	0.156 ± 0.024			No autokinase activity	40
Hamster BiP	T229S (sd)	T204	Not reported	0.217 ± 0.026			11.6% w.t. autokinase activity	40
E. coli DnaK	T199A (sd)	T204	0.003–0.007	0.03			No autokinase activity	35
E. coli DnaK	T199D (sd)	T204	0.009–0.011	0.07			No autokinase activity	35

E. coli DnaK	T199V (sd)	T204	0.001	0.007			No autokinase activity	35
Bovine Hsc70	T204E, E543K (sd/ca)		1.5 ± 0.3		97 ± 30		Reference to Hsc70 E543K mutant below	58
Bovine Hsc70	T204V, E543K (sd/ca)		0.07 ± 0.02		78 ± 34		Reference to Hsc70 E543K mutant below	58
E. coli DnaK	D201N (sel)	D206	0.31	10	8	8	Dominant; does not support λ replication; ≤ 1% activity in mini F′ replication viable up to 34°C; ATPase insensitive to peptide	23
E. coli Dnak756	G32D, G455D, G468D (sel)	G34D, G457D, G470D	12–18	~ 90			Temperature-sensitive; does not support λ replication	69
Hamster BiP	K446M, K447M (sd)	K423, Q424	"w.t. levels"	"w.t. levels"				75
Bovine Hsc70	E543K (ca)		0.072 ± 0.001 ($T = 25$°C)				Authors were not aware of mutation	17
Bovine Hsc70	E543K (ca)		0.2 ± 0.1		0.7 ± 0.5		Authors were not aware of mutation	58

[a]See Table 2 for assay conditions and comparable wild-type (w.t.) rates.
Origin: sd, site-directed; sel, selected in genetic screen; ca, cloning artifact.

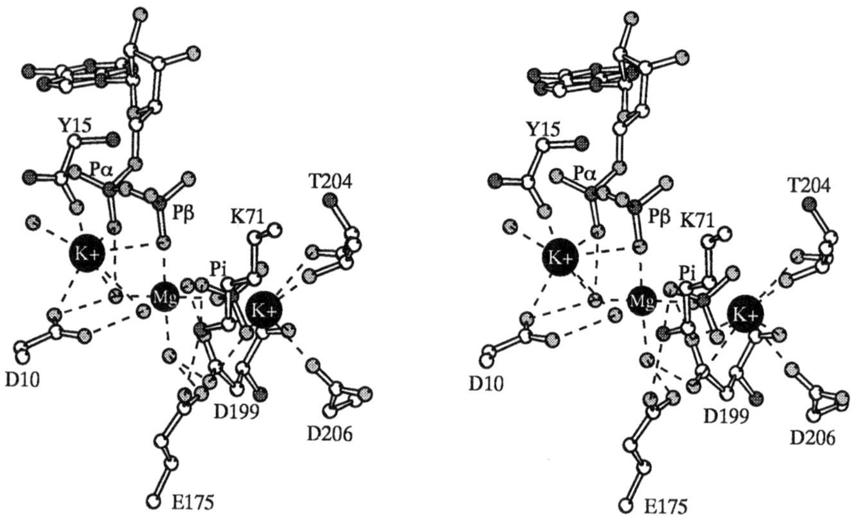

Figure 5 Stereo drawing showing selected features of the active site regions of the ATPase fragment, including MgADP, P_i, two K^+ ions, neighboring amino acid residues and H_2O molecules.

(substitution of nonhydrolyzable analogues for ATP) or monovalent ion (e.g., substitution of Na^+ for K^+), it can be suggested that MgATP must bind Hsp70 proteins in a complex with precise geometric constraints to trigger a conformational change, and that small perturbations of this complex defeat the coupling mechanism. The specific nature of this complex is not yet well-defined, because structures of representative fragments that undergo the change (either a full-length Hsp70 or a 60-kDa fragment with deletion of the COOH-terminal approximate 100 residues) are not available, and the structures of the ATPase and peptide-binding fragments have so far shown only a single overall conformation. A candidate for the nucleotide state that induces the change is a β,γ-bidentate MgATP complex that is a reaction intermediate state on the pathway of ATP hydrolysis, similar to the proposed intermediate shown in Figure 4. Consistent with this suggestion is the observation that ATP binding is a two-step process, and the conformational change appears to be predicated on the second step, after initial ATP binding [27].

V. STRUCTURAL ALTERATIONS: POSTTRANSLATIONAL MODIFICATION

Several posttranslational modifications have been documented for Hsp70 proteins. In vivo phosphorylation of serine and threonine has been seen in both prokaryotic DnaK [37] and eukaryotic Hsp70s [38–40]; the level of phosphorylation increases under conditions of cell stress or (for BiP) glucose starvation. In

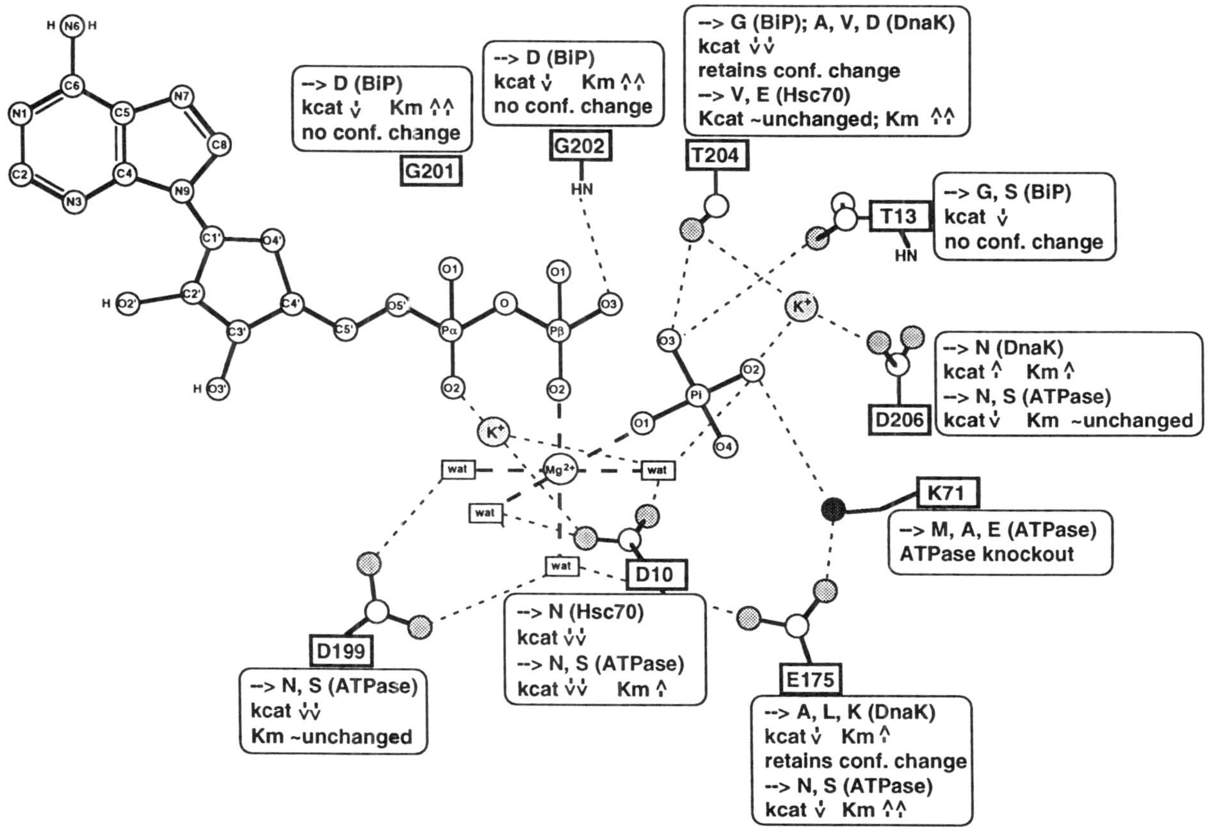

Figure 6 Schematic drawing summarizing the effects of mutagenesis of residues in the nucleotide-binding cleft of Hsp70 proteins. Refer to Tables 1 and 4 for details. “ATPase” refers to the ATPase fragment of bovine Hsc70; “Hsc70" refers to rat Hsc70: "DnaK" refers to *E. coli* DnaK: “BiP” refers to hamster BiP. Numbering corresponds to bovine Hsc70.

vitro Ca^{2+}-dependent phosphorylation of Thr-199 of DnaK (and the structurally equivalent residue of BiP) has been reported [35,41–43]; however, this targets a different residue from the in vivo phosphorylation and appears to be a distinct activity. ADP ribosylation of BiP has been documented; this modification has not been seen with other Hsp70 proteins [39,44]. In vivo, ADP ribosylated and phosphorylated BiP is localized in the oligomeric pools of the protein [39]. Methylation of lysine and arginine has been observed for eukaryotic Hsp70s, including both BiP and cytoplasmic proteins [45]. Differences between natural bovine BiP and recombinant mouse BiP in their propensity to self-aggregate, as well as in their basal ATPase activities, have been reported; the native bovine protein oligomerizes readily and has a lower basal ATPase rate, whereas the recombinant mouse protein is predominantly monomeric in vitro [46]. Posttranslational modification of the natural protein has been suggested as the source of the differences in behavior between these two forms of the proteins. However, although there appear to be intriguing correlations between different levels of posttranslational modification, on the one hand, and different oligomeric states of the Hsp70 proteins or different cellular circumstances (stressed versus nonstressed), on the other hand, the specific functional consequences of the known modifications are not yet understood.

VI. SELF-ASSOCIATION

It has been known for some time that Hsp70 proteins self-associate to form dimers and other low-order oligomers, and that these oligomers can be induced to dissociate to monomers by addition of ATP, but not by addition of nonhydrolyzable ATP analogues [47,48]. (Additionally, Hsp70 proteins form high molecular weight aggregates that cannot be dissociated by ATP; these appear to be inactive, dead-end aggregates [49]). The self-association parallels the interactions of Hsp70 proteins with target polypeptides and denatured proteins in many ways: it can be reversed by addition of short peptides (suggesting competition for the same peptide binding site [46]) and by addition of ATP [50]; stable complexes are formed in the presence of ADP. The region of Hsp70 proteins that is responsible for intramolecular association is not known with certainty; one candidate is the glycine-rich COOH-terminal 30- to 50-amino acid residues. Consistent with this suggestion is the observation that the 60-kDa fragments of Hsc70 and DnaK, which lack this COOH-terminal region, are substantially less prone to intramolecular aggregation [22,30]. The significance of the self-peptide-binding activity of Hsp70 proteins for their overall function is not yet clear; it may be a mechanism of self-regulating polypeptide-binding activity in vivo.

VII. SUMMARY

The activities with which Hsp70 proteins were first associated, including disassembly of a bacteriophage λ replication initiation complex by DnaK [51], disassembly of clathrin cages into triskelions by Hsc70 [52,53], facilitation of transmembrane translocation by Hsc70 [54,55], binding to immunoglobulin heavy chains (the consequence of which was nebulous) by BiP [56], renaturation of heat-denatured RNA polymerase by DnaK [3], were so diverse that it seemed inevitable that the Hsp70 proteins would be complex, multifunctional entities. However, both the structural studies summarized in this chapter and the biochemical studies summarized in other chapters of this volume appear to converge toward a relatively simple, unifying molecular mechanism. Hsp70 proteins employ a single nucleotide-binding site and a single peptide-binding site. The kinetics of peptide binding and release is regulated through interactions with ATP–ADP at the nucleotide-binding site; this is accomplished by nucleotide-induced switching between (to a first approximation) two conformations, one of which forms stable, long-lived complexes with peptides, and the other of which binds peptides only transiently. The peptide-binding domain has a unique structure that is adapted to its function of recognizing exposed hydrophobic segments of polypeptides.

However, describing the mechanism of Hsp70 proteins with such an elemental model fails to do justice to the diversity of functions ascribed to this protein family. Many questions still remain: What are the roles of the accessory proteins, particularly the DnaJ homologues, in modulating and targeting Hsp70 activities? What are the effects of posttranslational modifications on the functions of Hsp70s? Why are several different Hsp70 proteins needed in the cytoplasm–nucleus of eukaryotic cells, each having its own characteristic pattern of regulation of expression—do they have subtle differences in function that we have not yet identified? As the basic theme of structure and activity of isolated Hsp70 proteins becomes clarified, the question of how this protein family accommodates the many variations on the theme reflected in the proteins' complex and multifaceted in vivo activities emerges.

REFERENCES

1. Beckmann, R. P., Mizzen, L. A., and Welch, W. J., Interaction of Hsp 70 with newly synthesized proteins: Implications for protein folding and assembly, *Science 248*:850 (1990).
2. Scherer, P. E., Krieg, U. C., Hwang, S. T., Vestweber, D., and Schatz, G., A precursor protein partly translocated into yeast mitochondria is bound to a 70 kd mitochondrial stress protein, *EMBO J. 9*:4315 (1990).
3. Skowyra, D., Georgopoulos, C., and Zylicz, M., The *E. coli dnaK* gene product, the

Hsp70 homolog, can reactivate heat-inactivated RNA polymerase in an ATP hydrolysis-dependent manner, *Cell 62*:939 (1990).
4. Freeman, B. C., Myers, M. P., Schumacher, R., and Morimoto, R. I., Identification of a regulatory motif in Hsp70 that affects ATPase activity, substrate binding and interaction with HDJ-1, *EMBO J. 14*:2281 (1995).
5. Flaherty, K. M., DeLuca-Flaherty, C., and McKay, D. B., Three-dimensional structure of the ATPase fragment of a 70 kilodalton heat shock cognate protein, *Nature 346*:623 (1990).
6. Zhu, X., Zhao, X, Burkholder, W. F., Gragerov, A., Ogata, C. M., Gottesman, M. E., and Hendrickson, W. A., Structural analysis of substrate binding by the molecular chaperone DnaK, *Science 272*:1606 (1996).
7. Kabsch, W., Mannherz, H. G., Suck, D., Pai, E. F., and Holmes, K. C., Atomic structure of the actin:DNase I complex, *Nature 347*:37 (1990).
8. Flaherty, K. M., McKay, D. B., Kabsch, W., and Holmes, K. C., Similarity of the three-dimensional structures of actin and the ATPase fragment of a 70-kDa heat shock cognate protein, *Proc. Natl. Acad. Sci. USA 88*:5041 (1991).
9. Flajnik, M. F., Canel, C., Kramer, J., and Kasahara, M., Which came first, MHC class I or class II? *Immunogenetics 33*:295 (1991).
10. Rippmann, F., Taylor, W. R., Rothbard, J. B., and Green, N. M., A hypothetical model for the peptide binding domain of hsp70 based on the peptide binding domain of HLA, *EMBO J. 10*:1053 (1991).
11. Morshauser, R. C., Wang, H., Flynn, G. C., and Zuiderweg, E. R. P., The peptide-binding domain of the chaperone protein Hsc70 has an unusual secondary structure topology, *Biochemistry 34*:6261 (1995).
12. Flynn, G. C., Pohl, J., Flocco, M. T., and Rothman, J. E., Peptide-binding specificity of the molecular chaperone BiP, *Nature 353*:726 (1991).
13. Flaherty, K. M., Wilbanks, S. M., DeLuca-Flaherty, C., and McKay, D. B., Structural basis of the Hsc70 ATP hydrolytic activity. II. Structure of the active site with ADP or ATP bound to wild-type and mutant ATPase fragment, *J. Biol. Chem. 269*:12899 (1994).
14. O'Brien, M. C., Flaherty, K. M., and McKay, D. B., Lysine 71 of the chaperone protein Hsc70 is essential for ATP hydrolysis, *J. Biol. Chem. 271*:15874 (1996).
15. Gao, B., Emoto, Y., Greene, L., and Eisenberg, E., Nucleotide binding properties of bovine brain uncoating ATPase, *J. Biol. Chem. 268*:8507 (1993).
16. McCarty, J. S., Buchberger, A., Reinstein, J., and Bukau, B., The role of ATP in the functional cycle of the DnaK chaperone system, *J. Mol. Biol. 249*:126 (1995).
17. Ha, J.-H. and McKay, D. B., ATPase kinetics of recombinant bovine 70 kilodalton heat shock cognate protein and its amino-terminal ATPase domain, *Biochemistry 33*:14625 (1994).
18. Wilbanks, S. M., DeLuca-Flaherty, C., McKay, D. B., Structural basis of the Hsc70 ATP hydrolytic activity. I. Kinetic analyses of active site mutants, *J. Biol. Chem. 269*:12893 (1994).
19. O'Brien, M. C. and McKay, D. B., How potassium affects the activity of the molecular chaperone Hsc70: I. Potassium is required for optimal ATPase activity, *J. Biol. Chem. 270*:2247 (1995).
20. Wilbanks, S. M. and McKay, D. B., How potassium affects the activity of the

molecular chaperone Hsc70: II. Potassium binds specifically in the ATPase active site, *J. Biol. Chem. 270*:2251 (1995).
21. Liberek, K., Skowyra, D., Zylicz, M., Johnson, C., and Georgopoulos, C., The *Escherichia coli* DnaK chaperone, the 70-kDa heat shock protein eukaryotic equivalent, changes conformation upon ATP hydrolysis, thus triggering its dissociation from a bound target protein, *J. Biol. Chem. 266*:14491 (1991).
22. Buchberger, A., Theyssen, H., Schroder, H., McCarty, J. S., Virgallita, G., Milkereit, P., Reinstein, J., and Bukau, B., Nucleotide-induced conformational changes in the ATPase and substrate binding domains of the DnaK chaperone provide evidence for interdomain communication, *J. Biol. Chem. 270*:16903 (1995).
23. Kamath-Loeb, A. S., Lu, C. Z., Suh, W. C., Lonetto, M. A., and Gross, C. A., Analysis of three DnaK mutant proteins suggests that progression through the ATPase cycle requires conformational changes, *J. Biol. Chem. 270*:30051 (1995).
24. Wei, J., Gaut, J. R., and Hendershot, L. M., In vitro dissociation of BiP-peptide complexes requires a conformational change in BiP after ATP binding but does not require ATP hydrolysis, *J. Biol. Chem. 270*:26677 (1995).
25. Palleros, D. R., Reid, K. L., McCarty, J. S., Walker, G. C., and Fink, A. L., DnaK, hsp73, and their molten globules. Two different ways heat shock proteins respond to heat, *J. Biol. Chem. 267*:5279 (1992).
26. Banecki, B. and Zylicz, M., Real time kinetics of the DnaK–DnaJ–GrpE molecular chaperone machine action, *J. Biol. Chem. 271*:6137 (1996).
27. Ha, J.-H. and McKay, D. B., Kinetics of nucleotide-induced changes in the tryptophan fluorescence of the molecular chaperone Hsc70 and its subfragments suggest the ATP-induced conformational change follows initial ATP binding, *Biochemistry 34*:11635 (1995).
28. Banecki, B., Zylicz, M., Bertoli, E., and Tanfani, F., Structural and functional relationships in DnaK and DnaK756 heat-shock proteins from *Escherichia coli*, *J. Biol. Chem. 267*:25051 (1992).
29. Shi, L., Kataoka, M., and Fink, A. L., Conformational characterization of DnaK and its complexes by small-angle X-ray scattering, *Biochemistry 35*:3297 (1996).
30. Wilbanks, S. M., Chen, L., Tsuruta, H., Hodgson, K. O., and McKay, D. B., Solution small-angle X-ray scattering study of the molecular chaperone Hsc70 and its subfragments, *Biochemistry 34*:12095 (1995).
31. Schmid, D., Baici, A., Gehring, H., and Christen, P., Kinetics of molecular chaperone action, *Science 263*:971 (1994).
32. Wawrzynów, A. and Zylicz, M., Divergent effects of ATP on the binding of the DnaK and DnaJ chaperones to each other, or to their various native and denatured protein substrates, *J. Biol. Chem. 270*:19300 (1995).
33. Palleros, D. R., Reid, K. L., Shi, L., Welch, W. J., and Fink, A. L., ATP-induced protein–Hsp70 complex dissociation requires K^+ but not ATP hydrolysis, *Nature 365*:664 (1993).
34. Wei, J. and Hendershot, L. M., Characterization of the nucleotide binding properties and ATPase activity of recombinant hamster BiP purified form bacteria, *J. Biol. Chem. 270*:26670 (1995).
35. McCarty, J. S. and Walker, G. C., DnaK as a thermometer: Threonine-199 is site of

autophosphorylation and is critical for ATPase activity, *Proc. Natl. Acad. Sci. USA 88*:9513 (1991).

36. Schuster, H.-P., Theyssen, H., Buchberger, A., Bukau, B., and Reinstein, J., The mutation E171A of the molecular chaperone DnaK is deficient in ATP hydrolysis but still couples the binding of ATP to peptide release [poster abstract: A90], Conference on Biology of Molecular Chaperones, Aghia Pelaghia, Greece, 1995.
37. Rieul, C., Cortay, J. C., Bleicher, F., and Cozzone, A. J., Effect of bacteriophage M13 infection on phosphorylation of dnaK protein and other *Escherichia coli* proteins, *Eur. J. Biochem. 168*:621 (1987).
38. Loomis, W., Wheeler, S., and Schmidt, J., Phosphorylation of the major heat shock protein of *Dictyostelium discoideum*, *Mol. Cell. Biol. 2*:484 (1982).
39. Freiden, P. J., Gaut, J. R., and Hendershot, L. M., Interconversion of three differentially modified and assembled forms of BiP, *EMBO J. 11*:63 (1992).
40. Gaut, J. R. and Hendershot, L. M., The immunoglobulin-binding protein in vitro autophosphorylation site maps to a threonine within the ATP binding cleft but is not a detectable site of in vivo phosphorylation, *J. Biol. Chem. 268*:12691 (1993).
41. Leustek, T., Amir, S. D., Toledo, H., Brot, N., and Weissbach, H., Autophosphorylation of 70 kDa heat shock proteins, *Cell. Mol. Biol. 38*:1 (1992).
42. Panagiotidis, C. A., Burkholder, W. F., Gaitanaris, G. A., Gragerov, A., Gottesman, M. E., and Silverstein, S. J., Inhibition of DnaK autophosphorylation by heat shock proteins and polypeptide substrates, *J. Biol. Chem. 269*:16643 (1994).
43. Dalie, B. L., Skaleris, D. A., Kohle, K., Weissbach, H., and Brot, H., Interaction of DnaK with ATP: Binding, hydrolysis and Ca^{+2}-stimulated autophosphorylation, *Biochem. Biophys. Res. Commun. 166*:1284 (1990).
44. Carlsson, L. and Lazarides, E., ADP-ribosylation of the M_r 83,000 stress-inducible and glucose-regulated protein in avian and mammalian cells: Modulation by heat shock and glucose starvation, *Proc. Natl. Acad. Sci. USA 80*:4664 (1983).
45. Wang, C., Lin, J. M., and Lazarides, E., Methylations of 70,000-Da heat shock proteins in 3T3 cells: alterations by arsenite treatment, by different stages of growth and by virus transformation, *Arch. Biochem. Biophys. 297*:169 (1992).
46. Blond-Elguindi, S., Fourie, A. M., Sambrook, J. F., and Gething, M. J., Peptide-dependent stimulation of the ATPase activity of the molecular chaperone BiP is the result of conversion of oligomers to active monomers, *J. Biol. Chem. 268*:12730 (1993).
47. Schmid, S. L. and Rothman, J. E., Enzymatic dissociation of clathrin cages in a two-stage process, *J. Biol. Chem. 260*:10044 (1985).
48. Ladjimi, M. M., Benaroudj, N., Batelier, G., and Triniolles, F., Self-association of the molecular chaperone HSC70 as assessed by analytical ultracentrifugation, *Prog. Colloid Polym. Sci. 99*:1 (1995).
49. Palleros, D. R., Welch, W. J., and Fink, A. L., Interaction of hsp70 with unfolded proteins: Effects of temperature and nucleotides on the kinetics of binding, *Proc. Natl. Acad. Sci. USA 88*:5719 (1991).
50. Schmid, S. L., Braell, W. A., and Rothman, J. E., ATP catalyzes the sequestration of clathrin during enzymatic uncoating, *J. Biol. Chem. 260*:10057 (1985).
51. Liberek, K., Georgopoulos, C., and Zylicz, M., Role of the *Escherichia coli* DnaK

and DnaJ heat shock proteins in the initiation of bacteriophage lambda DNA replication, *Proc. Natl. Acad. Sci. USA 85*:6632 (1988).
52. Schlossman, D. M., Schmid, S. L., Braell, W. A., and Rothman, J. E., An enzyme that removes clathrin coats: Purification of an uncoating ATPase, *J. Cell Biol. 99*:723 (1984).
53. Chappell, T. G., Welch, W. J., Schlossman, D. M., Palter, K. B., Schlesinger, M. J., and Rothman, J. E., Uncoating ATPase is a member of the 70 kilodalton family of stress proteins, *Cell 45*:3 (1986).
54. Deshaies, T. J., Koch, B. D., Werner-Washburne, M., Craig, E. A., and Schekman, R., A subfamily of stress proteins facilitates translocation of secretory and mitochondrial precursor polypeptides, *Nature 332*:800 (1988).
55. Chirico, W. J., Waters, M. G., and Blobel, G., 70K heat shock related proteins stimulate protein translocation into microsomes, *Nature 332*:805 (1988).
56. Munro, S. and Pelham, H. R., An Hsp70-like protein in the ER: Identity with the 78 kd glucose-regulated protein and immunoglobulin heavy chain binding protein, *Cell 46*:291 (1986).
57. Kraulis, P., MOLSCRIPT: A program to produce both detailed and schematic plots of protein structures, *J. Appl. Cryst. 24*:946 (1991).
58. O'Brien, M. C. and McKay, D. B., Threonine 204 of the chaperone protein hsc70 influences the structure of the active site but is not essential for ATP hydrolysis, *J. Biol. Chem. 268*:24323 (1993).
59. Huang, S., Tsai, M.-Y., Tzou, Y.-M., Wu, W., and Wang, C., Aspartyl residue 10 is essential for ATPase activity of rat hsc70, *J. Biol. Chem. 268*:2063 (1993).
60. Braell, W. A., Schlossman, D. M., Schmid, S. L., and Rothman, J. E., Dissociation of clathrin coats coupled to the hydrolysis of ATP: Role of an uncoating ATPase, *J. Cell Biol. 99*:734 (1984).
61. Flynn, G. C., Chappell, T. G., and Rothman, J. E., Peptide binding and release by proteins implicated as catalysts of protein assembly, *Science 245*:385 (1989).
62. DeLuca-Flaherty, C., McKay, D. B., Parham, P., and Hill, B. L., Uncoating protein (hsc70) binds a conformationally labile domain of clathrin light chain LC_a to stimulate ATP hydrolysis, *Cell 62*:875 (1990).
63. Sadis, S. and Hightower, L. E., Unfolded proteins stimulate molecular chaperone Hsc70 ATPase by accelerating ADP/ATP exchange, *Biochemistry 31*:9406 (1992).
64. Wang, C. and Lee, M. R., High-level expression of soluble rat hsc70 in *Escherichia coli*: Purification and characterization of the cloned enzyme, *Biochem. J. 294*:69 (1993).
65. Gao, B., Greene, L., and Eisenberg, E., Characterization of nucleotide-free uncoating ATPase and its binding to ATP, ADP, and ATP analogues, *Biochemistry 33*:2048 (1994).
66. Greene, L. E., Zinner, R., Naficy, S., and Eisenberg, E., Effect of nucleotide on the binding of peptides to 70-kDa heat shock protein, *J. Biol. Chem. 270*:2967 (1995).
67. Kassenbrock, C. K. and Kelly, R. B., Interaction of heavy chain binding protein (BiP/GRP78) with adenine nucleotides, *EMBO J. 8*:1461 (1989).
68. Zylicz, M., LeBowitz, J. H., McMacken, R., and Georgopoulos, C., The DnaK protein of *Escherichia coli* possesses an ATPase and autophosphorylating activity and is

essential in an in vitro DNA replication system, *Proc. Natl. Acad. Sci. USA 80*:6431 (1983).
69. Liberek, K., Marszalek, J., Ang, D., Georgopoulos, C., and Zylicz, M., *Escherichia coli* DnaJ and GrpE heat shock proteins jointly stimulate ATPase activity of DnaK, *Proc. Natl. Acad. Sci. USA 88*:2874 (1991).
70. Buchberger, A., Valencia, A., McMacken, R., Sander, C., and Bukau, B., The chaperone function of DnaK requires the coupling of ATPase activity with substrate binding through residue E171, *EMBO J. 13*:1687 (1994).
71. Ziegelhoffer, T., Lopez, B. P., and Craig, E. A., The dissociation of ATP from hsp70 of *Saccharomyces cerevisiae* is stimulated by both Ydj1p and peptide substrates, *J. Biol. Chem. 270*:10412 (1995).
72. Takeda, S. and McKay, D. B., Kinetics of peptide binding to the bovine 70 kDa heat shock cognate protein, a molecular chaperone, *Biochemistry 35*:4636 (1996).
73. Skowyra, D. and Wickner, S., GrpE alters the affinity of DnaK for ATP and Mg^{2+}. Implications for the mechanism of nucleotide exchange, *J. Biol. Chem. 270*:26282 (1995).
74. Buchberger, A., Schröder, H., Büttner, M., Valencia, A., and Bukau, B., A conserved loop in the ATPase domain of the DnaK chaperone is essential for stable binding of GrpE, *Stuct. Biol. 1*:95 (1994).
75. Gaut, J. R. and Hendershot, L. M., Mutations within the nucleotide binding site of immunoglobulin-binding protein inhibit ATPase activity and interfere with release of immunoglobulin heavy chain, *J. Biol. Chem. 268*:7248 (1993).

5

The Hsp70 Reaction Cycle and Its Role in Protein Folding

Anthony L. Fink
University of California, Santa Cruz, California

I. INTRODUCTION

The focus of this chapter is on the role of the Hsp70 chaperones in protein folding and, in particular, the molecular details of the Hsp70 reaction cycle. Aspects of the structure, ATPase activity, and peptide specificity may be found in Chapters 4 and 6. The current picture for the overall role of Hsp70 in protein folding is that it binds unfolded or partially folded intermediate states of proteins (including nascent polypeptides) and prevents their aggregation. Thus, folding of a nascent polypeptide probably involves spontaneous folding at low concentrations of intermediates or through interaction with the Hsp60 chaperone system [1; see, however, 2] after release from Hsp70 (i.e., Hsp70 sequesters most of the nonnative molecules preventing their aggregation). In a similar fashion, nascent polypeptides are maintained in a translocation-competent (nonnative) state through being bound to Hsp70. Under conditions of stress, resulting in denatured proteins, the Hsp70s are thought to bind to the damaged proteins and either facilitate their subsequent renaturation or earmark them for degradation (see Chap. 20). Several recent reviews summarize the role of Hsp70 molecular chaperones in protein folding [1,3–10].

II. THE NEED FOR MOLECULAR CHAPERONES IN PROTEIN FOLDING

Many studies, both in vivo and in vitro, have shown that the transition of a protein from unfolded to folded state frequently results in the formation of partially folded intermediate states that have a very strong propensity to aggregate [11–14; see Chap. 1]. In vivo this may lead to formation of inclusion bodies when overexpression occurs, in vitro to soluble aggregates or visible precipitation; in both situations, reduced yields of soluble, native protein result. In response to the aggregation and misfolding of newly synthesized proteins, cells have developed molecular machinery, the chaperones, to circumvent this problem. Members of the Hsp60 and Hsp70 molecular chaperone families and their cofactors (including Hsp40) seem to be most directly, and most generally, involved. For example, repression of the Hsp70 chaperones in *Escherichia coli* leads to a dramatic increase in insoluble, newly synthesized protein, which can be eliminated by overexpression of the Hsp60 and Hsp70 chaperone machinery [15]. Direct evidence for the in vivo interaction of Hsp70 with nascent polypeptides was initially reported using coimmunoprecipitation [16]. Subsequent investigations have also implicated Hsp40 in the initial interactions with the nascent chain [17–20].

The Hsp70s are both constitutive and stress-induced. The constitutive roles appear to be mostly concerned with nascent polypeptides and are better understood than the role of Hsp70 in aiding cells to cope with stress. It is likely that only some newly synthesized proteins require the assistance of chaperones: in the coimmunoprecipitation studies of Welch [16] it was observed that smaller proteins were disproportionately absent, suggesting that they may fold more rapidly, either with or without the assistance of Hsp70 (and it has been calculated that there is only enough GroEL available for about 10% of nascent polypeptides [21]). Although most globular proteins will refold spontaneously in vitro in the absence of additional agents (with the exception of certain ones that involve posttranslational modifications), successful refolding may require very low protein concentrations to avoid aggregation. In contrast, in vivo the concentrations of nascent and folding proteins may be much higher. There is also some controversy about the role of misfolded or kinetically trapped partially folded intermediates during refolding; it is not yet clear how common these are, nor what the role of chaperones may be relative to them. It is certainly possible that misfolded intermediates are quite common in folding, and that they represent the main slow-folding species observed in in vitro experiments: the question thus arises of whether it is only, or predominantly, misfolded intermediates that interact with chaperones. At this point there is no reason to believe that this is true. A major difference between in vivo and in vitro folding in the presence of chaperones is that the former involves the vectorial presentation of the growing polypeptide to the intracellular chaperones, starting with the NH_2-terminal region, whereas in vitro folding allows the complete polypeptide chain to interact with the chaperones. There is some suggestion that this may result in significantly different chaperone effects [22]. Figure 1 illustrates some of the competing pathways for a nascent polypeptide.

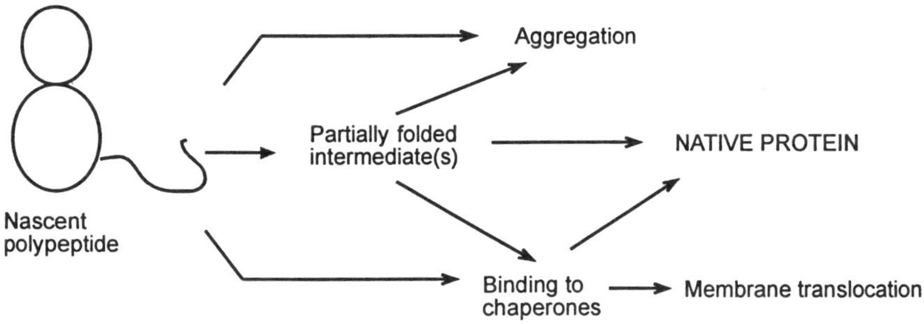

Figure 1 Overview of the role of molecular chaperones in the in vivo folding of newly synthesized proteins.

III. THE Hsp70 CHAPERONE FAMILY

The Hsp70 family of molecular chaperones is involved in a variety of cellular processes, including binding to nascent polypeptides [16], protein transport [23], disassembly of clathrin-coated vesicles [24], regulation of the heat-shock response [25], and bacteriophage DNA replication [26]. Thus, Hsp70s are involved in both constitutive, "housekeeping" functions in cells under nonstress conditions, as well as induced activities following a variety of stresses [3,5,7,8,27]. In *E. coli,* the main Hsp70 is DnaK, which requires two protein cofactors, DnaJ (an Hsp40) and GrpE, in at least some, if not all, of its functions. At least nine Hsp70 genes have been identified in humans and 14 in yeast (the cytosolic Ssa1p is the best studied yeast Hsp70). In mammalian systems, well-studied examples are the constitutive cytosolic members Hsc70 (also known as Hsp73), and the inducible form, usually referred to as Hsp70 or Hsp72; BiP, found in the endoplasmic reticulum (ER) and mitochondrial Hsp70. The basic paradigm of the function of Hsp70 is that it binds unfolded proteins or peptides to form a complex, which is dissociated on binding and hydrolysis of ATP. Thus, the two features common to all the functions associated with Hsp70 are (a) the binding to unfolded polypeptides [or (unfolded) regions of certain native proteins, see Sec. IV.A), and (b) their release on ATP binding and hydrolysis [28–33].

Multiple Hsp70s may be found in the same cell, but in different cellular locations, and they have somewhat different functions and specificity. Major sites are the cytosol, ER, and mitochondria in eukaryotes. The *E. coli* system seems to differ from that of eukaryotes in that two cofactors, DnaJ ($M_r = 41{,}000$; a member of the Hsp40 family) and GrpE ($M_r = 23{,}000$) are associated with Hsp70 function [34–36], whereas in eukaryotes only the DnaJ homologues have been found, except for mitochondria [37]. All three protein components also bind to the *E. coli* heat-shock–specific σ^{32} subunit of RNA polymerase [6,38,39].

GrpE functions as a nucleotide-exchange factor [40]; the role of DnaJ is less clear. DnaJ and its homologues stimulate the ATPase activity of Hsp70, and there is evidence that DnaJ enhances the interaction of DnaK with unfolded substrates, possibly by initially binding to them. The effects of the cofactors on Hsp70 are discussed in more detail later, and in Chapter 11.

The Hsp70 family members consist of two functional domains. A highly conserved NH_2-terminal domain (44 kDa) that contains the nucleotide-binding site (and ATPase activity) and a slightly less conserved COOH-terminal domain containing the substrate (unfolded proteins and peptides)-binding domain. The three-dimensional structure of the NH_2-terminal fragment of bovine brain Hsc70 is known at 2.2-Å resolution [41; see Chap. 4], and that the COOH-terminal region of DnaK at 2.0-Å resolution [42].

IV. STEPS IN THE Hsp70 REACTION CYCLE

The overall reaction cycle of Hsp70 must involve the following components: (a) binding of substrate protein or peptide, (b) binding of ATP, (c) hydrolysis of ATP, (d) release of substrate protein, and (e) the release of ADP (to allow another round of binding by ATP). Ascertainment of the order of these processes in the cycle has been just one of many challenges associated with elucidating the reaction cycle. In addition, substantial conformational changes are expected in connection with at least some of these steps, because communication between the two structural domains is clearly necessary for proper function. Furthermore, the interaction and regulatory effect of the Hsp70-associated factors, such as Hsp40, as well as nucleotides and metal ions must be factored into the reaction cycle mechanism. In the following subsections, various features associated with the main steps of the cycle are reviewed.

A. Binding of Substrate Protein

The peptide-binding region of Hsp70 was first localized to an 18-kDa region at the beginning of the COOH-terminal domain [43]. Because Hsp70s appear to bind to all, or at least most, unfolded or denatured proteins, the chaperones must be quite promiscuous in their affinity for polypeptides in their extended conformation. This has been demonstrated in several different ways, ranging from investigations of binding specificity toward peptides [28,44–49], to demonstrations that such bound peptides are in an extended conformation [50]. The recently determined structure of a complex between the peptide-binding domain of DnaK and a peptide [42] has confirmed earlier conclusions; namely, that the substrate polypeptide binds in an extended conformation, and that the maximum affinity will be for a five- to seven-residue stretch centered on a leucine, or related nonpolar residue, with hydrophobic residues on either side, and with weak preference for positively charged residues adjacent to these. The central-binding site appears to contribute most to the affinity for the substrate protein. Thus, Hsp70 shows preferential affinity for unfolded polypeptide segments involving significant hydrophobicity. There is a strong negative correlation with negatively charged side chains in the binding region, which has a structural basis in that unfavorable electrostatic interactions would occur with the surrounding chaperone residues [42].

The structure of the peptide-binding domain of DnaK, complexed with peptide, reveals that the peptide is bound in a novel, and unusual, way [42]. The major features of the binding domain consist of a β-sandwich subdomain and an α-helical subdomain. The loops and sheets of the β-subdomain form the binding sites for seven contiguous residues in a polypeptide. However, the α-helical subdomain appears to play a critical role by forming a flap that can "trap" the substrate peptide or protein, thereby preventing its departure. Hence, the structure of the substrate (peptide)-binding domain of Hsp70, particularly the α-helical lid, suggests that major conformational changes are required, both on binding and release of the substrate polypeptide.

As noted, Hsp70 binds to several native proteins (such as clathrin [24,29], the heat-shock factor σ^{32} [51], p53 [52], and phage λ replication proteins O and P [53]), presumably to extended surface loop segments, for peptides that bind to Hsp70 inhibit these interactions [54]. DnaJ is reported to also bind to at least some of these proteins [55].

Under typical physiological conditions, cellular ATP levels are several-fold higher than those of ADP: because Hsp70 binds both nucleotides very tightly (nanomolar K_d) [56–62], most of Hsp70 will be complexed with ATP under normal (nonstress) conditions, and the remainder will be in a complex with ADP. Thus under in vivo conditions, the reaction cycle will not involve free Hsp70, but rather, a complex with nucleotide. Size exclusion–high-performance liquid chromatography (SEC–HPLC) experiments show that stable complexes with substrate protein are formed only with the ADP complex, reflecting the much slower dissociation rate of the substrate protein from the Hsp70–ADP complex (and the nucleotide-free complex), compared with that of the Hsp70–ATP complex [31,54,60,63].

Comparison of the rates of binding of a fluorescence-labeled peptide to DnaK in the presence and absence of ATP revealed that binding (and dissociation) were very much faster in the presence of ATP than with ADP or no nucleotide [64]. Subsequent studies on the binding of a different fluorescence-labeled peptide to bovine Hsc70 showed biphasic kinetics for Hsc70–ADP, whereas binding to Hsc70–ATP was monophasic [65]. The association and dissociation rates for the first step of the binding of the peptide with Hsc70 are similar for the complexes with either nucleotide, whereas it is the slow-dissociation rate in the second phase of the interaction with Hsc70–ADP that accounts for the formation of a stable complex with peptide and the Hsc70–ADP, but not with the Hsc70–

ATP complex. The kinetics of peptide–substrate-binding to, and dissociation from, DnaK, especially the latter, are very strongly dependent on temperature [63,66]. Hsp70s' affinity for substrate polypeptides varies substantially depending on the polypeptide sequence and chaperone specificity: reported K_d ranges from submicromolar to millimolar.

B. ATP Hydrolysis and Nucleotide Exchange

Noteworthy features of the intrinsic ATPase activity of Hsp70 are that the turnover rates are so slow (on the order of 5×10^{-4} s^{-1}), and that the ATPase activity is stimulated a fewfold by substrate proteins or peptides. In the absence of the COOH-terminal domain, the ATPase domain exhibits maximum ATPase activity; therefore, it is the interactions between the domains that result in a depressed activity in the intact protein [60]. Detailed studies of the ATPase cycle of Hsc70 and DnaK in the absence of cofactors have been reported [33,60,65], involving measurements of both steady-state and pre–steady-state kinetics, as well as determination of the dissociation constants. For a fuller description, see Chapter 4.

There has been some controversy about the rate-limiting step in the Hsp70 reaction cycle. It has been suggested that the rate-limiting step for DnaK is different from that of mammalian cytosolic Hsp70 [67]. In addition, reported effects of ATP analogues may often be in error owing to the presence of contaminating nucleotides, such as ADP [61]. Another source of potentially apparent experimental discrepancies arises from the presence of bound nucleotide in supposedly nucleotide-free samples of Hsp70.

Initially, the rate-limiting step for the Hsc70 cycle in the absence of cofactors was reported to be the release of ADP, which was stimulated twofold by the presence of substrate protein (57): subsequent investigations, however, question this conclusion [68–69]. Several recent investigations indicate that, for both Hsc70 and DnaK, the rate-limiting step in the ATPase cycle is hydrolysis of ATP (i.e., cleavage of the γ-phosphate bond) [33,58,60,62,67,70]. Dissociation of ADP from Hsp70 is considerably faster than hydrolysis of the γ-phosphate, and it appears to be somewhat faster with human Hsc70 than with DnaK; however, ATP binds more tightly than ADP. In addition, P_i also binds significantly, especially to the complex with ADP [60,63]. ATP binds in a two-step process, in which the second step leads to a conformational change resulting in substrate polypeptide release (discussed later) [33,71].

For the ATP hydrolysis cycle, the minimum steps must be binding of ATP, hydrolysis of the γ-phosphate, leading to bound ADP and, finally, release of the ADP to start another round. Given that the rate-limiting step with DnaK is hydrolysis of ATP [33,70], rather than release of ADP, the question then arises of the need for the nucleotide exchange factor GrpE (see later). The answer could be either or both of the following reasons in which ADP release becomes rate-limiting: (a) the presence of DnaJ leads to the rapid formation of a complex involving DnaK, substrate protein, and ADP; and (b) other pathways leading to a DnaK–ADP–substrate protein complex. Ultimately, the basic reason is presumably a difference in rate-limiting step between DnaK and eukaryotic Hsp70s (except mitochondiral Hsp70, which also uses a GrpE homologue) under normal physiological conditions. There is some evidence to suggest that, in eukaryotes, the Hsp40 may also increase nucleotide-release rates [68].

Those functions of Hsp70 that involve ATP require the presence of both Mg^{2+} and K^+ for maximal effect [31,72]; this is because of the structural requirements in which the two ions are essential for proper binding and orientation of ATP [73; see Chap. 4].

Initial clues to the role of the cofactors DnaJ and GrpE on the molecular function of DnaK came from the observation that the combination of both cofactors substantially stimulated the ATPase acitivity of DnaK (in the absence of substrate polypeptides), whereas individually they had little effect [40]. Both Hsc70 and DnaK ATPase rates are slightly stimulated by Hsp40 or DnaJ [67,74]. However, the stimulating effect of DnaJ on DnaK ATPase activity occurs only in the absence of substrate polypeptide, whereas that of GrpE occurs only in the presence of substrate [62]. The human Hsp40 homologues of DnaJ also stimulate the ATPase activity of human Hsc70 by increasing the rate of hydrolysis of ATP, but have no effect on the rate of ADP release [74]; binding of Hsp40 induces a conformational change in Hsp70. In yeast, the ATPase activity of Ssa1p, the predominant cytosolic Hsp70, is enhanced by Ydj1, the yeast homologue of DnaJ: both Ydj1 and substrate polypeptides increased the rate of ATP dissociation from Hsp70 [68].

Several critical residues in the ATPase activity of Hsp70 have been identified, including Asp-10 [72,75], Thr-199 in DnaK (Thr-204 in Hsc70) [76,77], and Lys-71 in Hsc70 [78]. Coupling between peptide binding and ATPase activity has been demonstrated, and involves a conserved glutamate residue (Glu-171 in DnaK, Glu-175 in Hsc70) [79]. Although relatively little is currently known about the structural aspects of the interaction between Hsp40 and Hsp70, the effect of DnaJ on the ATPase activity of Hsp70 is due to interactions of the NH_2-terminal J-domain of DnaJ [67,74,80].

The observed polypeptide-stimulated ATPase activity of Hsp70 can be accounted for as follows. Under in vitro experimental conditions in which Hsp70, substrate protein, or peptide, and ATP are present, most of the Hsp70 will be complexed with ATP; therefore the substrate will bind to this complex, and also dissociate rapidly. However, if binding of substrate protein leads to a conformational change that increases the ATP hydrolysis rate [65], then substrate-stimulated ATPase enhancement will be observed.

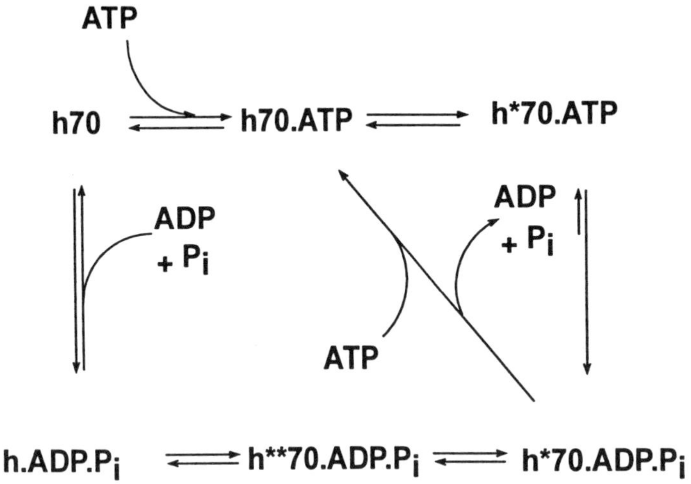

Figure 2 The ATPase cycle of Hsp70 (no cofactors or substrate): h70, h*70 and h**70 represent three different conformations of Hsp70. (From Ref. 33.)

Currently, the ATP hydrolysis cycle of Hsp70 in the absence of substrate protein and cofactors is best described by the scheme shown in Figure 2.

C. ATP-Induced Conformational Changes

It is now well established that binding of ATP (and some of its nonhydrolyzable analogues) to Hsp70 leades to a conformational change. Binding of substrates also stimulates the ATPase activity, and is consistent with several observations indicating a substrate-induced conformational change in Hsp70 as well.

The lines of evidence supporting a conformational change in Hsp70 on ATP binding include small-angle X-ray scattering [81,82], changes in tryptophan fluorescence emission (a blue shift in the λ_{max} on ATP binding) [33,71,83–85], and changes in proteolytic sensitivity (changes in the fragmentation pattern and peptides released on proteolysis) [13,32,83,86].

Interestingly, nucleotide binding to the NH_2-terminal ATP-binding domain also leads to conformational changes in the peptide-binding domain, as revealed by changes in proteolytic patterns in the latter domain [32]. This investigation also revealed that at least part of the tryptophan fluorescence change in DnaK on ATP binding was due to the peptide-binding domain. In both DnaK and yeast Ssa1p, three distinct conformational states have been demonstrated, corresponding to the absence of nucleotide, and the presence of ADP or ATP [32,87]. Given the changes in the sites of proteolysis, the binding of both nucleotides and peptides induced conformational changes in *both* the ATP-binding domain and the COOH-terminal domain, indicating transfer of effects of ligand binding in one domain to the other. These results clearly indicate that the conformations of the nucleotide- and polypeptide-binding domains are coupled.

That additional conformations of Hsp70 are important is suggested by the observation that there are three different conformational states for Hsp70 with bound substrate, depending on the nucleotide bound [65]. The weakest affinity for substrates is observed with bound ATP; the two ADP complexes both show higher affinity for polypeptides [54,65].

Studies on mutant DnaKs indicate that substitutions E171K and D201N abolish the ATP-induced conformational change, and also the DnaJ- and GrpE-enhanced ATPase activity [86]. The DnaK mutant A174T also exhibitied unusual properties, most notably differing from the wild-type in that it failed to exhibit the synergistic stimulation in ATPase activity induced by the *combination* of DnaJ and GrpE [86].

A regulatory motif of many Hsp70s, EEVD, which is conserved at the extreme COOH-terminus among nearly all cytosolic eukaryotic Hsp70s, is involved in determining the conformational state of Hsp70 and is critical for the ATPase activity, binding of substrates, and interaction with Hsp40 [88].

The propensity of Hsp70s to self-associate is abolished by ATP, indicating that the conformation induced by ATP has low affinity for itself, whereas those conformations existing in the absence of nucleotide or in the presence of ADP have strong affinity for self-association [30].

D. Dissociation of Substrate Protein

Several observations support the suggestion that the conformational change associated with binding of ATP to DnaK leads to release of the substrate protein [2,31,65]. This implies that the nucleotide binding, which occurs in the HN_2-terminal domain, is communicated to the substrate-binding COOH-terminal domain and leads to significantly weaker binding of the substrate protein. The rates of substrate polypeptide dissociation from Hsp70 complexes with ATP are much faster than those from complexes with ADP: for example, for Hsc70 and the peptide faf1, the dissociation rates are 10–15 times faster from the complex with ATP [65] and, for peptide c and clathrin from Hsc70, the rates for the ATP complex are more than 100 times faster [61]. A fluorescent peptide, has been reported to dissociate from the DnaK–ATP complex 1000 times faster than from the corresponding complex with ADP [33].

Comparison of the rates at which the substrate protein was released from the Hsp70 complex on addition of ATP with the rate at which ATP was hydrolyzed, demonstrated that substrate protein dissociation is much faster than ATP hydrolysis [31,65]. This is especially noticeable for the T199A mutant of DnaK, which has very low ATPase activity, and suggested that ATP binding was responsible for substrate dissociation [31]. The observation that the addition of the nonhydrolyzable ATP analogue, AMP-PNP results in release of the substrate protein from the DnaK–DnaJ–ADP–U complex is strong support that ATP binding is the trigger for release of substrate polypeptide [89].

E. Conformation of Bound and ATP-Released Substrate Protein

An investigation of the structure of substrate protein bound to DnaK, using a thermally unstable mutant of staphylococcal nuclease as substrate, revealed that the substrate is in a substantially unfolded conformation, as evidenced by circular dichroism and fluorescence spectra under conditions in which the substrate protein would have been native in the absence of the chaperone [63]. With a permanently unfolded substrate protein, this investigation also indicated that no major changes in the secondary structure of DnaK occurred on complex formation.

Further support that Hsp70 does not actively participate in protein folding comes from the observation that the conformation of the substrate protein after ATP-triggered release from a complex with DnaK, is substantially unfolded and, in fact, shows the same time-dependent tryptophan fluorescence changes as those observed from the refolding of denatured substrate protein under similar conditions [63].

F. Interactions Between Hsp70 and Its Cofactors

The essential requirement for DnaJ and GrpE, in addition to DnaK, for the efficient renaturation of firefly luciferase and rhodanese was demonstrated by Hartl and co-workers [90–91]. Maximum efficiency in the rhodanese system was demonstrated in the presence of GroE [90]; however, the exact role of the different chaperones in the folding process is still controversial: recent studies suggest that the earlier interpretation that unfolded substrate proteins were obligatorily transferred from the DnaK system to the GroE system for folding is oversimplified [2]. Several eukaryotic homologues of DnaJ [92] and GrpE [37,93] have been identified. The yeast [94] and human [67] systems of Hsp70 and their DnaJ homologues also faciliate the refolding of denatured rhodanese and firefly luciferase. Mitochondrial DnaJ is critical in the mitochondrial Hsp70-facilitated refolding of denatured firefly luciferase [95]. In fact, DnaJ and GrpE are necessary for the maximal effect of DnaK in all its various functions, including protein secretion [96] and λ replication [34]. Earlier studies on DnaJ and its Hsp40 homologues have been summarized [92]. Recent advances are discussed in Chapter 11.

Whereas the critical requirement for DnaJ and its Hsp40 homologues is clear in Hsp70-facilitated protein folding, the exact role of Hsp40 has been the subject of considerable discussion. For example, it has been suggested that the presence of DnaJ may increase the affinity of DnaK for substrates [62,97], that DnaJ can stabilize complexes between substrate protein and Hsp70–ATP [70,90], and that the role of DnaJ is to help deliver substrate proteins to the Hsp70-binding site and to trigger a conformational change resulting in stabilization of the Hsp70–substrate-protein complex [98].

It has been proposed that stabilization of Hsp70–ATP complexes with substrate is accomplished through the effects of DnaJ, which could increase the rate of ATP hydrolysis such that a significant fraction of the ATP in the complex is converted to ADP before substrate dissociation occurs [70]. Although this is certainly a possibility, such a mechanism has been questioned [74]. For this to happen, the rates of ATP hydrolysis in the presence of DnaJ must be faster than the rates of dissociation of the polypeptide substrate: although there are few quantitative measurements available, the data of McCarty et al. [70] suggest it is possible. An important observation is that DnaJ binds tightly only to DnaK when it is in the complex with ATP (i.e., DnaJ has low affinity for free DnaK or the ADP complex).

It is the NH_2-terminal region (the conserved J-region) of DnaJ that is essential for the enhanced activity of DnaK in λ-DNA replication and for stimulating ATP hydrolysis [99,100]. A conserved His-Pro-Asp sequence in the vicinity of residue 32–34 in the J-domain of DnaJ and its homologues is critical for binding to Hsp70 [101]. Studies with denatured rhodanese have shown that the zinc-finger (Gly–Phe-rich) region of DnaJ is required for binding substrate proteins [102]. A mutant of DnaK lacking the COOH-terminal 94 amino acids does not interact with DnaJ (in spite of having peptide-binding and ATPase activities), indicating that the site of interaction between DnaJ and DnaK may involve the COOH-terminal region of DnaK [55]. The ability to activate Hsc70 ATPase activity is partially preserved in a deletion mutant containing the J-domain and the G/F region of Hsp40, but not in a deletion mutant that contains the J-domain alone [67].

DnaJ itself has been reported to bind to at least some unfolded proteins (e.g., reduced carboxymethylated α-lactalbumin [55], rhodanese [27,103], and firefly luciferase [89,91,102]. DnaJ binds to σ^{32} in a mode different from that of DnaK, and forms a complex that is stable to ATP [38]. DnaJ catalytically activates DnaK for binding to σ^{32} in a process involving the NH_2-terminal J-domain [104]. Interestingly, the affinity of DnaJ alone for different substrates appears to correlate well with the affinity of DnaK in the presence of DnaJ and ATP for these same protein substrates [105].

GrpE, in the absence of substrate polypeptides, has very weak stimulation of the ATPase activity of DnaK [40,106]; however, the stimulation is dramatically enhanced in a sequence-dependent fashion in the presence of substrate polypeptides [62]. This is in contrast with the effect of DnaJ that stimulates the ATPase activity of DnaK in the absence of peptides [62].

GrpE alone increased the rate of nucleotide dissociation, but did not increase the rate of ATP hydrolysis. Whereas GrpE binds tightly to the ATPase domain of DnaK in the absence of ATP, or in the DnaK–ADP complex, no DnaK–GrpE complex is observed in the presence of ATP [13,107]. Binding of GrpE to DnaK leads to a conformational change, detectable by tryptophan fluorescence, that is probably responsible for facilitating nucleotide exchange [13,82]. A conserved loop in the vicinity of Gly-32 in DnaK is essential for the binding of GrpE [108]; deletion of this loop prevents stable GrpE binding, ATPase stimulation by GrpE, and DnaK chaperone activity. Conservation of this loop within the Hsp70 family suggests that cooperation between Hsp70 and GrpE-like proteins may be a general feature of this class of chaperone. Interactions between Mg^{2+} and GrpE that affect the ATPase activity of DnaK have been demonstrated, further complicating the picture: GrpE increases the K_d of DnaK for nucleotides more in the absence of Mg^{2+} than in its presence [106].

V. THE Hsp70 REACTION CYCLE

Several models have been proposed for the reaction cycle of Hsp70 [2,31,57,67,70,89,109]. The general outline of the reaction cycle is clear: (a) the chaperone binds the unfolded protein or partially folded intermediate or peptide; (b) binding of ATP leads to a conformational change causing release of the substrate protein; and (c) hydrolysis of the ATP leads to a return to the original conformation. However, complications arise owing to the involvement of various cofactors (nucleotides and proteins) in the Hsp70 reaction cycle. That there may be differences between the effects of chaperones under controlled in vitro conditions, compared with the in vivo situation, should not be surprising at this stage of our understanding of the role of molecular chaperones in protein folding, especially because of the likelihood that there are additional, as yet undiscovered, agents present in the in vivo situation. Some evidence for this has been recently reported [22].

One of the confounding factors in analysis of various conflicting observations on the mechanisms of chaperones results from the effects of low concentrations of critical contaminants: for example, samples of Hsp70 and Hsp90 are often contaminated with low levels of DnaJ or Hsp40, which may profoundly affect the experimental observations [110].

A. Models for the Hsp70 Reaction Cycle in the Absence of Cofactors

In the absence of cofactors, the simplest Hsp70 cycle is shown in Figure 3 [31]. This simplest cycle is unphysiological, in that it does not involve the cofactors, but does allow investigation of some of the basic molecular aspects of Hsp70

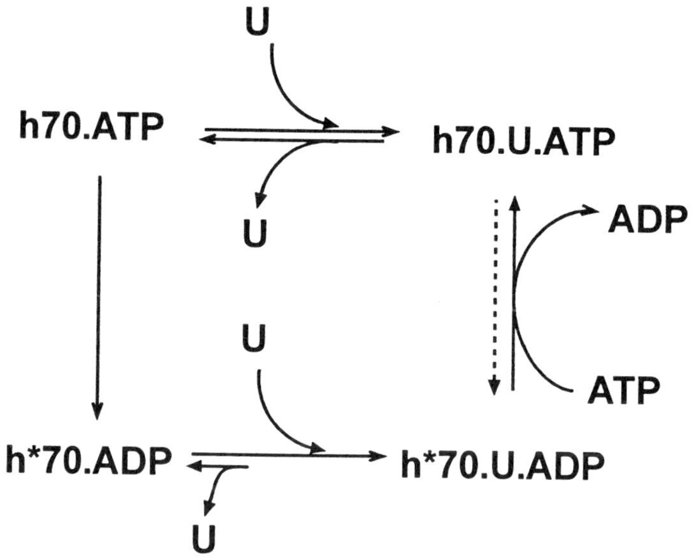

Figure 3 The minimum Hsp70 reaction cycle, in the absence of cofactors: U represents the unfolded substrate protein or polypeptide, h70 and h*70 represent two different conformations of Hsp70: U binds and dissociates rapidly from the h70 form, which consequently has low affinity for the substrate protein. h*70 binds and releases the substrate protein slowly and thus has high affinity for the substrate protein. (From Ref. 31.)

interactions. The first step in the mechanism involves the binding of substrate proteins or peptides (U) to either an Hsp70–ATP or Hsp70–ADP complex. Binding of denatured substrates to an Hsp70-ADP complex is faster than that to free Hsp70, although still quite slow, and the resulting Hsp70–ADP–substrate complex is more stable than the Hsp70–substrate complex [63]. Binding to an Hsp70–ATP complex is much faster, but so is dissociation [64]; consequently, no significant complex of Hsp70–ATP with denatured substrate is found [30,31]. Thus, substrate protein will cycle on and off the ATP complex, and will accumulate in the ADP complex. It is likely that the rate of ATP hydrolysis is faster in the complex with unfolded substrate than with ATP alone. This would be consistent with the suggestion that binding of denatured substrate to preexisting Hsp70–ATP complexes is responsible for the observed stimulation of the low intrinsic ATPase activity of Hsp70 [70].

The exchange of ADP by ATP in the Hsp70–U–ADP complex causes a conformational change in Hsp70 [12,13,71,82–84] that leads to reduced affinity for the substrate protein, which thus dissociates [31]. The data are consistent with the Hsp70–ADP complex having "high" affinity for denatured substrates, whereas the Hsp70–ATP complex has low affinity.

There are conflicting reports on the rate-limiting step in the intrinsic Hsp70–ATPase activity: currently the evidence is strongly in favor of the cleavage of the γ-phosphate of ATP being the rate-limiting step [70]. This also appears to be true in the presence of a peptide substrate and in the presence of DnaJ. However, that the presence of GrpE significantly stimulates the rate of ATP hydrolysis by DnaK in the presence of substrate protein suggests that ADP release is probably rate-limiting under such conditions.

That substrate protein dissociation precedes ATP hydrolysis was demonstrated by comparison of the corresponding rates: the rate for ATP hydrolysis being significantly slower than for substrate dissociation [31]. Until this report, it was assumed that ATP hydrolysis was concurrent with, or preceded, release of substrate protein. This mechanism (see Fig. 3) operates under physiological concentrations of potassium ion; lack of this ion results in a dramatic change in the binding properties of Hsp70 owing to the critical role of K^+ (and Mg^{2+}) in ATP binding [31,72,73]. The conformational state of the substrate protein bound to Hsp70 is substantially unfolded [63], and the conformational state of the substrate protein on release by ATP is also substantially unfolded [63].

Although this simple model does not correspond to the physiological situation, it made three important contributions to our understanding of Hsp70 function: namely, ATP binding caused a conformational change, substrate protein dissociation was faster than ATP hydrolysis and, therefore, preceded it (indicating that it was the ATP-induced conformational change that led to release of the substrate protein); and the Hsp70–ADP complex with substrate protein was relatively stable, whereas that with ATP was not, thereby suggesting that the ADP complex conformation had high affinity for the substrate protein and the ATP complex had low affinity for the substrate protein [31].

An alternative model for the cycle in the absence of cofactors has been proposed in which the substrate protein binds only to the ATP complex of Hsp70, forming a ternary complex in which the ATP rapidly undergoes hydrolysis to form a stable complex of Hsp70, ADP, and substrate protein [54]. This model would only be applicable to substrate proteins that dissociate very slowly from the Hsp70–ADP–U complex, because for most substrates, the rate of ATP hydrolysis is markedly slower than the rate of substrate dissociation from the Hsp70–ATP complex in the absence of the cofactors. In the presence of ATP and the absence of the cofactors, there is rapid cycling of substrate protein binding and releasing from the Hsp70–ATP complex. Simultaneously, there is slow ATP hydrolysis, leading to the formation of increasing amounts of ADP and consequently, Hsp70–ADP complexes. These, along with any complexes of Hsp70–ATP that bind substrate protein and undergo ATP hydrolysis before substrate release, lead to the gradual accumulation of stable complexes of Hsp70–ADP and substrate protein.

B. Models for the Hsp70 Reaction Cycle in the Presence of Cofactors

The involvement of additional cofactors in the function of Hsp70 was reported by Georgopoulos and co-workers [34,40,111,112]. The involvement of the cofactors DnaJ and GrpE have led to several proposed reaction cycles for DnaK [1,2,19,67,70,89,109,113]. Currently, there is no consensus on the exact details of the cycle and, in fact, there are major differences in some of the proposed schemes. Much of the controversy stems from conflicting, and unclear, roles of DnaJ or Hsp40 in the reaction cycle. Many observations indicate that the maximal functional effect of Hsp70 requires the presence of DnaJ and GrpE [38,91,96,97,109,114–116]. For example, all three components DnaK, DnaJ, and GrpE, are required in vivo and in vitro for efficient renaturation of firefly luciferase [19,89,91]. Interestingly, DnaJ alone was reported to bind denatured luciferase and prevent its aggregation. DnaJ has been reported to bind to nascent polypeptide chains and to require DnaK and GrpE to allow their subsequent folding [18,91,103,117]. In general, the features of the Hsp70 cycle in yeast (e.g., with Ssa1p and Ydj1) seem very similar to those of the corresponding mammalian (Hsc70) and bacterial (DnaK) systems [68].

There are conflicting reports concerning the effects of DnaJ on Hsp70–substrate protein interactions: Ssa1p–reduced-carboxymethylated α-lactalbumin complexes are dissociated by addition of ATP and Ydj1 [118], whereas a stable complex involving DnaK and reduced carboxymethylated α-lactalbumin is reported in the presence of ATP and DnaJ [105]. Possibly the reason for this apparent discrepancy is the experimental conditions or differences in rate constants between the yeast and *E. coli* systems: if hydrolysis of Hsp70-bound ATP is faster in the presence of DnaJ than substrate protein dissociation, then a stable complex of Hsp70–ADP and substrate protein is anticipated. For a given unfolded substrate protein, there will be several potential Hsp70- (and Hsp40?)-binding sites along the polypeptide chains, with different affinity for the chaperones.

The reaction cycle of Hartl [1] is outlined in Figure 4 and is derived from studies on the renaturation of firefly luciferase [89]. This model for the reaction cycle postulates that DnaJ (Hsp40) first interacts with an unfolded (nascent) polypeptide, targeting it for binding to DnaK (Hsp70). DnaK, in its ATP-bound state, then binds the unfolded target–DnaJ complex. The presence of the DnaJ in this complex stimulates the hydrolysis of the ATP to yield the more stable complex of DnaK, ADP, and unfolded substrate. The nucleotide-exchange factor GrpE, then facilitates the release of ADP from DnaK. The binding of ATP to DnaK leads to release of the substrate polypeptide, thereby completing the

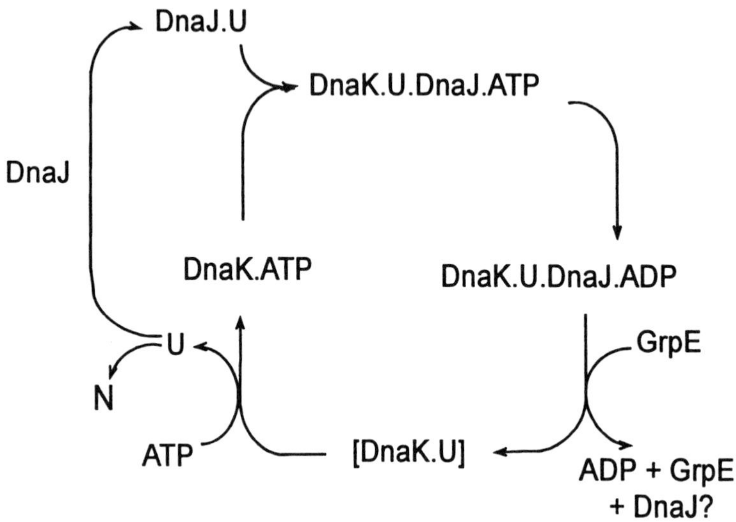

Figure 4 The DnaK reaction cycle in the presence of cofactors, as proposed by Hartl [1]: U and N represent unfolded and native protein. The basic cycle is that DnaJ binds unfolded substrate protein and interacts with the DnaK–ATP complex. The resulting complex undergoes hydrolysis of the ATP and then GrpE catalyzes the release of the nucleotide. DnaJ is also probably released at this stage of the cycle (although the exact point is not known; hence, the question-mark). Subsequent binding of ATP leads to release of the substrate protein and completion of one round of the cycle.

reaction cycle. The released substrate protein may then fold spontaneously, rebind to DnaJ or DnaK, or bind to another chaperone (e.g., Hsp60). A single cycle of binding and release results in the folding of only a fraction of luciferase molecules; thus, several rounds of ATP-dependent interaction with DnaK and DnaJ are required for complete folding [89]. This model is essentially the same as that proposed by Bukau and co-workers for the DnaK cycle involving the heat-shock transcription factor σ^{32} [109]. The apparent lack of a GrpE homologue in eukaryotic cytosol is accounted for by the assumption that the rate-limiting step is ATP hydrolysis in this system, compared with ADP dissociation in the prokaryotes (in the presence of DnaJ and substrate) [67].

In a related cycle (Fig. 5) Bukau and co-workers [70] proposed that it is only DnaK in the ATP complex state that interacts with the unfolded substrate. A relatively stable complex is then formed by the hydrolysis of the ATP to ADP. This may be accomplished either by the binding of DnaJ to this initial complex (free in solution or bound to an adjacent site on the polypeptide), or less efficiently through the unfolded substrate-stimulated ATPase effect. The action of

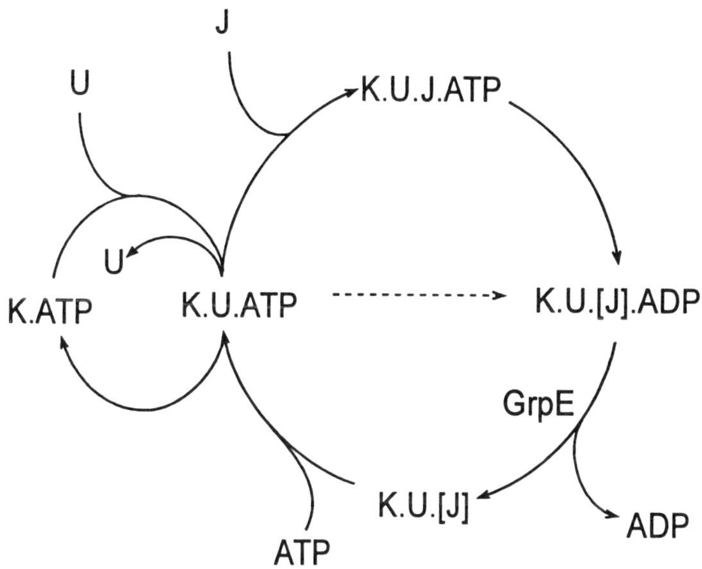

Figure 5 The DnaK reaction cycle in the presence of cofactors, as proposed by Bukau and co-workers [70]: The essence of this cycle is that unfolded substrate protein binds to the DnaK–ATP complex, which can either interact with DnaJ to form a quaternary complex, or undergo ATP hydrolysis to directly form a complex between DnaK, substrate protein, and ADP. GrpE-catalyzed release of the ADP and subsequent binding of ATP leads to release of the substrate.

GrpE then triggers the release of ADP, leading to the binding of ATP and release of the substrate polypeptide.

A substantially different cycle has been proposed by Zylicz [55,105,113], in which DnaJ binds to the DnaK–ADP complex, resulting in a DnaK–DnaJ–ADP complex, that then binds substrate polypeptide. Subsequent GrpE-facilitated nucleotide exchange leads to release of the polypeptide and DnaJ and hydrolysis of ATP. Major problems with this proposal are that several groups have oberved that DnaJ binds to DnaK only in the presence of ATP, and that the major Hsp70 nucleotide complex is with ATP, not ADP.

Thus, the exact details of the Hsp70 reaction cycle in the presence of the cofactors still remain to be sorted out. The puzzle, or problem, is to ascertain the mechanism given the following key observations: rapid substrate binding requires the Hsp70–ATP complex, but this leads to rapid substrate dissociation; "stable" Hsp70 complexes with substrate require the ADP complex; DnaJ (or Hsp40) binds significantly to only the Hsp70–ATP complex; DnaJ–Hsp40 stimulates the hydrolysis of ATP, thereby potentially converting Hsp70–ATP complexes into

Hsp70–ADP. In the cell, most of the Hsp70 will be present as the binary complex with ATP, which has low affinity for substrate protein owing to the fast dissociation rate. The binding of peptide substrates to Hsp70–ATP is fast, but is very much slower to Hsp70–ADP [64,65]. This leads to the apparent paradox that, under normal cellular conditions, Hsp70 is mostly in the form of a complex with ATP, which binds substrate proteins rapidly, but also, because of rapid dissociation, does not lead to the formation of significant Hsp70–substrate protein complex. It is likely, therefore, that one of the roles of DnaJ or Hsp40 is to circumvent this situation. Because DnaJ binds to both unfolded proteins and to DnaK–ATP, a major role of DnaJ is likely to be to allow the complexes of Hsp70 and ATP to form a productive (long-lived) complex with substrate protein. This may occur by initial interaction of Hsp40 with the substrate protein and then Hsp70–ATP, thereby acting as an anchor for both Hsp70–ATP and the substrate protein, and leading to formation of a relatively stable Hsp70–substrate polypeptide complex, or by interaction of Hsp40 with DnaK–ATP to form a ternary complex that both binds substrate protein rapidly and forms a stable complex. Several reports suggest that the ATP in a complex with Hsp70 and Hsp40 is "rapidly" hydrolyzed to ADP (although there is a paucity of data to support this). Thus, the initial stage of the "productive" reaction cycle,* starting with the Hsp70–ATP complex leads to formation of the Hsp70–Hsp40–U–ADP complex. Such a complex is undoubtedly unstable and rapidly dissociates the Hsp40, leaving a complex of Hsp70–ADP and substrate protein that is stable.

From the preceding discussion, we propose the following scheme (Fig. 6) for the Hsp70 reaction cycle in the presence of Hsp40: in *E. coli* and mitochondria, the cycle will differ in that GrpE will be present and facilitate the nucleotide exchange step. This scheme is similar in many ways to those of Hartl and Bukau and their co-workers (see Figs. 4 and 5), but it is more general. The main drawback with this scheme is that, until we know the relative rates of the various possible pathways, we still do not know which particular path a given combination of Hsp70, Hsp40, and substrate protein will take. It is quite likely that, in fact, different systems will involve different preferred paths. The most likely major routes for entry of the unfolded substrate protein into the cycle are through interaction with the Hsp40–Hsp70–ATP and Hsp70–ATP complexes, and probably also by initial binding to Hsp40.

*It is possible that the very short-lived complex of DnaK–ATP with substrate protein could be important physiologically for some systems. However, this seems unlikely, given our current understanding of the role of Hsp70 in protein folding and membrane translocation, because in both longlived complexes appear desirable.

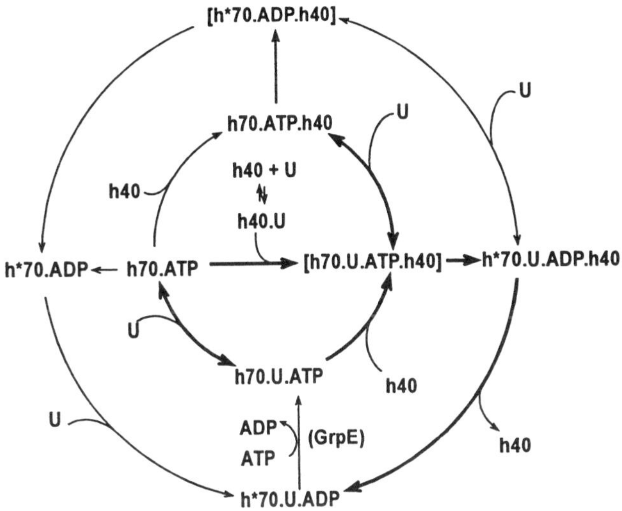

Figure 6 Proposed general Hsp70 reaction cycle, where h40 represents Hsp40, h70 and h*70 represent low- and high-affinity (for substrate protein) states of Hsp70. Square brackets indicate very short-lived complexes. As described in the text, different chaperones, substrate proteins, cofactors, and experimental conditions will affect the relative rates of the different paths, leading to different preferred routes. The most likely major paths for entry of the unfolded substrate protein into the cycle are through interaction with the Hsp40–Hsp70–ATP and Hsp70–ATP complexes, and by initial binding to Hsp40. The thicker lines represent the anticipated major pathways under optimal conditions.

VI. Hsp70 AND IN VIVO PROTEIN FOLDING

Several reports have implicated Hsp70 and Hsp40 in the earliest stages of folding in protein biosynthesis (see Chap. 3) [16–20]. The interaction of DnaJ with nascent ribosome-bound polypeptide chains as short as 55 residues has been demonstrated using firefly luciferase and chloramphenicol acetyltransferase in cross-linking experiments [18]. These investigations showed that both folding and subsequent mitochondrial translocation required DnaK, DnaJ, and GrpE, and led to the proposal that DnaJ protects nascent polypeptide chains from aggregation and, in cooperation with Hsp70, controls their productive folding once a complete polypeptide or a polypeptide domain has been synthesized. Analysis of the synthesis and folding of firefly luciferase in a rabbit reticulocyte lysate indicated that Hsp70 and Hsp40 interacted with the growing polypeptide as it emerged from the ribosome, and that subsequent interactions involved Hsp60 (TriC) [19].

DnaK and DnaJ are associated with nascent polypeptide chains in translating ribosomes, whereas GroEL, although transiently associated with newly synthesized proteins, was absent from the ribosomes, suggesting that Hsp70 and Hsp40 play an early role in protein maturation, whereas GroEL acts at a later stage [119]. Similar studies have been reported for the in vitro folding of the UmuC protein [3]. Recent studies on the interaction of BiP with immunoglobulin light chains during folding in the ER suggest that light chains undergo both Bip-dependent and BiP-independent folding steps, and that BiP must release the light chains before disulfide bond formation can occur in them [120].

Investigations using fluorescent-labeled rhodanese (by the cotranslational incorporation of a coumarin derivative at the NH_2-terminus of the nascent protein) demonstrated the accumulation of full-length, but enzymatically inactive, polypeptides accumulated as peptidly-tRNA on the ribosomes. These polypeptides could be activated and released by subsequent incubation with the bacterial chaperones (DnaJ, DnaK, GrpE, GroEL, GroES, and ATP) and release factor. Changes in fluorescence indicate that DnaJ binds to the nascent proteins and appears to be essential for folding of ribosome-bound rhodanese into the native conformation (27,103,117,121]. The results support the involvement of the chaperones in folding of nascent proteins while still associated with the ribosome, as well as after release from the ribosome.

A. The Hsp70 Reaction Cycle and In Vivo Protein Folding

The Hsp70 reaction cycles shown in Figures 4–6 can readily account for the interaction of Hsp70 and Hsp40 under nonstress conditions with nascent polypeptide chains emerging from the ribosome (possibly with additional cofactors or chaperones). After release from the ribosome, the substrate protein presumably remains bound to the Hsp70–ADP complex until exchange of ATP for Hsp70-bound ADP leads to dissociation and release of the substrate protein. The released substrate protein may fold spontaneously, or bind to Hsp60. Under in vivo stress conditions, when the levels of ATP may be quite low, relative to ADP, it is likely that the denatured proteins bind directly to Hsp70–ADP complexes, and may remain mostly bound until ATP levels recover or the cell dies.

Most models of chaperone-mediated protein folding have proposed a pathway in which Hsp70 and Hsp60 systems act in succession [1,90]. A recent study on the refolding of firefly luciferase denatured by guanidinium chloride and heat, however, found that DnaK and GroEL compete with each other for binding to nonnative luciferase [2]. Interestingly, only a small fraciton of luciferase molecules released from GroEL reached the native state, instead, the released luciferase bound repeatedly to the DnaK system and, only then, was it able to fold to the native state. It was suggested that this "lateral" chaperone network is of particular importance in the folding of proteins that are too large to allow folding inside the substrate-binding cavity of the GroEL ring underneath GroES.

The potential involvement of additional cofactors in vivo is supported by the discovery of Hip, the Hsc70-interacting protein [69]. Hip is involved in the regulation of Hsc70 and stabilizes the ADP state of Hsc70 that has a high affinity for substrate protein. A mechanism of regulation of eukaryotic Hsc70, distinct from that of bacterial Hsp70 has been proposed [69].

REFERENCES

1. Hartl, F.-U., Molecular chaperones in cellular protein folding, *Nature 381*:571–579 (1996).
2. Buchberger, A., Schroder, H., Hesterkamp, T., Schonfeld, H. J., and Bukau, B., Substrate shuttling between the DnaK and GroEL systems indicates a chaperone network promoting protein folding, *J. Mol. Biol. 261*:328–333 (1996).
3. Morimoto, R. I., Tissières, A., and Georgopoulos, C., Progress and perspectives on the biology of heat shock proteins and molecular chaperones, *The Biology of Heat Shock Proteins and Molecular Chaperones* (Morimoto, R. I., Tissières, A., and Georgopoulos, C., eds.), Cold Spring Harbor Laboratory Press, Cold Spring Harbor, NY, 1994, pp. 1–30.
4. Gething, M.-J. and Sambrook, J., Protein folding in the cell, *Nature 355*:33–45 (1992).
5. Hartl, F.-U., Martin, J., and Neupert, W., Protein folding in the cell: The role of molecular chaperones hsp70 and hsp60, *Annu. Rev. Biophys. Biomol. Struct. 21*:293–322 (1992).
6. Bukau, B., Regulation of the *E. coli* heat-shock response, *Mol. Microbiol. 9*:671–680 (1993).
7. Hendrick, J. P. and Hartl, F.-U., Molecular chaperone functions of heat-shock proteins, *Annu. Rev. Biochem. 62*:349–384 (1993).
8. Georgopoulos, C. and Welch, W. J., Role of the major heat shock proteins as molecualr chaperones, *Annu. Rev. Cell Biol. 9*:601–634 (1993).
9. Craig, E. A., Gambill, B. D., and Nelson, R. J., Heat shock proteins: Molecualr chaperones of protein biogenesis, *Microbiol. Rev. 57*:402–414 (1993).
10. Landry, S. J. and Gierasch, L. M., Polypeptide interactions with molecualr chaperones and their relationship to in vivo protein folding, *Annu. Rev. Biophys. Biomol. Struct. 23*:645–669 (1994).
11. DeYoung, L. R., Fink, A. L., and Dill, K. A., Aggregation of globular proteins, *Acc. Chem. Res. 26*:614–620 (1993).
12. Naylor, D. J., Hoogenraad, N. J., and Hoj, P. B., Isolation and characterisation of a cDNA encoding rat stress-inducible nucleotide-exchange factor of ubiquitous mammalian organs, *FEBS Lett. 396*:181–188 (1996).
13. Reid, K. L. and Fink, A. L., Physical interactions between members of the DnaK chaperone machinery: Characterization of the DnaK.GrpE complex, *Cell Stress Chaperones 1*:127–137 (1996).
14. Wu, B., Wawrzynow, A., Zylicz, M., and Georgopoulos, C., Structure–function analysis of the *E. coli* GrpE heat shock protein, *EMBO J. 15*:4806–4816 (1996).
15. Gragerov, A., Nudler, E., Komissarova, N., Gaitanaris, G. A., Gottesman, M. E., and Nikiforov, V., Cooperation of GroEL/GroES and DnaK/DnaJ heat shock proteins in

preventing protein misfolding in *E. coli*, *Proc. Natl. Acad. Sci. USA 89*:10341–10344 (1992).

16. Beckmann, R. P., Mizzen, L. E., and Welch, W. J., Interaction of hsp 70 with newly synthesized proteins: Implications for protein folding and assembly, *Science 248*:850–854 (1990).
17. Nelson, R. J., Ziegelhoffer, T., Nicolet, C., Werner-Washburne, M., and Craig, E. A., The translation machinery and 70 Kd heat shock protein cooperate in protein synthesis, *Cell 7*:97–105 (1992).
18. Hendrick, J. P., Langer, T., Davis, T. A., Hartl, F.-U., and Wiedmann, M., Control of folding and membrane translocation by binding of the chaperone DnaJ to nascent polypeptides, *Proc. Natl. Acad. Sci. USA 90*:10216–10220 (1993).
19. Frydman, J., Nimmesgern, E., Ohtsuka, K., and Hartl, F.-U., Folding of nascent polypeptide chains in a high molecular mass assembly with molecular chaperones, *Nature 370*:111–117 (1994).
20. Kudlicki, W., Odom, O. W., Merrill, G., Kramer, G., and Hardesty, B., Inhibition of the release factor-dependent termination reaction on ribosomes by DnaJ and the N-terminal peptide of rhodanese, *J. Bacteriol. 177*:5517–5522 (1995).
21. Lorimer, G. H., A quantitative assessment of the role of the chaperonin proteins in protein folding in vivo, *FASEB J. 10*:5–9 (1996).
22. Frydman, J. and Hartl, F.-U., Principles of chaperone-assisted protein folding: Differences between in vitro and in vivo mechanisms, *Science 272*:1497–1502 (1996).
23. Neupert, W., Hartl, F.-U., Craig, E. A., and Pfanner, N., How do polypeptides cross the mitochondrial membranes, *Cell 63*:447–450 (1990).
24. Schmid, S. L., Braell, W. A., and Rothman, J. E., ATP catalyzes the sequestration of clathrin during enzymatic uncoating, *J. Biol. Chem. 260*:10057–10062 (1985).
25. Yura, T., Nagai, H., and Mori, H., Regulation of the heat-shock response in bacteria, Annu. Rev. Microbiol. 47:321–350 (1993).
26. Zylicz, M., The *E. coli* chaperones involved in DNA replication, *Philos. Trans. R. Soc. Lond. Ser. B. 339*:271–277; discussion 277–278 (1993).
27. Craig, E. A., Baxter, B. K., Becker, J., Halladay, J., and Ziegelhoffer, T., Cytosolic hsp70s of *Saccharomyces cerevisiae*: Roles in protein synthesis, protein translocation, proteolysis and regulation, The Biology of Heat Shock Proteins and Molecular Chaperones (Morimoto, R. I., Tissières, A., and Georgopoulos, C., eds.), Cold Spring Harbor Laboratory Press, Cold Spring Harbor, NY, 1994, pp. 31–52.
28. Flynn, G. C., Chappell, T. G., and Rothman, J. E., Peptide binding and release by proteins implicated as catalysts of protein assembly, *Science 245*:385–390 (1989).
29. Greene, L. E. and Eisenberg, E., Dissociation of clathrin from coated vesicles by the uncoating ATPase, *J. Biol. Chem. 265*:6682–6687 (1990).
30. Palleros, D. R., Welch, W. J., and Fink, A. L., Interaction of hsp70 with unfolded proteins: Effects of temperature and nucleotides on the kinetics of binding, *Proc. Natl. Acad. Sci. USA 88*:5719–5723 (1991).
31. Palleros, D. R., Reid, K. L., Shi, L., Welch, W. J., and Fink, A. L., ATP-induced protein–hsp 70 complex dissociation requires K^+ but not ATP hydrolysis, *Nature 365*:664–666 (1993).
32. Buchberger, A., Theyssen, H., Schroder, H., McCarty, J. S., Virgallita, G., Milkereit,

P., Reinstein, J., and Bukau, B., Nucleotide-induced conformational changes in the ATPase and substrate binding domains of the DnaK chaperone provide evidence for interdomain communication, *J. Biol. Chem. 270*:16903–16910 (1995).

33. Theyssen, H., Schuster, H. P., Packschies, L., Bukau, B., and Reinstein, J., The second step of ATP binding to DnaK induces peptide release, *J. Mol. Biol. 263*:657–670 (1996).
34. Zylicz, M., Ang, D., Liberek, K., and Georgopoulos, C., Initiation of λ DNA replication with purified host- and bacteriophage-encoded proteins: The role of the DnaK, DnaJ and GrpE heat shock proteins, *EMBO J. 8*:1601–1608 (1989).
35. Georgopoulos, C., Ang, D., Liberek, K., and Zylicz, M., Properties of the *E. coli* heat shock proteins and their role in bacteriophage λ growth, *Stress Proteins in Biology and Medicine* (Morimoto, R. I., Tissières, A., and Georgopoulos, C., eds.), Cold Spring Harbor Laboratory Press, Cold Spring Harbor, NY, 1990, pp. 191–221.
36. Gross, C. A., Straus, D. B., Erickson, J. W., and Yura, T., The function and regulation of heat shock proteins in *E. coli*, *Stress Proteins in Biology and Medicine* (Morimoto, R. I., Tissières, A., and Georgopoulos, C., eds.), Cold Spring Harbor Laboratory Press, Cold Spring Harbor, NY, 1990, pp. 167–189.
37. Bolliger, L., Deloche, O., Glick, B. S., Georgopoulos, C., Jeno, P., Kronidou, N., Horst, M., Morishima, N., and Schatz, G., A mitochondrial homolog of bacterial GrpE interacts with mitochondrial hsp70 and is essential for viability, *EMBO J. 13*:1998–2006 (1994).
38. Gamer, J., Bujard, H., and Bukau, B., Physical interaction between heat shock proteins DnaK, DnaJ, and GrpE and the bacterial heat shock transcription factor sigma 32, *Cell 69*:833–842 (1992).
39. Liberek, K. and Georgopoulos, C., Autoregulation of the *E. coli* heat shock response by the DnaK and DnaJ heat shock proteins, *Proc. Natl. Acad. Sci. USA 90*:11019–11023 (1993).
40. Liberek, K., Marszalek, J., Ang, D., Georgopoulos, C., and Zylicz, M., *Escherichia coli* DnaJ and GrpE heat shock proteins jointly stimulate ATPase activity of DnaK, *Proc. Natl. Acad. Sci. USA 88*:2874–2878 (1991).
41. Deluca-Flaherty, C., McKay, D. B., Parham, P., and Hill, B. L., Uncoating protein (hsc70) binds a conformationally labile domain of clathrin light chain LCA to stimulate ATP hydrolysis, *Cell 62*:875–887 (1990).
42. Zhu, X., Zhao, X., Burkholder, W. F., Gragerov, A., Ogata, C. M., Gottesman, M. E., and Hendrickson, W. A., Structural analysis of substrate binding by the molecular chaperone DnaK, *Science 272*:1606–1614 (1996).
43. Wang, T. F., Chang, J. H., and Wang, C., Identification of the peptide binding domain of hsc70. 18-kilodalton fragment located immediately after ATPase domain is sufficient for high affinity binding, *J. Biol. Chem. 268*:26049–26051 (1993).
44. Flynn, G. C., Pohl, J., Flocco, M. T., and Rothman, J. E., Peptide-binding specificity of the molecular chaperone BiP, *Nature 353*:726–730 (1991).
45. Blond-Elguindi, S., Cwirla, S. E., Dower, W. J., Lipshutz, R. J., Sprang, S. R., Sambrook, J. F., and Gething, M.-J., Affinity panning of a library of peptides displayed on bacteriophages reveals the binding specificity of BiP, *Cell 75*:717–728 (1993).
46. Fourie, A. M., Sambrook, J. F., and Gething, M.-J., Common and divergent peptide

binding specificities of hsp70 molecular chaperones, *J. Biol. Chem. 269*:30470–30478 (1994).
47. Gragerov, A. and Gottesman, M. E., Different peptide binding specificities of hsp70 family members, *J. Mol. Biol. 241*:133–135 (1994).
48. Takenaka, I. M., Leung, S. M., McAndrew, S. J., Brown, J. P., and Hightower, L. E., Hsc70-binding peptides selected from a phage display peptide library that resemble organellar targeting sequences, *J. Biol. Chem. 270*:19839–19844 (1995).
49. McCarty, J. S., Rudiger, S., Schonfeld, H. J., Schneider-Mergener, J., Nakahigashi, K., Yura, T., and Bukau, B., Regulatory region c of the *E. coli* heat shock transcription factor, sigma32, constitutes a DnaK binding site and is conserved among eubacteria, *J. Mol. Biol. 256*:829–837 (1996).
50. Landry, S. J., Jordan, R., McMacken, R., and Gierasch, L. M., Different conformations for the same polypeptide bound to chaperones DnaK and GroEL, *Nature 355*:455–457 (1992).
51. Liberek, K., Galitski, T. P., Zylicz, M., and Georgopoulos, C., The DnaK chaperone modulates the heat shock response of *E. coli* by binding to the sigma 32 transcription factor, *Proc. Natl. Acad. Sci. USA 89*:3516–3520 (1992).
52. Hupp, T. R., Meek, D. W., Midgley, C. A., and Lane, D. P., Regulation of the specific DNA binding function of p53, *Cell 71*:875–886 (1992).
53. Liberek, K., Osipiuk, J., Zylicz, M., Ang, D., Skorko, J., and Georgopoulos, C., Physical interactions between bacteriophage and *E. coli* proteins required for initiation of λ DNA replication, *J. Biol. Chem. 265*:3022–3029 (1990).
54. Greene, L. E., Zinner, R., Naficy, S., and Eisenberg, E., Effect of nucleotide on the binding of peptides to 70-kDa heat shock protein, *J. Biol. Chem. 270*:2967–2973 (1995).
55. Wawrzynow, A. and Zylicz, M., Divergent effects of ATP on the binding of the DnaK and DnaJ chaperones to each other, or to their various native and denatured protein substrates, *J. Biol. Chem. 270*:19300–19306 (1995).
56. Dalie, B. L., Skaleris, D. A., Kohle, K., Weissbach, H., and Brot, N., Interaction of DnaK with ATP: Binding, hydrolysis and Ca^{2+}-stimulated autophosphorylation, *Biochem. Biophys. Res. Commun. 166*:1284–1292 (1990).
57. Sadis, S. and Hightower, L. E., Unfolded proteins stimulate molecular chaperone hsc70 ATPase by accelerating ADP/ATP exchange, *Biochemistry 31*:9406–9412 (1992).
58. Gao, B., Emoto, Y., Greene, L., and Eisenberg, E., Nucleotide binding properties of bovine brain uncoating ATPase, *J. Biol. Chem. 268*:8507–8513 (1993).
59. Gao, B., Greene, L., and Eisenberg, E., Characterization of nucleotide-free uncoating ATPase and its binding to ATP, ADP, and ATP analogues, *Biochemistry 33*:2048–2054 (1994).
60. Ha, J. H. and McKay, D. B., ATPase kinetics of recombinant bovine 70 kDa heat shock cognate protein and its amino-terminal ATPase domain, *Biochemistry 33*:14625–14635 (1994).
61. Gao, B., Eisenberg, E., and Greene, L., Interaction of nucleotide-free hsc70 with clathrin and peptide and effect of ATP analogues, *Biochemistry 34*:11882–11888 (1995).
62. Jordan, R. and McMacken, R., Modulation of the ATPase activity of the molecular

chaperone DnaK by peptides and the DnaJ and GrpE heat shock proteins, *J. Biol. Chem. 270*:4563–4569 (1995).

63. Palleros, D. R., Shi, L., Reid, K. L., and Fink, A. L., Hsp70-protein complexes. Complex stability and conformation of bound substrate protein, *J. Biol. Chem. 269*:13107–13114 (1994).
64. Schmid, D., Baici, A., Gehring, H., and Christen, P., Kinetics of molecular chaperone action, *Science 263*:971–973 (1994).
65. Takeda, S. and McKay, D. B., Kinetics of peptide binding to the bovine 70 kDa heat shock cognate protein, a molecular chaperone, *Biochemistry 35*:4636–4644 (1996).
66. Farr, C. D., Galiano, F. J., and Witt, S. N., Large activation energy barriers to chaperone–peptide complex formation and dissociation, *Biochemistry 34*:15574–15582 (1995).
67. Minami, Y., Hohfeld, J., Ohtsuka, K., and Hartl, F.-U., Regulation of the heat-shock protein 70 reaction cycle by the mammalian DnaJ homolog, hsp40, *J. Biol. Chem. 271*:19617–19624 (1996).
68. Ziegelhoffer, T., Lopez-Buesa, P., and Craig, E. A., The dissociation of ATP from hsp70 of *Saccharomyces cerevisiae* is stimulated by both Ydj1p and peptide substrates, *J. Biol. Chem. 270*:10412–10419 (1995).
69. Hohfeld, J., Minami, Y., and Hartl, F.-U., Hip, a novel cochaperone involved in the eukaryotic hsc70/hsp40 reaction cycle, *Cell 83*:589–598 (1995).
70. McCarty, J. S., Buchberger, A., Reinstein, J., and Bukau, B., The role of ATP in the functional cycle of the DnaK chaperone system, *J. Mol. Biol. 249*:126–137 (1995).
71. Ha, J. H. and McKay, D. B., Kinetics of nucleotide-induced changes in the tryptophan fluorescence of the molecular chaperone hsc70 and its subfragments suggest the ATP-induced conformational change follows initial ATP binding, *Biochemistry 34*:11635–11644 (1995).
72. O'Brien, M. C. and McKay, D. B., How potassium affects the activity of the molecular chaperone hsc70. I. Potassium is required for optimal ATPase activity, *J. Biol. Chem. 270*:2247–2250 (1995).
73. Wilbanks, S. M. and McKay, D. B., How potassium affects the activity of the molecular chaperone hsc70. II. Potassium binds specifically in the ATPase active site, *J. Biol. Chem. 270*:2251–2257 (1995).
74. Cheetham, M. E., Jackson, A. P., and Anderton, B. H., Regulation of 70-kDa heat-shock-protein ATPase activity and substrate binding by human DnaJ-like proteins, Hsj1a and Hsj1b, *Eur. J. Biochem. 226*:99–107 (1994).
75. Huang, S. P., Tsai, M. Y., Tzou, Y. M., Wu, W. G., and Wang, C., Aspartyl residue 10 is essential for ATPase activity of rat hsc70, *J. Biol. Chem. 268*:2063–2068 (1993).
76. McCarty, J. S. and Walker, G. C., DnaK as a thermometer—threonine-199 is site of autophosphorylation and is critical for ATPase activity, *Proc. Natl. Acad. Sci. USA 88*:9513–9517 (1991).
77. O'Brien, M. C. and McKay, D. B., Threonine 204 of the chaperone protein hsc70 influences the structure of the active site, but is not essential for ATP hydrolysis, *J. Biol. Chem. 268*:24323–24329 (1993).
78. O'Brien, M. C., Flaherty, K. M., and McKay, D. B., Lysine 71 of the chaperone

protein hsc70 is essential for ATP hydrolysis, *J. Biol. Chem. 271*:15874–15878 (1996).

79. Buchberger, A., Valencia, A., McMacken, R., Sander, C., and Bukau, B., The chaperone function of DnaK requires the coupling of ATPase activity with substrate binding through residue E171, *EMBO J. 13*:1687–1695 (1994).
80. Szyperski, T., Pellecchia, M., Wall, D., Georgopoulos, C., and Wuthrich, K., NMR structure determination of the *E. coli* DnaJ molecular chaperone: Secondary structure and backbone fold of the N-terminal region (residues 2–108) containing the highly conserved J domain, *Proc. Natl. Acad. Sci. USA 91*:11343–11347 (1994).
81. Wilbanks, S. M., Chen, L., Tsuruta, H., Hodgson, K. O., and McKay, D. B., solution small-angle X-ray scattering study of the molecular chaperone hsc70 and its subfragments, *Biochemistry 34*:12095–12106 (1995).
82. Shi, L., Kataoka, M., and Fink, A. L., Conformational characterization of DnaK and its complexes by small-angle X-ray scattering, *Biochemistry 35*:3297–3308 (1996).
83. Liberek, K., Skowyra, D., Zylicz, M., Johnson, C., and Georgopoulos, C., The *E. coli* DnaK chaperone, the 70-kDa heat shock protein eukaryotic equivalent, changes conformation upon ATP hydrolysis, thus triggering its dissociation from a bound target protein, *J. Biol. Chem. 266*:14491–14496 (1991).
84. Palleros, D. R., Reid, K. L., McCarty, J. S., Walker, G. C., and Fink, A. L., DnaK, hsp73, and their molten globules. Two different ways heat shock proteins respond to heat, *J. Biol. Chem. 267*:5279–5285 (1992).
85. Li, W. W., Sistonen, L., Morimoto, R. I., and Lee, A. S., Stress induction of the mammalian Grp78/BiP protein gene: In vivo genomic footprinting and identification of p70 core from human nuclear extract as a DNA-binding component specific to the stress regulatory element, *Mol. Cell Biol. 14*:5533–5546 (1994).
86. Kamath-Loeb, A. S., Lu, C. Z., Suh, W. C., Lonetto, M. A., and Gross, C. A., Analysis of three DnaK mutant proteins suggests that progression through the ATPase cycle requires conformational changes, *J. Biol. Chem. 270*:30051–30059 (1995).
87. Segal, M. S., Bye, J. M., Sambrook, J. F., and Gething, M.-J., Disulfide bond formation during the folding of influenza virus hemagglutinin, *J. Cell Biol. 118*:227–244 (1992).
88. Freeman, B. C., Myers, M. P., Schumacher, R., and Morimoto, R. I., Identification of a regulatory motif in hsp70 that affects ATPase activity, substrate binding and interaction with Hdj-1, *EMBO J. 14*:2281–2292 (1995).
89. Szabo, A., Langer, T., Schroder, H., Flanagan, J., Bukau, B., and Hartl, F.-U., The ATP hydrolysis-dependent reaction cycle of the *E. coli* hsp70 system DnaK, DnaJ, and GrpE, *Proc. Natl. Acad. Sci. USA 91*:10345–10349 (1994).
90. Langer, T., Lu, C., Echols, H., Flanagan, J., Hayer, M. K., and Hartl, F.-U., Successive action of DnaK, DnaJ and GroEL along the pathway of chaperone-mediated protein folding, *Nature 356*:683–689 (1992).
91. Schroder, H., Langer, T., Hartl, F.-U., and Bukau, B., DnaK, DnaJ and GrpE form a cellular chaperone machinery capable of repairing heat-induced protein damage, *EMBO J. 12*:4137–4144 (1993).
92. Caplan, A. J., Cyr, D. M., and Douglas, M. G., Eukaryotic homologues of *E. coli*

DnaJ: A diverse protein family that functions with hsp70 stress proteins, *Mol. Biol. Cell 4*:555–563 (1993).

93. Nakai, M., Kato, Y., Ikeda, E., Toh-E, A., and Endo, T., Yge1p, a eukaryotic GrpE homolog, is localized in the mitochondrial matrix and interacts with mitochondrial hsp70, *Biochem. Biophys. Res. Commun. 200*:435–442 (1994).
94. Levy, E. J., McCarty, J., Bukau, B., and Chirico, W. J., Conserved ATPase and luciferase refolding activities between bacteria and yeast hsp70 chaperones and modulators, *FEBS Lett. 368*:435–440 (1995).
95. Prip-Buus, C., Westerman, B., Schmitt, M., Langer, T., Neupert, W., and Schwarz, E., Role of the mitochondrial DnaJ homologue, Mdj1p, in the prevention of heat-induced protein aggregation, *FEBS Lett. 380*:142–146 (1996).
96. Wild, J., Rossmeissl, P., Walter, W. A., and Gross, C. A., Involvement of the DnaK–DnaJ–GrpE chaperone team in protein secretion in *E. coli*, *J. Bacteriol. 178*:3608–3613 (1996).
97. Osipiuk, J., Georgopoulos, C., and Zylicz, M., Initiation of λ DNA replication. The *E. coli* small heat shock proteins, DnaJ and GrpE, increase DnaK's affinity for the λ P protein, *J. Biol. Chem. 268*:4821–4827 (1993).
98. Wall, D., Zylicz, M., and Georgopoulos, C., The conserved G/F motif of the DnaJ chaperone is necessary for the activation of the substrate binding properties of the DnaK chaperone, *J. Biol. Chem. 270*:2139–2144 (1995).
99. Wall, D., Zylicz, M., and Georgopoulos, C., the NH_2-terminal 108 amino acids of the *E. coli* DnaJ protein stimulate the ATPase activity of DnaK and are sufficient for λ replication, *J. Biol. Chem. 269*:5446–5451 (1994).
100. Karzai, A. W. and McMacken, R., A bipartite signaling mechanism involved in DnaJ-mediated activation of the *E. coli* DnaK protein, *J. Biol. Chem. 271*:11236–11246 (1996).
101. Tsai, J. and Douglas, M. G., A conserved HPD sequence of the J-domain is necessary for Ydj1 stimulation of hsp70 ATPase activity at a site distinct from substrate binding, *J. Biol. Chem. 271*:9347–9354 (1996).
102. Szabo, A., Korszun, R., Hartl, F.-U., and Flanagan, J., A zinc finger-like domain of the molecular chaperone DnaJ is involved in binding to denatured protein substrates, EMBO J. 15:408–417 (1996).
103. Kudlicki, W., Odom, O. W., Kramer, G., and Hardesty, B., Binding of an N-terminal rhodanese peptide to DnaJ, *J. Biol. Chem. 271*:31160–31165 (1996).
104. Liberek, K., Wall, D., and Georgopoulos, C., The DnaJ chaperone catalytically activates the DnaK chaperone to preferentially bind the sigma 32 heat shock transcriptional regulator, *Proc. Natl. Acad. Sci. USA 92*:6224–6228 (1995).
105. Wawrzynow, A., Banecki, B., Wall, D., Liberek, K., Georgopoulos, C., and Zylicz, M., ATP hydrolysis is required for the DnaJ-dependent activation of DnaK chaperone for binding to both native and denatured protein substrates, *J. Biol. Chem. 270*:19307–19311 (1995).
106. Skowyra, D. and Wickner, S., GrpE alters the affinity of DnaK for ATP and Mg^{2+}. Implications for the mechanism of nucleotide exchange, *J. Biol. Chem. 270*:26282–26285 (1995).
107. Schonfeld, H. J., Schmidt, D., Schroder, H., and Bukau, B., The DnaK chaperone

system of *E. coli*: Quaternary structures and interactions of the DnaK and GrpE components, *J. Biol. Chem. 270*:2183–2189 (1995).

108. Buchberger, A., Schroder, H., Buttner, M., Valencia, A., and Bukau, B., A conserved loop in the ATPase domain of the DnaK chaperone is essential for stable binding of GrpE, *Nature Struct. Biol. 1*:95–101 (1994).
109. Gamer, J., Multhaup, G., Tomoyasu, T., McCarty, J. S., Rudiger, S., Schonfeld, H. J., Schirra, C., Bujard, H., and Bukau, B., A cycle of binding and release of the DnaK, DnaJ and GrpE chaperones regulates activity of the *E. coli* heat shock transcription factor sigma32, *EMBO J. 15*:607–617 (1996).
110. Schumacher, R. J., Hansen, W. J., Freeman, B. C., Alnemri, E., Litwack, G., and Toft, D. O., Cooperative action of hsp70, hsp90, and DnaJ proteins in protein renaturation, *Biochemistry 35*:14889–14898 (1996).
111. Ang, D., Chandrasekhar, G. N., Zylicz, M., and Georgopoulos, C., *E. coli* GrpE gene codes for heat shock protein b25.3, essential for both λ DNA replication at all temperatures and host growth at high temperature, *J. Bacteriol. 167*:25–29 (1986).
112. Liberek, K., Georgopoulos, C., and Zylicz, M., Role of the *E. coli* DnaK and DnaJ heat shock proteins in the initiation of bacteriophage λ DNA replication, *Proc. Natl. Acad. Sci. USA 85*:6632–6636 (1988).
113. Banecki, B. and Zylicz, M., Real time kinetics of the DNA/DnaJ/GrpE molecular chaperone machine action, *J. Biol. Chem. 271*:6137–6143 (1996).
114. Straus, D., Walter, W., and Gross, C. A., DnaK, DnaJ, and GrpE heat shock proteins negatively regulate heat shock gene expression by controlling the synthesis and stability of sigma 32, *Genes Dev. 4*:2202–2209 (1990).
115. Shi, W., Zhou, Y., Wild, J., Adler, J., and Gross, C. A., DnaK, DnaJ, and GrpE are required for flagellum synthesis in *E. coli*, *J. Bacteriol. 174*:6256–6263 (1992).
116. Hoffmann, H. J., Lyman, S. K., Lu, C., Petit, M. A., and Echols, H., Activity of the hsp70 chaperone complex—DnaK, DnaJ, and GrpE—in initiating phage λ DNA replication by sequestering and releasing λ P protein, *Proc. Natl. Acad. Sci. USA 89*:12108–12111 (1992).
117. Gething, M.-J., Blond-Elguindi, S., Mori, K., and Sambrook, J. F., Structure, function, and regulation of the endoplasmic reticulum chaperone BiP, *The Biology of Heat Shock Proteins and Molecular Chaperones* (Morimoto, R. I., Tissières, A., and Georgopoulos, C., eds.), Cold Spring Harbor Laboratory Press, Cold Spring Harbor, NY, 1994, pp. 111–135.
118. Cyr, D. M. and Douglas, M. G., Differential regulation of hsp70 subfamilies by the eukaryotic DnaJ homologue Ydj1, *J. Biol. Chem. 269*:9798–9804 (1994).
119. Gaitanaris, G. A., Vysokanov, A., Hung, S., Gottesman, M. E., and Gragerov, A., Successive action of *E. coli* chaperones *in vivo*, *Mol. Microbiol. 14*:861–869 (1994).
120. Hendershot, L., Wei, J., Gaut, J., Melnick, J., Aviel, S., and Argon, Y., Inhibition of immunoglobulin folding and secretion by dominant negative BiP ATPase mutants, *Proc. Natl. Acad. Sci. USA 93*:5269–5274 (1996).
121. Hightower, L. E., Sadis, S. E., and Takenada, I. M., Interactions of vertebrate hsc70 and hsp70 with unfolded proteins and peptides, *The Biology of Heat Shock Proteins and Molecular Chaperones* (Morimoto, R. I., Tissières, A., and Georgopoulos, C., eds.), Cold Spring Harbor Laboratory Press, Cold Spring Harbor, NY, 1994, pp. 179–207.

6

Substrate-Binding Specificity of the Hsp70 Family

Lawrence E. Hightower
and Sau-Mei Leung
The University of Connecticut, Storrs, Connecticut

I. INTRODUCTION

The three members of the Hsp70 family for which substantial information on peptide–unfolded protein-binding specificities have been collected are *Escherichia coli* DnaK, mammalian BiP, and mammalian Hsc70. The cytoplasmic members of the Hsp70 family or their accessory proteins are probably the first chaperones to interact with newly synthesized polypeptide chains in cells, possibly even as nascent chains in polyribosomes. There is evidence that DnaJ and its eukaryotic homologues which are Hsp70 family regulatory proteins and chaperones in their own right may bind nascent chains initially and both recruit and "present" polypeptides to Hsp70 family members in a complex. Whether Hsp70 family proteins are first or second on the scene, it is clear that their binding specificities and affinities for polypeptide chains emerging in a linear fashion from a ribosome will play significant roles in determining how many Hsp70 family proteins bind per independently folding domain and for how long. A broadly accepted hypothesis about the role of binding to nascent chains by Hsp70 family proteins is that they bind to hydrophobic sequences to prevent them from aggregating and to delay the initiation of folding of the polypeptide chain until the chain is completed and all of the primary sequence information required for folding is present. As we will describe in the following, the binding rules for peptides interacting with Hsp70 family proteins are fairly relaxed, so it is reasonable to assume that virtually all polypeptide chains associate at least transiently with these proteins. It is not yet clear what proportion of newly synthesized proteins are released immediately into the cytoplasm to fold, what proportion are transferred to cytoplasmic Cpn60 complexes in tandem chaperoning pathways, and what proportion are ferried to sites of translocation across membranes into the endoplasmic reticulum (ER), mitochondria, and chloroplasts. Another unresolved issue is how the interaction of Hsp70 family proteins with essentially folded proteins differ from their interactions with unfolded proteins. Recently, two different peptide-binding motifs were identified for bovine Hsc70, raising the possibility that sequence-binding specificity may play an important role in making this distinction. It is likely that the cochaperones and regulatory proteins of the Hsp70 family will also play important roles in both trafficking and substrate interactions. For vertebrate Hsc70, we will briefly describe a new accessory protein, a member of the nucleoside diphosphate kinase family (NDP kinase) and suggest how it may regulate the Hsc70–polypeptide interaction. Reviews have appeared covering peptide and unfolded protein binding to DnaK [1,2], BiP [3], and the mammalian Hsp70 family [4]. Here we will concentrate on more recent observations.

II. THE STRUCTURE OF THE PEPTIDE OR UNFOLDED PROTEIN-BINDING DOMAIN

The long-awaited molecular structure of an Hsp70 family peptide-binding domain is now available for DnaK [5]. Zhu and co-workers [5] crystallized the substrate-binding domain (residues 389–607) without the ATPase domain and the COOH-terminal sequence in the presence of a synthetic peptide (NRLLLTG). The structure was determined at 2.0-Å resolution using multiwavelength anomalous diffraction phasing applied to a selenomethionyl analogue. The structure consists of two domains, a β-strand sandwich subdomain containing eight β-strands followed by an α-helical subdomain containing five segments. The peptide is bound in an extended conformation, in agreement with previous nuclear magnetic resonance (NMR) studies [6], and in a channel that is created by loops from the β-sandwich. The channel crosses the shortest dimension of the structure, which is approximately 60 × 40 × 15 Å. The helix αA and the first half of αB form an arm with an elbow joint and are integrated with the same hydrophobic core as the first subdomain, whereas the rest of the helical elements are folded around a separate hydrophobic core. Three of the helices form a triple-helical bundle, the existence of which was predicted previously by Hightower and co-workers based on computerized analysis of Hsc70 [4]. Helix αB thus spans the two cores. The helical domain forms a lid over the peptide-binding domain, but it does not interact directly with the bound peptide. In a second crystal, the lid is rotated, suggesting that it may control access to the binding site and that a conformational change in the lid may be associated with a change in the affinities or rate constants for association–dissociation for peptides. The authors noted that the five-helix domain of DnaK has a topology and general structure similar to the helical domain of the Gα subunits of trimeric G proteins. Interestingly, the helical lid on Gα participates in a conformational change that regulates GDP–GTP exchange [7], a role similar to that proposed for the lid covering the peptide-binding site of DnaK.

The authors carried out a secondary structure-based sequence alignment of DnaK, bovine Hsc70, and human BiP. As expected from the high level of amino acid sequence conservation, the alignment predicts that the peptide-binding domain of the latter two proteins will be similar to that of DnaK. Sadis and Hightower [8] previously suggested that the COOH-terminal region of Hsc70 preceding the flexible terminal segment was composed entirely of α-helical elements and proposed three helices. These computer-aided predictions were supported by circular dichroism (CD) measurements. Morshauser and co-workers [6] identified the same β-strands, followed by an α-helix by NMR of Hsc70, as those found in the crystal structure of DnaK, except that the Hsc70 homologue of αA was run together with αB; however, the possibility of a kink, at about the position of the bend between the two DnaK helices, was noted. Thus, it is very likely that all of the Hsp70 family peptide-binding domains will be similar, but with some subtle chemical differences that will explain the differences in peptide-binding preferences that have been revealed by biochemical studies. There are also likely to be some differences in how the peptide-binding and ATPase domains communicate, because DnaK ends in a different COOH-terminal sequence than the eukaryotic Hsps, and this segment has been implicated in interdomain interactions [9].

III. INTERACTIONS OF Hsp70 FAMILY CHAPERONES WITH PEPTIDES AND UNFOLDED PROTEINS

A. Chemical Nature of the Binding Site

The synthetic peptide that was cocrystallized with DnaK (NRLLLTG) was identified in a screen of a phage display library, as described later. Leucine-4 is completely buried in a deep hydrophobic pocket, and most side-chain contacts between other residues and DnaK are also hydrophobic. Peptide binding occurs through numerous van der Waals interactions from its side chains and through seven main-chain hydrogen bonds: Leu-3 and Leu-4 contribute most of the contacts, the remainder coming from Arg-2, Leu-5, and Thr-6. The binding is predominately due to a five-residue core. The side chain of Leu-4 is down in the pocket and the side chains of the adjacent leucines point up and away from the β-sandwich. Selectivity at the Leu-4–binding site is predicted to be the greatest, but even here, methionine or isoleucine could also fit. Sites outside of the central three are even less restricted and may be selected primarily on the basis of the electrostatic potential of the protein surface, which is mainly negative. The dominant influence of the Leu-4 site, which the authors designated position 0, creates the interesting possibility that some peptides may bind in a polarity opposite the one in the crystal. The chemical nature of the binding site would best accommodate the sets of peptides selected by screens of phage display libraries using DnaK and Hsc70. And, in fact, a similar conclusion was reached based on a comparison of the types of peptides selected by DnaK, Hsc70, and BiP. The prediction is that the peptide-binding site of BiP will be the most chemically distinct.

The binding of bovine Hsc70 to peptides and unfolded proteins has been the focus of most of the biochemical studies of this family: therefore, we will concentrate on Hsc70 and compare other family members with it. Bovine Hsc70 is a constitutively expressed member of the Hsp70 family of peptide or unfolded protein-stimulated ATPases. It can distinguish between native and unfolded forms of the same protein [10,11]. Biochemical studies of a Hsc70 fragment generated using recombinant DNA technology have located the domain that binds unfolded proteins between residues 384 and 543 [12]. This localization agrees well with the molecular structure of the peptide-binding region of DnaK described earlier.

Pelham [13] suggested that Hsp70 family proteins may distinguish unfolded from folded proteins based on the display of aliphatic amino acid residues normally buried inside folded proteins. Studies of peptides selected from a pool of random sequences by BiP (Grp78), the family member located in the endoplasmic reticulum, indicate that the peptide-binding site holds aliphatic residues in an environment energetically equivalent to the hydrophobic interior of folded proteins and that the minimum sequence length for effective binding to BiP is seven amino acid residues [14]. Analyses of two-dimensional NMR spectra indicate that DnaK, the bacterial homologue of Hsp70, binds peptides in an extended conformation, with two large aliphatic residues having the strongest interactions with DnaK [15]; this work agrees well with the molecular structure obtained by X-ray crystallography.

B. Sequences from Phage Peptide-Display Libraries

Three groups have used affinity panning of f1 phage peptide-display libraries to obtain more detailed information on the binding specificity of Hsp70 family members. Gragerov et al. [16] used DnaK to screen a library and found that peptides such as NRLLLTG containing internal aliphatic residues are preferred substrates, inclusion of a basic residue is beneficial, and peptides with acidic residues are disfavored. The high-affinity BiP-binding peptides selected from display libraries by Gething and co-workers [3] contained a minimum of seven residues and had the motif Hy(W/X)HyXHyXHy where Hy is a large hydrophobic residue, W is tryptophan, and X is any amino acid except charged residues, which are not favored in BiP-binding peptides [17]. Subsequently, both groups compared the binding of DnaK, Hsc70, and BiP to sets of peptides. Gragerov and Gottesman [18] found that DnaK and Hsc70 are more similar to each other in their relative affinities for a set of three peptides than to BiP. They have the highest affinities for a peptide (NLLRLTG) containing two large hydrophobic internal residues, followed by a basic residue. Of the three Hsp70s, BiP was most tolerant of acidic residues. Fourie and co-workers [19] extended this comparison by showing that these three Hsp70 family proteins can have both common and exclusive binding specificities, depending on the peptide sequence. A convincing example of this was a comparison of the peptide KYLMFKT, which binds all three proteins with similar relative affinities, and peptide KKLMFKT, which binds with affinities in the order DnaK > Hsc70 >> BiP.

Takenaka and co-workers [20] screened a 15-mer phage display random peptide library with purified bovine Hsc70, and nucleotide sequence analysis of the selected clones showed a large enrichment for peptides containing basic sequences with at least KK, KR, or RR. Binding affinity for Hsc70 of representative peptides increased dramatically for heptamers compared with hexamers. The peptide NIVRKKK had the highest affinity for Hsc70, and substitution analyses showed that hydrophobic residues followed by basic residues play important roles in maintaining this affinity. In contrast, NIVRKKK was a weaker stimulator of the Hsc70 ATPase activity compared with pigeon cytochrome *c* peptide and FYQLALT, a peptide optimized for binding to Hsc70. FYQLALT effectively blocked the binding of NIVRKKK to Hsc70, possibly by causing a conformational change that masked Hsc70's binding site for the basic peptide.

The fact that all three Hsp70 family proteins require substrates containing large hydrophobic internal residues fits well with their proposed roles as molecular chaperones capable of distinguishing unfolded and native conformations of the same protein. However, several kinds of chaperoning have been proposed that may require these proteins to recognize different binding sites. For example, Hsp70 family proteins may very transiently bind hydrophobic sequences in nascent polypeptides to delay folding until the entire polypeptide is synthesized, or to prevent aggregation of a denatured protein. A different function, the chaperoning of proteins from cytoplasmic sites of synthesis to organelles, such as the endoplasmic reticulum, mitochondria, lysosomes, or nuclei, may require more stable associations, with Hsp70 family proteins requiring binding to different amino acid sequences. Indeed, mitochondrial Hsp70 (mtHsp70) can bind to both the mitochondrial-targeting sequence and to mature segments of unfolded precursor proteins during membrane translocation [21]. And apart from chaperoning, the binding sites for Hsp70 family proteins on folded proteins such as p53 are not hydrophobic and must involve a different type of interaction.

Thirteen known peptide sequences have now been reported that interact with Hsc70 (Table 1). Their binding constants are usually in the micromolar range; however, there are considerable differences both in their sequences and

Table 1 Peptides Known to Bind Hsc70

Peptides sequences	Comments
FYQLALT	Hydrophobic sequence
+++ -+ - + + I**F**AGIKKKAERADLIA**Y**LKQATAK	P*c*; pigeon cytochrome *c* peptide
+ + + + MVLPRL**Y**TATSRAA**F**KAAKQSAPLLS ++ + –TS**W**KRCMASAAQSTPITGKVTAVIG	F1β(1–51) of F1-ATPase β-subunit precursor protein (a.a. 1–51), mitochondrial-targeting signal-containing peptide
++++ NIVRKKK	Hydrophobic–basic sequence resembles organellar-targeting signal
+- + -+ + - KETAAAK**F**ERQHMDSSTSAA	S-peptide of ribonuclease A, lysosomal-targeting signal (KFERQ)-containing peptide
+++++ - CQPPKKKRKVE	NLS of SV40 large T-antigen
++ ++ ++++ - **Y**AVKRPAATKKAGQAKKKKLDC	NLS of nucleoplasmin
++ ++ - ARKSLKRLTG**W**ADVC	GT4; peptide derived from glucose transporter
- + - - - **F**AILDGGAPGPQPHGEPPGGPDAVD	Pa3; clathrin light-chain peptide
+ -- PLSQET**F**SGL**W**KLLPPEDG	Pro-17-Gly; peptide derived from p53 protein
+ + ++ ++++ + - - - SSHLKSKKGQSTSRHKKLM**F**KTEGPDSD	Peptide derived from p53 protein
+ + + KLIGVLSSL**F**RPK	Peptide C; derived from vesicular stomatitis virus glycoprotein
++ - - + + KRQI**Y**TDLEMNRLGK	Peptide A; derived from vesicular stomatitis virus glycoprotein

Highly hydrophobic amino acids (I, V, L, F, M) are underlined. Aromatic amino acids (F, Y, W) are bold. Basic amino acids (K, R, H) are indicated by plus signs, and acidic amino acids (E, D) are indicated by minus signs.

their ability to stimulate Hsc70 ATPase activity (Table 2). For example, the 7-mer peptides FYQLALT and NIVRKKK both bind with relatively high affinity to Hsc70, but they differ dramatically in their abilities to stimulate Hsc70 ATPase activity. Stimulation of ATPase activity by a hydrophobic peptide, FYQLALT occurred in the micromolar range, whereas stimulation by a hydrophobic and basic peptide, NIVRKKK occurred in the millimolar range [20]. Although the functional significance of this difference is still unclear, one possible explanation

Table 2 Kinetic Characteristics of Hsc70-Binding Peptides

Peptides	K_d (affinity to Hsc70) [Ref.]	K_m for Hsc70 ATPase [Ref.]
FYQLALT	5 μM (RCMLA)[a][19]	(0.03 mM, 2-fold increase) [20]
	14 and 4.3 μM	(presence of MgADP, 2-step process) [22]
	40–50 μM	(presence of MgATP) [22]
P*c*	8 μM [19]	(0.04 mM, 2-fold increase) [20]
	7 μM [25]	0.2 mM [25]
NIVRKKK	2 μM (P*c*)[a][20]	(2 mM, 2-fold increase) [20]
S-peptide	8 μM [31]	
	7.6 μM [36]	2 mM [36]
	4.2 μM [67]	
	7 μM[b][38]	(1 mM, 1.5-fold increase)[b][38]
	150 μM [25]	(1 mM gives weak stimulation) [25]
GT4	100 μM [36]	0.17 mM [36]
	300 μM[b][38]	
Pa3	13 μM[b][38]	0.56 mM[b][38]
	13 μM [36]	0.67 mM [36]
		0.3–0.7 mM [37]
Peptide C		0.25 mM [25]
		0.055 mM [6]
Peptide A	200–500 μM (RCMLA)[a][19]	0.77 mM [6]

[a]Competition binding using RCMLA or P*c*.
[b]Using 60-kDa fragment of Hsc70.

is that different types of substrates may interact differently with Hsc70. For example, Hsc70 binding to unfolded proteins (represented by FYQLALT; see later discussion) may require substantial hydrolysis of ATP to refold protein or prevent aggregation. However, for Hsc70-assisted protein translocation to organelles, active ATP hydrolysis may disrupt the necessary stable interaction between Hsc70 and transport-competent protein substrates (represented by NIVRKKK; see later) required for them to reach their target organelles.

C. Peptide Sequences Representing Unfolded Proteins

Gething and colleagues [19] studied the substrate specificities of three molecular chaperones (Hsc70, BiP, and DnaK) by measuring the ability of 53 synthetic peptides to inhibit the formation of Hsc70– BiP– and DnaK–RCMLA complexes. They demonstrated that peptides that bound with highest affinity to all three molecular chaperones contained stretches of at least seven residues that included large hydrophobic and basic amino acids, but few or no acidic residues. One synthetic peptide, FYQLALT, contains hydrophobic and aromatic residues and is a relatively good stimulator of Hsc70 ATPase activity. The FYQLALT-like peptide motifs may represent the hydrophobic amino acid residues normally buried inside folded proteins and probably represent sites recognized by Hsc70 to distinguish native from unfolded conformations. Takeda and McKay [22] used a fluorescently labeled version of FYQLALT to measure the kinetics of peptide binding to and release from recombinant Hsc70. Peptide binding is a two-step process in the presence of MgADP. A low-affinity peptide–Hsc70 complex is created in the first step ($K_d = 14$ μM) and the peptide is locked into a higher affinity complex ($K_d = 4.3$ μM) during the second step. In contrast, peptide binding is a one-step process in the presence of MgATP ($K_d = 40$–50 μM). A model was suggested for domain interactions in which the ATPase domain with bound ATP constrains the peptide-binding domain to a lower-affinity conformation. After ATP hydrolysis, the peptide-binding domain is proposed to be less constrained by the ATPase domain, allowing a higher-affinity interaction with the bound peptide. A two-step model for peptide binding was proposed previously for Hsc70–peptide interactions to account for the contributions of both ionic and hydrophobic residues [4].

P*c*, a fragment of pigeon cytochrome *c* (residues 81–104; IFAGIKKKAER ADLIAYLKQATAK) was first shown to bind to Pbp74, a constitutive 70-kDa heat-shock protein family member thought to be involved in antigen presentation on cell surfaces [23,24]. Later, P*c* was also shown to bind strongly to Hsc70 and stimulate Hsc70 ATPase activity [20,25]. Because Hsc70 binds to the unfolded protein apocytochrome *c*, but not its native form cytochrome *c* [11], and P*c*, similar to FYQLALT, contains hydrophobic and aromatic residues, P*c* probably contains a sequence motif that is normally buried inside the folded native protein and is recognized by Hsc70 to distinguish native from unfolded conformations. However, a highly basic region near the amino terminus (IFAGIKKK) which resembles the NIVRKKK motif (see following discussion) may also be a binding site for Hsc70.

D. Peptide Sequences Resembling Organellar-Targeting Signals

Takenaka and colleagues [20] have selected and identified hydrophobic and basic sequences that resemble organellar-targeting signals by affinity panning of f1 phage peptide-display libraries. One peptide, NIVRKKK, has the highest affinity for Hsc70. Subsequent substitution analyses of this peptide showed that hydrophobic residues followed by basic residues play important roles in maintaining this affinity. NIVRKKK contains the basic part of the nuclear localization signal (e.g., PKKKRKV of SV40 large T-antigen) and it may also be related to the mitochondrial-targeting signal that characteristically forms a basic amphiphilic α-helix. Therefore, it is proposed that Hsc70 recognizes two different amino acid sequence motifs in its dual roles of chaperoning proteins to organelles (NIVRKKK-like sequences) and facilitating protein folding (FYQLALT-like sequences). In addition, the NIVRKKK-like sequences have been observed in proteins, such as Hsc70, p53, and Dnaj2, that interact with Hsc70. Therefore, the NIVRKKK motif may be a recognition site used by Hsc70 to interact with these proteins in their folded states.

F1β(1–51) is a mitochondrial-targeting signal-containing peptide that contains the amino acids 1–51 (MVLPRLYTATSRAAFKAAKQSAPLLSTSWK RCMASAAQSTPI TGKVTAVIG) of the F1-ATPase β-subunit precursor protein. The first 20 residues form an amphiphilic α-helical structure with basic residues cluster on one face of the helix. This basic amphiphilic structure is proposed to be required for mitochondrial import [26–28]. Cyr and Douglas [29] demonstrated that F1β(1–51) forms stable complexes with Hsc70, BiP, Ssa1, and Ssa2. The ATPase activity of Ssa1 and Ssa2 (0.45 μM) was stimulated approximately twofold in the presence of 100 μM of F1β(1–51). This basic part of the signal sequence has also been proposed to be the binding site for Hsc70 [20].

In studying lysosomal degradation of intracellular proteins, such as RNase, during serum withdrawal, Dice and colleagues demonstrated that the RNase S-peptide (KETAAAKFERQHMDSSTSAA) binding protein prp73 is Hsc70 [30,31]. Hsc70 also binds to the lysosomal-targeting sequence KFERQ, although with less affinity. Hsc70 stimulates lysosomal uptake and degradation of KFERQ-containing proteins in vitro; therefore Hsc70 may help to selectively target these proteins to the lysosome for degradation during starvation [32]. Interestingly, similar to NIVRKKK, the S-peptide binds strongly to Hsc70, but is a weak Hsc70 ATPase stimulator (see Table 2). This observation further supports the idea that protein translocation to organelles may not require active ATP hydrolysis of Hsc70.

There is evidence that Hsc70 binds to the nuclear localization signal of both SV40 large T-antigen (CQPPKKKRKVE) and nucleoplasmin (YAVKRPAAT KKAGQAKKKKLDC) [33–35]. Hsc70 binds to a nucleoplasmin-Sepharose column and can be eluted by the NLS of both nucleoplasmin and SV40 large T-antigen. Therefore, Hsc70 has been implicated in importing proteins to the nucleus through binding to their nuclear localization signal.

E. Binding Sequences on Native Proteins

GT4, (ARKSLKRLTGWADVC), a 15-mer peptide derived from the insulin-regulated glucose transporter, was randomly chosen to study the peptide-stimulated ATPase activity of recombinant rat Hsc70 along with S-peptide and clathrin light- chain peptide, Pa3 [36]. Surprisingly, GT4 was a better ATPase stimulator than S-peptide and Pa3, even though it binds to Hsc70 10–20 times more weakly than S-peptide (see Table 2). The functional significance of these differences is still unclear.

Pa3 (FAILDGGAPGPQPHGEPPGGPDAVD), a 25-mer peptide derived from clathrin light chain (LCa), is the sequence recognized by Hsc70 [37]. Pa3 is a relatively good binder and ATPase stimulator for Hsc70 [36,38]. Interestingly, unlike the hydrophobic and hydrophobic–basic peptides (FYQLALT and NIVRKKK), described earlier, Pa3 is enriched for glycine and proline residues. These enriched residues may form a connecting loop or turn structure where Hsc70 may interact.

Lam and Calderwood [39] demonstrated that a peptide, Pro-17-Gly (PLSQ ETFSGLWKLLPPEDG), derived from the NH_2-terminal domain I of p53 tumor antigen, blocks the binding of mutant p53 to human Hsp70. Interestingly, similar to Pa3, Pro-17-Gly is also enriched for glycine and proline residues. However, Pro-17-Gly is not the only peptide sequence in p53 that interacts with Hsc70. Hainaut and Milner [40] reported that complex formation between mutant p53 and murine Hsp70 requires the carboxyl 28 amino acids of p53 (SSHLKSKK GQSTSRHKKLMFKTEGPDSD). This peptide is highly basic and contains a NIVRKKK-like sequence in the underlined portion of the peptide. Therefore it is possible that there are two different binding motifs for Hsc70 in mutant p53 protein.

Rothman and colleagues [41] investigated the ability of several randomly chosen peptides to bind Hsc70 and BiP and to stimulate their ATPase activity. Two peptides, peptide A (KRQIYTDLEMNRLGK) and peptide C (KLIGVLSSL FRPK), derived from the vesicular stomatitis virus glycoprotein, stimulate Hsc70 ATPase activity. These two peptides contain many hydrophobic residues in the middle and basic residues at both COOH- and NH_2-termini that further supports the idea that hydrophobic and basic residues are important for binding Hsc70.

Taken together, it is clear that the basis of substrate recognition for Hsc70 is not a specific amino acid sequence. The characteristics of Hsc70 substrate specificity can be summarized as follows: (a) Hsc70 recognizes at least seven-amino acid residues; (b) the recognition motif contains either hydrophobic or hydrophobic–basic residues, whereas acidic residues are disfavored for binding Hsc70; (c) for native proteins, the recognition motif may form an extended loop or turn structure.

IV. THERMAL ACTIVATION OF PEPTIDE BINDING TO Hsc70

Few studies have addressed the temperature dependence of members of the Hsp70 family binding to substrates. Binding of the unfolded substrate, reduced-carboxymethylated α-lactalbumin, to DnaK is temperature-dependent. It increases from 20°C, to a maximum at 40°C, followed by a sharp decrease as temperature increases to 50°C. The calculated activation energy of 12.4 kcal

mol^{-1} is relatively high [42]. When using a thermolabile mutant of staphylococcal nuclease with a t_m of 30°C as a substrate, Paleros and co-workers [10] found that binding to Hsc70 was activated between 25° and 30°C and increased up to 40°C. Interpretation of this experiment is complicated because both thermal unfolding of the substrate and thermal activation of Hsc70 probably occurred simultaneously. Others have noticed that activation of steroid hormone receptors by Hsc70 does not occur below 25°C. Formation of activated steroid hormone receptor requires bound steroid, a temperature of at least 25°C, ATP, and Hsc70 [43,44]. It is possible that this minimum temperature requirement is for conformational activation of Hsc70 (see later discussion).

Leung and co-workers [45] have shown that binding of peptides and unfolded proteins to Hsc70 increases substantially above 30°C. Binding of both the hydrophobic peptide FYQLALT and the basic peptide NIVRKKK, which is likely to bind at a different site, increases with increasing temperature, indicating that more than the hydrophobic peptide-binding domain of Hsc70 changes conformation. The model unfolded protein, apo *c*, contains sequences similar to both of these peptides, so apo *c* may bind to Hsc70 at more than one site. A temperature-dependent conformational change in Hsc70 was demonstrated by increased sensitivity to chymotryptic cleavage with increasing temperature, and this change was reversible. Previous studies have shown that several sensitive chymotryptic cleavage sites are located in the COOH-terminal region of Hsc70. The 60-kDa fragment detected in our study, is generated from a cleavage of an approximately 10-kDa segment from the COOH-terminus. Further degradation of the 60-kDa fragment to a 44-kDa fragment results in loss of substrate-binding activity, but not ATPase activity [46], indicating that this fragment represents the NH_2-terminal domain. Given the chymotryptic fragment patterns of Hsc70 incubated at 40°C and the activation of peptide-binding activity, we concluded that at least the COOH-terminal part of Hsc70 undergoes a temperature-dependent conformational change.

An exothermic transition detected by differential scanning calorimetry is associated with the conformational change in Hsc70 that activates peptide binding [45]. We hypothesize that this negative change in excess heat capacity may be due to hydration of polar and charged residues in the COOH-terminus. Other potential explanations, such as oligomerization and aggregation of Hsc70, have been ruled out.

It has been hypothesized that a homeostatic mechanism based on cellular levels of free Hsc70 and DnaK constitutes a molecular thermometer for sensing temperature changes [47,48]. In this model, these proteins are negative regulators of the heat-shock response that bind heat-shock transcription factors [49,50]. Higher temperatures would result in larger amounts of denatured proteins present in cells to bind free Hsc70 and DnaK, leaving more of the transcription factors in an unbound, activated state. McCarty and Walker [51] showed that DnaK autophosphorylation and ATPase activity are stimulated in the 30°–50°C range and proposed that these changes in DnaK activity may serve as a thermometer, sensing temperature changes and activating the heat-shock response of *E. coli.* Our studies suggest that the concept of Hsc70 as a thermal sensor, an analogy that we prefer over molecular thermometer, may be usefully applied to a much broader temperature range in mammalian cells than previously considered. The model described in the foregoing would apply to temperatures above the mammalian core body temperature of 37°C. However, if body extremities are considered, the normal temperature of mammalian tissues is more accurately described as a range of approximately 30–37°C. We propose that over this range, unfolded protein-binding activity of Hsc70 is matched to demand for chaperoning. As tissue temperatures increase or decrease, demand, including amounts of nascent polypeptides, amounts of partially folded proteins in folding pathways, participating in off-pathway interactions, or in transit to cellular compartments, and the levels of active Hsc70 chaperones may change roughly in parallel. The proportion of inactive and active Hsc70 in free pools would be determined by the reversible, temperature-dependent conformational change demonstrated herein (i.e., the Hsc70 thermal sensor). Above 37°C, a new chaperoning demand, the accumulation of denatured proteins, is placed on Hsc70 in addition to further increases in production of nascent polypeptides. It may be more than a coincidence that the maximum rate of conversion of inactive to active forms of Hsc70 in vitro occurs at 41°C, based on the T_m of the exothermic transition, and close to the upper limit for a survivable heat shock by mammalian cells.

The discovery of inactive and active chaperoning states of Hsc70 may bear on the altered cellular state known as acquired thermotolerance. Two states of thermotolerance have been described in mammalian cells, yeasts, and snails [52–54]. One of these states depends on the synthesis of heat-shock proteins, whereas the other can be established in the presence of cycloheximide. We suggest the following explanation for the latter type of thermotolerance. The inhibition of protein synthesis by cycloheximide blocks the temperature-dependent increase in one of the targets of Hsc70 activity, nascent polypeptide chains. Simultaneously, a rise in temperature conformationally activates more Hsc70, and the combined effect of this activation and the cycloheximide block is to rapidly increase the pools of active Hsc70 (i.e., rapidly increase chaperoning capacity). More active Hsc70 would then be available to bind increasing amounts of thermally denatured cellular proteins as the temperature increased. In the absence of cycloheximide, increasing temperature would result in increased rates of nascent polypeptide production and accumulation of denatured protein, rapidly exhausting active pools of Hsc70 and hastening the induction of the heat-shock response.

V. NDP KINASE (Hup) REGULATES PEPTIDE BINDING TO Hsc70 IN VITRO

Hsc70 is a multifunctional molecular chaperone, and it is likely that accessory proteins are used to specify its diverse chaperone activities. We recently found a 16-kDa cytoplasmic protein (p16) associated with Hsc70 during its purification from a fish hepatocyte cell line, PLHC-1. Hsc70 also coimmunoprecipitated with p16 from PLHC-1 cells. p16 was identified as a member of the Nm23/nucleoside diphosphate (NDP) kinase family based on its amino acid sequence similarity, NDP kinase activity, and recognition by antihuman NDP kinase-A antibody. p16 monomerized Hsc70 and released Hsc70 from pigeon cytochrome *c* peptide (P*c*), but not from FYQLALT peptide. Therefore, p16 may modulate Hsc70 function by maintaining Hsc70 in a monomeric state and dissociating unfolded proteins either through protein–protein interaction or by supplying ATP indirectly through phosphate transfer. p16 did not affect basal or unfolded protein-stimulated ATPase of bovine brain Hsc70 in in vitro assays. Interestingly, bovine liver NDP kinase did not dissociate the Hsc70–P*c* peptide complex. Therefore, p16 may be a unique Nm23/NDP kinase that functions as an accessory protein for cytoplasmic Hsc70 in eukaryotes. We suggest the name Hup (Hsc 70-unbinding protein) for this protein [68].

NDP kinase is generally considered to be one of the "housekeeping" enzymes necessary for maintaining the nucleotide pools of the cell. One of the remarkable features of this enzyme seems to be its association with proteins that require guanine nucleotides for their functions. These proteins include initiation factor eIF2 [55], and GTP-binding (G) proteins [56,57]. The ability of NDP kinase to convert GDP to GTP has led to the proposal that NDP kinase plays a role in the activation of G proteins in signal transduction by directly or indirectly providing GTP [58–60]. Interestingly, plant NDP kinases have been proposed as a part of the signal transduction pathway during stress [61]. Evidence that NDP kinases may bind to both Hsc70 and G proteins is interesting in light of the discussion in Section II on the structural similarity between the helical "lid" domains of these two proteins. The lid would appear to be the most likely binding site for NDP kinase, and this would be consistent with the effects that NDP kinase has on both peptide and guanosine nucleotide binding (i.e., regulation of access to the binding sites).

NDP kinase from cultured sugarcane cells exhibits heat-shock stimulated autophosphorylation activity [62] and in tomato plant, its mRNA level is up-regulated in response to wounding [63]. Because one of the most common signal transduction events during heat shock is protein phosphorylation, the up-regulated NDP kinase activity and protein levels may allow the plant to perceive and transduce signals related to high-temperature stress [61]. Because the protein levels of p16 increased when PLHC-1 cells were exposed to cadmium ions, it is possible that p16, similar to the plant NDP kinases, may also play a role in signal transduction pathways during stress. Furthermore, the direct association between p16 and Hsc70, combined with similar dose-dependent increases in both p16 and Hsc70/Hsp70 levels in response to Cd^{2+} that we have observed, demonstrate for the first time a direct link between NDP kinase and the stress response in animal cells.

The relation between NDP kinase and the stress response is further demonstrated by the identification of NDP kinase as a plant flavonoid cromoglycate-binding protein [64]. Cromoglycate is a widely used antiasthmatic drug that is structurally related to quercetin, a potent NDP kinase inhibitor in vitro (IC_{50} is 28.6 uM) [65]. More importantly, quercetin down-regulates the heat-shock response in human HeLa cells by reducing both HSF1 phosphorylation and protein levels. As a result, HSF1–heat-shock element complex declines [66]. A possible explanation of this inhibitory mechanism by quercetin is that quercetin may bind to NDP kinase and disrupt its association with Hsc70. Hsc70 is thought to transiently associate with HSF1 to keep it in an inactive state. This free pool of Hsc70 may then bind to HSF1 to suppress its activity during heat shock. Binding of quercetin to NDP kinase may also inhibit its ability to release Hsc70 from HSF1. As a result, HSF1 cannot be activated. Another possibility is that NDP kinase may modulate the phosphorylation state of HSF1 by using its kinase activity. Hsc70 may act as a bridge to allow the physical interaction between NDP kinase and HSF1 necessary for these effects. As a result, the inhibition of NDP kinase activity by quercetin may reduce the levels of phosphorylated HSF1.

ACKNOWLEDGMENTS

The authors wish to thank the members of the Hightower laboratory who have contributed to the studies of the structure and functions of the Hsp70 family: Peter T. Guidon, Jr., Seth Sadis, Ivone Takenaka, and Judy Boice.

Work in the author's laboratory was funded by the NSF, NIH, NIEHS, and the University of Connecticut Research Foundation.

REFERENCES

1. Landry, S. J. and Gierasch, L. M., Polypeptide interactions with molecular chaperones and their relationship to in vivo protein folding, *Annu. Rev. Biophys. Biomol. Struct. 23*:645 (1994).
2. Gierasch, L. M., Wang, Z., Hunt, J., Landry, S. J., Weaver, A., and Deisenhofer, J., Chaperone–substrate interactions, *Protein Eng. 8*:14 (1995).
3. Gething, M.-J., Blond-Elguindi, S., Mori, K., and Sambrook, J. F., Structure, function and regulation of the endoplasmic reticulum chaperone, BiP, *The Biology of Heat Shock Proteins and Molecular Chaperones* (Morimoto, R. I., Tissieres, A., and

Georgopoulos, C., eds.), Cold Spring Harbor Laboratory Press, Cold Spring Harbor, NY, 1994, p. 111.

4. Hightower, L., Sadis, S., and Takenaka, I., Interactions of vertebrate hsc70 and hsp70 unfolded proteins and peptides, *Biology of Heat Shock Proteins and Molecular Chaperones* (Morimoto, R. I., Tissieres, A., and Georgopoulos, C., eds.), Cold Spring Harbor Laboratory Press, Cold Spring Harbor, NY, 1994, p. 179.
5. Zhu, X., Zhao, X., Burkholder, W. F., Gragerov, A., Ogata, C. M., Gottesman, M. E., and Hendrickson, W. A., Structural analysis of substrate binding by the molecular chaperone DnaK, *Science 272*:1606 (1996).
6. Morshauser, R. C., Wang, H., Flynn, G. C., and Zuiderweg, E. R. P., The peptide-binding domain of the chaperone protein hsc70 has an unusual secondary structure topology, *Biochemistry 34*:6261 (1995).
7. Noel, J. P., Hamm, H. E., and Sigler, P. B., The 2.2 angstrom crystal structure of transducin-α complexed with GTPγS, *Nature 366*:654 (1993).
8. Sadis, S., Raghavendra, K., and Hightower, L. E., Secondary structure of the mammalian 70-kilodalton heat shock cognate protein analyzed by circular dichroism spectroscopy and secondary structure prediction, *Biochemistry 29*:8199 (1990).
9. Freeman, B. C., Myers, M. P., Schumacher, R., and Morimoto, R. I., Identification of a regulatory motif in Hsp70 that affects ATPase activity, substrate binding and interaction with HDJ-1, *EMBO J. 14*:2281 (1995).
10. Palleros, D., Welch, W., and Fink, A., Interaction of Hsp70 with unfolded proteins: Effects of temperature and nucleotides on the kinetics of binding, *Proc. Natl. Acad. Sci. USA 88*:5719 (1991).
11. Sadis, S. and Hightower, L. E., Unfolded proteins stimulate molecular chaperone Hsc70 ATPase by accelerating ADP/ATP exchange, *Biochemistry 31*:9406 (1992).
12. Wang, T. F., Chang, J. H., and Wang, C., Identification of the peptide binding domain of hsc70—18-kilodalton fragment located immediately after ATPase domain is sufficient for high affinity binding, *J. Biol. Chem. 268*:26049 (1993).
13. Pelham, H. R. B., Speculations on the functions of the major heat shock and glucose-regulated proteins, *Cell 46*:959 (1986).
14. Flynn, G. C., Pohl, J., Flocco, M. T., and Rothman, J. E., Peptide-binding specificity of the molecular chaperone Bip, *Nature 353*:726 (1991).
15. Landry, S., Jordan, R., McMacken, R., and Gierasch, L., Different conformations for the same polypeptide bound to chaperones DnaK and GroEL, *Nature 355*:455 (1992).
16. Gragerov, A., Li, Z., Xun, Z., Burkholder, W., and Gottesman, M. E., Specificity of DnaK-peptide binding, *J. Mol. Biol. 235*:848 (1994).
17. Blond-Elguindi, S., Cwirla, S., Dower, W., Lipshutz, R., Sprang, S., Sambrook, J., and Gething, M., Affinity panning of a library of peptides displayed on bacteriophages reveals the binding specificity of BiP, *Cell 75*:717 (1993).
18. Gragerov, A. and Gottesman, M. E., Different peptide binding specificities of hsp70 family members, *J. Mol. Biol. 241*:133 (1994).
19. Fourie, A. M., Sambrook, J. F., and Gething, M. J. H., Common and divergent peptide binding specificities of hsp70 molecular chaperones, *J. Biol. Chem. 269*:30470 (1994).
20. Takenaka, I. M., Leung, S.-M., McAndrew, S. J., Brown, J. P., and Hightower, L. E.,

Hsc70-binding peptides selected from a phage display peptide library that resemble organellar targeting sequences, *J. Biol. Chem. 270*:19839 (1995).
21. Ungermann, C., Neupert, W., and Cyr, D. M., The role of hsp70 in conferring unidirectionality on protein translocation into mitochondria, *Science 266*:1250 (1994).
22. Takeda, S. and McKay, D. B., Kinetics of peptide binding to the bovine 70 kDa heat shock cognate protein, a molecular chaperone, *Biochemistry 35*:4636 (1996).
23. Vanbuskirk, A., Crump, B. L., Margoliash, E., and Pierce, S. K., A peptide binding protein having a role in antigen presentation is a member of the hsp70 heat shock family, *J. Exp. Med. 170*:1799 (1989).
24. Domanico, S. Z., Denagel, D. C., Dahlseid, J. N., Green, J. M., and Pierce, S. K., Cloning of the gene encoding peptide-binding protein-74 shows that it is a new member of the heat shock protein-70 family, *Mol. Cell. Biol. 13*:3598 (1993).
25. Greene, L. E., Zinner, R., Naficy, S., and Eisenberg, E., Effect of nucleotide on the binding of peptides to 70-kDa heat shock protein, *J. Biol. Chem. 270*:2967 (1995).
26. Cyr, D. M. and Douglas, M. G., Early events in the transport of proteins in mitochondria: Import competition by a mitochondrial presequence, *J. Biol. Chem. 266*:21700 (1991).
27. Hoyt, D. W., Cyr, D. M., Gierasch, L. M., and Douglas, M. G., Interaction of peptides corresponding to mitochondrial presequences with membranes, *J. Biol. Chem. 266*:21693 (1991).
28. Kiebler, M., Keil, P., Schneider, H., van der Klei, I. J., Pfanner, N., and Neupert, W., The mitochondrial receptor complex: A central role of MOM22 in mediating preprotein transfer from receptors to the general insertion pore, *Cell 74*:483 (1993).
29. Cyr, D. M. and Douglas, M. G., Differential regulation of Hsp70 subfamilies by the eukaryotic DnaJ homologue YDJ1, *J. Biol. Chem. 269*:9798 (1994).
30. Chiang, H. L., Terlecky, S. R., Plant, C. P., and Dice, J. F., A role for a 70-kilodalton heat shock protein in lysosomal degradation of intracellular proteins, *Nature 246*:382 (1989).
31. Terlecky, S., Chiangs, H., Olson, T., and Dice, J., Protein and peptide binding and stimulation of in vitro lysosomal proteolysis by the 73-kDa heat shock cognate protein, *J. Biol. Chem. 267*:9202 (1992).
32. Dice, J. F., Agarraberes, F., Kirven-Brooks, M., Terlecky, J., and Terlecky, S. R., Heat shock 70-kDa proteins and lysosomal proteolysis, *The Biology of Heat Shock Proteins and Molecular Chaperones* (Morimoto, R. I., Tissieres, A., and Georgopoulos, C., eds.), Cold Spring Harbor Laboratory Press, Cold Spring Harbor, NY, 1994, p. 137.
33. Imamoto-Sonobo, N., Matsuoka, Y., Semba, T., Okada, Y., Uchida, T., and Yoneda, Y., A protein recognized by antibodies to Asp-Asp-Glu-Asp shows specific binding activity to heterogeneous nuclear transport signals, *J. Biol. Chem. 265*:16504 (1990).
34. Imamoto, N., Matsuoka, Y., Kurihara, T., Kohno, K., Miyagi, M., Sakiyama, F., Okada, Y., Tsunasawa, S., and Yoneda, Y., Antibodies against 70-kD heat shock cognate protein inhibit mediated nuclear import of karyophilic proteins, *J. Cell. Biol. 119*:1047 (1992).
35. Okuno, Y., Imamoto, N., and Yoneda, Y., 70-kDa heat-shock cognate protein colocalizes with karyophilic proteins into the nucleus during their transport in vitro, *Exp. Cell Res. 206*:134 (1993).

36. Wang, C. and Lee, M. R., High-level expression of soluble rat hsc70 in *Escherichia coli*—purification and characterization of the cloned enzyme, *Biochem. J. 294*:69 (1993).
37. DeLuca-Flaherty, C., McKay, D. B., Parham, P., and Hill, B. L., Uncoating protein (hsc70) binds a conformationally labile domain of clathrin light chain LCa to stimulate ATP hydrolysis, *Cell 62*:875 (1990).
38. Tsai, M. Y. and Wang, C., Uncoupling of peptide-stimulated ATPase and clathrin-uncoating activity in deletion mutant of Hsc70, *J. Biol. Chem. 269*:5958 (1994).
39. Lam, K. T. and Calderwood, S. K., Hsp70 binds specifically to a peptide derived from the highly conserved domain (I) region of p53, *Biochem. Biophys. Res. Commun. 184*:167 (1992).
40. Hainant, P. and Milner, J., Interaction of heat-shock protein 70 with p53 translated in vitro: Evidence for interaction with dimeric p53 and for a role in the regulation of p53 conformation, *EMBO J. 11*:3513 (1992).
41. Flynn, G. C., Chappell, T. G., and Rothman, J. E., Peptide binding and release by proteins implicated as catalysts of protein assembly, *Science 245*:385 (1989).
42. Palleros, D., Reid, K., McCarty, J. S., Walker, G. C., and Fink, A., DnaK, hsp73 and their molten globules: Two different ways heat shock proteins cope with heat, *J. Biol. Chem. 267*:5279 (1992).
43. Smith, D., Stensgard, B., Welch, W., and Toft, D., Assembly of progesterone receptor with heat shock proteins and receptor activation are ATP mediated events, *J. Biol. Chem. 267*:1350 (1992).
44. Bohen, S. P., Kralli, A., and Yamamoto, K. R., Hold'em and fold'em: Chaperones and signal transduction, *Science 268*:1303 (1995).
45. Leung, S.-M., Senisterra, G., Ritchie, K. P., Sadis, S. E., Lepock, J. R., and Hightower, L. E., Thermal activation of the bovine Hsc70 molecular chaperone at physiological temperatures: Physical evidence of a molecular thermometer, *Cell Stress Chaperones 1*:78 (1996).
46. Chappell, T. G., Konforti, B. B., Schmid, S. L., and Rothman, J. E., The ATPase core of a clathrin uncoating protein, *J. Biol. Chem. 262*:746 (1987).
47. DiDomenico, B. J., Bugaisky, G. E., and Lindquist, S., The heat shock response is self-regulated at both the transcriptional and posttranslational levels, *Cell 31*:593 (1982).
48. Craig, E. A. and Gross, C. A., Is hsp70 the cellular thermometer? *Trends Biochem. Sci. 16*:135 (1991).
49. Baler, R., Guettouche, T., and Voellmy, R., On the model of feedback regulation of heat shock gene expression by heat shock proteins: Demonstration of heat shock protein 70-containing complexes of unactivated heat shock transcription factor 1, *Protein Eng. 8*:35 (1995).
50. Nunes, S. L. and Calderwood, S. K., Heat shock factor-1 and the heat shock cognate 70 protein associate in high molecular weight complexes in the cytoplasm of NIH-3T3 cells, *Biochem. Biophys. Res. Commun. 213*:1 (1995).
51. McCarty, J. S. and Walker, G. C., DnaK as a thermometer—threonine-199 is site of autophosphorylation and is critical for ATPase activity, *Proc. Natl. Acad. Sci. USA 88*:9513 (1991).

52. Hall, B. G., Yeast thermotolerance does not require protein synthesis, *J. Bacteriol. 157*:1363 (1983).
53. Boon-Niermeijer, E. K., Tuyl, M., and Van de Scheur, H., Evidence for two states of thermotolerance, *Int. J. Hyperthermia 2*:93 (1986).
54. Laszlo, A., Evidence for two states of thermotolerance in mammalian cells, *Int. J. Hyperthermia 4*:513 (1988).
55. Walton, G. M. and Gill, G. N., Nucleotide regulation of a eukaryotic protein synthesis initiation complex, *Biochim. Biophys. Acta 390*:231 (1975).
56. Kimura, N. and Shimada, N., Direct interaction between membrane associated nucleoside diphosphate kinase and GTP-binding protein (Gs), and its regulation by hormones and guanine nucleotides, *Biochem. Biophys. Res. Commun. 151*:248 (1988).
57. Kimura, N. and Shimada, N., Evidence for complex formation between GTP binding protein (Gs) and membrane associated nucleoside diphosphate kinase, *Biochem. Biophys. Res. Commun. 168*:99 (1990).
58. Bominaar, A. A., Molijin, A. C., Pestel, M., Veron, M., and Haastert, P. J. M. V., Activation of G-proteins by receptor-stimulated nucleoside diphosphate kinase in *Dictyostelium*, *EMBO J. 12*:2275 (1993).
59. Klinker, J. F., et al., Mastoparan may activate GTP hydrolysis by Gi-proteins in HL-60 membranes indirectly through interaction with nucleoside diphosphate kinase, *Biochem. J. 304*:377 (1994).
60. Klinker, J. F. and Seifert, R., Synthetic lipopeptides activate nucleoside diphosphate kinase in HL-60 membranes, *Biochem. Biophys. Res. Commun. 209*:575 (1995).
61. Harrington, H. M., Dharmasiri, N., and Dharmasiri, S., Heat-shock proteins: A search for functions, *Aust. J. Plant Physiol. 21*:843 (1994).
62. Moisyadi, S., Harrington, H. M., and Lukas, T. J., Characterization of a low molecular mass autophosphorylating protein in cultured sugarcane cells and its identification as nucleoside diphosphate kinase, *Plant Physiol. 104*:1401 (1994).
63. Harris, N., Taylor, J. E., and Roberts, J. A., Isolation of a mRNA encoding a nucleoside diphosphate kinase from tomato that is up-regulated by wounding, *Plant Mol. Biol. 25*:739 (1994).
64. Hemmerich, S., and Pecht, I., A cromoglycate binding protein from rat mast cells of a leukemia line is a nucleoside diphosphate kinase, *Biochemistry 31*:4574 (1992).
65. Martin, M. W., Osullivan, A. J., and Gomperts, B. D., Inhibition by cromoglycate and some flavonoids of nucleoside diphosphate kinase and of exocytosis from permeabilized mast cells, *Br. J. Pharmacol. 115*:1080 (1995).
66. Nagai, N., Nakai, A., and Nagata, K., Quercetin suppresses heat shock response by down regulation of HSF1, *Biochem. Biophys. Res. Commun. 208*:1099 (1995).
67. Huang, S. P., Tsai, M. Y., Tzou, Y. M., Wu, W. G., and Wang, C., Aspartyl residue -10 is essential for ATPase activity of rat hsc 70, *J. Biol. Chem. 268*:2063 (1993).
68. Leung, S.- M. and Hightower, L. E., A 16-kDa protein functions as a new regulatory protein for Hsc 70 molecular chaperone and is identified as a member of the Nm 23/nucleoside diphosphate kinase family, *J. Biol. Chem. 272*:2607 (1997).

7

Major Heat-Shock Chaperones in Protein Folding and Inclusion Body Formation

Alexander Gragerov
NCI-Frederick Cancer Research and Development Center, Frederick, Maryland

I. INTRODUCTION

The purpose of this chapter is to give a short general overview of the chaperone–inclusion bodies connection, rather than to cover extensively the fields of molecular chaperones and protein overproduction. Consequently, the list of references plays merely an illustrative role. I apologize to those colleagues whose work is not appropriately cited.

Normally, protein is synthesized on ribosomes, folds, assembles in homo- or heterooligomeric tertiary structures, and becomes stable (as a rule, with a notable exception of some regulatory proteins) and functional. Molecular chaperones may assist a protein on this route. A rare mistake may cause protein misfolding or misassembly. In this case a protein may try to fold and assemble correctly again, once more with help from chaperones, and if this attempt fails, it will be targeted for proteolysis. Involvement of chaperones in these processes has been revealed by using bacterial mutants unable to induce or synthesize them. In these mutants, a portion of newly formed protein accumulates as insoluble aggregates. Such aggregation results from the deficiency of both chaperone functions and proteolysis of abnormal proteins. In wild-type *Escherichia coli*, the increased demand for molecular chaperones associated with the massive overproduction of a protein is usually met by an appropriate induction of chaperones that either allows formation of large quantities of the folded soluble protein, or aids in proteolytic destruction of the bulk of misfolded protein. If both folding and proteolytic machineries are overwhelmed by protein overproduction, the excessive protein forms inclusion bodies.

II. INCLUSION BODIES FORMATION ON SYNTHESIS OF ABNORMAL OR MUTANT PROTEINS

Protein that is synthesized by cells in the presence of nonnatural amino acid analogues or puromycin is unstable and prone to aggregation and formation of dense intracellular inclusions, called inclusion bodies (IB) [1]. Such behavior was interpreted as an indication of massive protein misfolding, with misfolded protein being susceptible to both proteolytic degradation and nonspecific aggregation. These early experiments set an important distinction between "normal" soluble and usually long-living proteins and misfolded, or "abnormal" unstable and aggregation-prone proteins. The nature of the abnormal state became clear with the advent of genetic engineering and overproduction of numerous foreign proteins in bacteria. Unlike canavanine or puromycin peptides, overproduced proteins were composed of the normal natural amino acids, still many proteins overproduced in bacteria formed insoluble aggregates [2]. These aggregates (IB) are clearly seen as homogeneous inclusions by electron microscopy (EM), or as bright refractile subcellular granules by phase-contrast light microscopy. The IB sediment very rapidly, which may serve as a useful purification step for an overproduced protein, especially when activity of the protein is not a concern (as in proteins overproduced as antigens for immunizations). Usually (but not always), proteins accumulating in IB are functionally (e.g., enzymatically) inactive; at the same time, if only a portion of the protein aggregates, the part that remains soluble is usually fully active. Aggregated protein from IB may be dissolved only in strong denaturants, such as urea or guanadine hydrochloride (GuHCl). After removal of the denaturant, a portion of the protein could be recovered in a soluble active form [3]. These observations with overproduced proteins made it clear that the main feature that distinguishes normal and abnormal (aggregating) proteins is their conformation. Moreover, some proteins remain active and soluble when overproduced, but it is possible to select for mutations causing aggregation of a portion of the overproduced protein, which may serve as an additional confirmation of the notion of altered conformation of abnormal proteins [4,5]. Mutations may also have the reverse effect: to increase the solubility of the overproduced protein [6]. Other factors affecting overproduced protein aggregation (besides the nature of the protein itself) are: extent of overproduction; growth conditions, such as medium composition and growth temperature; and bacterial strain (see following).

A conceptually important step in the study of IB came with the demonstration that aggregation may be strongly temperature-dependent, even for extremely thermostable folded protein [7,8]. Moreover, at least some mutations affecting either the extent or temperature-sensitivity of the aggregation, are without any influence on stability or on denaturation temperature of the soluble folded protein [9]. The most logical conclusion from these data is that aggregation is a property of certain folding or of assembly intermediates, and that these intermediates are more temperature-sensitive and more susceptible to adverse effects of amino acid substitutions, than the native state of the corresponding proteins.

III. INCLUSION BODY ACCUMULATION INDUCES HEAT-SHOCK PROTEINS

Elevated temperature and some other adverse conditions induce synthesis of a set of proteins known as heat-shock proteins (Hsps) [reviewed in 10]. These include the molecular chaperones GroEL, GroES, DnaK, DnaJ, and GrpE, the proteases Lon and ClpP, and many others [for reviews see 11,12]. Some of the Hsps are indispensable. A certain quantity of these proteins is always present in cells; however, under heat-shock or other conditions causing accumulation of unfolded polypeptides, a steady-state level of Hsps in *E. coli* increases significantly. This indicates that Hsps have important normal cellular functions, and that a heat-shock environment does not create, but merely increases the demand for these proteins. The biochemistry of some of these proteins is described elsewhere in this book, so we will concentrate here mainly on their in vivo functions.

Overproduction of proteins in bacteria has also been noted to induce the heat-shock proteins [13]. Accumulation of insoluble aggregates, rather than the overexpression itself, correlates with Hsp induction [14]. These data were among several lines of evidence that eventually converge to the notion that one of the major functions of Hsps is to deal with intracellular protein misfolding. From this perspective, the effect of temperature itself is to cause protein misfolding, which will increase the cellular need for Hsps. These data are also relevant to a feedback mechanism regulating the heat-shock response according to a demand for Hsps' function. Specific mechanisms of such feedback are partially understood and are briefly described later in this chapter.

IV. HEAT-SHOCK CHAPERONES IN PROTEIN DEGRADATION AND AGGREGATION

One way for a cell to deal with abnormal proteins is to degrade them. Misfolded polypeptides, such as amber fragments, or some foreign proteins are, indeed, very unstable in *E. coli*. Several cellular components are responsible for this instability. First of all, heat-shock–induced proteases, including the ATP-dependent protease *lon* (protease La) and *clpP* proteases, are designed to recognize and destroy misfolded proteins [for reviews see 15,16]. In addition, the heat-shock chaperones are also involved in abnormal protein degradation. For example, mutations in *dnaK*, *dnaJ*, or *grpE* stabilize both amber fragments and canavanine peptides [17,18]. Misfolded proteins form transient complexes with chaperones before they are degraded, these complexes are believed to be degradation intermediates [19,20]. Not surprisingly, mutations of the transcriptional activator of Hsp genes, σ^{32}, that cause down-regulation of all heat-shock proteins, including chaperones and proteases (discussed later), are known to stabilize abnormal proteins. A speculative model of the involvement of chaperones in protein degradation is given elsewhere in this chapter.

Another side of the chaperone involvement in protein biosynthesis was revealed in experiments following protein behavior in vivo under conditions of Hsp scarcity. Normally, bacterial cytoplasmic proteins are soluble and do not aggregate to any significant extent. This was thought to be their natural property, despite early indications that normal protein assembly may require dedicated protein cofactors, now known as chaperones [21]. We demonstrated that solubility of the bulk of cellular proteins results from the constant action of Hsps on newly formed polypeptides. σ^{32} mutants unable to induce Hsps at elevated temperature, including an *rpoH* deletion mutant and its revertants [22], continue to synthesize proteins at a virtually normal rate; however, significant portion of proteins aggregated, forming typical inclusion bodies [23]. This aggregation was strictly dependent on protein synthesis, and pulse–chase experiments demonstrated that aggregates consisted mostly of newly synthesized polypeptides. This means that folded proteins (not surprisingly) are resistant to a moderate temperature increase (42°C), but folding intermediates of the same proteins require molecular chaperone function of Hsps to stay soluble and fold properly. Five Hsps—DnaK, DnaJ, GrpE, GroEL, and GroES—are necessary and sufficient for the chaperoning of the bulk of intracellular folding. Strong mutant alleles of either the DnaK chaperone machine or the GroE chaperone machine cause intracellular protein misfolding and aggregation, similar to mutants unable to synthesize all Hsps [24,25], and even weak alleles of *dnaK* or *groE* caused aggregation when combined in a single cell [26]. On the other hand, the aggregation phenotype of σ^{32} mutants unable to produce Hsps may be completely reversed by moderate, close to physiological amounts of DnaK, DnaJ, GroEL, and GroES. That neither pair of chaperones alone is able to efficiently prevent intracellular aggregation, indicates that these two classes of Hsps either act in the same folding pathway, or have complementary functions [26], in good agreement with in vitro data [27] and studies of the mitochondial protein-import machinery [reviewed in 28]. A possible order of chaperone action in normal protein folding was established in experiments demonstrating association of DnaK and DnaJ with nascent chains on translating ribosomes [29], similar to what was found for yeast and mammalian Hsp70s [30,31]. In contrast, GroEL was completely absent from ribosomal fractions and, instead, bound transiently to full-length newly formed polypeptides [29,32].

V. MECHANISTIC ROLE OF GroEL AND DnaK CHAPERONES IN PROTEIN FOLDING

The biochemistry of GroEL, DnaK, and their cochaperones is described in detail in other chapters of this volume. Here I will give a very brief (and biased) view of their mechanistic properties for the sake of this discussion. Biochemical studies indicate quite distinct substrate specificities [33] and different functions of the GroEL and DnaK chaperone machines in protein folding [reviewed in 34]. GroEL and GroES at certain stages in their ATP-driven reaction cycle first capture and then encapsulate a single molecule of a protein-folding intermediate in their hydrophilic cavity, thus isolating it from other folding molecules and allowing intramolecular folding without interference from intermolecular off-pathway interactions [35,36]. The exact mechanism of DnaK (and other Hsp70 chaperones) is not yet known; however, the picture emerging from studies of different systems, including bacterial DnaK and eukaryotic Hsp70s from the cytosol, endoplasmic reticulum, and mitochondria, suggests that DnaK, DnaJ, and substrate polypeptide may form a ternary complex in which DnaK binds both DnaJ and a completely unfolded segment of the substrate protein chain [12,28,34,37]. A conformational change occurring somewhere in the DnaK ATPase cycle may pull the unfolded segment from the rest of the substrate protein bound to DnaJ, thereby providing for its unfolding [38]. If this, or a similar mechanism is correct, the function of DnaK is very different from GroEL: to unfold or disassemble any proteins or protein aggregates having DnaK and DnaJ recognition sites, such as loosely hanging polypeptide segments. In this model an important target of DnaK- and DnaJ-mediated unfolding would be small off-pathway aggregates. The chaperones are apparently unable to dissolve large IB-type aggregates; however, disassembly of specific protein complexes composed of only few polypeptides is well documented [for review see 12]. Unfolded products of such a reaction may be handed over to GroEL for an attempt at folding, or to proteases for degradation. Such a view is consistent with the ability of DnaK and cochaperones to repair heat-induced protein damage in vivo [39]. Unfolding may be important for abnormal protein degradation, because at least some proteolytic enzymes specific for such substrates require threading of the polypeptide through the central cavity containing proteolytic sites [16,40]. DnaK, DnaJ, and GrpE are not the only chaperones "preparing" polypeptides for proteolysis. ATPase subunits of ClpP protease—ClpA and ClpX—fulfill similar functions in protein disassembly and proteolysis [41–43].

Figure 1 summarizes these ideas about likely functions of major *E. coli* heat-shock chaperones in normal protein folding and in IB formation. The position of DnaK in this scheme makes it ideally situated to sense abnormalities in the folding process and to regulate the heat-shock response.

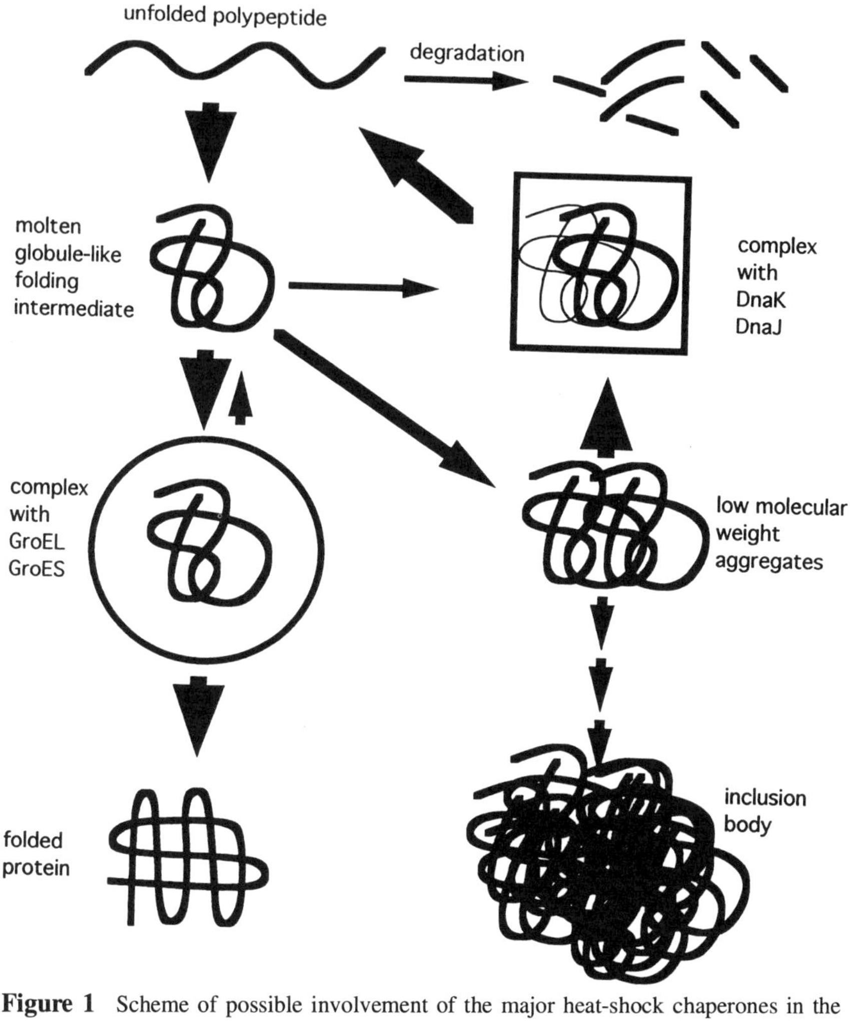

Figure 1 Scheme of possible involvement of the major heat-shock chaperones in the folding of soluble cytoplasmic proteins, their degradation, and aggregation. Polypeptide chain enters the folding pathway in this "flow diagram" either as unfolded polypeptide, or as a complex with DnaK and DnaJ [29]. Unfolded polypeptide collapses very rapidly into compact molten globule-like folding intermediate. They may fold correctly into mature protein; however, a high concentration of such intermediates makes them highly aggregation-prone. GroEL is designed to trap such molten globule-like polypeptides [34] and, together with GroES, allows them to fold in an isolated environment [36]. GroEL–GroES releases folded protein, or returns the intermediate to the pool if the folding attempt failed. Some intermediates may escape from this straightforward path and form aggregates. This is where the well-documented ability of DnaK and DnaJ to disassemble protein complexes [12] may work to retrieve polypeptide chains back to the mainstream folding pathway.

VI. REGULATION OF THE HEAT-SHOCK RESPONSE

The heat-shock response is regulated mainly at the transcriptional level. Genes for the Hsps are either under the control of specialized heat-shock promoters, or have the heat-shock promoter in addition to a constitutive one responsible for the basal level of expression. The heat-shock promoters are recognized by RNA polymerase having the heat-shock σ-subunit, σ^{32}, the product of the *rpoH* gene [11,44,45]. Unlike the bulk of cellular proteins, σ^{32} is rapidly proteolyzed, resembling in this respect misfolded proteins. Interestingly, the major Hsps DnaK and DnaJ are involved in recognition of σ^{32} as abnormal and facilitate its proteolysis [46]. Mutations inactivating these proteins dramatically stabilize σ^{32} and induce expression of all Hsps [47,48]. In addition, the same proteins, DnaK and DnaJ inhibit translation of the σ^{32} RNA template [48,49]. Intracellular accumulation of other targets for these chaperones (e.g., aggregation-prone folding intermediates) distract them from σ^{32}, causing its immediate stabilization and stimulation of its synthesis. Both effects act to rapidly increase the amount of σ^{32} in the cell and induce Hsps transcription. After Hsps accumulate to a level saturating all their targets, an excess of DnaK and DnaJ will again destabilize σ^{32} and inhibit its translation. Such a feedback mechanism maintains the amount of Hsps sufficient for treating all the necessary protein targets; that is, the amount either promoting proper protein folding and assembly, or leading to the degradation of irreversibly misfolded protein.

VII. FOLDING AND DEGRADATION OF OVERPRODUCED PROTEINS

How do the principles described in the foregoing relate to IB formation during protein overproduction? And why do Hsps promoting proper folding of proteins and degradation of misfolded proteins not prevent IB formation? First of all, if the overproduced protein stays soluble, it does so because of the action of Hsp chaperones. In our expriments with *E. coli* strain unable to induce Hsps, β-galactosidase and tumor necrosis factor-α (TNF-α), normally soluble on overproduction, formed typical IB even at permissive temperature at which the bulk of the cellular proteins did not significantly aggregate. Interestingly, mutant TNF-α that did not accumulate in wild-type cells at all, apparently because of very rapid

---

Aggregates that reach a certain threshold size become refractory to disassembly. One reason for this may be an unfavorable ratio of the volume of big aggregates (IB) to their surface area that is accessible to chaperones. A block in the formation of the folded protein (absence of the required subunit, destabilizing mutation, or deficiency in GroE function) will redirect the protein toward two other irreversible final states: IB aggregates, or proteolytic destruction.

degradation, readily formed IB in the same Hsps-defective cells [M. Paktoris, A. Vysokanov, and A. Gragerov, unpublished data]. In other words, the same set of Hsps provided for the normal folding of native TNF and for the rapid destruction of the mutant form of the protein. The deficiency of Hsp function caused accumulation of the misfolded protein as IB in both instances. Similar results were also obtained with several other overproduced proteins. Different proteins may differ in the requirements for particular chaperones from the Hsps cocktail: for example, modest overproduction of GroEL and GroES completely suppressed TNF IB formation in Hsps-defective *E. coli*, but was ineffective against β-galactosidase IB [M. Paktoris, A. Vysokanov, and A. Gragerov, unpublished data].

We acquired further insight into the relations between chaperone function, protein degradation, and IB formation by studying overproduction of RNA polymerase β- and β′-subunits. The high molecular weight of these proteins, separating them in a gel from the rest of bacterial protiens, allows direct pulse–chase experiments. In wild-type *E. coli*, β- and β′-subunits stay soluble, and the extent of their overproduction is very moderate owing to rapid degradation of all the excess protein. If the same plasmid construct is used to transform mutant cells defective in different Hsps (e.g., GroEL, DnaJ, or Lon protease), large quantities of the subunits accumulate as IB. Pulse–chase experiments showed that most of the newly formed polypeptides are soluble, and only a small portion aggregates immediately after (or during) synthesis. Soluble and insoluble fractions are, however, dramatically different in stability: soluble overproduced subunits degrade with somewhat varying rates in different Hsp mutants, but always slower than in the wild-type, whereas the IB fraction is not susceptible to degradation at all [K. Monastyrskaya, A. Vysokanov, and A. Gragerov, unpublished data]. It looks as though in mutant strains some excessive protein escapes from the proteolytic pathway into IB and becomes stable. In this respect an IB acts as a trap in which otherwise unstable, misfolded proteins, or folding intermediates, become nondegradable and, therefore, accumulate. These data also shed some light on the role of Hsp chaperones in protein degradation. On the one hand, chaperones, as has been discussed earlier, may participate in protein unfolding for proteolysis, and may even be involved in the processive threading of a polypeptide into the active center of the protease; equally important, however, is their role in preventing the aggregation of misfolded proteins that would render them completely protease-resistant. In the next paragraph I will try to stretch these conclusions to IB formation in wild-type bacteria.

When a protein is overproduced in *E. coli*, it may have several fates, depending on the protein itself, media composition, and temperature. One possibility is that the protein is generally able to fold and (if necessary) assemble correctly in the environment of the bacterial cytoplasm, which will make it soluble. Molecular chaperones play a role in achieving this solubility, similar to their function in folding normal cellular proteins. The elevated load on the chaperone machinery caused by overproduction will be met by an appropriate (usually insignificant) induction of Hsps. It may happen, however, that a particular protein requires an unusually high level of chaperones, and because of overexpression exceeds the ability of Hsps to be induced, resulting in IB formation. Most likely, lowering growth temperature will solve the problem by reducing the demand for Hsps both by the overproduced protein and host proteins. Another scenario is possible if one is trying to overproduce protein unable to fold correctly. A very common cause of such inability is expression of a subunit of a multisubunit assembly that has a chance to fold only in the presence of its counterparts. Occasionally, coexpression of more than one subunit may solve the problem [50]. Otherwise, a protein unable to fold is destined either for rapid degradation or for IB formation. In this manner, accumulation of the protein in IB saves it from proteolysis. Most likely, attempts to solubilize the protein, for example, by coexpression of chaperones, will result in loss of the material to proteolysis. Numerous anecdotal data indicate that in such situations a high rate of protein synthesis will result in IB formation, whereas lower rates of synthesis yield no protein at all. The "no-protein" mode may switch to the "IB mode" when a mutation inactivating one of the heat-shock proteases (e.g., *lon*) is introduced. This means that the absence of protein results from its rapid degradation. The IB mode of unstable protein overproduction is always accompanied by very strong Hsp induction, indicating that the heat-shock regulatory system is trying to control the presence of misfolded protein in the cells, but is overwhelmed by the extent of overproduction. Such prolonged induction of Hsps is toxic for cells, which may be an explanation for the usual instability of many overproducer strains. If instability is a problem and protein aggregates anyway, it makes sense to try overproduction in strains with impaired ability to induce Hsps, such as strains with the classic *rpoH* (σ^{32}) mutations, or *rpoH* mutations selected as suppressors of the harmful effects of Hsps induction by *dnaK* mutations [51].

Does the previous discussion mean that it is useless to try coexpression of molecular chaperones to overproduce proteins in their properly folded soluble form? Not necessarily. Chaperone overproduction increases yield of active proteins in at least some instances [52]. It is true that for the reasons discussed earlier chaperone coexpression cannot serve as universal cure for the problem of aggregation of overproduced proteins. Certain proteins, however, may have specific requirements for particular chaperones (see, for example, the differences between TNF and β-galactosidase described earlier). The normal folding pathway involves complementary action of different groups of chaperones, and the rate-limiting step may depend on the particular folding protein. It is conceivable that overproduction of the group of chaperones required for the limiting step may improve chances of the protein to fold. Note, that the normal induction of Hsps affects uniformly all proteins regulated by the heat-shock σ factor, including numerous heat-shock proteases. That is why coexpression of either DnaK and

DnaJ, or GroEL and GroES sometimes helps improve the yield of particular soluble proteins.

ACKNOWLEDGMENTS

I thank Bill Burkholder and Michael Sherman for helpful comments.

REFERENCES

1. Prouty, W. F., Karnovsky, M. J., and Goldberg, A. L., Degradation of abnormal proteins in *Escherichia coli*. Formation of protein inclusions in cells exposed to amino acid analogs, *J. Biol. Chem. 250*:1112 (1975).
2. Kane, J. F. and Hartley, D. L., Properties of recombinant protein-containing inclusion bodies in *Escherichia coli*, *Bioprocess Technol. 12*:121 (1991).
3. Marston, F. A., The purification of eukaryotic polypeptides synthesized in *Escherichia coli*, *Biochem. J. 240*:1 (1986).
4. Krueger, J. K., Stock, A. M., Schutt, C. E., and Stock, J. B., Inclusion bodies from proteins produced at high levels in *Escherichia coli*, *Protein Folding* (Gierasch, L. N. and King, J., eds.), AAAS, Washington, DC, 1990, p. 136.
5. Villafane, R. and King, J., Nature and distribution of sites of temperature-sensitive folding mutations in the gene for the P22 tailspike polypeptide chain, *J. Mol. Biol. 204*:607 (1988).
6. Wetzel, R., Perry, L. J., and Veilleux, C., Mutations in human interferon gamma affecting inclusion body formation identified by a general immunochemical screen, *Biotechnology 9*:731 (1991).
7. Goldenberg, D. P., Berget, P. B., and King, J., Maturation of the tail spike endorhamnosidase of salmonella phage P22, *J. Biol. Chem. 257*:7864 (1982).
8. Yu, M. H. and King, J., Surface amino acids as sites of temperature-sensitive folding mutations in the P22 tailspike protein, *J. Biol. Chem. 263*:1424 (1988).
9. Sturtevant, J. M., Yu, M.-H., Haase-Pettingell, C., and King, J., Thermostability of temperature-sensitive folding mutants of the P22 tailspike protein, *J. Biol. Chem. 264*:10693 (1989).
10. Neidhardt, F. C., VanBogelen, R. A., and Vaughn, V., The genetics and regulation of heat shock proteins, *Annu. Rev. Genet. 18*:295 (1984).
11. Gross, C. A., Straus, D. B., Erickson, J. W., and Yura, T., The function and regulation of heat shock proteins in *Escherichia coli*, *Stress Proteins in Biology and Medicine* (Morimoto, R. I., Tissieres, A., and Georgopoulos, C., eds.), Cold Spring Harbor Laboratory Press, Cold Spring Harbor, NY, 1990, p. 167.
12. Georgopoulos, C., Liberek, K., Zylicz, M., and Ang, D., Properties of the heat shock proteins of *Escherichia coli* and the autoregulation of the heat shock response, *The Biology of Heat Shock Proteins and Molecular Chaperones* (Morimoto, R. I., Tissieres, A., and Georgopoulos, C., eds.), Cold Spring Harbor Laboratory Press, Cold Spring Harbor, NY, 1994, p. 209.
13. Goff, S. A. and Goldberg, A. L., Production of abnormal protein in *E. coli* stimulates transcription of *lon* and other heat shock genes, *Cell 41*:587 (1985).

14. Parsell D. A. and Sauer, R. T., Induction of heat shock-like response by unfolded protein in *Escherichia coli*: Dependence on protein level not protein degradation, *Genes Dev. 3*:1226 (1989).
15. Gottesman, S. and Maurizi, M. R., Regulation by proteolysis: Energy-dependent proteases and their targets, *Microbiol. Rev. 56*:592 (1992).
16. Goldberg, A. L., The mechanism and functions of ATP-dependent proteases in bacterial and animal cells, *Eur. J. Biochem. 203*:9 (1992).
17. Strauss, D., Walter, W., and Gross, C. A., *Escherichia coli* heat shock gene mutants are defective in proteolysis, *Genes Dev. 2*:1851 (1988).
18. Kelly, J. and Simon, L., Divergent effects of *dnaK* mutation on abnormal protein degradation in *Escherichia coli*, *Mol. Microbiol. 2*:31 (1988).
19. Sherman, M. Y. and Goldberg, A. L., Formation in vitro of complexes between an abnormal fusion protein and the heat shock proteins from *Escherichia coli* and yeast mitochondria, *J. Bacteriol. 173*:7249 (1991).
20. Sherman, M. Y. and Goldberg, A. L., Involvement of the chaperonin DnaK in the rapid degradation of a mutant protein in *Escherichia coli*, *EMBO J. 11*:71 (1992).
21. Georgopoulos, C., Hendrix, R., Kaiser, A., and Wood, W., Role of the host cell in bacteriophage morphogenesis: Effects of a bacterial mutation on T4 heat assembly, *Nature New Biol. 239*:38 (1972).
22. Kusukawa, N. and Yura, T., Heat shock protein GroE of *Escherichia coli*: Key protective roles against thermal stress, *Genes Dev. 2*:874 (1988).
23. Gragerov, A., Martin, E., Krupenko, M., Kashlev, M., and Nikiforov, V., Protein aggregation and inclusion body formation in *Escherichia coli rpoH* mutant defective in heat shock protein induction, *FEBS Lett. 291*:222 (1991).
24. Vysokanov, A., Synthesis of chloramphenicol acetyltransferase in a coupled transcription–translation in vitro system lacking the chaperones DnaK and DnaJ, *FEBS Lett. 375*:211 (1995).
25. Horwich, A. L., Low, K. B., Fenton, W. A., Hirshfield, I. N., and Furtak, K., Folding in vivo of bacterial cytoplasmic proteins: Role of GroEL, *Cell 74*:909 (1993).
26. Gragerov, A., Nudler, E., Komissarova, N., Gaitanaris, G. A., Gottesman, M. E., and Nikiforov, V., Cooperation of GroEL/GroES and DnaK/DnaJ heat shock proteins in preventing protein misfolding in *Escherichia coli*, *Proc. Natl. Acad. Sci. USA 89*:10341 (1992).
27. Langer, T., Lu, C., Echols, H., Flanagan, J., Hayer, M. K., and Hartl, F. U., Successive action of DnaK, DnaJ, and GroEL along the pathway of chaperon-mediated protein folding, *Nature 356*:683 (1992).
28. Langer, T. and Neupert, W., Chaperoning mitochondrial biogenesis, *The Biology of Heat Shock Proteins and Molecular Chaperones* (Morimoto, R. I., Tissieres, A., and Georgopoulos, C., eds.), Cold Spring Harbor Laboratory Press, Cold Spring Harbor, NY, 1994, p. 53.
29. Gaitanaris, G. A., Vysokanov, A., Hung, S.-C., Gottesman, M. E., and Gragerov, A., Successive action of *Escherichia coli* chaperones in vivo, *Mol. Microbiol. 14*:861 (1994).
30. Beckmann, R. P., Mizzen, L. A., and Welch, W., Interaction of hsp70 with newly synthesized proteins: Implication for protein folding and assembly, *Science 248*:850 (1990).

31. Nelson, R. J., Ziegelhoffer, T., Nicolet, C., Werner-Washburne, M., and Craig, E. A., The translation machinery and 70kd heat shock protein cooperate in protein synthesis, *Cell 71*:97 (1992).
32. Bochkareva, E., Lissin, N., and Girshovich, A., Transient association of newly synthesized unfolded proteins with heat shock GroEL protein, *Nature 336*:254 (1988).
33. Landry, S. J., Jordan, R., McMacken, R., and Gierasch, L., Different conformations for the same polypeptide bound to chaperones DnaK and GroEL, *Nature 355*:455 (1992).
34. Frydman, J. and Hartl, F. U., Molecular chaperone functions of hsp70 and hsp60 in protein folding, *The Biology of Heat Shock Proteins and Molecular Chaperones* (Morimoto, R. I., Tissieres, A., and Georgopoulos, C., eds.), Cold Spring Harbor Laboratory Press, Cold Spring Harbor, NY, 1994, p. 251.
35. Agard, D. A., To fold or not to fold . . . , *Science 260*:1903 (1993).
36. Weissman, J. S., Hohl, C. M., Kovalenko, O., Kashi, Y., Chen, S., Braig, K., Saibil, H. R., Fenton, W. A., and Horwich, A. L., Mechanism of GroEL action: Productive release of polypeptide from a sequestered position under GroES, *Cell 83*:577 (1995).
37. Brodsky, J. L. and Schekman, R., Heat shock cognate proteins and polypeptide translocation across the endoplasmic reticulum membrane, *The Biology of Heat Shock Proteins and Molecular Chaperones* (Morimoto, R. I., Tissieres, A., and Georgopoulos, C., eds.), Cold Spring Harbor Laboratory Press, Cold Spring Harbor, NY, 1994, p. 85.
38. Zhu, X., Zhao, X., Barkholder, W. F., Gragerov, A., Ogata, C. M., Gottesman, M. E., and Hendrickson, W. A., Structural analysis of substrate binding by the molecular chaperone DnaK, *Science 272*:1606 (1996).
39. Schroder, H., Langer, T., Hartl, F. U., and Bukau, B., DnaK, DnaJ, GrpE form a cellular chaperone machinery capable of repairing heat-induced protein damage, *EMBO J. 12*:4137 (1993).
40. Goldberg, A. L., Functions of the proteasome: The lysis at the end of the tunnel, *Science 268*:522 (1995).
41. Wickner, S., Gottesman, S., Skowyra, D., Hoskins, J., McKenney, K., and Maurizi, M. R., A molecular chaperon, ClpA, functions like DnaK and DnaJ, *Proc. Natl. Acad. Sci. USA 91*:12218 (1994).
42. Wawrzynow, A., Wojtkowiak, D., Marszalek, J., Banecki, B., Jonsen, M., Graves, B., Georgopoulos, C., and Zylicz, M., The ClpX heat shock protein of *Escherichia coli*, the ATP-dependent substrate specificity component of the ClpP–ClpX protease, is a novel molecular chaperon, *EMBO J. 14*:1867 (1995).
43. Levchenko, I., Luo, L., and Baker, T. A., Disassembly of Mu transposase tetramer by the ClpX chaperon, *Genes Dev. 9*:2399 (1995).
44. Grossman, A. D., Erickson, J. W., and Gross, C. A., The *htpR* gene product of *E. coli* is a σ factor for heat shock promoters, *Cell 38*:383 (1984).
45. Cowing, D. W., Bardwell, J. C. A., Craig, E. A., Woolford, C., Hendrix, R. W., and Gross, C. A., Consensus sequence of *Escherichia coli* heat shock gene promoters, *Proc. Natl. Acad. Sci. USA 82*:2679 (1985).
46. Gamer, J., Bujard, H., and Bukau, B., Physical interaction between heat shock

proteins DnaK, DnaJ, and GrpE and the bacterial heat shock transcription factor σ^{32}, *Cell 69*:833 (1992).

47. Tilly, K., Spence, J., and Georgopoulos, C., Modulation of stability of the *Escherichia coli* heat shock regulatory factor σ^{32}, *J. Bacteriol. 171*:1585 (1989).
48. Strauss, D., Walter, W., and Gross, C. A., DnaK, DnaJ, and GrpE heat shock proteins negatively regulate heat shock gene expression by controling the synthesis and stability of σ^{32}, *Genes Dev. 4*:2202 (1990).
49. Nagai, H., Yuzawa, H., Kanemori, M., and Yura, T., A distinct segment of the σ^{32} polypeptide is involved in DnaK-mediated negative control of the heat shock response in *Escherichia coli*, *Proc. Natl. Acad. Sci. USA 91*:10280 (1994).
50. Samuelsson, E., Jonasson, P., Viklund, F., Nilsson, B., and Uhlén, M., Affinity assisted in vivo folding of a secreted human peptide hormone in *Escherichia coli*, *Nature Biotechnol. 14*:??? (1996).
51. Bukau, B. and Walker, G. C., Mutations altering heat shock specific subunit of RNA polymerase suppress major cellular defects of *E. coli* mutants lacking the DnaK chaperone, *EMBO J. 9*:4027 (1990).
52. VanDyk, T. K., Gatenby, A. A., and LaRossa, T. A., Demonstration by genetic suppression of interaction of GroE products with many proteins, *Nature 342*:451 (1989).

8

Structure and Function of the 90-kDa Stress Protein Hsp90

Ichiro Yahara
The Tokyo Metropolitan Institute of Medical Science, Tokyo, Japan

I. MEMBERS OF THE Hsp90 FAMILY

Hsp90 and its homologues are ubiquitously distributed among all organisms from bacteria to human. Hsp90 is a major soluble cytosolic protein at least in eukaryotes. Amino acid sequences of these proteins are well conserved during evolution. For instance, the HtpG protein of *Escherichia coli* is 42% identical in amino acid sequence with human Hsp90, although two regions of eukaryotic Hsp90 are missing in HtpG [1]. One is a hydrophilic stretch from the 218- to 276-amino acid residues of human Hsp90β and the other is the COOH-terminal 35-amino acid residues of eukaryotic Hsp90. The *htpG* gene is not essential for *E. coli*, although *hspG*-deletion mutants showed temperature-sensitive growth [2].

Similar to other major stress proteins, such as Hsp70 (DnaK), Hsp90 is essential for growth of cells in *Saccharomyces cerevisiae* [3]. The lethality of yeast cells associated with the deletion was rescued by expressing mammalian Hsp90 [4], indicating that functions of Hsp90 are well conserved among eukaryotic cells. Mammals and the budding yeast have two cytosolic Hsp90 genes per haploid genome, whereas *Drosophila* has only one Hsp90 gene (*Hsp83*) [5].

Hsp90 of eukaryotes is expressed both under normal and stressful conditions [6]. The expression of Hsp90 of cultured mammalian cells is enhanced severalfold by increasing the temperature from 37° to 43°C.

Eukaryotes have a homologue of Hsp90 in endoplasmic reticulum (ER) [6]. In mammals, they are called Grp94 (glucose-regulated protein 94), Hsp100, or endoplasmin. Grp94 has a signal sequence at its NH_2-terminus, which is processed after translocation into the ER, and the ER-retention signal sequence KDEL at its COOH-terminus [7]. Although Grp94 is expressed at a basal level under normal conditions, the expression is enhanced by glucose starvation or by treating with inhibitory agents of glycosylation, such as tunicamycin.

A chromosomal gene of rye leaves (*Secale cereale*) encoding plastid Hsp90 homologue has been identified [8]. The encoded protein has an NH_2-terminal transit peptide to plastids. The expression of the plastid-specific Hsp90 is constitutive and is enhanced by exposure to 42°C. No member of the Hsp90 family has yet been identified in mitochondria.

II. THE NATIVE FORM OF Hsp90

Results obtained by physicochemical studies of purified mouse Hsp90 indicated that Hsp90 is present in a dimeric form [9]. A relatively large frictional ratio (f/f_0) indicated that Hsp90 has a nonspherical structure. Indeed, electron microscopic observation, with the aid of the low-angle rotary shadowing technique, revealed that native Hsp90 appeared to be composed of two spherical domains connected with each other by a flexible junction.

Mammalian Hsp90 has two isoforms, Hsp90α and Hsp90β, which are encoded by separate genes [10]. Native polyacrylamide gel electrophoresis followed by isoform analysis revealed that most Hsp90 exists as Hsp90α/α and Hsp90β/β homodimers [11]. Only a minor part of Hsp90β is present as a monomer. The COOH-terminal region of Hsp90α is necessary for dimerization [12]. The COOH-terminally truncated human Hsp90α neither formed dimers in yeast cells nor rescued the lethality of yeast haploid cells caused by the disruption of the two Hsp90-related genes, although the expression of the full-length human Hsp90α did. This result supports the idea that the dimerization is required for Hsp90 to elicit its functions in intact cells [12].

Grp94 exists also as a dimeric form [9]. Proteolytic removal of the COOH-terminal 10-kDa peptide from Grp94 caused dissociation of the dimeric form to monomer (Y. Miyata, unpublished results). This suggests that Grp94 forms dimers in a fashion similar to Hsp90.

Exposure of purified Hsp90 to elevated temperatures induces oligomerization of the protein [13]. The oligomerization is associated with induction of a potent chaperone activity of Hsp90 (see following section).

III. Hsp90 AS A CONVENTIONAL CHAPERONE

One of the major functions of molecular chaperones is to prevent unproductive aggregation of unfolded polypeptides and to facilitate their refolding. Major stress proteins, including Hsp70 (DnaK) and chaperonin (GroEL), are well known to exert such chaperone activities over various substrate proteins in an ATP-dependent manner [14]. Hsp90 has been reported to have a similar chaperone activity, although the activity is weak and is not dependent on ATP [15].

In test tubes, purified Hsp90 prevents aggregation of denatured citrate synthetase and the antibody Fab fragment, probably by stoichiometrically binding to the substrates, and enhances their reactivation [15]. The effect of Hsp90 is very weak; even under the optimal conditions, Hsp90 stimulated refolding of these substrate proteins 1.5- to 2.0-fold, compared with spontaneous refolding of the same proteins. Thus, this in vitro effect of Hsp90 appears similar to that elicited by GroEL in the absence of GroES and ATP, raising the possibility that Hsp90 elicits its chaperone activity in a cooperative manner together with other chaperones or cofactors. In fact, Hsp90 binds to denatured β-galactosidase and can maintain it in a refoldable state over extended periods [16]. Interestingly, trapped β-galactosidase on Hsp90 molecules can be dissociated and reactivated with the aid of Hsp70 and Hdj-1 in an ATP-dependent manner.

Hsp90 acquires a potent chaperone activity when incubated at temperatures higher than 46°C, which is coupled with its self-ologomerization [17]. Heated Hsp90 binds unfolded dihydrofolate reductase (DHFR) and inhibits spontaneous refolding of the enzyme. Trapped DHFR can be dissociated and fully reactivated when incubated with GroEL and GroES in the presence of ATP. Heated Hsp90 also traps unfolded firefly luciferase and maintains it in a folding-competent structure. Partially folded luciferase trapped on heated Hsp90 can be released by incubation with reticulocyte lysate with an ATP-regenerating system. The GroE chaperone does not effectively work for refolding of luciferase trapped on Hsp90. Active components contained in reticulocyte lysate in the refolding of luciferase remain to be identified. Because Hsp90 is abundant in unstressed cells, this inducible chaperone activity may have a pivotal role in protecting cells from thermal damage [17].

IV. SPECIFIC SUBSTRATE PROTEINS OF Hsp90

Hsp90 forms complexes with functionally key proteins involved in signal transduction pathways [18]. These substrates include transcription factors, nonreceptor tyrosine–protein kinases, receptor tyrosine–protein kinases, serine/threonine–protein kinases, subunits of G proteins [19], and cytoskeletal proteins [9].

A. Actin Filaments

Hsp90 binds and cross-links actin filaments [20]. Under saturable-binding conditions, one Hsp90 dimer binds ten actin monomeric units in polymerized form. The K_d of binding was estimated to be 2×10^{-6} M.

Tropomyosin and Hsp90 compete in binding to actin filaments. This explains why Hsp90 is not distributed on actin stress fibers in fibroblastic cells to which tropomyosin preferentially binds. Calcium–calmodulin binds Hsp90 and inhibits the binding of Hsp90 to actin filaments. Calmodulin-binding sites of Hsp90α and Hsp90β were identified to be the sequence of Lys-500 to Ile-520 of Hsp90α and that of Lys-491–Val-511 of Hsp90β, respectively [21]. These calmodulin-binding sequences can be predicted to form positively charged amphiphilic α-helices. These sequences bind calmodulin in a Ca^{2+}-dependent manner. Surprisingly, synthetic peptides corresponding to the calmodulin-binding sequence of Hsp90 also bind Hsp90, suggesting that the calmodulin-binding sites in Hsp90 might interact with other sites on the same molecules [17]. The Hsp90 peptides augment their self-oligomerization and induction of the chaperone activity that are induced at elevated temperatures. These results suggest that Ca^{2+}–calmodulin might modulate functions of Hsp90 in vivo.

B. Steroid Hormone Receptors and Other Transcription Factors

A body of evidence has indicated that cytosolic steroid receptors are present in complex forms with Hsp90 and other chaperone-related proteins in the absence of steroid hormones [22]. Cytosolic receptor complexes bind actin filaments in vitro by their Hsp90 moiety, suggesting that the complexes might be anchored on the cellular actin cytoskeleton [23]. The complexes do not have the ability to bind steroid-responsive elements and, therefore, do not function as transcription-regulating factors. The binding of hormones to the receptor complexes induces dissociation of the complexes into active receptors. In the glucocorticoid receptors, the dissociation induced by ligands is followed by translocation of the receptors into the nucleus. The hormone-binding domain and Hsp90-binding domain of glucocorticoid receptors overlap [22]. Truncation of this region induced a constitutively active glucocorticoid receptor. This appears to suggest that the region negatively regulates functions of the receptor. However, in the yeast system, Hsp90 is required for signal transduction pathways involving steroid receptors [4]. The receptor complexes consist of one steroid receptor, dimeric Hsp90, and other proteins, including three immunophillins (FKBP54, FKBP52, and CyP-40), p23, Hsp70, P60, and p49 [24]. However, the binding of these proteins, other than Hsp90, to steroid receptors is substoichiometric. When purified glucocorticoid receptors, free from associated proteins, were incubated with reticulocyte lysate, active receptor complexes, as measured by ligand binding, could be reconstituted [25]. The assembly is Mg^{2+}/ATP- and K^{+}-dependent. In vitro reconstitution experiments using fractionated proteins from reticulocyte lysate and recombinant proteins revealed that four proteins, Hsp70, p60, Hsp90, and p23, are required for the reconstitution, although optimal complex assembly might require other proteins [25]. A genetic approach has revealed that a DnaJ homologue might also be involved in regulation of steroid receptor complexes [26].

A transient interaction between MyoD1, a basic helix–loop–helix protein, and Hsp90 has been suggested as necessary for its activation [27]. The activation does not necessarily require the whole Hsp90 molecule, but it is induced by interacting of MyoD1 with the COOH-terminal 194-amino acid residues of murine Hsp84. It is also suggested that the dioxin receptor interacts with Hsp90 through the basic helix–loop–helix and a Per-Arnt-Sim (PAS) domains [28].

C. Protein–Serine/Threonine Kinases

Casein kinase II (CKII) is a second-messenger–independent protein kinase that is ubiquitously distributed and phosphorylates various substrate proteins. Purified CKII is labile and tends to form self-aggregates in association with loss of its activity. When CKII is present with Hsp90, the kinase is significantly stabilized (Fig. 1) [29]. CKII in cytosols prepared from cultured cells forms 8S complexes with Hsp90. The K_d of the binding is approximately 10^{-7} M. CKII is a hetero-tetrameric structure, $\alpha\alpha/\beta\beta$, and Hsp90 interacts with CKIIα. An Hsp90-binding site on CKIIα was identified to be the same as that for heparin binding, the lysine-rich region of the 62–83 residues [30], consistent with the observation that heparin inhibits the binding of CKII to Hsp90 [29]. Positive charges of this region may interact with two clusters of negatively charged amino acid residues on Hsp90α, the 227–249 and 251–273 residues in the case of human Hsp90α. Interestingly, each of these regions contains a serine residue (Ser-211 and Ser-263, respectively) to be phosphorylated by CKII [31].

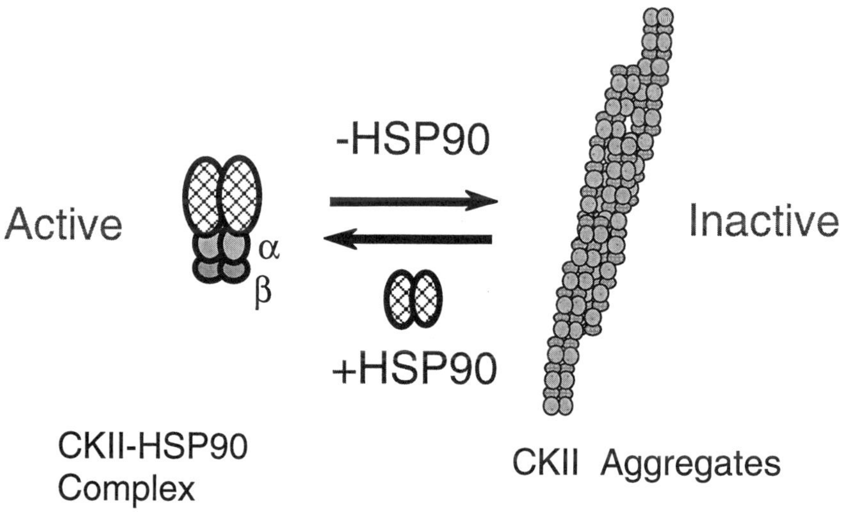

Figure 1 Chaperone activity of Hsp90 on casein kinase II.

Raf-1, another serine/threonine kinase, exists in cytosol as a high molecular weight complex with Hsp90 [32]. Benzoquinone ansamycins, such as geldanamycin and herbimycin A, are anti-Hsp90 antibiotics [33] that disrupt the Raf-1–Hsp90 complex and destabilize the Raf-1 kinase [32]. Hsp90 is a key factor that ensures targeting of Raf-1 to plasma membrane where the kinase elicits its function.

D. Protein–Tyrosine Kinases

Newly synthesized pp60$^{v\text{-}src}$, the product of the viral oncogene v-*src*, is transiently associated with Hsp90 and p50 [34]. After myristyration at the NH_2-terminus of pp60$^{v\text{-}src}$, the tyrosine kinase is anchored on the inner surface of the plasma membrane and elicits its kinase activity. During this processing, pp60$^{v\text{-}src}$ is dissociated from the associated proteins. By using a mutant of the budding yeast in which the level of Hsp90 expression was reduced, it has been demonstrated that Hsp90 is required for stabilization of pp60$^{v\text{-}src}$ and expression of its activity [35]. In the fission yeast *Schizosaccharomyces pombe*, Wee1 tyrosine kinase negatively regulates the Cdc2–cyclin B protein kinase by phosphorylating Cdc2, the catalytic subunit of the kinase, and causes G_2 arrest. A gene encoding a member of the Hsp90 family was isolated as a suppressor of G_2 arrest caused by overexpression of Wee1. In addition, Hsp90 coimmunoprecipitates with Wee1, suggesting that Wee1 kinase activity requires a complex formation with Hsp90 [36]. Studies on a *Drosophila* system revealed that two receptor tyrosine kinases, sevenless and Torso, genetically interact with Hsp90 [5].

V. MUTATIONS AFFECTING FUNCTIONS OF Hsp90

By taking advantage of the fact that Hsp90 is essential for yeast cells to survive, a number of mutations in the Hsp90 gene have been isolated. Point mutations were created by treating the *HSP82* or *HSC82* DNA with hydroxylamine and were mapped [37–39]. These mutations with temperature-sensitive growth, deficient phenotypes in expression of functional steroid receptors, and viral oncogenic tyrosine kinase were scattered along the entire *hsp82* gene (Fig. 2). So far tested, *trans*-complementation between different mutation alleles have not been observed [38]. This suggests that functional domains of Hsp90 structurally or functionally depend on each other. Alternatively, the integrity of the whole Hsp90 molecule may be necessary for the function of its domains. Probably for the same reason, no multicopy suppressor has been isolated that suppresses any of the temperature-sensitive mutations (Y. Kimura, unpublished results). Instead, a synthetic lethal second mutation was isolated that causes cell lethality when combined with any allele of ts-*hsp82* mutations [26]. The synthetic lethal mutation was identified as a *ydj1* mutation; the wild-type *YDJ1* encodes a homologue

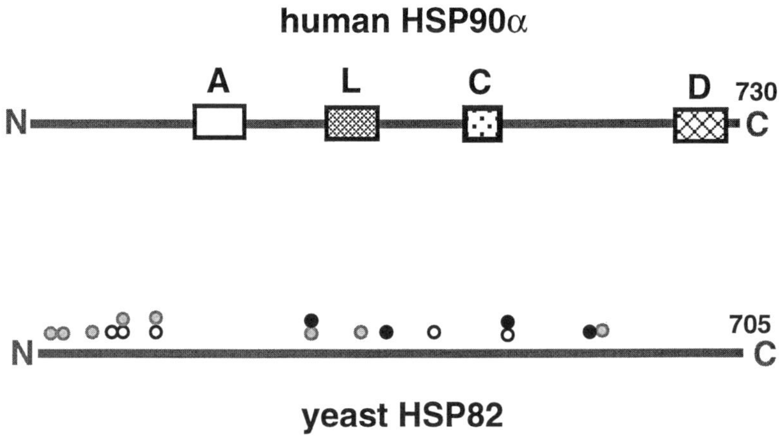

Figure 2 Functional domains of Hsp90: human Hsp90α: A, hydrophilic domain rich in acidic amino acid residues; L, a putative leucine-zipper domain; C, calmodulin-binding domain; D, a region involved in Hsp90 dimerization. Yeast Hsp82: Point mutations of yeast Hsp82. (Data from Refs. 37–39.)

of DnaJ, suggesting that two major Hsp chaperones function in vivo in a cooperative manner.

Two functional domains, dimerization domain [12,40,41] and calmodulin-binding domain [21], were identified (see Fig. 2). Two characteristic regions are present in eukaryotic Hsp90; one region (A) is rich in charged amino acids and the other (L) is a putative leucine-zipper sequence. The region A is missing in HtpG, the *E. coli* homologue of Hsp90 [1] and is dispensable in *S. cerevisiae* [41]. The binding site of Hsp90 to casein kinase IIα has been mapped in the A region [30]. Several deletion mutants of mammalian and avian Hsp90 have been created and examined for their functions, such as binding to steroid receptors [41–43]. However, the results and their interpretations are controversial. For instance, a deletion (residues 221–290) of the A region of chicken Hsp90 caused defect in binding to estrogen, mineralocorticoidsteroid, and glucocorticoid receptors [43,44] whereas a similar deletion (residues 226–287) of the A region did not affect binding to estrogen receptor [41]. Moreover, NH_2-terminal deletion mutants up to the 380th residue of chicken Hsp90 were reported to form functional complexes with progesterone receptor [42]. Reasons for these discrepancies remain to be solved. On the basis of deletion analysis, it was also suggested that the properties of Hsp90, including dimerization, interaction with target proteins, and the ability to rescue yeast *Δhsp90* mutants are not necessarily interdependent [41]. Elucidation of tertiary structure of Hsp90 must clarify controversial issues concerning the functional domains of Hsp90.

REFERENCES

1. Bardwell, J. C. A. and Craig, E. A., Eukaryotic M_r 83,000 heat shock protein has a homologue in *Escherichia coli*, *Proc. Natl. Acad. Sci. USA 84*:5177–5181 (1987).
2. Bardwell, J. C. A. and Craig, E. A., Ancient heat shock gene is dispensable, *J. Bacteriol. 170*:2977–2983 (1988).
3. Borkovich, K. A., Farrelly, F. W., Finkelstein, D. B., Taulien, J., and Lindquist, S., hsp82 is an essential protein that is required in higher concentration for growth of cells at higher temperatures, *Mol. Cell. Biol. 9*:3919–3930 (1989).
4. Picard, D., Khursheed, B., Garabedian, M. J., Fortin, M. G., Lindquist, S., and Yamamoto, K. R., Reduced levels of hsp90 compromise steroid receptor actin in vivo, *Nature 348*:166–168 (1990).
5. Cutforth, T. and Rubin, G. M., Mutations in Hsp83 and Cdc37 impair signaling by the sevenless receptor tyrosine kinase in *Drosophila*, *Cell 77*:1027–1036 (1994).
6. Lindquist, S. and Craig, E. A., The heat-shock proteins, *Annu. Rev. Genet. 22*:631–677 (1988).
7. Sorger, P. K. and Pelham, H. R. B., The glucose-regulated protein grp94 is related to heat shock protein hsp90, *J. Mol. Biol. 194*:341–344 (1987).
8. Schmitz, G., Schmidt, M., and Feierabend, J., Characterization of a plastid-specific HSP90 homologue: Identification of a cDNA sequence, phylogenetic descendence and analysis of its mRNA and protein expression, *Plant Mol. Biol. 30*:479–492 (1996).
9. Koyasu, S., Nishida, S., Kadowaki, T., Matsuzaki, F., Iida, K., Harada, F., Kasuga, M., Sakai, H., and Yahara, I., Two mammalian heat shock proteins, HSP90 and HSP100, are actin-binding proteins, *Proc. Natl. Acad. Sci. USA 83*:8054–8058 (1986).
10. Moore, S. K., Kozak, C., Robinson, E. A., Ullrich, S. J., and Appella, E., Murine 86- and 84-kDa heat shock proteins, cDNA sequences, chromosomal assignments, and evolutionary origins, *J. Biol. Chem. 264*:5343–5351 (1989).
11. Minami, Y., Kawasaki, H., Miyata, Y., Suzuki, K., and Yahara, I., Analysis of native forms and isoform compositions of the mouse 90-kDa heat shock protein, HSP90, *J. Biol. Chem. 266*:10099–10103 (1991).
12. Minami, Y., Kimura, Y., Kawasaki, H., Suzuki, K., and Yahara, I., The carboxy-terminal region of mammalian HSP90 is required for its dimerization and function in vivo, *Mol. Cell. Biol. 14*:1459–1464 (1994).
13. Lanks, K. W., Temperature-dependent ologomerization of hsp85 in vitro, *J. Cell. Physiol. 140*:601–607 (1989).
14. Hendrick, J. P. and Hartl, F.-U., Molecular chaperone functions of heat-shock proteins, *Annu. Rev. Biochem. 62*:349–384 (1993).
15. Wiech, H., Buchner, J., Zimmermann, R., and Jakob, U., Hsp90 chaperones protein folding in vitro, *Nature 358*:169 (1992).
16. Freeman, B. C. and Morimoto, R. I., The human cytosolic molecular chaperones hsp90, hsp70 (hsc70) and hdj-1 have distinct roles in recognition of a nonnative protein and protein refolding, *EMBO J. 15*:2969–2979 (1996).
17. Yonehara, M., Minami, Y., Kawata, Y., Nagai, J., and Yahara, I., Heat-induced chaperone activity of HSP90, *J. Biol. Chem. 271*:2641–2645 (1996).

18. Rutherford, S. L. and Zuker, C. S., Protein folding and the regulation of signaling pathways, *Cell 79*:1129–1132 (1994).
19. Inanobe, A., Takahashi, K., and Katada, T., Association of the βγ subunits of trimeric GTP-binding proteins with 90-kDa heat shock protein, hsp90, *J. Biochem. 115*:486–492 (1993).
20. Nishida, E., Koyasu, S., Sakai, H., and Yahara, I., Calmodulin-regulated binding of the 90-kDa heat shock protein to actin filaments, *J. Biol. Chem. 261*:16033–16036 (1986).
21. Minami, Y., Kawasaki, H., Suzuki, K., and Yahara, I., The calmodulin-binding domain of the mouse 90-kDa heat shock protein, *J. Biol. Chem. 268*:9604–9610 (1993).
22. Bohen, S. P. and Yamamoto, K. R., Modulation of steroid receptor signal transduction by heat shock proteins, *The Biology of Heat Shock Proteins and Molecular Chaperones* (Morimoto, R. I., Tissieres, A., and Georgopoulos, C., eds.), Cold Spring Harbor Laboratory Press, Cold Spring Harbor, NY, 1994, pp. 313–334.
23. Miyata, Y. and Yahara, I., Cytoplasmic 8S-glucocorticoid receptor binds actin filaments through the HSP90 moiety, *J. Biol. Chem. 266*:8779–8783 (1991).
24. Johnson, J., Carbisier, R., Stenggard, S., and Toft, D., The involvement of p23, hsp90, and immunophillins in the assembly of progesterone receptor complexes, *J. Steroid Biochem. Mol. Biol. 56*:31–37 (1996).
25. Dittmar, K. D., Hutchison, K. A., Owens-Grillo, J. K., and Pratt, W. B., Reconstitution of the steroid receptor–hsp90 heterocomplex assembly system of rabbit reticulocyte lysate, *J. Biol. Chem. 271*:12833–12839 (1996).
26. Kimura, Y., Yahara, I., and Lindquist, S., Role of the protein chaperone YDJ1 in establishing Hsp90-mediated signal transduction pathways, *Science 268*:1362–1365 (1995).
27. Shaknovich, R., Shue, G., and Kohtz, D. S., Conformational activation of a basic helix–loop–helix protein (MyoD1) by the C-terminal region of murine HSP90 (HSP84), *Mol. Cell. Biol. 12*:5059–5068 (1992).
28. Antonsson, C., Whitelaw, M. L., McGuire, J., Gustafsson, J.-A., and Poellinger, L., Distinct roles of the molecualr chaperone hsp90 in modulating dioxin receptor function via the basic helix–loop–helix and PAS domains, *Mol. Cell. Biol. 15*:756–765 (1995).
29. Miyata, Y. and Yahara, I., The 90-kDa heat shock protein, HSP90, binds and protects casein kinase II from self-aggregation and enhances its kinase activity, *J. Biol. Chem. 267*:7042–7047 (1992).
30. Miyata, Y. and Yahara, I., Interaction between casein kinase II and the 90-kDa stress protein, HSP90, *Biochemistry 34*:8123–8129 (1995).
31. Lees-Miller, S. P. and Anderson, C. W., Two human 90-kDa heat shock proteins are phosphorylated in vivo at conserved serines that are phosphorylated in vitro by casein kinase II, *J. Biol. Chem. 264*:2431–2437 (1989).
32. Schulte, T. W., Blagosklonny, M. V., Ingui, C., and Neckers, L., Disruption of the Raf-1–Hsp90 molecular complexes results in destabilization of Raf-1 and loss of Raf-1–Ras association, *J. Biol. Chem. 270*:24585–24588 (1995).
33. Whitesell, L., Mimnaugh, E. G., Costa, B. D., Myers, C. E., and Neckers, L. M., Inhibition of heat shock protein Hsp90–pp60$^{v\text{-}src}$ heteroprotein complex formation by

benzoquinone ansamycins: Essential role for stress proteins in oncogenic transformation, *Proc. Natl. Acad. Sci. USA 91*:8324–8328 (1994).
34. Brugge, J. S., Interaction of the Rous sarcoma virus protein pp60$^{v\text{-}src}$ with the cellular proteins p50 and p90, *Curr. Top. Microbiol. Immunol. 123*:1–22 (1986).
35. Xu, Y. and Lindquist, S., Heat-shock protein hsp90 governs the activity of pp60$^{v\text{-}src}$ kinase, *Proc. Natl. Acad. Sci. USA 90*:7074–7078 (1993).
36. Aligue, R., Akhavan-Niak, H., and Russell, P., A role for Hsp90 in cell cycle control: Weel tyrosine kinase activity requires interaction with Hsp90, *EMBO J. 13*:6099–6106 (1994).
37. Bohen, S. P. and Yamamoto, K. R., Isolation of Hsp90 mutants by screening for decreased steroid receptor function, *Proc. Natl. Acad. Sci. USA 90*:11424–11428 (1993).
38. Kimura, Y., Matsumoto, S., and Yahara, I., Temperature-sensitive mutants of *hsp82* of the budding yeast, *Saccharomyces cerevisiae*, *Mol. Gen. Genet. 242*:517–527 (1994).
39. Nathan, D. F. and Lindquist, S., Mutational analysis of Hsp90 function: Interaction with a steroid receptor and a protein kinase, *Mol. Cell. Biol. 15*:3917–3925 (1995).
40. Nemoto, T., Ohara-Nemoto, Y., Ota, M., Takagi, T., and Yokoyama, K., Mechanism of dimer formation of the 90-kDa heat-shock protein, *Eur. J. Biochem. 233*:1–8 (1995).
41. Meng, X., Devin, J., Sullivan, W. P., Toft, D., Baulieu, E.-E., and Catelli, M.-G., Mutational analysis of Hsp90α dimerization and subcellular localization: Dimer disruption does not impede "in vivo" interaction with estrogen receptor, *J. Cell Sci. 109*:1677–1687 (1996).
42. Sullivan, W. P. and Toft, D. O., Mutational analysis of hsp90 binding to the progesterone receptor, *J. Biol. Chem. 268*:20373–20379 (1993).
43. Cadepond, F., Binart, N., Chambraud, B., Jibard, N., Schweizer-Groyer, G., Segard-Maurel, I., and Baulieu, E. E., Interaction of glucocorticoid receptor and wild-type or mutated 90-kDa heat shock protein coexpressed in baculovirus-infected Sf9 cells, *Proc. Natl. Acad. Sci. USA 90*:10434–10438 (1993).
44. Binart, N., Lombes, M., and Baulieu, E.-E., Distinct functions of the 90-kDa heat shock-protein (hsp90) in oestrogen and mineralocorticoidsteroid receptor activity: Effects of hsp90 deletion mutants, *Biochem. J. 311*:797–804 (1995).

9

Hsp104

John R. Glover, Eric C. Schirmer,
Mike A. Singer, and Susan L. Lindquist
The University of Chicago, Chicago, Illinois

I. INTRODUCTION

A readily observable consequence of heat stress is aggregation of proteins within the cell [1,2]. Heat stress induces the expression of heat-shock proteins (Hsps), which appear to contribute to prevention and repair of heat damage by functioning as molecular chaperones [3]. During heat shock, they bind to and block the irreversible aggregation of thermally unfolded proteins and promote refolding and reactivation when the heat stress is relieved. During extreme or prolonged stress, the sheer numbers of unfolded substrates may overwhelm the capacity of chaperone systems to prevent aggregation. Under these circumstances, survival would be enhanced by the ability to separate proteins that have already started to aggregate. *Saccharomyces cerevisiae* Hsp104, the first Hsp for which thermotolerance function is uniquely associated with resistance to extreme stress, exhibits the capacity to help resolubilize and reactivate heat-damaged proteins. Through the agency of Hsp104 or other Hsp100–Clp family members, many proteins that have been diverted into "off-pathway" aggregation reactions are given a second chance at life.

II. Hsp104 IS A MEMBER OF THE Hsp100–Clp FAMILY

Although the Hsp100–Clp family of proteins was first described only fairly recently [4], in the past few years many new members have been identified. With these additional sequences, criteria have emerged for grouping the HSP100–Clp proteins into two major classes that provisionally are further divided into eight

Amino-terminus: Signature Sequence I

Con:	xxFTxxhxxhhxxAxxxAxxLxHxxhxxxHLLLGLh
A:	xxxxxxLxxxhxxAxAxAxxxRHExxTVEHLLLALh
B:	xKFTxxxxxALAxAxxLAxxxxHxxhxPhHLAxALh
C1:	ERFTEKAxKVIxLAQEEARRLGHNFVGTEQILLGLI
C2:	IPFTPRAKRVLELSxEEARQLGHNYIGTEHLLLGLL
D:	ERFTERAxRAIxxSQxEAKxLGKxxVxTxHLLLGLI

ATP-1: Walker consensus distinctions

	A	B1	B2
Classical:	GX_4GKT	$RX_{6-8}h_4D$	$RX_{6-8}h_4D$
A:	GX_4GKT	KX_9h_4D	$RX_6AIDLhD$
B:	GX_4GKT	$KX_{6-10}h_4D$	$RX_6AIDLhD$
C, D:	GX_4GKT	R/KX_6h_4D	$RX_6AIDLhD$

Middle: Signature Sequences

II		III	
Con:	RxxDxxxAxELRxxxxx	**Con:**	xWTGIPVxKh
Bp:	RxGDLARxSELQYGxIP	**A:**	RIARIPxRxV
Be:	RxxDLxxAADLRYxxIP	**B:**	RWTGIPVxKh
C:	RxQDFEKAGxLRDxExx	**C:**	SWTGIPVxKh
		D:	VWSGIPVQQh

ATP-2: Walker consensus distinctions

	A	B
Classical	GX_4GKT	$RX_{6-8}h_4D$
A, B, C, D:	GX_4GKT	RX_6h_4D
M, N, X:	GX_4GKT	$RX_{4-12}h_4D$
Y:	GX_4GKT	$RX_{4-12}h_4D$

Carboxyl-terminus: Signature Sequences

IV		V	
Con:	FRPEFLNRLDEIIVFxxL	**Con:**	GARPLRRxI
A:	FTPEFRNRLDxIIxFxxL	**Others:**	GARPLRRxI
Others:	FRPEFLNRLDEIIVFxxL	**X:**	GARGLRRxI
X, Y:	LRPEFxGRLPIxVxLxxL	**Y:**	GARRLRRxI

subfamilies [5]. The primary sequence elements characteristic for each of these subfamilies are listed in Figure 1 and the overall domain structure of each family is represented in diagram form in Figure 2.

Class 1 Hsp100s contain two nucleotide-binding domains (ATP-1 and ATP-2), each with canonical Walker A and Walker B elements, similar to those found in P-type transporters and the mitochondrial F_1-ATPase [6]. A general topology has been proposed for these domains based on the crystal structures of several proteins that contain these elements and catalyze ATP-triggered reactions [7]. The Walker A motif, or P-loop [8], occurs at the beginning of the central α-helix. The side chain of a conserved lysine (K218 in ATP-1 and K620 in ATP-2 of Hsp104) coordinates the β- and γ-phosphates of the bound ATP molecule; the next amino acid side chain in the primary sequence—a threonine—forms one coordinate of the active site Mg^{2+} through a water molecule. The conserved aspartic acid side chain of the Walker B motif is located at the end of one of six overlying β-strands and forms a second coordinate for the active site Mg^{2+} ion.

In other protein families, the members of which typically contain two nucleotide-binding domains, the domains are very similar to one another. In contrast, the ATP-1 and ATP-2 domains of the Hsp100 class 1 proteins are very distinct, suggesting that class 1 arose by gene fusion, rather than gene duplication. The occurrence of two putative Walker B motifs in the ATP-1 of class 1 proteins makes this domain unique. It is not currently known whether only one of these is involved in binding nucleotides, or if class 1 proteins toggle between conformational states, requiring utilization of alternate motifs. Hsp104 derivatives with alanine substituted for the conserved aspartic acid residue in either B1 or B2 exhibit reduced function in vivo [E. C. Schirmer, unpublished observations].

The ATP-1 and ATP-2 domains of class 1 proteins are flanked by amino(NH_2)-terminal and carboxy(COOH)-terminal regions and are separated by a middle region. Class 1 proteins can be divided into four subfamilies based on the size of the middle region combined with signature sequences throughout the protein, designated I through V. A-type proteins have the shortest middle region (about

Figure 1 ATP-binding motifs and signature sequences that define HSP100-Clp subfamily distinctions: Signature sequence I in the NH_2-terminal region and sequences II and III of the middle region are present in class1 proteins only. Signature sequences IV and V in the COOH-terminal region are present in all subfamilies. The signature sequence I in the NH_2-terminal region of C-type Hsp100s is duplicated and is designated C1 and C2. The class 1 proteins have two Walker B consensus motifs that are designated B1 and B2. Signature sequence II in the middle region of B-type Hsp100s is distinct in proteins from prokaryotic and eukaryotic sources and are designated Bp and Be, respectively. The overall consensus sequence Hsp100 is listed above each set of subfamily consensus sequences as well as the classic consensus sequences for Walker A and B motifs. x = nonconserved residues; h = hydrophobic residues. (From Ref. 5.)

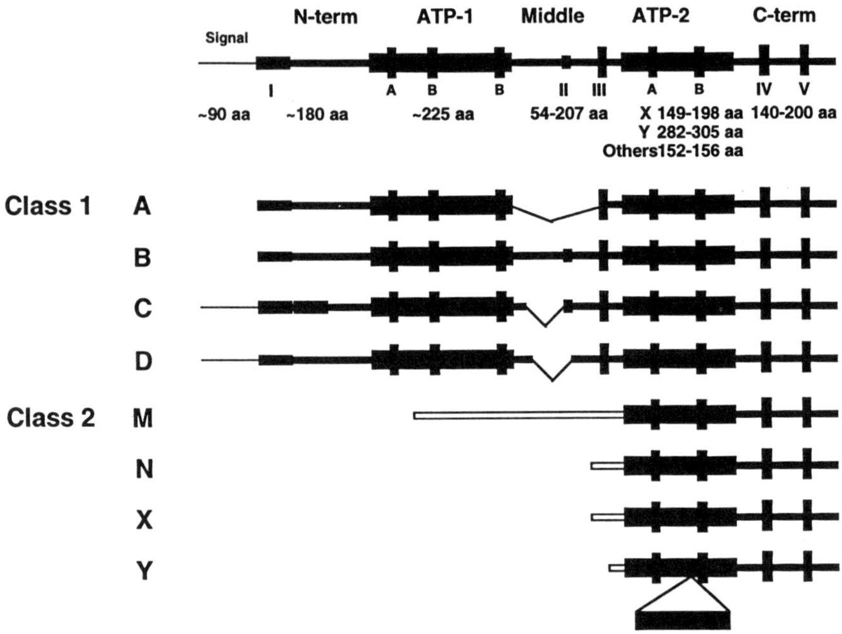

Figure 2 Diagrammatic representation of HSP100–Clp subfamilies: (top) a prototypical HSP100–Clp protein indicating the main regions of the protein, the typical length of each region, and the positions of signature sequences I–V. The degree of conservation in each diagram is indicated by the thickness of the bar. Open bars in class 2 proteins represent NH_2-terminal sequences that are not shared by class 1 proteins. An approximately 140-amino acid residue insertion unique to the ATP-2 region of Y subfamily members is also indicated. (Adapted from Ref. 5.)

54-amino acid residues), whereas those of C- and D-types are intermediate in length (about 101–118 residues), and those of the B-types are long (about 172–207). The C- and D-type proteins are further distinguished by a duplication in the C-type proteins of the NH_2-terminal signature sequence I. Some class 1 proteins are localized in plastids or mitochondria and so contain targeting signal peptides that are presumably cleaved on import.

Hsp104 and Hsp78, two of the three Hsp100–Clp proteins present in *S. cerevisiae*, are both B-type proteins based on the lengths of their middle regions. Hsp78, a heat-inducible mitochondrial matrix protein [9] and another B-type protein from *Mycoplasma genitalium* (Genbank accession number U39719) are distinct from other full-length HSP100s in that they lack signature sequence I of Hsp104 and other class 1 proteins.

Class 2 proteins share only the ATP-2 domain and the COOH-terminal region in common with class 1 proteins. The ATP-2 domains of the M and N subfamily most closely align with the class 1 proteins; the X and, to a greater extent, the Y subfamilies are more divergent. Members of the Y subfamily carry an insertion of approximately 140-amino acid residues between the Walker A and B elements in their ATP-2 domains. The M-type HSP100s are distinguished from the N-type by their long NH_2-terminal extensions that, in mouse SKD3 protein, contains ankyrin-like repeats [10].

III. FUNCTIONAL DIVERSITY OF Hsp100–Clp PROTEINS

The known functions of Hsp100–Clp proteins in biological systems are summarized in Table 1. One of the most intriguing features of this family is that some of its members can either restructure inactive protein complexes, or promote their proteolysis through specific association with a partner protease [for reviews see 11,12]. *Escherichia coli* ClpA embodies this seeming paradox. ClpA was named for its capacity to promote the ATP-dependent proteolysis of casein (caseinolytic protease) when complexed with the unrelated ClpP (Ti) protease in vitro and is involved in the turnover of abnormal proteins in vivo [13–15]. However, in the absence of ClpP, ClpA can substitute for the DnaK–DnaJ chaperone system in activating the DNA-binding capacity of the P1 protein, RepA by disassembling the inactive dimer and releasing active monomers [16,17].

The B-type Hsp100–Clp proteins are primarily involved in mitigating damage incurred by a variety of stresses [18–25]. The C-type proteins also function in stress tolerance [26,27]. In addition to its role in stress tolerance, the *Bacillus subtilis* C-type protein is involved in the regulation of gene expression associated with metabolic stress and entry into stationary phase [28–30]. The inability to establish strains with reduced ClpC in either cyanobacteria or in plant plastids suggests that this protein is essential in photosynthetic organisms [31,32]. The only D-type Hsp100 identified so far is the product of the *Arabidopsis Erd1* gene and is induced by desiccation, rather than by heat stress [33].

The M- and N-type proteins are postulated to serve as ATPase components of two-component membrane transporters [10,34], although there is currently insufficient functional characterization to establish a definitive role for members of these subfamilies.

Proteins in the X and Y subfamilies function in the regulation of gene expression and phage DNA replication, by determining the turnover rates of specific proteins, in conjunction with a protease partner protein [35–42]. ClpX from *E. coli*, like ClpA, appears to have biological functions that involve disassembling protein complexes independently of protein turnover [43–45].

We have proposed that a common mechanism underlies each of these diverse biological activities [5]. Recent work on Hsp104, ClpA, and ClpX indicate that all three proteins employ ATP to promote changes in the folding and

Table 1 Functional Diversity of Hsp100–Clp Proteins[a]

Type	Name	Organism	Function
A	ClpA	*E. coli*	ATPase subunit of Ti protease [13]
B	ClpB	*E. coli*	Required for growth at high temperatures [19]
	Hsp104	*S. cerevisiae*	Tolerance to extreme temperature and other stress [18,20]; regulation of [PSI+] genetic element [93,113]
	Hsp78	*S. cerevisiae*	Thermotolerance in mitochondria [25]
	Hsp101	*Arabadopsis thaliana*, *Glycine max*	Complementation of thermotolerance defect in *hsp104* mutant yeast [21,22]
	ClpB	*Leischmania*	Required for rapid growth at high temperature [23]
	ClpB	*Synechococcus*	Thermotolerance [24]
C	MecB/ ClpC	*Bacillus subtilis*	Heat and salt tolerance [26]; regulation of competence specific gene expression [28,29]; regulation of expression of sigma factor *sigD* [30]
	ClpC	*Synechococcus*	Essential [32]
	ClpC	Chloroplasts	Essential, functions with *E. coli* ClpP in proteolysis in vitro [31]
	ClpC	*Listeria monocytogenes*	Stress tolerance and intracellular survival factor [27]
	TlyB	*Serpulina hyodistenteriae*	Hemolytic activity [120]
M	SKD3	Mouse	Suppressor of potassium transport defect in yeast [10]
N	AmiB	*Pseudomonas aeruginosa*	ATPase component of putative two-component transporter [34]
X	ClpX	*E. coli*	Alternative ATPase of Ti protease [92]; turnover of starvation sigma factor in log phase [42]; turnover of UmuD′ in regulation of SOS mutagenesis [41]
Y	ClpY	*E. coli*	ATPase component of ClpYQ protease [121,122]
	LapA	*Pasteurella haemolytica*	Inhibition of leukotoxin expression [36]

[a]The functions of some Hsp100–Clp proteins are summarized. In some cases further details are offered in the text.

assembly of other proteins. Because these proteins belong to different classes and subfamilies, it seems likely that all members of the family will share this fundamental mechanism. Thus, the Hsp100–Clp proteins constitute a new type of molecular chaperone.

IV. ATPase ACTIVITY AND OLIGOMERIZATION OF Hsp104

Each of the Hsp100–Clp proteins tested, including ClpA [46], ClpB [47], and ClpX [44] from *E. coli*, *Pseudomonas aeruginosa* AmiB [34], and Hsp104 [48], are ATPases. Each has a weak intrinsic ATPase activity that is either stimulated or inhibited by specific proteins or peptides [44,46,47; E. C. Schirmer, unpublished observations]. Nucleotides also play a role in the assembly of oligomers of ClpA and Hsp104. In keeping with the distinct sequence characteristics of the ATP-1 and ATP-2 domains suggested by sequence comparisons, identical mutations in the conserved lysine residues (e.g., G/AX$_4$GKT/S to G/AX$_4$GTT/S) within each domain have different effects on the ability of the mutant protein to hydrolyze ATP or to form oligomers, suggesting that the two domains contribute to protein function in different ways. In the yeast Hsp104 protein, a K218T mutation in the ATP-1 domain affects ATPase activity more so than oligomerization, whereas the equivalent K620T mutation in the ATP-2 domain compromises oligomerization more than ATPase activity [48,49]. Curiously, the same amino acid substitutions in the *E. coli* ClpA protein suggest that the relative contributions to ATPase activity and oligomerization of ATP-1 and ATP-2 domains in that protein are reversed relative to Hsp104 [50,51].

More recent data demonstrate that subunit contacts within the assembled oligomer of Hsp104 are essential for full ATPase activity, and that ATP hydrolysis within the Hsp104 hexamer is concerted. First, the oligomerization defect of the K620T derivative of Hsp104 can be overcome at high protein concentrations [48]. At low protein concentrations, at which K620T is monomeric, the protein hydrolyzes ATP at a rate that is three- to fourfold lower than at high concentrations at which the mutant protein is mostly hexameric, and its ATPase activity approaches that of the wild-type enzyme. In contrast, the observed rates of ATP hydrolysis by wild-type Hsp104 is independent of protein concentration, and no detectable ATPase activity is associated with the K218T mutant, even at high protein concentrations. Second, a deficit in ATP hydrolysis, comparable with that observed with the K620T mutant, was also observed when wild-type protein was assayed in the presence of 10 mM CHAPS, which inhibits oligomerization. Finally, the ATPase activity of wild-type enzyme in hetero-oligomers containing K218T mutant subunits is reduced, suggesting that interactions among subunits within the Hsp104 hexamer are required for full activity.

Electron microscopy reveals that both ClpA and Hsp104 assemble into a ring-like structure with sixfold symmetry [49,52]. Images of ClpA and Hsp104 are reminiscent of those of the chaperonin proteins GroEL (Hsp60) and CCT (also known as TriC or TCP-1), but the chaperonin proteins share no sequence homology with the Hsp100 family [53]. When the GroEL oligomer is complexed with GroES, the encompassed cavity can readily accommodate an average-sized protein in a compact conformation [54]. Within this protected environment, nonnative proteins are shielded from unfavorable interactions that might lead to aggregation during folding and may acquire the native state before being released from the chaperonin complex. Although the central channels of Hsp104 or ClpA may open in association with partner proteins, it seems unlikely that they are used to enclose whole polypeptides because their diameters are much smaller. CCT has a pore size of approximately 6 nm and GroEL, even in its compact form (without GroES), has a channel diameter of 4.5 nm [53], whereas Hsp104 has a pore of only 2.5 nm [49], and the center of the ClpA oligomer not complexed with ClpP appears to be occluded [52].

Nonetheless, the similar tertiary structures of these unrelated proteins may reflect a shared mechanistic constraint. The toroidal arrangement of subunits allows the display of active surfaces at a high local concentration, permitting multiple simultaneous interactions between the chaperone and its substrates. Such interactions, combined with coordinated utilization of ATP, may be required to overcome the forces stabilizing collapsed or assembled conformations of substrates to promote disaggregation, unfolding, or dissociation of their targets. For Hsp100 proteins, whether the final outcome is reactivation or degradation of the substrate likely depends on distinct properties of the individual chaperone, together with its ability to associate with other folding catalysts or the proteolytic machinery of the cell.

V. Hsp104 IS A CRUCIAL THERMOTOLERANCE FACTOR IN YEAST

A. Stress-Inducible Expression of Hsp104

Although the promoter region of *Hsp104* has not been examined in detail, it is known to contain at least one HSE element required for HSF-mediated regulation in response to heat, and three STREs (stress response elements [55]). The latter are binding sites for at least one transcription factor, Msn2p [56]. In general, STREs regulate the transcription of numerous genes in response to an array of stresses, including heat shock, high osmolarity, oxidative stress, exposure to ethanol, and agents that cause DNA damage [for reviews see 57,58]. Thus, control of Hsp104 expression is mediated by at least two overlapping stress-response pathways.

A key aspect of Hsp104 regulation is its induction by heat. Hsp104 is barely detectable in cultures during log-phase growth in rich medium at 25°C. A shift to

39°C for 1 h induces a 50-fold increase in Hsp104 accumulation [18]. Induction of Hsp104 is also conferred by exposure to ethanol [20], which is thought to promote protein denaturation and invoke a pattern of cellular response similar to that of heat shock [for review see [59].

The basal or uninduced level of Hsp104 varies, depending on the carbon source and growth phase of the culture. In general, cells growing on carbon sources requiring respiratory metabolism (acetate and galactose) have substantially higher basal levels of Hsp104 [20]. As cells enter stationary phase, which is accompanied by a switch from glucose fermentation to respiratory ethanol metabolism, Hsp104 accumulates [20]. Stress resulting from the formation of reactive oxygen species during respiration may induce many antioxidant and general stress-responsive proteins [for review see 60]. Induction of Hsp104 by arsenite, cadmium, and copper may also involve the response to the intracellular formation of reactive oxygen species [for review see 61].

The levels of Hsp104 and other stress proteins are also elevated in response to the loss of the signal recognition particle Srp54 [62]. Absence of this function leads to accumulation of mislocalized (misfolded?) precursor proteins in the cytosol, which may trigger a general stress response.

B. Hsp104 in Thermotolerance

Cells challenged with a moderate heat stress survive subsequent exposure to a lethal heat shock much better than cells shifted directly to the lethal temperature. This phenomenon, known as induced thermotolerance, is universally observed and is correlated with the induction of heat shock proteins [63] as well as with the accumulation of osmolytes [64]. In yeast, Hsp104 is a strong determinant of survival in thermotolerant cells during extreme stress. When cells with a wild-type *Hsp104* gene and cells with the gene disrupted are given identical mild-heat treatments and then shifted to an extreme temperature, 100- to 1000-fold more cells survive if Hsp104 is present [18].

Another observation that highlights the pivotal role of Hsp104 in thermotolerance comes from experiments in which expression of Hsp104 is uncoupled from the heat-shock response and placed under the control of a hormone-inducible promoter [65]. In hormone-treated cells, expression of Hsp104 is sufficient to confer a level of basal (uninduced) thermotolerance nearly equal to cells given a preconditioning heat shock. Nonetheless, the contribution to thermotolerance of hormone-induced Hsp104 is enhanced in cells given an additional heat stress, indicating that Hsp104 is either further activated by heat, or that its function is enhanced by other stress tolerance factors induced by preconditioning heat treatments.

Hsp104 not only contributes to thermotolerance, but also to ethanol tolerance and, to a lesser extent (tenfold), to that of arsenite, and to long-term storage of stationary-phase cells at cold temperatures [20]. Although Hsp104 is induced by copper or cadmium, it does not appear to play a significant role in tolerance to these stresses [20]. The fatal lesions produced by toxic metals are linked to direct interaction with sensitive amino acid side chains and to oxidative damage to DNA, proteins, and lipids [for review see 60]. The prevention and repair of such damage, most of which involves covalent modification, is less likely to involve chaperone-mediated protein refolding as a rate-limiting step. Likewise, Hsp104 does not confer tolerance to the accumulation of abnormal proteins containing the arginine analogue canavanine [66]. Rather, canavanine-containing proteins appear to be degraded by a ubiquitin-dependent pathway in yeast [67,68].

Hsp104 is critical for survival during brief exposure to extreme temperature (50°C) or high concentrations of ethanol (20%), but of reduced importance during less severe treatments that take several hours to kill cells [20]. In the wild, yeasts are more likely to encounter a multitude of environmental signals and stresses, in contrast with a typical laboratory experiment during which cells are exposed to a single, severe challenge, with all other conditions remaining optimal. Mild stresses (40°C or 6% ethanol), which are not lethal when applied separately, become toxic when combined [69]. Under such combined conditions, Hsp104 plays a vital role in survival. The up-regulation of Hsp104 in response to growth conditions requiring respiration or entry into stationary phase also enhances the basal thermotolerance of yeast [20]. For wild yeasts, the nonstress response to respiratory metabolism may anticipate a gradual increase in ethanol and heat in the local environment and help increase survival as conditions worsen. Thus, even though Hsp104 has no housekeeping function under optimal growth conditions, its function in stress tolerance is likely under consistent selective pressure in wild populations.

C. Interactions in Thermotolerance

The critical components of thermotolerance vary from organism to organism. For example, in *Drosophila* [70,71] and mammalian cells [72], which are not currently known to have a functional homologue of Hsp104, expression of members of the Hsp70 family of heat-shock proteins contributes to thermotolerance. In *S. cerevisiae*, the cytosolic Hsp70s are encoded by six genes—*SSA1–4* and *SSB1–2* [73]. When the accumulation of Ssa proteins is limited by disruption of the genes encoding heat-inducible forms, cells become temperature-sensitive for growth, but little effect on the acquisition of thermotolerance is observed [74,75]. However, in yeast lacking Hsp104, reducing Hsp70 abolishes much of the residual thermotolerance exhibited by cells during the first few minutes of exposure to extreme heat, whereas overexpression of Ssa1 partially compensates in thermotolerance for the loss of Hsp104 [75].

As a corollary, no role for Hsp104 can be detected over the entire temperature range permissive for yeast growth, except when the constitutively expressed Ssa1 and Ssa2 Hsp70s are deficient [75]. Because yeast require at least one intact *SSA* gene for growth, and the expression of these genes cannot be entirely eliminated without perturbing both the heat-shock response [74] and protein translocation [76], the functional overlap of Hsp104 with Hsp70 cannot be rigorously tested. It can be concluded, however, that Hsp104 is increasingly important when Hsp70 is limiting and vise versa, and both probably function either in parallel pathways or in different steps of the same pathway, leading to stress tolerance.

In addition to Hsps, low molecular weight osmolytes accumulate to high levels in stressed organisms [64]. In yeast, heat stress induces a 100-fold increase in the accumulation of the disaccharide trehalose [77]. Comparison of thermotolerance in yeast strains differing in their capacity to accumulate trehalose indicates that it contributes to survival during exposure to extreme heat [78]. Trehalose accumulation is also associated with thermotolerance that has been induced in *hsp104* mutant strains [79].

Whereas the genetic interactions between Hsp70 and Hsp104 suggest an overlapping or parallel function in thermotolerance, a potent synergism in the loss of thermotolerance in yeast is observed when both Hsp104 expression and trehalose accumulation are blocked [80; M. A. Singer, unpublished observation]. The most likely explanation of this synergism is that trehalose and Hsp104 function by separate mechanisms and even at different times. Trehalose acts as a thermostabilizer of proteins in vitro [81], suggesting that it serves as thermoprotectant in vivo by a similar mechanism. Trehalose most likely functions during heat stress by preventing thermally induced protein denaturation and, possibly, by altering the tendency of already-unfolded protein to form irretrievable aggregates, whereas Hsp104 appears to function exclusively during recovery to resolubilize and reactivate aggregated proteins [82].

VI. STRESS TOLERANCE FUNCTIONS OF OTHER Hsp100–Clp PROTEINS

The expression patterns and stress-related phenotypes of other B- and C-type Hsp100 proteins are similar to those of Hsp104 and suggest similar functions. ClpB of the cyanobacterium *Synechococcus* is also heat inducible and like Hsp104 in yeast, it is critical for the development of induced thermotolerance [24]. *Escherichia coli* ClpB expression is induced by heat, and mutations in the gene reduce growth at high temperatures [19], but it contributes only a fivefold enhancement of survival of thermotolerant bacteria [K. Uno and S. Lindquist, unpublished observation]. In the gram-positive bacterium *B. subtilis*, a C-type Hsp100 (mecB) also functions in stress tolerance [26]. This protein is strongly induced by exposure to heat, ethanol, or high salt levels, and loss of its function reduces survival in cells exposed to these stresses. The B-type Hsp100 of *Leishmania* is strongly induced by a temperature shift similar to that which the parasite experiences when it is transferred from the fly host to the much warmer mammalian blood stream; deletion of the Hsp100 gene significantly reduces leishmania's virulence [23; J. Clos, personal communication]. In *Listeria* a similar role is played by a C-type Hsp100 [27]. Moreover, the heat-inducible B-type Hsp100s from plants exhibit a conserved stress-tolerance function: both the *Arabidopsis* and the soybean proteins restore thermotolerance in yeast cells lacking Hsp104 [21,22].

Internal translational initiation sites are present in the genes encoding some prokaryotic Hsp100 proteins, although the lower molecular weight translation products derived from them represent only a small fraction of the total gene product. In vitro analysis of purified proteins corresponding to these shortened gene products shows that the shortened form of ClpB is not responsive to protein or peptide stimulation [83] and that the shortened form of ClpA exhibits normal ATPase activity, but no longer promotes proteolysis of substrates by ClpP [84]. Inhibition of the enzyme activities of the full-length proteins in mixtures containing the shortened versions suggests that the latter could hypothetically play a role in the regulation of Hsp100 activity in vivo under conditions for which their accumulation is favored.

In yeast, the mitochondrial ClpB Hsp78 [9] is reminiscent of the prokaryotic forms derived from internal initiation, because it lacks a large portion of the NH_2-terminal region shared by most B-type Hsp100s. Unlike the shortened prokaryotic forms, Hsp78 is encoded by a gene that is distinct from *HSP104* and is functionally competent in thermotolerance, both in mitochondria and in the cytosol where it can partially substitute for Hsp104 [25]. In mitochondria, Hsp78 can partially compensate for a *ts*-allele of mitochondrial Hsp70 in the import of precursor proteins [85]. Hsp78 also contributes to the maintenance of ts-Hsp70 protein solubility in the mitochondrial matrix [86], possibly by stabilizing the ts-protein or helping it refold. Similar to its cytosolic counterpart Hsp104, Hsp78 does not prevent heat-induced damage (assessed by loss of translational competence within the mitochondrial matrix), but is essential for the restoration of translation during recovery [25].

VII. DISASSEMBLY OF PROTEIN AGGREGATES UNDERLIES THE STRESS-TOLERANCE FUNCTION

Members of the best-studied Hsp families (e.g., Hsp70 and Hsp60) inhibit aggregation of unfolded proteins by binding to hydrophobic surfaces exposed by denaturation. Except in a few special cases [87], Hsp60 and Hsp70 are ineffective in resolubilizing aggregates once they have formed. A unique property of the Hsp100 family is the ability to promote the resolubilization of protein aggregates, particularly those induced by stress. When yeast cells expressing a temperature-sensitive bacterial luciferase protein are subjected to a moderate heat stress (44°C), which does not affect cell viability, luciferase activity is lost, and the protein aggregates [82]. After return to normal temperatures, luciferase is re-solubilized and reactivated in wild-type cells, but not in cells lacking Hsp104.

Separate experiments indicate that the general stress-tolerance function of Hsp104 is to promote the resolubilization of protein aggregates. For example, when cells not expressing the luciferase reporter protein are heat-shocked and examined by electron microscopy (Fig. 3), wild-type and *hsp104* mutant cells both display large electron-dense material (presumably protein aggregates) in their nuclei and cytoplasms [82]. During recovery from heat shock, these deposits disappear from wild-type cells, but not from the mutants. Because no difference in proteolysis is detected in wild-type or *hsp104* mutant cells [66], the disappearance of these aggregates is most likely attributable to resolubilization. Similarly, when RNA splicing is damaged in vitro by heating cell-free lysates lacking Hsp104, subsequent addition of purified Hsp104 partially restores splicing activity [88]. Thus, Hsp104 appears to mediate the repair of substrates already damaged by heat, rather than to protect them from damage.

Our understanding of the detailed molecular mechanism by which Hsp104 accomplishes its function in promoting resolubilization and refolding is at a formative stage. The results of in vitro experiments using unfolded firefly luciferase as a substrate suggest that purified Hsp104 does not promote refolding by itself. However, when urea-denatured luciferase is refolded in the presence of Ydj1 and Ssa1, which together function as an ATP-dependent folding catalyst [89], Hsp104 greatly enhances the yield of the reaction (Fig. 4) [J. R. Glover and S. Lindquist, in preparation]. The K218T *hsp104* mutant, which is defective in ATP hydrolysis, does not increase folding in these reactions indicating that Hsp104-mediated enhancement of luciferase folding is dependent on ATP [J. R. Glover, unpublished observation].

The functional interaction between Hsp104 and Ydj1–Ssa provides a biochemical explanation for some of the genetic interactions between Hsp104 and the Ssas [75] and for the ability of both Hsp70 and Hsp104 to mediate the in vitro repair of mRNA splicing [88]. Although still uncertain, it is plausible that Hsp104 functions at the beginning of a "pathway" for the refolding of heat-damaged proteins in vivo. In this scenario, Hsp104 simply helps to pull apart kinetically trapped and partially aggregated proteins, providing other folding catalysts access to suitable substrates. The apparent inability of Hsp104 to promote refolding or to provide a protected environment for stabilizing nonnative intermediates suggests that productive release of unfolded substrate from Hsp104 should be intimately coupled to the acquisition of the substrate by the factors that do promote refolding. The observation that Hsp104 does not, by itself, mediate refolding of the limited number of test substrates analyzed in vitro does not

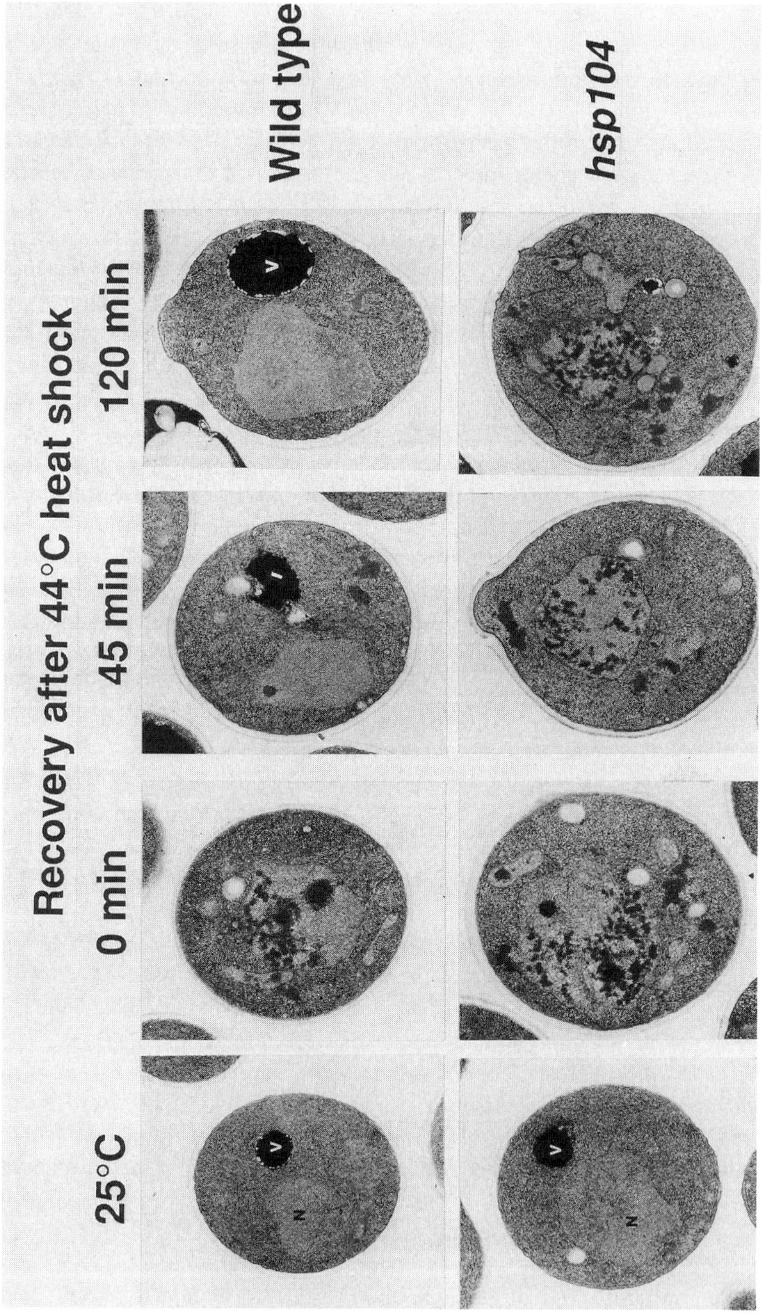
Recovery after 44°C heat shock
25°C
0 min
45 min
120 min
Wild type
hsp104
V
N
I

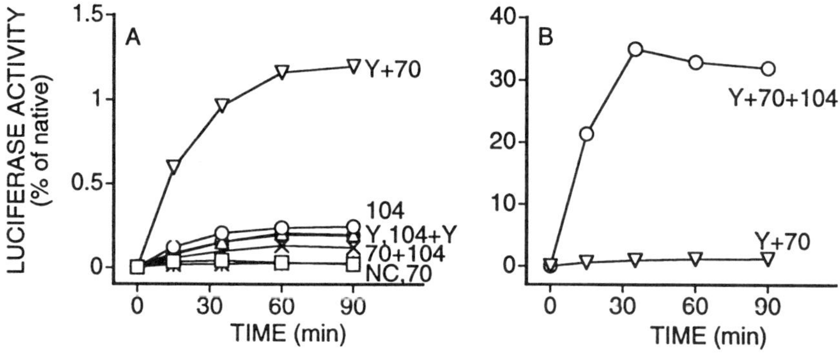

Figure 4 Hsp104 mediates the refolding of chemically denatured firefly luciferase in conjunction with Ydj1 and Hsp70 of the Ssa subfamily: Urea-denatured luciferase (20 nM) was diluted directly into buffer alone (25 mM HEPES-KOH pH 7.6, 150 mM potassium acetate, 10 mM magnesium acetate, 10 mM DTT, 5 mM ATP) or buffer supplemented with 1 μM purified Hsp104 (104), Ydj1(Y), and Ssa1 (70), either each alone or in various combinations and incubated at 25°C. Duplicate aliquots of the refolding reaction were assayed for luciferase activity at the indicated times and expressed as a percentage of an equivalent amount of native enzyme assayed in the presence of 1 mg mL^{-1} BSA. (A) Under these conditions luciferase refolding by Ydj1 and Ssa1 (Y+70) was inefficient, but severalfold better than spontaneous refolding in buffer alone (B) or any single chaperone (Y, 70, 104) or any other combination of two (Y+104, 104+70). (B) Refolding of luciferase by Ydj1 and Ssa1 (Y+70) is dramatically enhanced by the addition of Hsp104 (Y+70+104).

Figure 3 Hsp104 is required for the resolution of heat-induced aggregation in vivo: Wild-type and an isogenic *hsp104* deletion mutant were cultured at 25°C. Midlog phase cultures were shifted to 37°C for 30 min, then heat-shocked at 44°C for 60 min and subsequently allowed to recover for the indicated period at 25°C. In electron micrographs, intracellular damage that is absent from non–heat-shocked cells (25°C), is indicated by the occurrence of electron-dense, morphologically irregular structures in the cytoplasms and nuclei of cells sampled at the end of the heat-shock treatment (0 min). During recovery in wild-type cells these structures disappear, whereas in the mutant yeast the damage persists. Cells from both strains were viable throughout the experiment. The large, dark structures present in some micrographs are vacuoles, and their occurrence in sections is not associated with the treatments given. (From Ref. 82.)

exclude the possibility that in vivo some stress-damaged proteins are refolded directly by Hsp104 or by Hsp104 in cooperation with folding catalysts other than Ydj1 and Ssa.

VIII. THE MOLECULAR FUNCTIONS OF OTHER Hsp100s ALSO DEPEND ON A "DISASSEMBLING" ACTIVITY

A. Proteolysis

Both ClpA and ClpX were identified in biochemical screens for proteins that promote the degradation of specific test substrates [35,90–92]. Neither has a proteolysis function on its own; rather, they promote proteolysis through association with the ClpP (Ti) protease. The first-identified targets of the ClpAP complex were casein and a ClpA–lacZ fusion protein, the latter presumably reflecting the capacity of ClpA to regulate its own intracellular concentrations through proteolysis [46,90,91]. Inactivation of ClpA alone does not cause a detectable proteolytic phenotype, but in the background of a *lon* protease mutation, it enhances defects in the degradation of abnormal, canavanine-containing proteins [91]. The ClpAP protease is also required for degradation of N-end rule substrates in *E. coli*; inactivation of ClpA results in almost complete stabilization of these substrates [15]. ClpX complexed with ClpP specifies the degradation of a different subset of proteins, including the *vir* group of repressor protein mutants of bacteriophage mu [38], the P1 plasmid PhD protein [39], and the starvation σ factor [42] in vivo.

In vitro studies demonstrate that ClpA and ClpX have two effects on the ClpP protease catalytic core: they enhance its proteolytic functions in response to ATP, and in agreement with their in vivo functions, determine its range of substrates. The ClpAP complex hydrolyzes casein, heat-shock transcription factor σ^{32}, plasmid origin-binding factor RepA, and the bacteriophage origin-binding protein λO [17,44]. The ClpXP complex hydrolyzes λO (at three times the rate of ClpAP), but not σ^{32} or casein [35,44,92]. Just as Hsp104-dependent thermotolerance is abolished by site-directed mutations in Hsp104 that compromise its capacity to hydrolyze ATP or to form oligomers [48,94], similar mutations in ClpA eliminate its ability to function in proteolysis [50,51].

The proteolysis function of Hsp100–Clp proteins is now thought to reflect their capacity to affect the higher-order structure of their targets, thereby providing better substrates for the ClpP protease. In the absence of ClpP, ClpA disassembles RepA dimers in vitro in an ATP-dependent fashion and activates them for DNA binding [17]. ClpX protects λO from heat-induced aggregation and dissociates such aggregates once formed, thereby restoring the ability of λO to bind DNA [44]. The dissociation of preformed aggregates, but not the prevention of aggregation, requires hydrolyzable ATP. It seems likely that the dissociation function is used to present an accessible conformation of the substrate protein to the protease.

ClpA, in association with the ClpP protease, exhibits a striking resemblance to the eukaryotic 26S proteasome [52]. The oligomeric nature of Hsp100 proteins leads to two hypotheses about the interface with the protease. In one, the Hsp100 chaperone provides a surface on which a denatured substrate can be maintained in its unfolded state through interactions with multiple subunits of the oligomer. In its folded state, the substrate would be protease-resistant, but when held in an unfolded conformation by the Hsp100 oligomer, the surfaces of the substrate would be exposed to the ClpP protease for hydrolysis. Alternatively, the small central pore of the oligomerized Hsp100 proteins, although not sufficient to accommodate folding of a substrate, may be large enough to serve as a channel through which the substrate may be unraveled and passed to ClpP for processive degradation. The extended conformation of the substrate would thus be resistant to aggregation, but sensitive to the protease. The mismatch in symmetry between ClpP (sevenfold) and ClpA (sixfold) precludes simple mechanisms involving one-to-one association between ClpA and ClpP subunits, and has led to the intriguing, but as yet unsubstantiated, conjecture that rotation of ClpA relative to ClpP results in a ratcheting of substrate from ClpA to ClpP [52].

Unlike ClpA and ClpX, ClpB does not associate directly with the ClpP protease. However, occasionally its disaggregating function may also lead to proteolysis. In *clpB*⁻ mutants, certain unstable variants of bacteriophage T4 lysozyme are stable and accumulate in inclusion bodies [95]. Presumably, in wild-type cells ClpB prevents these lysozyme variants from accumulating in an aggregated state, thereby enhancing their degradation. Native conformations of proteins are relatively protease-resistant, and unfolded proteins are protease-sensitive. But at high concentrations, unfolded proteins are likely to aggregate into large protease-resistant complexes. The Hsp100 proteins may commonly employ a disassembly function to maintain unstable proteins in a protease-sensitive state. The absence of proteolysis defects in *hsp104* deletion mutants at high temperatures indicates that this is not Hsp104's essential stress-tolerance function—presumably because these heat-damaged substrates can be degraded by factors that do not require Hsp104 function, or can refold into their active conformations after resolubilization by Hsp104 and cofactors.

B. Regulation

Several Hsp100 proteins have been identified as regulators of important biological functions, such as the induction of genes mediating competence to take up exogenous DNA [29], and the control of toxin expression [36]. Little is known about the role of Hsp100 proteins in some of these phenomena, or the nature of their substrates. Strikingly, in the few examples examined in detail, the Hsp100s regulate specific pathways by promoting the dissociation of key components.

The best-understood example involves the lytic phase of the bacteriophage Mu life cycle, a complex, highly regulated process involving numerous protein–protein and protein–DNA interactions [96]. When activated, the Mu genome resident on the host chromosome is amplified by repeated cycles of replication and transposition. Insertion into the host chromosome is mediated by a DNA-binding complex consisting of several proteins, including a tetramer of the MuA transposase. This complex must be dislodged from Mu DNA ends for the host replication machinery to synthesize a complementary strand of the phage DNA for the next round of transposition.

In bacterial host strains lacking ClpX, the replication process is blocked at the stage requiring removal of the MuA tetramer [38]. Replication is relatively unaffected in *clpP* mutants, indicating that the role of ClpX in this process is distinct from its role as a component of the ClpXP protease. In vitro, MuA is released by ClpX, in an ADP-driven reaction, from a complex composed of MuA, MuB, and host HU protein bound to DNA. This complex (type II transpososome) is otherwise stable at high temperatures and high concentrations of urea. MuA is released as a monomer that retains its native capacity to associate into protein–DNA complexes [43,45].

In some circumstances, ClpX plays an additional regulatory role at the level of the Mu repressor protein. Certain virulent mutants of bacteriophage Mu (*vir*) carry frameshift mutations in the Mu repressor gene. The half-life of the altered repressor proteins is significantly reduced, resulting in derepression of Mu transposition. The *vir* mutants can be suppressed by null mutations in *clpP* and *clpX*, which restore longevity to the repressor [38,97]. Thus, the mutant repressor protein is a proteolytic substrate of the ClpXP protease; *clpB* and *clpA* mutants have no effect. The absence of an effect of *clp* mutations on the stability of the wild-type repressor protein suggest that it is not a substrate of the ClpXP protease. However, the lability of *vir* repressors is transdominant—the presence of the mutant protein promotes degradation of the wild-type repressor protein. The most likely explanation for this observation is that mutant repressors form hetero-oligomers with wild-type repressors, and ClpX targets both components of the complex for degradation.

IX. Hsp104 CONTROLS THE AGGREGATION STATE OF Sup35

A. [PSI$^+$] is a Yeast Prion

[PSI$^+$] was identified over 30 years ago as an extrachromosomal genetic element that causes omnipotent suppression of nonsense codons in certain yeast strains [98]. [PSI$^+$] is dominant and cytoplasmically inherited, but could not be attributed to any similarly inherited nucleic acid. When [psi$^-$] cells are exposed to the cytoplasm of cells carrying the [PSI$^+$] element through mating or cytoduction, the progeny of such encounters are uniformly [PSI$^+$]. [PSI$^+$] is dependent on a nuclear gene, *SUP35* [99]. The Sup35 protein normally functions in a complex with Sup45 in translation termination [100] and consists of a highly conserved COOH-terminal domain that is essential for growth and provides the translation termination function, and an NH_2-terminal domain that is dispensable for growth, but is required for the formation of the [PSI$^+$] element [101]. [PSI$^+$] is now thought to be a conformationally altered form of Sup35 which forms aggregates that are analogous to those formed by mammalian prion proteins and other amyloidogenic precursors [93,102].

Wickner [103,104] first proposed that parallels could be drawn between current notions of mammalian prion infectivity and the curious mode of inheritance of both [PSI$^+$] and another yeast extrachromosomal element, [URE3]. Indeed, the idea that a self-propagating conformational change in Sup35 is the basis for the inheritance of the [PSI$^+$] element lends an astonishing coherence to several disparate observations that had long puzzled investigators.

1. Mutations of the *SUP35* gene, which cause a reduction in translation termination efficiency, mimicking the phenotype associated with [PSI$^+$], are inherited in a simple mendelian fashion [99]. In contrast, the ability of [PSI$^+$] cytoplasm to "transmit" the [PSI$^+$] element to [psi$^-$] strains is analogous to the transmission of prion disease to healthy animals by exposure to prion particles derived from diseased animals.
2. Overexpression of Sup35 induces the formation of [PSI$^+$] [105], presumably by increasing the probability of spontaneous conformational change that becomes self-propagating in the absence of continued overexpression. Likewise, overexpression of PrP induces neuromuscular degeneration associated with prion–protein deposition in transgenic mice [106].
3. Cells containing a deletion of the NH_2-terminal domain of Sup35 are resistant to [PSI$^+$] [101,107] just as deletion of *PrP* gene renders mice resistant to prion infection [108]. Furthermore, the NH_2-terminal region of Sup35 contains tandem repeats [109,110] strikingly similar to those in the mammalian prion protein (mouse PrP = PHGGGWGQ; Sup35 = PQGGYQQYN) which, when expanded, lead to increased susceptibility to inherited spontaneous prion disease [for review of inherited prion diseases see 111].

The prion hypothesis of [PSI$^+$] genesis is borne out by physical evidence. In [psi$^-$] cells, Sup35 is mostly soluble, whereas in isogenic [PSI$^+$] strains, Sup35 is found in high molecular weight, protease-resistant aggregates [93,102]. The simplest mechanistic explanation for the suppression of nonsense codons afforded by these results is that aggregation of Sup35 reduces the level of freely diffusable translation termination factor to the extent that read-through of stop codons occurs with increased frequency.

Amyloid fibrils are postulated to grow by the capture of amyloidogenic protomers undergoing folding transitions and stabilizing the "abnormal" conformation by incorporating the protomer at the end of the growing fibril [112]. The ability of preexisting amyloid structures to stabilize the aberrant conformation explains the apparent self-propagating nature of prions and other amyloids. The self-propagating nature of the Sup35 aggregate is supported by the observation that Sup35 derived from lysates of [PSI$^+$] but not [psi$^-$] strains can bind to an immobilized Sup35–GST fusion protein in vitro [102]. Likewise, a protein consisting of the NH_2-terminal prion-determining domain (NPD) of Sup35 fused to the green fluorescent protein (NPD–GFP) coalesces into foci in [PSI$^+$] cells while remaining diffusely distributed in [psi$^-$] cells [93]. Presumably, daughter cells inherit [PSI$^+$] elements through the transfer of Sup35 aggregates as cytoplasm from the mother cell flows into the bud.

B. Hsp104 Is Required for [PSI$^+$] Propagation and Maintenance

An exciting development in our understanding of the [PSI$^+$] element came with the identification of *Hsp104* as an extracopy "antisuppressor" of [PSI$^+$] [113]. Overexpression of Hsp104, even transiently, not only suppresses the [PSI$^+$] phenotype but actually "cures" the cells of [PSI$^+$]—cells become heritably [psi$^-$]. However, if Hsp104 concentration is reduced to its normal constitutive level, these cells can be reconverted to [PSI$^+$] by overexpressing Sup35. Elimination of Hsp104 expression by gene disruption also cures cells of [PSI$^+$], but now they become "psi-no-more"—[PSI$^+$] cannot be propagated even by overexpression of Sup35. Furthermore, when Sup35 solubility is examined under various regimens of Hsp104 expression, changes in the [PSI$^+$] status of cells coincide with heritable changes in the aggregation state of Sup35 [93,102].

Current information about the propagation, maintenance, and curing of [PSI$^+$] can be synthesized into a coherent model (Fig. 5). In this model, de novo [PSI$^+$] formation depends on an intermediate level of Hsp104 expression, reflecting a requirement for a chaperone-mediated transition between a stably folded conformation of Sup35 that is active in translation termination and a metastable, aggregation-prone conformer. In the absence of preexisting prion particles, the probability that aggregation actually proceeds is low and the metastable conformer readily reverts to its normal conformation. Overexpression of Sup35 increases the concentration of the prion intermediate, raising the overall probability that "seed" aggregates will be formed. Once the aggregation process is nucleated, newly synthesized Sup35 is rapidly integrated into the [PSI$^+$] element. Extension of this hypothesis argues that elimination of Hsp104 decreases the chances for the formation of the metastable prion protomer and eventually eliminates preexisting [PSI$^+$] elements by attrition and by preventing the formation of new [PSI$^+$] elements.

It is not yet clear how increasing the level of Hsp104 relative to that of Sup35 may contribute to the curing of [PSI$^+$]. Consistent with the dissaggregating function of Hsp104 in recovery from thermal stress, overexpression of Hsp104 may simply promote the dissolution of the Sup35 aggregate. The relatively short duration of Hsp104 overexpression required for the curing of [PSI$^+$] seems to favor this possibility. However, Hsp104 has an extended half-life that may be sufficient to allow other processes to play a role. For example, Hsp104 at an elevated concentration may compete with the [PSI$^+$] element for the binding of Sup35 conformers. Whether the curing of [PSI$^+$] occurs by a rapid or slow mechanism is difficult to determine, for assessment of the [PSI$^+$] status of cells requires several generations of growth on indicator plates. Because the aggregation state of the GFP–NPD fusion protein corresponds to the [PSI$^+$] status of the cells in which it is expressed, the observation of changes in the distribution of GFP-tagged prion particles in living cells during treatments that either induce the formation or loss of [PSI$^+$] may help distinguish between rapid and slow processes.

The acute dependence of [PSI+] propagation on a relatively narrow range of Hsp104 concentrations suggests that Hsp104 may play a more direct role in the assembly of aggregation prone intermediates into prion particles. When the ratio of Sup35 to Hsp104 is favorable, the effective local concentration of these intermediates would be increased by simultaneous binding of two or more destabilized monomers on an individual Hsp104 hexamer and, by their proximity to one another, promote their aggregation on release. Overexpression of Hsp104 would create a Sup35/Hsp104 ratio unfavorable to these encounters and effectively eliminate the propagation and maintenance of [PSI+] elements. The dominant character of ATP-binding domain mutants of Hsp104 relative to the suppression and curing of [PSI+] [113] can be explained by this model if the assembly of mutant and wild-type Hsp104 subunits into mixed oligomers has the effect of reducing the overall number and altering the distribution of productive-binding surfaces within individual Hsp104 rings. Under such circumstances the effect of promoting aggregation by bringing intermediates into proximity is lost, and newly synthesized Sup35 molecules tend to remain soluble.

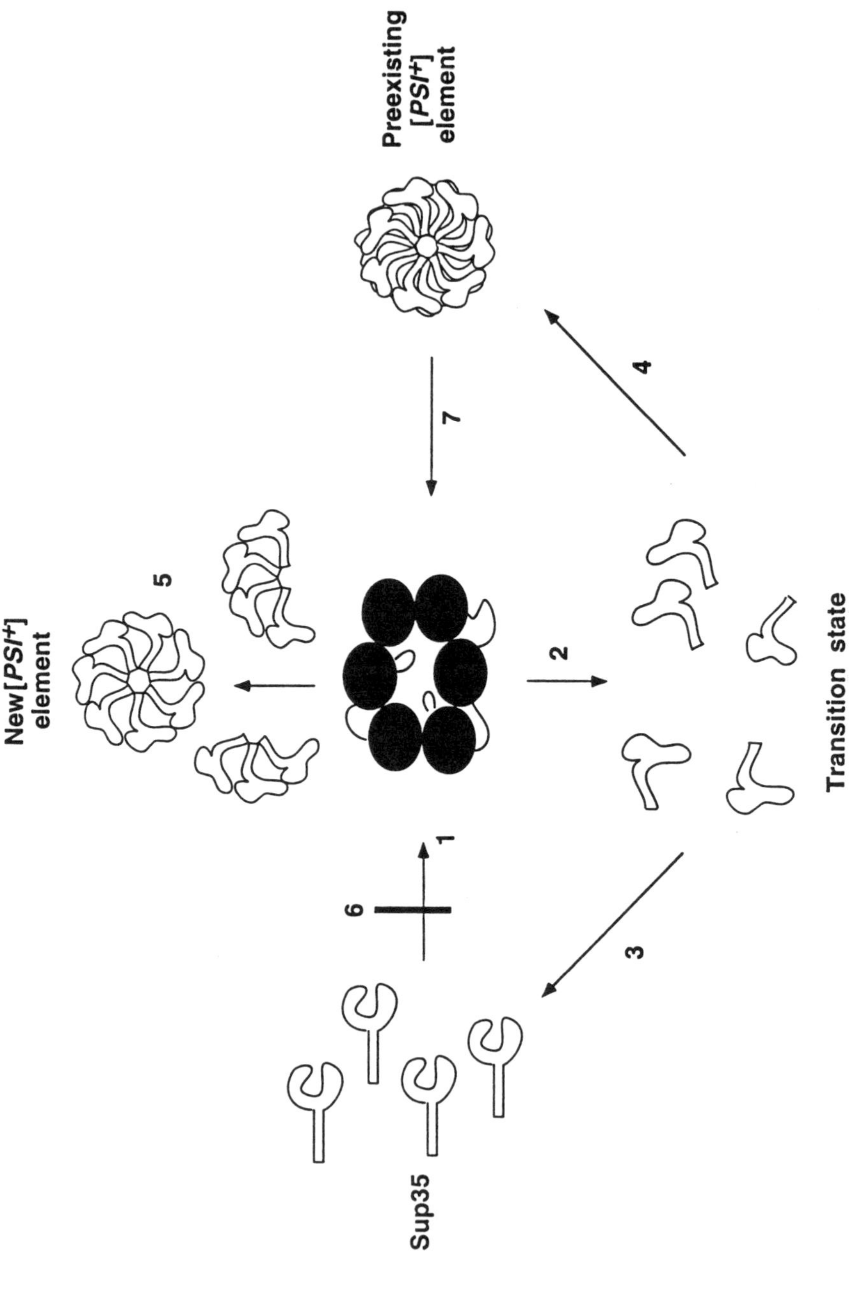
Preexisting
[PSI+]
element
New[PSI+]
element
Transition state
Sup35
1
2
3
4
5
6
7

X. FUTURE PROSPECTS

We are still at an early stage in understanding how Hsp104 and other Hsp100–Clp proteins accomplish their various functions. A unique capacity of the Hsp100 protein is to alter the conformation of proteins that are already stably assembled or aggregated to the extent that they become accessible to other molecular chaperones or proteases. The ability to overcome multiple stabilizing forces within these assemblages suggests an active, coordinated pulling or tugging motion applied to substrate molecules that contrasts sharply with the simple binding and release mechanism suggested for other folding catalysts. It is clear that ATP is required for this activity, and it is possible that motion within the Hsp100 oligomer may be coordinated by ATP hydrolysis. The key steps in unraveling the molecular mechanism of Hsp100–Clp chaperone function will likely involve a detailed description of the atomic structure of the proteins themselves and collection of complementary data concerning the conformational dynamics of Hsp100s in response to nucleotide binding, hydrolysis, and exchange, and in their interactions with substrates and partner proteins.

The ability to modulate the stress tolerance of organisms through the introduction of new genes or the harnessing restorative power of existing genes has important implications for biotechnology, agriculture, and medicine. A crucial first step toward this goal is to identify genes that would be most effective in achieving dramatic gains in stress resistance, without perturbing normal growth and development. The impressive role that Hsp104 plays in yeast thermotolerance, even when only modestly expressed in the absence of other heat-inducible factors [65], suggests that Hsp100–Clp proteins may be ideal candidates. The function of plant cytosolic Hsp100s in complementing the thermotolerance defect caused by *hsp104* mutations in yeast indicates that the intrinsic capacity of Hsp100s to interact with substrates and cofactors is highly conserved [21,22]. Engineering heat tolerance by increasing the complement of endogenous Hsp100 genes, or heterologously expressing yeast Hsp104 in transgenic plant varieties,

Figure 5 Hypothetical model of the interaction of Hsp104 and Sup35 in the formation of the yeast prion-like element [PSI$^+$]: (1) Newly synthesized Sup35 protomers interact with Hsp104 hexamers. (2) Sup35 polypeptides are released from Hsp104 in a metastable transition state. (3) When concentration of Sup35 is low and in the absence of [PSI$^+$] Sup35 spontaneously reverts to its functional state. (4) In presence of preexisting [PSI$^+$] transition state conformers are incorporated into aggregates. (5) When Sup35 is overexpressed the concentration of transition state conformers is sufficient to induce formation of new [PSI$^+$] elements. (6) In the absence of Hsp104 conversion to transition state conformation is blocked and [PSI$^+$] cannot be propagated or maintained. (7) Overexpression of Hsp104 may disaggregate [PSI$^+$] prions and cause reversion to [psi$^-$]. See text for expanded discussion of various aspects of this model. (From Ref. 93.)

may allow the expansion of crop production into areas of marginal climate as well as provide a buffer against the incertitudes of weather in temperate zones.

There has been considerable thought given to the notion that chaperones may facilitate the conformational transitions required for the conversion of the normal form of the mammalian prion protein (PrP^{c}) to the abnormal aggregated form (PrP^{Sc}) [114]. Experimental evidence supporting a direct role of chaperones in this process is lacking, but tantalizing hints have emerged from a few studies that have shown that the pattern of distribution and induction of heat-shock proteins is altered in scrapie-infected tissue culture cells [115], that the development of prion disease in transgenic mice is dependent on a proposed factor X [116], and that segments of PrP interact with chaperonin Hsp60 both in vivo and in vitro [117]. The linkage of prion formation and "curing" with molecular chaperones has been firmly established in yeast [93,102]. It is clear that Sup35 and Hsp104 can undergo a biologically relevant interaction because they occur in the same subcellular compartment. Although it is also clear that at various stages of their existence, amyloidogenic proteins are likely exposed to interactions with molecular chaperones in one compartment or another, there is no clear evidence that these potential interactions are involved in amyloid pathogenesis.

Even though the number of Hsp100–Clp proteins that have been sequenced has exploded over the last year, only a single mammalian homologue has been sequenced in its entirety [10]. Several partial sequences of apparent human homologues have appeared in databases, courtesy of the Human Genome Sequencing Project. Furthermore, a human ClpP homologue has been cloned [118], suggesting that human homologues to the ATPase components of prokaryotic Ti/Clp protease also exist. What roles these proteins might play in normal and stressed cells and tissues are, for now, simply fodder for speculation. For example, there are some promising indications that manipulation of Hsp70 levels may be used as a means of minimizing damage to tissues caused by ischemic injury [119]. Hypothetically, the positive effects of Hsp70s might be further enhanced by concomitant expression of an Hsp104-like functionality. Exploration of the fundamental properties of the members of the Hsp100 family and how these might be exploited represents a new frontier in the molecular chaperone field, and we hope, one with many more surprises yet to come.

REFERENCES

1. Kabakov, A. E. and Gabai, V. L., Protein aggregation as a primary and characteristic cell reaction to various stresses, *Experientia 49*:706 (1993).
2. Kampinga, H. H., Thermotolerance in mammalian cells: Protein denaturation and aggregation, and stress proteins, *J. Cell Sci. 104*:11 (1993).
3. Ellis, Proteins as molecular chaperones, *Nature 328*:378 (1987).
4. Gottesman, S., Squires, C., Pichersky, E., Carrington, M., Hobbs, M., Mattick, J. S.,

Dalrymple, B., Kuramitsu, H., Shiroza, T., and Foster, T., Conservation of the regulatory subunit for the Clp ATP-dependent protease in prokaryotes and eukaryotes, *Proc. Natl. Acad. Sci. USA 87*:3513 (1990).

5. Schirmer, E. C., Glover, J. R., Singer, M. A., and Lindquist, S., HSP100/Clp proteins: A common mechanism explains diverse functions, *Trends Biochem. Sci. 21*:289 (1996).
6. Walker, J. E., Saraste, M., Runswick, M. J., and Gay, N. J., Distantly related sequences in the α- and β-subunits of ATP synthase, myosin, kinases and other ATP-requiring enzymes and a common nucleotide binding fold, *EMBO J. 1*:945 (1982).
7. Yoshida, Y. and Amano, R., A common topology of proteins catalyzing ATP-triggered reactions, *FEBS Lett. 359*:1 (1995).
8. Saraste, M., Sibbald, P. R., and Wittinghofer, A., The P-loop—a common motif in ATP- and GTP-binding proteins, *Trends Biochem. Sci. 15*:430 (1990).
9. Leonhardt, S. A., Fearson, K., Danese, P. N., and Mason, T. L., *HSP78* encodes a yeast mitochondrial heat shock protein in the Clp family of ATP-dependent proteases, *Mol. Cell. Biol. 13*:6304 (1993).
10. Perier, F., Radeke, C. M., Raab-Graham, K. F., and Vandenberg, C. A., Expression of a putative ATPase suppresses the growth defect of a yeast potassium transport mutant: Identification of a mammalian member of the Clp/HSP104 family, *Gene 152*:157 (1995).
11. Horwich, A. L., Resurrection or destruction, *Curr. Biol. 5*:455 (1995).
12. Clarke, A. K., Variations on a theme: Combined molecular chaperone and proteolysis functions in Clp/Hsp100 proteins, *J. Biosci. 21*:161 (1996).
13. Katayama-Fujimura, Y., Gottesman, S., and Maurizi, M. R., A multiple-component, ATP-dependent protease from *Escherichia coli*, *J. Biol. Chem. 262*:4477 (1987).
14. Gottesman, S., Clark, W. P., and Maurizi, M. R., The ATP-dependent Clp protease of *Escherichia coli*: Sequence of *clpA* and identification of a Clp-specific substrate, *J. Biol. Chem. 265*:7886 (1990).
15. Tobias, J. W., Shrader, T. E., Rocap, G., and Varshavsky, A., The N-end rule in bacteria, *Science 254*:1374 (1991).
16. Wickner, S., Hoskins, J., and McKenney, K., Monomerization of RepA dimers by heat shock proteins activates binding to DNA replication origin, *Proc. Natl. Acad. Sci. USA 88*:7903 (1991).
17. Wickner, S., Gottesman, S., Skowyra, D., Hoskins, J., McKenney, K., and Maurizi, M. R., A molecular chaperone, ClpA, functions like DnaK and DnaJ, *Proc. Natl. Acad. Sci. USA 91*:12218 (1994).
18. Sanchez, Y, and Lindquist, S. L., HSP104 is required for induced thermotolerance, *Science 248*:1112 (1990).
19. Squires, C. L., Pedersen, S., Ross, B. M., and Squires, C., ClpB is the *Escherichia coli* heat shock protein F84.1, *J. Bacteriol. 173*:4254 (1991).
20. Sanchez, Y., Taulien, J., Borkovich, K. A., and Lindquist, S., Hsp104 is required for tolerance to many forms of stress, *EMBO J. 11*:2357 (1992).
21. Lee, Y.-R. J., Nagao, R. T., and Key, J. L., Soybean 100-kD heat shock protein complements a yeast *hsp104* deletion mutant in the acquisition of thermotolerance, *Plant Cell 6*:1889 (1994).

22. Schirmer, E. C., Lindquist, S., and Vierling, E., An *Arabidopsis* heat shock protein complements a thermotolerance defect in yeast, *Plant Cell 6*:1899 (1994).
23. Hubel, A., Brandau, S., Dresel, A., and Clos, J., A member of the ClpB family of stress proteins is expressed during heat shock in *Leishmania* spp., *Mol. Biochem. Parasitol. 70*:107 (1995).
24. Eriksson, M.-J. and Clarke, A. K., The heat shock protein ClpB mediates the development of thermotolerance in the cyanobacterium *Synechococcus* sp. strain PCC 7924, *J. Bacteriol. 178*:4839 (1996).
25. Schmitt, M., Neupert, W., and Langer, T., The molecular chaperone Hsp78 confers compartment specific thermotolerance to mitochondria, *J. Cell Biol. 134*:1375 (1996).
26. Kruger, E., Volker, U., and Hecker, M., Stress induction of ClpC in *Bacillus subtilis* and its involvement in stress tolerance, *J. Bacteriol. 176*:3360 (1994).
27. Rouquette, C., Ripio, M.-T., Pellegrini, E., Bolia, J.-M., Tascon, R. I., Vazquez-Boland, J.-A., Berche, P., Identification of a ClpC ATPase required for stress tolerance and in vivo survival of *Listeria monocytogenes*, *Mol. Microbiol. 21*:977 (1996).
28. Kong, L. and Dubnau, D., Regulation of competence-specific gene expression by Mec-mediated protein–protein interaction in *Bacillus subtilis*, *Proc. Natl. Acad. Sci. USA 91*:5793 (1994).
29. Msadek, T., Kunst, F., and Rapoport, G., MecB of *Bacillus subtilis*, a member of the ClpC ATPase family, is a pleiotropic regulator controlling competence gene expression and growth at high temperature, *Proc. Natl. Acad. Sci. USA 91*:5788 (1994).
30. Rashid, M. H., Tamakoshi, A., and Sekiguchi, J., Effects of *mecA* and *mecB* (*clpC*) mutations on the expression of *sigD*, which encodes an alternative sigma factor, and autolysin operons and on flagellin synthesis in *Bacillus subtilis*, *J. Bacteriol. 178*:4861 (1996).
31. Shanklin, J., DeWitt, N. D., and Flanagan, J. M., The stroma of higher plants plastid contain ClpP and ClpC, functional homologues of *Escherichia coli* ClpP and ClpC: An archetypal ATP-dependent protease, *Plant Cell 7*:1713 (1995).
32. Clarke, A. K. and Eriksson, M.-J., The cyanobacterium *Synechococcus* sp. PCC 7942 possesses a close homologue to the chloroplast ClpC protein of higher plants, *Plant Mol. Biol. 31*:721 (1996).
33. Kiyosue, T., Yamaguchi-Shinozaki, K., and Shinozaki, K., Characterization of cDNA for a dehydration-inducible gene that encodes a Clp A, B-like protein in *Arabidopsis thaliana* L, *Biochem. Biophys. Res. Commun. 196*:1214 (1993).
34. Wilson, S. A., Williams, R. J., Pearl, L. H., and Drew, R. E., Identification of two new genes in the *Pseudomonas aeruginosa* amidase operon, encoding an ATPase (AmiB) and a putative integral membrane protein (AmiS), *J. Biol. Chem. 270*:18818 (1995).
35. Gottesman, S., Clark, W. P., de Crecy-Lagard, V., and Maurizi, M. R., ClpX, an alternative subunit for the ATP-dependent Clp protease of *Escherichia coli*: Sequence and in vivo activities, *J. Biol. Chem. 268*:22618 (1993).
36. Highlander, S. K., Wickersham, E. A., Garza, O., and Weinstock, G. M., Expression of the *Pasteurella haemolytica* leukotoxin is inhibited by a locus that encodes a ATP-binding cassette homolog, *Infect. Immun. 61*:3942 (1993).

37. Slack, F. J., Mueller, J. P., and Sonenshein, A. L., Mutations that relieve nutritional repression of the *Bacillus subtilis* dipeptide permease operon, *J. Bacteriol. 175*:4605 (1993).
38. Mhammedi, A. A., Pato, M., Gama, M. J., and Toussaint, A., A new component of bacteriophage Mu replicative transposition machinery: The *Escherichia coli* ClpX protein, *Mol. Microbiol. 11*:1109 (1994).
39. Lehnherr, H. and Yarmolinsky, M. B., Addiction protein Phd of plasmid prophage P1 is a substrate of the ClpXP serine protease of *Escherichia coli*, *Proc. Natl. Acad. Sci. USA 92*:3274 (1995).
41. Frank, E. G., Ennis, D. G., Gonzalez, M., Levine, A. S., and Woodgate, R., Regulation of SOS mutagenesis by proteolysis, *Proc. Natl. Acad. Sci. USA 93*:10291 (1996).
42. Schweder, T., Lee, K.-H., Lomovskaya, O., and Matin, A., Regulation of *Escherichia coli* starvation sigma factor (σ^s) by ClpXP protease, *J. Bacteriol. 178*:470 (1996).
43. Levchenko, I., Luo, L., and Baker, T. A., Disassembly of the Mu transposase tetramer by the ClpX chaperone, *Genes Dev. 9*:2399 (1995).
44. Wawrzynow, A., Wojtkowiak, D., Marszalek, J., Banecki, B., Jonsen, M., Graves, B., Georgopoulos, C., and Zylicz, M., The ClpX heat-shock protein of *Escherichia coli*, the ATP-dependent substrate specificity component of the ClpP–ClpX protease, is a novel molecular chaperone, *EMBO J. 14*:1867 (1995).
45. Kruklitis, R., Welty, D. J., and Nakai, H., ClpX protein of *Escherichia coli* activates bacteriophage Mu transposase in the strand transfer complex for initiation of Mu DNA synthesis, *EMBO J. 15*:935 (1996).
46. Maurizi, M. R., Thompson, M. W., Singh, S. K., and Kim, S.-H., Endopeptidase Clp: ATP-dependent Clp protease from *Escherichia coli*, *Methods Enzymol. 244*:314 (1994).
47. Woo, K. M., Kim, K. I., Goldberg, A. L., Ha, D. B., and Chung, C. H., The heat-shock protein ClpB in *Escherichia coli* is a protein-activated ATPase, *J. Biol. Chem. 267*:20429 (1992).
48. Schirmer, E. C., Queitsch, C., Kowal, A. S., and Lindquist, S., Evidence for concerted hydrolysis in Hsp104 hexamers (submitted).
49. Parsell, D. A., Kowal, A. S., and Lindquist, S., *Saccharomyces cerevisiae* Hsp104 protein: Purification and characterization of ATP-induced structural changes, *J. Biol. Chem. 269*:4480 (1994).
50. Singh, S. K. and Maurizi, M. R., Mutational analysis demonstrates different functional roles for the two ATP-binding sites in ClpAP protease from *Escherichia coli*, *J. Biol. Chem. 269*:29537 (1994).
51. Seol, J. H., Baek, S. H., Kang, M. S., Ha, D. B., and Chung, C. H., Distinctive roles of the two ATP-binding sites in ClpA, the ATPase component of protease Ti in *Escherichia coli*, *J. Biol. Chem. 270*:8087 (1995).
52. Kessel, M., Maurizi, M. R., Kim, B., Kocsis, E., Trus, B. L., Singh, S. K., and Steven, A. C., Homology in structural organization between *E. coli* ClpAP protease and the eukaryotic 26 S proteasome, *J. Mol. Biol. 250*:587 (1995).
53. Kim, S., Willison, K. R., and Horwich, A. R., Cytosolic chaperonin subunits have

a conserved ATPase domain but divergent polypeptide-binding domains, *Trends Biochem. Sci. 19*:543 (1994).
54. Langer, T., Pfeifer, G., Martin, J., Baumeister, W., and Hartl, F.-U., Chaperonin mediated protein folding: GroES binds to one end of the GroEL cylinder which accommodates the protein substrate in its central cavity, *EMBO J. 11*:4757 (1992).
55. Kobayashi, N. and McEntee, K., Identification of *cis* and *trans* components of a novel heat shock stress regulatory pathway in *Saccharomyces cerevisiae*, *Mol. Cell Biol. 13*:248 (1993).
56. Schmitt, A. P. and McEntee, K., Msn2p, a zinc finger DNA-binding protein, is the transcriptional activator of the multistress response in *Saccharomyces cerevisiae*, *Proc. Natl. Acad. Sci. USA 93*:5777 (1996).
57. Mager, W. H. and De Kruijff, A. J. J., Stress-induced transcriptional activation, *Microbiol. Rev. 59*:506 (1995).
58. Ruis, H. and Schuller, C., Stress signalling in yeast, *BioEssays 17*:959 (1995).
59. Piper, P. W., The heat shock and ethanol responses of yeast exhibit extensive similarity and functional overlap, *FEMS Microbiol. Lett. 134*:121 (1995).
60. Moradas-Ferreira, P., Costa, V., Piper, P., and Mager, W., The molecular defences against reactive oxygen species in yeast, *Mol. Microbiol. 19*:651 (1996).
61. Stohs, S. J. and Bagchi, D., Oxidative mechanisms in the toxicity of metal ions, *Free Radical Biol. Med. 18*:321 (1995).
62. Arnold, C. E. and Wittrup, K. D., The stress response to loss of signal recognition particle function in *Saccharomyces cerevisiae*, *J. Biol. Chem. 269*:30412 (1994).
63. Parsell, D. A. and Lindquist, S., Heat shock proteins and stress tolerance, *The Biology of Heat Shock Proteins and Molecular Chaperones* (Morimoto, R. I., Tissieres, A., and Georgopoulos, C., eds.), Cold Spring Harbor Laboratory Press, Cold Spring Harbor, NY, 1994, p. 457.
64. Yancey, P. H., Clark, M. E., Hand, S. C., Bowlus, R. D., and Somero, G. N., Living with water stress: Evolution of osmolyte systems, *Science 217*:1214 (1982).
65. Lindquist, S. and Kim, G., Heat-shock protein 104 expression is sufficient for thermotolerance in yeast, *Proc. Natl. Acad. Sci. USA 93*:5301 (1996).
66. Parsell, D. A., Taulien, J., and Lindquist, S., The role of heat-shock proteins in thermotolerance, *Philos. Trans. R. Soc. Lond. 339*:279 (1993).
67. Finley, D., Sadis, S., Monia, B. P., Boucher, P., J., E. D., Crooke, S. T., and Chau, V., Inhibition of proteolysis and cell cycle progression in a multiubiquitination-deficient yeast mutant, *Mol. Cell. Biol. 14*:5501 (1994).
68. Chen, Y. and Piper, P. W., Consequences of the overexpression of ubiquitin in yeast: Elevated tolerances of osmostress, ethanol and canavanine, yet reduced tolerances of cadmium, arsenite and paromomycin, *Biochim. Biophys. Acta 1268*:59 (1995).
69. Lindquist, S., Patino, M. M., Chernoff, Y. O., Kowal, A. S., Singer, M. A., Liebman, S. W., Lee, K.-H., and Blake, T., The role of Hsp104 in stress tolerance and [PSI^+] propagation in *Saccharomyces cerevisiae*, *Cold Spring Harbor Symp. Quant. Biol.*, p. 451 (1995).
70. Welte, M. A., Tetrault, J. M., Dellavalle, R. P., and Lindquist, S. L., A new method for manipulating transgenes: Engineering heat tolerance in a complex, multicellular organism, *Curr. Biol. 3*:824 (1993).
71. Feder, M. E., Cartano, N. V., Milos, L., Krebs, R. A., and Lindquist, S. L., Effect of

engineering *hsp70* expression and tolerance of ecologically relevant heat shock in larvae and pupae of *Drosophila melanogaster*, *J. Exp. Biol. 199*:1837 (1996).

72. Li, G. C., Li, L. G., Liu, Y. K., Mak, J. Y., Chen, L. L., and Lee, W., Thermal response of rat fibroblasts stably transfected with human 70-kDa heat shock protein-encoding gene, *Proc. Natl. Acad. Sci. USA 88*:1681 (1991).
73. Craig, E. A., Baxter, B. K., Becker, J., Halladay, J., and Ziegelhoffer, T., Cytosolic hsp70s of *Saccharomyces cerevisiae*: Roles in protein synthesis, protein translocation, proteolysis, and regulation, *The Biology of Heat Shock Proteins and Molecular Chaperones* (Morimoto, R. I., Tissieres, A., and Georgopoulos, C., eds.), Cold Spring Harbor Laboratory Press, Cold Spring Harbor, NY, 1994, p. 31.
74. Werner-Washburne, M., Stone, D. E., and Craig, E. A., Complex interactions among members of an essential subfamily of *hsp70* genes in *Saccharomyces cerevisiae*, *Mol. Cell. Biol. 7*:2568 (1987).
75. Sanchez, Y., Parsell, D. A., Taulien, J., Vogel, J. L., Craig, E. A., and Lindquist, S., Genetic evidence for a functional relationship between Hsp104 and Hsp70, *J. Bacteriol. 175*:6484 (1993).
76. Deschaies, R. J., Koch, B. D., Werner,-Washburne, M., Craig, E. A., and Scheckman, R., A subfamily of stress proteins facilitates translocation of secretory and mitochondrial proteins, *Nature 332*:800 (1988).
77. De Virgilio, C., Hottiger, T., Dominguez, J., Boller, T., and Wiemken, A., The role of trehalose for the acquisition of thermotolerance in yeast. I. Genetic evidence that trehalose is a thermal protectant, *Eur. J. Biochem. 219*:179 (1994).
78. Attfield, P. V., Kletsas, S., and Hazell, B. W., Concomitant appearance of intrinsic thermotolerance and storage of trehalose in *Saccharomyces cerevisiae* during early respiratory phase of batch-culture is CIF1-dependent, *Micobiology 140*:2625 (1994).
79. De Virgilio, C., Piper, P., Boller, T., and Wiemken, A., Acquisition of thermotolerance in *Saccharomyces cerevisiae* without heat shock protein hsp104 and in the absence of protein synthesis, *FEBS Lett. 288*:86 (1991).
80. Elliot B., Haltiwanger, R. S., and Futcher, B., Synergy between trehalose and Hsp104 for thermotolerance in yeast, *Genetics 144*:923 (1996).
81. Hottiger, T., De Virgilio, C., Hall, M. N., Boller, T., and Wiemken, A., The role of trehalose for the acquisition of thermotolerance in yeast. II. Physiological concentrations of trehalose increase the thermostability of proteins in vitro, *Eur. J. Biochem. 219*:187 (1994).
82. Parsell, D. A., Kowal, A. S., Singer, M. A., and Lindquist, S., Protein disaggregation mediated by heat-shock protein Hsp104, *Nature 372*:475 (1994).
83. Park, S. K., Kim, K. I., Woo, K. M., Seol, J. H., Tanaka, K., Ichihara, A., Ha, D. B., and Chung, C. H., Site-directed mutagenesis of the dual translational initiation sites of the *clpB* gene of *Escherichia coli* and characterization of its gene products, *J. Biol. Chem. 268*:20170 (1993).
84. Seol, J. H., Yoo, S. J., Kim, K. I., Kang, M.-S., Ha, D. B., and Chung, C. H., The 65-kDa protein derived from the internal translational initiation site of the ClpA gene inhibits the ATP-dependent protease Ti in *Escherichia coli*, *J. Biol. Chem. 269*:29468 (1994).
85. Schmitt, M., Neupert, W., and Langer, T., Hsp78, a Clp homologue within mito-

chondria, can substitute for chaperone functions of mt-hsp70, *EMBO J. 14*:3434 (1995).

86. Moczko, M., Schonfisch, B., Voos, W., Pfanner, N., and Rassow, J., The mitochondrial ClpB homolog Hsp78 cooperates with matrix Hsp70 in maintenance of mitochondrial function, *J. Mol. Biol. 254*:538 (1995).
87. Skowyra, D., Georgopoulos, C., and Zylicz, M., The *E. coli dnaK* gene product, the hsp70 homolog, can reactivate heat-inactivated RNA polymerase in an ATP hydrolysis-dependent manner, *Cell 62*:939 (1990).
88. Vogel, J. L., Parsell, D. A., and Lindquist, S., Heat-shock proteins Hsp104 and Hsp70 reactivate mRNA splicing after heat inactivation, *Curr. Biol. 5*:306 (1995).
89. Bardwell, J. C. and Craig, E. A., Major heat shock gene of *Drosophila* and the *Escherichia coli* heat-inducible *dnaK* gene are homologous, *Proc. Natl. Acad. Sci. USA 81*:848 (1984).
90. Hwang, B. J., Park, W. J., Chung, C. H., and Goldberg, A. L., *Escherichia coli* contains a soluble ATP-dependent protease (Ti) distinct from protease La, *Proc. Natl. Acad. Sci. USA 84*:5550 (1987).
91. Katayama, Y., Gottesman, S., Pumphery, J., Rudikoff, S., Clark, W. P., and Maurizi, M. R., The two-component, ATP-dependent Clp protease of *Escherichia coli*, *J. Biol. Chem. 263*:15226 (1988).
92. Wojtkowiak, D., Georgopoulos, C., and Zylicz, M., Isolation and characterization of ClpX, a new ATP-dependent specificity component of the Clp protease of *Escherichia coli*, *J. Biol. Chem. 268*:22609 (1993).
93. Patino, M. M., Liu, J.-J., Glover, J. R., and Lindquist, S., Support for the prion hypothesis for inheritance of a phenotypic trait in yeast, *Science 273*:622 (1996).
94. Parsell, D. A., Sanchez, Y., Stitzel, J. D., and Lindquist, S., Hsp104 is a highly conserved protein with two essential nucleotide-binding sites, *Nature 353*:270 (1991).
95. Inoue, I. and Rechsteiner, M., On the relationship between the metabolic and thermodynamic stabilities of T4 lysozymes, *J. Biol. Chem. 269*:29241 (1994).
96. Baker, T. A., Protein-DNA assemblies controlling lytic development of bacteriophage Mu, *Curr. Opin. Genet. Dev. 3*:708 (1993).
97. Geuskens, V., Mhammedi-Alaoui, A., Desmet, L., and Toussaint, A., Virulence in bacteriophage Mu: A case of *trans*-dominant proteolysis by the *Escherichia coli* Clp serine protease, *EMBO J. 11*:5121 (1992).
98. Cox, B. S., ψ, a cytopolasmic suppressor or super-suppressor in yeast, *Heredity 20*:505 (1965).
99. Chernoff, Y. O., Inge-Vechtomov, S. G., Derkach, I. L., Ptyushkina, M. V., Tarunina, O. V., Dagkesamanskaya, A. R., and Ter-Avanesyan, M. D., Dosage dependent translational suppression in yeast *Saccharomyces cerevisiae*, *Yeast 8*:489 (1992).
100. Stansfield, I., Jones, K. M., Kushnirov, V. V., Dagkesamanskaya, A. R., Poznyakovski, A. I., Paushkin, S. V., Nierras, C. R., Cox, B. S., Ter-Avanesyan, M. D., and Tuite, M. F., The products of the *SUP35* (eRFl) and *SUP45* genes interact to mediate translation termination in *Saccharomyces cerevisiae*, *EMBO J. 14*:4365 (1995).
101. Ter-Avanesyan, M. D., Kushmirov, V. V., Dagkesamanskaya, A. R., Didichenko, S. A., Chernoff, Y. O., and Inge-Vechtomov, S. G., Deletion analysis of the *sup35* gene

of the yeast *Saccharomyces cerevisiae* reveals two nonoverlapping functional regions in the encoded protein, *Mol. Microbiol. 7*:683 (1993).

102. Paushkin, S. V., Kushnirov, V. V., Smirnov, V. N., and Ter-Avanesyan, M. D., Propagation of the yeast prion-like [*Psi*$^+$] determinant is mediated by oligomerization of the *SUP35*-encoded polypeptide release factor, *EMBO J. 15*:3127 (1996).
103. Wickner, R. B., [URE3] as an altered URE2 protein: Evidence for a prion analog in *Saccharomyces cerevisiae*, *Science 264*:566 (1994).
104. Wickner, R. B., Masison, D. C., and Edskes, H. K., [PSI] and [URE3] as yeast prions, *Yeast 11*:1671 (1995).
105. Chernoff, Y. O., Derkach, I. L., and Inge-Vechtomov, S. G., Multicopy *SUP35* gene induces de novo appearance of *psi* like factors in the yeast *Saccharomyces cerevisiae*, *Curr. Genet. 24*:268 (1993).
106. Westaway, D., Dearmond, S. J., Cayetano-Canlas, J., Groth, D., Foster, D., Yang, S. L., Torchia, M., Carlson, G. A., and Prusiner, S. B., Degeneration of skeletal muscle, periferal nerves, and the central nervous system in transgenic mice overexpressing wild-type prion proteins, *Cell 76*:117 (1994).
107. Doel, S. M., McCready, S. J., Nierras, C. R., and Cox, B. S., The dominant *PNM2*–mutation that eliminates the [PSI] factor of *Saccharomyces cerevisiae* is the result of a missense mutation in the *SUP35* gene, *Genetics 137*:659 (1994).
108. Prusiner, S. B., Groth, D., Serban, A., Koehler, R., Foster, D., Torchia, M., Burton, D., Yang, S. L., and DeArmond, S. J., Ablation of the prion protein (PrP) gene in mice prevents scrapie and facilitates production of anti-PrP antibodies, *Proc. Natl. Acad. Sci. USA 90*:10608 (1993).
109. Cox, B., Prion-like factors in yeast, *Curr. Biol. 4*:744 (1994).
110. Tuite, M. F. and Lindquist, S., The maintenance and inheritance of yeast prions, *Trends Genet. 12*:467 (1996).
111. Pruisner, S. B., Inherited prion diseases, *Proc. Natl. Acad. Sci. USA 91*:4611 (1994).
112. Lansbury, P. T. and Caughey, B., The chemistry of scrapie infection: Implications of the "ice 9" metaphor, *Chem. Biol. 2*:1 (1995).
113. Chernoff, Y. O., Lindquist, S. L., Ono-B-I., Inge-Vechtomov, S. G., and Liebman, S. W., Role of the chaperone protein Hsp104 in propagation of the yeast prion-like factor [*psi*$^+$], *Science 268*:808 (1995).
114. Kenwald, N., Landon, M., Laszlo, L., and Mayer, R. J., Heat shock proteins, molecular chaperones and the prion encephalopathies, *Cell Stress Chaperones 1*:18 (1996).
115. Tatzelt, J., Zuo, J., Voellmt, R., Scott, M., Hartl, F.-U., Puisiner, S. B., and Welch, W. J., Scrapie prions selectively modify the stress response in neuroblastoma cells, *Proc. Natl. Acad. Sci. USA 92*:2944 (1995).
116. Telling, G. C., Scott, M., Mastriani, J., Gabizon, R., Torchia, M., Cohen, F. E., Dearmond, S. J., Prusiner, S. B., Prion propagation in mice expressing human and chimeric PrP transgenes implicates the interaction of cellular PrP with another protein, *Cell 83*:79 (1995).
117. Edenhofer, F., Rieger, R., Famulok, M., Wendler, W., Weiss, S., and Winnacker, E.-L., Prion protein PrPc interacts with molecular chaperones of the Hsp60 family, *J. Virol. 70*:4724 (1996).
118. Bross, P., Andersen, B. S., Knudsen, I., Kruse, T. A., and Gregersen, N., Human

ClpP protease: cDNA sequence, tissue-specific expression and chromosomal assignment of the gene, *FEBS Lett. 377*:249 (1995).
119. Plumier, J.-C. L. and Currie, W. L., Heat shock induced myocardial protection against ischemic injury: A role for Hsp70? *Cell Stress Chaperones 1*:13 (1996).
120. ter Huurne, A. A. H. H., Muir, S., Van Houten, M., Van der Zeijst, B. A. M., Gaastra, W., and Kusters, J. G., Characterization of three putative *Serpulina hyodysenteriae* hemolysins, *Microb. Pathog. 16*:269 (1994).
121. Gottesman, S., Wu, W.-F., and Missiakas, D., *Escherichia coli* ClpY, *Guidebook to Molecular Chaperones and Protein Folding Factors* (M.- J. Gething ed., Sambrook and Tooze at Oxford University Press, Oxford), in press.
122. Missiakas, D., Schwager, F., Betton, J. M., Georgopoulus, C. Raina, S. Identification and characterization of HslV HslU (ClpQClpY) proteins involved in overall proteolysis of misfolded proteins in *Eschericia coli, EMBO J. 15*:6899 (1996).

10

Involvement of Hsp47 in the Folding and Processing of Procollagen in the Endoplasmic Reticulum

Kazuhiro Nagata, Mamoru Satoh, and Nobuko Hosokawa

Kyoto University, Kyoto, Japan

Andrew D. Miller

Imperial College of Science, Technology and Medicine, London, United Kingdom

I. INTRODUCTION

Newly synthesized membrane and secretory proteins are translocated across the endoplasmic reticulum (ER) membrane and are then transported to the Golgi apparatus for subsequent distribution to the cell surface, lysosomes, and secretory vesicles. Many of these proteins are not exported from the ER until they are folded and assembled correctly [1–3]. The ER contains several proteins that are termed either *molecular chaperones* or *folding enzymes*. The molecular chaperones are involved in the processing and maturation of secretory and membrane proteins. They are thought to associate with folding intermediates or misfolded proteins to prevent nonproductive side reactions, such as irreversible aggregation, and they may accelerate the slower steps in the folding process [1,2,4,5]. Of these proteins, certain stress proteins, such as glucose-regulated protein 78 (Grp78 or immunoglobulin heavy chain-binding protein [BiP]) and GRP94 are thought to facilitate translocation into the ER and then to assist in folding, oligomeric assembly, and sorting in the ER [6,7]. The folding enzymes, which include prolyl *cis–trans*-isomerase and protein disulfide isomerase (PDI), are involved in the posttranslational modification of proteins such as procollagen [8,9]. The sequential formation of disulfide bonds in the ER is required for normal folding and for subsequent transport to the Golgi apparatus and beyond [1,10]. These soluble ER-resident proteins have the sequence Lys-Asp-Glu-Leu (KDEL) at their C-terminus, which acts as an ER-retention signal [11]. Similarly, some ER-resident type I transmembrane proteins contain an ER-targeting motif consisting of two lysine residues at the C-terminus [12,13].

Hsp47 is another ER-resident protein, first identified as the major collagen-binding heat-inducible glycoprotein in fibroblasts [14]. It is characterized by its substrate specificity for collagen, and plays a major role in collagen processing and quality control under stress conditions, by preventing the secretion of procollagen with abnormal conformations [15].

II. MOLECULAR CHARACTERIZATION AND LOCALIZATION OF Hsp47

Hsp47 is a 47-kDa heat-inducible protein found in collagen-producing cells, where it functions as a collagen-binding glycoprotein [14,16–18]. Complementary DNA encoding Hsp47 was cloned from the cDNA library of chick cells [19]. Chick Hsp47 contains signal sequences at the NH_2-terminus and two glycosylation sites (Asn-X-Thr), both of which are glycosylated with high mannose-type sugar moieties [19–21]. The amino acid sequence of Hsp47 revealed that it has a homology with members of the serpin protein super-family, including α_1-antitrypsin, α_1-antichymotrypsin, antithrombin III, and ovalbumin. The serpin super-family comprises a group of proteins with similar tertiary structures and functional domains [22]. Hsp47 had the highest homology (31%) with human protein C inhibitor (hPCI) [23,26]. Unlike other serpin super-family proteins, Hsp47 contains an Arg-Asp-Glu-Leu (RDEL) sequence at its COOH-terminus. The RDEL sequence functions as the ER-retention signal, for the deletion of this sequence caused the secretion of the mutant Hsp47 out of the cells [21].

Immunofluorescence staining of chick embryo fibroblasts with anti-Hsp47 antibody revealed that the perinuclear network was stained but not the nucleus and the Golgi apparatus [17]. Typically, strip-like structures were seen by immune electronmicroscopy [17]. Thus, Hsp47 was shown to be an ER-luminal protein with an ER-retention signal at the COOH-terminus. Similar collagen-binding proteins were also reported as colligin and gp46 in the mouse teratocarcinoma cell line F9 and rat L6 myoblasts, respectively, and were later established to be identical with Hsp47 [24,25].

III. MOLECULAR STRUCTURE OF Hsp47

Because Hsp47 has the highest homology (31% sequence identity and 70% sequence similarity) hPCI [19], molecular models of two intact hPCI forms, the inhibitory and latent states, were generated using the X-ray crystal structure coordinates of proteolytically cleaved hPCI, and were used as templates for homology modeling of the inhibitory and latent states of Hsp47 [26]. Such homology modeling is an appropriate and effective method for determining the three-dimensional structure of a protein, provided that the three-dimensional structure of another protein that is related by at least a 25% amino acid sequence identity is available as a template [27]. As Hsp47 did not have inhibitory activities against several serine proteases including trypsin, thrombin, elastase, kallikrein, etc., the latent state model seemed to be more suitable for explaining the structure of Hsp47 [26].

The latent state model of Hsp47 was found to possess a long, deep cleft (seen in side view [Fig. 1a] and top view [Fig. 1b]) that appears to offer an explanation for its ability to bind specifically and tightly to procollagen and collagen peptides at standard physiological pH [28,29]. This cleft is a very promising procollagen–collagen peptide-binding groove for two main reasons. First, two deletion mutants of Hsp47, NΔ1 and CΔ3, which were unable to bind to collagen, have been prepared (Yokota et al., unpublished data). NΔ1 is missing the first NH_2-terminal 32 amino acids (a portion of α-helix in Fig. 1b) and CΔ3 the last COOH-terminal 34 amino acids (a portion of β-sheet in Fig. 1b). Since both deletion mutants of Hsp47 have no collagen-binding activity, this suggests the importance of the cleft in the interaction with collagen, for the deleted peptide lengths constitute substantial parts of this putative collagen-binding groove. Second, this putative-binding groove bears reasonable similarity to the peptide-binding regions of human class I histocompatibility glycoprotein HLA-Aw68 [30] and the human class II histocompatibility protein HLA-DR1 [31].

IV. ASSOCIATION OF Hsp47 WITH PROCOLLAGEN

To understand the function of Hsp47 in the ER, it is important to establish where it binds to procollagen and where it dissociates from procollagen. Pulse label and chase experiments, combined with immunoprecipitation by anticollagen and anti-Hsp47 antibodies, were adapted to examine the sites of its association with procollagen. Immediately after the α-chains of procollagen enter the ER, Hsp47 binds to these nascent α-chains [21]. When polysome fractions were isolated and immunoprecipitated with anti-Hsp47 antibody, nascent procollagen chains with various lengths were coprecipitated with Hsp47, which again suggested that it binds to nascent polypeptides of procollagen [32,33]. Hsp47 also binds to the

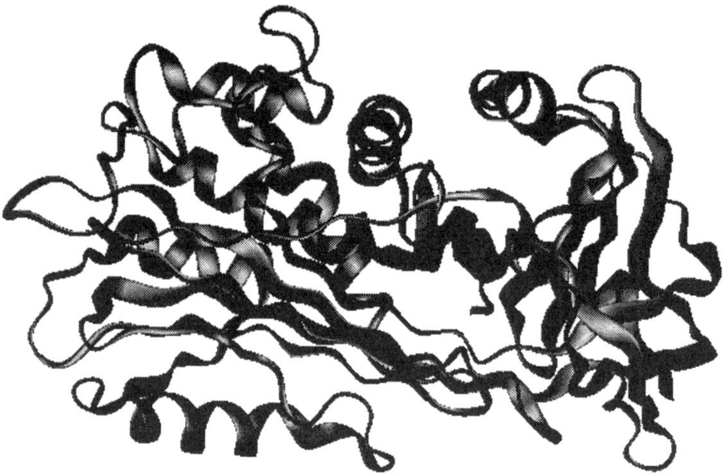

(a)

Figure 1 (a) Side view of latent-state homology model of Hsp47. α-Helices are represented by thick helices, β-sheets by ribbons, and loops by thin lines. (b) Close-up top view of the putative procollagen–collagen-binding groove. The peptide segment deleted in the NΔ1 mutant are extending from the N-term arrow to the second arrow. The segment deleted in the CΔ3 mutant is the light-shaded section of β-sheet.

triple-helix form of procollagen, as demonstrated by the resistance of immunoprecipitated collagen chains to trypsin digestion [21].

In vitro-binding assays using purified mature collagen also indicated that Hsp47 binds to the triple-helix form of mature collagen [29]. Dissociation constants as well as association and dissociation rate constants were determined using purified collagens (types I–V) and recombinant mouse Hsp47 (rmHsp47) by the optical sensor system BIAcore, which is based on surface plasmon resonance spectroscopy. The dissociation constants of rmHsp47 with types I–V collagens are 10^{-6} to 10^{-7} M, which is relatively weak for molecular interactions and can be accounted for by the rapid dissociation rate constant ($k_d > 10^{-2}$ s^{-1}) and high association rate constant (k_a of about 2×10^4 M^{-1} s^{-1}) [29].

The binding of Hsp47 with collagen is pH-sensitive both in vivo and in vitro. When the pH of the buffer is lowered to near 6.3, no binding is observed and, indeed, Hsp47 can be purified from the cell lysate by eluting the collagen–Sepharose-binding proteins with low-pH buffer [17]. When the intracellular pH was lowered by treating the cells with the proton ionophore carbonylcyanide-*m*-chlorophenylhydrazone (CCCP) in low-pH buffer (pH < 6.3), coprecipitation of procollagen with Hsp47 by immunoprecipitation with anti-Hsp47 antibody was no longer observed [18].

By circular dichroism and fluorescence spectroscopy, Hsp47 undergoes

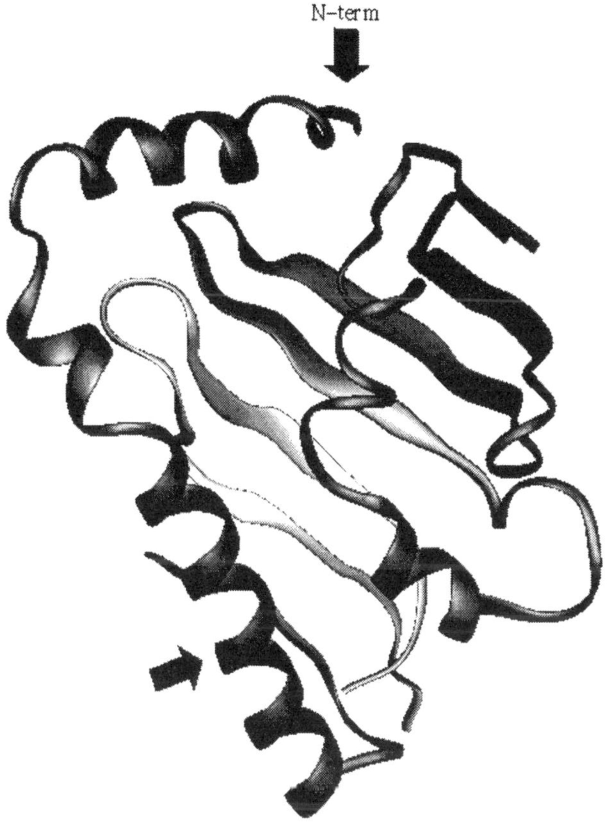

(b)

Figure 1 Continued

reversible pH-induced conformational changes that could account for its pH-dependent binding behavior [34]. These experiments showed a two-step conversion of the secondary and tertiary structures of Hsp47 from a high pH (or alkali) state to a low pH (or acidic) state, through an intermediate state. This transition from alkali (> pH 6.4) to intermediate then acidic (< pH 5.7) states was characterized according to standard curve-fitting procedures by a substantial drop in α-helical character (from 25 to 11%), paralleled by a substantial rise in β-sheet character (from 36 to 57%) [34]. When a site-specific histidine mutant (H335L) of recombinant mouse Hsp47 was bound to collagen–Sepharose, a lower pH was required to cause the dissociation of this mutant Hsp47 from collagen [S. Yokota et al., unpublished data]. This suggests that the pH-induced structural changes in Hsp47 are driven initially by histidine protonation and deprotonation.

V. DISSOCIATION OF Hsp47 FROM PROCOLLAGEN

To assess the question of where Hsp47 dissociates from procollagen in the central secretory pathway, we treated the cells with various inhibitors of secretion, and then analyzed the dissociation of Hsp47 from procollagen by pulse label–chase experiments. Confocal microscopic observation after simultaneous staining of the cells with anti-Hsp47 and anticollagen antibodies was also adopted. α, α′-Dipyridyl, an iron-chelating agent, inhibits the proryl and lysyl hydroxylation of procollagen, preventing triple-helix formation and of the secretion of procollagen from cells [18,35]. Brefeldin A releases coat proteins and an ADP ribosylation factor from the Golgi stack and, thereby inhibiting the transport of secretory proteins from the ER to the Golgi [36]. Monensin is an ionophore for Na^+, K^+, and protons, and it inhibits intracellular transport within the Golgi complex, preferentially between the *cis*- or *medial*-Golgi and *trans*-Golgi compartments [37].

In normal fibroblasts, newly synthesized procollagen is secreted within 20 min, and Hsp47 dissociates from procollagen in a little under 20 min. However, when cells were treated with α, α′-dipyridyl or brefeldin A, the secretion of procollagen was completely inhibited [18,21]. The pulse–chase and immunoprecipitation experiments revealed that the Hsp47 that was bound to newly synthesized procollagen, did not dissociate within 2 h (Fig. 2). Immunostaining observations using confocal microscopy showed that both Hsp47 and procollagen colocalized in the ER and in the intermediate compartment between the ER and the Golgi, in the α, α′-dipyridyl-treated and brefeldin A-treated cells, respectively [21]. These results suggest that Hsp47 does not dissociate from procollagen either in the ER or in the intermediate compartment.

When the cells were treated with monensin, the dissociation of Hsp47 from procollagen occurred within 15 min of pulse labeling, whereas the secretion of procollagen out of the cells was totally inhibited [21]. Double staining of Hsp47 and procollagen with specific antibodies also supported the foregoing result: in the cells treated with monensin, the Golgi area was stained with anticollagen antibody, but not with anti-Hsp47 antibody, whereas the ER-staining was observed with anti-Hsp47 antibody, but not with anticollagen antibody [21]. These results indicate that Hsp47 dissociates from the procollagen when it enters the *cis*-Golgi compartment. Thus, Hsp47 binds to the procollagen immediately after the nascent polypeptides of procollagen enter the ER and may dissociate from the procollagen in the *cis*-Golgi network (Fig. 3) [15,21].

VI. Hsp47 AS A MOLECULAR CHAPERONE

Procollagen α-chains first associate with each other at the C-terminal propeptide region, resulting in formation of trimers in the ER. Once α-chains form trimers

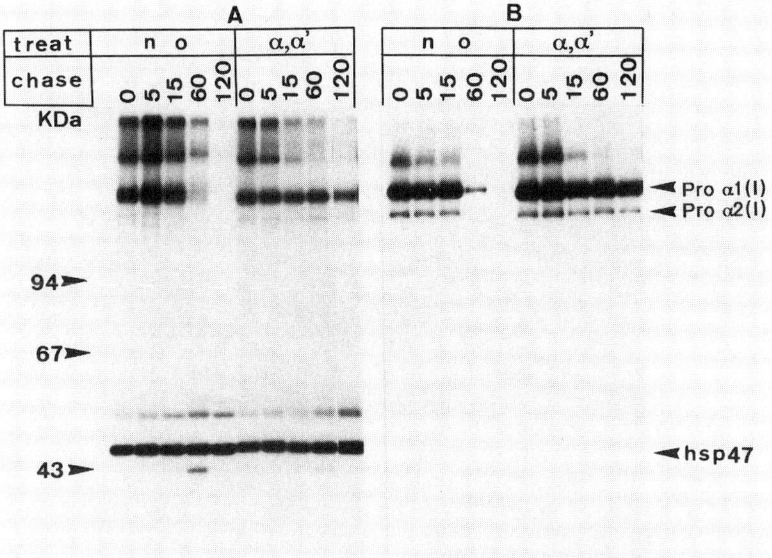

Figure 2 Immunoprecipitation of Hsp47 and procollagen in the cells treated with α, α′-dipyridyl after pulse label and during chase periods. Chick embryo fibroblasts were pulse-labeled for 10 min with [^{35}S]methionine and chased in the medium containing excess methionine in the presence (indicated as α, α′ on the top of the panel) or absence (indicated as *no*) of α, α′-dipyridyl for various periods. Immune complexes precipitated with anti-Hsp47 antibody (A) or anticollagen antiserum (B) were applied to SDS–8% polyacrylamide gel. The numbers on the top of the each lane show the chase periods in minutes.

(two α_1(I) chains and one α_2(I) chain in the type I collagen) at the C-terminal propeptides, triple-helix formation proceeds from the C-terminal propeptide toward the N-terminal propeptide. This implies that nascent polypeptide α-chains should be maintained unfolded until they form trimers, otherwise triple-helix formation would be prevented. Hsp47 may prevent nascent α-chains of procollagen from being randomly folded or from aggregation in the ER.

Recently, we have obtained several stable transfectants in Balb c/3T3 mouse fibroblastic cells, in which the cDNA encoding the antisense of mouse Hsp47 was transfected and constitutively expressed (M. Satoh et al., unpublished data). In these cells, the level of Hsp47 was prominently decreased. The initial rate of procollagen secretion was lower in antisense-transfected cells than in control mock-transfected cells. When α_1-chains of type I collagen were transiently overexpressed in these Hsp47-deficient cells, α_1-chains were recovered in a detergent-insoluble fraction, whereas they were recovered in the detergent-soluble fraction when transfected into normal Balb c/3T3 cells. These observa-

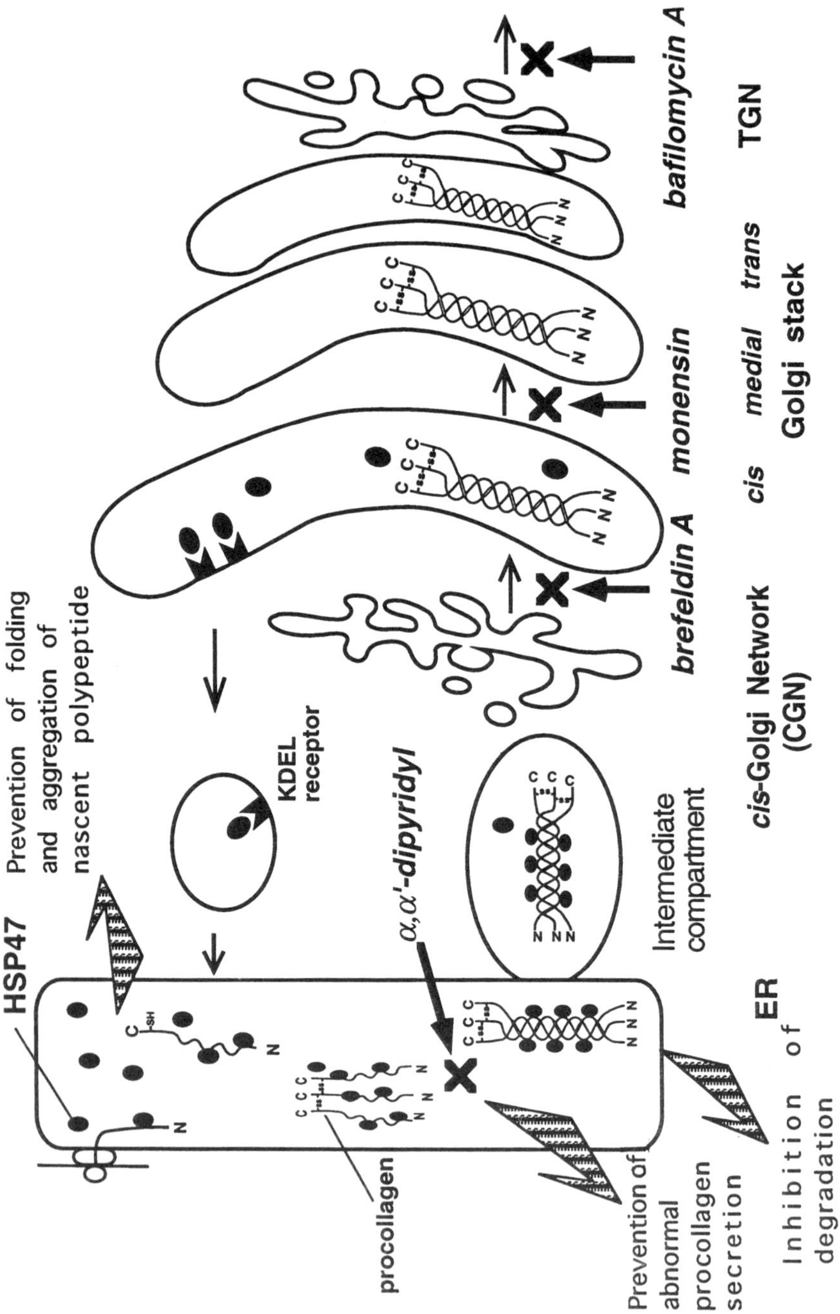
HSP47
Prevention of folding and aggregation of nascent polypeptide
KDEL receptor
α,α'-dipyridyl
procollagen
Intermediate compartment
ER
Prevention of abnormal procollagen secretion
Inhibition of degradation
cis-Golgi Network (CGN)
brefeldin A
monensin
bafilomycin A
cis
medial
trans
Golgi stack
TGN

tions also support the suggestion that Hsp47 prevents newly synthesized procollagens from forming aggregates. These possible functions of Hsp47 have also been attributed to other molecular chaperones.

When cells were heat shocked or treated with α, α′-dipyridyl, Hsp47 did not dissociate from procollagen during the chase period for up to 2 h, whereas the dissociation occurred within 15 min in nontreated control cells (see Fig. 2) [18,21]. These results suggest that the function of Hsp47 under stress conditions may be similar to the quality control mechanism such as BiP/Grp78 [2,15]. Consistent with this, Hsp47 has a higher affinity for heat-denatured collagen than for mature normal collagen (T. Natsume et al., unpublished data).

By incorporating these observations, Figure 3 shows a schematic representation of the interactions of Hsp47 during the folding and assembly of procollagen, the behavior of which is consistent with Hsp47 being a molecular chaperone that is specific to collagen. The possible functions of Hsp47 in collagen processing and secretion may be (a) to prevent the incorrect folding and aggregation of newly synthesized procollagen polypeptide chains until the correct C-terminal associations have been made and until the triple-helix formation proceeds, and (b) to prevent the secretion of abnormal procollagen caused by heat shock or the exposure to α, α′-dipyridyl. Hsp47 has been reported to prevent the degradation of procollagen in the ER [38]. Further experiments would be required to determine whether the inhibition of procollagen degradation is due to serpin-like activity of Hsp47, because our experiments to date suggest that Hsp47 has no inhibitory activity against serine proteases [26].

When the Hsp47–procollagen complex enters the *cis*-Golgi network, Hsp47 rapidly dissociates from procollagen. In vitro kinetic analysis of the binding of rmHsp47 with purified mature collagens revealed that dissociation rate constants were high enough for Hsp47 to be dissociated very rapidly from collagen when the concentration of free Hsp47 becomes low in the *cis*-Golgi compartment.

Figure 3 Schematic representation of the interaction of Hsp47 with procollagen and of its putative functions. Hsp47 binds to nascent procollagen polypeptides immediately after they cotranslationally enter the ER. The triple helix of type I procollagen is then formed: two α_1(I) and one α_2(I) chains of procollagen make contact at their C-termini and then make disulfide bonds among them. The formation of a triple helix then proceeds from the C- to the N-terminus, which provides the appropriate alignment and orientation to ensure correct helix formation. The triple helix then proceeds through the Golgi complex, where it dissociates from the Hsp47 in the *cis*-compartment, before continuing on through the *trans*-Golgi network to be secreted at the cell surface. Hsp47 molecules are cycled back to the ER when bound to the receptors that recognize the RDEL peptide. The sites of action of inhibitors used to study the passage of procollagen are indicated. Possible functions of Hsp47 are also indicated.

When the small number of Hsp47 molecules left is transported through the Golgi compartment with procollagen, Hsp47 may be dissociated from procollagen in the *trans*-Golgi compartment or *trans*-Golgi network because the pH in this compartment falls to as low as pH 5.7 [39,40]. Thus, the dissociation of Hsp47 from procollagen is thought to occur mainly in a pH-independent, concentration-dependent manner in the *cis*-Golgi network as well as possibly by a pH-dependent manner when it reaches the *trans*-Golgi.

It is likely that Hsp47 recycles back to the ER through the Golgi. Hsp47 probably becomes trapped by receptors in the Golgi membrane that recognize the RDEL peptide (perhaps KDEL receptors [41]), accounting for the low concentration of free Hsp47 in the Golgi compartment which, in turn, allows the rapid dissociation of Hsp47 from procollagen in the *cis*-Golgi compartment. Hsp47 is then free to shuttle back to the ER (see Fig. 3).

VII. CORRELATIONAL EXPRESSION OF Hsp47 WITH COLLAGEN GENES

Cloned Hsp47 genes reveal a well-conserved heat-shock element (Hse) consisting of three tandem repeats of five base-pair units (nGAAn, where *n* is any nucleotide) at the promoter region (about –80 from the transcription start site) [42,43]. This suggests that, under stressful conditions, Hsp47 is induced by an Hse-heat-shock factor (Hsf) system as with other stress proteins. Nuclear runon assays demonstrated that a more than tenfold increase in the transcription rate of Hsp47 gene occurred after 2 h of the heat treatment of chick embryo fibroblasts at 45°C, which rapidly returned to basal levels during the recovery period at 37°C [19]. The collagen synthesis decreases under heat-shock conditions similar to other proteins, whereas Hsp47 synthesis is up-regulated.

On the other hand, the expression of Hsp47 in various cell lines and tissues is always closely correlated with the expression of various types of collagens under nonstressed conditions [15].

The synthesis of type I collagen, a major collagen secreted by fibroblasts, decreases after malignant transformation of these cells, and this decrease in synthesis is concomitant with that of other extracellular matrix proteins, including fibronectin and laminin. When fibroblasts are transformed with Rous sarcoma virus, simian virus 40, or the oncogenic protein *c-Ha-ras*, synthesis of both type I collagen and Hsp47 decreases by the same extent [28,44,45]. In contrast, both type IV collagen and Hsp47 (colligin) are dramatically induced during the differentiation of the mouse teratocarcinoma cell line F9 after treatment with retinoic acid alone or with retinoic acid plus dibutylyl cAMP (Fig. 4) [24,46,47]. Interestingly, Hsp47 synthesis is not observed in cells in which collagen synthesis is not detected, such as mouse myeloid leukemic M1 cells and mouse pheo-

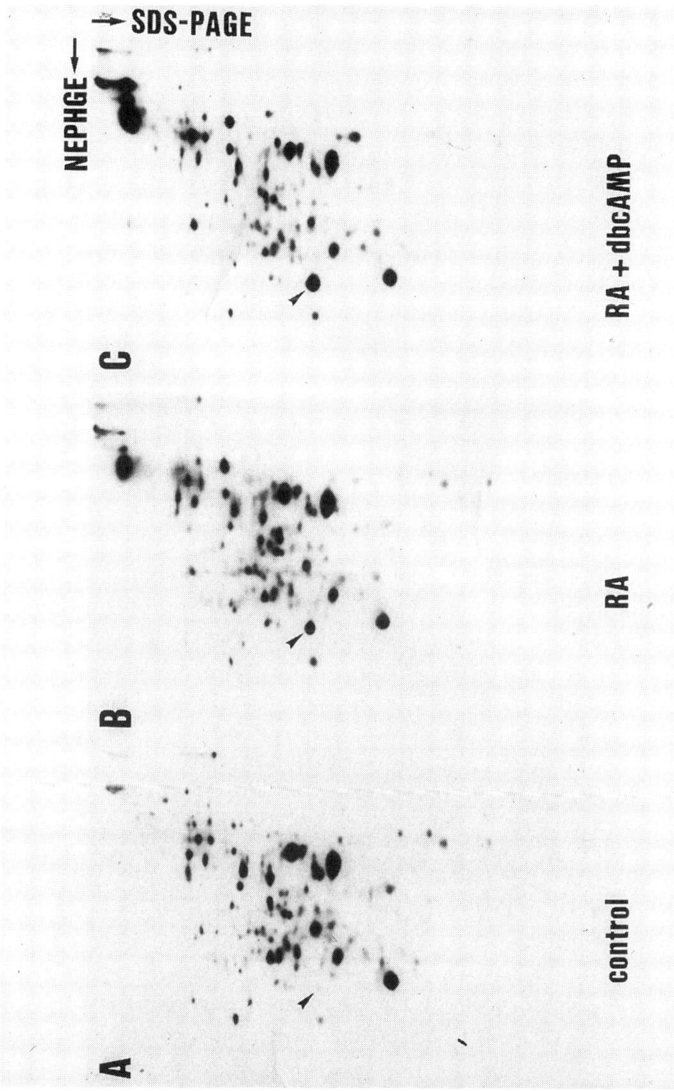

Figure 4 Two-dimensional analysis, by NEPHGE and SDS–PAGE, of the cell extract of mouse teratocarcinoma cell line F9: (A) F9 stem cells without treatment; (B) F9 cells treated with 1 μM retinoic acid for 3 days; (C) F9 cells treated with 1 μM retinoic acid and 1 mM $(But)_2cAMP$ for 3 days. Cells were labeled with [^{35}S]methionine for 1 h, and cell extracts were analyzed by NEPHGE as the first dimension and SDS–PAGE as the second dimension. Arrowheads indicate Hsp47 spots and arrows indicate collagen spots.

chromocytoma PC12 cells [45]. Such a correlation between the expression of Hsp47 and several types of collagen has also been reported in rat cells, including hepatocytes in which neither Hsp47 nor collagen were detected [48,49]. Clarke et al. [50] reported the parallel induction of Hsp47 with collagens after treatment of the myoblastic cell lines with transforming growth factor β-1 (TGFβ-1), again suggesting that the constitutive expression of Hsp47 is always accompanied by that of collagen under nonstressed conditions.

We also observed a strong correlational relation in the regulation of Hsp47 with collagen under pathological conditions in vivo: the synthesis of Hsp47, as well as type I and III collagens, is dramatically increased during the progression of rat liver fibrosis induced by carbon tetrachloride [48]. In situ hybridization analysis shows that only those cells adjacent to the collagen fibrils in the carbon tetrachloride-treated rat liver expressed large amounts of Hsp47 mRNA, whereas cells far from the collagen fibrils did not. The Hsp47-producing cells in rat liver were identified as Itoh cells (vitamin A-storing cells) and myofibroblast cells, which also actively produce type I and III collagens after treatment with carbon tetrachloride [48,49].

Such a correlation in the expression between Hsp47 and various collagens is also observed during the development of the mouse embryo. For example, Hsp47 is strongly expressed in cartilaginous regions, particularly in chondrocytes in the proliferation and maturation zones of epiphyseal growth plates, in which the expression of type II collagen is also prominent. In contrast, both type I collagen and Hsp47 were localized to the extracellular matrix and osteoblasts in the vascularization zone during ossification (H. Masuda et al., unpublished data). Kambe et al. [51] also reported that Hsp47 was expressed in the epiphyseal cartilage and in cultured chondrocytes from chick embryos. Coexpression of the genes encoding both Hsp47 and collagen is also reported in the development of murine femurs and molars [52] and various chick tissues [53]. Thus, Hsp47 is a substrate-specific stress protein, the regulation of which is closely correlated with its substrate, collagen.

The regulatory mechanism of the correlational expression of Hsp47 with that of various types of collagens is one of the most interesting projects to be examined in the future. One candidate for such a regulatory mechanism is the GATA family of transcription factors, which are reported to function during the differentiation of F9 cells. The Hsp47 gene has GATA-binding motifs in the promoter region and these elements are occupied by GATA-4 protein after the differentiation of F9 cells with retinoic acid [54,55]. However, whether collagen induction during F9 differentiation is also regulated by the same factors, and also whether this parallel induction of Hsp47 and collagen is regulated by the same mechanism in other cell lines or systems, is not yet clear.

REFERENCES

1. Hurtley, S. M. and Helenius, A., Protein oligomerization in the endoplasmic reticulum, *Annu. Rev. Cell Biol. 5*:277 (1989).
2. Gething, M. J. and Sambrook, J., Protein folding in the cell, *Nature 355*:33 (1992).
3. Bergerron, J. J., Brenner, M. B., Thomas, D. Y., and Williams, D. B., Calnexin: A membrane-bound chaperone of the endoplasmic reticulum, *Trends Biochem. Sci. 19*:124 (1994).
4. De Silva, A. M., Balch, W. E., and Helenius, A., Quality control in the endoplasmic reticulum: Folding and misfolding of vesicular stomatitis virus G protein in cells and in vitro, *J. Cell Biol. 111*:857 (1990).
5. De Silva, A., Braakman, I., and Helenius, A., Posttranslational folding of vesicular stomatitis virus G protein in the ER: Involvement of noncovalent and covalent complexes, *J. Cell Biol. 120*:647 (1993).
6. Kozutsumi, Y., Segai, M., Normington, K., Gething, M. J., and Sambrook, J., The presence of malfolded proteins in the endoplasmic reticulum signals the induction of glucose-regulated proteins, *Nature 332*:462 (1988).
7. Melnick, J., Aviel, S., and Argon, Y., The endoplasmic reticulum stress protein grp94, in addition to BiP, associates with unassembled immunoglobulin chains, *J. Biol. Chem. 267*:21303 (1992).
8. Chessler, S. D. and Byers, P. H., BiP binds type I procollagen proα chains with mutations in the carboxyl-terminal propeptides synthesized by cells from patients with osteogenesis imperfecta, *J. Biol. Chem. 268*:18226 (1993).
9. Chessler, S. D., Wallis, G. A., and Byers, P. H., Mutations in the carboxyl-terminal propeptide of the proα1(I) chain to type I collagen result in defective chain association and produce lethal osterogenesis imperfecta, *J. Biol. Chem. 268*:18218 (1993).
10. Koivu, J. and Myllyla, R., Interchain disulfide bond formation in type I and II procollagen, *J. Biol. Chem. 262*:6159 (1987).
11. Munro, S. and Pelham, H. R. B., A C-terminal signal prevents secretion of luminal ER proteins, *Cell 48*:899 (1987).
12. Nilsson, T., Jackson, M. R., and Peterson, P. A., Short cytoplasmic sequences serve as retention signals for transmembrane proteins in the endoplasmic reticulum, *Cell 58*:707 (1989).
13. Jackson, M. R., Nilsson, T., and Peterson, P. A., Retrieval of transmembrane proteins to the endoplasmic reticulum, *J. Cell Biol. 121*:317 (1993).
14. Nagata, K., Saga, S., and Yamada, K. M., A major collagen-binding protein of chick embryo fibroblasts is a novel heat shock protein, *J. Cell Biol. 103*:223 (1986).
15. Nagata, K., HSP47, a collagen-specific molecular chaperone, *Trends Biochem. Sci. 21*:23 (1996).
16. Nagata, K., Hirayoshi, K., Obara, M., Saga, S., and Yamada, K. M., Biosynthesis of a novel transformation-sensitive heat-shock protein that binds to collagen: Regulation by mRNA levels and in vitro synthesis of a functional precursor, *J. Biol. Chem. 263*:8344 (1988).
17. Saga, S., Nagata, K., Chen, W.-T., and Yamada, K. M., pH-dependent function,

purification, and intracellular location of a major collagen-binding glycoprotein, *J. Cell Biol. 105*:517 (1987).

18. Nakai, A., Satoh, M., Hirayoshi, K., and Nagata, K., Involvement of the stress protein HSP47 in procollagen processing in the endoplasmic reticulum, *J. Cell Biol. 117*:903 (1992).
19. Hirayoshi, K., Kudo, H., Takechi, H., Nakai, A., Iwamatsu, A., Yamada, K. M., and Nagata, K., HSP47: A tissue-specific, transformation-sensitive, collagen-binding heat shock protein of chicken embryo fibroblasts, *Mol. Cell Biol. 11*:4036 (1991).
20. Hughes, R. C., Tayler, A., Sage, H., and Hogan, B. L. M., Distinct patterns of glycosylation of colligin, a collagen-binding glycoprotein, and SPARC (osteonectin), a secreted Ca^{2+}-binding glycoprotein. Evidence for the localisation of colligin in the endoplasmic reticulum, *Eur. J. Biochem. 163*:57 (1987).
21. Satoh, M., Hirayoshi, K., Yokota, S., Hosokawa, N., and Nagata, K., Intracellular interaction of collagen-specific stress protein HSP47 with newly synthesized procollagen, *J. Cell Biol. 133*:469 (1996).
22. Carrell, R. W. and Boswell, D. R., Serpins: The superfamily of plasma serine proteinase inhibitors, *Proteinase Inhibitors* (Barrett, A. and Salvesen, G., eds.), Elsevier/North-Holland Biomedical, Amsterdam, 1986, p. 403.
23. Suzuki, K., Deyashiki, Y., Nishioka, J., Kurachi, K., Akira, M., Yamamoto, S., and Hashimoto, S., Characterization of a cDNA for human protein C inhibitor, *J. Biol. Chem. 262*:611 (1987).
24. Kurkinen, M., Tayler, A., Garrels, J. I., and Hogan, B. L., Cell surface-associated proteins which bind native type IV collagen or gelatin, *J. Biol. Chem. 259*:5915 (1984).
25. Cates, G. A., Nandan, D., Brickenden, A. M., and Sanwal, B. D., Differentiation defective mutants of skeletal myoblasts altered in a gelatin-binding glycoprotein, *Biochem. Cell. Biol. 65*:767 (1987).
26. Davids, J. W., El-Thaher, T. S. H., Nakai, A., Nagata, K., and Miller, A. D., Modeling the three-dimensional structure of serpin/molecular chaperone HSP47, *Bioorg. Chem. 23*:427 (1995).
27. Blundell, T. L., Sibanda, B. L., Sternberg, M. J. E., and Thornton, J. M., *Nature 326*:347 (1987).
28. Nagata, K. and Yamada, K. M., Phosphorylation and transformation sensitivity of a major collagen-binding protein of fibroblasts, *J. Biol. Chem. 261*:7531 (1986).
29. Natsume, T., Koide, T., Yokota, S., Hirayoshi, K., and Nagata, K., Interactions between collagen-binding stress protein HSP47 and collagen. Analysis of kinetic parameters by surface plasmon resonance biosensor, *J. Biol. Chem. 269*:31224 (1994).
30. Silver, M. L., Guo, H.-C., Strominger, J. L., and Wiley, D. C., Atomic structure of a human MHC molecule presenting an influenza virus peptide, *Nature 360*:367 (1992).
31. Stern, L. J., Brown, J. H., Jardetsky, T. S., Gorga, J. C., Urban, R. G., Strominger, J. L., and Wiley, D. C., Crystal structure of the human class II MHC protein HLA-DR1 complexed with an influenza virus peptide, *Nature 368*:215 (1994).
32. Sauk, J. J., Smith, T., Norris, K., and Ferreira, L., Hsp47 and the translation–translocation machinery cooperate in the production of alpha 1(I) chains of type I procollagen, *J. Biol. Chem. 269*:3941 (1994).

33. Hu, G., Gura, T., Sabsay, B., Sauk, J., Cixit, S. N., and Veis, A., Endoplasmic reticulum protein HSP47 binds specifically to the N-terminal globular domain of the amino-propeptide of the procollagen I a1(I)-chain, *J. Cell. Biochem. 59*:350 (1995).
34. El-Thaher, T. S. H., Drake, A. F., Yokota, S., Nakai, A., Nagata, K., and Miller, A. D., The pH-dependent, ATP-independent interaction of collagen specific serpin/stress protein HSP47, *Prot. Pept. Lett. 3*:1 (1996).
35. Kao, W. W., Prockop, D. J., and Berg, R. A., Kinetics for the secretion of nonhelical procollagen by freshly isolated tendon cells, *J. Biol. Chem. 254*:2234 (1979).
36. Lippincott-Schwartz, J., Yuan, L. C., Bonifacino, J. S., and Klausner, R. D., Rapid redistribution of Golgi proteins into the ER in cells treated with brefeldin A: Evidence for membrane cycling from Golgi to ER, *Cell 56*:801 (1989).
37. Tartakoff, A. M., Perturbation of vesicular traffic with the carboxylic ionophore monensin, *Cell 32*:1026 (1983).
38. Jain, N., Brickenden, A., Ball, E. H., and Sanwal, B. D., Inhibition of procollagen I degradation by colligin: A collagen-binding serpin, *Arch. Biochem. Biophys. 314*:23 (1994).
39. Anderson, R. G. and Orci, L., A view of acidic intracellular compartments, *J. Cell Biol. 160*:539 (1988).
40. Wilson, D. W., Lewis, M. J., and Pelham, H. R. B., pH-dependent binding of KDEL to its receptor in vitro, *J. Biol. Chem. 268*:7465 (1993).
41. Semenza, J. C., Hardwick, K. G., Dean, N., and Pelham, H. R., *ERD2*, a yeast gene required for the receptor-mediated retrieval of luminal ER proteins from the secretory pathway, *Cell 29*:1349 (1990).
42. Wang, S. Y., Structure of the gene and its retinoic acid-regulatory region for murine J6 serpin. An F9 teratocarcinoma cell retinoic acid-inducible protein, *J. Biol. Chem. 267*:15362 (1992).
43. Hosokawa, N., Takechi, H., Yokota, S., Hirayoshi, K., and Nagata, K., Structure of the gene encoding the mouse 47-kDa heat-shock protein (HSP47). *Gene 126*:187 (1993).
44. Nakai, A., Hirayoshi, K., and Nagata, K., Transformation of BALB/3T3 cells by simian virus 40 causes a decreased synthesis of a collagen-binding heat-shock protein (hsp47), *J. Biol. Chem. 265*:992 (1990).
45. Nagata, K., Nakai, A., Hosokawa, N., Kudo, H., Takechi, H., Satoh, M., and Hirayoshi, K., Interaction of HSP47 with newly synthesized procollagen, and regulation of HSP expression, *Heat Shock* (Maresca, B. and Lindquist, S., eds.), Springer-Verlag, New York, 1991, p. 105.
46. Wang, S. Y. and Gudas, L. J., A retinoic acid-inducible mRNA from F9 teratocarcinoma cells encodes a novel protease inhibitor homologue, *J. Biol. Chem. 265*:15818 (1990).
47. Takechi, H., Hirayoshi, K., Nakai, A., Kudo, H., Saga, S., and Nagata, K., Molecular cloning of a mouse 47-kDa heat-shock protein (HSP47), a collagen-binding stress protein, and its expression during the differentiation of F9 teratocarcinoma cells, *Eur. J. Biochem. 206*:323 (1992).
48. Masuda, H., Fukumoto, M., Hirayoshi, K., and Nagata, K., Coexpression of the collagen-binding stress protein HSP47 gene and the alpha 1(I) and alpha 1(III)

collagen genes in carbon tetrachloride-induced rat liver fibrosis, *J. Clin. Invest. 94*:2481 (1994).

49. Kawada, N., Kuroki, T., Kobayashi, K., Inoue, M., Nakatani, K., Kaneda, K., and Nagata, K., Expression of heat-shock protein 47 in mouse liver, *Cell Tissue Res. 284*:341 (1996).
50. Clarke, E. P., Jain, N., Brickenden, A., Lorimer, I. A., and Sanwal, B. D., Parallel regulation of procollagen I and colligin, a collagen-binding protein and a member of the serine protease inhibitor family, *J. Cell Biol. 121*:193 (1993).
51. Kambe, K., Yamamoto, A., Yoshimori, T., Hirayoshi, K., Ogawa, R., and Tashiro, Y., Preferential localization of heat shock protein 47 in dilated endoplasmic reticulum of chicken chondrocytes, *J. Histochem. Cytochem. 42*:833 (1994).
52. Shroff, B., Smith, T., Norris, K., Pileggi, R., and Sauk, J. J., Hsp47 is localized to regions of type I collagen production in developing murine femurs and molars, *Connect. Tissue Res. 29*:273 (1993).
53. Miyaishi, O., Sakata, K., Matsuyama, M., and Saga, S., Distribution of the collagen binding heat-shock protein in chicken tissues, *J. Histochem. Cytochem. 40*:1021 (1992).
54. Wang, S. Y., A retinoic acid-inducible GATA-binding protein binds to the regulatory region of J6 serpin gene, *J. Biol. Chem. 269*:607 (1994).
55. Bielinska, M. and Wilson, D. B., Regulation of J6 gene expression by transcription factor GATA-4, *Biochem. J. 307*:183 (1995).

11

DnaJ Proteins

Ulrich Zuber
Universität Greifswald, Greifswald, Germany

Alexander Buchberger and
***Thomas Laufen**
Universität Heidelberg, Heidelberg, Germany

Bernd Bukau
Universität Freiburg, Freiburg, Germany

I. INTRODUCTION

The first discovery of a member of the DnaJ family, the DnaJ heat-shock protein of *Escherichia coli,* dates back to 1971 [Georgopoulos and Herskowitz, 1971]. Two decades later the first eukaryotic homologue was identified [Sadler et al., 1989], and by now, the family of DnaJ proteins (also referred to as Hsp40) comprises more than 100 members (Table 1). DnaJ homologues have been found in a large variety of prokaryotic and eukaryotic cells, as well as some viruses, with the noticeable exception of specific archaeabacteria that lack the entire DnaK chaperone system including DnaJ [Bult et al., 1996; K. Willison and A. Horwich, personal communication]. DnaJ proteins are multidomain proteins that share and are defined by a highly conserved domain of approximately 78 amino acids, the J-domain [Bork et al., 1992; Cyr et al., 1994; Silver and Way, 1993]. For the *E. coli* homologue the J-domain allows DnaJ to interact with DnaK, the prokaryotic homologue of Hsp70 chaperones and, thereby, to control its functional cycle. Therefore, it is conceivable that DnaJ family members share the

*Current affiliation: Universität Freiburg, Freiburg, Germany

Table 1 All Protein Sequences Homologous to *E. coli* DnaJ that were Available by the End of 1996

Origin[a]	Name(s)[b]	J[c]	G/F[c]	Zn[c]	Ct[c]	aa[d]	Acc. Nr.[e]
Prokaryotes							
Agrobacterium tumefaciens	DnaJ (fragment)	+	?	?	?	66	P50018
Bacillus stearothermophilus	DnaJ	+	+	+	+	380	Q45552
B. subtilis	DnaJ	+	+	+	+	372	P17631
Borrelia burgdorferi	DnaJ	+	+	+	+	352	P28616
B. burgdorferi	DnaJ*	+	+	+	+	295	*BBGRPEPLS_3*
Brucella ovis	DnaJ	+	+	+	+	375	Q05980
Caulobacter crescentus	DnaJ (fragment)	+	+	½	?	190	P22305
Clostridium acetobutylicum	DnaJ	+	+	+	+	374	P30725
Coxiella burnetii	DnaJ	+	+	+	+	367	P42381
C. burnetii	*	+	–	–	–	270	*CBMUCZ_1*
Erysipelothrix rhusiopathiae	DnaJ	+	+	+	+	370	Q05646
Escherichia coli	DnaJ	+	+	+	+	376	P08622
E. coli	CBPA	+	–	–	+	297	P36659
E. coli	YFHE-hscB	+	–	–	–	171	P36540
E. coli	YABH-DjlA	+	–	–	–	271	P31680
Francisella tularensis	DnaJ	+	+	+	+	371	P48207
Haemophilus actinomycetemcomitans	DnaJ*	+	+	+	+	290	*HAD753_2*
H. ducreyi	DnaJ	+	+	+	+	377	P48208
H. influenzae	DnaJ/HI1238	+	+	+	+	382	P43735
H. influenzae	DnaJ*	+	+	+	+	394	*C64112*
H. influenza	HI0721*	+	–	–	–	288	P44607
Lactococcus lactis	DnaJ	+	+	+	+	379	P35514
Legionella pneumophila	DnaJ	+	+	+	+	379	U15010
Lyme disease spirochette	DnaJ*	+	+	+	+	352	*A49210*
Methanosarcina mazei	DnaJ	+	+	+	+	389	P35515
Mycoplasma genitalium	DnaJ-MG019	+	+	+	+	389	P47265
M. genitalium	MG200	+	–	–	–	601	P47442
M. genitalium	MG002	+	–	–	–	310	P47248
M. hyopneumoniae	*	+	–	–	–	145	*MH50211_1*
M. pneumoniae	*	+	–	–	–	309	*MP34816_2*
Mycobacterium leprae	DnaJ	+	+	+	+	389	Q02605
M. tuberculosis	DnaJ	+	+	+	+	356	P07881
Pseudomonas denitrificans	*	+	–	–	–	171	P29944
Rhizobium fredii	NOLC	+	+	+	+	392	P26508

Table 1 Continued

Origin[a]	Name(s)[b]	J[c]	G/F[c]	Zn[c]	Ct[c]	aa[d]	Acc. Nr.[e]
Rhodobacter capsulatus	DnaJ	+	+	+	+	384	*RCNAKJGE_2*
Salmonella typhimurium	DnaJ*	+	+	+	+	379	*ST58360_2*
Solanum tuberosum	DnaJ	+	+	+	+	419	*X94301*
Staphylococcus aureus	DnaJ	+	+	+	+	379	P45555
Streptomyces coelicor	DnaJ*	+	+	+	+	387	*S41948*
S. coelicor	DnaJ	+	+	+	+	399	P40170
Synechocystis sp.	DNJH*	+	+	–	+	377	*SSSLLLH_82*
Synechocystis sp.	DnaJ*	+	–	–	+	332	P50027
Synechococcus sp.	DnaJ* (fragment)	+	?	?	?	189	P50026
Thermus thermophilus	T.DnaJ	+	+	–	–	189	Motohashi et al., 1996
Eukaryotes							
Allium porrum	DnaJ1	+	+	+	+	397	Q03363
A. porrum	DnaJ2	+	+	+	+	418	P42824
Arabidopsis thaliana	DnaJ homologue	+	+	+	+	420	*AT22340_1*
A. thaliana	ATJ	+	+	+	–	419	P42825
A. thaliana	ATJ1	+	–	+	+	368	*AT16246_1*
A. thaliana	J-domain protein	+	–	–	–	539	*ATD3MR_1*
Atriplex nummularia	DNJH	+	+	+	+	417	P43644
Bos taurus	p58	+	–	–	–	504	A56534
B. taurus	(csp)	+	–	–	–	167	*BTCYSPRT2_1*
B. taurus	(csp)	+	–	–	–	198	*BTCYSPRT1_1*
B. taurus	Auxilin	+	–	–	–	910	Q27974
Caenorhabditis elegans	DnaJ*	+	+	+	+	943	P34408
C. elegans	T15H9.1*	+	+	–	+	355	*CET15H9_1*
C. elegans	F54D5.8*	+	+	–	+	331	*CEF54D5_7*
C. elegans	*	+	+	–	–	457	*CEW03A5_3*
C. elegans	F22B7.5.*	+	–	–	–	943	P34408
C. elegans	B0035.14*	+	–	–	–	401	*CEB0035_13*
C. elegans	*	+	–	–	–	378	*CER74_2*
C. elegans	B0035.2*	+	–	–	–	337	*CEB0035_5*
C. elegans	T04A8.9*	+	–	–	–	249	*CET04A8_7*
C. elegans	T24H10.3*	+	–	–	–	242	*CET24H10_3*
C. elegans	*	+	–	–	–	215	*CEK02G10_7*
C. elegans	csp	+	–	–	–	239	(Buchner et al., 1997)
Cucumis sativus	DnaJ1	+	+	+	+	413	Q04960
Drosophila melanogaster	Tid(56)	+	+	+	+	518	*S42091*
D. melanogaster	DROJ1	+	+	–	–	333	DM34904_1
D. melanogaster	csp32	+	–	–	–	249	Q03751

Table 1 Continued

Origin[a]	Name(s)[b]	J[c]	G/F[c]	Zn[c]	Ct[c]	aa[d]	Acc. Nr.[e]
D. melanogaster	csp29	+	–	–	–	223	*DMCSPAA_1*
D. virilis	Tid	+	+	+	+	529	*DMEL2131_1*
H. sapiens	Hdj2/DnaJ2	+	+	+	+	397	P31689
H. sapiens		+	+	½	–	189	*S34632*
H. sapiens		+	+	–	+	339	*S20062*
H. sapiens	p58	+	+	–	–	504	*HS2842410_1*
H. sapiens	hsj1	+	–	–	–	351	P25686
H. sapiens	DnaJ1/hdj1	+	+	–	+	340	P25685
H. sapiens		+	–	–	–	277	*S23509*
H. sapiens		+	–	–	–	178	*HS409921_1*
H. sapiens	csp1	+	–	–	–	201	(Buchner et al., 1997)
H. sapiens	csp2	+	–	–	–	198	(Buchner et al., 1997)
Mus musculus	MIDA1	+	–	–	–	621	*MMMIDA1_1*
M. musculus	MTJ1	+	–	–	–	552	*JC4030*
M. musculus	ZRF1	+	–	–	–	514	P54103
M. musculus	p58	+	–	–	–	504	*MM284231_1*
Nicotiana tabacco	(fragment)	+	?	?	?	88	*S56704*
Pacific electric ray		+	–	–	–	195	*JH0719*
Picea glauca		+	–	–	–	158	*PGHSPA_1*
Plasmodium falciparum Papua n. Guenea	RESA	+	–	–	–	1073	P13830
P. falciparum Ghana	RESA (fragment)	+	?	?	?	760	P13831
P. falciparum	RESA	+	–	–	–	755	*PFRESAG1_1*
Rattus norvegicus	DnaJ2-RDJ1	+	+	+	+	397	P54102
R. norvegicus	(csp)	+	–	–	–	198	P54101
Saccharomyces cerevisiae	YDJ1 or MAS5-YNL064C/ N2418/ YNL248C	+	+	+	+	409	P25491
S. cerevisiae	XDJ1-YLT090W	+	+	+	–	459	P39102
S. cerevisiae	SCJ1-YMR214W	+	+	+	–	404	P25303
S. cerevisiae	MDJ1-YFL016C	+	+	+	+	511	P35191
S. cerevisiae	SIS1-YNL007C/ N2679	+	+	–	+	352	P25294
S. cerevisiae	HLJ1-YMR161W/ YM8530.10	+	+	–	–	224	P48353
S. cerevisiae	ZOU1-YGR285C	+	–	–	–	433	P32527

Table 1 Continued

Origin[a]	Name(s)[b]	J[c]	G/F[c]	Zn[c]	Ct[c]	aa[d]	Acc. Nr.[e]
S. cerevisiae	SEC63 or - NPL1 or PTL1/ YOR254C	+	–	–	–	663	P14906
S. cerevisiae	CAJ1-YER048C	+	–	–	–	391	P39101
S. cerevisiae	YIS4-YIR004W/ YIB4W*	+	–	–	–	432	P40564
S. cerevisiae	YFL1-YFR041C*	+	–	–	–	295	P43613
S. cerevisiae	YJL073W/J1083/ HRC558*	+	–	–	–	692	P40358
S. cerevisiae	YJL162C/ J0549*	+	–	–	–	482	P46997
S. cerevisiae	YP9499.16c*	+	–	–	–	301	*S54082*
S. cerevisiae	YNL227C*	+	–	+	–	590	P53863
S. cerevisiae	YNL077W*	+	+	–	–	528	*SCYNL077W_1*
S. cerevisiae	YGL128C/ G2861*	+	–	–	–	347	P52868
S. cerevisiae	YGL018C*	+	–	–	–	184	P53193
Schizosaccharomyces pombe	PSI	+	+	–	+	379	Q09912
S. pombe	*	+	–	–	–	392	*SPAC4H3_1*
S. pombe	*	+	–	–	–	270	*SPAC4G9_19*
Torpedo c.	csp	+	–	–	–	195	(Buchner et al., 1997)
Trypanosoma cruzi	TCJ2	+	+	+	+	399	*TCTCJ2G_1*
T. cruzi	TCJ3	+	+	+	+	390	*TCTCJ3G_1*
T. cruzi	TCJ4	+	+	+	+	441	*TCTCJ4G_1*
T. cruzi	TJC1	+	–	–	–	307	*TCTCJ1G_1*
T. cruzi		+	–	–	–	324	*TCDNAJ_1*
Viruses							
Budgerigar fledgling disease virus	Large T antigen	+	–	–	–	587	P13894
Budgerigar fledgling disease virus	Small T antigen	+	–	–	–	145	P13895
Polyomavirus BA	Large T antigen	+	–	–	–	691	P14999
Polyomavirus BK	Large T antigen	+	–	–	–	695	P03071
Polyomavirus BK	Small T antigen	+	–	–	–	172	P03082
Polyomavirus BK strain AS	Small T antigen	+	–	–	–	172	P15000
Polyomavirus bovine	Small T antigen	+	–	–	–	124	P24852
Polyomavirus hamster	Large T antigen	+	–	–	–	751	P03075
Polyomavirus hamster	Middle T antigen	+	–	–	–	401	P03079

Table 1 Continued

Origin[a]	Name(s)[b]	J[c]	G/F[c]	Zn[c]	Ct[c]	aa[d]	Acc. Nr.[e]
Polyomavirus hamster	Small T antigen	+		–	–	194	P03080
Polyomavirus JC	Large T antigen	+	–	–	–	688	P03072
Polyomavirus JC	Small T antigen	+	–	–	–	172	P03083
Polyomavirus lymphotrophic	Large T antigen	+	–	–	–	597	P04008
Polyomavirus lymphotrophic	Small T antigen	+	–	–	–	189	P04009
Polyomavirus mouse	Large T antigen	+	–	–	–	777	P12905
Polyomavirus mouse strain Crawford	Middle T antigen	+	–	–	–	421	P12906
Polyomavirus mouse strain Crawford	Middle T antigen	+	–	–	–	421	P12906
Polyomavirus mouse strain Crawford	Small T antigen	+	–	–	–	158	P24598
Polyomavirus mouse strain 2	Middle T antigen	+	–	–	–	421	P03077
Polyomavirus strain 2 and 3	Small T antigen	+	–	–	–	195	P03078
Polyomavirus mouse strain 3	Large T antigen	+	–	–	–	782	P03074
Polyomavirus mouse strain 3	Middle T antigen	+	–	–	–	440	P03076
Simian virus 40	Large T antigen	+	–	–	–	708	P03070
Simian virus 40	Small T antigen	+	–	–	–	174	P03081

This table was generated by performing a homology search with the program "Blast" [Altschul et al., 1990] o the "nr" databank. The string for which homology was searched was from *E. coli* DnaJ, consisting of residue 2–108 (J-domain and G/F-motif) and 143–376 (Zn-finger motif and substrate-binding domain).

To appear in the table, the protein sequences found had to possess a J-domain, including the HPD motif Exceptions are all RESA proteins (*Plasmodium falciparum*), for which this motif is changed to YPY; and CAJ (*Sacharomyces cerevisiae*), for which the HPD motif is repeated twice. Homology for the individual domain was considered to be significant and is marked with a + when $P(N) \leq 0.05$, where $P(N)$ is a likelihood valu describing the significance of a hit. It cannot be excluded that more than one name and Acc. Nr. are assigned t the same sequence. For the *S. cerevisiae* homolouges duplicates were excluded by pairwise sequence alignments

[a]Origin, organism in which the homologue protein sequence is found.

[b]Name(s), name(s) found in the databank for the homologous protein sequence (An asterisk behind the nam indicates that the found protein sequence is hypothetical), csp, protein sequence belongs to the csp family; (csp) protein sequence has no name, but has homology to the csp family; (Fragment), protein sequence is fragmental ?, unassigned homologies.

[c]Four columns denote homology of the identified protein sequence to the individual domains of *E. coli.* DnaJ: J J-domain; G/F, G/F-motif; Zn, Zn-finger; Ct, COOH-terminus; 1/2 in the Zn-finger column means one, rathe than two, Zn-fingers in the protein sequence.

[d]aa, length of the identified protein sequence in amino acid residues.

[e]Acc Nr, accession number of the identified protein sequence. Protein sequences with numbers in regular typ can be obtained from Swissprot (http://expasy.hcuge.ch/sprot/sprot-top.html); protein sequences with numbers i italics can be obtained from SRS (http://www.ebi.ac.uk/srs/wgetz) with all databanks checked. In the few in stances in which more than one entry was found the PIR entry contains the protein sequence shown in this table

ability to cooperate with Hsp70 chaperones in assisting protein folding, and there is increasing evidence that this cooperation is an obligatory step for the chaperone activity of Hsp70. DnaJ proteins exhibit strong variability in their other domains. This diversity of DnaJ proteins is likely to provide an important basis for the functional diversity of Hsp70 chaperone systems. This chapter summarizes our current knowledge of the functions and mechanism of action of DnaJ proteins, with a focus on the *E. coli* DnaJ homologue.

II. *ESCHERICHIA COLI* DnaJ PROTEINS

Escherichia coli has at least four genes encoding DnaJ family members, *dnaJ* [Georgopoulos and Herskowitz, 1971], *cbpA* [Ueguchi et al., 1994], *djlA or YABH* [Clarke et al., 1996; Yura et al., 1992] and *hscB* or *YFHE* [Lelivelt and Kawula, 1995]. Of these homologues, the *dnaJ* gene product appears to be the most important for metabolism of growing *E. coli* cells and is mechanistically the best studied of all DnaJ homologues in prokaryotes and eukaryotes.

A. Cellular Functions

1. DnaJ

Gene *dnaJ* is the distal gene of the *dnaK dnaJ* operon [Bardwell and Craig, 1984] and is expressed in a stress-dependent manner as part of the heat-shock regulon [Bukau 1993; Gross 1996; Yura et al., 1993]. It was discovered as a point mutation, *dnaJ259,* which blocks the lytic growth of bacteriophage λ [Saito and Uchida, 1977; Sunshine et al., 1977]. This phenotype was originally designated GroP because it can be suppressed by secondary mutations in the λP gene [Sunshine et al., 1977]. Later work established that the DnaJ259 mutant protein has a H33E exchange in the J-domain that prevents its assistance in the initiation of λ DNA replication [Alfano and McMacken, 1989; Echols, 1986; Wall et al., 1994]. DnaJ targets DnaK to a nucleoprotein complex formed by the λO and λP proteins and *E. coli* DnaB helicase at the origin of replication of λ. This activity involves an interaction of DnaJ with DnaK and with λP. In presence of ATP and GrpE, the nucleotide-exchange factor for DnaK [Jordan and McMacken, 1995; Liberek et al., 1991a; McCarty et al., 1995], the activity of the DnaK chaperone system allows DnaB to dissociate from the nucleoprotein complex and bi-directionally unwind λ DNA.

The *dnaJ* knockout mutant cells grow at 30° and 42°C, though slightly (30°C) or much (42°C) slower than *dnaJ*$^+$ cells, indicating that DnaJ plays a role in metabolism at a variety of temperatures but is not strictly essential for growth at intermediate temperatures [Georgopoulos et al., 1990; Sell et al., 1990]. At temperatures higher than 43°C *dnaJ* knockout mutants fail to grow [Georgopoulos et al., 1994], suggesting roles for DnaJ in heat stress-related processes. This increase in physiological importance of DnaJ with increase in temperature is also reflected by the induction of its synthesis after heat-shock treatment as part of the *E. coli* heat-shock response. The levels of DnaJ (< 0.02% of cellular protein; < 200 copies per cell at 37°C) [Neidhardt and VanBogelen, 1987] increase severalfold after heat-shock treatment of the cells.

Further genetic and biochemical analysis identified cellular processes involving the activity of DnaJ. In virtually all of these processes, DnaJ acts as part of the DnaK chaperone system, as evidenced genetically by the similarity of the phenotypes of *dnaJ, dnaK,* and *grpE* mutants [Bukau and Walker, 1989a,b; Georgopoulos et al., 1990; Sell et al., 1990; Ueguchi et al., 1995] and biochemically by in vitro reconstitution of some of these activities. This picture is complicated somewhat by the recent finding that CbpA and DnaJ can partially complement each other in vivo (see later discussion). Demonstrated and postulated functions for DnaJ are as follows:

Folding of Newly Synthesized Proteins. It has been proposed that the Hsp70 chaperone system assists the folding of newly synthesized proteins in prokaryotes and eukaryotes by blocking premature folding through association with nascent chains [Hartl, 1996; Hendrick and Hartl, 1993a; Mayhew and Hartl, 1996]. For a subset of newly synthesized proteins, this interaction is proposed to be followed by association of ring-shaped chaperonins (e.g., GroEL), which assist late-folding steps until the native state is reached [Bochkareva et al., 1988; Frydman and Hartl, 1996; Frydman et al., 1994]. The role of DnaJ proteins in this scenario is to block premature folding of nascent chains, to keep them in a folding competent state, and to target the Hs70 chaperone to them.

There is no firm experimental evidence for such a proposed role of the *E. coli* DnaK system [Bukau et al., 1996]. DnaJ cross-links to nascent chains when added at a high concentration to a cell-free wheat germ translation system [Hendrick et al., 1993b]. However, another study failed to demonstrate cross-links of DnaJ to nascent chains after addition of lysates from wild-type or DnaJ-overproducing cells [Valent et al., 1995], suggesting that detectable cross-links occur only when DnaJ is present in large excess. Furthermore, no cross-links of DnaJ and DnaK to nascent chains were observed in a cell-free *E. coli* translation system [J. Luirink, personal communication]. Finally, deletion of the *dnaJ* and *dnaK* genes is not lethal for *E. coli* at intermediate growth temperatures (e.g., 30°C). [Bukau and Walker, 1989a,b; Georgopoulos et al., 1994] indicating that a potential role of the DnaK system in assisting the cotranslational folding of newly synthesized proteins is not strictly essential. For most *E. coli* proteins, including β-galactosidase, as well as the model protein firefly luciferase produced in *E. coli,* the DnaK chaperone system is dispensable for in vivo folding to native structure at 30°C and higher temperatures [T. Hesterkamp and B. Bukau, manuscript submitted], whereas it is essential for refolding of misfolded luciferase after heat denaturation [Schröder et al., 1993]; however, these findings are valid only for *E. coli.* The role of Hsp70 chaperones in protein folding of newly synthesized proteins in the cytosol of eukaryotic cells may be different [Bukau et al., 1996; Hartl, 1996].

In view of the recent findings for the DnaK–chaperone system of *E. coli* it was proposed [Bukau et al., 1996] that the folding of cytosolic proteins occurs via multiple, flexible pathways. An ordered chaperone pathway involving the DnaK and GroEL systems probably accounts for folding of only a minority of nascent chains. Most newly synthesized proteins may fold to native state without cotranslational assistance by these chaperones. According to this model a variable network of folding pathways exists that comprises alternative routes and backup systems.

Refolding of Nonnative Proteins. *dnaJ259* mutants (as well as various *dnaK* and *grpE* mutants) are defective in the refolding of proteins denatured by stress treatment, including heat-denatured firefly luciferase [Schröder et al., 1993] and λ cI857 repressor [Gaitanaris et al., 1990]. A direct role for the DnaK chaperone system in assisting protein refolding has been demonstrated in vitro for several proteins, including heat-inactivated luciferase [Schröder et al., 1993] RNA polymerase [Blaszczak et al., 1995; Skowyra et al., 1990; Ziemienowicz et al., 1993], chemically unfolded rhodanese [Langer et al., 1992], and luciferase [Buchberger et al., 1994; Szabo et al., 1994]. This activity involves the prevention of protein aggregation and, for RNA polymerase, the disassembly of aggregated substrate [Skowyra et al., 1990; Ziemienowicz et al., 1993]. The activity to refold misfolded proteins is likely to be a major physiological role for DnaJ in *E. coli* metabolism, which becomes essential for *E. coli* growth under stress conditions. The extent of protein misfolding occurring in vivo can be quite substantial. This is suggested by the high amount of aggregated protein accumulating in mutant cells lacking most of the activities of the GroEL and DnaK chaperone systems [Gragerov et al., 1992]. However, the misfolded substrates that require the refolding activity of DnaJ are largely unknown.

Denatured firefly luciferase can shuttle between the DnaJ–DnaK and GroEL chaperones [Buchberger et al., 1996]. Only a small fraction of luciferase molecules released from GroEL can reach the native state. Instead, the released luciferase must bind repeatedly to the DnaK system and, only then, is it able to fold to the native state. These experiments indicate that the DnaK and GroEL chaperone systems do not obligatorily act in succession by promoting earlier- and later-folding steps, but instead, can form a laterally acting chaperone network.

Degradation of Unstable Proteins. *dnaJ259* mutants, as well as the *dnaK756* and *grpE280* missense mutants, are defective in degradation of unstable proteins, including mutant proteins (X90 fragment of β-galactosidase; the alkaline phosphatase mutant protein encoded by the *phoA84* allele), puromycyl fragments, and the *E. coli* heat-shock transcription factor, σ^{32} [Sherman and Goldberg, 1992; Straus et al., 1988, 1990). The mechanism by which DnaJ, together with DnaK and GrpE, promotes the degradation of unstable proteins is unclear. It is conceivable that these chaperones keep protein substrates in a degradation-competent status through transient association, thereby preventing aggregation. In such a scenario, they would act rather indirect by affecting the kinetic partitioning of the substrate between folding to native state (if possible); off-pathway folding, leading to aggregation; and proteolytic degradation. It is also possible, although not shown experimentally, that DnaJ and DnaK play more direct roles in proteolysis by handing over substrates to proteases through formation of ternary substrate–chaperone–protease complexes [Sherman and Goldberg, 1991, 1992]. Furthermore, it is not excluded that occasionally they may induce conformational changes in substrates that render them susceptible to proteolysis.

A puzzling result suggests a role for DnaJ in protein stabilization that is opposite of that just described. Mutations in *dnaJ* cause the destabilization of the catabolite activator protein, CAP (CRP), of *E. coli* [Ohki et al., 1992]. This finding suggests that DnaJ may be required to alter the CAP protein such that it is resistant to proteolytic attack in vivo. It is unfortunate that this finding has not been followed up experimentally.

Protein Translocation. At least some of the conditional protein translocation defects of *secB::Tn5* mutants lacking the SecB chaperone [Kumamoto and Beckwith, 1985] can be suppressed by overexpression of *dnaJ* together with *dnaK* [Altman et al., 1991; Wild et al., 1992, 1996]. It is reasonable to assume that the DnaK chaperone system, including DnaJ, constitutes a chaperone backup system for SecB, which acts to keep precursors in a translocation-competent status.

Autoregulation of Heat-Shock Genes. The stress-dependent regulation of heat-shock gene expression is altered in *dnaJ* (as well as *dnaK* and *grpE*) mutants [Bukau, 1993; Georgopoulos et al., 1994; Gross, 1996; Yura et al., 1993]. The basal level of heat-shock gene transcription is increased two- to fivefold at 30°C and the shutoff phase of the heat-shock response is prolonged [Straus et al., 1990]. These defects are due to a failure of these mutants to mediate the degradation and to repress the activity of σ^{32} [Straus et al., 1989, 1990]. Furthermore, genetic data indicate roles for DnaJ in the repression of translation of the *rpoH* mRNA encoding σ^{32} during the shutoff phase of the heat-shock response [Grossman et al., 1987; Straus et al., 1990].

Biochemical evidence supports that the activity control of σ^{32} relies on the reversible association of DnaJ and DnaK with σ^{32} [Gamer et al., 1992, 1996; Liberek and Georgopoulos, 1993]. Regulation models predict that this association is in direct competition with the association of these chaperones with misfolded proteins accumulating under stress conditions. Accordingly, the sequestering of the chaperones away from σ^{32} during stress induction allows the rapid induction of the heat shock response in *E. coli* [Bukau, 1993; Craig and Gross, 1991; Gamer et al., 1996]. The roles of DnaJ together with DnaK and GrpE in the control of σ^{32} stability and synthesis remain unclear.

DNA Replication. A role for DnaJ in replication has been suggested for chromosomal DNA and demonstrated for DNA of phage λ (see foregoing) and low–copy-number plasmids. The role in chromosomal replication is suggested by the finding that a shift in growth temperature from 30° to 42°C leads to rapid inhibition of DNA synthesis in *dnaJ* mutant cells (Wada et al., 1982). In *dnaK* mutants similar defects exist that are at the level of initiation [Sakakibara, 1988]; thus, it is likely that DnaJ also acts at this stage of replication. DnaK interacts in vitro with DnaA, the initiator protein for *E. coli* chromosomal replication and, thereby, allows the conversion of DnaA from a phospholipid-bound inactive state into an active state [Hwang et al., 1990; Hupp and Kaguni, 1993; Hwang and Kaguni, 1991]. This raises the possibility that the DnaK chaperone system acts at an early step in chromosomal replication by activating DnaA. Such a function may be essential for replication at high growth temperature, but nonessential, although of relevance for optimal replication [Bukau and Walker, 1989b], at low temperature.

DnaJ has essential roles in the replication of P1- and F-derived plasmids. For both plasmids this role was indicated by the demonstration that mutations in *dnaJ,* as well as *dnaK* and *grpE,* block their replication [Bukau and Walker, 1989b; Kawasaki et al., 1992; Tilly et al., 1989]. Biochemical analysis revealed that DnaJ, in cooperation with DnaK and GrpE, is required to activate the RepA initiator protein of P1-plasmid DNA replication for binding to the origin of replication [Wickner et al., 1990; 1991b] by mediating the monomerization of inactive RepA dimers [Wickner et al., 1991a] or, according to an alternative proposal, by activating RepA monomers [DasGupta et al., 1993]. The activation reaction involves the direct association of DnaJ with RepA, probably as part of the regular substrate-binding–release cycle of the DnaK system (discussed later). Similarly, for the F-plasmids, DnaJ increases the ability of the RepE initiator protein for binding to the replication origin of F-derived plasmids [Kawasaki et al., 1992]. Interestingly, a point mutation in the *repE* gene (*repE54*) allows F-derived plasmids to replicate in *dnaJ* mutant cells. The RepE54 mutant protein is monomeric, in contrast with the dimeric $RepE^+$ protein, suggesting that the DnaK system activates RepE by assisting the monomerization of RepE dimers [Ishiai et al., 1994].

Cell Division. A role for DnaJ in cell division is indicated by the finding that *dnaJ* insertion mutants form unseptated short filaments at intermediate (30°C) and high growth temperatures [Sell et al., 1990; B. Bukau, unpublished observation]. For *ΔdnaK52* mutants cells the formation of long filaments, occurring even at 30°C, can be significantly suppressed by overproduction of FtsZ [Bukau and Walker, 1989]. FtsZ is a tubulin-like protein that initiates septum formation and assembles at leading edges of growing septa in ring-like structures [Lutkenhaus, 1992, 1993]. Although traces of DnaK copurify with FtsZ [Mukherjee et al., 1993], it remains unclear whether a physiological relevant in vivo interaction exists between FtsZ and DnaK and, possibly, between the DnaJ and GrpE cochaperones.

Synthesis of Flagella. Cells deleted for *dnaJ* or *dnaK,* or carrying specific *dnaK* or *grpE* mutant alleles, lack motility, because they fail to produce flagella [Shi et al., 1992], the absence of which in these mutants is due to a 10- to 20-fold decrease in the rate of flagellin synthesis. This defect is caused by reduced transcription of the *flhD* master operon and the *fliA* operon encoding the flagellum gene-specific sigma factor σ^F (or σ^{28}). There is no further evidence on the mechanism by which DnaJ interferes with *flhD* expression.

2. *Escherichia coli DnaJ Homologues*

One DnaJ homologue, the CbpA protein (58% identity in the J-domain, 39% overall identity, no Zn-finger), was discovered in a biochemical screen for proteins with affinity for curved DNA (CbpA; curved DNA-binding protein) [Ueguchi et al., 1994]. The *cbpA* gene is controlled by σ^S as part of the general stress regulon and, thus, is induced under nutrient starvation (e.g., in stationary growth phase) [Yamashino et al., 1994].

A knockout mutation of *cbpA* does not result in apparent growth defects [Ueguchi et al., 1994], indicating that CbpA is not essential for viability. There is, however, evidence that CbpA and DnaJ have complementary functions in vivo. First, cellular defects of *dnaJ259* mutants, including temperature-sensitive growth and block of λ DNA replication, can be suppressed by multiple copies of *cbpA* [Ueguchi et al., 1994; Wegrzyn et al., 1996]. Second, *cbpA dnaJ* double-knockout mutants have more severe growth defects than *dnaJ* single mutants, including a very narrow temperature range for growth from 16° to 37°C, defects in cell division, and susceptibility to killing by carbon starvation [Ueguchi et al., 1995]. The phenotypes of these double mutants are almost identical with those reported for *dnaK* knockout mutants [Bukau and Walker, 1989a,b; Spence et al., 1990], suggesting that the CbpA and DnaJ proteins are functionally related and act to cooperate with DnaK, although it remains unclear whether they have identical biological activities. Importantly, the ability of the *cbpA dnaJ* double-null mutants to grow at 30°C, albeit with cellular defects, indicates that the products of both genes are not strictly essential for viability at that temperature.

A second homologue of the *dnaJ* gene was identified in the course of both the *E. coli* genome project (open-reading frame termed *YABH*) [Yura et al., 1992] and as a multicopy suppressor of a mutation conferring resistance to the anti-calmodulin drug W7 [Clarke et al., 1996]. This suppressor gene was referred to as *dj1A* (DnaJ-like). The *dj1A* null mutants have no striking phenotype except an increased spectinomycin sensitivity, indicating cell envelope alterations. Even moderate overexpression of this gene is toxic for cells growing in a minimal medium and results in growth–medium-dependent pleiotropic defects in cell division, chromosomal segregation, and in an altered ability to form mucoid colonies. Dj1A is a type III inner membrane protein that has a single transmem-brane-spanning domain, with a small NH_2-terminal segment localized in the periplasm, and a large COOH-terminal cytosolic domain that shares homology with the J-domain. *dj1A* is unable to complement *dnaJ* null mutants [W. L. Kelley and C. Georgopoulos, unpublished data], suggesting that this family member has distinct functions in vivo. These findings are consistent with the hypothesis that Dj1A has a specialized function in processes correlated with the cell envelope.

A fourth homologue of *dnaJ* (*YFHE* or *hscB*) has been identified by sequencing the chromosomal region near the gene encoding the *dnaK* homologue *hscA* (also *hsc66*) [Lelivelt and Kawula, 1995; Seaton and Vickery, 1996]. The *hscB* gene is expressed together with *hscA* as part of an operon and encodes a rather small homologue of 171 residues. The homology of HscB to *E. coli* DnaJ is confined to the J-domain. The expression of the *hscAB* operon is not induced after heat-shock treatment of the cells, but instead, increases three- to fourfold on cold shock from 37° to 10°C. The physiological roles of this DnaJ homologue are unknown.

B. Role of DnaJ in the Functional Cycle of the DnaK System

The available data indicate that the physiological roles of DnaJ result from its activity as cofactor of the DnaK chaperone system. This role involves stimulation of ATPase activity of DnaK, as well as the binding of substrates and their transfer to DnaK.

1. Control of the ATPase Activity of DnaK

The binding of polypeptide to the substrate-binding domain of DnaK is regulated by the nucleotide state of the ATPase domain [Liberek et al., 1991b; Palleros et al., 1993; Schmid et al., 1994; McCarty et al., 1995] (Fig. 1). The DnaK–ATP conformation exhibits unstable substrate binding, whereas the DnaK–ADP conformation exhibits stable substrate binding. These two affinity states of DnaK are mainly the result of drastic changes in the dissociation constants, with those of DnaK–ADP–substrate complexes being at least two orders of magnitude lower than those of DnaK–ATP–substrate complexes [Schmid et al., 1994]. The intrinsic ATPase activity of DnaK is very low ($k_{cat} = 0.04$ M ATP · M DnaK^{-1} · min^{-1}), with γ-phosphate cleavage being the rate-limiting step [McCarty et al., 1995; Theyssen et al., 1996]. The ATPase activity of DnaK is tightly regulated by DnaJ and GrpE. DnaJ acts as hydrolysis factor [Liberek et al., 1991a; McCarty et al., 1995] that efficiently stimulates the cleavage of γ-phosphate. An apparent dissociation rate constant of 0.2–0.3 μM for activation of the DnaK ATPase by DnaJ has been reported [Karzai and McMacken, 1996]; DnaK and DnaJ do not form stable complexes, and only the transient ATP-dependent interactions can be detected [Wawrzynów and Zylicz, 1995].

In presence of DnaJ, the dissociation of ADP ($k_{off} = 2.1$ min^{-1}) becomes the

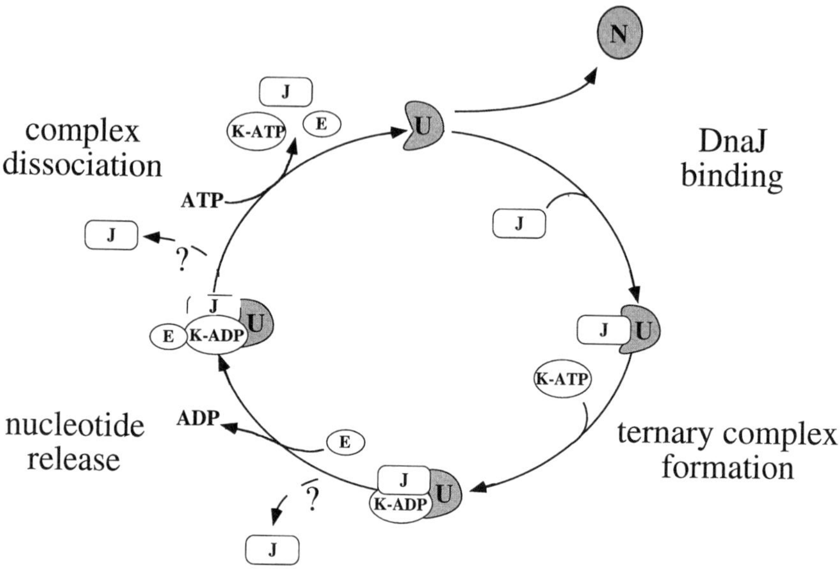

Figure 1 Model of the functional cycle of the DnaK chaperone system: The cycle starts with the rapid association of DnaJ (J) with a protein substrate that is typically unfolded (U). DnaJ, associated with this substrate, interacts with DnaK (K) in the ATP conformation and rapidly stimulates ATP hydrolysis, which results in the formation of a complex of DnaK, ADP, substrate, and DnaJ. DnaJ may dissociate from this complex before DnaK. GrpE (E) can associate with DnaK complexed to the substrate (and possibly DnaJ), thereby triggering nucleotide release and forming nucleotide-free complexes of DnaK, GrpE, and substrate, with or without DnaJ. Spontaneous nucleotide release in the absence of GrpE is considered less important because GrpE is essential for virtually all chaperone functions of the DnaK system. The binding of ATP induces dissociation of the substrate from DnaK. The dissociated substrate partitions between productive folding (N) and rebinding to DnaJ.

rate-limiting step of the ATPase cycle [McCarty et al., 1995]. This limitation is overcome by the nucleotide-exchange factor GrpE [Liberek et al., 1991a; McCarty et al., 1995] that maximally stimulates ADP dissociation 5000-fold ($k_{off} = 10{,}000\ min^{-1}$) [Packschies et al., 1997] and thus allows ATP to bind to DnaK rapidly ($k_{on} = 90\ min^{-1}$) [Theyssen et al., 1996]. In summary, there are two checkpoints for the ATPase cycle of DnaK; ATP hydrolysis and nucleotide exchange, which are tightly controlled by DnaJ and GrpE, respectively. The availability of DnaJ and GrpE, therefore, will critically influence the kinetics of the cycle and, hence, the activity of the DnaK system in protein folding. The efficiency of the chaperone-assisted in vitro refolding of unfolded firefly luciferase is indeed highly sensitive to changes in the concentration of DnaJ and GrpE, relative to DnaK [Szabo et al., 1996; Packschies et al., 1997; T. Laufen, H. Schröder, and B. Bukau, unpublished data].

Recently, the role of DnaJ as a hydrolysis factor for DnaK–ATP has been challenged [Banecki and Zylicz, 1996]. A kinetic correlation between conformational changes within the ATPase domain of DnaK, as measured by a rapidly occurring ATP-induced quench and blueshift of DnaK's tryptophan fluorescence emission maximum [Banecki et al., 1992; Palleros et al., 1993; Buchberger et al., 1995], and a pre–steady-state burst in ATPase activity was observed by these authors, suggesting that (a) the fluorescence changes result from hydrolysis of ATP and (b) hydrolysis of ATP is rapid. Addition of DnaJ reversed the ATP-induced fluorescence changes. This led the authors to conclude that DnaJ interacts with a rapidly formed "high-energy" DnaK*–ADP–P_i conformation, rather than with DnaK–ATP [Banecki and Zylicz, 1996]. However, several lines of evidence disfavor this model: (a) the initial ATPase burst is not generally observed with all DnaK preparations [T. Barthel, G. C. Walker, and J. Reinstein personal communications; J. McCarty, A. Buchberger and B. Bukau, unpublished data]; (b) the absence of an observable initial burst does not affect the ability of DnaJ to alter DnaK's tryptophan fluorescence in the presence of ATP [J. Reinstein, personal communication], which excludes a strong correlation of ATP hydrolysis with ATP-induced fluorescence changes; (c) DnaK mutant proteins deficient in ATP hydrolysis exhibit the same ATP-induced tryptophan fluorescence changes as $DnaK^+$ [T. Laufen and B. Bukau, unpublished data], which demonstrates that the tryptophan fluorescence changes reflect binding, rather than hydrolysis of ATP.

2. Binding to Substrates

DnaJ itself can associate with substrates and, as will be outlined later, this allows the targeting of DnaK to substrates. DnaJ association with substrates has been demonstrated for nonnative conformers of luciferase [Schröder et al., 1993], rhodanese [Langer et al., 1992], and the folded proteins σ^{32} [Gamer et al., 1992, 1996; Liberek et al., 1993], λP [Alfano and McMacken, 1989; Liberek et al., 1990], and RepA [Wickner et al., 1991a,b]. The association of DnaJ with folding rhodanese and luciferase at least partially prevents the aggregation of these substrates, even in the absence of DnaK, thus demonstrating a chaperone activity of DnaJ on its own [Langer et al., 1992; Schröder et al., 1993]. In contrast, DnaK alone, with or without ATP, cannot prevent aggregation of these proteins, indicating the essential role for DnaJ in this process. The ability of DnaJ to prevent the aggregation of luciferase is stimulated by the presence of DnaK and, furthermore, the refolding of luciferase requires the concerted action of DnaJ, DnaK, and GrpE [Schröder et al., 1993]. Therefore, the efficient prevention of aggregation and refolding of this substrate is the product of a cooperative action of both chaperones and, for refolding, of GrpE. In addition to these physiological substrates, association of DnaJ with peptides [A. Fink, personal communication], carboxymethylated α-lactalbumin [Wawrzynów and Zylicz, 1995], and casein [Langer et al., 1992] has been reported.

The association of DnaJ with substrates is apparently rapid enough that it can compete with aggregation reactions. For the native σ^{32} substrate, the binding constants were determined by plasmon surface resonance spectroscopy [Gamer et al., 1996], taking advantage of the fact that this substrate is not aggregating in the absence of chaperones. The association and dissociation rate constants of DnaJ–σ^{32} complexes are high, 3.3×10^5 M^{-1} s^{-1} and 6.2×10^{-3} s^{-1}, respectively. These values are 5900- and 20-fold higher, respectively, than those of DnaK–σ^{32} complexes in the absence of ATP. DnaJ thus interacts only transiently, although with high affinity ($K_d = 20$ nM), with this substrate. It remains to be determined whether DnaJ interactions with other substrates exhibit similar kinetic features.

3. *Role in the Substrate-Binding–Release Cycle*

With its dual role as hydrolysis factor for DnaK and chaperone, possessing its own substrate-binding site, DnaJ plays a key role in the functional cycle of the DnaK–chaperone system by allowing the efficient formation of DnaK–substrate complexes (see Fig. 1) [Gamer et al., 1996; Liberek et al., 1991a; Szabo et al., 1994]. A two-step mechanism for chaperone association with substrates has been proposed [McCarty et al., 1995; Gamer et al., 1996]. Initially, it is the ATP-bound form of DnaK that binds with a high-association rate either to substrate or to DnaJ–substrate complexes. This initial binding is followed by rapid conversion of DnaK–ATP to DnaK–ADP through stimulation by DnaJ bound to substrate, resulting in a lowering of the dissociation rate of the DnaK–substrate complex. This DnaJ-mediated mechanism of stable complex formation leads to an increased affinity of DnaK for substrate (lower k_{off}), thereby allowing the DnaK chaperone system to associate with substrates even at low levels of available chaperone and substrate and providing a kinetic advantage to these chaperones in the competition with aggregation reactions.

Following the formation of DnaK–substrate complexes, GrpE triggers ADP release, rebinding of ATP, and consequently, the release of DnaK-bound substrate. This regulated cycle of substrate binding and release provides a means for repetitive interactions of the DnaK system with aggregation-prone folding intermediates.

This functional cycle may not be entirely conserved within the prokaryotic world. Analysis of the *Thermus thermophilus* DnaK chaperone system revealed the existence of a stable complex, consisting of three molecules each of the *T. thermophilus* DnaK, DnaJ, and a cofactor DafA, which is required for assembly and stability of this complex [Motobashi et al., 1994]. The entire complex possesses an ATPase activity that is lower than the basal ATPase activity reported for *E. coli* DnaK; furthermore, it cannot be disassembled by MgATP, DafA is encoded by a gene that is part of an operon encoding DnaK, DnaJ, and GrpE. The mechanism of action of DafA and the functional consequences of its activity are unclear.

C. Structure–Function Relation of DnaJ

The *dnaJ* gene codes for a basic protein of 376 amino acids that has the NH_2-terminal methionine removed posttranslationally in vivo [Bardwell et al., 1986; Ohki et al., 1986]. The pH value of DnaJ is somewhat unclear, for it has been reported to be pH 7.5 [Schönfeld et al., 1995] and 8.5 [Zylicz et al., 1985]. The processed protein has a molecular mass of 40,973 Da, but migrates as a 37-kDa species under denaturing conditions [Zylicz et al., 1985]. DnaJ was proposed to be a homodimeric protein, as based on its elution profile after glycerol gradient centrifugation [Zylicz et al., 1985]. Analytical ultracentrifugation under equilibrium sedimentation conditions, however, revealed a pH-dependent broad distribution of oligomers, ranging between monomers and 20 mer at pH 7.7, which could not be resolved into single components [Schönfeld et al., 1995]. This behavior is likely to result from a rapid equilibrium between different oligomeric forms of DnaJ. The functionally active form of DnaJ is unclear.

1. Domain Structure

Sequence alignment of DnaJ proteins suggests four regions of homology [Bork et al., 1992; Caplan and Douglas, 1991; Luke et al., 1991]: a highly conserved "J-domain" (at least 50% identity) often located near the NH_2-terminus (residues 2–78 of *E. coli* DnaJ); a G/F motif, possibly acting as a flexible linker (79–105); an adjacent region (106–143), with unknown function, linking the G/F-motif to the "zinc-binding domain" (144–200); and a low-homology region, generally found at the COOH-terminus (201–376; Fig. 2). In contrast with the J-domain, which constitutes the signature for this family, the three other regions are frequently missing in DnaJ homologues (see Table 1) [Bork et al., 1992]. Therefore, DnaJ proteins appear to be mosaics containing different modules with distinct functions.

The existence of domains in *E. coli* DnaJ was demonstrated experimentally by analysis of DnaJ fragments generated by proteolysis and *dnaJ* deletion mutagenesis [Banecki et al., 1996; Karzai and McMacken, 1996; Szabo et al., 1996; Wall et al., 1994, 1995]. Papain digestion of DnaJ generated two stable fragments, with approximate molecular weights of 9 and 30 kDa, that constitute the J-domain (residue 2–75) and COOH-terminal regions (104–376), respectively [Karzai and McMacken, 1996]. The G/F-motif is highly unstable during proteolysis and probably represents an unstructured flexible linker connecting the J-domain with the Zn-binding region. Our current knowledge of the functional and structural characteristics of DnaJ domains is summarized in the following:

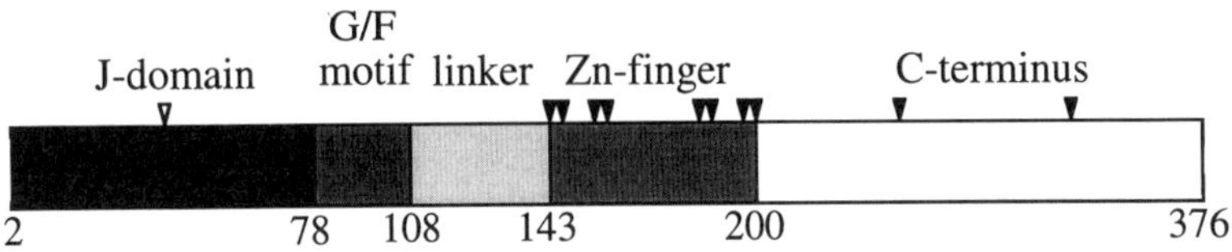

Figure 2 Domain structure of *E. coli* DnaJ: Individual domains are represented by differentially colored boxes. The numbers indicate COOH-terminal domain borders. Closed arrows indicate the cystein residues, the open arrow indicates the HPD motif. The definition of domains is based on homology searches of sequences of DnaJ family members and slightly differs from the domain boundaries identified by proteolytic digestion of DnaJ [Karzai and McMacken, 1996]. For further assignment of the different domains, see Table 1.

2. Activities of Domains of DnaJ

Stimulation of the DnaK ATPase Activity. The J-domain (residues 1–78) is essential for stimulation of the DnaK ATPase by DnaJ. This is evidenced by the findings that deletion of the J-domain (Δ1–73) or single-amino acid alterations within this domain, in particular, within the entirely conserved HPD motif (see following for details), result in a failure of the mutant proteins to stimulate the ATPase activity [Karzai and McMacken, 1996; Wall et al., 1994]. However, the J-domain is insufficient for stimulation of the DnaK ATPase. In single-turnover experiments, fragments of DnaJ comprising the J-domain (1–75) completely fail to stimulate the ATPase activity, even when added at high concentrations [Karzai and McMacken, 1996]. In contrast, a fragment comprising both the J-domain and the adjacent hydrophobic G/F-motif (1–106) stimulates the ATPase of DnaK, although with significantly lower efficiency than wild-type (> 5-fold lower stimulation of the ATPase activity and 20-fold increased K_A for stimulation). In overall agreement with this result are studies showing, by steady-state ATPase assays, that fragments comprising the J-domain and the G/F-motif (1–108; 1–104) stimulate the DnaK ATPase activity [Liberek et al., 1995; Szabo et al., 1996; Wall et al., 1994]. One discrepancy with the former study is that stimulation can reach wild-type levels when the fragments are present in a large molar excess. It is thus safe to conclude that the J-domain, together with the G/F-motif, provides the minimal segment of DnaJ required for efficient although possibly not maximal, stimulation of the DnaK ATPase.

It was speculated that the hydrophobic G/F-motif mimics a DnaK substrate and associates with the substrate-binding pocket of DnaK during DnaJ-mediated stimulation of the ATPase activity [Karzai and McMacken, 1996]. This hypothesis is based on the finding that a J-domain fragment (1–75) becomes about as efficient as full-length DnaJ in stimulating the DnaK ATPase activity when a peptide substrate for DnaK (peptide C of vesicular stomatitis virus [vsv] glycoprotein) is added at saturating concentration [Karzai and McMacken, 1996]. The maximal stimulation of the ATPase activity of DnaK is proposed to occur as result of a "bipartite-signaling mechanism" involving two separate interactions: the J-domain, with an unknown-binding site on DnaK, and the G/F-motif, with the substrate-binding pocket of DnaK. Notably, peptides derived from the G/F-motif fail to cooperate with the J-domain in stimulation of the ATPase activity which, on the basis of this model, would reflect a rather weak affinity of the G/F-motif for DnaK. This limited affinity for DnaK and the structural flexibility may be a prerequisite for the G/F-motif to allow DnaK to switch from an association with the G/F-motif to an association with a protein substrate-bound DnaJ [Karzai and McMacken, 1996]. The failure of a DnaJ mutant protein lacking the G/F-motif (Δ77–107) to target DnaK to substates in presence of ATP [Wall et al., 1995] is in agreement with this model.

It is possible that the proposed bipartite-signaling mechanism acting to stimulate the ATPase of DnaK is conserved within the Hsp70 family. The J-domain of human Hsp40 is also essential, but insufficient, for stimulation of the human Hsc70 ATPase activity, and the G/F-motif in this homologue partially increases the activity of Hsp40 [Minami et al., 1996]. Interestingly, several members of the DnaJ family, including the Sec63 homologue of yeast that is part of the translocation apparatus of the endoplasmic reticulum (ER) membrane and interacts with the luminal Hsp70 homologue BiP, lack the G/F-motif [Feldheim et al., 1992]. In Sec63, it might be the translocating chain itself that provides the second signal required for stimulation of the BiP ATPase and, thereby, couples the import of proteins with the association of BiP.

Clearly, the model of a bipartite-signaling mechanism lacks experimental support for key predictions, such as the switching of DnaK from the G/F-motif to a folding substrate. Furthermore, there are discrepancies in the literature that need clarification relative to the role of the G/F-motif. Thus, it has been reported that an internal deletion removing the G/F-motif (Δ77–107) does not affect the ability of DnaJ to stimulate the ATPase activity of DnaK [Wall et al., 1995], in apparent contradiction with results described earlier. To explain this contradiction, possibly the missing G/F-motif is functionally replaced by sequences that are COOH-terminal to it [Karzai and McMacken, 1996]. However, such potential replacement does not extend to the role of the G/F-motif in targeting DnaK to substrates.

Substrate Binding. The role of different domains of DnaJ in the binding to substrates has been tested using enzyme-linked immunosorbent assay (ELISA) and aggregation assays. The association of DnaJ with σ^{32} is not affected by deletion of the G/F-motif [Wall et al., 1995]. The ability of DnaJ to prevent aggregation of unfolded [^{3}H]rhodanese on dilution from denaturant is preserved in fragments carrying residues 1–209 and 121–376 of DnaJ, but is lost with fragments that comprise only the J-domain and G/F-motif [Szabo et al., 1996].

Furthermore, an internal deletion of the Zn-binding region (Δ144–200) decreases the affinity of DnaJ for some substrates (RepA and λP, but not σ^{32}) [Banecki et al., 1996]. These data are consistent with a role of the region between residues 121 and 209 in substrate binding. This region contains eight cysteines in a spacing (twofold repetition of $C–X_2–C–X_{10–13}–C–X_2–C$) similar to that of cysteines in Zn-finger domains [Cyr et al., 1994]. Two Zn-binding modules have been proposed (close to residues 144–170 and 184–204). Atomic absorption spectroscopy analysis and titration of Zn^{2+} with a chelator did indeed demonstrate binding of two Zn^{2+} ions per DnaJ monomer [Banecki et al., 1996; Szabo et al., 1996]. Zinc binding stabilizes the tertiary structure of DnaJ [Banecki et al., 1996]. X-ray absorption fine structure spectroscopy revealed that the structure of the Zn-binding domain resembles that of Zn-finger proteins [Szabo et al., 1996]. Site-specific mutational alteration of individual cysteine residues of DnaJ decreases the number of Zn atoms bound to DnaJ and the activity of the DnaJ mutant proteins in prevention of rhodanese aggregation, as well as association with, and refolding of, nonnative luciferase. Furthermore, by using a sulfhydryl-specific heterobifunctional cross-linking reagent bound to cysteine residues located in the two cysteine clusters, denatured luciferase can be cross-linked to DnaJ [Szabo et al., 1996]. Together, these findings indicate a direct or indirect involvement of the Zn-binding domain in substrate recognition.

Activity in Protein-Folding Reactions. Several studies testing the activity of DnaJ fragments in DnaJ-assisted–folding reactions showed that such activity requires the complete structure of DnaJ. In the replication of λ DNA in vitro and in vivo, DnaJ fragments were either completely (1–75, 73–376) or largely (1–106/108) inactive [Karzai and McMacken, 1996; Wall et al., 1994]. In refolding of unfolded firefly luciferase, none of several DnaJ fragments tested (1–78; 1–104; 1–127; 1–209; 121–367; 209–376) had detectable chaperone activity, even when present in large excess [Szabo et al., 1996].

It is conceivable that NH_2-terminal fragments of DnaJ (e.g., 1–106) fail to efficiently support chaperone activities because of their failure to associate with substrates. The remaining low activity of such DnaJ fragments in mediating replication of λ DNA [Karzai and McMacken, 1996; Wall et al., 1994] and targeting of DnaK to σ^{32} [Liberek et al., 1995] is likely to result from their ability to stimulate the ATPase activity of DnaK, without being associated with substrates. Here, the conversion of DnaK into an ADP-bound form capable of stable substrate binding is an untargeted process that is likely to be inefficient. Consistent to this hypothesis is that folding reactions requiring the rapid association of the DnaK–DnaJ chaperones (e.g., the refolding of luciferase that is endangered by aggregation) are blocked by truncation of DnaJ.

Three-Dimensional Structure of the J-Domain and G/F-Motif. A nuclear magnetic resonance (NMR)-derived structural model of an NH_2-terminal DnaJ fragment (residues 1–78), comprising the J-domain [Hill et al., 1995], and the complete NMR solution structure of a fragment (residues 2–108), constituting the J-domain and the G/F-motif [Pellecchia et al., 1996; Szyperski et al., 1994] were recently reported. Furthermore, the NMR structure of the J-domain of the human DnaJ homologue Hdj-1, was solved and is highly similar to the corresponding structure of *E. coli* DnaJ [Qian et al., 1996]. The J-domain comprises four helices (Fig. 3), of which the two amphiphilic antiparallel helices 2 and 3 form a coiled coil, connected by a flexible turn accessible to solvent to the J-domain fragment, and the two helices 1 and 4 are oriented perpendicular to it. The structure has a

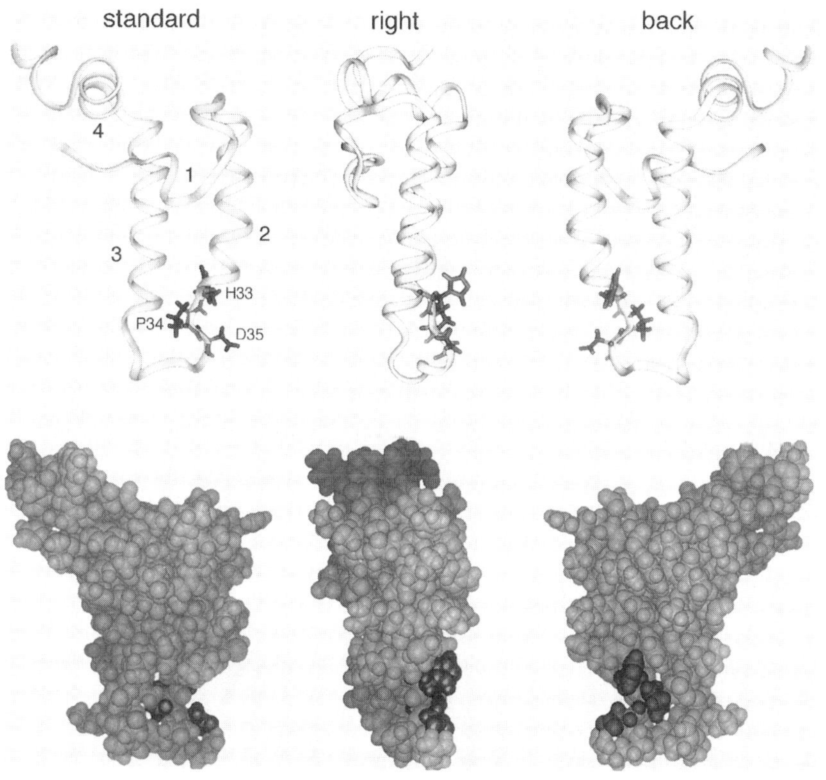

Figure 3 Nuclear magnetic resonance (NMR) structure of the J-domain of *E. coli* DnaJ: One of several similar NMR structures (termed "1" in the pdb-file containing the J-domain coordinates) of the J-domain [Pellecchia et al., 1996] is shown as ribbon (top) and space-filling (bottom) presentation in standard view and rotated rightward by 90° and 180°. The helices are numbered according to the reference, and the conserved HPD motif is shown as sticks (top; side chains in black, backbone in grey) or marked in black (bottom).

hydrophobic core formed by contacts between helices 2 and 3 and additional hydrophobic contacts of these helices with helix 1. The positioning of helix 4 in the NMR structure is less well defined than those of helices 1–3 (Fig. 4). Furthermore, the loop connecting helices 2 and 3 is highly mobile, which may be important for the process of association of DnaJ with DnaK (see later discussion). Even more unstructured in the DnaJ fragment is the G/F-motif located adjacent to the J-domain [Pellecchia et al., 1996; Szyperski et al., 1994] (see Fig. 4).

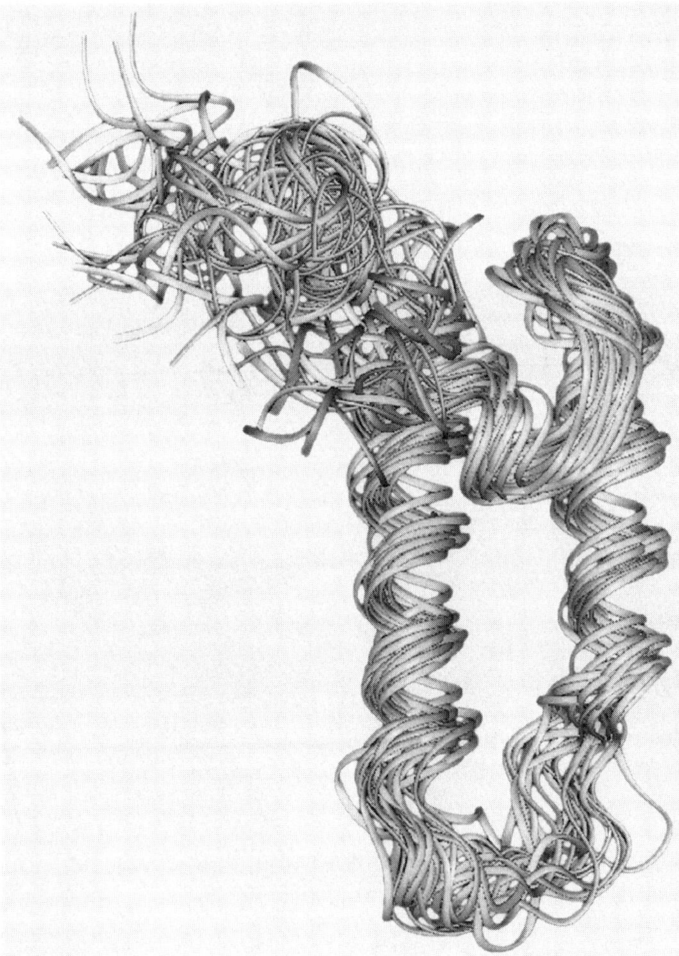

Figure 4 Ribbon blot of the backbones of 20 NMR structures of the J-domain of *E. coli* DnaJ [Pellecchia et al., 1996] in standard view.

Although this is consistent with a role for this region as flexible linker connecting the J-domain with the substrate-binding domain of DnaJ, it is likely that this region has increased rigidity in the context of the full-length molecule.

The highest degree of conservation within the J-domain is found for residues forming helix 2 and the connecting mobile loop. In particular, this loop contains the entirely conserved sequence motif H33–P34–D35 [Bork et al., 1992], alterations in which have strong effects on the activity of DnaJ proteins. The exchanges H33Q in DnaJ259 and corresponding H34Q in yeast Ydj1p completely abolish the ability of these mutant proteins to stimulate the ATPase activity of their Hsp70 partner proteins and, as shown for DnaJ259, to cooperate with DnaK in folding reactions [Tsai and Douglas, 1996; Wall et al., 1994]. The P156N and D157A alterations in the HPD loop of Sec63p inactivates this homologue in protein translocation functions [Feldheim et al., 1992]. Together, these phenotypes suggest a conserved key role for the HPD motif in the interaction between DnaJ and Hsp70 proteins that is required for stimulation of the ATPase activity.

III. EUKARYOTIC DnaJ PROTEINS

A. Occurrence and Functional Diversity

The DnaJ proteins exhibit a large variability for their functions in cell metabolism. This is indicated by their broad distribution within the cell. DnaJ homologues are found in a variety of compartments, including the cytosol, mitochondria, ER, and nucleus. They exist as soluble proteins, in association with membranes through farnesylation modification [Caplan et al., 1992] and as integral membrane proteins [Feldheim et al., 1992]. Furthermore, the different family members show strong variation in their molecular weights, ranging from 167 (Csp of *Bos taurus*) to 1073 (RESA of *Plasmodium falciparum*) residues, and 40% of the DnaJ proteins listed in Table 1 share homology with other family members only in the J-domain.

Eukaryotic DnaJ proteins have been implicated in various cellular processes, including folding of newly synthesized proteins [Bukau et al., 1996; Craig et al., 1993; Hartl, 1996], refolding of misfolded proteins [Hartl, 1996], import of proteins into the ER and mitochondria [Hartl, 1996; Schatz and Dobberstein, 1996], intracellular vesicle traffic [Buchner and Gundersen, 1997; Ungewickell et al., 1995], activity control of regulatory proteins [Kimura et al., 1995], degradation of proteins [Lee et al., 1996; Sherman and Goldberg, 1996], promotion and inhibition of transcription initiatin [Zhong and Arndt, 1993; Lee et al., 1994], tumor suppression [Gateff, 1994], and possibly transformation of cells after viral infection [Kelley and Landry, 1994]. The common feature of these functions appears to be that DnaJ proteins can bind to protein substrates and target them to Hsp70 proteins.

B. Role in the Hsp70 Functional Cycle

The evolutionary conservation of components of the Hsp70 chaperone system suggests that key features of the Hsp70 functional cycle, in particular the role of DnaJ proteins in this cycle, are also conserved. This is supported by the findings that Hsp70 proteins act in protein folding in cooperation with DnaJ homologues [Minami et al., 1996; Cyr, 1995; Freeman and Morimoto, 1996; Freeman et al., 1994], and that there is a significant degree of functional complementation among members of the Hsp70 and DnaJ families from different species. The yeast Ssa1p and Ydj1p proteins, for example, efficiently complement their corresponding *E. coli* homologues in the refolding of denatured firefly luciferase [Levy et al., 1995].

Eukaryotic and prokaryotic DnaJ homologues most likely act at the same stage in the ATPase cycle of Hsp70 proteins. γ-Phosphate cleavage is rate-limiting in the ATPase cycle of eukaryotic Hsp70 members [Ziegelhoffer et al., 1995; Gao et al., 1993; Ha and McKay, 1994; Cheetham et al., 1994]. The ability of DnaJ homologues to stimulate the ATPase activity of Hsp70 has been demonstrated for several eukaryotic pairs, including yeast Ssa1p and Ydj1p [Cyr et al., 1992; Levy et al., 1995; Ziegelhoffer et al., 1995], Ssa1p and Sis1p [James et al., 1997], Ssa2p and Ydj1p [Cyr, 1995], bovine and human Hsp70 and human Hsp40 [Höhfeld et al., 1995; Minami et al., 1996], bovine Hsc70 and rat Csp [Braun et al., 1996], porcine Hsc70 and Hsj1a/b [Cheetham et al., 1994], maize cytosolic Hsp70 and *Arabidopsis* mitochondrial ATJ1 [Kroczynska et al., 1996], human Hsp70 and Hdj1 [Freeman et al., 1994]. Interestingly, human Hsp40 is slower than *E. coli* DnaJ in stimulating the ATP hydrolysis by human Hsc70 [Minami et al., 1996].

Although the cooperation between Hsp70 and DnaJ proteins is a hallmark of the Hsp70 chaperone system, other components of this system are less well conserved. In particular, GrpE homologues appear to exist only in prokaryotes and in mitochondria of eukaryotes. The cochaperone requirement for the refolding of unfolded proteins indeed differs between prokaryotic DnaK and eukaryotic cytosolic Hsp70. Efficient refolding of firefly luciferase by *E. coli* DnaK requires DnaJ as well as GrpE [Schröder et al., 1993; Szabo et al., 1994], whereas its efficient refolding by human Hsp70 and yeast Ssa1p requires only human Hsp40 and yeast Ydj1p, respectively [Freeman et al., 1994; Levy et al., 1995]. Furthermore, the refolding of unfolded β-galactosidase and rhodanese is efficiently assisted by pairs of Hsp70 and DnaJ from human and yeast [Cyr, 1995; Freeman and Morimoto, 1996].

Although the foregoing experiments do not exclude that additional cochaperones may further stimulate the activity of eukaryotic Hsp70 proteins (see later discussion of Hip), they suggest that stimulation of nucleotide exchange is not a strict requirement for the chaperone activity of eukaryotic Hsp70 proteins located in the cytosol and the ER. This is suggested by the finding that stimulation of the yeast Ssa1p and bovine Hsc70 ATPase activities by yeast Ydj1p and human Hsp40, respectively, does not lead to accumulation of Hsp70–ADP complexes [Höhfeld et al., 1995; Ziegelhoffer et al., 1995]. Release of ADP may not be rate-limiting under these conditions and, consistent with this hypothesis, the rate of ADP release by human Hsc70 has been determined to be faster than ADP release by *E. coli* DnaK [Minami et al., 1996]. The second checkpoint of the ATPase cycle of Hsp70 proteins—nucleotide exchange—may thus be controlled differently in prokaryotic and eukaryotic Hsp70 homologues. Rather than stimulation of nucleotide exchange by a GrpE homologue, it may be required that some eukaryotic Hsp70 proteins stabilize the ADP state to maintain productive Hsp70 interactions with substrates. The identification of the Hsp70 interacting protein Hip [Höhfeld et al., 1995], provides a possible solution to this problem. Hip interacts with the ATPase domain of bovine Hsp70, stabilizes the Hsp70–ADP conformation in the presence of Hsp40; thus, it acts as an antagonist to a nucleotide-exchange factor. Hip has been reported to stimulate the activity of bovine Hsp70 and human Hsp40 in the refolding of denatured luciferase [Höhfeld et al., 1995].

In contrast with the results for yeast and bovine homologues, however, the combination of porcine Hsp70 and human Hsj1a/b leads to formation of stable Hsp70–ADP complexes, reminiscent of the prokaryotic system [Cheetham et al., 1994]. It remains to be determined whether these differences between eukaryotic Hsp70 systems reflect variability of the functional cycle of eukaryotic Hsp70 chaperones, or whether they result from differences in the experimental approaches and conditions used to investigate these features.

C. Role of DnaJ Proteins in Providing Functional Specificity in Hsp70 Systems: Possible Mechanisms

Which determinants allow DnaJ proteins to provide functional specificity to Hsp70 proteins remain unclear. The yeast *Saccharomyces cerevisiae* is an excellent example to illustrate this point. Sequencing of the genome identified open-reading frames for 14 Hsp70 (E. Craig, personal communication) and approximately 17 DnaJ homologues (see Table 1). For a subset of the DnaJ open-reading frames, the gene products have been identified and are localized to the cytosol, nucleus, ER, and mitochondria.

It seems conceivable that the cellular localization of DnaJ homologues is one major determinant for functional diversity of Hsp70 systems. Thus, the Sec63p homologues, which is part of the protein translocation machinery involved in the import of precursors into the ER, allows it to target its Hsp70 partner, the luminal Kar2p, to the site of entry of nascent polypeptides. Similarly, the Sis1p is associated with ribosomes, where it is involved in the initiation of translation [Zhong and Arndt, 1993].

A second determinant may be provided by restrictions in the interaction between Hsp70 and DnaJ proteins, as suggested by elegant genetic experiments in yeast. The swapping of the J-domain from the luminal ER homologue Scj1p for a similar region in Sec63p results in a chimeric Sec63p that has retained its activity in protein translocation [Schlenstedt et al., 1995]. This indicates that Kar2p can interact with both DnaJ homologues in the ER. In contrast, the swapping of the J-domains from the cytosolic Sis1p or the mitochondrial Mdj1p for the J-domain in Sec63p results in inactivation of the chimeric Sec63p. Three-amino acid changes in the swapped J-domain of Sis1p (Q13R; K17S; and K42V) are sufficient to render the chimeric Sec63p protein active [Schlenstedt et al., 1995]. These three residues thus appear to be involved in providing specificity of these DnaJ homologue for their Hsp70 partner proteins.

A third determinant may be that DnaJ homologues differ in their substrate specificity and the kinetics of substrate binding. Such a mechanism can apply only to DnaJ homologues containing a substrate-binding domain (e.g., Ydj1 and Sis1 of yeast, and DnaJ and CbpA of *E. coli*). However, there is no information on the substrate specificity of DnaJ homologues. It should also be considered that, irrespective of potential variations in the substrate specificity of DnaJ homologues, these proteins will have to transfer the substrates to an Hsp70 partner protein. Given the high conservation of the architecture of the substrate-binding site of Hsp70 proteins [Rüdiger et al., 1997a,b; Zhu et al., 1996], it seems unlikely that there is room for strong variations in their substrate specificity.

Taken together, the picture of how DnaJ proteins act to specify the function of Hsp70 proteins in cellular processes is only beginning to emerge. It is clear, though, that the highly diverse family of DnaJ proteins provides a key to understanding the cellular roles of the Hsp70 chaperone system.

ACKNOWLEDGMENTS

We thank K. Wüthrich for the coordinates of the J-domain of *E. coli* DnaJ, and E. Deuerling, T. Hesterkamp, and S. Rüdiger for critical comments on the manuscript.

REFERENCES

Alfano, C. and McMacken, R., Ordered assembly of nucleoprotein structures at the bacteriophage λ replication origin during the initiation of DNA replication, *J. Biol. Chem. 264*:10699 (1989).

Altman, E., Kumamoto, C. A., and Emr, S. D., Heat-shock proteins can substitute for SecB function during protein export in *Escherichia coli, EMBO J. 10*:239 (1991).

Altschul, S. F., Gish, W., Miller, W., Myers, E. W., and Lipman, D. J., Basic local alignment search tool. *J. Mol. Biol. 215*:403 (1990).

Banecki, B., Liberek, K., Wall, D., Wawrzynow, A., Georgopoulos, C., Bertoli, E., Tanfani, F., and Zylicz, M., Structure–function analysis of the zinc finger region of the DnaJ molecular chaperone. *J. Biol. Chem. 271*:14840 (1996).

Banecki, B. and Zylicz, M., Real time kinetics of the DnaK/DnaJ/GrpE molecular chaperone machine action, *J. Biol. Chem. 271*:6137 (1996).

Banecki, B., Zylicz, M., Bertoli, E., and Tanfani, F., Structural and functional relationships in DnaK and DnaK756 heat-shock proteins from *Escherichia coli, J. Biol. Chem. 267*:25051 (1992).

Bardwell, J. C., Tilly, K., Craig, E., King, J., Zylicz, M., and Georgopoulos, C., The nucleotide sequence of the *Escherichia coli* K12 *dnaJ*$^+$ gene. A gene that encodes a heat shock protein, *J. Biol. Chem. 261*:1782 (1986).

Bardwell, J. C. A. and Craig, E. A., Major heat shock gene of *Drosophila* and the *Escherichia coli* heat-inducible *dnaK* gene are homologous, *Proc. Natl. Acad. Sci. USA 81*:848 (1984).

Blaszczak, A., Zylicz, M., Georgopoulos, C., and Liberek, K., Both ambient temperature and the DnaK chaperone machine modulate the heat shock response in *Escherichia coli* by regulating the switch between σ^{70} and σ^{32} factors assembled with RNA polymerase, *EMBO J. 14*:5085 (1995).

Bochkareva, E. S., Lissin, N. M., and Girshovich, A. S., Transient association of newly synthesized unfolded proteins with the heat-shock GroEL protein, *Nature 336*:254 (1988).

Bork, P., Sander, C., Valencia, A., and Bukau, B., A module of the DnaJ heat shock proteins found in malaria parasites, *Trends Biochem. Sci. 17*:129 (1992).

Braun, J. E. A., Wilbanks, S. M., and Scheller, R. H., The cysteine string secretory vesicle protein activates Hsc70 ATPase, *J. Biol. Chem. 271*:25989 (1996).

Buchberger, A., Schröder, H., Hesterkamp, T., Schönfeld, H. -J., and Bukau, B., Substrate shuttling between the DnaK and GroEL systems indicates a chaperone network promoting protein folding, *J. Mol. Biol. 261*:329 (1996).

Buchberger, A., Theyssen, H., Schröder, H., McCarty, J. S., Virgallita, G., Milkereit, P., Reinstein, J., and Bukau, B., Nucleotide-induced conformational changes in the ATPase and substrate binding domains of the DnaK chaperone provide evidence for interdomain communication, *J. Biol. Chem. 270*:16903 (1995).

Buchberger, A., Valencia, A., McMacken, R., Sander, C., and Bukau, B., The chaperone function of DnaK requires the coupling of ATPase activity with substrate binding through residue E171, *EMBO J. 13*:1687 (1994).

Buchner, E. and Gundersen, C. B., The DnaJ-like cysteine string protein and exocytotic neurotransmitter release, *Trends Neurol. Sci.* (in press; 1997).

Bukau, B., Regulation of the *E. coli* heat shock response, *Mol. Microbiol. 9*:671 (1993).

Bukau, B., Hesterkamp, H., and Luirink, J., Growing up in a dangerous environment: A network of multiple targeting and folding pathways for nascent polypeptides in the cytosol, *Trends Cell. Biol. 6*:480 (1996).

Bukau, B. and Walker, G. C., Cellular defects caused by deletion of the *Escherichia coli dnaK* gene indicates roles for heat shock protein in normal metabolism, *J. Bacteriol. 171*:2337 (1989a).

Bukau, B. and Walker, G. C., Δ*dnaK* mutants of *Escherichia coli* have defects in chromosomal segregation and plasmid maintenance at normal growth temperatures, *J. Bacteriol. 171*:6030 (1989b).

Bult, C. J., White, O., Olsen, G. J., et al., Complete genome sequence of the methanogenic archaeon, *Methanococcus jannaschii, Science 273*:1058 (1996).

Caplan, A. J. and Douglas, M. G., Characterization of YDJ1: A yeast homologue of the bacterial DnaJ protein, *J. Cell Biol. 114*:609 (1991).

Caplan, A. J., Tsai, J., Casey, P. J., and Douglas, M. G., Farnesylation of YDJ1p is required for function at elevated growth temperatures in *Saccharomyces cerevisiae, J. Biol. Chem. 267*:18890 (1992).

Cheetham, M. E., Jackson, A. P., and Anderton, B. H., Regulation of 70-kDa heat-shock-protein ATPase activity and substrate binding by human DnaJ-like proteins, HSJ1a and HSJ1b, *Eur. J. Biochem. 226*:99 (1994).

Clarke, D. J., Jacq, A., and Holland, I. B., A novel DnaJ-like protein in *Escherichia coli* inserts the cytoplasmic membrane with a type III topology, *Mol. Microbiol. 20*:1273 (1996).

Craig, E. A., Gambill, B. D., and Nelson, R. J., Heat shock proteins: Molecular chaperones of protein synthesis, *Microbiol. Rev. 57*:402 (1993).

Craig, E. A. and Gross, C. A., Is hsp70 the cellular thermometer? *Trends Biol. Sci. 16*:135 (1991).

Cyr, D., Langer, T., and Douglas, M., DnaJ-like proteins: Molecular chaperones and specific regulators of Hsp70, *Trends Biochem. Sci. 19*:176 (1994).

Cyr, D. M., Lu, X., and Douglas, M. G., Regulation of Hsp70 function by a eukaryotic DnaJ homolog, *J. Biol. Chem. 29*:20927 (1992).

Cyr, M. D., Cooperation of the molecular chaperone Ydj1 with specific Hsp70 homologs to suppress protein aggregation, *FEBS Lett. 359*:129 (1995).

DasGupta, S., Mukhopadhyay, G., Papp, P. P., Lewis, M. S., and Chattoraj, D. K., Activation of DNA binding by the monomeric form of the P1 regulation initiator RepA by heat shock proteins DnaJ and DnaK, *J. Mol. Biol. 232*:23 (1993).

Echols, H., Multiple DNA-protein interactions governing high-precision DNA transactions, *Science 233*:1050 (1986).

Feldheim, D., Rothblatt, J., and Schekman, R., Topology and functional domains of Sec63p, an endoplasmic reticulum membrane protein required for secretory protein transloction, *Mol. Cell. Biol. 12*:3288 (1992).

Freeman, B. C. and Morimoto, R. I., The human cytosolic molecular chaperones hsp90, hsp70 (hsc70) and hdj-1 have distinct roles in recognition of a non-native protein and protein refolding, *EMBO J. 15*:2969 (1996).

Freeman, B. C., Myers, M. P., Schumacher, R., and Morimoto, R. I., Identification of a regulatory motif in Hsp70 that affects ATPase activity, substrate binding and interaction with HDJ-1, *EMBO J. 14*:2281 (1994).

Frydman, J. and Hartl, F., Principles of chaperone-assisted protein folding: Differences between in vitro and in vivo mechanisms, *Science 272*:1497 (1996).

Frydman, J., Nimmesgern, E., Ohtsuka, K., and Hartl, F. U., Folding of nascent polypeptide chains in a high molecular mass assembly with molecular chaperones, *Nature 370*:111 (1994).

Gaitanaris, G. A., Rubock, A. G. P., Silverstein, S. J., and Gottesman, M. E., Renaturation of denatured λ repressor requires heat shock proteins, *Cell 61*:1013 (1990).

Gamer, J., Bujard, H., and Bukau, B., Physical interaction between heat shock proteins DnaK, DnaJ, GrpE and the bacterial heat shock transcriptional factor σ^{32}, *Cell 69*:833 (1992).

Gamer, J., Multhaup, G., Tomoyasu, T., McCarty, J. S., Rüdiger, S., Schönfeld, H. -J., Schirra, C., Bujard, H., and Bukau, B., A cycle of binding and release of the DnaK, DnaJ and GrpE chaperones regulates activity of the *E. coli* heat shock transcription factor σ^{32}, *EMBO J. 15*:607 (1996).

Gao, B., Yumiko, E., Greene, L., and Eisenberg, E., Nucleotide binding properties of bovine brain uncoating ATPase, *J. Biol. Chem. 268*:8507 (1993).

Gateff, E., Tumor suppressor and overgrowth suppressor genes of *Drosophila melanogaster:* Developmental aspects, *Int. J. Dev. Biol. 38*:565 (1994).

Georgopoulos, C., Ang, D., Liberek, K., and Zylicz, M., Properties of the *Escherichia coli* heat shock proteins and their role in bacteriophage λ growth, in *Stress Proteins in Biology and Medicine* (Morimoto, R., Tissieres, A., and Georgopoulos, C., eds.), Cold Spring Harbor Press, Cold Spring Harbor, NY, 1990, p. 191.

Georgopoulos, C., Liberek, K., Zylicz, M., and Ang, D., Properties of the heat shock proteins of *Escherichia coli* and the autoregulation of the heat shock response, in *The Biology of Heat Shock Proteins and Molecular Chaperones* (Morimoto, R. I., Tissières, A., Georgopoulos, C., ed.), Cold Spring Harbor Press, Cold Spring Harbor, NY, 1994, p. 209.

Gragerov, A., Nudler, E., Komissarova, N., Gaitanaris, G., Gottesman, M., and Nikiforov, V., Cooperation of GroEL/GroES and DnaK/DnaJ heat shock proteins in preventing protein misfolding in *Escherichia coli, Proc. Natl. Acad. Sci. USA 89*:10341 (1992).

Grossman, A. D., Straus, D. B., Walter, W. A., and Gross, C. A., sigma 32 Synthesis can regulate the synthesis of heat shock proteins in *Escherichia coli, Genes Dev. 1*:179 (1987).

Ha, J. -H. and McKay, D. B., ATPase kinetics of recombinant bovine 70 kDa heat shock cognate protein and its amino-terminal ATPase domain, *Biochemistry 33*:14625 (1994).

Hartl, F. U., Molecular chaperones in cellular protein folding, *Nature 381*:571 (1996).

Hendrick, J. P. and Hartl, F. -U., Molecular chaperone functions of heat-shock proteins, *Annu. Rev. Biochem. 62*:349 (1993a).

Hendrick, J. P., Langer, T., Davis, T. A., Hartl, F. U., and Wiedmann, M., Control of folding and membrane translocation by binding of the chaperone DnaJ to nascent polypeptides, *Proc. Natl. Acad. Sci. USA 90*:10216 (1993b).

Hill, R., Flanagan, J., and Prestegard, J., ^{1}H and ^{15}N magnetic resonance assignments, secondary structure, and teriary fold of *Escherichia coli* DnaJ(1–78), *Biochemistry 34*:5587 (1995).

Höhfeld, J., Minami, Y., and Hartl, F. U., Hip, a novel cochaperone involved in the eukaryotic Hsc70/Hsp40 reaction cycle, *Cell 83*:589 (1995).

Hupp, T. and Kaguni, J., Activation of mutant forms of DnaA protein of *Escherichia coli* by DnaK and GrpE proteins occurs prior to DNA replication, *J. Biol. Chem. 268*:13143 (1993).

Hwang, D. S., Crooke, E., and Kornberg, A., Aggregated DnaA protein is dissociated and activated for DNA replication by phospholipase or DnaK protein, *J. BIol. Chem. 265*:19244 (1990).

Hwang, D. S., and Kaguni, J. M., DnaK protein stimulates a mutant form of DnaA protein in *Escherichia coli* DNA replication, *J. Biol. Chem. 266*:7537 (1991).

Ishiai, M., Wada, C., Kawasaki, Y., and Yura, T., Replication initiator protein RepE of mini-F plasmid: Functional differentiation between monomers (initiator) and dimers (autogenous repressor), *Proc. Natl. Acad. Sci. USA 91*:3839 (1994).

James, P., Pfund, C., and Craig, E. A., Functional specificity among Hsp70 molecular chaperones, *Science 275*:387 (1997).

Jordan, R. and McMacken, R., Modulation of the ATPase activity of the molecular chaperone DnaK by peptides and the DnaJ and GrpE heat shock proteins, *J. Biol. Chem. 270*:4563 (1995).

Karzai, A. W. and McMacken, R., A bipartite signaling mechanism involved in DnaJ-mediated activation of the *Escherichia coli* DnaK protein, *J. Biol. Chem. 271*:11236 (1996).

Kawasaki, Y., Wada, C., and Yura, T., Roles of *Escherichia coli* heat shock proteins DnaK, DnaJ and GrpE in mini-F plasmid replication, *Mol. Gen. Genet. 220*:277 (1990).

Kawasaki, Y., Wada, C., and Yura, T., Binding of RepE initiator protein to mini-F DNA origin (ori2). Enhancing effects of repE mutations and DnaJ heat shock protein, *J. Biol. Chem. 267*:11520 (1992).

Kelley, W. L. and Landry, S. J., Chaperone power in a virus? *Trends Biochem. Sci. 19*:277 (1994).

Kimura, Y., Yahara, I., and Lindquist, S., Role of the protein chaperone YDJ1 in establishing Hsp90-mediated signal transduction pathways, *Science 268*:1362 (1995).

Kroczynska, B., Zhou, R., Wood, C., and Miernyk, J. A., AtJ1, a mitochondrial homologue of the *Escherichia coli* DnaJ protein, *Plant Mol. Biol. 31*:619 (1996).

Kumamoto, C. A. and Beckwith, J., Evidence for specificity at an early step in protein export in *Escherichia coli, J. Bacteriol. 163*:267 (1985).

Langer, T., Lu, C., Echols, H., Flanagan, J., Hayer, M. K., and Hartl, F. U., Successive action of DnaK, DnaJ and GroEL along the pathway of chaperone-mediated protein folding, *Nature 356*:683 (1992).

Lee, D. H., Sherman, M. Y., and Goldberg, A. L., Involvement of the molecular chaperone Ydj1 in the ubiquitin-dependent degradation of short-lived and abnormal proteins in *Saccharomyces cerevisiae, Mol. Cell. Biol. 16*:4773 (1996).

Lee, T. G., Tang, N., Thompson, S., Miller J, and Katze, M. G., The 58,000-dalton cellular inhibitor of the interferon-induced double-stranded RNA-activated protein kinase (PKR) is a member of the tetratricopeptide repeat family of proteins, *Mol. Cell. Biol. 14*:2331 (1994).

Lelivelt, M. and Kawula, T., Hsc66, an Hsp70 homolog in *Escherichia coli,* is induced by cold shock but not by heat shock, *J. Bacteriol. 177*:4900 (1995).

Levy, E. J., McCarty, J., Bukau, B., and Chirico, W. J., Conserved ATPase and luciferase refolding activities between bacteria and yeast Hsp70 chaperones and modulator, *FEBS Lett. 368*:435 (1995).

Liberek, K., Osipiuk, J., Zylicz, M., Ang, D., Skorko, J., and Georgopoulos, C., Physical interactions between bacteriophage and *Escherichia coli* proteins required for initiation of λ DNA replication, *J. Biol. Chem. 265*:3022 (1990).

Liberek, K. and Georgopoulos, C., Autoregulation of the *Escherichia coli* heat shock response by the DnaK and DnaJ heat shock proteins, *Proc. Natl. Acad. Sci. USA 90*:11019 (1993).

Liberek, K., Marszalek, J., Ang, D., Georgopoulos, C., and Zylicz, M., *Escherichia coli* DnaJ and GrpE heat shock proteins jointly stimulate ATPase activity of DnaK, *Proc. Natl. Acad. Sci. USA 88*:2874 (1991a).

Liberek, K., Skowyra, D., Zylicz, M., Johnson, C., and Georgopoulos, C., The *Escherichia coli* DnaK chaperone, the 70-kDa heat shock protein eukaryotic equivalent, changes conformation upon ATP hydrolysis, thus triggering its dissociation from a bound target protein, *J. Biol. Chem. 266*:14491 (1991b).

Liberek, K., Wall, D., and Georgopoulos, C., The DnaJ chaperone catalytically activates the DnaK chaperone to preferentially bind the sigma 32 heat shock transcriptional regulator, *Proc. Natl. Acad. Sci. USA 92*:6224 (1995).

Luke, M. M., Sutton, A., and Arndt, K. T., Characterization of SIS1, a *Saccharomyces cerevisiae* homologue of bacterial dnaJ proteins, *J. Cell Biol. 114*:623 (1991).

Lutkenhaus, J., Isolation and characterization of *ftsZ* alleles that affect morphology, *J. Bacteriol. 174*:5414 (1992).

Lutkenhaus, J., FtsZ ring in bacterial cytokinesis, *Mol. Microbiol. 9*:403 (1993).

Mayhew, M. and Hartl, F. U., Molecular chaperone proteins, in Escherichia coli and Salmonella, (Neidhardt, F. C., ed.), ASM Press, Washington, DC, 1996, p. 922.

McCarty, J. S., Buchberger, A., Reinstein, J., and Bukau, B., The role of ATP in the functional cycle of the DnaK chaperone system, *J. Mol. Biol. 249*:126 (1995).

Minami, Y., Höhfeld, J., Ohtsuka, K., and Hartl, F. -U., Regulation of the heat-shock protein 70 reaction cycle by the mammalian DnaJ homolog, Hsp40, *J. Biol. Chem. 271*:19617 (1996).

Motohashi, K., Taguchi, H., Ishii, N., and Yoshida, M., Isolation of the stable hexameric DnaK–DnaJ complex from *Thermus thermophilus, J. Biol. Chem. 269*:27074 (1994).

Motohashi, K., Yohda, M., Endo, I., and Yoshida, M., A novel factor required for the assembly of the DnaK and DnaJ chaperones of *Thermus thermophilus, J. Biol. Chem. 271*:17343 (1996).

Mukherjee, A., Dai, K., and Lutkenhaus, J., *Escherichia coli* cell division protein FtsZ is a guanine nucleotide binding protein, *Proc. Natl. Acad. Sci. USA 90*:1053 (1993).

Neidhardt, F. C. and vanBogelen, R. A., Heat shock response, in Escherichia coli and Salmonella typhimurium: *Cellular and Molecular biology* (Neidhard, F. C., ed.), ASM Press, Washington, DC, 1987, p. 1334.

Ohki, M., Tamura, F., Nishimura, S., and Uchida, H., Nucleotide of the *Escherichia coli dnaJ* gene and purification of the gene product, *J. Biol. Chem. 261*:1778 (1986).

Ohki, R., Kawamata, T., Katoh, Y., Hosoda, F., and Ohki, M., *Escherichia coli dnaJ* deletion mutation results in loss of stability of a positive regulator, CRP, *J. Biol. Chem. 267*:13180 (1992).

Packschies, L., Theyssen, H., Buchberger, A., Bukau, B., Goody, R. S., and Reinstein, J., The molecular chaperone GrpE accelerates nucleotide exchange of DnaK by an associative displacement mechanism, *Biochemistry 36*:3417 (1997).

Palleros, D. R., Reid, K. L., Shi, L., and Fink. A. L., DnaK ATPase activity revisited, *FEBS Lett. 336*:124 (1993).

Pellecchia, M., Szyperski, T., Wall, D., Georgopoulos, C., and Wüthrich, NMR structure of the J-domain and the Gly/Phe-rich region of the *Escherichia coli* DnaJ chaperone, *J. Mol. Biol. 260*:236 (1996).

Qian, Y. Q., Patel, D., Hartl, F. U., and D. J. McColl, Nuclear magnetic resonance solution structure of the human Hsp40 (HDJ-1) J-domain, *J. Mol. Biol. 260*:224 (1996).

Rüdiger, S., Buchberger, A., and Bukau, B., Principles governing the interaction of Hsp70 chaperones with substrates, *Nature Struct. Biol. 7*:1501 (1997a).

Rüdiger, S., Germeroth, L., Schneider-Mergener, J., and Bukau, B., Substrate specificity of the DnaK chaperone determined by screening of cellulose-bound peptide libraries, *EMBO J.* (in press; 1997b).

Sadler, I., Chiang, A., Kurihara, T., Rothblatt, J., Way, J., and Silver, P., A yeast gene important for protein assembly into the endoplasmic reticulum and the nucleus has homology to DnaJ, an *Escherichia coli* heat shock protein, *J. Cell BIol. 109*:2665 (1989).

Saito, H. and Uchida, H., Initiation of the DNA replication of bacteriophage lambda in *Escherichia coli* K12, *J. Mol. Biol. 113*:1 (1977).

Sakakibara, Y., The *dnaK* gene of *Escherichia coli* functions in initiation of chromosome replication, *170*:972 (1988).

Schatz, G. and Dobberstein, B., Common principles of protein translocation across membranes, *Science 271*:1519 (1996).

Schlenstedt, G., Harris, S., Risse, B., Lill, R., and Silver, P. A., A yeast DnaJ homologue, Scj1p, can function in the endoplasmic reticulum with BiP/Kar2p via a conserved domain that specifies interactions with Hsp70s, *J. Cell Biol. 129*:979 (1995).

Schmid, D., Baici, A., Gehring, H., and Christen, P., Kinetics of molecular chaperone action, *Science 263*:971 (1994).

Schönfeld, H. J., Schmidt, D., and Zulauf, M., Investigation of the molecular chaperone DnaJ by analytical ultracentrifugation, *Prog. Colloid Polym. Sci. 99*:7 (1995).

Schröder, H., Langer, T., Hartl, F. -U., and Bukau, B., DnaK, DnaJ, GrpE form a cellular chaperone machinery capable of repairing heat-induced protein damage, *EMBO J. 12*:4137 (1993).

Seaton, B. L. and Vickery, L. E., A gene encoding Dnak/hsp70 homologue in *Escherichia coli, Proc. Natl. Acad. Sci. U.S.A. 91*:2066 (1994).

Sell, S. M., Eisen, C., Ang, D., Zylicz, M., and Georgopoulos, C., Isolation and characterization of *dnaJ* null mutants of *Escherichia coli, J. Bacteriol. 172*:4827 (1990).

Sherman, M. and Goldberg, A. L., Involvement of the chaperon DnaK in the rapid degradation of a mutant protein in *Escherichia coli, EMBO J. 11*:71 (1992).

Sherman, M. Y. and Goldberg, A. L., Formation in vitro of complexes between an abnormal fusion protein and the heat shock proteins from *Escherichia coli* and yeast mitochondria, *J. Bacteriol. 173*:7249 (1991).

Sherman, M. Y. and Goldberg, A. L., Involvement of molecular chaperones in intracellular protein breakdown, *EXS 77*:57 (1996).

Shi, W., Zhou, Y., Wild, J., Adler, J., and Gross C. A., DnaK, DnaJ, and GrpE are required for flagellum synthesis in *Escherichia coli, J. Bacteriol. 174*:6256 (1992).

Silver, P. A. and Way, J. C., Eucaryotic DnaJ homologs and the specificity of Hsp70 activity, *Cell 74*:5 (1993).

Skowyra, D., Georgopoulos, C., and Zylicz, M., The *E. coli dnaK* gene product, the Hsp70 homolog, can reactivate heat-inactivated RNA polymerase in an ATP hydrolysis-dependent manner, *Cell 62*:939 (1990).

Spence, J., Cegielska, A., and Georgopoulos, C., Role of *Escherichia coli* heat shock proteins DnaK and HtpG (C62.5) in response to nutritional deprivation, *J. Bacteriol. 172*:7157 (1990).

Straus, D., Walter, W., and Gross, C., DnaK, DnaJ, and GrpE heat shock proteins negatively regulate heat shock gene expression by controlling the synthesis and stability of σ^{32}, *Genes Dev. 4*:2202 (1990).

Straus, D. B., Walter, W. A., and Gross, C. A., *Escherichia coli* heat shock gene mutants are defective in proteolysis, *Genes Dev. 2*:1851 (1988).

Straus, D. B., Walter, W. A., and Gross, C. A., The activity of σ^{32} is reduced under conditions of excess heat shock protein production in *Escherichia coli, Genes Dev. 3*:2003 (1989).

Sunshine, M., Feiss, M., Stuart, J., and Yochem, Y., A new host gene (*groPC*) is necessary for lambda DNA replication, *Mol. Gen. Genet. 151*:27 (1977).

Szabo, A., Korzun, R., Hartl, F. U., and Flanagan, J., A zinc finger-like domain of the molecular chaperone DnaJ is involved in binding to denatured protein substrates, *EMBO J. 15*:408 (1996).

Szabo, A., Langer, T., Schröder, H., Flanagan, J., Bukau, B., and Hartl, F. U., The ATP hydrolysis-dependent reaction cycle of the *Escherichia coli* Hsp70 system—DnaK, DnaJ and GrpE, *Proc. Natl. Acad. Sci. USA 91*:10345 (1994).

Szyperski, T., Pellecchia, M., Wall, D., Georgopoulos, C, and Wüthrich, K., NMR structure determination of the *Escherichia coli* DnaJ molecular chaperone: Secondary structure and backbone fold of the N-terminal region (residues 2–108) containing the highly conserved J domain, *Proc. Natl. Acad. Sci. USA 91*:11343 (1994).

Theyssen, H., Schuster, H. -P., Bukau, B., and Reinstein, J., The second step of ATP binding to DnaK induces peptide release, *J. Mol. Biol. 263*:657 (1996).

Tilly, K. and Yarmolinsky, M., Participation of *Escherichia coli* heat shock proteins DnaJ, DnaK, and GrpE in P1 plasmid replication, *J. Bacteriol. 171*:6025 (1989).

Tsai, J. and Douglas, M. G., A conserved HPD sequence of the J-domain is necessary for YDJ1 stimulation of Hsp70 ATPase activity at a site distinct from substrate binding, *J. Biol. Chem. 271*:9347 (1996).

Ueguchi, C., Kakeda, M., Yamada, H., and Mizuno, T., An analogue of the DnaJ molecular chaperoine in *Escherichia coli, Proc. Natl. Acad. Sci. USA 91*:1054 (1994).

Ueguchi, C., Shiozawa, T., Kakeda, M., Yamada, H., and Mizuno, T., A study of the double mutation of *dnaJ* and *cbpA,* whose gene products function as molecular chaperones in *Escherichia coli, J. Bacteriol. 177*:3894 (1995).

Ungewickell, E., Ungewickell, H., Holstein, S. E., Lindner, R., Prasad, K., Barouch, W., Martin, B., Greene, L. E., and Eisenberg, E., Role of auxilin in uncoating clathrin-coated vesicles, *Nature 378*:632 (1995).

Valent, Q. A., Kendall, D. A., High, S., Kusters, R., Oudega, B., and Luirink, J., Early events in preprotein recognition in *E. coli:* Interaction of SRP and trigger factor with nascent polypeptides, *EMBO J. 14*:5494 (1995).

Wada, M., Kadokami, Y., and Itikawa, H., Thermosensitive synthesis of DNA and RNA in DnaJ mutants of *Escherichia coli* K-12, *Jpn. J. Genet. 57*:407 (1982).

Wall, D., Zylicz, M., and Georgopoulos, C., The NH_2-terminal 108 amino acids of the *Escherichia coli* DnaJ protein stimulate the ATPase activity of DnaK and are sufficient for λ replication, *J. Biol. Chem. 269*:5446 (1994).

Wall, D., Zylicz, M., and Georgopoulos, C., The conserved G/F motif of the DnaJ chaperone is necessary for the activation of the substrate binding properties of the DnaK chaperone, *J. Biol. Chem. 270*:2139 (1995).

Wawrzynów, A. and Zylicz, M., Divergent effects of ATP on the binding of the DnaK and DnaJ chaperones to each other, or to their various native and denatured protein substrates, *J. Biol. Chem. 270*:19300 (1995).

Wegrzyn, A., Taylor, K., and Wegrzyn, G., The *cbpA* chaperone gene function compensates for *dnaJ* in lambda plasmid replication during amino acid starvation of *Escherichia coli, J. Bacteriol. 178*:5847 (1996).

Wickner, S., Hoskins, J., and McKenney, K., Monomerization of RepA dimers by heat shock proteins activates binding to DNA replication origins, *Proc. Natl. Acad. Sci. USA 88*:7903 (1991a).

Wickner, S., Hoskins, J., and McKenny, K., Function of DnaJ and DnaK as chaperones of origin specific DNA binding by RepA, *Nature 350*:165 (1991b).

Wickner, S. H., Three *Escherichia coli* heat shock proteins are required for P1 plasmid replication: Formation of an active complex between *E. coli* DnaJ protein and the P1 initiator protein, *Proc. Natl. Acad. Sci. USA 87*:2690 (1990).

Wild, J., Altman, E., Yura, T., and Gross, C. A., DnaK and DnaJ heat shock proteins participate in protein export in *Escherichia coli, Genes Dev. 6*:1165 (1992).

Wild, J., Rossmeissl, P., Walter, W. A., and Gross, C. A., Involvement of the DnaK–DnaJ–GrpE chaperone team in protein secretion in *Escherichia coli, J. Bacteriol. 178*:3608 (1996).

Yamashino, T., Kakeda, M., Ueguchi, C., and Mizuno, T., An analogue of the DnaJ molecular chaperone whose expression is controlled by σ^S during the stationary phase and phosphate starvation in *Escherichia coli, Mol. Microbiol. 13*:475 (1994).

Yura, T., Mori, H., Hiroki, N., Nagat, T., Ishihama, A., Fujita, N., Isono, K., Mizobuchi, K., and Nakata, A., Systematic sequencing of the *Escherichia coli* genome: Analysis of the 0–2.4 min. region, *Nucleic Acids Res. 20*:3305 (1992).

Yura, T., Nagai, H., and Mori, H., Regulation of the heat-shock response in bacteria, *Annu. Rev. Microbiol. 47*:321 (1993).

Zhong, T. and Arndt, K. T., The yeast SIS1 protein, a DnaJ homolog, is required for the initiation of translation, *Cell 73*:1175 (1993).

Zhu, X., Zhao, X, Burkholder, W. F., Gragerov, A., Ogata, C. M., Gottesman, M., and Hendrickson, W. A., Structural analysis of substrate binding by the molecular chaperone DnaK, *Science 272*:1606 (1996).

Ziegelhoffer, T., Lopez-Buesa, P., and Craig E., The dissociation of ATP from hsp70 of *Saccharomyces cerevisiae* is stimulated by both Ydj1p and peptide substrates, *J. Biol. Chem. 270*:10412 (1995).

Ziemienowicz, A., Skowyra, D., Zeilstra-Ryalls, J., Fayet, O., Georgopoulos, C., and Zylicz M., Both the *Escherichia coli* chaperone systems GroEL/Gro/ES and DnaK/DnaJ/GrpE, can activate heat treated RNA polymerase. Different mechanisms for the same activity, *J. Biol. Chem. 268*:25425 (1993).

Zylicz, M., Yamamoto, T., McKittrick, N., Sell, S., and Georgopoulos, C., Purification and properties of the dnaJ replication protein of *Escherichia coli, J. Biol. Chem. 260*:7591 (1985).

12

Some Structural Aspects of Chaperonin-Assisted Folding by GroEL and GroES

Paul M. Horowitz
The University of Texas Health Science Center, San Antonio, Texas

I. OVERVIEW OF THE MOLECULAR CHAPERONES GroEL AND GroES

The essential *Escherichia coli* proteins GroEL and GroES are ring-shaped, oligomeric molecular chaperones that have been suggested to be involved in functions including protein folding, protein turnover [1], and RNA processing and degradation [1–3]. All of these basic functions may be related to the ability to make interactive surfaces on GroEL appear and disappear in an ATP-dependent manner, thereby permitting controlled binding, sequestration and release of nonnative conformers of proteins for folding, degradation, modification, or assembly. These proteins, which belong to a class of ubiquitous and conserved prokaryotic proteins called chaperonins, are by far the most widely studied for their ability to facilitate the folding and assembly of many diverse proteins. GroEL is homologous with a mitochondrial matrix heat-shock protein Hsp60, which is required for import, folding, and maturation of mitochondrial matrix proteins. GroEL is a tetradecamer (14 mer) consisting of 14 identical 58-kDa subunits with ATPase activity, and it is one of the most negatively charged proteins in the cell. The active protein consists of two stacked rings of seven subunits each. Its full activity in assisting folding requires MgATP, K^+, and a second protein, GroES, which forms a heptamer (7 mer) composed of seven identical 10-kDa subunits. In the presence of ATP, the GroEL and GroES rings bind to each other, and this binding inhibits the ATPase activity of GroEL. The GroEL–GroES system is one of the most extensively reviewed in biochemistry [4 (especially Chaps. 7 and 9), 5–7]. These reviews represent excellent sources, and they should be consulted directly, especially to appreciate the beautiful structures, and to find details concerning the effects of ligands that are only peripherally treated here. In addition, recent publications dealing with chaperonins, especially GroEL, along with a wealth of additional information are available on the Internet at the Chaperonin Home Page [8]. In this review, we will focus on some conformational aspects of chaperonin-assisted protein folding, with an emphasis on GroEL. In view of the rich sources available, the material presented here will be selective, so some interesting topics will be treated only lightly.

GroEL is the central player in assisting the folding of other proteins and, although we will concentrate on the complete system including GroES, GroEL alone can assist the folding of appropriate proteins. Both GroEL and GroES are encoded within the same operon of the *E. coli* genome with GroES positioned first. Thus, it might be expected that these proteins would be produced in a 1:1 ratio of polypeptide chains.

The basic folding function is exerted by binding interactive, nonnative protein structures, thereby limiting aggregation or misfolding that would compete with the acquisition of the native structure of a great variety of target proteins. This interaction has the effect of limiting aggregation and, at the same time, retarding or even arresting any folding of the target protein. Subsequent steps, which in the most efficient processes involve ATP hydrolysis and the co-chaperonin GroES, lead to protein release and folding. These features by themselves imply that, under some circumstances, GroEL would be expected to unfold proteins or to dissolve aggregates. GroEL does not bind tightly to native proteins, and it shows very little specificity for the sequences of target proteins, e.g., at least half of all the proteins in *E. coli* can be bound to GroEL [9], although there appears to be only enough GroEL to facilitate the folding of 5% of the proteins in *E. coli* [10]. These facts imply that, although the interactions involving GroEL are general, they are not universal, so some elements of specificity must be operating. Therefore, it is expected that the recognition of folding intermediates must involve general physicochemical signals present on folding intermediates that are not expressed on native proteins and are not necessarily contained within specific amino acid sequences.

This review will emphasize the protein-folding aspects of the GroEL/ES system, and it will present an overview of the structural aspects of the process that must be accounted for in any mechanism that captures the essence of the assisted-folding process. Overall, the interactions that can be responsible for the essential events are so general and the substrates are so diverse, that the detailed mechanism for a particular system is likely to be a consensus pathway chosen from the repertoire of interactions and kinetics for a particular substrate. Thus, there are potentials inherent in the structure of GroEL that must interact with surfaces on the target protein within a temporal window, defined by the rate at which the target folds to a form that does not interact with the chaperonin. In addition, the interaction with the substrate protein can influence the structure of GroEL; hence, the nature of the substrate can also affect the behavior of the GroEL.

II. PROTEIN MISFOLDING

It is accepted that the information for protein folding is encoded in the one-dimensional sequence of the genetic code that is translated into the amino acid sequence of the protein in question [11]. In this view, proteins know how to fold, so they do not need to be instructed by GroEL. Thus, with the exception of influences of posttranslational modifications or special interactions, the three-dimensional structure of a protein is thermodynamically predestined and ordained by its sequence. However, many proteins do not refold efficiently after they are unfolded. This raises the questions: What information is in the code for the sequence, and what information is in the mechanism or environment in which the code is expressed? That is, how does the biological machinery affect folding [12]? Among the many reasons for the apparent discrepancy between expected and observed protein folding is that competing processes kinetically limit renaturation. Thus, although the native state may be the most stable, it is not accessible in a biologically reasonable time, and is characterized, even in vitro, as being "irreversible." Aggregation is one of the major kinetic traps that competes with protein folding. For example, inappropriate association of partially folded proteins has been suggested as leading to the inclusion bodies that limit the production of correctly folded recombinant proteins in *E. coli.* This picture of kinetic traps involving off-pathway interactions is in keeping with recent discussions of the basic features of protein folding [13,14]. Thus, proteins fold to their native states directly over an energy landscape that may be rough, but folding is limited or slowed by kinetic barriers representing the results of a rough energy landscape during folding or competition from off-pathway events. In this picture, then, the job of GroEL is to prevent untoward interactions and to permit the target protein to express the conformational information that is inherent in its sequence. Proteins that misfold or aggregate by association of hydrophobic surfaces can often be protected by general strategies. Thus, rhodanese refolding can be assisted by micellar amounts of nondenaturing detergents or by nonchaperonin proteins that have some degree of exposure of hydrophobic surfaces, such as bovine serum albumin [15,16]. GroEL binds partially folded species and prevents both further folding and off pathway interactions. There are several points that follow from this effect. First, folding, in general, is expected to be slowed. Second, GroEL is expected to be able to unfold some proteins to a nonnative state. This state, depending on the protein, may be native-like, or it may be closer to a fully unfolded conformation.

III. STRUCTURAL STARTING POINTS AND PHENOMENOLOGIC MODELS

All models for GroEL function take as starting points the structures of GroEL and GroES, both of which are considered to function as oligomeric proteins, although evidence has been presented that, at least, a significant fraction of GroES can be monomeric under conditions used for many refolding assays [17]. These structures are based on electron microscopic and X-ray crystallographic analyses. The subunits of the GroEL 14 mer are arranged in two seven-membered rings that are stacked on top of each other to form a hollow, cylindrical structure, with a height of 146 A and a diameter of 137 A [18]. The central channel has a diameter of 45 A, and it is blocked at the equatorial junction of the two rings by portions of each monomer. The rings are arranged back-to-back so that the two ends of the cylinder facing the solution are the same. These ends are the regions that interact with both GroES and the target proteins that are to be refolded. The cylindrical space within each ring is large enough to accommodate the molten globule form of a protein of about 40 kDa. Somewhat larger proteins may be accommodated because the space expands during the folding cycle, and the space within the GroEL–GroES complex is further increased consequent to the juxtaposition of the two oligomers. In the presence of appropriate nucleotides, the single-ring GroES 7 mer can form complexes with GroEL that represent either one GroES ring per GroEL 14 mer (1:1 complex or asymmetric complex) or two GroES rings per GroEL 14 mer (2:1 complexes or symmetric complexes). There is some controversy over the functional primacy of these 1:1 and 2:1 complexes [19], but it has been demonstrated that both forms can be involved in GroEL function (discussed later). The GroEL 14 mer is a weak ATPase, and it will cooperatively hydrolyze seven ATP molecules, presumably by binding ATP to seven symmetrically located sites on one of the rings of GroEL. In evaluating various studies, one notes the physiological concentrations of nucleotides: ATP = 2.7 mM and ADP = 0.38 mM [20]. GroEL displays positive cooperativity for binding a nucleotide within one ring, whereas there is negative cooperativity for binding to the associated ring. This so-called nested cooperativity has been well described [21].

A typical phenomenologic mechanism involving both symmetric and asymmetric complexes can be invoked to serve as a basis for detailed discussions of GroEL and GroES function. This model would be most appropriate for proteins that bind tightly to GroEL, for slow-folding proteins, and for proteins that are relatively small (< 40 kDa). The mechanism could be considered to begin with a 1:1 complex in which seven ADP molecules are bound to the ring of GroEL containing the bound GroES 7 mer. This state is the acceptor complex, and it can bind polypeptides on the open GroEL ring opposite the GroES. The mnemonic rules that guide the binding of components are (a) each GroEL ring cooperatively binds seven ATP or seven ADP molecules; (b) GroES can bind tightly to a ring containing ADP; (c) binding of GroES to form a 2:1 complex in the presence of ADP is weak, so that asymmetric complexes are greatly favored under these conditions; (d) GroES can form 2:1 complexes in the presence of ATP, and fairly stable 2:1 complexes are formed when one ring contains bound ATP and the second contains ADP. A typical refolding cycle involving GroEL and GroES would then be as follows:

The acceptor complex binds seven ATP molecules and the unfolded polypeptide chain to the ring *trans* to the bound ADP and GroES. This complex could then bind a second GroES ring to form the 2:1 complex, with the protein under the GroES cap. The binding of GroES and the protein are linked, so that binding of GroES loosens the binding of the protein [22], and it tends to be released into the space defined by the GroES–GroEL interaction. The 2:1 complex initially contains an asymmetric distribution of nucleotides. Cooperative hydrolysis of seven ATP molecules induces a protein conformational response that leads to dissociation of GroES from the ring opposite that in which hydrolysis occurred. This ATP hydrolysis has the effect of introducing directionality into the release and binding of GroES, and the hydrolysis rate acts as a timer for binding and release. This leaves a complex that is similar to the acceptor complex, with the exception that the target protein is under the cap provided by the bound GroES. In one scenario, the target protein can fold in the protected space formed between the GroES cap and the cavity in that ring of the GroEL oligomer. It is possible for a second protein to bind to the open side of the complex to produce a species with the stoichiometry: [$GroES_7$:protein:ADP_7:$GroEL_{14}$:ATP_7:protein:$GroES_7$]. The formation of this species would depend on the concentration of partially folded protein during the lifetime of the open side of the GroEL. A repetition of the cycle of GroES binding to the open side of the oligomer, ATP hydrolysis and GroEL dissociation from the opposite end of the cylinder completes the cycle and permits exit of the protein from the end now freed. The system is now able to bind repetitively other molecules of unfolded protein or molecules of previously bound protein, if it was misfolded to an appropriately interactive state. Once the cycle is initiated, if both GroEL rings are occupied by bound protein, only one round of ATP hydrolysis is needed to weaken the interactions between the protein and GroEL which, with some proteins, leads to discharge of protein states that are folded or committed to fold. The simple fact that nonnative proteins bind whereas native conformers do not, indicates that it is expected that GroEL would have an "unfoldase" activity in that it would bind the nonnative conformer of a particular protein for which the interaction energy with GroEL would be greater than the folding free energy of that form. The general mechanism here would also lead to the expectation of some type of "iterative annealing" in which less stable forms of the protein, if released, could rebind in competition with folding to the native state. Use of special solution conditions or of specific substrate proteins leads to results that emphasize portions of this mechanism or particular properties of GroEL. Because the mechanism is complex, and because the properties of target proteins are so diverse, there are apparent discrepancies in observations that have led to differences in detailed mechanisms that have been proposed for GroEL and GroES function.

There are several generally agreed on characteristics of the interactions between GroEL and target proteins that form tight isolatable complexes. Proteins in their native conformations are bound only weakly, if at all. In addition, such a large number of diverse proteins are bound that it is unlikely that specific sequences can be expected as the recognition signal. There are several proteins whose folding is influenced by GroEL and, although no complexes can be isolated, there is clear evidence from the effects on the refolding kinetics that interactions between GroEL and the protein do occur. Different proteins differ in their requirements for completion of the cycle. Some proteins, such as barnase, can be influenced by GroEL alone. Some proteins, depending on the conditions, can form a tight complex with GroEL and require the complete complement of known accessory components for successful release (i.e., ATPMg binding and hydrolysis, K^+, and the cochaperonin GroES). It appears that this complete system is required when the target protein folds under nonpermissive conditions (i.e., if the protein would not spontaneously fold in the absence of the GroEL system) [23]. It is the complete system that has attracted considerable attention. In any event, if one considers the spectrum of proteins that can be folded, together with the requirements for the process, it seems that, because GroEL is always the same, the differences must rest in a property of the target protein itself or a target protein-dependent property of the complex induced by the interaction. Thus, although the target proteins are presumed to have some general properties in common, the specific nature of the target proteins must influence the interactions.

A. X-Ray Structure of GroEL and GroES

There are many important research articles and reviews that present details of chaperonin structure derived from both X-ray or electron microscopic examinations. To appreciate the elegance of the work and the beauty of the structures, these sources should be consulted directly [18,24]. In addition, several Web sites can be contacted for recent views of the chaperonin structures, and one of them even displays a "movie" based on a sequence of cryoelectron microscopic images. These sites can be accessed through the Chaperonin Home Page [8]. Here, we will present a brief summary of the structural information that is relevant to the function of GroEL.

The GroEL structure has been reported at 2.8-A resolution [18]. Each of the

14 identical 58-kDa subunits that make up the GroEL oligomer contains 523 amino acids and is folded into three domains: equatorial, intermediate, and axial. The fold of the monomer is somewhat unusual, and the topology can be best appreciated by picturing the polypeptide chain to be folded back on itself as an elongated letter U, with the termini together. The domains can then be imagined to be formed from this state. The equatorial domain is the largest, and it consists of residues 6–133 together with residues 409–523 for a total of 243 residues. This domain is largely helical, appears well ordered, and it provides most of the contacts between the subunits within a ring and all of the contacts that hold the rings together across the equatorial plane. The intermediate domain consists of residues 134–190 together with residues 377–408 (total of 89 residues). This domain connects the apical and intermediate domains, and contains structural elements that contact the apical domain on an adjacent subunit. It was suggested that the connections of this domain to the other two could serve as hinges for allosteric adjustments. The apical domain is formed from a continuous region in the center of the sequence, and it is composed of residues 191–376 (186 residues). This domain is the least well structured, and it contains those segments in which mutational changes can be made that will disrupt the binding of both GroES and polypeptide chains. It is this domain that appears to be flexible, and part of the mechanism of GroEL action might involve this flexibility or en bloc movements of the entire domain about hinge connections with the intermediate domain.

The 45-A–diameter central cavity in the cylindrical structure of the 14 mer does not traverse the entire cylinder, because there are apparently disordered sequences from the NH_2- and COOH-termini of the GroEL subunits that project into the channel and block direct access between the two rings. Thus, the 14 mer contains two separate, outward-facing cylindrical spaces. Each ring of the structure contains seven side openings that connect the solvent to the central channel, which might provide access for nucleotides and metal ions.

Refinement of the X-ray structure of GroEL resulted in a model that was consistent with deviations from exact sevenfold symmetry, and it was suggested that the conformational variability exists owing to rigid body movements between the apical and intermediate domains of GroEL. The regions that have been suggested to be involved in polypeptide and GroES binding (see later discussion) show high B factors, which would be consistent with conformational malleability [25].

A structure of GroEL, at a resolution of 2.4 A, has been reported with bound ATPγS [26]. The conditions used for nucleotide binding produced a structure that contained 14 molecules of ATPγS bound per molecule of GroEL, so that the binding does not reflect the negative cooperativity observed in solution experiments. However, these X-ray studies [26], as well as the preceding studies on the unliganded GroEL [26], were performed on crystals of GroEL containing the double mutation, R13G/A26V. In solution studies, this form of GroEL has been demonstrated to lack the negative cooperativity for nucleotide binding that is observed with the wild-type protein [27], although the crystallographic study reported that this mutant functioned normally. Even if several aspects of the allosteric effects of nucleotide binding remain unclear, there are interesting conclusions that can be drawn from these structures. The ATPγS interacts with residues confined to the equatorial domain, and it binds in a pocket located at the top surface of the equatorial domain facing toward the inside of the cylinder. The Mg^{2+} ion interacts with a nonbridging oxygen from each of the three phosphate groups. Other coordination sites of Mg^{2+} are provided by Asp-87 and two water molecules. In addition, the α-phosphate is linked to Thr-30 and Lys-51 by what is interpreted to be a bound metal ion, which could be K^+, an ion involved in ATP binding and cooperative hydrolysis [28]. When GroEL subunit structures are compared with and without bound ATPγS, there are local changes in the equatorial domain at the sites of contact with the nucleotide and the neighboring subunits. The largest change is a modest translational movement of two helices. The tip of a stem loop between two helices that is involved in intersubunit contacts is pushed away from the α-phosphate of ATPγS. These changes in the polar interactions between the domains of the subunits could influence their conformational stability in a nucleotide-dependent manner. Although the structures do not immediately explain the functional response to ATP, several points are clear. The binding of ATP is in a cavity composed of residues from the equatorial domain, and the nucleotide makes no direct contacts with the apical or intermediate domains. Because the bound polypeptides and GroES interact with only the apical domains, it is clear that the ATP influence is allosteric. Each of the ATP phosphates form hydrogen bonds to GroEL, mostly between the β- and γ-phosphates and a phosphate-binding loop of GroEL that is highly conserved and unique to the chaperonins. The nucleotide-induced shifts of one of the helices across the central cavity to the polar residues on the under surface of the apical domain could relay conformational changes induced by ATP binding to the apical domain [29].

The GroES structure has been reported at a resolution of 2.8 A by Hunt et al. [30], and a structure at 3.5 A was reported by Mande et al. [31] at about the same time. The structures are in general agreement, and the higher-resolution structure shows that the GroES heptamer forms a dome approximately 75 A in diameter and 30 A high. There is an 8-A hole in the center of the roof that is suggested to have considerable conformational plasticity. Thus, weak structural interactions and the juxtaposition of considerable negative charge suggests that the roof of the GroES dome might be a metastable structure. The authors raise the speculative possibility that a conformational change in the roof opening could enlarge the hole to 30A and, thereby, present a means for escape of nonnative polypeptide chains. No direct evidence for this suggestion has been presented, and all current models for GroES function assume that protein trapping and release is related to binding of intact 7 mers. Each GroES subunit contains a mobile loop previously identified from nuclear magnetic resonance (NMR) studies as a GroEL-binding determinant. The structure of GroES is perfectly compatible with a model in which the mobile loops interact with GroEL, resulting in an enclosed cavity, the walls and floor of which are provided by the GroEL cylinder that enlarges on interaction, and for which the roof is made up of the GroES dome.

B. Electron Microscopic Structure of GroEL and Its Complexes with GroES

Electron microscopic structures have been reported by both negative-staining and cryomicroscopic methods that are fully compatible with the X-ray results. At the level that comparisons can be made (30 A), there is good agreement between the electron microscopic structures and the crystal structure. The electron microscopic images add greatly to understanding the GroEL system, because various distinct structures can be visualized that are related to stages in the functional cycle of chaperonin-assisted folding [7,24,99]. Electron microscopic images of frozen and hydrated GroEL complexes show the characteristic ring-like end views and rectangular side views. The side views have four layers of density, corresponding to the two-stacked rings of subunits, with each subunit containing the two major domains (the intermediate domains are not obvious). The end views show the expected sevenfold symmetry. In the presence of ADP, the structure is elongated and radially expanded at the ends. When both GroES and ADP are present, asymmetric, bullet-shaped complexes are formed. With GroES and AMPPNP, both asymmetric and some symmetric, football-shaped complexes are formed. All of the GroEL nucleotide complexes show vertical opening and twisting of the apical domains. The ADP structure shows the most opening. Although the equatorial domains stay relatively constant, there are differences in the interring contacts between the ATP structures and all of the others. In the presence of AMPPNP, the upper ring looks similar to the ADP-bound structure, and the lower ring is close to an ATP-bound structure. When images are arranged in the sequence GroEL, GroEL ADP, GroEL ATP, it appears that the apical domains pivot about the upper hinge regions between the intermediate and apical domains. This causes the substrate-binding sites on the apical domains to be rotated out of the central channel toward the intersubunit interface. In ADP, the lower apical domains pivot counterclockwise, but the main movement in ATP is radially inward. The domain movements that are suggested to occur during the GroEL ATPase cycle are best appreciated by viewing the succession of forms as a movie, available on the Website http://www.cell.com.

Electron microscopic images have also been evaluated with a view to understand the cooperative interactions in solution studies of the GroEL ATPase. This cooperative mechanism of GroEL ATPase has been analyzed in terms of each ring being in either a T or R allosteric state, and it has been extended to take into account positive cooperativity within rings and negative cooperativity between rings [21,32,33]. Unliganded GroEL is mainly in the TT state, ATP binding converts it to the TR state, and at high ATP concentrations, to RR. The R state favors GroES binding, so the RR state is capable of forming a symmetric complex. The ADP state is a distinct R-like state, intermediate between the T and R states, and it only slowly binds GroES.

The electron microscopic images show a considerable range of domain movements in GroEL. Allosteric R states are characterized by flexibility of the hinge region between the intermediate and apical domains. Roseman et al. [7] present a schematic model for the mechanism mainly based on rigid body rotations of the equatorial and apical domains that are coupled by twisting or other concerted changes through the intermediate domain. The authors suggest that nucleotide binding controls the allosteric state by transmitting movements in two directions. First, nucleotide binding causes rotations around the equatorial–intermediate domain hinge, leading to reorientation of the equatorial domain. Changes in the contacts are transmitted to the opposite ring by interactions between equivalent helices at the equatorial boundary, thereby mediating negative cooperativity. Second, to explain specific effects of different nucleotides on the observed twists of the apical domains, it is proposed that changes at the equatorial–intermediate hinge are propagated by the intermediate domain inducing large rotations at the intermediate–apical hinge, thereby twisting the apical domains and controlling substrate and GroES binding. In this way, changes in conformation can be cooperatively propagated from the GroEL equatorial domain to the apical domains in both rings, an effect that modulates the accessibility of the substrate-binding sites.

C. Information from Site-Directed Mutagenesis

Coupled with the X-ray crystallographic study, site-directed mutagenesis has been used to map functional sites within the structure. Among the sites that have been probed are the nucleotide-binding site, the substrate-binding site, and GroES-binding site. Although, additional results from mutagenesis studies will be mentioned in other sections, several points can be made here.

The X-ray structure of the ATPγS-bound form of GroEL shows the nucleotide in a pocket formed by a helix bundle in the equatorial domain. This region is bordered by a highly conserved sequence motif that includes Asp-87. Mutations of this residue abolish ATPase activity, although ATP analogues can still bind. In addition, mutations in the region of the intermediate–equatorial hinge region near residues 150, 151, 152, 405, and 406, also abolish ATPase activity. These results are consistent with the suggestion that hinge movements are coupled with ATP hydrolysis [34]. Mutational analysis has also been used in a different way to investigate domain movement [29]. In an interesting study, two mutationally introduced cysteine residues replaced Asp-83 and Lys-327. When these sulfhydryl groups were oxidized, the resulting GroEL was locked in a closed conformation by an interdomain disulfide between the equatorial and apical domains. This closed conformation could bind six to seven molecules of ATP, which remained bound. However, there was no hydrolysis. Furthermore, this ATP-bound, oxidized GroEL could bind to the stable nonnative substrate protein, isopropylmalate dehydrogenase, whereas, the nucleotide-free, oxidized protein bound this substrate with weaker affinity. This oxidized GroEL failed to proceed with further reaction steps, such as ATP hydrolysis, GroES binding, and dissociation of the substrate protein. When the disulfide bonds were reduced, the GroEL regained the ability to process all of the reactions, just as the wild-type GroEL. These results are consistent with the hypothesis that hydrolysis of ATP is coupled to opening movements of the apical domains, and the reciprocal expectation that ATP hydrolysis can cause the opening of the apical domains. GroES would then be able to bind to the open form of GroEL, and the substrate polypeptide could be released from GroEL coupled with ATP hydrolysis or opening of the apical domains.

Many of the same mutations to residues lining the surfaces of the apical channel interfere with binding of both polypeptides and GroES, suggesting that the binding sites for GroES and substrate overlap, or at least that they occur in the same region of the structure. It is evident from the comparison of cryoelectron microscopy pictures that a large rotation of at least the hinge region between the apical and intermediate domains in the GroEL subunits is required to reorient the apical domains by 50°–60°, which places the substrate-binding surfaces close to the bases of the β-barrel domains of the GroES subunits.

Substrate proteins bound to GroEL have been suggested to be partially folded forms with native secondary structure, but with expanded and incomplete tertiary structure (a so-called molten globule). There is general agreement that GroEl binds folding intermediates with exposed hydrophobic surfaces, and, in this way, suppresses aggregation that often competes with proper folding in solution. In the crystal structure, Arg-197 in the apical domain is involved in a contact with Glu-386 in the neighboring intermediate domain within the same ring. Mutation of this arginine to alanine reduces both the positive cooperativity of the ATPase within the rings and its negative cooperativity between rings, indicating that this residue is involved in both in intra- and intersubunit allosteric interactions. These residues are near the apical hinge, whereas ATP binding is next to the equatorial hinge [33]. Several site-directed mutations have indicated that binding and release can be blocked by very different mutations, overall it might be thought that binding may be relatively easy to understand, but release is a much more complex phenomenon, involving the ATPase and GroES mechanisms. The wide distribution of sites affecting release suggests that global rearrangements accompany the release step.

The interactions between the individual rings of GroEL are particularly interesting, and isolated single rings are capable of performing several partial reactions (i.e., they can bind nonnative proteins). However, these forms tend to be deficient in release. The properties of a quadruple mutant (R452E/E461A/S463A/V464A), SR1, have been studied. This SR1 mutant forms a single ring. In the presence of GroES, the ATPase activity of SR1 is completely inhibited, and a stable, asymmetric half-football is formed. This mutant form cannot complete the catalytic cycle, which involves dissociation of the ligands ADP and GroES [35]. These properties have led to the use of SR1 as a trapping agent for partially folded polypeptide chains in studies of the forms of proteins released from functional GroEL.

IV. THE ROLE OF HYDROPHOBIC INTERACTIONS IN GroEL FUNCTION

A. What Is the Nature of the Interactions Between Bound Proteins and GroEL?

The very general nature of the interactions between GroEL and the numerous and diverse target proteins that can be bound suggest a more general recognition motif than normally expected for a biological system. The specific sequence of the bound protein is unlikely to contain the appropriate binding signal, and it is some more general characteristics of the nonnative states that are involved. Hydrophobic interactions are most often considered to be the primary forces operating between the target protein and GroEL [36,37]. This idea is attractive, because the appropriate surfaces would be available on a variety of interactive, nonnative states of proteins, that are hidden in the fully folded, noninteractive conformers. For hydrophobic interactions to dominate, there must be appropriate hydrophobic surfaces on both GroEL and on the target proteins.

Site-directed mutagenesis and X-ray results suggest the importance of hydrophobic interactions (see foregoing). Single mutations of hydrophobic residues within a region of the apical domain, shown to be flexible in the X-ray structure, abolish peptide binding (Y199E, Y203E, F204E, L234E, L259S, V263S, and V264S) [34]. These mutagenesis studies have been supplemented by several solution studies involving the binding of proteins and by the use of fluorescent-probes such as bisanilinonaphthalene sulfonate (bisANS), the fluorescence intensity of which increases dramatically when it binds noncovalently to hydrophobic surfaces. In addition, bisANS can be covalently photoincorporated into GroEL, and the major site of incorporation is within a peptide that maps to the apical domain (residues 203–249), which is located in the most hydrophobic portion of the sequence [38]. Thermodynamic parameters derived from the temperature dependence of the binding of various substrates, including a variant of subtilism

BPN′, α-casein [39], and β-lactamase [40] gave results consistent with an important role for hydrophobic interactions as the driving force for the association of these proteins with GroEL. An interesting further example is provided by studies of the interaction of the dimeric malate dehydrogenases (MDH) with GroEL [41]. Mitochondrial MDH is an interesting system, because at 36°C, dilution from guanidine HCl gives a stable folding intermediate that is not prone to aggregate and that can be refolded by GroEL–GroES [42]. Mitochondrial MDH is considerably more hydrophobic than the cytoplasmic MDH, based on their amino acid compositions [43], and it is bound more tightly to GroEL, as measured by the requirements for release of the bound proteins (i.e., release of the more hydrophobic protein requires both ATPMg and GroES, whereas the less hydrophobic protein can be released by ATPMg alone). The homologous lactate dehydrogenase from *Bacillus stearothermophilus* has a hydrophobicity and a release requirement intermediate between the MDH molecules. The involvement of hydrophobic interactions is further suggested by studies in which GroEL interacts preferentially with the side chains of hydrophobic amino acids and more weakly with polar or charged amino acids [44]. Interestingly, this strong specificity is reduced by binding of GroES [45], which is associated with polypeptide chain release. Again, these results suggest that the modulation of hydrophobic surfaces can be important in protein-binding and release steps that are central to the mechanism of chaperonin-assisted refolding.

B. What Is the Role of Substrate Protein in the Interaction with GroEL?

Charge, in addition to hydrophobic exposure, appears to be important in the interactions of proteins with GroEL. The binding site on human granulocyte ribonuclease possesses three positive and no negative charges within 17 residues [46]. Similarly, the rhodanese NH_2-terminal 22-residue sequence is positively charged, and it can interact with GroEL (see following). Barley chymotrypsin inhibitor studies show that reducing net charge weakens interaction with GroEL [47]. Binding of α-lactalbumin, as for the previous examples, is entropy-driven at room temperature; the heat capacity change for binding is negative, indicating that binding involves hydrophobic interactions. Furthermore, the effects of monovalent cations and ionic strength on binding of α-lactalbumin led to the idea that binding is strengthened by electrostatic screening [48]. Ding et al. formed complexes of GroEL with gp5, a net-negatively charged, 42-kDa major head protein of the bacteriophage HK97 [49]. Interactions of this protein, its fragments, and several variants, the size of which was varied by internal duplications, were studied by gel electrophoresis, taking advantage of the fact that interactions with some of these proteins produced a gel shift in which the complex showed decreased mobility that could be ascribed to a decreased net negative charge on the complex. The results indicated that only peptides above 31 kDa could cause a band shift, and that the apparently anomolous migration (slower anodic migration, with a more negatively charged complex that was not due to the larger size of the complex) suggested that binding of a substrate protein to GroEL could be accompanied by binding of cations from the solution. These results suggest that protein and cation binding are coupled, which is consistent with the idea that charge interactions could modulate the binding of substrate proteins through hydrophobic interactions.

The binding of peptides is instructive for the potential role of charge in modulating interactions. It has been proposed that the secondary structure of regions of target proteins can be important for binding to GroEL. For example, transferred nuclear Overhauser enhancement NMR experiments were used to study the binding of a peptide based on the NH_2-terminal sequence of rhodanese [50–52]. The binding was extremely weak ($K_d = 1$ mM), but the peptide, which was a random coil in solution, appeared, when bound, to adopt a helical conformation. This sequence forms an amphiphilic helix in the native structure of rhodanese, and it was suggested that the binding could be along the hydrophobic face. This system was further studied using hybrid peptides that were synthesized to contain portions of both the bee venom peptide, apamin, and the sequence from rhodanese that forms the amphiphilic helix [52]. The apamin sequence contains disulfides that constrain the rhodanese peptide to a helical conformation, with one unobstructed face of the helix facing the solution. Hybrid peptides could be constructed such that the unobstructed face could be hydrophobic or hydrophilic. Only the disulfide-bonded structure with the hydrophobic face exposed could bind significantly to GroEL. Reduction to produce the random-coil form of the peptide abolished measurable binding. In addition, binding the peptide to GroEL appeared to induce more extensive exposure of hydrophobic surfaces in the complex than were present in the initial components. Although amphiphilic structures appear to be bound most easily, the specific secondary structure is not likely to be the determinant for the recognition by GroEL, for GroEL can capture an all β-sheet Fab fragment [53]. It does follow, though, that if hydrophobic surfaces are preferentially bound, and if these surfaces are formed by elements of structure, then the binding will favor the structured form and the interactions with GroEL will appear to have an organizing effect on the protein. Because interactions with charges and hydrophobic surfaces appear to be linked, it is tempting to speculate that amphiphiles are preferentially bound. Thus, if amphiphilic structures are bound, and those structures are amphiphilic because of conformation, then the interaction must favor the structured form, and the interaction with GroEL can have an organizing effect.

Hydrophobic interactions require that both partners have appropriate hydrophobic surfaces. Although mutations of hydrophobic residues give results consistent with their importance, quantitative bisANS-binding studies indicate very few available sites on unperturbed GroEL for such binding [36]. However, hydrophobic surface exposure on GroEL can be modulated by polar and ionic interactions. Cations, especially polyvalent ones such as spermidine, are effective in increasing hydrophobic exposure on GroEL without dissociating the oligomer [38]. The bisANS-binding studies with the conformationally constrained apamin-based peptides (see foregoing) suggested that peptide–GroEL complexes contained more exposed hydrophobic surfaces than was expected, based on the sum of the components [52]. In evaluating the effects of ions, it should be recognized that GroEL has an extremely high negative charge density. Ions could influence hydrophobic binding by several mechanisms in addition to conformational changes: (a) the known effect of ions on strengthening hydrophobic interactions; (b) electrostatic screening to reduce repulsion between GroEL and target proteins; or (c) counterion binding to GroEL or targets to increase hydrophobic effectiveness.

It may be that binding to GroEL is favored by structures on the target protein that organize and expose hydrophobic surfaces, a feature found in, but not exclusive to, amphiphilic helices. Furthermore, it is possible that the binding of a hydrophobic surface to GroEL can induce further exposure of complementary surfaces on GroEL complexes, thus amplifying interactions available for target proteins. The major point would be that amphiphilic structures that present hydrophobic surfaces within a charged context would be particularly effective in binding to GroEL. The organization of the interactive surfaces on GroEL (sevenfold symmetry within a ring) would permit multiple weak interactions to contribute to a tight, net binding of target proteins while allowing individual binding sites to bind only weakly. This could account for the finding that, generally, peptides bind in a significantly weaker manner than large proteins. This is in keeping with the idea that sequences are not the major determinant of binding.

Further evidence for modulation of hydrophobic surfaces of GroEL by mechanistically relevant ligands is provided by studies of hydrophobic exposure on GroEL during its cycle of ATP hydrolysis [54]. In the presence of ATP, but not ADP or the nonhydrolyzable ATP analogue, AMPPNP, there was a time-dependent increase in bisANS fluorescence that correlated with the hydrolysis of ATP. However, the rate of fluorescence increase was slower than the rate of ATP hydrolysis. These results suggest that a transient conformational state of GroEL is formed consequent to ATP hydrolysis, possibly resulting from electrostatic repulsion between the bound products of ATP hydrolysis.

V. WHAT FORM OF THE TARGET PROTEIN IS BOUND TO GroEL?

It is difficult to assess the conformations of proteins bound to GroEL because of the great disparity between the mass of GroEL (840 kDa) and that of the target proteins (typically 30 kDa or less) in the complexes. Thus, the properties of

GroEL dominate most measurements of complexes. An additional difficulty, is that properties of the complex are typically taken to be the sum of the properties of the separated components, and it is most often assumed that the properties of GroEL in the complexes are the same as free GroEL under the same solution conditions. Furthermore, influencing protein folding does not necessarily require the formation of tight, stable complexes; yet these are the only ones that can be measured with any facility. Any mode of tight binding of a nonnative conformer would lead to the observed result that folding is arrested and aggregation is inhibited. The major methods that have been used to monitor the conformations of bound proteins rely on optical signals that are assumed to be from the bound protein, such as tryptophan (Trp) fluorescence or fluorescent probe binding [56]. One technical problem in the early studies is that, although tryptophan is absent from the sequence of GroEL, it is almost always found in highly purified samples, presumably from contaminating peptides that copurify with the GroEL, and the fluorescence spectra often display a peak in the same region as that of many protein–GroEL complexes (e.g., the rhodanese–GroEL complex). Recent advances in GroEL purification have made it possible to ameliorate this problem, and future investigations should not suffer from this ambiguity [57,58]. In any event, spectral results must be interpreted with caution. The tryptophan and ANS environments in complexes could be provided by surfaces on GroEL or by regions formed on interaction, rather than existing solely on the bound protein itself. In a similar vein, the methods used commonly assume that the properties of the complexes are additive, and changes in GroEL as a result of complexation are routinely ignored. Thus, the conclusion from ANS fluorescence studies that rhodanese is bound to GroEL as a molten globule rests on the assumption that hydrophobic exposure in the complex is entirely from the bound protein. For example, there is the possibility that increased hydrophobic exposure in complexes may be partly due to changes in GroEL, rather than resulting entirely from the bound proteins [56]. Finally, as a general caveat, the use of probes must be interpreted with caution because they are capable of perturbing the protein conformation [59]. Having said that, there are various interesting studies that have sought to determine the states of bound proteins. Different proteins lead to somewhat different conclusions depending on the nature of the bound polypeptide [55].

A. What Is the Nature of Bound Proteins?

The environmental sensitivity of Trp fluorescence was used to suggest that rhodanese bound to GroEL is in a molten globule state, because the Trp in the complex has an apparent solvent exposure between that of the native protein and that of a fully solvent exposed Trp. In addition, that complex binds the hydrophobic probe, anilinonaphthalene sulfonate (ANS) with an increase in fluorescence intensity, indicating that increased hydrophobic surfaces are present in the complex compared with native rhodanese and GroEL itself [56]. Limited proteolysis of bound rhodanese was interpreted as indicating that individual domains of rhodanese are folded and separately bound to GroEL [60]. The individual-folded domains of rhodanese have extensive hydrophobic surfaces that interact to form the native structure [61]. Binding to GroEL would keep the domains from interacting to form aggregates and could even facilitate domain formation by selecting conformers with appropriately organized surfaces. These results are consistent with structured intermediates being able to bind to GroEL. However, other studies suggested that several intermediates along the folding path of rhodanese could be bound by GroEL [62].

Experiments with LDH led to a somewhat different picture from that suggesting that structured intermediates are bound. When LDH was unfolded as far as the molten globule state, it did not form a tight complex with GroEL, although, when the protein was allowed to collapse from a fully denatured state, a tight complex was formed [41,63]. These experiments with LDH, and others with barnase [64], are consistent with binding of more unfolded states to GroEL and, in these cases, states further from the native state are bound to GroEL. Cyclophilin is a small protein that does not form a tight complex, but it interacts reversibly with GroEL. The NMR studies of cyclophilin indicate that all amide protons in cyclophilin bound to GroEL are exchanged with solvent, although the exchange does not occur in free cyclophilin. The result was taken to indicate that the complete secondary structure of cyclophilin is disrupted on complexation [65]. These results indicate that unfolded states are preferentially bound, and they lead to the expectation that interactions with GroEL would be capable of extensively unfolding bound proteins.

Studies with dihydrofolate reductase (DHFR) extend the picture of the form of the bound protein [57,66]. DHFR is an example of a protein that readily folds in the absence of chaperonins, and it does not display significant misfolding or aggregation. At most, DHFR requires ATP, but not GroES, for its assisted folding. DHFR from *E. coli* (ecDHFR) makes only a transient complex with GroEL, whereas murine DHFR (muDHFR) forms a tight, stable complex, even though the two proteins are structurally homologous. They differ by the presence of surface loops in the structure of muDHFR. Although ecDHFR binding is transient, it can form complexes with GroEL that influence folding. Either of two surface loops of eukaryotic DHFR (seven or nine amino acids) can be inserted into corresponding regions of ecDHFR, which lacks them, and the resulting constructs can now be bound tightly to DHFR. The DHFR molecules in the stable complexes with GroEL appear to be the same whether they are prepared from the native DHFR or from DHFR denatured in urea. Detailed analysis of the kinetics of GroEL-assisted folding are consistent with a picture in which native DHFR is in a preexisting equilibrium with a late-folding intermediate, and it is that form that is bound. Thus, in this example, the bound protein is close to the native form, and because the state of the bound species is independent of its mode of preparation indicates that some folding can occur within the complex. A further example is provided by the studies of interactions of GroEL with an Fab fragment [67]. It is shown that GroEL is able to bind both an early-folding intermediate as well as with a quaternary-structured–folding intermediate that is populated late in the folding process. The interaction does not lead to unfolding of this late intermediate.

B. How Does GroEL Respond to Protein Binding?

In addition to changes in GroEL hydrophobic exposure, other studies have indicated that GroEL has different conformational responses that depend on the nature of the protein. For example, the interaction of GroEL with DHFR-folding intermediates gives rise to complexes with different properties, depending on which of two common protocols is used to make the complex [68]. If the complex is prepared by diluting urea-denatured DHFR, a complex is formed with DHFR on the GroEL ring *trans* to bound GroES. If DHFR is heated together with GroEL to 45°C, then a complex can be formed with DHFR on the same ring as GroES and protected by it from proteolysis [69]. Because GroEL is reported to undergo a thermally induced conformational change [70], and the refolding mechanism is similar whether DHFR is spontaneously refolded from urea or after thermal unfolding, it was suggested that differences in GroEL conformations could contribute to the differences in the behavior of the complexes. On a different note, a very interesting study reported that, although the proteolytic fragmentation patterns of GroEL in the presence of ligands were similar, the rates of formation and degradation of the fragments depended on the ligands' nature [71]. Significantly, a fragment of approximately 25 kDa, resulting from cleavage within the apical domain of GroEL, was generated with GroEL alone or with bound rhodanese or DHFR, whereas this fragment was not prominent with bound citrate synthase. The data support the idea that changes in trypsin-sensitive regions of GroEL occur that are influenced by binding of chaperonin ligands and substrate polypeptides.

The major conclusions from various studies lead one to suspect that the form of the protein that is bound, the tightness of the complexes, and the position of the bound intermediate along the folding pathway depend on the surface characteristics and relative stabilities of the intermediates in relation to the rates of their interconversion. These factors will determine, for example, whether interactions with GroEL can substantially unfold a native protein. In this picture, and consistent with observations, there is a common set of potentials from which a consensus mechanism can be selected for an individual protein. This behavior of the chaperonin stems from different properties of folding intermediates and different responses of GroEL to the interactions.

VI. STOICHIOMETRY OF GroEL–GroES COMPLEXES

When proteins fold under nonpermissive conditions (i.e., when they cannot easily fold without assistance), they require GroES, ATPMg, and K^+ in addition to GroEL [72]. GroES, in the presence of nucleotides, binds to the apical domains of the initially symmetric GroEL oligomer by a highly mobile and accessible polypeptide loop present in each monomer of the GroES 7 mer [30,73]. This interaction forms a protected space between the overlying dome-like cap provided by GroES and the cylindrical region within one ring of GroEL that is expanded by the interaction. A partially folded protein could initially bind to the apical domains opposite the bound GroES, and successive rounds of GroES binding and release would first encapsulate and displace the bound protein [74], permitting it to reside in the protected space, perhaps fold, and then be released when the GroES is released [35,69,75,76]. In principal, a second GroES could bind to the opposite end of the GroEL molecule to form a symmetric 2:1 complex. In most instances, there is apparent anticooperativity, and the binding of the second GroES oligomer is less favorable. Although there is some controversy over the required intermediacy of the symmetric complex [19], it appears that symmetric complexes (2:1 complexes) can be comfortably accommodated within the most stringent mechanisms for GroEL function under nonpermissive conditions [20,77–81]. These symmetric complexes provide a simple way to cycle the GroES oligomers; a process required for completion of the folding cycle. Part of the controversy stems from the transient nature of the symmetric complexes under some conditions, and part stems from the fact that conditions can be arranged involving only a portion of the complete cycle so that protein folding can proceed without requiring the symmetric complexes.

Negative-stain electron microscopy has demonstrated that asymmetric bullet-shaped complexes are exclusively formed in the presence of ADP, whereas a significant number of symmetric, football-shaped complexes can be observed in the presence of ATP [78,82]. Because the symmetric complexes were considerably more populated with ATP than with the nonhydrolyzable ATP analogues, it can be suggested that ATP hydrolysis favors the formation of the symmetric complexes [82,83].

Cross-linking studies, coupled with electron microscopy, have been valuable in defining the roles of various complexes, although potential artifacts and distortion of distributions of species should be considered. It was possible to correlate the presence of symmetric and asymmetric complexes with bands on sodium dodecyl sulfate (SDS) gels run after cross-linking. In this way, it was possible to define the distribution of particles under a wide range of solution conditions by quantifying the SDS gel patterns [79,83]. These studies led to the conclusion that the symmetric complexes acted as highly efficient intermediates in the protein-folding activity of GroEL. These conclusions were supported by direct quantitation of particle distributions from electron microscopic images, which led to a correlation between the presence of symmetric complexes and the rhodanese-folding activity [81].

Fluorescence studies used GroEL labeled with I-AEDANS, a fluorescent probe that is designed to react with protein sulfhydryl groups and for which fluorescence is sensitive to structural changes within the protein [84]. The fluorescence response after ATP addition in the presence of GroES corresponded to the sequential binding of two GroES 7 mers. Binding of thermally denatured MDH could be followed by changes in AEDANS fluorescence. The fluorescence changes led to the suggestion of the following sequence of events: The asymmetric complex containing ADP and MDH could bind ATP; a second molecule of GroES could bind to form a symmetric complex containing the bound MDH; breakdown of the symmetric complex led to release of the MDH. The results were consistent with the notion that the symmetric complex is an intermediate in the folding of MDH in this system.

Fluorescence has been used in another way to investigate the distributions of complexes in the presence of nucleotides [85]. Fluorescence anisotropy, which measures the rate of molecular tumbling, was used to quantify the binding to GroEL of labeled GroES (70 kDa), which tumbles more slowly when it is part of a complex with GroEL (> 900 kDa). In the presence of MgATP and 10 mM KCl, two GroES rings bind strongly per 14 mer. In the presence of ADP, only one GroES ring is tightly bound to GroEL, although a second GroES can bind weakly, and analysis of the binding yields a Hill coefficient of less than 1, which is consistent with negative cooperativity. The ability to form some symmetric complexes in the presence of ADP is supported by studies of GroES exchange reactions in which it has been demonstrated that GroES in asymmetric complexes can be exchanged with excess free GroES in the absence of prior dissociation of the bound GroES [86]. AMPPNP gives results similar to those with ADP, although there is a somewhat larger population of 2:1 complexes. Strikingly, addition of AMPPNP to a solution containing asymmetric complexes formed in the presence of ADP results in binding of a second GroEL; therefore, 2:1 complexes can be more easily formed in the presence of mixed nucleotides than they can with either of the individual nucleotides in the absence of hydrolysis. These results suggest that ADP strongly favors asymmetric complexes, whereas symmetric complexes are favored by asymmetric occupancy of nucleotides in which one ring contains seven bound ATP molecules, and the second ring contains seven molecules of ADP.

VII. FOLDING AND OLIGOMERIZATION OF GroEL AND GroES

In considering the role of GroEL in folding a diverse group of proteins, the question is often raised about how GroEL itself is folded and assembled. For some time, attempts to refold GroEL after complete denaturation were reported to be unsuccessful [87–89]. Because native GroEL and GroES could assist the assembly of monomeric GroEL, it was suggested that folding of GroEL might require the intervention of preexisting chaperones [89–91]. GroEL contains all the amino acids encoded by the gene except the initiating methionine, so it would appear to have all the sequence information required for folding and assembly. Furthermore, the GroEL monomer presents an interesting folding problem, because the first biosynthesized sequence must extensively interact with the last biosynthesized portion to form the largest and most structured equatorial domain. Studies of GroEL unfolding led to the suggestion that GroEL, although largely unfolded in 3–4 M urea, contains a region of hydrophobic residual structure within the apical domain that could serve to nucleate folding and put the chain ends into register for the observed fold [92]. From these results, it was possible to devise a protocol that led to the refolding and reassembly of active GroEL after extensive denaturation, although a fraction of the GroEL misfolded and formed high molecular weight aggregates [58]. Related studies demonstrated that GroES could be fully refolded and reassembled after denaturation, and that GroES could assist the folding and assembly of GroEL and reduce the heterogeneity observed earlier [93]. These results demonstrate that the GroEL–GroES sequences have the ability to refold spontaneously and suggest that the *groE* operon may be organized to provide a structural role in GroES–GroEL assembly as well as a functional role.

VIII. CONCLUDING REMARKS

The diversity of substrates for GroEL, together with the lack of strong specificity for the interactions that characterize the system, suggest that there can be a number of detailed mechanisms, each of which represents a consensus of interactions involving the substrates and the chaperonins. Because GroEL brings the same potentials to every interaction, and the target proteins represent the elements of diversity, they must represent the major source of different detailed requirements for the folding mechanisms that have been observed. These considerations lead to a relatively simple expectation. The interaction of a protein with GroEL fundamentally depends on the competition among several equilibria, and in consequence, on the rates of interconversion among species: first, there are the equilibria among states of the protein, including monomeric intermediate forms, the native protein; the denatured state, and associated or aggregated species; second, those equilibria describing the binding of the intermediates to GroEL; third, equilibria involving binding of nucleotides and GroES to GroEL; and finally, those equilibria involving interactions of elements of the GroEL structure itself, such as those interactions among the domains of the monomers. For example, the form of the protein that binds to GroEL is likely to be the one with sufficiently exposed, interactive hydrophobic surfaces under the conditions of the interaction. So, for example, if a compact, near native conformation can expose sufficient surfaces with exposed loops or terminal segments, it could be bound in a nonnative state. If domain interfaces can be exposed, the protein may be bound by an intermediate with extensive, although nonnative, structure, and if the protein is sufficiently unstable, it may be significantly unfolded to permit interaction [40,94]. In the simplest version, the balance between the stability of the protein and its ability to expose appropriate interactive surfaces may determine its ability to bind to GroEL and determine the subsequent release requirements (i.e., the tighter the binding, the more stringent the requirements for release). Furthermore, these considerations lead to the possibilities of finding different apparent mechanisms for the folding of different substrates of different conditions are used. Thus, in the presence of GroES, a weakly bound protein, or one that can fold rapidly (folding under permissive conditions), may interact with the *trans*-ring of a 1:1 complex and fold without the necessity of GroES capping [20], a type of mechanism that may also hold for large proteins that could not be accommodated under a GroES cap. Among these proteins would be those for which folding could be influenced and for which aggregation could be suppressed without the necessity of forming complexes with GroEL stable enough to be isolated. For proteins that bind tightly (Note: proteins that extensively aggregate in competition with folding may be expected to have those surfaces that lead to tight interactions) or fold slowly, the requirements for release may be more extensive and may require nucleotides alone, ATP hydrolysis, or the complete chaperonin system, including GroES [95]. The most loosely bound proteins, with a relatively strong tendency to fold spontaneously may have their folding influenced by transient complexes with GroEL alone.

Multiple, cooperative interactions can form stable complexes. Changing the cooperativity could then permit release. The sevenfold symmetry of the interactive, apical domains of GroEL raises the possibility that stable complexes between a protein and GroEL might result from multiple weak binding interactions. This would allow formation of tight net binding, while still permitting flexibility from weak binding equilibria at individual sites. In light of these considerations, it is interesting that the intact GroEL 7 mer binds rhodanese under nonpermissive conditions, arrests its spontaneous folding, and requires GroES for release. On the other hand, single, folded, recombinantly prepared apical domains [96] or monomeric proteolytic fragments of GroEL containing the apical region [97,98] can facilitate the folding of rhodanese without ATP or GroES. In intact GroEL, the increased separation among the apical domains after interaction with GroES may reduce the cooperative component of protein binding, and contribute to the release of bound proteins into the protected space within the capped complex. These results with individual domains are reminiscent of the "buffering" effect of detergent micelles when they assist rhodanese refolding by forming transient complexes with interactive intermediates and favor folding in competition with aggregation [15].

REFERENCES

1. Kandror, O., Busconi, L., and Goldberg, A. L., *J. Biol. Chem. 269*:23575–23582 (1994).
2. Sohlberg, B., Lundberg, U., Hartl, F. U., and von Gabain, A., *Proc. Natl. Acad. Sci. USA 90*:277–281 (1993).
3. Ybarra, Y. and Horowitz, P. M., *J. Biol. Chem. 271*:25063–25066 (1996).
4. Ellis, R. J. (ed.), *The Chaperonins*, Academic Press, New York, 1996.
5. Ellis, R. J., *Philos. Trans. R. Soc. Lond. B. 339*:257–261 (1993).
6. Ellis, R. J. *Folding Design 1*:R9–R15 (1996).
7. Roseman, A. M., Chen, S., White, H., Braig, K., and Saibil, H. R., *Cell 87*:241–251 (1996).
8. http://bioc09.uthscsa.edu/~seale/Chap/chap.html
9. Viitanen, P. V., Gatenby, A. A., and Lorimer, G. H., *Protein Sci. 1*:363–369 (1992).
10. Lorimer, G. H., *FASEB J. 10*:1–5 (1996).
11. Anfinsen, C. B., *Science 181*:223–230 (1973).
12. Ellis, R. J. and Hartl, F. U., *FASEB J. 10*:20–26 (1996).
13. Sosnick, T. R., Mayne, L., Hiller, R., and Englander, S. W., *Nature Struct. Biol. 1*:149–156 (1994).
14. Wolynes, P. G., Onuchic, J. N., and Thirumalai, *Science 267*:1619–1620 (1995).
15. Horowitz P. M., *Protein Folding: In Vivo and In Vitro* (J. Cleland, ed.), ACS Symp. *526*:156–163 (1993).
16. Jarabak, R., Westley, J., Dungan, J. M., and Horowitz, P. M., *J. Biochem. Toxicol. 8*:41–48 (1993).
17. Zondlo, J., Fisher, K. E., Lin, Z., Ducote, K. R., and Eisenstein, E., *Biochemistry 34*:10334–10339 (1995).
18. Braig, K., Otwinowski, Z., Hedge, R., Boisvert, D. C., Joachimiak, A., Horwich, A. I., and Sigler, P. B., *Nature 371*:578–586 (1994).
19. Engel, A., Hayer-Hartl, M. K., Goldie, K. N., Pfeifer, G., Hegerl, R., Muller, S., da Silva, A. C. R., Baumeister, W., and Hartl, F. U., *Science 269*:832–836 (1995).
20. Corrales, F. J. and Fersht, A. R., *Proc. Natl. Acad. Sci. USA 93*:4509–4512 (1996).
21. Yifrach, O. and Horovitz, A., *Biochemistry 34*:5303–5308 (1995).
22. Hayer-Hartl, M. K., Weber, F., and Hartl, F. U., *EMBO J. 15*:6111–6121 (1996).
23. Lorimer, G. H., Todd, M. J., and Viitanen, P. V., *Philos. Trans. R. Soc. Lond. B. 339*:297–304 (1993).
24. Saibil, H. R. *The Chaperonins* (Ellis, R. J., ed.), Academic Press, New York., 1996, pp. 245–266.
25. Braig, K., Adams, P. D., and Brunger, A. T., *Nature Struct. Biol. 2*:1083–1094 (1995).
26. Boisvert, D. C., Wang, J., Otwinowski, Z., Horwich, A. L., and Sigler, P. B., *Nature Struct. Biol. 3*:170–177 (1996).
27. Aharoni, A. and Horovitz, A., *J. Mol. Biol. 258*:732–735 (1996).
28. Todd, M. J., Viitanen, P. V., and Lorimer, G. H., *Biochemistry 32*:8560–8567 (1993).

29. Murai, N., Makino, Y., and Yoshida, M., *J. Biol. Chem. 271*:28229–28234 (1996).
30. Hunt, J. F., Weaver, A. J., Landry, S. J., Gierasch, L., and Deisenhofer, J., *Nature 379*:37–45 (1996).
31. Mande, S. C., Mehra, V., Bloom, B. R., and Hol, W. G. J., *Science 271*:203–207 (1996).
32. Kovalenko, O., Yifrach, O., and Horovitz, A., *Biochemistry 33*:14974–14978 (1994).
33. Yifrach, O. and Horovitz, A., *J. Mol. Biol. 243*:397–401 (1994).
34. Fenton, W. A., Kashi, Y., Furtak, K., and Horwich, A. L., *Nature 371*:614–619 (1994).
35. Weissman, J. S., Hohl, C. M., Kovalenko, O., Kashi, Y., Chen, S., Braig, K., Saibil, H. R., Fenton, W. A., and Horwich, A. L., *Cell 83*:577–587 (1995).
36. Mendoza, J. A., Rogers, E., Lorimer, G. H., and Horowitz, P. M., *J. Biol. Chem. 266*:13044–13049 (1991).
37. Ellis, R. J., van deer Vies, S. M., and Hemmingsen, S. M., *Biochem. Soc. Symp. 55*:145–153 (1990).
38. Horowitz, P. M., S. Hua, and D. L. Gibbons, *J. Biol. Chem. 270*:1535–1542 (1995).
39. Lin, Z., Schwarz, F. P., and Eisenstein, E., *J. Biol. Chem. 270*:1011–1014 (1995).
40. Zahn, R. and Plukthun, A., *J. Mol. Biol. 242*:165–174 (1994).
41. Staniforth, R. A., Cortes, A., Burston, S. G., Atkinson, T., Holbrook, J. J., and Clarke, A. R., *FEBS Lett. 344*:129–135 (1994).
42. Peralta, D., Hartman, D. J., Hoogenraad, N. J., and Hoj, P. B., *FEBS Lett. 339*:45–49 (1994).
43. Nozaki, Y. and Tanford, C., *J. Biol. Chem. 246*:2211–2217 (1971).
44. Richarme, G. and Kohiyama, M., *J. Biol. Chem. 269*:7095–7098 (1994).
45. de Crouy-Chanel, A., El Yaagoubi, A., Kohiyama, M., and Richarme, G., *J. Biol. Chem. 270*:10571–10575 (1995).
46. Rosenberg, H. F., Ackerman, S. J., and Tenin, D. G., *J. Biol. Chem. 268*:4499–4503 (1993).
47. Izhaki, L. S., Otzen, D. E., and Fersht, A. R., *Biochemistry 34*:14581–14587 (1995).
48. Katsumata, K., Okazaki, A., Tsurpa, G. P., and Kuwajima, K., *J. Biol. Chem.* (in press, 1997).
49. Ding, Y.-H., Duda, R. L., Hendrix, R. W., and Rosenberg, J. M., *Biochemistry 34*:14918–14931 (1995).
50. Landry, S. J. and Gierasch, L. M., *Biochemistry 30*:7359–7362 (1991).
51. Landry, S. J. and Gierasch, L. M., *Annu. Rev. Biophys. Biomol. Struct. 23*:645–669 (1994).
52. Brazil, B. T., Cleland, J., McDowell, R. S., Skelton, N. J., Paris, K., and Horowitz, P. M., *J. Biol. Chem. 272*:5105–5111 (1997).
53. Schmidt, M. and Buchner, J., *J. Biol. Chem. 267*:16829–16833 (1992).
54. Gorovits, B. M., Ybarra, J., and Horowitz, P. M., *J. Biol. Chem. 272*:6842–6845. (1997).
55. Clarke, A. R. and Lund, P. A., *The Chaperonins* (Ellis, R. J., ed.), Academic Press, New York, 1996, pp. 167–212.
56. Martin, J., Langer, T., Boteva, R., Schramel, A., Horwich, A. L., and Hartl, F. U., *Nature 352*:36–42 (1991).
57. Clark, A. C., Hugo, E., and Frieden, C., *Biochemistry 35*:5893–5901 (1996).
58. Ybarra, J. and Horowitz, P. M., *J. Biol. Chem. 270*:22113–22115 (1995).

59. Shi, L., Palleros, D. R., and Fink, A. L., *Biochemistry 33*:7536–7546 (1994).
60. Hlodan, R., Tempst, P., and Hartl, U., *Nature Struct. Biol. 2*:587–594 (1995).
61. Ploegman, J. H., Drent, G., Kalk, K. H., Hol, W. G. J., Heinrikson, R. L., Keim, P., Weng, L., and Russell, J., *Nature 273*:124–129 (1978).
62. Smith, K. E. and Fisher, M. T., *J. Biol. Chem. 270*:21517–21523 (1995).
63. Badcoe, I. G., Smith, C. J., Wook, S., Hallall, D. J., Holbrook, J. J., Lund, P., and Clarke, A. R., *Biochemistry 30*:9195–9200 (1991).
64. Gray, T. E. Fersht, A. R., *J. Mol. Biol. 232*:1197–1207 (1993).
65. Zahn, R., Spitzfaden, C., Ottiger, M., Wuthrich, K., and Plukthun, A., *Nature 368*:261–265 (1994).
66. Clark, A. C. and Frieden, C., *J. Mol. Biol.* (in press, 1997).
67. Lilie, H. and Buchner, J., *Proc. Natl. Acad. Sci. USA 92*:8100–8104 (1995).
68. Gorovits, B. M. and Horowitz, P. M., *J. Biol. Chem. 272*:32–35 (1997).
69. Mayhew, M., da Silva, A. C. R., Martin, J., Erdjument-Bromage, H., Tempst, P., and Hartl, F. U., *Nature 379*:420–427 (1996).
70. Hansen, J. E. and Gafni, A., *J. Biol. Chem. 269*:6286–6289 (1994).
71. Mendoza, J. A. and Del Campo, G., *J. Biol. Chem. 271*:16344–16349 (1996).
72. Schmidt, M., Buchner, J., Todd, M. J., Lorimer, G. H., and Viitanen, P. V., *J. Biol. Chem. 269*:10304–10311 (1994).
73. Landry, S. J., Zellstra-Ryalls, J., Fayet, O., Georgopoulos, C., and Gierasch, L. M., *Nature 364*:255–258 (1993).
74. Murai, N., Taguchi, H., and Yoshida, M., *J. Biol. Chem. 270*:19957–19963 (1995).
75. Weissman, J. S., Rye, H. S., Fenton, W. A., Beechem. J. M., and Horwich, A. L., *Cell 84*:481–490 (1996).
76. Ranson, N. A., Burston, S. C., and Clarke, A. R., *J. Mol. Biol. 266*:656–664 (1997).
77. Todd, M. J., Viitanen, P. V., and Lorimer, G. H., *Science 265*:659–666 (1994).
78. Schmidt, M., Rutkat, K., Rachel, R., Pfeifer, G., Jaenkcke, R., Viitanen, P., Lorimer, G., and Buchner, J., *Science 265*:656–659 (1994).
79. Azem, A., Diamant, S., Kessel, M., Weiss, C., and Goloubinoff, P., *Proc. Natl. Acad. Sci. USA 92*:12021–12025 (1995).
80. Corrales, F. J. and Fersht, A. R., *Folding Design 1*:265–273 (1996).
81. Llorca, O., Carrascosa, J. L., and Valpuesta, J. M., *J. Biol. Chem. 271*:68–76 (1996).
82. Llorca, O., Marco, S., Carrascosa, J. L., and Valpuesta, J. M., *FEBS Lett. 345*:181–186 (1994).
83. Azem, A., Kessel, M., and Goloubinoff, P., *Science 265*:653–658 (1994).
84. Torok, Z., Vight, L., and Goloubinoff, P., *J. Biol. Chem. 271*:16180–16186 (1996).
85. Gorovits, B. M. and Horowitz, P. M., (in preparation, 1997).
86. Ybarra, J. and Horowitz, P. M., (in preparation, 1997).
87. Price, N. C., Kelly, S. M., Thomson, G. J., Coggins, J. R., Wood, S., and auf der Mauer, A., *Biochim. Biophys. Acta 1161*:52–58 (1993).
88. Mizobata, T. and Kawata, Y., *Biochim. Biophys. Acta 1209*:83–88 (1994).
89. Mendoza, J. A., Martinez, J. L., and Horowitz, P. M., *Biochim. Biophys. Acta 1247*:209–214 (1994).
90. Lissen, N. M., Venyaminov, S. Y., and Girshovich, A. A., *Nature 348*:339–342 (1990).
91. Lissin, N. M. and Hemmingsen, S. M., *FEBS Lett. 324*:41–44 (1993).

92. Gorovits, B. M., Seale, J. W., and Horowitz, P. M., *Biochemistry 34*:13928–13933 (1995).
93. Seale, J. W., Gorovits, B. M., Ybarra, J., and Horowitz, P. M., *Biochemistry 35*:4079–4083 (1996).
94. Zahn, R. and Plukthun, A., *J. Mol. Biol. 242*:150–164 (1994).
95. Schmidt, M., Buchler, U., Kaluza, B., and Buchner, J., *J. Biol. Chem. 269*:27964–27972 (1994).
96. Zahn, R., Buckle, A. M., Perrett, S., Johnson, C. M., Corrales, F. J., Golbik, R., and Fersht, A. R., *Proc. Natl. Acad. Sci. USA 93*:15024–15029 (1996).
97. Makino, Y., Taguchi, H., and Yoshida, M., *FEBS Lett. 336*:363–367 (1993).
98. Taguchi, H., Makino, Y., and Yoshida, M., *J. Biol. Chem. 269*:8529–8534 (1994).
99. Llorca, O., Marco, S., Carrascosa, J. L., and Valpuesta, J. M., *J. Struct. Biol. 118*:31–42 (1997).

13

A Chaperonin from a Thermophilic Bacterium, *Thermus thermophilus*

Mitsuko Yamakoshi, Hideki Taguchi, and Masasuke Yoshida

Tokyo Institute of Technology, Yokohama, Japan

Noriyuki Ishii*

National Institute of Bioscience and Human Technology, Tsukuba, Japan

I. INTRODUCTION

Chaperonins are ubiquitous, abundant protein in bacteria, mitochondria, and chloroplasts that facilitate the proper folding of proteins by preventing aggregations in an ATP-, K^+-dependent manner [1–8]. Elucidation of the molecular mechanism of chaperonin activity on folding may contribute a great deal to our understanding about the general pathway of protein folding. Because most studies of chaperonin-dependent folding of proteins have been carried out for *Escherichia coli* chaperonins, we have developed another system to study the general features of chaperonin function using a thermophilic eubacteria, *Thermus thermophilus. T. thermophilus* grows optimally at 75°C and cannot grow at temperatures lower than 50°C. Although amino acid sequences of the chaperonin from *T. thermophilus* (*Thermus* chaperonin; *Th* - cpn 0) have a significant similarity to that of *E. coli* chaperonin, the *Thermus* chaperonin has characteristic features such as thermostability; therefore, we have studied the molecular mechanism of chaperonins by taking advantage of these features.

*Current affiliation: New York University Medical Center, New York, New York

II. STRUCTURAL FEATURES OF *THERMUS* CHAPERONIN

A. Molecular Size and Shape of *Thermus* Chaperonin

Chaperonins comprise two proteins, a 57-kDa protein chaperonin 60 (cpn60) and a 10-kDa protein chaperonin 10 (cpn10) [9–11]. The *E. coli* cpn60 (called GroEL) is purified as $GroEL_{14}$, a tetradecameric oligomer, with an approximate M_r of 840 kDa that shows a heptagonal shape from a top view and a rectangular shape with four clear stripes from a side view by electron micrograph [12,13]. According to X-ray crystallography, $GroEL_{14}$ is composed of two heptameric rings stacked back to back [14]. The *E. coli* cpn10 (GroES) is purified separately as $GroES_7$, a heptameric oligomer, with an approximate M_r of 70 kDa, which shows a ring-like shape from a top view [11–15]. $GroEL_{14}$ and $GroES_7$ associate with each other in the presence of ADP (or ATP) and a bullet-shaped particle that has a round cap on one side of the rectangular complex ($GroEL_{14}$–$GroES_7$) is formed [16]. Usually, cpn60 and cpn10 are purified separately, similar to GroE, even though both of them are required for function. However, only for *Thermus* chaperonin, are they copurified as a functional complex [17]. Under the electron microscope, the purified *Thermus* chaperonin has a sevenfold rotational symmetry when viewed from the top. The diameter of the particles is approximately 13–14 nm [18]. The side view appears to be a bullet-like shape, with four stripes being less obvious than chaperonins from other sources [17–20] (Fig. 1A). It is composed of two stacked heptamer rings of cpn60 ($cpn60_{14}$) and a heptamer ring of cpn10 ($cpn10_7$) that are associated at one axial end of $cpn60_{14}$ (see Fig. 1D). This was directly proved by immunoelectron micrography using a specific antibody to *Th*-cpn10 [19]. Two chaperonin molecules are connected by antibodies through only the round tops of the bullet-shaped molecules (see Fig. 1B). *Th*-$cpn10_7$ cannot be easily separated from *Th*-$cpn60_{14}$ by the purification procedure.

It has been suggested that binding sites of nonnative substrate proteins are located in the central cavity of tetradecameric cpn60 [21]. Our previous observation by immunoelectron microscopy, using antibody against substrate protein, showed that the folding intermediate binds to the bottom end of bullet-shaped chaperonin, which is opposite the end where $cpn10_7$ resides (*trans* side; see Fig. 1C) [22]. However, it was recently proposed for the *E. coli* chaperonin that the folding intermediate can bind not only to the *trans* side, but also to the *cis* side (the side where GroES resides), and folding proceeds more effectively in the cavity [23–25]. The same function might be observed for the *Thermus* chaperonin.

B. Sequencing of *Thermus* Chaperonin

The gene coding *Thermus* chaperonin was sequenced, and the amino acid sequence was determined (accession number D45880) [26]. Th-cpn60 and *Th*-cpn10 are composed of 543- and 101-amino acid residues, respectively. The

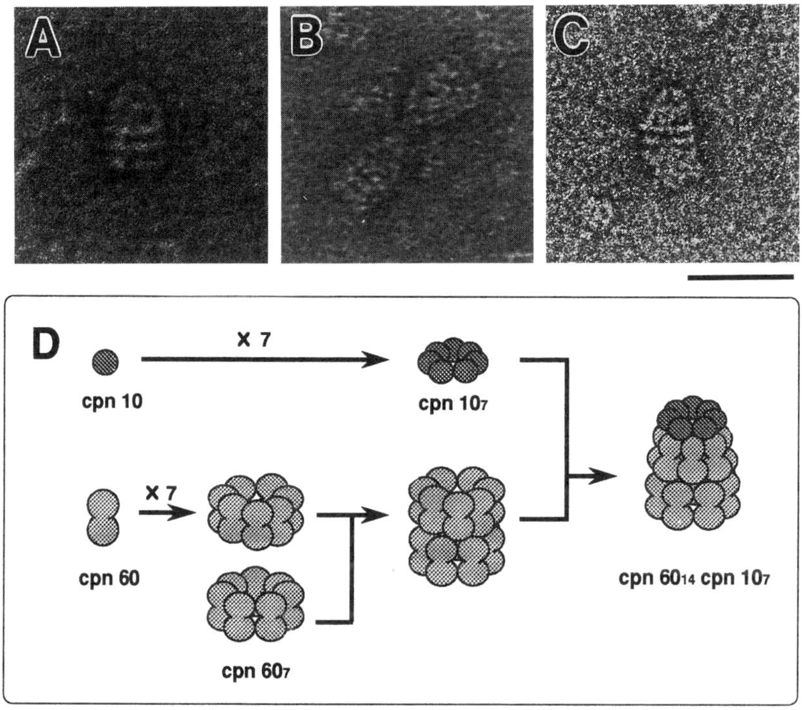

Figure 1 Structure of Thermus chaperonin: Electron micrographs of (A) purified *Thermus* chaperonin; (B) intact *Thermus* chaperonin treated with antibody against cpn10. Two *Thermus* chaperonin molecules are connected by antibodies, and antibodies bind only a round top of the bullet-shaped chaperonin. (C) *Thermus* chaperonin with the bound folding intermediate of IPMDH decorated with anti-IPMDH IgG at first, and second antibody against IgG next. Samples were stained with uranyl acetate. Scale bar represents 25 nm. Detailed protocol is described under "Experimental Procedures" of reference [19,22]. (D) Image of the oligomeric structure of *Thermus* chaperonin.

amino acid sequences of both subunits are highly homologous with chaperonins from other sources. When *Th*-cpn60 was compared with cpn60 from *E. coli* (GroEL), *Bacillus stearothermophilus,* thermophilic *Bacillus* PS3, and *Saccharomyces cerevisiae* mitochondria, 52.0, 67.9, 66.5, and 50.1%, respectively, were identical. When *Th*-cpn10 was compared with cpn10 from *E. coli* (GroES), *B. stearothermophilus,* thermophilic *Bacillus* PS3, and *S. cerevisiae* mitochondria, 54.3, 64.5, 61.7, and 40.7%, respectively, were identical.

Thermus chaperonin showed higher scores of homology to chaperonins from other thermophilic bacteria, such as *B. stearothermophilus* and thermophilic *Bacillus* PS3, than to those from *E. coli* and *S. cerevisiae* mitochondria. This high homology among thermophilic chaperonins could be a reflection of thermostability, which is a special feature of these chaperonins. There is no cysteine residue in either *Th*-cpn60 or *Th*-cpn10, as is generally seen for many proteins from thermophilic bacteria. Although most cpn60s, including GroEL, have a GGM repeat in the COOH-terminal region, it is not found in *Th*-cpn60.

C. Purified *Thermus* Chaperonin Contains Endogenous ADP

For *T. thermophilus,* cpn60 and cpn10 were copurified as a large complex [17–20]. $GroEL_{14}$ and $GroES_7$ associate with each other to form a binary complex in the presence of ADP (or ATP) and dissociation of $GroES_7$ from $GroEL_{14}$ occurs when bound adenine nucleotides are released [16,27,28]. Therefore, we examined whether the purified *Thermus* chaperonin (binary complex) contained endogenously bound adenine nucleotide. The *Thermus* chaperonin contains endogenous ADP, but stoichiometry of bound ADP ranged from 0.3 to 2.8 mol mol^{-1} of the *Thermus* chaperonin (binary complex), dependent on its preparation [20]. The reason for this fluctuation is unknown but it is at least obvious that the stability of binary *Thermus* chaperonin complex cannot be attributed to the bound ADP.

D. ATPase Activity and Thermostability of *Thermus* Chaperonin

Thermus chaperonin is stable up to 80°C judging from temperature dependency of ATPase activity and the decay of the circular dichroism spectrum (Fig. 2) [17]. The ATPase activity was strongly temperature-dependent: 0.1 U mg^{-1} at 80°C and very little activity below 40°C. These results are consistent with the fact that *T. thermophilus* grows optimally at 75°C and cannot grow at a temperature lower than 50°C.

III. *THERMUS* CHAPERONIN-DEPENDENT FOLDING OF SEVERAL PROTEINS

A. Folding Process by Chaperonin

The *Thermus* chaperonin can promote folding of various proteins, not only thermophilic, but also mesophilic, in an ATP-dependent manner [17,29]. As observed for the *E. coli* chaperonin system, *Thermus* chaperonin-dependent protein folding proceeds through two major steps: the association step and the dissociation one. *Thermus* chaperonin binds to the folding intermediate* of the

*The term *folding intermediate* used in this chapter does not mean any defined conformational state of the proteins. It only means a protein in a nonnative structure that appears during the folding process.

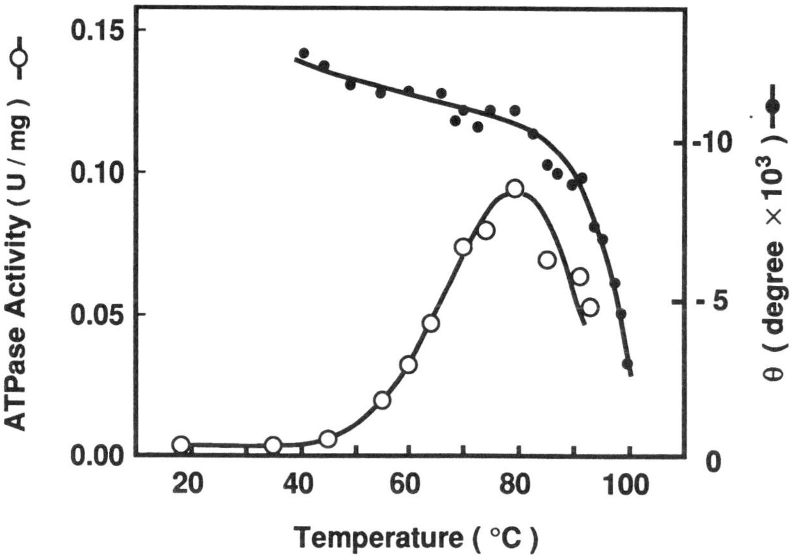

Figure 2 Temperature dependency of ATPase activity and stability of secondary structure of Thermus chaperonin: 40 μg of the purified *Thermus* chaperonin was incubated in an ATPase assay mixture for 15 min at indicated temperatures and liberated P_i was measured. A magnitude of circular dichroism at 220 nm of the purified *Thermus* chaperonin (200 μg ml^{-1} in 50 mM Tris HCl, pH 7.5) was monitored as the temperature was elevated at a constant rate (0.5°C min^{-1}). A cuvette with a 5-mm light path was used. (Adapted from Ref. 17.)

target protein to form the complex. Then coupling, with ATP hydrolysis, the intermediate is released from the complex [10,17,27,29]. Basically, only cpn60 is required to arrest the folding intermediate. We purified recombinant *Th*-cpn60$_{14}$ by gene expression, and found that the chaperonin can arrest the folding intermediate in the absence of cpn10 [26].

B. *Thermus* Chaperonin Acts Differently at Three Temperature Ranges

The effect of *Thermus* chaperonin on the folding of target proteins is varied, being dependent on the temperature range of the experiments (Fig. 3) [17]. At temperatures lower than 25°C (see Fig. 3I), *Thermus* chaperonin-dependent folding in the presence of MgATP was very poor, even though spontaneous folding occurred very efficiently. ATPase activity of *Thermus* chaperonin is very weak at this temperature range (see Fig. 2), whereas its ability to bind the folding intermediate remains intact. It appears that the *Thermus* chaperonin is active in the association

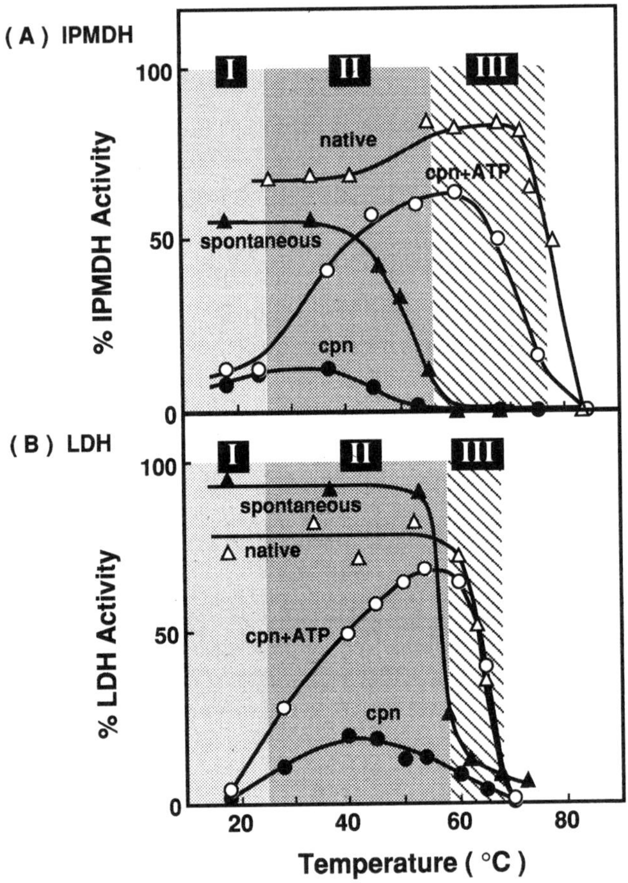

Figure 3 Temperature dependency of spontaneous and chaperonin-dependent folding of (A) *T. thermophilus* IPMDH and (B) *B. stearothermophilus* LDH: The enzyme solution containing guanidine HCl-denatured IPMDH or LDH was diluted into the dilution buffer preincubated at the indicated temperature. Incubation was continued another 10 min (IPMDH) or 30 min (LDH), and the recovered activities were then measured [17]. The dilution buffer contained *Thermus* chaperonin and MgATP (open circles); *Thermus* chaperonin (filled circles); or neither chaperonin nor MgATP (filled triangles). As a control (open triangles), native IPMDH or LDH was treated with the same procedures except that guanidine HCl was contained not in the enzyme solution, but in the dilution buffer at a concentration that gave the same final concentration of guanidine HCl as in the previous experiments. Temperature range I, II, and III are shown by different shades. (Adapted from Ref. 20.)

step at all temperatures lower than 75°C, but is not so in the dissociation step at less than 30°C. At middle temperatures 25°–55°C (see Fig. 3II), spontaneous folding of the proteins occurs and the folding is inhibited by *Thermus* chaperonin in the absence of MgATP. The folding is initiated at the time MgATP is added as supplement to the solution. The chaperonin is dispensable for folding at this temperature range, for spontaneous folding proceeds efficiently. At temperatures higher than 55°C (see Fig. 3III), spontaneous folding fails, but once-refolded (native) protein is sufficiently stable. Both *Thermus* chaperonin and MgATP are absolutely required for productive folding. The chaperonin alone, in the absence of MgATP, interfered with the spontaneous folding at all temperatures. The border of the temperature range II and III (see Fig. 3) varies from one protein to another and is dependent on experimental conditions. For example, isopropylmalate dehydrogenase (IPMDH) from *T. thermophilus,* the temperature range III (see Fig. 3) covers 55°–75°C (see Fig. 3A), whereas lactate dehydrogenase (LDH) from *B. stearothermophilus* has only a narrow temperature range (58°–65°C) because of the limited thermal stability of the native LDH (see Fig. 3B). Spontaneous folding of proteins appears to undergo an intermediate form that is generally more heat-labile than the native proteins and is easily irreversibly denatured, even at the optimal growth temperature of the cell. Chaperonin protects the intermediates from heat denaturation. This may provide an answer to why the chaperonin is expressed abundantly at normal physiological temperature for the cell in the absence of heat shock.

C. *Thermus* Chaperonin Increases Heat Stability of Other Proteins

Chaperonin binds folding intermediates that occur during folding process from chemically denatured state, and the complex between chaperonin and folding intermediate is stabilized, to some extent, in the absence of MgATP. Because chaperonin is a heat-shock protein, it is also expected that it can protect proteins from heat denaturation [30]. *Thermus* chaperonin is stable and functional up to 80°C (see Fig. 2). Heterologous combination of *Thermus* chaperonin and relatively heat-labile enzymes provides a suitable in vitro assay system to study the effect of chaperonin on heat denaturation. A mixture of *Thermus* chaperonin and the enzyme is incubated at various high temperatures for 30 min, it is subsequently incubated at a fixed moderate temperature in the presence of MgATP, and the residual enzyme activity is then assayed. In general, the critical temperature at which irreversible denaturation of the enzymes occurs is elevated 8–15°C by inclusion of *Thermus* chaperonin in the solution (Table 1) [20,29]. Addition of MgATP in the solution during the first incubation abolishes the effect of chaperonin. The addition of *Thermus* chaperonin to the completely heat-denatured

Table 1 Temperature of 50% Loss of Activity (°C)

	LDH (°C)	Rhodanese (°C)	IPMDH (°C)
+BSA	65	53	58
+BSA, then ATP	65	52	58
+cpn	65	52	58
+cpn, then ATP	73	67	67

Protection of relatively heat-labile proteins from irreversible heat denaturation by *Thermus* chaperonin. The experiment consisted of three steps. (1) incubation: the enzyme solution (0.1 mg ml^{-1}) containing *B. stearothermophilus* LDH or bovine rhodanese was diluted 50-fold (LDH) or 25-fold (rhodanese) into the dilution buffer preincubated at various high temperatures, and incubation was continued for 30 min. The dilution buffer contained 5 mM dithiothreitol, 50 mM potassium phosphate, pH 7.8, and, when indicated, bovine serum albumin or *Thermus* chaperonin (0.4 mg mL^{-1} for LDH, 0.2 mg mL^{-1} for rhodanese system). In the rhodanese system, the dilution buffer also contained 50 mM sodium thiosulfate in addition of the foregoing components. (2) incubation: the solution was transferred to a water bath thermostated at 55°C (LDH) or 37°C (rhodanese) with concomitant addition of 50 μM MgATP (final concentration), and it was incubated for 30 min (LDH) or 60 min (rhodanese). Assay of enzyme activity: the enzyme activity in the solution was measured. This activity is the sum of survived activity at the first incubation and recovered activity at the second incubation. Temperatures of the first incubation that cause 50% loss of enzyme activity are shown.

protein does not restore the activity of the protein. Thus, heat denaturation of a protein proceeds through an intermediate, the structure of which is more or less common to the folding intermediate from the chemically denatured state, and is recognizable by chaperonin. As expected, the complex formed between *Thermus* chaperonin and the folding intermediate of chemically denatured proteins is highly thermostable [20,29]. It seems that thermostability of the chaperonin–intermediate complex is limited only by the thermostability of *Thermus* chaperonin itself. The results are very clear, as shown in Figure 4 [29]. heat stabilities of folding intermediates of three enzymes captured by *Thermus* chaperonin were all the same. This temperature is close to the denaturation temperature of *Thermus* chaperonin itself, 80°–85°C. Whether the direction of process is folding or unfolding, once the intermediate is captured by the chaperonin, the intermediate can retain its ability to resume productive folding, even after exposure to high temperatures that otherwise would cause the protein's rapid denaturation. Chaperonin can act as a "shelter" for labile proteins from transient exposure to high temperature or protein-destabilizing chemicals.

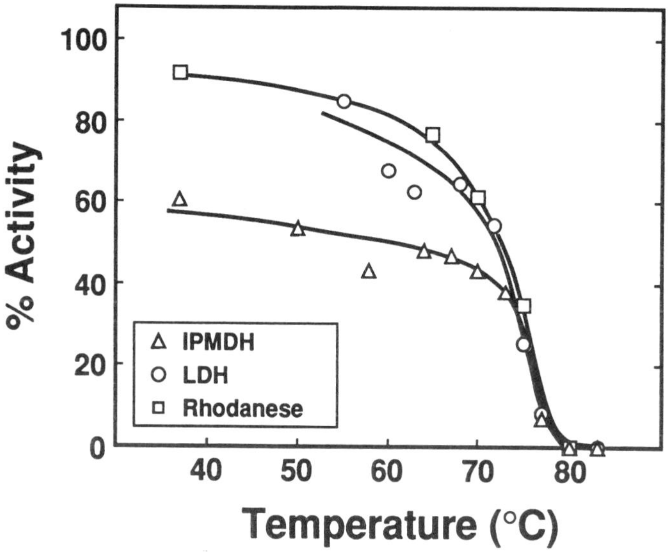

Figure 4 Heat stability of a folding intermediate captured by *Thermus* chaperonin: Stability of the folding intermediate was defined as retention of the ability to resume folding at a moderate temperature when MgATP is supplied. Experiments consisted of four steps. (1) "cpn-1" formation: the enzymes, *B. stearothermophilus* LDH (○), *E. coli* IPMDH (△), or bovine rhodanese (□) were dissolved in 6.4 M guanidine HCl that was diluted into the buffer containing *Thermus* chaperonin preincubated at 55°C (LDH) or at 37°C (IPMDH and rhodanese). (2) Exposure to high temperature: The complex formed between the folding intermediate and chaperonin was exposed to the indicated temperatures for 30 min. (3) folding: The solution was shifted again to 37°C (rhodanese and IPMDH) or to 55°C (LDH) with concomitant addition of 500 μM MgATP. Incubation was continued for 60 min (rhodanese and IPMDH) or 30 min (LDH). (4) Assay of recovered enzyme activity: recovered enzyme activity in the solution was measured at 37°C (rhodanese and IPMDH) or at 55°C (LDH). Details of experimental conditions are described in "Experimental Procedures" of cited reference. (Adapted from Ref. 29.)

IV. OTHER FEATURES OF *THERMUS* CHAPERONIN

A. Monomeric *Thermus* Chaperonin-60 Can Promote Protein Foldings Without MgATP

1. *Background*

It has been suggested that binding sites of nonnative substrate proteins are located in the central cavity of tetradecameric cpn60 [21], and oligomeric forms of cpn60 are believed to be essential for the function of chaperonins [16,23–25,31].

However, the cpn60 monomer isolated from *T. thermophilus* (*Th*-cpn60_m) possesses the ability to promote protein folding without MgATP by suppressing the aggregation of nonnative substrate proteins under certain conditions [32,33].

2. Preparation of Monomeric Chaperonin

We prepared monomer of cpn60 (cpn60_m) from purified *Thermus* chaperonin as follows: The purified *Thermus* chaperonins were denatured by acid and two subunits, cpn60 and cpn10, were separated by reversed-phase high-performance liquid chromatography (HPLC). The purified cpn60 and cpn10 spontaneously refolded in a buffer at neutral pH. The refolded cpn60 existed as monomers and did not assemble into oligomers in the absence of nucleotides. The refolded cpn10 assembled into oligomers, probably heptamers. The 50-kDa fragment was prepared from *Th*-cpn60_m by treatment with TPCK-trypsin for 3 h. The NH_2-terminal sequence of the 50-kDa fragment was determined to start from Thr-79 in the whole sequence of *Th*-cpn60. The peptide segment 1–79 corresponds to a part of an equatorial domain that interacts with equatorial domains of two adjacent cpn60 molecules, one in the same ring and the other in an opposite ring, in the tetradecamer chaperonin [14]. The segment also contains the residues that contribute to form the ATP or ADP-binding site of chaperonin [14]. Therefore, it is not surprising that the 50-kDa fragment of *Th*-cpn60_m did not assemble into tetradecamers even in the presence of MgATP or *Th*-cpn10. The ATP hydrolyzing activities of *Th*-cpn60_m and the 50-kDa fragment of *Th*-cpn60_m were null or very low [33].

3. Monomeric cpn60 Suppresses Aggregation of Rhodanese

Dilution of bovine mitochondrial rhodanese from 6-M guanidine HCl into buffer alone resulted in aggregation [34], but when the dilution buffer contained intact *Thermus* chaperonin, aggregation was completely suppressed (Fig. 5). When the dilution buffer contained *Th*-cpn60_m instead of intact *Thermus* chaperonin,* aggregation of rhodanese was also suppressed. Furthermore, the 50-kDa fragment of *Th*-cpn60_m was also effective in suppression of aggregation. The degree of the suppression increased in parallel with the amount of added *Th*-cpn60_m. These effects were not due to a nonspecific effect of increasing concentration of proteins because inclusion of bovine serum albumin in the dilution buffer caused no effect on aggregation of rhodanese. These observations indicate that monomeric cpn60 can interact with rhodanese molecules with nonnative structure and prevent them from aggregation [33].

---

*The term *intact* Thermus *chaperonin* in this chapter means a bullet-shaped chaperonin ($cpn60_{14}$–$cpn10_7$) from *T. thermophilus.*

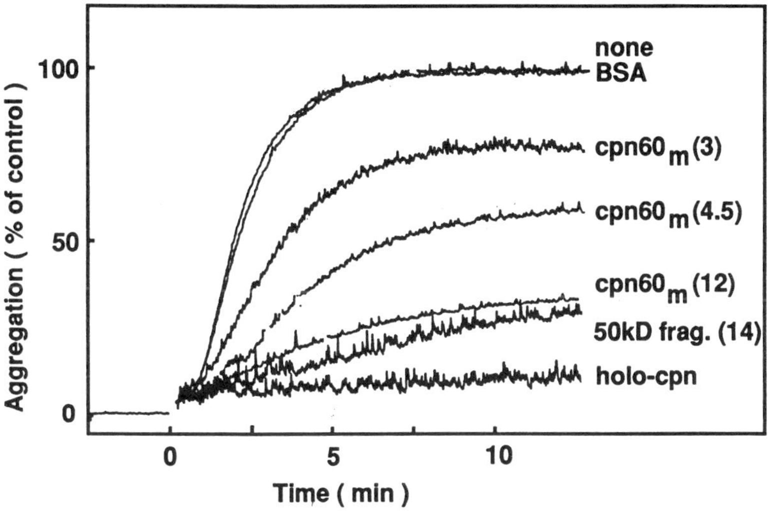

Figure 5 Suppression of aggregation during the folding of denatured rhodanese by *Th*-cpn60$_m$: The 6-M guanidine HCl-denatured rhodanese was diluted 100-fold (0.14 μM final concentration) at 30°C into the dilution buffer, 50 mM Tris-Cl, pH 7.5, containing the following components; none; 120 μg mL^{-1} bovine serum albumin (BSA); 2.0 μM (calculated in terms of cpn60 monomer) intact *Thermus* chaperonin (holo-cpn); *Th*-cpn60$_m$ at 3-, 4.5-, and 12-fold molar excess to rhodanese; or the 50-kDa fragment of *Th*-cpn60$_m$ at 14-fold molar excess to rhodanese. The time courses of aggregation were measured by light scattering using a Jasco FP-777 spectofluorometer as absorbance at 320 nm. (Adapted from Ref. 33.)

4. *Chaperone Activity of* Th-*cpn60*$_m$ *Is Independent of MgATP*

When the folding of rhodanese was monitored by measuring the reactivation of rhodanese activity [34,35], promotion of functional folding by *Th*-cpn60$_m$ was clearly observed (Fig. 6) [33]. When the denatured rhodanese was diluted into a buffer (0.13 μM rhodanese, final concentration) containing *Th*-cpn60$_m$, the reactivation yield reached to 25% in a few minutes, whereas the yield of spontaneous reactivation of rhodanese was as low as approximately 5–7%. Further inclusion of MgATP in the dilution buffer had little effect on the yield of reactivation. The similar amount (weight) of intact *Thermus* chaperonin arrested folding completely in the absence of MgATP and promoted reactivation—up to 50%—in the presence of MgATP. The time course of reactivation of rhodanese promoted by *Th*-cpn60$_m$ was much faster than that promoted by intact *Thermus* chaperonin plus MgATP, indicating that the folding by intact *Thermus* chaperonin occurred in a more modulated manner. The 50-kDa fragment of *Th*-cpn60$_m$ was even more effective in reactivating rhodanese than *Th*-cpn60$_m$ (Table 2, part A). The presence of cpn10 had no effect on the yield of reactivation in any experiments tested.

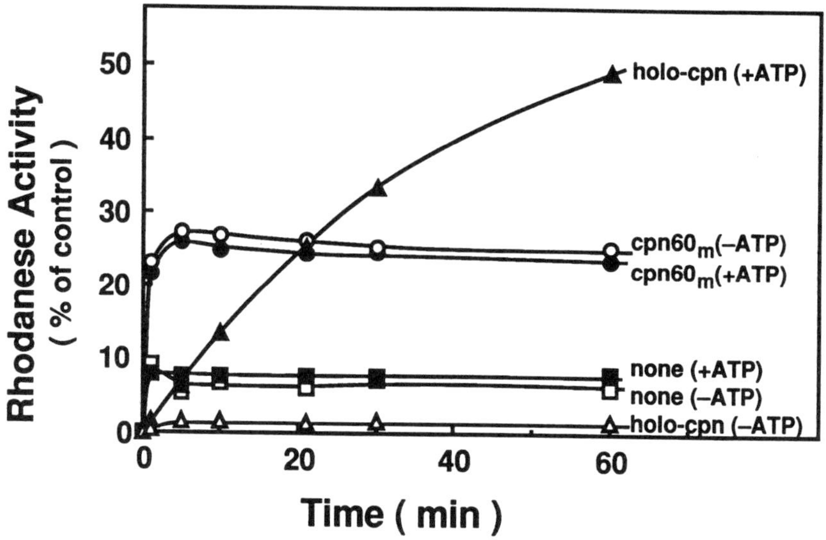

Figure 6 The *Th*-cpn60$_m$-dependent, MgATP-independent reactivation of denatured rhodanese: The denatured rhodanese was diluted 25-fold (0.13-μM final concentration) at 37°C into 50 mM of potassium phosphate buffer, pH 7.8, 50 mM dithiothreitol, 50 mM $Na_2S_2O_3$, and the following components: none; 3.3 μM *Th*-cpn60$_m$, or 3.3 μM (calculated in terms of cpn60 monomer) intact *Thermus* chaperonin (holo-cpn). Each sample was examined with or without addition of 1 mM MgATP. After the indicated times of incubation, an aliquot was withdrawn and the rhodanese activities were assayed at 20°C for 15 min [33].

When the final concentration of rhodanese was increased up to 1.5 μM, the effect of *Th*-cpn60$_m$ and the 50-kDa fragment on rhodanese reactivation became more evident. At this concentration, denatured rhodanese absolutely cannot fold spontaneously whereas *Th*-cpn60$_m$ and the 50-kDa fragment can promote the reactivation (the reactivation yield reached close to 15% [33].

We then immobilized *Th*-cpn60$_m$ onto the solid surface of resins through covalent bonds to prevent possible transient formation of oligomers during the folding assays. The immobilized *Th*-cpn60$_m$ promoted reactivation of rhodanese at a significant yield (see Table 2, part B) [33]. Such a promotion was not observed for the immobilized bovine serum albumin.

The promotion of functional folding by *Th*-cpn60$_m$ was not restricted to rhodanese. When lactate dehydrogenase (LDH) from *B. stearothermophilus* and IPMDH from *T. thermophilus* were used as substrate proteins, the reactivation yields increased about double those of the spontaneous ones by *Th*-cpn60$_m$ [17]. Thus, the cpn60$_m$-dependent folding seems to be indiscriminate for a variety of proteins.

Table 2 Effects of Various Chaperonin Species on the Reactivation Yields of Rhodanese

Addition	Reactivation (%)
A	
None	4.5
Th-cpn60$_m$	24.7
Th-cpn60$_m$ + *Th*-cpn10	24.6
Th-cpn60$_m$ + *Th*-cpn10 + MgATP	25.2
50-kDa fragment	35.0
B	
Immobilized BSA	0.0
Immobilized *Th*-cpn60$_m$	10.0

The activity of reactivated rhodanese was assayed at 30 min after starting folding at 37°C. When indicated, following concentrations of components were contained in the dilution buffer: 3.3 μM of *Th*-cpn60$_m$; 0.28 μM of *Th*-cpn10 (calculated as heptamer); 3.3 μM of the 50-kDa fragment of *Th*-cpn60$_m$; 1 mM $MgCl_2$ and 1 mM ATP. Under the same conditions, intact *Thermus* chaperonin (3.3 μM, calculated as cpn60 monomer) plus MgATP gave 34% of reactivation of rhodanese. For experiments of immobilized chaperonin, 6 mg of resins supposed to have 20 μg of immobilized *Th*-cpn60$_m$ or 20 μg of immobilized bovine serum albumin (BSA) were contained in the dilution buffer and the mixtures were shaken gently during incubation. It was confirmed that resins without immobilized proteins had no effect on reactivation of rhodanese.

The dependency of reactivation yield on the concentration of *Th*-cpn60$_m$ is also very different from that of intact *Thermus* chaperonin. Optimal reactivation of rhodanese activity was observed when the molar ratio of *Th*-cpn60$_m$/rhodanese was about 10:1, and the reactivation yield leveled off or even slightly decreased when more *Th*-cpn60$_m$ was included in the dilution mixture. Conversely, the reactivation yield increased as the amount of intact *Thermus* chaperonin increased up to molar ratio, expressed in terms of monomer cpn60 and monomer substrate proteins, reached 100:1 [33].

5. Functional Meaning of the Oligomeric Structure of Chaperonin

If monomeric cpn60 has the primitive ability to promote folding of other proteins, then one might ask: what is the functional meaning of tetradecameric structure of native chaperonin? The intact *Thermus* chaperonin requires MgATP for promotion of folding [17,36,37], but monomeric cpn60 does not. Similarly, ATP-independent promotion of folding has been observed for some other molecular chaperones, such as Hsp90 [38] and small heat-shock proteins [39]. Transient and reversible bindings of proteins with nonnative structure to these molecular chaperones should result in lowering the concentration of nonnative proteins in the solution. Because the aggregation is highly dependent on the concentration of nonnative proteins, this favors strongly productive folding in kinetic competition between folding and aggregation [20,40]. When monomeric cpn60s are organized into tetradecamers, the affinity of cpn60 for nonnative proteins becomes very strong, and the binding of nonnative proteins is almost irreversible. Then, the releasing step of nonnative proteins captured on intact *Thermus* chaperonin is subjected to regulation, such that MgATP accelerates the releases. Thus, oligomerization of cpn60 adds a regulatory function to the intrinsic chaperone activity of monomeric cpn60.

B. Tetradecamer–Heptamer Transition of *Thermus* Chaperonin

1. Background

Thermus chaperonin is purified as a bullet-shaped particle. However, when MgATP and high concentration of K^+ are present in solution, it splits along its equatorial plane into two components. When MgATP was eliminated from the solution, the split chaperonins slowly associated with each other to regenerate the bullet-shaped chaperonin. Furthermore, football-shaped chaperonins ($cpn10_7$–$cpn60_{14}$–$cpn10_7$) appeared when MgATP and excess $cpn10_7$ were included in the solution. Thus, *Thermus* chaperonin can undergo reversible tetradecamer–heptamer transition of cpn60, even though its functional significance is as yet unclear.

2. Equatorial Split of Thermus *Chaperonin Requires ATP Hydrolysis*

Purified *Thermus* chaperonin splits at the plane between two cpn60 rings into two halves in the presence of ATP and K^+ (Fig. 7b) [41]. Because the split reaction is not induced by ADP plus K^+, AMP–PNP plus K^+, or ATP plus Na^+ and ATPase activity of chaperonin is known to be K^+-dependent [27,42], it is likely that ATP hydrolysis is required for the split reaction of *Thermus* chaperonin. Magnesium was essential for the split reaction, because when Mg^{2+} was removed by prior addition of a strong metal chelator, cyclohexyldiaminetetraacetic acid, to the chaperonin solution, no split reaction was induced by the addition of ATP. This split was observed at a wide range of temperatures, 0°–60°C [41]. At low temperature, although the rate is slow [17], intact *Thermus* chaperonin can still hydrolyze ATP; for example, one *Thermus* chaperonin hydrolyzed three ATP molecules per minute at 18°C. A small number of (or even a single) catalytic

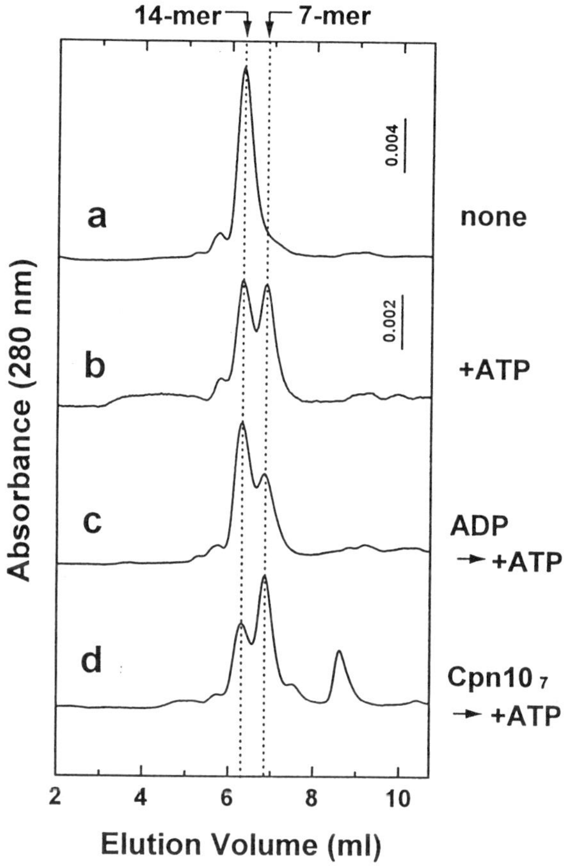

Figure 7 Yield of the split chaperonin monitored with gel permeation HPLC: Intact *Thermus* chaperonin (710 μg mL^{-1}) was dissolved in (a) 40 mM sodium phosphate buffer (pH 7.5); (b) 40 mM sodium phosphate buffer (pH 7.5); (c) 25 mM sodium phosphate buffer (pH 7.5); and 200 μM ADP; or (d) 40 mM sodium phosphate buffer (pH 7.5) in the presence of *Th*-cpn10_7 (180 μg mL^{-1}). After the solutions were incubated for 10 min at 37°C, except for (a), ATP, $MgCl_2$, and KCl at 2 mM, 2 mM, and 200 mM at final concentrations, respectively, were added to the solutions. For (a), KCl was added at the foregoing concentrations, but MgATP was not added. The solutions were incubated at 37°C for another 1 min. Then, 50 μL of the solution was applied to a gel permeation HPLC column (Tosoh G3000SWxl) equilibrated with 50 mM Tris-Cl buffer (pH 6.8) containing 200 mM Na_2SO_4 (buffer A). The column was eluted at 0.5 mL min^{-1} with buffer A at 25°C and was monitored with absorbance at 280 nm. The elution volumes corresponding to the tetradecamer chaperonin* and the split chaperonin* are indicated in the figure as 14-mer and 7-mer, respectively. The absorbance scales (shown by a scale bar) of b–d are the same, whereas the scale of a is half the others.

turnovers of ATP hydrolysis seem to be sufficient to induce the dissociation of *Thermus* chaperonin.

3. Electron Microscopic Observation of the Split Chaperonin

As shown in Figure 7, a peak of the split chaperonin emerges at 6.9 mL only in the presence of MgATP and K^+ with gel permeation HPLC. Then the fraction was separated and observed by electron microscopy. Images of molecules in the fractions at 6.9 mL were as shown in Figure 8A. It is easy to distinguish between a side-on view of "half a football" (cone-like particles) and the end-on view looking down the sevenfold axis of symmetry. Similar images were observed for the 6.9-mL fractions at 60°C. Two intact *Thermus* chaperonin molecules were connected by anti-cpn10 IgGs, the solution was then incubated for

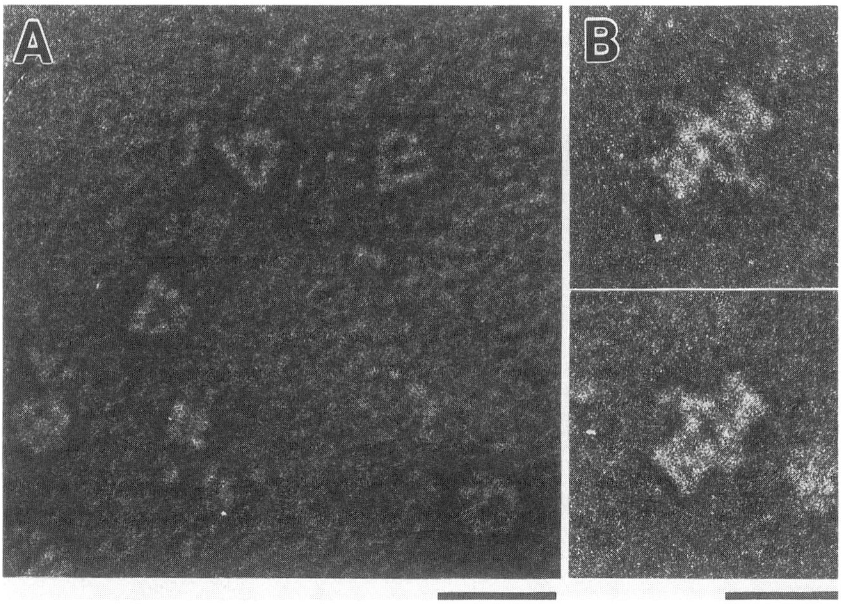

Figure 8 Electron micrographs of the split chaperonin: (A) *Thermus* chaperonin that was eluted at 6.9 mL on gel filtration HPLC (see trace b in Fig. 7) stained with methylamine tungstic acid. Cone-like particles and round particles with a clear central cavity and with an obscured central cavity are seen. (B) Two intact *Thermus* chaperonin molecules connected by anti-cpn10 IgGs, and those after the incubation for 30 min at 22°C with 50 mM of Tris-Cl buffer (pH 7.51) containing 200 mM NaCl, 90 mM KCl, and 1.5 mM MgATP; particles have lost lower halves of connected intact *Thermus* chaperonin. Scale bars in each of the figures represent 25 nm. (Detailed experimental conditions are described in Materials and Methods in reference [41].)

30 min with MgATP plus K^+ at 22°C, and the sample was observed with an electron microscope (see Fig. 8B). The bullet-shaped chaperonin disappeared and two cone-like particles connected by anti-cpn10 IgGs through round tops were observed. Irregular aggregates that seemed to be connected by IgGs were also observed. Thus, cpn10s reside at the round top of the cone-like particle that is derived from the upper half of the bullet-shaped chaperonin. From these observations, it appears that intact *Thermus* chaperonin splits into two particles in the presence of MgATP plus K^+ and the split occurs at the boundary plane ("equatorial" split) between two rings of $cpn60_7$ in the *Thermus* chaperonin molecule [41].

4. The Effect of ATP and KCl Concentrations on the Split Reaction

The split reaction of *Thermus* chaperonin is well monitored in real time by light scattering; therefore, the effect of ATP and KCl concentrations on the split reaction was examined by this method. The magnitude of scattering change increased as the concentrations of ATP and KCl increased, as shown in Figure 9, and it was saturated at 20 μM ATP and 150 mM KCl (see Fig. 9A and B, insets). Consistent with the previous result with HPLC [41], NaCl could not substitute for KCl to induce the split reaction. This indicates that the effect of KCl is not due to mere ionic strength, because the total concentrations of KCl and NaCl in the chaperonin solutions were the same in all experiments in Figure 9. In any event, the split reaction occurred rapidly; the scattering transition finished in less than 5 s.

5. The Split Reaction Is Inhibited by ADP and Stimulated by Excess $Cpn10_7$

To estimate relative population of the split chaperonin, *Thermus* chaperonin was analyzed with gel permeation HPLC. As shown in Figure 7, preincubation of intact *Thermus* chaperonin with ADP before the addition of ATP resulted in a decrease of 39% in the yield of the split chaperonin (see Fig. 7c). Interestingly, when the order of nucleotide addition was reversed (i.e., ATP was added first and ADP next), the yield was only slightly decreased. It seems that the preloading nucleotide-binding sites of chaperonin with ADP before the addition of ATP prevents the chaperonin from splitting. Probably related to this observation, the ATPase activity of the chaperonin was inactivated when the empty nucleotide-binding sites were occupied by ADP [42,43]. Moreover, the population of the split chaperonin was increased significantly to 63% by inclusion of 3.3 M excess of the *Thermus* $cpn10_7$ in the incubation mixture (see Fig. 7d). A peak at 8.7 mL in Fig. 7d corresponds to the free $cpn10_7$ [20].

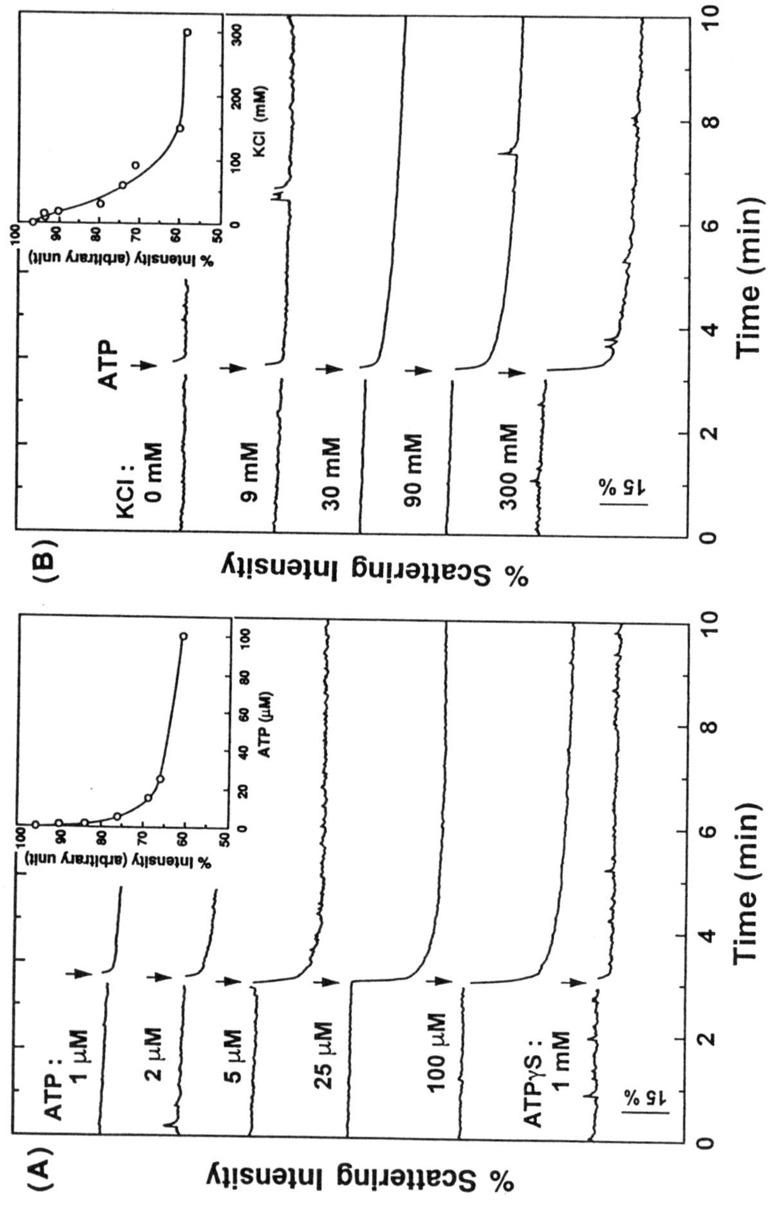
(A)
% Scattering Intensity
ATP : 1 μM
2 μM
5 μM
25 μM
100 μM
ATPγS : 1 mM
15 %
Time (min)
% Intensity (arbitrary unit)
ATP (μM)
(B)
% Scattering Intensity
KCl : 0 mM
9 mM
30 mM
90 mM
300 mM
ATP
15 %
Time (min)
% Intensity (arbitrary unit)
KCl (mM)

6. Tetradecamer–Heptamer Transition of Thermus *Chaperonin*

To know whether all, or a fixed population of *Thermus* chaperonins undergo the split reaction and whether the split chaperonins can reassociate with each other, we isolated the tetradecamer chaperonin* and the split chaperonin* by a gel permeation HPLC column. The bullet-shaped *Thermus* chaperonins were incubated with the $cpn10_7$, ATP, and KCl for 10 min, applied to the column, and fractions of the tetradecamer chaperonins (6.3 mL) and those of the split chaperonins (6.9 mL) were pooled separately (Fig. 10a). The isolated tetradecamer chaperonin fraction was aged for 170 min and was again applied to the column before (see Fig. 10b) and after (see Fig. 10c) another 1 min incubation with ATP and KCl. The elution profile of Figure 10c shows the appearance of a peak of the split chaperonin at 6.9 mL, ensuring that the tetradecamer chaperonin of the first HPLC (see Fig. 10a) does not represent a fixed population of chaperonin that is unable to split. The isolated split chaperonin was applied to the column after aging for 15, 95, and 170 min at 25°C (see Fig. 10d–f). With each increasing aging period, the peak height of the split chaperonin at 6.9 mL decreased, with a reciprocal increase in the peak at 6.3 mL. Electron microscopy showed that the peak fraction at 6.3 mL of Figure 10f contained the bullet-shaped chaperonin as a dominant species (see Fig. 10g). Essentially no football-shaped chaperonin was found. Only very few cylinder-shaped tetradecamer chaperonins without cpn10 were found in the side view (Table 3, line 6). Thus, the split chaperonins can reassociate with each other and the bullet-shaped chaperonins are regenerated. There can be several pathways to form the bullet-shaped chaperonin from two

---

Figure 9 The split reaction of *Thermus* chaperonin induced by various concentrations of (A) ATP or (B) KCl, monitored with light scattering: (A) Intact *Thermus* chaperonin (74 μg) in 1 mL of 25 mM Tris-Cl buffer (pH 7.5) containing 300 mM KCl and 5 mM $MgCl_2$, was incubated at 40°C and the indicated concentrations of ATP (final concentration) were added to the solution. (Lowest trace) 1 mM (final concentration) of ATPγS was added instead of ATP. (Inset) Dependency of the scattering decrease on the ATP concentrations. (B) Intact *Thermus* chaperonin (74 μg) in 1 mL of 25 mM Tris-Cl buffer (pH 7.5) containing 5 mM $MgCl_2$, and (from the top trace) 300 mM NaCl, 9 mM KCl, and 291 mM NaCl; 30 mM KCl and 270 mM NaCl; 90 mM KCl and 210 mM NaCl; or 300 mM KCl, was incubated at 40°C. ATP at 1 mM (final concentration) was added to the solution at the time indicated. (Inset) Dependency of the scattering decrease on the KCl concentrations. The magnitude of the scattering intensity at 320 nm before addition of ATP was taken as 100%.

---

*The term *tetradecamer chaperonin* used in this chapter includes all the chaperonin species that have 14 cpn60 molecules, such as the bullet-shaped, the football-shaped, and the cylinder-shaped chaperonins. Similarly, the term *split chaperonin* includes all the chaperonin species that have 7 cpn60 molecules.

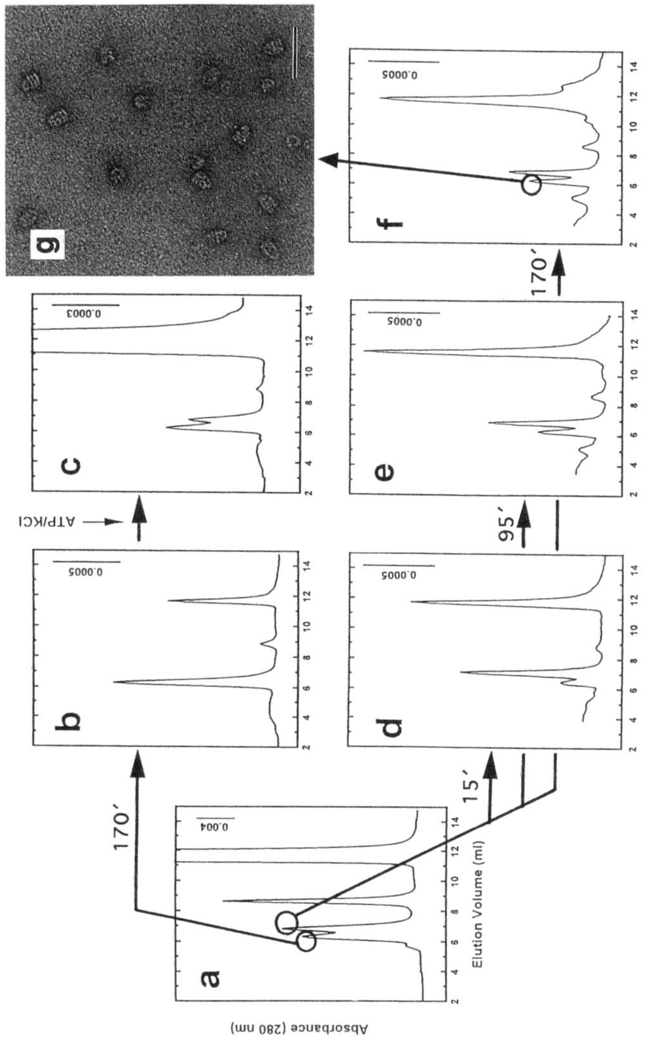
a
b
c
d
e
f
g
170′
15′
95′
170′
ATP/KCl
Absorbance (280 nm)
Elution Volume (ml)
0.004
0.0005
0.0003
0.0005
0.0005
0.0005

cone-shaped chaperonins, including transient formation of cylinder-shaped chaperonin, but there are as yet no available experimental results to determine the real pathway. The height of a peak at near 8.7 mL, which corresponded to the $cpn10_7$ and the $cpn60_m$ [20], remained small throughout the aging period; therefore, $cpn60_m$ might not appear during the reassociation reaction, even if we cannot exclude the possibility of rapid transient dissociation.

Reassociation of the split chaperonin was also monitored by light scattering. As seen in Figure 11, the ATP-induced split reaction was partly reverted by deprivation of ATP from the solution. ATP was converted to ADP by addition of hexokinase and glucose (see Fig. 11A, trace a), or hydrolyzed to AMP by addition of apyrase (see Fig. 11A, trace b). These procedures resulted in the gradual increase of the scattering signal, which finally reached the intermediate intensity. It was confirmed that the addition of the hexokinase and apyrase alone, at the concentrations used in the experiment, did not affect the scattering signal and that the increase of the signal was not due to the aggregation of the hexokinase or apyrase. Similar results were obtained from analysis with gel permeation HPLC. Intact *Thermus* chaperonin was at first incubated with ATP, KCl, and $cpn10_7$ to induce the split reaction (see Fig. 11B, trace a) and then ATP was converted to ADP by the addition of hexokinase and glucose (see Fig. 11B, trace b), or hydrolyzed to AMP by the addition of apyrase (see Fig. 11B, trace c). After a 5-min incubation, samples were analyzed with HPLC. When compared with an elution profile of the control sample (see Fig. 11B, trace a), the relative populations of the split chaperonin (the peak at 6.9 mL) of hexokinase-treated and apyrase-treated samples were clearly decreased and that of the tetradecamer

---

Figure 10 Gel permeation HPLC of the isolated tetradecamer and the split chaperonins: (a) Intact *Thermus* chaperonin (1.0 mg mL^{-1}) and *Thermus* $cpn10_7$ (0.39 mg mL^{-1}) were incubated in 40 mM sodium phosphate buffer (pH 7.5) containing 2 mM ATP, 2 mM $MgCl_2$, and 200 mM KCl (final concentration) for 10 min at 37°C, and 100 μL of the solution was applied to the gel permeation HPLC as described in Fig. 7. (b) The peak fractions at 6.3 mL (tetradecamer chaperonins) and 6.9 mL (the split chaperonins) were pooled separately. The tetradecamer fraction was aged for 170 min at 25°C. Then, an aliquot (100 μL) of the solution was taken out and applied to the HPLC column. (c) To the rest of the solution, 2 mM ATP, 2 mM $MgCl_2$, and 200 mM KCl (final concentrations) were added. After the solution was incubated for 1 min at 37°C, an aliquot (100 μL) was injected onto the HPLC column. The fraction of the split chaperonin (100 μL each) was applied to the column after aging for (d) 15 min, (e) 95 min, and (f) 170 min at 25°C. (g) Electron micrograph of the tetradecamer chaperonins produced by reassociation of the split chaperonins. The peak fraction of the tetradecamer chaperonins at 6.3 mL of f was immediately used for electron microscopy. The scale bar represents 50 nm.

Table 3 Distribution of Negatively Stained *Thermus* Chaperonins Identified by Electron Micrography

	Side views								
Conditions	Total	Bullet (%)	Football (%)	Cone (%)	Cylinder (%)	7-mer (%)	Top views	Unclassified	Particle analyzed
1. ATP + KCl (0.1 mM) + cpn10	357	15.4	67.8	16.5	0.3	—	387	74	818
2. ATP + KCl (100 mM) + cpn10	389	21.6	43.4	34.7	—	0.3	359	86	834
3. ADP + KCl (0.1 mM) + cpn10	108	98.1	1.9	—	—	—	755	72	935
4. ADP + KCl (100 mM) + cpn10	180	98.9	0.6	0.6	—	—	716	111	907
5. ATP + KCl (0.5 mM)	143	63.6	28.7	7.7	—	—	474	68	685
6. Reassociated fractions	256	92.2	0.8	3.9	2.7	0.4	366	60	682

Chaperonin particles were inspected and classified manually in a large sample of fields from Figure 12A (line 1); 12B (line 2); 12C (line 3); 12D (line 5); and 10g (line 6). Line 4; The KCl concentration of the solution used for Figure 12C was subsequently increased to 100 mM, the resultant solution was incubated for 10 min at 60°C, and an aliquot was observed by electron microscopy. Distributions of identified side views of particles are presented in percentage.

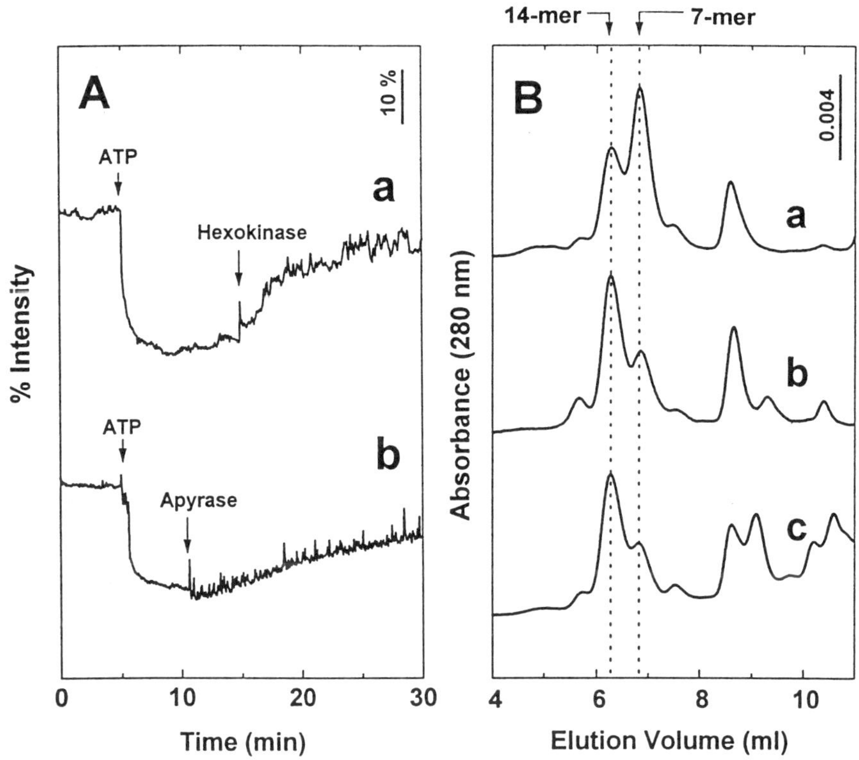

Figure 11 The effect of ATP deprivation on the reassociation of the split chaperonin monitored by (A) light scattering or (B) gel permeation HPLC: (A) The split reaction was initiated by addition of 200 μM ATP and 200 μM $MgCl_2$ (final concentrations) to 1 mL of the solution containing 100 μg mL^{-1} of intact *Thermus* chaperonin, 40 mM sodium phosphate (pH 7.5), 200 mM KCl and, for A, trace a, 500 μM glucose. After the scattering intensity fell down completely, (A, trace a) hexokinase (20 μg) or (A, trace b) apyrase (40 μg) was added at the time indicated by an arrow. (B) Intact *Thermus* chaperonin (680 μg mL^{-1}) and *Thermus* $cpn10_7$ (170 μg mL^{-1}) in 30 mM sodium phosphate buffer (pH 7.5) containing 2 mM ATP, 2 mM $MgCl_2$, and 200 mM KCl was incubated for 1 min at 37°C. Then, (B, trace a) nothing; (B, trace b) hexokinase and glucose (final concentrations, 80 μg mL^{-1} and 20 mM, respectively); or (B, trace c) apyrase (final concentration, 160 μg mL^{-1}) were added and further incubated for 5 min. The sample solutions (50 μL each) were applied to the gel permeation HPLC column as described in Figure 7.

chaperonin (the peak at 6.3 mL) was reciprocally increased, even in the presence of excess amount of the $cpn10_7$. These observations suggest that the deprivation of ATP drags the equilibrium between the split and the tetradecamer chaperonin in the direction of the latter.

7. Physiological Meaning of the Split Reaction

Conditions required for the split reaction are nearly physiological. It occurred at 20 μM MgATP, 150 mM KCl, neutral pH, and at a wide range of temperatures, including optimum growth temperature of *T. thermophilus* (60°–75°C) [41]. It is tempting to speculate that tetradecamer–heptamer transition of *Th*-cpn60 constitutes a part of chaperonins functional cycle. If this is really true, the presence of the single-ring form of cpn60 ($cpn60_7$) in bovine mitochondrial cpn60 [6] and in *Thermoacetobacter brockii* cpn60 [44,45] is well explained; they are an intermediate of the functional cycle of active chaperonin. That they can facilitate the protein folding in an ATP-, cpn10-dependent manner [6,44,45] seems to be consistent with this speculation. However, the kinetics of the tetradecamer–heptamer transition does not support it at present. Although the rate of the split reaction is fairly fast (< 5 s) (see Fig. 9), the reassociation proceeds only slowly (see Figs. 10 and 11). Actually, the slow rate of reassociation allowed us to isolate the split chaperonin as a stable molecular species by gel permeation HPLC. We have not been able to observe the similar quaternary structure transition for the *E. coli* GroEL system. Altogether, more studies are needed to draw a definite conclusion on the functional meaning of the tetramer–heptamer transition of *Th*-cpn60.

8. Formation of the Football-Shaped Chaperonin Under Physiological Conditions

It has been reported that *E. coli* chaperonin forms a football-shaped symmetrical complex ($GroES_7$–$GroEL_{14}$–$GroES_7$) when ATP occupies the nucleotide-binding sites of GroEL [46–52]. However, the functional significance of the football-shaped complex is in controversy [28,43,51–56]. In *Thermus* chaperonin, the football-shaped chaperonin ($cpn10_7$–$cpn60_{14}$–$cpn10_7$) was formed at 1.7 mM $MgCl_2$ and pH 7.25. Because the concentration of ATP was also 1.7 mM in that experiment and almost all Mg^{2+} existed as MgATP, the concentration of free Mg^{2+} in the solution should be very low. Even though high concentrations of free Mg^{2+} could have a stabilizing effect on the football-shaped chaperonin, it is not obligatory for *Thermus* chaperonin to form this shape.

As seen in Figure 12A, at a low concentration of KCl (0.1 mM), at which the split reaction was suppressed, the football-shaped chaperonins ($cpn10_7$–$cpn60_{14}$–$cpn10_7$), which have a shape very similar to those reported for the *E.*

coli equivalent complex [46–50], appeared, and it became a dominant species when excess cpn10_7 was included in the solution. The concentration of KCl of the solution was subsequently increased from 0.1 to 100 mM by the addition of KCl, at which the split reaction could occur, and the chaperonins in the solution were observed by electron microscopy (see Fig. 12B). The population of the cone-shaped chaperonins increased from 17 to 35% of the total side-view particles along with the decrease from 68 to 44% of the football-shaped chaperonins (see Table 3). Populations of top-view particles remained almost unchanged before and after the addition of KCl. ADP was unable to substitute for ATP to produce the football-shaped chaperonin at 0.1 mM KCl (see Fig. 12C) and at 100 mM KCl (see Table 3, line 4). Similar to the untreated *Thermus* chaperonin (see Fig. 1A), only the bullet-shaped chaperonins were observed under both conditions in side-view particles. The population of top-view particles was high. We could not distinguish the football-shaped chaperonin from the bullet-shaped one in top-view particles. However, according to Engel et al. [56], top-view particles are mostly bullet-shaped chaperonins (or cylinder-shaped chaperonin), rather than the football-shaped ones. It appears that binding of ATP (but not ADP) triggers the conformational change of the "bottom" end of the bullet-shaped chaperonin so that the second cpn10_7 can bind to form the football-shape.

The football-shaped chaperonins were generated from the bullet-shaped chaperonins even in the absence of added cpn10_7 (see Fig. 12D; Table 3, line 5). Probably, after binding of ATP, the cpn10_7 bound to cpn60_{14} in the bullet-shaped chaperonin becomes more mobile, and a small pool of the free cpn10_7 in equilibrium with the bullet-shaped one is generated.

9. *Transition of Quaternary Structure of* Thermus *Chaperonin*

On the basis of these results, it was established that there is a change of chaperonin quaternary structure. *Thermus* chaperonin is purified in the bullet-shaped form [20]. If the empty nucleotide-binding sites are occupied by ATP, the lower cpn60_7 ring of the bullet-shaped chaperonin becomes able to bind the free cpn10_7, and the football-shaped chaperonin is formed (see Fig. 12A). Even when excess free cpn10_7 is unavailable, a small population of the free cpn10_7 may be generated by the mobilization of the cpn10_7 of the bullet-shaped chaperonin (see Fig. 12D). If the empty sites are occupied by ADP instead of ATP, the bullet-shaped chaperonin cannot bind the cpn10_7, and the football-shape is not formed (see Fig. 12C).

The cone-shaped chaperonins appears when ATP and K^+ are present in the solution. However, it is still unclear which structure of the chaperonin is the

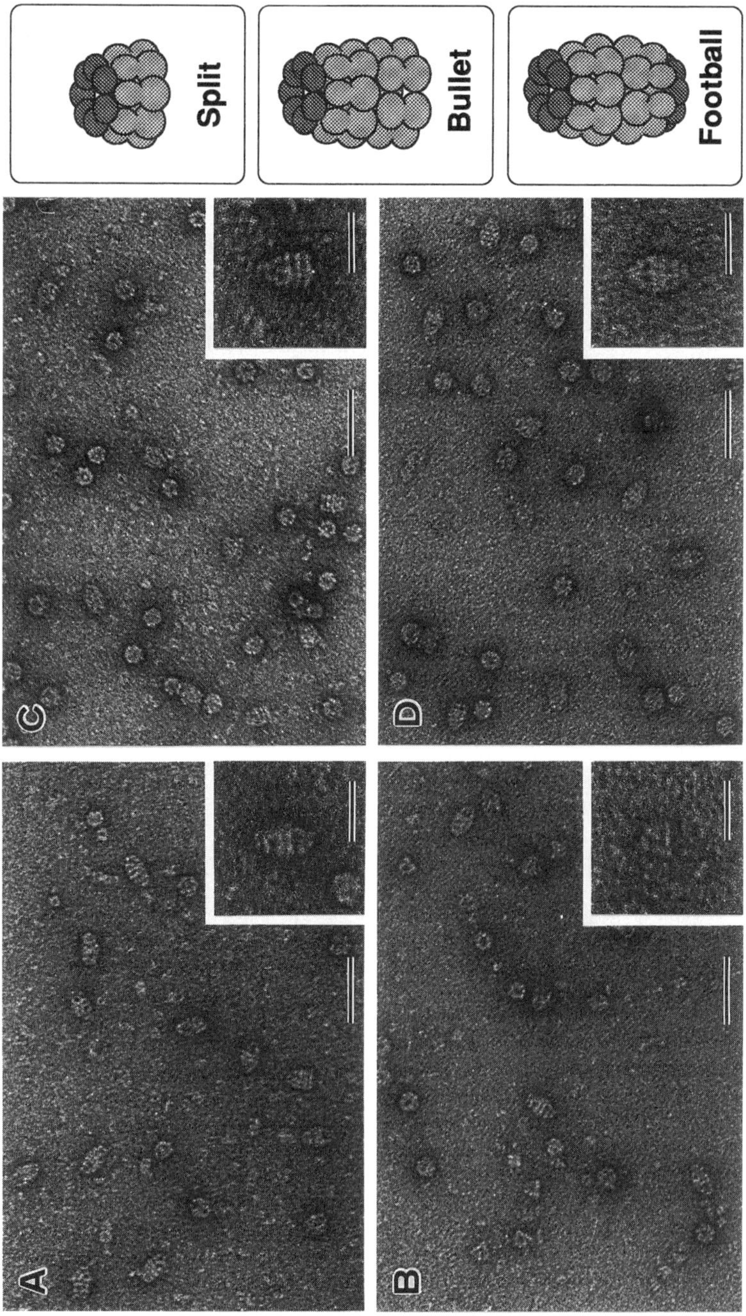
Split
Bullet
Football
A
B
C
D

precursor of the split chaperonin. We can just mention that the cone-shaped chaperonin splits from tetradecameric chaperonin after hydrolysis of the bound ATP. Then the cone-shaped chaperonins reassociate with each other to form the bullet-shaped one. There are several possible orders of release and binding of $cpn10_7$ between the cone- and the bullet-shaped chaperonins.

V. CONCLUSIONS

Compared with *E. coli* chaperonin (GroEL and GroES), the *T. thermophilus* chaperonin has several unique features that sometimes, from a different aspect, shed light on the common mechanism of chaperonin in protein folding. The heat stability of *T. thermophilus* chaperonin enabled us to examine the role of chaperonin as a shelter for the heat-labile proteins in a hot environment. The stability of the monomer cpn60 of *T. thermophilus* chaperonin provides us an opportunity to know the primitive ability of isolated monomer cpn60 to help the folding of other proteins. The split of chaperonin at the equatorial plane into heptamers has as yet been observed only for *T. thermophilus* chaperonin. It occurs under physiological conditions, suggesting functional significance, but further evidence will be required to conclude so.

Figure 12 Electron micrographs of *Thermus* chaperonin under various conditions: Electron micrographs of *Thermus* chaperonin (bullet-shaped) incubated with (A) *Th*-$cpn10_7$, MgATP, and 0.1 mM KCl; (B) *Th*-$cpn10_7$, MgATP, and 0.1 mM KCl and then 100 mM KCl; (C) *Th*-$cpn10_7$, MgADP, and 0.1 mM KCl; (D) MgATP and 0.5 mM KCl. *Thermus* chaperonin (170 μg mL^{-1}) and *Th*-$cpn10_7$ (65 μg mL^{-1}) in 40 mM Tris-Cl (pH 7.25 at 60°C) were mixed and incubated in 0.1 mM KCl, 1.7 mM $MgCl_2$, and 1.7 mM ATP for 10 min at 60°C. After the incubation, half of the solution was taken out for electron microscopic observation (see panel A). KCl was added to the rest of the solution to a final concentration of 100 mM. After a 10-min incubation at 60°C, the solution was subjected to electron microscopic observation (see panel B). The experimental procedures for panel C were the same as those for A except that ADP was used instead of ATP. (D) *Thermus* chaperonin (170 μg mL^{-1}) was incubated in 30 mM Tris-Cl (pH 7.25 at 60°C) containing 0.5 mM KCl, 1.7 mM $MgCl_2$, and 1.7 mM ATP for 10 min at 60°C and an aliquot was taken out for investigation by electron microscopy. Enlarged images of typical (A) football-, (B) cone-, (C) bullet-, and (D) football-shaped chaperonins found in each of the specimens are attached. Scale bars in each of the Figures represent 50 nm and 25 nm (enlarged images). Models of three oligomeric states of *Thermus* chaperonin are also illustrated on the right.

REFERENCES

1. Ellis, R. J. and van der Vies, S. M. (1991), *Annu. Rev. Biochem. 60*:321–347.
2. Gething, M. -J. and Sambrook, J. (1992), *Nature 355*:33–45.
3. Horwich, A. L. and Willison, K. R. (1993), *Philos. Trans. R. Soc. Lond. B 339*:313–326.
4. Gatenby, A. A. and Viitanen, P. V. (1994), *Annu. Rev. Plant Physiol. Plant Mol. Biol. 45*:469–491.
5. Horwich, A. L., Low, K. B., Fenton, W. A., Hirshfield, I. N., and Furtak, K. (1993), *Cell 74*:909–917.
6. Viitanen, P. V., Lorimer, G. H., Seetharam, R., Gupta, R. S., Oppenheim, J., Thomas, J. O., and Cowan, N. J. (1992), *J. Biol. Chem. 267*:695–698.
7. Pushkin, A. V., Tsuprun, V. L., Solovjeva, N. A., Shubin, V. V., Evstigneeva, Z. G., and Kretovich, W. L. (1982), *Biochim. Biophys. Acta 704*:379–384.
8. Hemmingsen, S. M. and Ellis, R. J. (1986) *Plant Physiol. 80*:269–276.
9. Hemmingsen, S. M., Woolford, C., van der Vies, S. M., Tilly, K., Dennis, D. T., Georgopoulos, C. P., Hendrix, R. W., and Ellis, R. J. (1988), *Nature 333*:330–334.
10. Goloubinoff, P., Christeller, J. T., Gatenby, A. A., and Lorimer, G. H. (1989), *Nature 342*:884–889.
11. Chandrasekhar, G. N., Tilly, K., Woolford, C., Hendrix, R., and Georgopoulos, C. (1986), *J. Biol. Chem. 261:12414–12419.*
12. Hohn, T., Hohn, B., Engel, A., Wurtz, M., and Smith, P. R. (1979), *J. Mol. Biol. 129*:359–373.
13. Hendrix, R. W. (1979), *J. Mol. Biol. 129*:375–392.
14. Braig, K., Otwinowski, Z., Hegde, R., Boisvert, D. C., Joachimiak, A., Horwich, A. L., and Sigler, P. B. (1994), *Nature 371*:578–586.
15. Hunt, J. F., Weaver, A. J., Landry, S. J., Gierasch, L., and Deisenhofer, J. (1996), *Nature 379*:37–45.
16. Langer, T., Pfeifer, G., Martin, J., Baumeister, W., and Hartl, F. -U. (1992), *EMBO J. 11*:4757–4765.
17. Taguchi, H., Konishi, J., Ishii, N., and Yoshida, M. (1991), *J. Biol. Chem. 266*:22411–22418.
18. Ishii, N., Taguchi, H., Yoshida, M., Yoshimura, H., and Nagayama, K. (1991), *J. Biochem. 110*:905–908.
19. Ishii, N., Taguchi, H., Sumi, M., and Yoshida, M. (1992), *FEBS Lett. 299*:169–174.
20. Yoshida, M., Ishii, N., Muneyuki, E., and Taguchi, H. (1993), *Philos. Trans. R. Soc. Lond. B 339*:305–312.
21. Fenton, W. A., Kashi, Y., Furtak, K., and Horwich, A. L. (1994), *Nature 371*:614–619.
22. Ishii, N., Taguchi, H., Sasabe, H., and Yoshida, M. (1994), *J. Mol. Biol. 236*;691–696.
23. Weissman, J. S., Hohl, C. M., Kovalenko, O., Kashi, Y., Chen, S., Braig, K., Saibil, H. R., Fenton, W. A., and Horwich, A. L. (1995), *Cell 83*:577–587.
24. Weissman, J. S., Rye, H. S., Fenton, W. A., Beechem, J. M., and Horwich, A. L. (1996), *Cell 84*:481–490.

25. Mayhew, M., da Silva, A. C. R., Martin, J., Erdjument-Bromage, H., Tempst, P., and Hartl, F. U. (1996), *Nature 379*:420–426.
26. Amada, K., Yohda, M., Odaka, M., Endo, I., Ishii, N., Taguchi, H., and Yoshida, M. (1995), *J. Biochem. 118*:347–354.
27. Viitanen, P. V., Lubben, T. H., Reed, J., Goloubinoff, P., O'Keefe, D. P., and Lorimer, G. H. (1990), *Biochemistry 29*:5665–5671.
28. Bochkareva, E. S., Lissin, N. M., Flynn, G. C., Rothman, J. E., and Girshovich, A. S. (1992), *J. Biol. Chem. 267*:6796–6800.
29. Taguchi, H. and Yoshida, M. (1993), *J. Biol. Chem. 268*:5371–5375.
30. Höll-Neugebauer, B., Rudolph, R., Schmidt, M., and Buchner, J. (1991), *Biochemistry 30*:11609–11614.
31. Braig, K., Simon, M., Furuya, F., Hainfeld, J. F., an Horwich, A. L. (1993), *Proc. Natl. Acad. Sci. USA 90*:3978–3982.
32. Makino, Y., Taguchi, H., and Yoshida, M. (1993), *FEBS Lett. 336*:363–367.
33. Taguchi, H., Makino, Y., and Yoshida, M. (1994), *J. Biol. Chem. 269*:8529–8534.
34. Martin, J., Langer, T., Boteva, R., Schramel, A., Horwich, A. L., and Hartl, F. -U. (1991), *Nature 352*:36–42.
35. Mendoza, J. A., Rogers, E., Lorimer, G. H., and Horowitz, P. M. (1991), *J. Biol. Chem. 266*:13044–13049.
36. Lorimer, G. H. (1992), *Curr. Opin. Struct. Biol. 2*:26–34.
37. Hartl, F. U., Martin, J., and Neupert, W. (1992), *Annu. Rev. Biophys. Biomol. Struct. 21*:293–322.
38. Wiech, H., Buchner, J., Zimmermann, R., and Jakob, U. (1992), *Nature 358*:169–170.
39. Jakob, U., Gaestel, M., Engel, K., and Buchner, J. (1993), *J. Biol. Chem. 268*:1517–1520.
40. Kiefhaber, T., Rudolph, R., Kohler, H. -H., and Buchner, J. (1991), *Biotechnology 9*:825 829.
41. Ishii, N., Taguchi, H., Sasabe, H., and Yoshida, M. (1995), *FEBS Lett. 362*:121–125.
42. Todd, M. J., Viitanen, P. V., and Lorimer, G. H. (1993), *Biochemistry 32*:8560–8567.
43. Todd, M. J., Viitanene, P. V., and Lorimer, G. H. (1994), *Science 265*:659–666.
44. Truscott, K. N., Hoj, P. B., and Scopes, R. K. (1994), *Eur J. Biochem. 222*:277–284.
45. Todd, M. J., Walke, S., Lorimer, G., Truscott, K., and Scopes, R. K. (1995), *Biochemistry 34*:14932–14941.
46. Llorca, O., Marco, S., Carrascosa, J. L., and Valpuesta, J. M. (1994), *FEBS Lett. 345*:181–186.
47. Azem, A., Kessel, M., and Goloubinoff, P. (1994), *Science 265*:653–656.
48. Schmidt, M., Rutkat, K., Rachel, R., Pfeifer, G., Jaenicke, R., Viitanen, P., Lorimer, G., and Buchner, J. (1994), *Science 265*:656–659.
49. Harris, J. R., Plückthun, A., and Zahn, R. (1994), *J. Struct. Biol. 112*:216–230.
50. Hayer-Hartl, M. K., Martin, J., and Hartl, F. U. (1995), *Science 269*:836–841.
51. Azem, A., Diamant, S., Kessel, M., Weiss, C., and Goloubinoff, P. (1995), *Proc. Natl. Acad. Sci. USA 92*:12021–12025.
52. Llorca, O., Carrascosa, J. L., and Valpuesta, J. M. (1996), *J. Biol. Chem. 271*:68–76.

53. Jackson, G. S., Staniforth, R. A., Halsall, D. J., Atkinson, T., Holbrook, J. J., Clarke, A. R., and Burston, S. G. (1993), *Biochemistry 32*:2554–2563.
54. Martin, J., Mayhew, M., Langer, T., and Hartl, F. U. (1993), *Nature 366*:228–233.
55. Weissmann, J. S., Kashi, Y., Fenton, W. A., and Horwich, A. L. (1994), *Cell 78*:693–702.
56. Engel, A., Hayer-Hartl, M. K., Goldie, K. N., Günter Pfeifer, Hegerl, R., Müller, S., Silva, A. C. R. d., Baumeister, W., and Hartl, F. U. (1995), *Science 269*:832–836.

14

Protein Disulfide Isomerase

Kenneth W. Walker and Hiram F. Gilbert
Baylor College of Medicine, Houston, Texas

I. INTRODUCTION

The realization that spontaneous disulfide formation was too slow to be kinetically competent in the cell stimulated the search for a catalyst of oxidative folding. This search led to the initial discovery of protein disulfide isomerase (PDI) [1], a component of the folding machinery of the eukaryotic endoplasmic reticulum (ER). This abundant, 55-kDa protein is a catalyst of disulfide formation and rearrangements [2–4]; a chaperone that inhibits aggregation [5–7]; a component of the ER quality control system [5]; and a binding protein that interacts with peptides [8,9], estrogen [10], thyroid hormone [11], and calcium [12]. The disulfide isomerase activity of PDI is essential for yeast viability [13,14]. In this chapter, we will discuss and review what is known about the diverse functions of this remarkable protein.

This chapter is organized by PDI function. After a brief disulfide primer, we will begin by discussing the general aspects of PDI catalysis of disulfide formation and rearrangement. The focus will then shift to the features of PDI's active sites—interaction with substrates, cooperation between active sites, and the mechanism of PDI-catalyzed disulfide formation and rearrangement. We will then turn our attention to the recently discovered chaperone activity, followed by a description of the unusual ability of PDI to facilitate protein aggregation, termed antichaperone activity. Finally, a summary of experiments that address the functions of PDI in the cell will be presented.

II. A DISULFIDE PRIMER

Disulfide formation between cysteine residues is a posttranslational modification that can affect the stability, structure, and activity of proteins [15]. The covalent cross-linking of residues that are far apart in primary sequence decreases the number of accessible conformations available in the unfolded polypeptide. This reduces the entropy of the unfolded state, thereby reducing the entropic penalty for folding [16]. Intermolecular disulfide bonds can also stabilize multimeric structures by tethering the subunits together. Finally, it is possible to control the biological activity of a protein by changing the enzyme's structure, depending on whether an essential cysteine residue is in the reduced or oxidized state. This allows the cellular thiol–disulfide redox environment to regulate biological activity [17].

In vivo, the primary mechanism of disulfide formation in proteins is by thiol–disulfide exchange, rather than by direct oxidation using nondisulfide oxidants such as oxygen [18]. Thiol–disulfide exchange occurs by nucleophilic attack of a thiolate anion on an existing disulfide bond [19].

$$R^*S^- + RS\text{–}SR \rightleftarrows R^*S\text{–}SR + RS^- \tag{1}$$

$$R^*S^- + R^*S\text{–}SR \rightleftarrows R^*S\text{–}S^*R + RS^- \tag{2}$$

Because of the requirement of a nucleophilic thiolate anion, disulfide exchange rates are sensitive to the pH and the pK_a of the involved cysteine residues. The pK_a of a typical cysteine is about 8.6; however, this value can vary widely, depending on the local environment within the protein [15]. For a nucleophilic cysteine [R*; see Eq. (1)], the optimum pK_a for maximum reaction rate is observed when the pK_a is equal to the pH of the medium. Lowering the pK_a farther will actually decrease the observed rate because the nucleophilicity of the attacking thiolate decreases with the pK_a [15,20]. For sterically accessible disulfides, nucleophilic attack can occur on either sulfur. In an asymmetrical disulfide (R*S–SR), the cysteine with the lower pK_a will be the favored leaving group because its thiolate anion is the most stable [19]. In contrast with the nucleophilic cysteine, as the pK_a of the leaving group decreases, the thiol–disulfide-exchange rate continues to increase, regardless of pH.

The formation of an intramolecular protein disulfide is a two-step reaction. The first step is a bimolecular reaction in which a protein thiolate reacts with a disulfide donor, as in Eq. (3).

$$RSSR + \text{protein}\begin{matrix}/\text{SH}\\ \backslash\text{SH}\end{matrix} \rightleftarrows RSH + \text{protein}\begin{matrix}/\text{SSR}\\ \backslash\text{SH}\end{matrix} \tag{3}$$

$$\text{Protein}\begin{matrix}/\text{SSR}\\ \backslash\text{SH}\end{matrix} \rightleftarrows RSH + \text{protein}\begin{matrix}/\text{S}\\ |\\ \backslash\text{S}\end{matrix} \tag{4}$$

The second step [see Eq. (4)] is an intramolecular attack of the remaining free cysteine on the mixed disulfide formed in the first step. Because the first step is bimolecular, the rate is dependent on the concentration of both the protein and the donor disulfide. The second step is unimolecular and, at higher concentrations of the donor disulfide, the overall rate may saturate as the intramolecular reaction becomes rate-limiting [21].

The proximity of the attacking cysteine to the target, mixed disulfide has a large effect on the reaction rate. The effective local concentration of the nucleophile can be as high as 10^8 M [22]. To achieve this high local concentration, the α-carbons of the two cysteine residues must be within 4–9 Å of each other in the folding intermediate before the reaction; therefore, intermediate structure can have a major effect on disulfide bond formation rates [23,24].

$$K_{OX} = \frac{[P_{SS}][RSH]^2}{[P_{(SH)_2}][RSSR]} \tag{5}$$

The stability of an intramolecular disulfide can be described by its K_{ox} [see Eq. (5)]; a high K_{ox} indicates a very stable disulfide, whereas a low K_{ox} indicates an unstable disulfide. For an intramolecular disulfide [see Eq. (3) and (4)], the K_{ox} can vary enormously. For PDI, K_{ox} is approximately 1 mM [25]. By contrast, thioredoxin has a K_{ox} of 2.0 M [15], whereas certain disulfides that are introduced to stabilize proteins can have K_{ox} values as high as 10^7 M [15]. Proteins, such as PDI, that are designed to serve as oxidants during the oxidative phase of protein folding have a relatively low K_{ox}.

III. CATALYTIC ACTIVITY

A. An Overview of Oxidative Folding

Observation of the rate of disulfide formation during the in vitro refolding of reduced, denatured RNase reveals that disulfide bonds are formed rapidly, followed by slow thiol–disulfide rearrangements [26–29]. This two-step process is apparent in the kinetics of the formation of native RNase from the reduced, denatured protein (Fig. 1), for which a significant lag is observed before the formation of native, active RNase (four disulfides). The lag represents the accumulation of numerous redox isomers of RNase that have incorrect disulfide bonds, consistent with the observation that oxidation is largely complete (more than 3.5 disulfides per RNase) before the detection of significant RNase activity [26–28]. This accumulation of oxidized folding intermediates is followed by disulfide rearrangements that lead to the more stable native RNase.

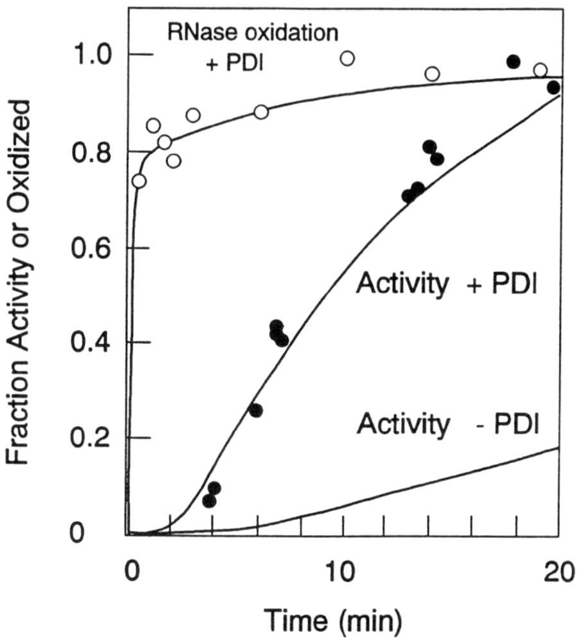

Figure 1 Gain of activity and disulfide formation during the oxidative folding of reduced RNase: Disulfide formation (○) and the regain of RNase activity (●) during the oxidative refolding of 8 μM reduced RNase in the presence of 1.4 μM PDI and a glutathione redox buffer. The fraction of oxidation was measured by observing the formation of GSH from the GSSG-dependent oxidation of reduced RNase; 8 mol GSH per mole of RNase is taken as 100% oxidation. The fraction of activity recovered is calculated relative to that expected from complete refolding of the RNase present. The curve through the solid points is the activity of RNase observed in a continuous assay. The solid line at the bottom of the figure is the amount of activity recovered in the absence of PDI. (From Ref. 36.)

B. Nonnative Disulfides

There has been some controversy over the role of nonnative disulfides during oxidative folding. The early work done by Creighton and co-workers [26,30] suggested that many quasistable intermediates with nonnative disulfide bonds form during the in vitro folding of bovine pancreatic trypsin inhibitor (BPTI). More recently, Weissman and Kim [31] repeated the BPTI folding experiments using a more effective acid quench to trap the intermediates. When using this system, most of the populated species of BPTI-folding intermediates contained native disulfides; however, owing to the mode of disulfide bond formation in BPTI, some of these intermediates must rearrange through the transient formation of nonnative disulfides to reach the native structure. Experiments with other proteins suggest that the dominance of nonnative intermediates depends on the protein [4]. For example, hirudin [32] initially forms random disulfides that rearrange to the native state; however, human chorionic gonadtropin proceeds to the native state by sequential formation of native disulfides [33].

C. Redox Optimum

Because of the requirement of a multistep reaction for the formation and isomerization of protein disulfide bonds, oxidative folding requires both an oxidant to support disulfide formation and a reductant to allow thiol-dependent rearrangements. Consequently, oxidative folding is usually more rapid and efficient when performed in the presence of a low molecular weight disulfide and its corresponding thiol (a redox buffer) [28,34,35]. At low concentrations of a disulfide oxidant, such as glutathione disulfide (GSSG), protein disulfide formation itself will be slow; however, if the GSSG concentration is too high, too many glutathione-mixed disulfides will result, preventing the intramolecular step in the oxidation from occurring. Suboptimal reductant (GSH) concentrations will slow disulfide rearrangements, which require a free thiol to initiate breaking the disulfides of nonproductive intermediates; however, excess GSH will inhibit the formation of productive disulfides. For RNase oxidative folding, both in the presence or absence of catalysis by PDI, the redox optimum occurs at a GSSG concentration of 0.2 mM and a GSH concentration of 1 mM [28] (Fig. 2).

Consequently, protein-folding compartments must carefully regulate the concentrations of their redox-active pair to maintain optimal oxidative-folding rates. Surprisingly, the optimal redox buffer for correct disulfide bond formation in eukaryotic proteins usually requires more reductant than oxidant [28,34,35]. Because GSH can break only disulfide bonds, this may indicate that rearrangement or reduction of incorrect disulfide bonds is a critical, rate-limiting step in oxidative protein folding.

D. Catalysis of Oxidative Folding

With reduced, denatured substrates, PDI catalyzes the initial formation of disulfides and their subsequent rearrangement into the native configuration [28,36]. With scrambled RNase, catalysis of disulfide isomerization is observed in the presence of a trace amount of thiol, presumably to keep PDI in the reduced state [37] that is required for catalysis of disulfide isomerization. PDI also catalyzes the formation of native protein from denatured protein in which the cysteines are modified to glutathione-mixed disulfides [38]. However, glutathione-mixed disulfides are not essential intermediates in the folding reaction, because stoichiometric amounts of PDI, in its disulfide form, can provide all of the disulfide equivalents required for substrate oxidation [36]. There is no

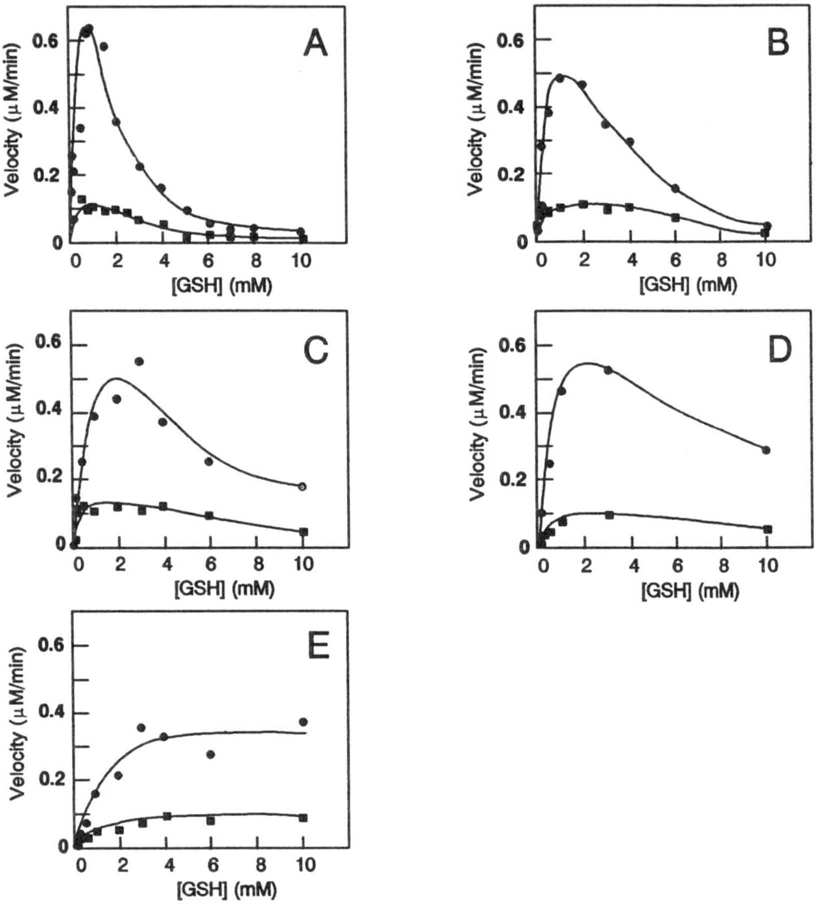

Figure 2 Redox optimum for catalysis of the oxidative-refolding of reduced RNase: The steady-state rates of PDI-dependent oxidative folding (●) and the uncatalyzed reaction (■) are shown as a function of the GSH concentrations at several, fixed levels of GSSG: (A) 0.2 mM; (B) 0.5 mM; (C) 1 mM; (D) 2 mM; (E) 4 mM. (From Ref. 28.)

evidence that PDI significantly alters the refolding mechanism from that observed in the absence of catalysis [28,33,39].

PDI is not a particularly good catalyst compared with most enzymes. With either RNase [37,40] or BPTI [41] as substrate, the turnover number is an unimpressive 0.5–1 min^{-1}. However, this low activity is all that is necessary, given the high local concentration of PDI found in cells (see later discussion), but it often makes it difficult or impossible to observe PDI activity at low concentrations of enzyme (less than 1 μM). Oxidation and rearrangement do not appear to be processive—there is no indication that multiple oxidations or rearrangements occur within the enzyme–substrate complex—but this question has not been carefully examined. The substrate K_m is available only for scrambled RNase [37], reduced RNase [28,40], and BPTI [41]. For all these protein substrates, the K_m is near 5–10 μM.

IV. ISOLATION, CELLULAR LOCATION, AND SEQUENCE

Because the spontaneous formation and rearrangement of disulfide bonds involves relatively slow covalent changes, catalysis of oxidative folding is essential for kinetically competent folding in the cell. PDI activity was first identified by Anfinsen and colleagues over 30 years ago as a component of microsomal extracts that accelerates the oxidative folding of reduced RNase [1,42]. More than 20 years later, the gene for PDI was cloned and sequenced; PDI has significant homology to the ubiquitous, redox-active protein, thioredoxin [43]. Since then, many more genes belonging to the PDI superfamily have been identified. This superfamily includes four subfamilies: PDI, Erp60, Erp72, and P5 [for review see Ref. 2]. However, of all the oxidoreductases now identified, PDI is the most effective catalyst of oxidative folding, accelerating reactions 100- to 6000-fold [28,41].

A. Cellular Location

The PDI gene is constitutively expressed but its concentration can be up-regulated 10- to 100-fold in cells, such as B cells, that are especially active in protein secretion [44]; however, the mechanism of tissue-specific regulation is still not understood [45,46]. In actively secreting tissues, the ER concentration of PDI can approach the millimolar range [44]. Almost all PDI is located in the lumen of the ER [47]. Its COOH-terminus is KDEL, a signal that confers ER localization on PDI and other ER proteins [48,49]. In the ER, PDI is an exceptionally stable protein, with a half-life of approximately 7 days. The high ER concentrations of PDI are due to both elevated expression levels and the unusual stability of the enzyme [50]. However, overproduction of PDI results in saturation of the ER retention system, leading to PDI secretion. This sets an upper limit on lumenal PDI concentration [51].

B. The Thioredoxin Superfamily

Sequence analysis of the PDI intron–exon boundaries and protease-cleavage patterns suggest that it is a multidomain protein [2]. The catalytic activity resides in two domains that are homologous to thioredoxin [52]. One is located near the NH_2-terminus and the other near the COOH-terminus (Fig. 3).

PDI possesses a conventional NH_2-terminal leader sequence that designates

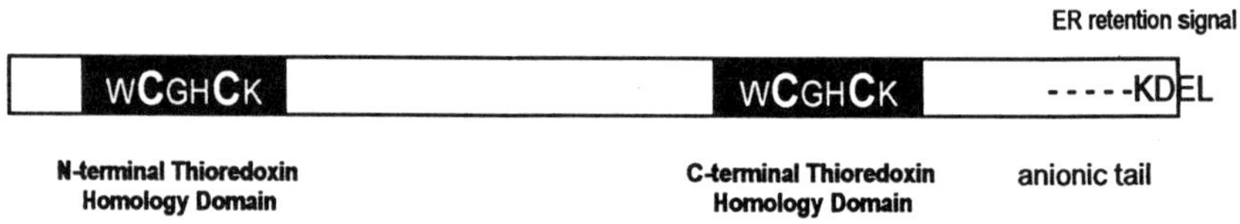

Figure 3 Domain structure of PDI: The two active sites are located in two regions of the sequence that are homologous to thioredoxin [2,43]. The COOH-terminal of the molecule is very acidic and terminates in KDEL, the ER-retention signal.

it for secretion into the ER, and it has a COOH-terminal KDEL sequence (HDEL in yeast), resulting in ER retention of the enzyme [48]. Homology is also found to the estrogen receptor; however, this homology does not appear to constitute an independent structural domain [25]. Interestingly, PDI binds estrogen; however, no physiological role has been assigned to the PDI–estrogen interaction [10]. The b- and b′-domains are internal repeats of unknown function, whereas the c domain is a highly acidic sequence that may function in Ca^{2+} binding [12].

V. CHEMISTRY OF THE ACTIVE SITES

A. Active-Site Cysteines

The two thioredoxin-like active sites of PDI are distinguished by a WCGHCK motif that contains the essential thiols. In the reduced enzyme, chemical modification of these thiols inactivates PDI's oxidoreductase activity [53]. It has been suggested that the structure of the PDI active sites closely resembles that of thioredoxin and the prokaryotic homologue of PDI, DsbA. In thioredoxin [54] and DsbA [55], high-resolution X-ray crystal structures indicate that the more NH_2-terminal cysteine in the active site CXXC motif is exposed on the surface, whereas the more COOH-terminal cysteine is buried in a pocket of the protein; thereby, sterically hindering the reaction of this cysteine residue with any disulfide bonds other than those that involve the NH_2-terminal cysteine [55].

Consistent with this, only two highly reactive cysteines are observed in wild-type PDI, even though it has four active-site cysteines and two internal cysteines. Thus, each active site has one exposed, reactive cysteine and one buried, unreactive cysteine [53]. A nuclear magnetic resonance (NMR) structure of the NH_2-terminal thioredoxin domain of PDI has been recently reported [56], and the isolated domain is similar in structure to thioredoxin. This truncated form of PDI is active as an oxidase, but it has no detectable isomerase activity.

B. Redox Properties and Chemical Reactivity

The thiol–disulfide redox properties of the active sites of the thioredoxin family members are in line with their biological function. Thioredoxin, which functions in the highly reducing environment of the cytoplasm and is an excellent catalyst of disulfide reduction, has the lowest reduction potential, –260 mV [57]. DsbA, a catalyst of disulfide formation found in the highly oxidizing *Escherichia coli* periplasmic space, is the best-oxidizing agent, with a reduction potential of –89 mV [58,59]. PDI, a catalyst of both oxidation and rearrangement, has a reduction potential of approximately –100 mV that is much closer to DsbA than thioredoxin [25,60]. The redox potential disparity is partly due to the difference in the central residues of the active sites of thioredoxin (CGPC), compared with PDI (CGHC) and DsbA (CPHC). Conversion of the NH_2-terminal active site of PDI to CGPC results in a loss of almost 50% of PDI's activity in the oxidative folding of RNase, but it results in less than a 10% decrease in its insulin reduction ability [61]. Conversion of the CGPC sequence of thioredoxin to that of PDI (CGHC) increases the reduction potential by 30 mV (tenfold) and increases its disulfide isomerase activity fivefold [60]. The role of the two residues between the active site cysteines can be seen in a series of random mutations of these two residues in DsbA. Depending on the identity of the intervening amino acids, the reduction potentials vary by as much as 100 mV [62].

The structural basis for the high-reduction potential of PDI active-site disulfides is not completely clear. Electrostatic effects, including backbone interactions, the intervening histadine residue, and a positive helix dipole may stabilize the thiolate anion of the reduced active site [63,64]. Stabilization of the thiolate anion of the reduced active site also lowers the pK_a of the more NH_2-terminal active-site cysteine. Hawkins and Freedman [53] estimated a pK_a value of 6.7 for the active-site cysteine, using the pH-dependence of the rate of alkylation by iodoacetamide. However, Darby and Creighton [25] conclude that the rate constants for reaction of the dithiol forms of the thioredoxin domains are more consistent with a pK_a that is substantially lower than 5. For DsbA, the low active-site pK_a (3.5) provides much of the stabilization of the reduced active site [62]. The pK_a of the more COOH-terminal active-site cysteine is not known accurately, but titration of Cys/Ser mutants of the active site suggest that it may be greater than 9.5 [64].

C. Binding and Specificity

There are no recognizable sequence cues for which two cysteine residues in the substrate should be paired to form a disulfide bond during oxidative folding [24]. As a result, PDI must be able to recognize protein cysteine residues in a variety of sequence contexts. The second-order rate constant for reaction of the active site with peptide [21,65] and protein substrates [66] appears to be too high to be accounted for simply by the chemical reactivity of PDI's active site. The implication is that there are likely to be noncovalent interactions of peptides and proteins with PDI. Indeed, PDI activity can be inhibited by peptides of various sequences and lengths [8,67]. There is no obvious sequence preference, and longer peptides bind more tightly in proportion to their length (Fig. 4). This is expected if a longer peptide can bind to PDI anywhere along its length. Peptides shorter than about five residues are much poorer inhibitors than expected for their length, suggesting that they may not completely fill the peptide–protein-binding site. Cysteine-containing peptides are about eightfold better inhibitors than peptides of the same length that do not contain cysteine. Overall, peptide binding is relatively weak with K_d values in the range of 0.05–1 mM [8]. A tripeptide photoaffinity probe labels a region of PDI near the COOH-terminus [9,68] and inactivates the isomerase activity. If PDI has only one peptide–protein-binding site, the two active sites must share access to it or one must function without it.

In addition to peptides and proteins, PDI interacts with an amazing variety of other molecules. It binds estrogen quite tightly, with a K_d in the nanomolar range [10]. Thyroid hormone binds with similar affinity and, in fact, PDI was identified using a thyroid hormone-based photoaffinity label on cells [11]. PDI binds tightly to some other folded proteins, including prolyl hydroxylase and

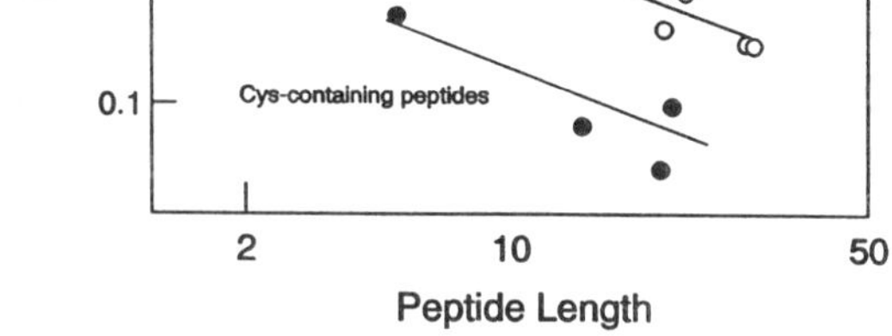

Figure 4 Inhibition of PDI by peptides: Peptides of different sequence were used to inhibit the reduction of insulin or the oxidative folding of reduced RNase. The inhibition constant (K_i) for competitive inhibition against substrate is plotted as a function of the length of the peptide. Note that cysteine-containing peptides (●) bind more tightly than peptides of the same length that do not contain cysteine (○). Short peptides (- - - -) or peptides with no side chains also bind less tightly than expected from their length (note the position of GGGGGG). (Data from Ref. 8.)

triglyceride transferase. Although its catalytic activity is not used, PDI is the β-subunit of prolyl hydroxylase and copurifies with the catalytic α-subunit [69,70]. It appears to serve a similar function in the microsomal triglyceride transferase complex [71]. Whether or not these sites overlap or represent exclusive binding sites is unknown. However, PDI appears to be capable of binding multiple molecules of the hydrophobic probe bis-anilinonaphthalene sulfonate [Primm and H. F. Gilbert, unpublished data], suggesting a rather large hydrophobic surface.

D. Cooperation Between Active Sites

Both of PDI's thioredoxin-like active sites are required for full disulfide bond formation and rearrangement activity; however, the two active sites are not equivalent. Vouri et al. [52] used site-directed mutagenesis of the active-site cysteines to demonstrate that the PDI active sites can act independently of each other. In more detailed study of the enzymatic properties of PDI active-site mutants [40], the active site that is closer to the NH_2-terminus ($\mathbf{N}_{SS}$:$\mathbf{C}_{OO}$)* has a higher k_{cat} than the active site that is closer to the COOH-terminus. A mutant in which both of the cysteines of the more COOH-terminal active site have been mutated to serine ($\mathbf{N}_{SS}$:$\mathbf{C}_{OO}$) exhibits a k_{cat} value similar to the wild-type enzyme; however, the K_m value is three-fold higher than the wild-type enzyme. Consequently, at high substrate concentration, the NH_2-terminal active site is capable of providing all of the catalytic activity that is normally observed from both active sites of the wild-type enzyme. By contrast, mutation of the more NH_2-terminal active site ($\mathbf{N}_{OO}$:$\mathbf{C}_{SS}$) has no effect on K_m but decreases k_{cat} threefold. Thus, at saturating substrate concentrations, the activity of the wild-type enzyme is approximately one-third lower than expected from the additive contribution of the two sites. This may reflect some internal competition between the two sites when acting on bound substrate [40].

E. Oxidation Mechanisms

Three mechanisms of PDI catalyzed oxidation have been proposed [see Ref. 2 for review]: One mechanism (Fig. 5A) involves the donation of an active-site disulfide directly from PDI to the substrate protein through a covalent intermedi-

*The nomenclature used to designate mutations of cysteines in the two thioredoxin-like active sites of PDI is as follows: The location of the domain is indicated as **N** or **C**, depending on whether it is the one closer to the NH_2-terminus (N) or the COOH-terminus (C) of PDI. The two cysteines at each of the two active sites are designated as S for sulfur or O for oxygen, depending on whether a cysteine or serine residue is present at a given position. The wild-type enzyme would be designated $\mathbf{N}_{SS}$:$\mathbf{C}_{SS}$, whereas a mutant with all active-site cysteines mutated to serine would be designated as $\mathbf{N}_{OO}$:$\mathbf{C}_{OO}$.

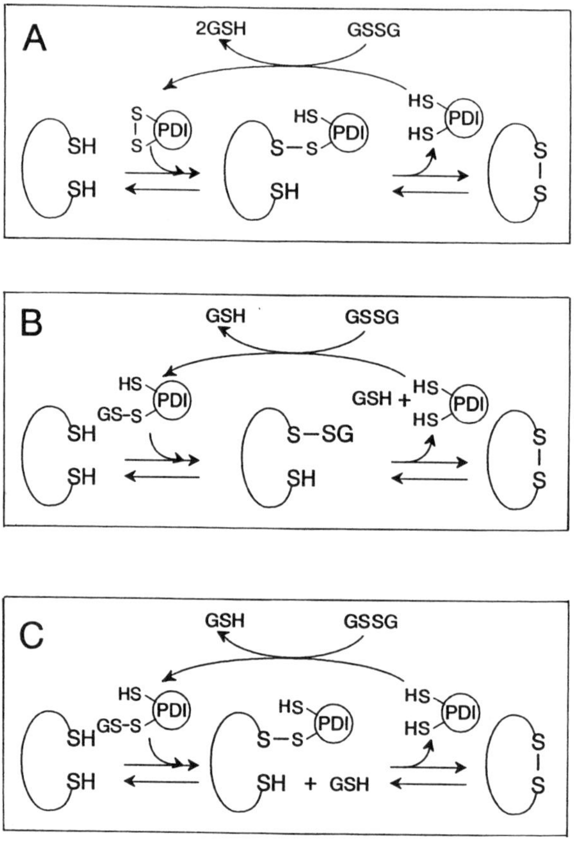

Figure 5 Possible mechanism for the PDI-catalyzed oxidation of substrate proteins to form protein disulfides: (A) PDI catalyzes direct transfer of its active site disulfide to the substrate and the reduced enzyme is recycled to the oxidized enzyme by reaction with GSSG; (B) involves the transfer of the disulfide from a donor disulfide, such as GSSG, to the protein without the involvement of a PDI–substrate mixed disulfide, (C) a mixed-disulfide between PDI and glutathione is displaced by a substrate thiol to yield a PDI-substrate covalent intermediate. The subsequent intramolecular displacement to form the substrate disulfide occurs with displacement of PDI.

ate, followed by re-formation of PDI's disulfide bond using a donor disulfide such as GSSG. Importantly, this mechanism requires PDI to possess a disulfide bond in its active site to maintain oxidase activity, and it requires PDI to form an intermolecular disulfide with its substrate. Because PDI can act as the sole oxidant in the folding of reduced RNase, it must be capable of direct oxidation of its substrates, consistent with mechanism A [28]. However, in vivo the oxidant

GSSG is present in relatively high concentrations in the ER, and PDI may catalyze substrate oxidation by using the donor disulfide GSSG, rather than a disulfide between the two active-site cysteines. PDI may catalyze the formation of a glutathione–substrate-mixed disulfide, followed by catalysis of an intramolecular displacement of GSH from the glutathione-mixed disulfide intermediate (see Fig. 5B), or a PDI–GSSG-mixed disulfide may function as the oxidant (see Fig. 5C). PDI catalysis, using the mechanisms of Fig. 5B and C, has been demonstrated in vitro using RNase T1 that has been preoxidized with GSSG to mixed disulfides [38]. In addition, the PDI-catalyzed oxidation of a model peptide by GSSG proceeds through a peptide–glutathione-mixed disulfide followed by catalysis of the intramolecular reaction [65]. Mechanisms B and C differ significantly from A in that they do not require PDI to form an intramolecular disulfide involving both of the active site disulfides. In support of mechanism A, mutants of PDI with a single active site cysteine at either active site ($\mathbf{N}_{SO}$:$\mathbf{C}_{OO}$, $\mathbf{N}_{OO}$:$\mathbf{C}_{SO}$, and $\mathbf{N}_{SO}$:$\mathbf{C}_{SO}$) are virtually inactive in catalyzing the oxidation of reduced RNase A in a glutathione redox buffer, although they exhibit significant isomerase activity [14,29]. At least for RNase oxidative-folding, direct oxidation by an intramolecular disulfide involving both cysteines of each thioredoxin domain is dominant (see Fig. 5A), even when alternative mechanisms involving GSSG are available.

F. Isomerization Mechanisms

The other major catalytic activity of PDI is disulfide isomerization, and two mechanisms have been proposed to explain how PDI catalyzes disulfide bond rearrangements (Fig. 6) [see Ref. 2 for review].

Disulfide isomerization converts one disulfide isomer to another with no net change in the redox state of the substrate and product. Disulfide isomerization requires a free thiol to initiate the rearrangement by thiol–disulfide exchange. This thiol could be provided by PDI, the substrate, or by the thiol component of the redox buffer. The initial attack of an active site thiol from PDI on one of the substrate disulfides could initiate such a rearrangement (see Fig. 6A). Rearrangement would be completed by attack of the resulting substrate thiol to form a different disulfide in the substrate and release PDI in the reduced state. An alternative mechanism for substrate rearrangement would involve reduction of substrate disulfides, followed by their rearrangement (see Fig. 6B). The mechanisms depicted in Figure 6 differ significantly in two respects. First, an internal rearrangement (see Fig. 6A) does not require the participation of both cysteines at a PDI active site; one is sufficient. Second, in the internal rearrangement (see Fig. 6A), PDI relies on a substrate thiol to release it from the intermediate. For mutants of PDI with only one active site cysteine ($\mathbf{N}_{SO}$:$\mathbf{C}_{OO}$; $\mathbf{N}_{OO}$:$\mathbf{C}_{SO}$; $\mathbf{N}_{SO}$:$\mathbf{C}_{SO}$), the enzyme can catalyze disulfide isomerization (during RNase oxidative folding) without being able to form a disulfide with its vicinal

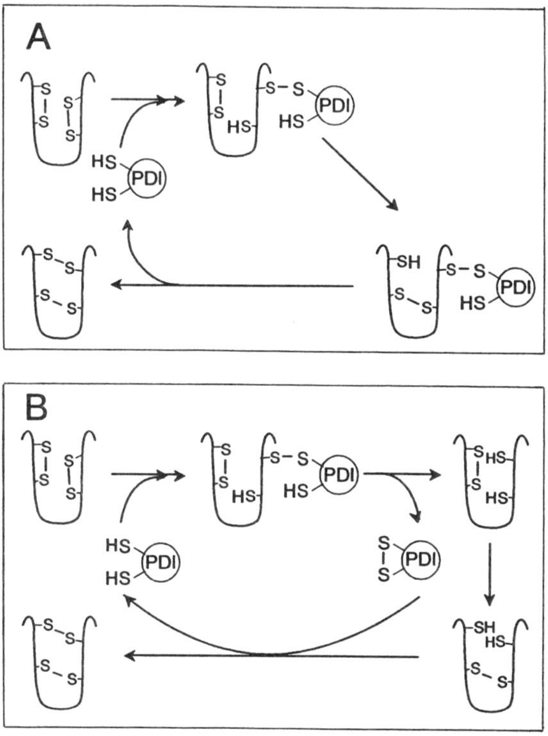

Figure 6 PDi catalysis of disulfide isomerization reactions: (A) a rearrangement of substrate that occurs with PDI covalently attached. Here a free thiol on the substrate is necessary to displace PDI from the complex. (B) The reduction of a substrate disulfide followed by a rearrangement and reoxidation to give the rearranged product. This mechanism also provides PDI with a way to escape from covalent complexes with substrates that rearrange slowly.

active-site cysteines; however, the rearrangement is slowed by the failure of the substrate to efficiently release from PDI. Although small amounts of PDI–RNase covalent complexes have been detected with wild-type PDI [72], the single-cysteine mutants accumulate much larger quantities of these intermediates [29], suggesting that both mechanisms contribute to PDI-catalyzed rearrangements.

In addition to catalyzing the formation of intramolecular disulfide bonds in its substrates, PDI also catalyzes the formation of intermolecular disulfides. For example, PDI catalyzes the formation of disulfides between monoclonal antibody chains in vitro [73] and monomers of the IgM pentamer [44]. PDI also catalyzes the formation of the procollagen trimer from its monomers; however, it was not able to catalyze the final step of triple-helix formation [74].

VI. UNFOLDING ACTIVITY

Weissman and Kim [41] reported that PDI was an extremely effective catalyst of the intramolecular rearrangements that are rate-limiting in the refolding of reduced BPTI to the native, three-disulfide protein. Compared with the uncatalyzed rearrangement, PDI increases the rate by up to 6000-fold. This is in contrast with the approximately 100-fold rate acceleration observed in the refolding of reduced RNase [28]. Because the BPTI rearrangement must involve buried cysteines and disulfides, one might suspect that the interaction of folding intermediates with PDI disrupts the structure of such intermediates. To investigate the mechanisms available to PDI for dealing with thiols or incorrect disulfides that become sterically trapped during folding, Walker and Gilbert [66] used reduced RTEM-β-lactamase as a model. Native β-lactamase has a single disulfide that is buried in the hydrophobic core of the molecule. However, the disulfide is not essential to stability, and the reduced protein folds into a relatively stable, catalytically active conformation [75]. The second-order rate constant for the oxidation of this folded, reduced β-lactamase by PDI is about 500-fold faster than the second-order rate constant for oxidation of β-lactamase by GSSG. However, this is very similar to the differences in chemical reactivity between the active-site disulfides of PDI and GSSG [21]. High concentrations of PDI do not disrupt the native structure of reduced β-lactamase, and saturation kinetics with a similar maximum rate are observed for uncatalyzed alkylation or oxidation of the buried cysteines. Thus, it appears that PDI must wait for the spontaneous opening of the reduced β-lactamase to permit access to the buried cysteines. However, the high chemical reactivity of PDI's active-site disulfides ensures that once the β-lactamase structure opens sufficiently to expose the buried cysteines, PDI can oxidize the cysteines faster than the protein refolds to bury them.

VII. CHAPERONE ACTIVITY

Protein disulfide isomerase has additional protein-folding activities other than disulfide bond formation and rearrangement. At high concentrations, PDI can act as a chaperone. For example, it increases the yield of properly folded lysozyme from denatured lysozyme by preventing aggregation [5]. For lysozyme oxidative folding, PDI's chaperone activity requires its ability to form disulfides, suggesting that rapid disulfide formation results in intermediate species that have a lower tendency to aggregate [76]. However, PDI's chaperone activity is not restricted to proteins that form disulfides. Stoichiometric concentrations of PDI inhibit the aggregation of glyceraldehyde-3-phosphate dehydrogenase and rhodanese, proteins that do not contain disulfides [6,7].

The interaction between PDI and the β-subunit of prolyl hydroxylase has also been suggested to involve a chaperone component. Because the α-subunit does not have an ER retention signal, it is believed that association with PDI prevents the α-subunit from being secreted out of the ER [77]. Furthermore, PDI may act as a chaperone to the α-subunit. When the α-subunit is expressed alone in a baculovirus system, it aggregates, but when coexpressed with PDI it forms the normal tetrameric $\alpha_2\beta_2$-complex [78,79]. PDI also has dehydroascorbate reductase activity that results in the production of ascorbic acid [80]. It has been proposed that not only does PDI act as a chaperone and ER retention anchor for the α-subunit, it also provides it with the necessary cofactor for proline hydroxylation, ascorbic acid [80].

PDI is also the β-subunit of the αβ-heterodimeric microsomal triglyceride transfer complex [81]. Again, PDI may serve as a chaperone in this system, because its removal from the αβ-complex results in α-subunit aggregation [82]. Finally, PDI has been implicated as a component of the system required for vitamin K reduction, which is required for γ-carboxylation of glutamate residues [83].

VIII. ANTICHAPERONE ACTIVITY

Molecular chaperones make folding more efficient by inhibiting aggregation, and folding catalysts, such as PDI, make folding more efficient by making it faster. Therefore, it is surprising to find that during the refolding of reduced, denatured lysozyme, low concentrations of PDI promote aggregation and decrease the yield of native lysozyme [5] (Fig. 7). The decrease in lysozyme yield that is observed at lower concentrations of PDI results from the formation of aggregates. At a ratio of only one PDI monomer per ten lysozyme monomers, all of the lysozyme and all of the PDI is found in large, insoluble aggregates. PDI incorporation into lysozyme aggregates is specific, for various other soluble proteins do not induce lysozyme aggregation and are not found in the insoluble aggregates induced in the presence of PDI [5]. Although the lysozyme–PDI aggregates are extensively disulfide cross-linked, disulfide formation is not required for aggregation. Low concentrations of PDI induce lysozyme aggregation in 10-mM DTT without forming disulfide cross-links, and an enzymatically inactive mutant of PDI still has antichaperone activity [5]. Antichaperone activity is not restricted to lysozyme aggregation. The aggregation of thermally or chemically denatured alcohol dehydrogenase is also stimulated by low concentrations of PDI, but is inhibited by higher, chaperone concentrations (Primm and H. F. Gilbert, submitted). However, antichaperone activity is not observed for all aggregating proteins. The aggregation of citrate synthesis is not enhanced by PDI; only chaperone activity is observed at higher PDI concentrations.

In many respects, the PDI-facilitated aggregation of lysozyme and alcohol dehydrogenase is analogous with classic immunoprecipitation by polyclonal antibodies. Bivalent (or multivalent) interactions between PDI and the aggregat-

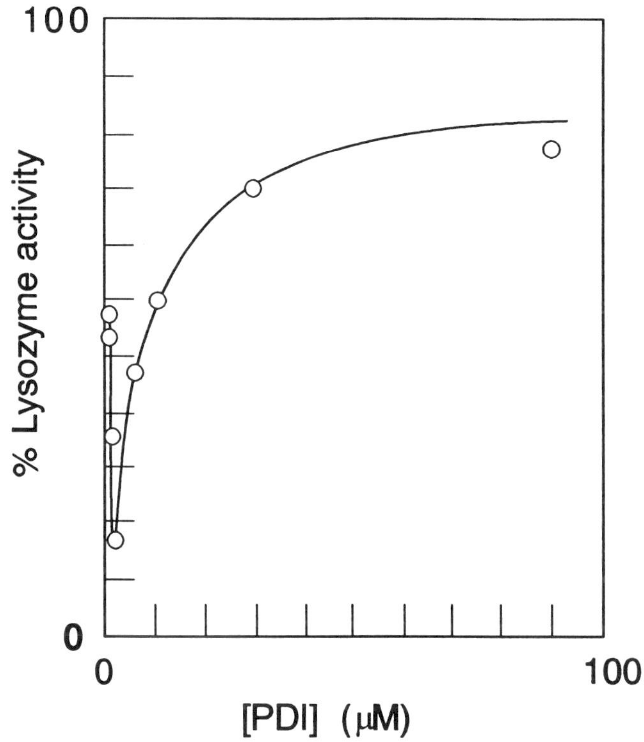

Figure 7 Effects of PDI concentration on the refolding of reduced hen egg white lysozyme: Various concentrations of PDI were present when reduced, denatured lysozyme (10 μM final concentration) was added to a redox buffer containing GSH and GSSG. The final yield of refolded lysozyme was measured after completion of the folding reaction. (From Ref. 76.)

ing protein may induce further growth of aggregates. However, at high concentrations of PDI (analogous with the antibody-excess region in an immunoprecipitation), aggregation is inhibited as PDI covers sites on the substrate protein that are necessary for aggregate growth [5,76].

It has been suggested that the transition between antichaperone–chaperone activity may serve to increase the capacity of the ER to retain malfolded proteins as chaperone-associated aggregates. Under conditions during which the concentration of unfolded protein begins to exceed the chaperone capacity of the ER, PDI and other ER antichaperones [5] could hold the aggregates in the ER until they are properly folded or degraded. The ability to facilitate aggregation under conditions of substrate excess would allow a single antichaperone to retain multiple molecules of unfolded protein. Unlike most of the irreversible aggregates observed in vitro, aggregates formed in the cell are not irreversible [84,85]. As a consequence, facilitated aggregation could play a role in the ER quality control system (discussed in the following).

IX. DISULFIDE FORMATION IN THE CELL

As a general rule, disulfide bonds are found in extracellular proteins [24]; however, extracellular thiols and intracellular disulfides do exist [86,87]. Oxidative folding occurs in specialized cellular compartments, the bacterial periplasm [88], or the eukaryotic ER [89]. The ER and the periplasm provide suitable redox environments and the presence of high concentrations of chaperones and folding catalysts. In eukaryotes, a quality control mechanism ensures that proteins do not leave the ER until they are correctly folded [90]. In prokaryotes, proteins that do not fold rapidly and correctly are degraded by periplasmic proteases [91].

A. Disulfide Formation in the Endoplasmic Reticulum

Recently, significant progress has been made in determining PDI's role in the cell. PDI-depleted microsomes are capable of translocation of newly synthesized peptides, but they are deficient in the formation of disulfide bonds unless PDI is reintroduced [92]. Furthermore, PDI interaction with its substrates in vivo has been observed by the addition of a bifunctional cross-linker to antibody-producing cells, resulting in cross-linking of PDI to nascent antibodies [93]. In *Saccharomyces cerevisiae,* PDI is an essential gene [13]. The *PDI1* gene is required for efficient secretion of disulfide-containing peptides, because depletion of *PDI1* results in a failure to secrete the disulfide-containing protein carboxypeptidase Y [94], and use of a truncated *PDI1* gene with poor isomerase activity results in the failure to secrete the intrachain disulfide-bonded killer toxin [95]. LaMantia and Lennarz [13] reported that a mutant of PDI that complements the *PDI1* deletion and contains only the most NH_2-terminal cysteines in each active site ($\mathbf{N}_{SO}$:$\mathbf{C}_{SO}$) is incapable of catalyzing disulfide isomerization but does rescue the PDI null strain. However, recent evidence shows that although this mutant of PDI ($\mathbf{N}_{SO}$:$\mathbf{C}_{SO}$) is much less efficient than the wild-type enzyme in catalyzing oxidation, it is almost as good as wild-type PDI at catalyzing isomerizations [14,29]. This would argue that the essential function of PDI lies in its ability to catalyze disulfide rearrangements.

The eukaryotic redox environment in vivo appears to be maintained by a glutathione redox buffer, composed of glutathione (GSH) and glutathione–disulfide (GSSG). Hwang et al. [89] introduced peptides with the sequence NYTC into murine hybridoma cells. These cells transferred the peptide to the ER, which ultimately led to the secretion of the glycosylated peptide into the medium. The glycosylated peptide was reisolated as a mixture of the reduced peptide and a mixed disulfide with glutathione [89]. If we assume equilibrium between the glutathione redox buffer and the redox state of the peptide, an estimate can be obtained for the GSH/GSSG ratio in the ER. Additional evidence for an ER glutathione redox buffer comes from *S. cerevisiae,* in which a mutant lysozyme, with one of its normally disulfide-bonded cysteines mutated to alanine, was secreted as a mixed disulfide with glutathione [96]. This mixed disulfide species is proposed to be the intermediate that normally forms during the course of oxidation. If the other cysteine were present, it would then react with the mixed disulfide to form the native disulfide-releasing GSH to complete the reaction. Although there appears to be a high K_m GSSG transporter in microsomes, the mechanism by which the redox environment of the ER is regulated is unknown [89].

B. Disulfide Formation in the Bacterial Periplasm

Disulfide formation in the bacterial periplasm is accomplished by an elaborate system of proteins that catalyze disulfide formation and rearrangement. DsbA is a member of the thioredoxin family and, similar to PDI, has an active-site sequence motif of CXXC [88]. As secreted proteins enter the periplasmic space, they are efficiently oxidized by DsbA. In turn, the reduced DsbA is reoxidized by DsbB, a transmembrane protein that transfers oxidizing equivalents to the periplasm by an unknown mechanism [97]. Unlike PDI, DsbA is not a particularly effective catalyst of disulfide isomerization. In *E. coli,* isomerization activity may be provided by a separate protein, DsbC, another periplasmic member of the thioredoxin family [98]. Yet another protein, DsbD, may be involved in maintaining the redox state of the bacterial periplasm by providing reducing equivalents [99]. There are significant differences between disulfide formation in prokaryotes and eukaryotes. Prokaryotic proteins contain fewer disulfides, generally one or two, and often fold without the necessity for disulfide rearrangement [66]. Consequently, the oxidative folding of bacterial proteins may require more efficient oxidation mechanisms and be less reliant on catalysis of rearrangements.

C. Folding Pathways in the Endoplasmic Reticulum

Because of its complexity, it has been very difficult to study disulfide formation in the ER in any great detail. For some proteins, such as IgG, disulfide formation occurs cotranslationally; disulfides can be isolated on nascent chains [100]. However, for other proteins, such as human chorionic gonadotropin (hCG), disulfide formation occurs posttranslationally. For hCG, Ruddon and his colleagues have mapped the pathway for disulfide formation both in the cell and in vitro [101,102]. The intermediates observed have native disulfide configurations and accumulate and disappear in the same order in vivo and in vitro. At least, for this protein, the ER-folding machinery does not influence the pathway of disulfide formation [33]. Disulfide formation in influenza virus hemagglutin has also been followed during its maturation, and it, too, folds posttranslationally [103]. Because of the high local concentration of PDI and the high reactivity of its active-site disulfides, it may be necessary for some proteins to invoke local folding or interaction with chaperones to prevent sequential, cotranslational disulfide formation [103].

D. Aggregation and Antichaperones in the Cell

Aggregation is not always the result of nonspecific hydrophobic association of unfolded proteins. The experiments of Brems [104] and King [105] and their co-workers have provided compelling evidence that, at least for many proteins, aggregation occurs through the specific self-association of folding intermediates. For example, aggregation of bovine growth hormone is maximal at intermediate concentrations of urea and can be inhibited by specific peptides, suggesting a high degree of specificity in the aggregation process [104]. In addition, mutants of the phage P22 tailspike protein form specific aggregates both in vivo and in vitro [105] that derive from specific interactions in folding intermediates, rather than by nonspecific interactions. Mutations that destabilize these aggregation-prone intermediates act as global suppressors of aggregation [106]. Protein aggregates may also have considerable secondary structure. Infrared spectra of inclusion bodies isolated from *E. coli,* show considerable secondary structure, implying that there may be specificity in their formation [107,108].

When conditions in the ER favor protein misfolding, the misfolded proteins accumulate in the ER, often as very large, chaperone-associated aggregates [84,109,110]. When lysozyme folding is disturbed by mutation of one of its disulfides, misfolded lysozyme accumulates in the ER and can be covalently cross-linked to PDI [111]. Similarly, when disulfide formation is inhibited by including dithiothreitol in the medium or when cellular ATP is depleted, Chinese hamster ovary (CHO) cells accumulate large aggregates in the ER. When conditions that permit normal folding are restored, proteins are recovered from these aggregates and productively folded and secreted [84]. Transient and reversible aggregation may also be a part of the normal secretion pathway for some proteins. The induction of thyroglobulin synthesis in HepG2 cells by thyroid hormone leads to the initial accumulation of thyroglobulin in the ER as large, disulfide-bonded aggregates that, with time, are solubilized and secreted [85]. In a yeast strain that constitutively overexpresses human granulocyte colony-stimulating factor, erythropoietin, or yeast acid phosphatase, virtually all of the PDI and BiP was present in insoluble aggregates; very little PDI or BiP remained soluble [112]. Overexpression of BiP in these cells results in no significant increase in soluble BiP or secretion of BiP, although higher levels of BiP are imported into the ER. Surprisingly, overexpression of BiP in CHO cells increases ER retention and decreased secretion of several proteins [113], suggesting that BiP was more involved with ER-retention than with behaving as a chaperone that promotes correct folding. Taken together these paradoxical effects of chaperone overexpression are consistent with an active role for ER chaperones, or foldases, in facilitating the formation of large, ER-retained aggregates as part of the ER-quality control machinery.

E. Expression of Recombinant Proteins

It is often difficult to achieve high-level synthesis and secretion of eukaryotic proteins in bacterial expression systems. One of the factors that may complicate bacterial expression is a mismatch between the requirements for the oxidative folding of eukaryotic proteins and the redox environment and folding catalysts of the bacterial periplasm. For the few examples that have been examined, bacterial proteins fold well in an oxidizing environment [114], whereas the oxidative folding of eukaryotic proteins often occurs faster under more reducing conditions [28,34]. Furthermore, PDI is an excellent catalyst of the disulfide isomerizations that often limit the rate of productive folding for eukaryotic proteins [28,37]; however, bacterial disulfide isomerases such as DsbA and DsbC are much less effective catalysts of isomerization [98].

Consistent with the mismatch between folding conditions and catalysts in the eukaryotic and prokaryotic systems, providing an appropriate glutathione redox buffer in the *E. coli* growth medium leads to a 14-fold increase in the expression levels of α-amylase or α-trypsin inhibitor [115]. However, the folding of bovine pancreatic trypsin inhibitor in the periplasm of *E. coli* depends on the presence of DsbA; without DsbA, oxidants such as GSSG must be added to the medium to support BPTI folding [116]. When expressed in the bacterial periplasm, eukaryotic PDI can complement the DsbA null mutation in *E. coli* and provides a severalfold increase in the amount of BPTI recovered in this expression system [117]. In yeast, a four- to tenfold increase in the expression of platelet-derived growth factor and yeast acid phosphatase by coexpressing high levels of PDI have been achieved [118]; however, coexpressing BiP provides no benefit in this system [112]. The complex interactions between folding intermediates, chaperones, antichaperones, and folding catalysts, coupled with differences in the oxidative folding mechanisms of various proteins, have made it very difficult to devise global solutions to achieve high-level expression of recombinant proteins in bacteria.

F. Extracellular PDI

Although most of the PDI found in cells is located in the lumen of the ER, there have been reports of PDI associated with the plasma membrane [47]. It has been suggested that extracellular PDI may play a role in reductive activation or disulfide isomerization of membrane proteins. Bacitracin, an inhibitor of PDI, and anti-PDI antibodies inhibit the cytotoxicity of diphtheria toxin, a disulfide-linked heterodimer that must be reductively cleaved to exhibit toxicity [119]. In addition, bacitracin and anti-PDI antibodies, as well as the thiol reagent DTNB, inhibit the interaction of human immunodeficiency virus with its extracellular receptor, suggesting that some disulfide isomerization may be required for this interaction [120].

REFERENCES

1. Goldberger, R. F., Epstein, C. J., and Anfinsen, C. B., Acceleration of reactivation of reduced bovine pancreatic ribonuclease by a microsomal system from rat liver. *J. Biol. Chem. 238*:628–635 (1963).
2. Freedman, R. B., Hirst, R. R., and Tuite, M. F., Protein disulphide isomerase: Building bridges in protein folding. *Trends Biol. Sci. 19*:331–336 (1994).
3. Noiva, R., Enzymatic catalysis of disulfide formation, *Protein Expr. Purif. 5*:1–13 (1994).
4. Gilbert, H. F., The formation of native disulfide bonds, *Protein Folding* (Pain, R., ed.), Oxford Press, Oxford, 1993, pp. 104–136.
5. Puig, A. and Gilbert, H. F., Protein disulfide isomerase exhibits chaperone and antichaperone activity in the oxidative refolding of lysozyme, *J. Biol. Chem. 269*:7764–7771 (1994).
6. Cai, H., Wang, C., and Tsou, C., Chaperone-like activity of protein disulfide isomerase in the refolding of a protein with no disulfide bonds, *J. Biol. Chem. 269*:7764–7771 (1994).
7. Song, J. L. and Wang, C. C., Chaperone-like activity of protein disulfide-isomerase in the refolding of rhodanese, *Eur. J. Biochem. 2*:312–316 (1995).
8. Morjana, N. A. and Gilbert, H. F., Effect of protein and peptide inhibitors on the activity of protein disulfide isomerase, *Biochemistry 30*:4985–4990 (1991).
9. Noiva, R., Freedman, R. B., and Lennarz, W. J., Peptide binding to protein disulfide isomerase occurs at a site distinct from the active sites, *J. Biol. Chem. 268*:19210–19217 (1993).
10. Tsibris, J. C. M., Hunt, L. T., Ballejo, G., Barker, W. C., Toney, L. J., and Spellacy, W. N., Selective inhibition of protein disulfide isomerase by estrogen, *J. Biol. Chem. 264*:13967–13970 (1989).
11. Cheng, S. Y., Gong, Q. H., Parkinson, C., Robinson, E. A., Appela E., Merlino, G. T., and Pastan, I., The nucleotide sequence of a human cellular thyroid hormone binding protein present in endoplasmic reticulum, *J. Biol. Chem. 262*:11221–11227 (1987).
12. Lebeche, D., Lucero, H. A., and Kaminer, B., Calcium binding properties of rabbit liver protein disulfide isomerase, *Biochem. Biophys. Res. Commun. 202,* 556–561 (1994).
13. LaMantia M. and Lennarz, W. J., The essential function of yeast protein isomerase does not reside in its isomerase activity, *Cell 74*:899–908 (1993).
14. Laboissière, M. C. A., Sturley, S. L., and Raines, R. T., The essential function of

protein–disulfide isomerase is to unscramble non-native disulfide bonds, *J. Biol. Chem. 270*:28006–28009 (1995).
15. Gilbert, H. F., Molecular and cellular aspects of thiol–disulfide exchange, *Adv. Enzymol. 63*:69–172 (1990).
16. Pace, C. N., Grimsley, G. R., Thomson, J. A., and Barnett, B. J., Conformational stability and activity of ribonuclease T1 with zero, one and two intact disulfide bonds, *J. Biol. Chem. 263*:11820–11825 (1988).
17. Gilbert, H. F., Redox control of enzyme activities by thiol/disulfide exchange, *Methods Enzymol. 107*:330–351 (1984).
18. Creighton, T. E., Disulfide bond formation in proteins, *Methods Enzymol. 107*:305–329 (1984).
19. Wilson, J. M., Bayer, R. J., and Hupe, D. J., Structure–reactivity correlations for the thiol–disulfide interchange reaction, *J. Am. Chem. Soc. 99*:7922–7926 (1977).
20. Shaked, Z., Szajewski, R. P., and Whitesides, G. M., Rates of thiol–disulfide interchange reactions involving proteins and kinetic measurements of thiol pK_a values, *Biochemistry 19*:4156–4161 (1980).
21. Gilbert, H. F., Catalysis of thiol/disulfide exchange: Single-turnover reduction of protein disulfide isomerase by glutathione and catalysis of peptide disulfide reduction, *Biochemistry 27*:7298–7305 (1989).
22. Creighton, T. E., Disulfide bonds as probes of protein folding pathways, *Methods Enzymol. 131*:83–106 (1986).
23. Richardson, J. A., The anatomy and taxonomy of protein structure, *Adv. Protein Chem. 34*:167–173 (1981).
24. Thornton, J. M., Disulphide bridges in globular proteins, *J. Mol. Biol. 151*:261–273 (1981).
25. Darby, N. J. and Creighton, T. E., Characterization of the active site cysteine residues of the thioredoxin-like domains of protein disulfide isomerase, *Biochemistry 34*:11725–11735 (1995).
26. Creighton, T. E., Intermediates in the folding of reduced ribonuclease A, *J. Mol. Biol. 129*:411–431 (1984).
27. Scheraga, H. A., Konishi, Y., and Ooi, T., Multiple pathways for regenerating ribonuclease A, *Adv. Biophys. 18*:21–41 (1984).
28. Lyles, M. M. and Gilbert, H. F., Catalysis of the oxidative folding of ribonuclease A by protein disulfide isomerase: Dependence of the rate on the composition of the redox buffer, *Biochemistry 30*:613–619 (1991).
29. Walker, K. W., Lyles, M. M., and Gilbert, H. F., Catalysis of oxidative protein folding by mutants of proteins disulfide isomerase with a single active-site cysteine, *Biochemistry 35*:1972–1980 (1996).
30. Creighton, T. E. and Goldenberg, D. P., Kinetic role of a meta-stable native-like two-disulphide species in the folding transition of bovine pancreatic trypsin inhibitor, *J. Mol. Biol. 179*:497–526 (1984).
31. Weissman, J. S. and Kim, P. S., Reexamination of the folding of BPTI: Predominance of native intermediates, *Science 253*:1386–1390 (1991).
32. Chang, J. -Y., The properties of scrambled hirudins, *J. Biol. Chem. 270*:25661–25666 (1995).
33. Huth, J. R., Perini, F., Lockridge, O., Bedows, E., and Ruddon, R. W., Protein

folding and assembly in vitro parallel intracellular folding and assembly. Catalysis of folding and assembly of the human chorionic gonadotropin alpha beta dimer by protein disulfide isomerase, *J. Biol. Chem. 268*:16472–16482 (1993).

34. Saxena, V. P. and Wetlaufer, D. B., Formation of three-dimensional structure in proteins. I. Rapid nonenzymic reactivation of reduced lysozyme, *Biochemistry 9*:5015–5022 (1970).
35. Konishi, Y., Ooi, T., and Sheraga, H. A., Regeneration of ribonuclease A from the reduced protein. Rate-limiting steps, *Biochemistry 21*:4734–4740 (1982).
36. Lyles, M. M. and Gilbert, H. F., Catalysis of the oxidative folding of ribonuclease A by protein disulfide isomerase: Pre-steady-state kinetics and the utilization of the oxidizing equivalents of the isomerase, *Biochemistry 30*:619–624 (1991).
37. Hillson, D. A. and Freedman, R. B., Resolution of protein disulphide–isomerase and glutathione–insulin transhydrogenase activities by covalent chromatography, *Biochem. J. 191*:373–388 (1980).
38. Ruoppolo M. and Freedman, R. B., Refolding by disulfide isomerization: The mixed disulfide between ribonuclease T1 and glutathione as a model refolding substrate, *Biochemistry 34*:9380–9388 (1995).
39. Creighton, T. E., Hillson, D. A., and Freedman, R. B., Catalysis by protein–disulphide isomerase of the unfolding and refolding of proteins with disulphide bonds, *J. Mol. Biol. 142*:43–62 (1980).
40. Lyles, M. M. and Gilbert, H. F., Mutations in the thioredoxin sites of protein disulfide isomerase reveal functional non-equivalence of the N- and C-terminal domains, *J. Biol. Chem. 269*:30946–30952 (1994).
41. Weissman, J. S. and Kim, P. S., Efficient catalysis of disulphide bond rearrangements by protein disulphide isomerase, *Nature 365*:185–188 (1993).
42. Venetainer, P. and Straub, R. B., The enzymic reactivation of reduced ribonuclease, *Biochim. Biophys. Acta 67*:166–168 (1963).
43. Edman, J. C., Ellis, L., Blacher, R. W., Roth, R. A., and Rutter, W. J., Sequence of protein disulphide isomerase and implications of its relationship to thioredoxin, *Nature 317*:267–270 (1985).
44. Roth, R. A. and Koshland, M. E., Role of disulfide interchange enzyme in immunoglobulin synthesis, *Biochemistry 20*:6594–6599 (1981).
45. Tasanen, K., Oikarinen, J., Kivirikko, K. I., and Pihlajaniem, T., Promoter of the gene for the multifunctional protein disulfide isomerase polypeptide. Functional significance of the six CCAAT boxes, *J. Biol. Chem. 267*:11513–11519 (1992).
46. Tasanen, K., Oikarinen, J., Kivirikko, K. I., and Pihlajaniem, T., Interaction of transcription factor *Sp1* with the promoter of the gene for the multifunctional protein disulphide isomerase polypeptide, *Biochem. J. 292*:41–45 (1993).
47. Akagi, S., Yamamoto, A., Yoshimori, T., Masaki, R., Ogawa R., and Tashiro, Y., Distribution of protein disulfide isomerase in rat hepatocytes, *J. Histochem. Cytochem. 36*:1533–1542 (1988).
48. Munro, S. and Pelham, H. R. B., A C-terminal signal prevents secretion of luminal ER proteins, *Cell 48*:899–907 (1987).
49. Pelham, H. R. B., Control of protein exit from the endoplasmic reticulum, *Annu. Rev. Cell Biol. 5*:1–23 (1989).

50. Ohba, H., Harano, T., and Omura, T., Biosynthesis and turnover of a microsomal protein disulfide isomerase in rat liver, *J. Biochem. 89*:901–907 (1981).
51. Dorner, A. J., Wasley, L. C., Raney, P., Haugejorden, S., Green, M., and Kaufman, R., The stress response in Chinese hamster ovary cells. Regulation of ERp72 and protein disulfide isomerase expression and secretion, *J. Biol. Chem. 265*:22029–22034 (1990).
52. Vuori, K., Myllyla, R., Pihlajaniemi, T., and Kivirikko, K. I., Expression and site-directed mutagenesis of human protein disulfide isomerase in *Eschericia coli, J. Biol. Chem. 267*:7211–7214 (1992).
53. Hawkins, H. C. and Freedman, R. B., The reactivities and ionization properties of the active-site dithiol groups of mammalian protein disulphide-isomerase, *Biochem. J. 275*:335–339 (1991).
54. Eklund, H., Gleason, F. K., and Holmgren, A., Structural and functional relations among thioredoxins of different species, *Proteins 11*:13–28 (1991).
55. Martin, J. L., Bardwell, J. C., and Kuriyan, J., Crystal structure of the DsbA protein required for disulphide bond formation in vivo, *Nature 365*:464–467 (1993).
56. Kemmink, J., Darby, N. J., Dijkstra, K., Scheek, R. M., and Creighton, T. E., Nuclear magnetic resonance characterization of the N-terminal thioredoxin-like domain of protein disulfide isomerase, *Protein Sci. 4*:2587–2593 (1995).
57. Holmgren, A., Enzymatic reduction–oxidation of protein disulfides by thioredoxin, *Methods Enzymol. 107*:295–300 (1984).
58. Zapun, A., Bardwell, J. C., Creighton, T. E., The reactive and destabilizing disulfide bond of DsbA, a protein required for protein disulfide bond formation in vivo, *Biochemistry 32*:5083–5092 (1993).
59. Wunderlich, M., Jaenicke, R., and Glockshuber, R., The redox properties of protein disulfide isomerase (DsbA) of *Escherichia coli* result from a tense conformation of its oxidized form, *J. Mol. Biol. 233*:559–566 (1993).
60. Lundström, J., Krause, G., and Holmgren, A., A Pro to His mutation in active site of thioredoxin increases its disulfide-isomerase activity 10-fold. New refolding systems for reduced or randomly oxidized ribonuclease, *J. Biol. Chem. 267*:9047–9052 (1992).
61. Lu, X., Gilbert, H. F., and Harper, W. J., Conserved residues flanking the thiol/disulfide centers of protein disulfide isomerase are not essential for catalysis of thiol/disulfide exchange, *Biochemistry* 31, 4205–4210 (1992).
62. Grauschopf, U., Winther, J. R., Korber, P. I., Zander, T., Dallinger, P., and Bardwell, J. C. A., Why is DsbA such an oxidizing disulfide catalyst? *Cell 83*:947–955 (1995).
63. Gane, P. J., Freedman, R. B., and Warwicker, J., A molecular model for the redox potential difference between thioredoxin and DsbA, based on electrostatics calculations, *J. Mol. Biol. 249*:376–387 (1995).
64. Kortemme T. and Creighton, T. E., Ionisation of cysteine residues at the termini of model alpha-helical peptides. Relevance to unusual thiol pK_a values in proteins of the thioredoxin family, *J. Mol. Biol. 253*:799–812 (1995).
65. Darby, N. J., Freedman, R. B., and Creighton, T. E., Dissecting the mechanism of protein disulfide isomerase: Catalysis of disulfide bond formation in a model peptide, *Biochemistry 33*:7937–7947 (1994).

66. Walker K. W. and Gilbert, H. F., Oxidation of kinetically trapped thiols by protein disulfide isomerase, *Biochemistry 34*:13642–13650 (1995).
67. Chandler, M. L. and Varandani, P. T., Kinetic analysis of the mechanism of insulin degradation by glutathione–insulin transhydrogenase (thiol: protein-disulfide oxidoreductase), *Biochemistry 14*:2107–2114 (1975).
68. Noiva, R. and Lennarz, W. J., Protein disulfide isomerase: A multifunctional protein resident in the lumen of the endoplasmic reticulum, *J. Biol. Chem. 267*:3553–3556 (1992).
69. Pihlajaniemi, T., Helaakoski, T., Tasanen, K., Myllylä, R., Huhtala, M., Koivu, J., and Kivirikko, K. I., Molecular cloning of the beta-subunit of human prolyl 4-hydroxylase. This subunit and protein disulphide isomerase are products of the same gene, *EMBO J. 6*:643–649 (1987).
70. Kivirikko, K. I., Myllylä, R., and Pihlajaniemi, T., Protein hydroxylation: Prolyl 4-hydroxylase, an enzyme with four cosubstrates and a multifunctional subunit, *FASEB J. 3*:1609–1617 (1989).
71. Wetterau, J. R., Combs, K. A., Spinner, S. N., and Joiner, B. J., Protein disulfide isomerase is a component of the microsomal triglyceride transfer protein complex, *J. Biol. Chem. 265*:9800–9807 (1990).
72. Hu, C. H. and Tsou, C. L., Formation of enzyme–substrate disulfide linkage during catalysis by protein disulfide isomerase, *FEBS Lett. 290*:87–89 (1991).
73. Park, S. H. and Ryu, D. D. Y., Protein disulfide isomerase reaction kinetics in endoplasmic reticulum for monoclonal antibody refolding and assembly, *Ann. N. Y. Acad. Sci. 63*:291–299 (1995).
74. Koivu, J. and Myllyla, R., Interchain disulfide bond formation in types I and II procollagen. Evidence for a protein disulfide isomerase catalyzing bond formation, *J. Biol. Chem. 262*:6159–6164 (1987).
75. Laminet, A. A. and Pluckthun, A., The precursor of β-lactamase: purification, properties and folding kinetics, *EMBO J. 8*:1469–1477 (1989).
76. Puig, A., Lyles, M. M., Noiva, R., and Gilbert, H. F., The role of the thiol/disulfide centers and peptide binding site in the chaperone and anti-chaperone activities of protein disulfide isomerase, *J. Biol. Chem. 269*:19128–19135 (1994).
77. Bassuk, J. A., Kao, W. W. Y, Herzer, P., Kedersha, N. L., Seyer, J., DeMartion, J. A., Daugherty, B. L., Mark, G. E., and Berg, R. A., Prolyl 4-hydroxylase: Molecular cloning and the primary structure of the alpha subunit from chicken embryo, *Proc. Natl. Acad. Sci. USA 86*:7382–7386 (1989).
78. Vuori, K, Pihlajaniemi, T., Marttila, M., and Kivirikko, K. I., Characterization of the human prolyl 4-hydroxylase tetramer and its multifunctional protein disulfide-isomerase subunit synthesized in a baculovirus expression system, *Proc. Natl. Acad. Sci. USA 89*:7467–7470 (1992).
79. Helaakoski, T., Annunen, P., Vuori, K., MacNeil, I., Pihlajaniemi, T., and Kivirikko, K. I., Cloning, baculovirus expression, and characterization of a second mouse prolyl 4-hydroxylase alpha-subunit isoform: Formation of an $\alpha_2\beta_2$ tetramer with the protein disulfide-isomerase/beta subunit, *Proc. Natl. Acad. Sci. USA 92*:4427–4431 (1995).
80. Wells, W. W., Xu, D. P., Yang, Y., and Rocque, P. A., Mammalian thioltransferase

(glutaredoxin) and protein disulfide isomerase have dehydroascorbate reductase activity, *J. Biol. CHem. 265*:15361–15364 (1990).
81. Wetterau, J. R., Combs, K. A., McLean, R. R., Spinner, S. N., and Aggerbeck, L. P., Protein disulfide isomerase appears necessary to maintain the catalytically active structure of the microsomal triglyceride transfer protein, *Biochemistry 30*:9728–9735 (1991).
82. Wetterau, J. R., Aggerbeck, L. P., Laplaud, M., and McLean, L. R., Structural properties of the microsomal triglyceride–transfer protein complex, *Biochemistry 30*:4406–4412 (1991).
83. Soute, B. A. M., Monique, M. C. L., Dooren, G., Holmgren, A., Lundstrom, J., and Vermeer, C., Stimulation of the dithiol-dependent reductases in the vitamin K cycle by the thioredoxin system. Strong synergistic effects with protein disulphide-isomerase, *Biochem. J. 281*:255–259 (1992).
84. Braakman, I., Helenius, J., and Helenius, A., Role of ATP and disulphide bonds during protein folding in the endoplasmic reticulum, *Nature 356*:260–262 (1992).
85. Kim, P. S., Bole, D., and Arvan, P., Transient aggregation of nascent thyroglobulin in the endoplasmic reticulum: Relationship to the molecular chaperone, BiP, *J. Cell. Biol. 118*:541–549 (1992).
86. Hammel, K. E., Cornwell, K. L., and Buchanan, B. B., Ferredoxin/flavoprotein-linked pathway for the reduction of thioredoxin, *Proc. Natl. Acad. Sci. USA 80*:3681–3685 (1983).
87. Benkoura, R., Guidoni, A. A., DeCaro, J. D., Bonicel, J. J., Desnuelle, P. A., and Rovery, M., Porcine pancreatic lipase. The disulfide bridges and the sulfhydryl groups, *Eur. J. Biochem. 128*:331–341 (1982).
88. Bardwell, J. C. A., Lee, J. -O., Jander, M., Martin, N., Belin, D., and Beckwith, J. A., A pathway for disulfide bond formation in vivo, *Proc. Natl. Acad. Sci. USA 90*:1038–1042 (1991).
89. Hwang, C., Sinskey, A. J., and Lodish, H. F., Oxidized redox state of glutathione in the endoplasmic reticulum, *Science 257*:1496–1502 (1992).
90. Gething, M. J. and Sambrook, J., Protein folding in the cell, *Nature* 355:33–45 (1992).
91. Meerman, H. J. and Georgiou, G., Construction and characterization of a set of *E. coli* strains deficient in all known loci affecting the proteolytic stability of secreted recombinant proteins, *Biotechnology 12*, 1107–1110 (1994).
92. Bulleid, N. J. and Freedman, R. B., Defective co-translational formation of disulphide bonds in protein disulphide-isomerase-deficient microsomes, *Nature 335*:649–651 (1988).
93. Roth, R. A. and Pierce, S. B., In vivo cross-linking of protein disulfide isomerase to immunoglobulins, *Biochemistry 26*:4179–4182 (1987).
94. Freiden, P. J., Gaut, J. R., and Hendershot, L. M., Interconversion of three differentially modified and assembled forms of BiP, *EMBO J. 11*:63–70 (1992).
95. Dunn, A., Luz, J. M., Natalia, D., Bamble, J. A., Freedman, R. B., and Tuite, M. F., Protein disulphide isomerase (PDI) is required for the secretion of a native disulphide-bonded protein from *Saccharomyces cerevisiae, Biochem. Soc. Trans. 23*:78S–79S (1995).
96. Taniyama, Y., Seko, C., and Kikuchi, M., Secretion in yeast of mutant human

lysozymes with and without glutathione bound to cysteine 95, *J. Biol. Chem. 265*:16767–16771 (1990).
97. Bardwell, J. C. A., Lee, J. -O., Jander, B., Martin, N., Belin, D., and Beckwith, J., A pathway for disulfide bond formation in vivo, *Proc. Natl. Acad. Sci. USA* 90:1038–1042 (1993).
98. Zapun, A., Missiakas, D., Raina, S., and Creighton, T. E., Structural and functional characterization of DsbC, a protein involved in disulfide bond formation in *Escherichia coli, Biochemistry 34*:5075–5089 (1995).
99. Missiakas, D., Schwager, R., and Raina, S., Identification and characterization of a new disulfide isomerase-like protein (DsbD) in *Escherichia coli, EMBO J. 14*, 3415–3424 (1995).
100. Bergman, L. W. and Kuehl, W. M., Formation of an intrachain disulfide bond on nascent immunoglobulin light chains, *J. Biol. Chem. 254*:8869–8876 (1979).
101. Huth, J. R., Mountjoy, K., Perini, F., and Ruddon, R. W., Intracellular folding pathway of human chorionic gonadotropin β subunit, *J. Biol. Chem. 267*:8870–8879 (1992).
102. Bedows, E., Huth, J. R., Suganuma, N., Bartels, C. R., Boime, I., and Ruddon, R. W., Disulfide bond mutations affect the folding of the human chorionic gonadotropin-beta subunit in transfected Chinese hamster ovary cells, *J. Biol. Chem. 268*:11655–11662 (1993).
103. Segal, M. S., Bye, J. M., Sambrook, J. F., and Gething, M. J., Disulfide bond formation during the folding of influenza virus hemagglutinin, *J. Cell Biol. 118*:227–244 (1992).
104. DeFelippis, M. R., Alter, L. A., Pekar, A. H., Havel, H. A., and Brems, D. N., Evidence for a self-associating equilibrium intermediate during folding of human growth hormone, *Biochemistry 32*:1555–1562 (1993).
105. King, J., Haase-Pettingell, C., Robinson, A. S., Speed, M., and Mitraki, A., Thermolabile folding intermediates: Inclusion body precursors and chaperonin substrates, *FASEB J. 10*:57–66 (1996).
106. Mitraki, A., Danner, M., King, J., and Seckler, R., Temperature-sensitive mutations and second-site suppressor substitutions affect folding of the P22 tailspike protein in vitro, *J. Biol. Chem. 268*:20071–20075 (1993).
107. Przybycien, T. M., Dunn, J. P., Valax, P., and Georgiou, G., Secondary structure characterization of β-lactamase inclusion bodies, *Protein Eng. 7*:131–136 (1994).
108. Oberg, K., Chrunyk, B. A., Wetzel, R. B., and Fink, A. L., Native-like secondary structure in interleukin 1β inclusion bodies by attenuated total reflectance FTIR, *Biochemistry 33*:2628–2634 (1994).
109. Tatu, U., Braakman, I., and Helenius, A., Membrane glycoprotein folding, oligomerization and intercellular transport: Effects of dithiothreitol in living cells, *EMBO J. 12*:2151–2157 (1993).
110. Braakman, I., Helenius, J., and Helenius, A., Manipulating disulfide bond formation and protein folding in the endoplasmic reticulum, *EMBO J. 11*:1717–1722 (1992).
111. Otsu, M., Omura, F., Yoshimori, T., and Kikuchi, M., Protein disulfide isomerase associates with misfolded human lysozyme in vivo, *J. Biol. Chem. 269*:6874–6877 (1994).
112. Robinson, A. S. and Wittrup, K. D., Constitutive overexpression of secreted hetero-

logous proteins decreases extractable BiP and protein disulfide isomerase levels in *Saccharomyces cerevisiae, Biotechnol. Prog. 11*:171–177 (1995).

113. Dorner, A. J., Wasley, L. C., and Kaufman, R. J., Overexpression of GRP78 mitigates stress induction of glucose regulated proteins and blocks secretion of selective proteins in Chinese hamster ovary cells, *EMBO J. 11*:1563–1571 (1992).
114. Walker, K. W. and Gilbert, H. F., Effect of redox environment on the in vitro and in vivo folding of RTEM-1 β-lactamase and *Escherichia coli* alkaline phosphatase, *J. Biol. Chem. 269*:28487–28493 (1994).
115. Wunderlich, M. and Glockshuber, R., In vivo control of redox potential during protein folding catalyzed by bacterial protein disulfide-isomerase (DsbA), *J. Biol. Chem. 268*:24547–24550 (1993).
116. Ostermeier, M. and Georgiou, G., The folding of bovine pancreatic trypsin inhibitor in the *Escherichia coli* periplasm, *J. Biol. Chem. 269*:21072–21077 (1994).
117. Ostermeier, M., De Sutter, K., and Georgiou, G., Eukaryotic protein disulfide isomerase complements *Escherichia coli* dsbA mutants and increases the yield of a heterologous secreted protein with disulfide bonds, *J. Biol. Chem 271*:10616–10622 (1996).
118. Robinson, A. S., Hines, V., and Wittrup, K. D., Protein disulfide isomerase over-expression increases secretion of foreign proteins in *Saccharomyces cerevisiae, Biotechnology 12*:381–384 (1994).
119. Mandel, R., Ryser, H. J., Ghani, F., Wu, M., and Peak, D., Inhibition of a reductive function of the plasma membrane by bacitracin and antibodies against protein disulfide-isomerase, *Proc. Natl. Sci. USA 90*:4112–4116 (1993).
120. Ryser, H. J., Levy, E. M., Mandel, R., and DiSciullo, G. J., Inhibition of human immunodeficiency virus infection by agents that interfere with thiol–disulfide interchange upon virus–receptor interaction, *Proc. Natl. Acad. Sci. USA 91*:4559–4563 (1994).

15

Catalysis of Protein Folding by Prolyl Isomerases

Franz Xaver Schmid
Universität Bayreuth, Bayreuth, Germany

I. INTRODUCTION

Unlike other peptide bonds, those preceding proline often occur in the *cis*-conformation in folded proteins [1,2]. The *cis* ⇌ *trans*-isomerizations about these bonds (prolyl isomerizations) are slow processes in oligopeptides as well as in proteins [3], and they frequently determine the overall rate of protein-folding reactions [4–8]. The first enzyme that could catalyze a prolyl isomerization in a tetrapeptide was isolated from porcine kidney in 1984 by G. Fischer and his co-workers [9]. According to its function they named this enzyme peptidyl prolyl *cis/trans* isomerase (prolyl isomerase). This enzyme was not only active toward small substrates, but also it accelerated slow protein-folding reactions, which were thought to involve prolyl isomerizations as rate-determining steps [10,11]. Since then many other prolyl isomerases have been discovered, and it is now clear that they are ubiquitous enzymes that occur in virtually all organisms and in all subcellular compartments [12–14]. Three unrelated families of prolyl isomerases are now known: the cyclophilins, the tacrolimus (FK 506)-binding proteins (FKBPs), and the parvulins [13]. The cyclophilins and FKBPs are also termed *immunophilins,* because they are the cellular receptors for the immunosuppressants cyclosporine (cyclosporin A; CsA) and tacrolimus, respectively. This function of prolyl isomerases as mediators of immunosuppression is described in several reviews and will not be considered here [12–14]. Recently, the trigger factor of *Escherichia coli* was identified as an efficient prolyl isomerase that can associate with the ribosome and with nascent protein chains [15–17].

This chapter will focus on the function of the prolyl isomerases in protein folding. First, mechanistic aspects of the enzymes and the catalyzed reactions are described, then the importance of prolyl isomerization reactions as slow, rate-limiting steps of folding and their catalysis in vitro will be discussed, followed by a description of recent results that point to a role for prolyl isomerases in cellular folding. It is still largely unknown whether the catalysis of slow steps in folding is, in fact, a major function of prolyl isomerases in the cell. If one considers the enormous diversity and the ubiquitous distribution of these enzymes, it is almost certain that they are involved in a multitude of functions. They might be similar to proteases, which function in processes as diverse as digestion, enzyme activation, blood clotting, and the immune response.

II. PROLYL ISOMERIZATION

A. Prolyl Peptide Bonds

Peptide bonds are planar and the two successive C_α atoms, which flank the peptide bond, can be either in the *trans*- or in the *cis*-conformation. These conformations are equivalent to dihedral angles ω of 180° and 0°, respectively. For peptide bonds preceding residues other than proline the *trans*-state is strongly favored. The *cis*-state is believed to be populated to less than 1 or even 0.1% in oligopeptides [18,19] and, in fact, very few nonproline *cis*-peptide bonds have been found in native, folded proteins by X-ray crystallography [1,2].

Unlike other peptide bonds, those between proline and its preceding amino acid (Xaa–Pro bonds) can exist in both the *cis*- and the *trans*-state in solution, unless structural constraints (such as in folded proteins) strongly stabilize one isomer relative to the other. The *trans*-isomer is usually favored slightly over the *cis*-isomer in the absence of ordered structure and in short linear peptides. Frequently, *cis* contents of 10–30% are found [20,21]. The *cis/trans* equilibrium depends to some extent on the chemical nature of the flanking amino acids and on the charge distribution around the Xaa–Pro bond.

In proteins with known three-dimensional structure the conformational state of each peptide bond is usually clearly defined. It is either *cis* or *trans* in every molecule, depending on the structural framework imposed by the folded protein chain, and about 7% of all prolyl peptide bonds in folded proteins are *cis* [1,2]. There are a few exceptions to this rule. In the native states of staphylococcal nuclease [22], insulin [23], and calbindin [24,25] *cis/trans* equilibria at particular Xaa–Pro bonds have been detected in solution by nuclear magnetic resonance (NMR). In staphylococcal nuclease, the *cis*-conformer of the Lys116–Pro117 bond can be selectively stabilized by binding of Ca^{2+} ions and the inhibitor thymidine 3′5′-phosphate [22,26,27]. It is conceivable that *cis* $\rightleftharpoons$ *trans* equilibria at Xaa–Pro bonds are more common for proteins with dynamic conformations (which cannot be crystallized) or in flexible chain regions (which are not well-defined in crystal structures).

Prolyl *cis* $\rightleftharpoons$ *trans*-isomerization involves the rotation about a partial double bond. Therefore, it is an intrinsically slow reaction (time constants in the range of 10–100 s are frequently observed at 25°C), with a high activation energy. The barrier to isomerization is enthalpic. The average activation enthalpy is about 80 kJ mol^{-1}, and the activation entropy is close to zero. The rate of isomerization is increased in organic solvents [28], which indicates that the transition state for amide rotation is less polar than the reactant state. Prolyl isomerization is also catalyzed by strong acids [29], probably by protonation at the imide nitrogen, which abolishes the partial double-bond character of the peptide bond [3].

Prolyl isomerization is a very simple reaction. Covalent bonds are neither formed nor broken, and reactant and product are chemically identical. Therefore, the mechanism of its catalysis by the prolyl isomerases is difficult to investigate. Available experimental evidence and theoretical calculations point to an enzymatic mechanism that involves catalysis by distortion [3,14,30]. In this mechanism the enzyme would stabilize a transition state with a partially rotated C–N bond, in which the partial double-bond character of this bond is diminished. Such a twisted amide could be stabilized by specific enzyme–substrate interactions and by the low polarity of the substrate-binding site. The prolyl isomerases with known three-dimensional structure are all hydrophobic in their active sites [31–34] and, thus, the transfer of the folding protein chain into this nonpolar environment could partly explain the observed high-catalytic efficiencies. It is not known whether some kind of acid catalysis also contributes to the enzymatic mechanism of prolyl isomerases.

B. Prolyl Isomerization in Protein Folding

In 1973, Garel and Baldwin discovered that unfolded ribonuclease A (RNase A) consists of a kinetically heterogeneous mixture of molecules [35]. The fast-folding U_F molecules could refold in less than a second, whereas the slow-folding U_S molecules required several minutes to fold to completion. Subsequently, similar U_F and U_S species have been detected in the folding of many other proteins [5,6,36]. A plausible molecular explanation for this phenomenon was provided by the proline hypothesis of Brandts and co-workers. They suggested that the fast- and slow-folding molecules differ in the *cis–trans*-isomeric state of one or more Xaa–Pro peptide bonds [4].

In the native protein (N) usually each prolyl peptide bond is in a defined conformation being either *cis* or *trans* in every molecule. After unfolding [$N \rightarrow U_F$; Eq. (1)], however, these bonds become free to isomerize slowly as in short oligopeptides [in the $U_F \rightarrow U_S^i$ reaction; see Eq. (1)].

$$N \rightarrow U_F \rightleftharpoons U_S{}^i \quad (1)$$

Thus, a mixture of a species is created, which consists of a single unfolded form with correct prolyl isomers (U_F), and one or more unfolded species with incorrect prolyl isomers ($U_S{}^i$). As just outlined, the U_F molecules can refold rapidly to the native conformation. $U_S{}^i$ molecules refold slowly, because refolding is coupled with the reisomerizations of the incorrect prolyl bonds. Nonnative isomers do not block refolding right at its beginning, and reisomerization is not required to be the first step of folding, as suggested initially. A heterogeneity at prolines can usually be well accommodated in partially folded intermediates, but not in native proteins. Conformational folding can thus start while some prolines are still in the nonnative state. The final folding steps, however, require correct prolines and, therefore, these crucial reactions are limited in rate by the slow prolyl isomerizations [7,36–38].

The extent of conformational folding before prolyl isomerization depends on the location of the nonnative prolyl isomers in the structure and on the folding conditions. Generally, incorrect prolyl bonds at the surface of the folded protein or in flexible-chain regions will not strongly interfere with conformational folding, and solvent conditions that strongly stabilize folded proteins will also stabilize partially folded structures in intermediates with incorrect isomers. Conformational-folding steps and prolyl isomerization are thus mutually interdependent. On the one hand, the presence of incorrect isomers in the chain can decelerate its folding and, on the other hand, rapid chain folding can affect the equilibrium and the kinetic properties of Xaa–Pro peptide bond isomerization. This close interrelation between structure formation and prolyl peptide bond isomerization is a key feature of slow-folding steps and is of central importance for understanding the role of prolyl isomerases in these processes.

III. PROLYL ISOMERASES

A. Assays

It took almost a decade from the formulation of the proline hypothesis for protein folding [4] to the discovery of the first prolyl isomerase. This discovery was possible because Fischer and his co-workers developed an ingenious assay for this class of enzymes [9]. With several minor modifications, their protease-coupled assay is routinely used to identify and characterize prolyl isomerases. The assay is based on the conformational specificity of chymotrypsin, which cleaves a chromogenic reporter group from a tetrapeptide, such as succinyl-Ala-Ala-Pro-Phe-4-nitroanilide, only when the Ala–Pro bond of this peptide is in the *trans*-conformation. In aqueous solution the assay peptide exists as a 90:10 mixture of molecules with the Ala–Pro bond in *trans* and in *cis,* respectively. Therefore, in the presence of a high concentration of chymotrypsin, 90% of the hydrolysis reaction occurs within the dead time of manual mixing. Hydrolysis of the remaining 10% is slow, because it is limited in rate by the *cis* → *trans* isomerization of the Ala–Pro bond. It can be followed directly by absorption spectroscopy.

This assay was improved by Kofron et al. [39,40]. They increased the fraction of the *cis*-isomer from 10 up to 70% by dissolving the assay peptide in an anhydrous mixture of trifluoroethanol and LiCl. In addition to chymotrypsin, several other isomer-specific proteases are suitable for the coupled assay, and thus a wider range of peptides can be used to investigate the activities and specificities of prolyl isomerases [13]. Alternative assays for prolyl isomerase activity are based on the use of fluorescing peptides [41,42] or of peptides with a modified amino acid incorporated before proline. Also, NMR methods are available for directly measuring prolyl isomerase activities without the need for an isomer-specific protease [43,44]. These assays require the synthesis of modified peptides or advanced instrumentation, but are the method of choice for assaying prolyl isomerases, which are sensitive to proteases.

B. Identity with Immunophilins

A surprising result emerged when the first prolyl isomerase was sequenced [45,46]. It was identical with cyclophilin (Cyp), which is the cellular receptor for the immunosuppressive drug CsA. Incidentally, this protein, of 18 kDa, had been discovered twice in 1984: as the binding protein for CsA [47] and as an efficient prolyl isomerase [9]. The cyclophilins are ubiquitous proteins. They seem to be present in all organisms and in all subcellular compartments. In yeast, seven different forms of cyclophilin exist, which are found in the cytosol, the mitochondria, the endoplasmic reticulum, and the nucleus [J. Heitman, personal communication].

The amino acid sequences and the three-dimensional structures of the cyclophilins are highly conserved. The cytosolic cyclophilins from humans and *E. coli* show about 25% sequence identity [34,46]. The backbone of the cyclophilins consists of a central eight-stranded antiparallel β-sheet structure and two α-helices. The core structures of the human [32,33] and the *E. coli* protein [34] are almost identical, and 75 Cα atoms within the β-sheet and the helices could be superimposed with a root-mean-square (rms) deviation of 0.61 Å. Stronger deviations are found in the loops at the protein periphery, caused primarily by several deletions and insertions. Structures of the enzyme complexed with cyclosporine and with proline-containing tripeptides were determined by a combination of X-ray crystallography and two-dimensional NMR spectroscopy [48–51]. The substrate peptide is bound in a long groove at the surface of the β-sheet. The Ala–Pro bond of the bound peptide is in the *cis*-conformation and the *cis*-proline ring is inserted into a hydrophobic pocket. Residues that were identified to participate in CsA binding by NMR cluster at the peptide-binding site, indicating that CsA binds at the prolyl isomerase active site as a competitive inhibitor.

Interestingly, members of a further class of prolyl isomerases, the FK 506-binding proteins (FKBPs) are also involved in immunosuppression. Human FKBP12, a 12-kDa protein, was discovered in 1989 [52,53]. The immunosuppressants FK 506 and rapamycin (but not CsA) bind and inhibit FKBP12. There is no similarity in sequence to the cyclophilins, but the FKBPs also catalyze prolyl isomerization in oligopeptides [53] and in protein folding [54]. The FKBPs share the ubiquitous distribution with the cyclophilins, and five FKBPs are found in yeast [J. Heitman, personal communication]. The spatial structures of FKBP12s were solved by NMR spectroscopy [55,56] and by X-ray crystallography [31] in the free form and in complexes with the inhibitors FK 506 and rapamycin [57]. FKBP12 is composed of a five-stranded antiparallel β-sheet, a short α-helix and a fairly high amount of aperiodic structure [31]. The pipecolic amide moiety of FK 506, which is thought to mimick the proline residue of the substrate, is also bound in a hydrophobic pocket of FKBP, presumably at the active site [31].

C. Substrate Specificities of Cyclophilin and FKBP

The FKPBs and the cyclophilins differ substantially in substrate specificity [3,13,58]. The cyclophilins are highly tolerant to the nature of the amino acids that flank the proline in a tetrapeptide, and the specificity constant k_{cat}/K_M varied by less than one order of magnitude when different amino acids were inserted at these positions in a tetrapeptide. In contrast, the specificity of the FKBPs is much higher, and a more than 1000-fold variation in k_{cat}/K_M was observed when the amino acid before proline was varied. The k_{cat}/K_M value was particularly high for hydrophobic residues, such as leucine and phenylalanine at this position ($k_{cat}/K_M = 620 \times 10^3$ $M^{-1}s^{-1}$) and very low for glutamate ($k_{cat}/K_M = 0.6 \times 10^3$ $M^{-1}s^{-1}$) [58].

There is good evidence now that the functions of Cyp18 and FKBP12 in immunosuppression do not involve their prolyl isomerase activity or the catalysis of a protein-folding reaction. Rather, the inhibitors CsA and FK 506 form tight 1:1 complexes with Cyp18 and FKBP12, respectively. These complexes bind to the protein phosphatase calcineurin and, thereby, inhibit the transport to the nucleus of a cytosolic component of the transcription factor NF-AT in T cells [12,13,59]. The molecular mechanisms of these immunusuppressive processes have been reviewed and will not be considered further.

D. FKBP and Cyclophilin as Domains of Larger Proteins

Domains that are homologous to Cyp18 or FKBP12 are found in many proteins. The NinA protein (26 kDa) from an insect eye contains in addition to the Cyp18 domain a hydrophobic sequence, which serves as a membrane anchor [60–62].

FKBP25 of *Legionella pneumophila* (also called the Mip protein) is localized at the outer membrane surface of this pathogenic organism. Its COOH-terminal domain is homologous with FKBP12, and its NH_2-terminal domain (residues 1–106) is probably required for attaching the protein to the cell membrane [63–66]. The infectivity of *L. pneumophila* is strongly reduced when FK 506 binds to this FKBP25. A 22-kDa FKBP, which is homologous with FKBP25, was also discovered recently in the periplasm of *E. coli* [67]. The cyclophilin Cyp40 binds to the heat-shock protein Hsp90 in glucocorticoid receptor complexes, and in this function resembles FKBP52 (also known as Hsp56), which is also found in these complexes [68,69]. Moreover, Cyp40 and FKBP52 share a related tetratricopeptide sequence that was suggested to function as a general association domain to allow protein–protein interactions. Probably Cyp40 and FKBP52 compete in their interaction with Hsp90. Their roles in the function of steroid receptor complexes are still unknown, but it seems unlikely that their prolyl isomerase activities are involved in folding events of the complex or its subunits [70]. Additional proteins with cyclophilin- or FKBP-related domains have been described [12,13,71]. Recently, the trigger factor of *E. coli* was added to the list of proteins with a FKBP domain. It is described further in Section III.F.

E. Parvulin

Recently, Rahfeld et al. discovered a novel, very small prolyl isomerase in *E. coli* that could neither be inhibited by CsA nor by FK 506 [72,73]. The authors called this enzyme "parvulin" (from the Latin word *parvulus*, which means very small). Parvulin and its gene were sequenced. With 92 amino acids, it is one of the smallest enzymes that are now known. It is not a fragment of a larger protein, and it shows no sequence homology with cyclophilins or FKBPs. Homologies were found with domains of PrsA from *Bacillus subtilis* [74,75], SurA from *E. coli* [76], and PrtM from *Lactococcus lactis* [77,78]. All these proteins seem to be involved in protein maturation, in particular of secreted proteins, and it has been suggested that mutations cause misfolding and defects in the function of other proteins. The SurA protein is located in the periplasm of *E. coli* and participates in early stages of the maturation of proteins of the outer membrane [79]. Also, it is involved in the pathway that signals the accumulation of unfolded proteins from the periplasm to the cytoplasm [80].

The prolyl isomerase activity of parvulin is high and, with a specificity constant k_{cat}/K_M of $1.69 \times 10^7\ M^{-1}s^{-1}$ for the substrate Suc-Ala-Leu-Pro-Phe-4-nitroanilide, it is as active as the cyclophilins of *E. coli* and much more active than the FKBPs [72]. Still it resembles the FKBPs, rather than the cyclophilins, in its strong preference for hydrophobic residues, such as leucine and phenylalanine, in the position preceding proline in the assay peptides [72,73].

F. Trigger Factor

The trigger factor is an abundant soluble protein of *E. coli* with an M_r of 48 kDa. It was discovered originally by Crooke and Wickner in 1987, when they searched for cytosolic components that are involved in the export of secretory proteins [81,82]. These authors found that trigger factor binds to and thus stabilizes the export-competent form of a (presumably unfolded or incompletely folded) precursor protein, proOmpA. In this function the trigger factor could be substituted by the canine signal recognition particle [83]. The trigger factor associated with the large subunit of the ribosome at low ionic strength and could be dissociated by adding salts, such as LiCl, at high concentration. The function of trigger factor in vivo remained unclear, for the export of proOmpA appeared to function normally in an *E. coli* strain in which the trigger factor was depleted to about 5% of its original level [84]. An important function of the trigger factor is suggested, however, by the finding that a homologue of the trigger factor gene exists in the extremely small, stream-lined genome of *Mycoplasma genitalium* [85].

Recently, the trigger factor was rediscovered by three groups, which used different experimental approaches to investigate early processes in cellular protein maturation. The groups of Bukau [17] and Luirink [16] incorporated photoactivatable chemical cross-linkers into nascent proteins and, after arresting translation, searched for proteins that possibly associate and are cross-linked with these newly formed protein chains while they are still associated with the ribosome. Both groups found cross-linking of presecretory proteins to only two proteins, the P48 component of the prokaryotic signal recognition particle and the trigger factor. Nonsecretory proteins were efficiently cross-linked to only the trigger factor.

Fischer and co-workers [15] searched for a prolyl isomerase activity at the *E. coli* ribosome. They discovered a prolyl isomerase that was specifically associated with the 50S subunit of the ribosome and could be dissociated by adding LiCl at concentrations higher than 1.5 M. Isolation and sequencing of proteolyic fragments identified this prolyl isomerase as the trigger factor. In its enzymatic activity this isomerase resembles the FKBPs. Although it does not seem to be inhibited by FK 506, it is reminiscent of FKBP in its subsite specificity toward the amino acid Xaa preceding proline in oligopeptides. Indeed, a weak sequence homology was noted between the 165–240 region of the trigger factor and human FKBP12 [17,86]. This homology is significant only for the residues that are necessary for substrate binding and for activity. Proteolytic fragments of the trigger factor, which encompass the putative FKBP domain (residues 132–247 and 145–252, respectively) and a recombinant form of the 142–251 fragment retained the full prolyl isomerase activity of the intact protein when assayed with proline-containing oligopeptides [87,88]. The trigger factor of *B. subtilis* is also a prolyl isomerase, and this protein is weakly inhibited by FK 506 [M. Marahiel, personal communication].

IV. CATALYSIS OF IN-VITRO PROTEIN FOLDING REACTIONS BY PROLYL ISOMERASES

A. Acceleration of Proline-Limited Folding

A catalysis of folding by a prolyl isomerase (porcine cytoplasmic cyclophilin 18) could first be demonstrated for the immunoglobulin light chain, porcine ribonuclease (RNase), and the S-protein fragment of bovine RNase A [10]. In all these the activity in protein folding was much lower than toward prolines in tetrapeptides. Initial attempts to catalyze the folding of intact RNase A and of thioredoxin were unsuccessful [11], although the folding reactions of both proteins involve proline-limited steps [29,89]. It is now clear that, during folding, prolyl peptide bonds can rapidly become inaccessible for prolyl isomerases because the onset of conformational folding is not blocked by an incorrect proline [6,7,37,38]. This is probably the major reason for the decreased activity of the cyclophilins in protein folding. Prolyl and hydroxyprolyl isomerizations are rate-limiting steps for the maturation of the collagen triple helix, and this folding reaction is also accelerated by cyclophilin 18 [90–92].

Meanwhile, a catalysis of proline-limited steps by various prolyl isomerases was observed in the folding of many proteins, including barnase [93], carbonic anhydrase [94,95], β-lactamase (A. Lejeune, unpublished results), chymotrypsin inhibitor CI2 [96], yeast iso-2-cytochrome *c* [97,98], the immunoglobulin light chain [10,99], and *trp* aporepressor [100]. Most of these proteins are small, and in the absence of prolyl isomerization they refold in the time range of milliseconds. Both *cis* $\rightarrow$ *trans*- and *trans* $\rightarrow$ *cis*-isomerizations are catalyzed equally well in the folding of these proteins. The role of the accessibility of the prolines was studied for iso-2-cytochrome *c*. In aqueous buffer the folding of this protein is barely catalyzed by cyclophilin. However, when GdnCl is added in increasing, but still nondenaturing concentrations, catalysis is markedly improved, presumably because the denaturant destabilizes folding intermediates and, thereby, improves the accessibility of the prolyl bonds [98].

Not all slow steps in protein folding are prolyl isomerizations. The very slow refolding rate of large proteins is often limited by other events, such as slow conformational rearrangements, domain-pairing reactions, or subunit associations. Whether prolyl isomerizations occur at earlier stages in the folding of such proteins remains to be demonstrated.

B. Ribonuclease T_1 as a Model Protein for Investigating Catalyzed Protein Folding

The RNase T_1 [101] proved to be a good model for investigating the mechanisms of proline-limited protein folding and of its catalysis by prolyl isomerases. Almost its entire folding involves prolyl isomerizations [102,103], both in the presence and in the absence of the two disulfide bonds [104,105], and it is a good substrate not only for prolyl isomerases, but also for other folding enzymes and chaperones [6,7]. The properties and the folding mechanism of this protein are briefly described in the next section.

RNase T_1 is a small single-domain protein of 104 amino acids [101,106, 107], with two disulfide bonds (Cys2–Cys10 and Cys6–Cys103). It contains four prolyl peptide bonds; two are *trans* (Trp59–Pro60 and Ser72–Pro73) and two are *cis* (Tyr38–Pro39 and Ser54–Pro55) in the native protein. RNase T_1 is most stable near pH 5 and is further stabilized when NaCl is added [108,109]. Importantly, in the absence of the disulfide bonds, RNase T_1 can still fold to a native-like conformation in the presence of 1 M NaCl. Variants of RNase T_1 have been designed to create simple models for investigating particular aspects of protein folding. These variants differ in the conformational stability and in the number and location of *cis*-prolyl bonds and of disulfide bonds [110–112].

The folding mechanism of the wild-type protein with the two *cis*-prolines and the two disulfide bonds intact has been studied most extensively [113].

In unfolding (Scheme 1) after the conformational unfolding step ($N \rightleftharpoons U^{39c}_{55c}$), the two *cis*-prolyl peptide bonds at Pro-39 and Pro-55 isomerize largely to the incorrect *trans*-conformation until both Pro-39 and Pro-55 are 80–90% *trans*. As a consequence four unfolded species are formed that differ in the isomeric states of Pro-39 and Pro-55.

The refolding of RNase T_1 initiates from these four unfolded species, its mechanism is shown in Scheme 2. The unfolded species U^{39c}_{55c} with both Pro-39

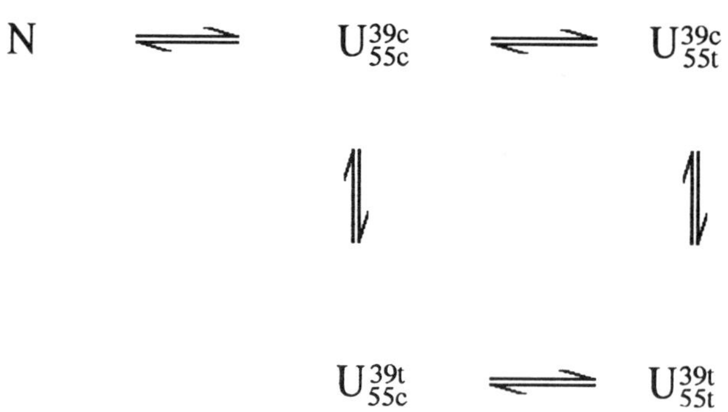

Scheme 1 Kinetic model for the unfolding and isomerization of RNase T_1. This model is valid for unfolding only. The superscript and the subscript indicate the isomeric states of prolines 39 and 55, respectively, in the correct, native-like *cis* (c)- and in the incorrect, nonnative *trans* (t)-isomeric states. As an example, U^{39t}_{55c} is an unfolded species with Pro-39 in the incorrect *trans*- and Pro-55 in the correct *cis*-state. In the denatured protein the two isomerizations are independent of each other; therefore, the scheme is symmetric with identical rate constants in the horizontal and vertical directions, respectively.

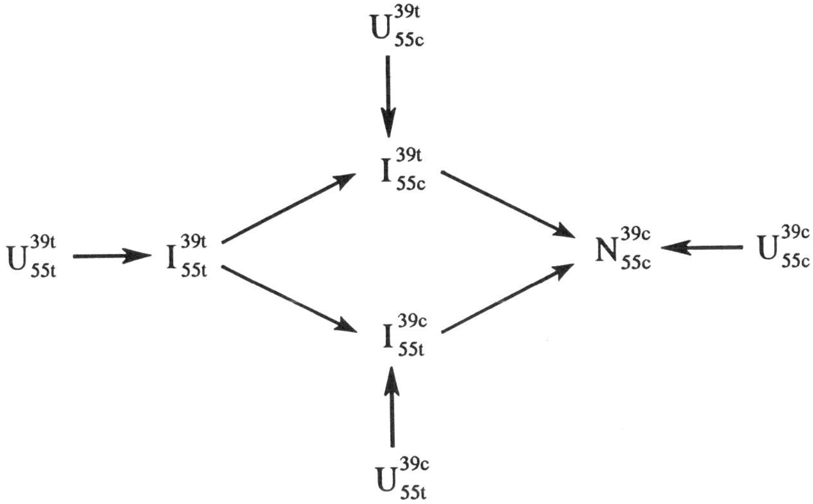

Scheme 2 Kinetic model for refolding of RNase T_1 under native conditions. U stands for unfolded species, I for intermediates of refolding, and N for native protein. The superscript and the subscript indicate the isomeric states of prolines 39 and 55, respectively, in the correct *cis* (c)- and the incorrect *trans* (t)-isomeric states.

and Pro-55 in the correct *cis*-conformation refolds to the native state N in less than a second [113]. The protein chains with incorrect prolyl isomers (U^{39c}_{55t}, U^{39t}_{55c}, U^{39t}_{55t}) also regain most of their secondary structure and, presumably, part of their tertiary structure early in refolding. The products of these reactions, however, are partially folded intermediates and still contain incorrect prolines (see the I_i species in Scheme 2). The subsequent slow-folding steps involve the reisomerizations of the incorrect prolyl isomers and are coupled with further folding. Scheme 2 is valid for folding only under strongly native conditions, for which intermediates are populated, and reverse reactions are not significant.

Conformational folding and the reisomerizations of the prolyl peptide bonds are tightly interrelated in the folding of RNase T_1. Very rapid partial folding is possible in the presence of nonnative proline isomers [114,115], but the correct isomers are required for the final events of folding, which therefore, are slow and rates limited by the prolyl isomerizations. Ordered structure in the intermediates affects the isomerization kinetics. Surprisingly, in RNase T_1, the *trans* → *cis* isomerization at Pro-39 (in the $I^{39t}_{55c} \rightarrow N$ step) is strongly decelerated by the folded structure that is already present in the intermediate I^{39t}_{55c} [114]. In the folding of pancreatic RNase A prolyl isomerization is accelerated in a folding intermediate [37,38], and in the folding of dihydrofolate reductase (DHFR) an intramolecular catalysis of a prolyl isomerization was proposed to occur [116].

The two disulfide bonds of RNase T_1 can be broken by reduction and subsequent carboxymethylation. The resulting reduced and carboxymethylated (RCM)-RNase T_1 is unfolded in aqueous buffer, but reversible folding to a native-like ordered conformation can be induced by adding 1 M, or higher, NaCl [117]. The RCM form of the S54G/P55N variant [(-P55)–RCM–T1] is a particularly valuable model protein for studying assisted protein folding. Its folding mechanism is very simple, because it involves only a single *trans* → *cis*-isomerization (of Pro-39) and because the access to this proline is not impaired by premature structure formation, as in the presence of the disulfide bonds. Because all RCM forms are unfolded in aqueous buffer, spontaneous as well as assisted and catalyzed unfolding and refolding can be studied in the absence of denaturants, simply by varying the NaCl concentration [118,119]. This is important because several prolyl isomerases are very sensitive to residual concentrations of denaturants, such as guanidinium chloride or urea.

C. Catalysis of RNase T_1 Folding by Cyclophilins and FKBPs

In the folding of RNase T_1 with intact disulfide bonds, all slow phases are catalyzed by prolyl isomerases, albeit with varying efficiency. Catalysis at Pro-55 is generally good, because this proline is solvent-exposed in the native protein and, presumably, also in folding intermediates. Pro-39 is buried in the native protein. Its isomerization is not well catalyzed; rapid conformational folding blocks the access of prolyl isomerases. Catalysis of Pro-39 isomerization is strongly improved when this rapid folding is suppressed by destabilizing mutations, or by breaking of the two disulfide bonds (as in the reduced and carboxymethylated RCM form of RNase T_1). Partially folded intermediates can no longer form in these variants; consequently, prolyl isomerases are not hindered in their access to the refolding protein molecules [117].

Refolding of RNase T_1 is catalyzed by prolyl isomerases of the cyclophilin and the FKBP families from a variety of organisms. In general, their activities were similar [7]. The best catalysis of Pro-39 isomerization in the refolding of RNase T_1 with intact disulfide bonds was observed with the cytoplasmic cyclophilin 18 from *E. coli*. In the presence of 29 μM cyclophilin 18, the rate of this folding reaction was 300-fold increased [120].

Prolyl isomerases function in protein folding as enzymes. They catalyze *cis* ⇌ *trans*-isomerization in either direction and do not determine the isomeric states of the prolyl peptide bonds in the protein substrates. Catalysis of prolyl isomerization in protein *unfolding* cannot be measured easily, because prolyl isomerases are inactivated by denaturants and at elevated temperatures. These problems could be overcome by using the RCM form of RNase T_1 as a substrate [117]. As outlined earlier, this protein is folded only in the presence of more than 1 M NaCl and can be unfolded by dilution to, for example, 0.4 M NaCl at pH 8 and 15°C. Prolyl isomerases are stable under these conditions and retain their full enzymatic activity. Unfolding, or to be more precise, prolyl *cis* ⇌ *trans* isomerization in the unfolding chains of RCM-RNase T_1 was indeed catalyzed and the efficiency of Cyp18 was the same in unfolding and refolding experiments under identical conditions near the midpoint of the unfolding transition. Similar results were also obtained for the catalysis of folding by FKBP12 [C. Scholz and F. X. Schmid, unpublished data]. The extents of unfolding in the kinetic experiments with and without prolyl isomerase were identical, indicating that these enzymes do not affect the equilibrium between native and unfolded protein. Clearly, prolyl isomerases have no directional information for folding. The isomeric states of the prolines are determined only by the structure and the conformational stability of the folding protein under the employed conditions.

D. Catalysis by Parvulins

An additional family of prolyl isomerases, the parvulins, was discovered by Rahfeld et al. [72]. These small proteins show no sequence homologies with the cyclophilins and the FKBPs, and they are not inhibited by the respective inhibitors CsA and FK 506. Parvulin from *E. coli* also catalyzes the folding and unfolding of our model protein (-P55)–RCM–T_1, and it shows an efficiency that is intermediate between the cyclophilins and the FKBPs [C. Scholz and F. X. Schmid, unpublished results].

E. Autocatalytic Folding of a Prolyl Isomerase

As folding enzymes the prolyl isomerases have the potential to catalyze their own folding. The folding of human cytosolic FKBP12 is indeed an autocatalytic process for the mature protein and is more pronounced for a variant with an NH_2-terminal extension of 16 residues [121,122]. Native FKBP contains seven *trans*-prolyl peptide bonds, and the *cis* → *trans*-isomerizations of some or all of them determine the rate of its folding. In an autocatalytic reaction the product catalyzes its own formation; therefore, the rate should increase with reactant concentration. This was indeed observed in the folding of the FKBP12 fusion protein: it was more than tenfold accelerated when the protein concentration was increased from 0.05 to 10 μM [121].

The seven *trans*-prolyl bonds of native FKBP12 are flanked by different residues [52]. They probably do not remain equally accessible during folding, and only two of them (F15–P16 and I91–P92) should be good substrates for the native FKBP molecules. Therefore, in its self-catalyzed refolding, FKBP12 probably accelerates the prolyl isomerizations with different efficiencies, and it is not possible to analyze the autocatalytic folding of FKBP in more detail. First results indicate that the refolding of parvulin is also an autocatalytic process [C. Scholz, unpublished observation].

As mentioned, many large proteins (including heat-shock proteins) contain FKBP-, cyclophilin-, or parvulin-like domains [13,73,123,124], but the functions of these prolyl isomerase modules are unknown. It is tempting to speculate that they serve a role as intermolecular or even intramolecular catalysts of the folding or of conformational rearrangements that are necessary for the biological functions of these proteins. However, there is as yet no experimental evidence in support of such a role.

V. THE TRIGGER FACTOR AS A FOLDING ENZYME

A. Prolyl Isomerase Activity

The trigger factor is located at the ribosome and can interact with newly formed protein chains. Therefore, it could be the prime candidate for a folding enzyme that accelerates prolyl isomerizations in the de novo folding of nascent proteins. In fact, trigger factor is much more efficient as a catalyst of RNase T_1 folding than all the other prolyl isomerases yet known. Typically, the prolyl isomerase activities of the cyclophilins and FKBPs are much lower in protein folding than toward exposed prolyl peptide bonds in short tetrapeptides. In contrast, trigger factor is more active in protein folding than in assays with proline-containing tetrapeptides. The addition of as low as 2.5 nM trigger factor leads to a doubling of the folding rate of (-P55)–RCM–T_1, and in the presence of 20 nM trigger factor this folding reaction is 14-fold accelerated. This remarkable catalytic efficiency of the trigger factor as a folding enzyme is reflected in a specificity constant k_{cat}/K_M of 1.1×10^6 M^{-1} s^{-1}, which is almost 100-fold higher than this value for human FKBP12 [125].

B. Enzyme Kinetics of Trigger–Factor-Catalyzed Folding

To find out why trigger factor is such an efficient folding enzyme, the K_M and k_{cat} of catalyzed folding were determined separately [125]. In these experiments, the trigger factor behaved like a classic enzyme. The initial rates of catalyzed folding showed saturation behavior and obeyed the Michaelis–Menten equation. The analysis of the kinetic data gave values of 0.7 μM for the K_M value and 1.3 s^{-1} for the catalytic rate constant k_{cat} of trigger–factor-catalyzed folding. Clearly, the high catalytic efficiency of trigger factor originates from the tight binding to its substrate, as reflected in the low K_M value. The k_{cat} value of 1.3 s^{-1} is very small. For the catalysis by cyclophilin 18 of the *trans* → *cis*-prolyl isomerization in a tetrapeptide K_M and k_{cat} values of 220 μM and 620 s^{-1}, respectively, were obtained [44]. This comparison indicates that the high activity of the trigger factor as a folding catalyst does not originate from a high turnover number, but from a good affinity for the protein substrate. For other prolyl isomerases such a kinetic analysis cannot be performed, because their binding to protein or peptide substrates is very weak.

C. Chaperone Properties of the Trigger Factor

Other unfolded proteins, such as the reduced and carboxymethylated (RCM) form of bovine α-lactalbumin (RCM-La), interfere with substrate binding to the trigger factor and inhibit the catalysis of folding. RCM-La, which is also a good substrate for the chaperone GroEL [126,127], competes efficiently with RCM-T_1 for binding to the trigger factor. This inhibition follows a saturation curve, and at 0.6 μM RCM-La the trigger factor is 50% inhibited. When the Michaelis–Menten kinetics of catalyzed folding was measured in the presence of 1.0 μM RCM-La, the observed curve was significantly shifted to higher substrate concentrations. This is indicative of competitive inhibition and suggests that the folding substrate RCM-T1 and the nonfolding inhibitor RCM-La compete for the same binding site on the trigger factor.

The high-affinity binding site for unfolded proteins is probably distinct from the catalytic site of the trigger factor. The isolated FKBP domain of the trigger factor is fully active as a prolyl isomerase toward a short tetrapeptide [87], but in protein folding its activity is about 800-fold reduced and, moreover, this low-residual activity of the FKBP domain is no longer inhibited by RCM-La. This suggests that the high-affinity binding site for unfolded substrate proteins extends over several domains of the intact trigger factor or requires the interaction of these domains.

The additional protein-binding site(s) on the intact trigger factor probably decelerate the dissociation of the protein substrate from the trigger factor, and the low k_{cat} value of 1.3 s^{-1} may reflect a change in the rate-limiting step from bond rotation (in tetrapeptide substrates) to product dissociation (in protein substrates). Additionally, it is quite likely that some of the binding events are nonproductive because the reactive prolyl peptide bonds are not positioned correctly within the prolyl isomerase site. Indeed, a lowering of both K_M and k_{cat}, as observed for the intact trigger factor, points to nonproductive binding of a substrate to an enzyme [128].

The trigger factor thus seems to have properties of a folding enzyme and of a chaperone. The cooperation of both functions is required for its very high-catalytic efficiency in protein folding. It is not yet known whether the chaperone properties of the trigger factor are also required for additional functions, such as accepting newly synthesized protein chains at the ribosome.

Prolyl isomerizations are presumed to be late steps in protein folding [36] and, therefore, it seems surprising that a prolyl isomerase binds very early, possibly cotranslationally, to a nascent protein chain. The trigger factor might, however, remain associated with the folding chains and could even accompany the folding protein chain to other chaperone systems, such as GroEL. Indeed, there is evidence that the trigger factor can bind to GroEL in a substrate-dependent manner [129]. It is also possible that not all newly formed proteins require the DnaK or the GroE chaperone systems and, consequently, might fold rapidly to completion right after their synthesis at or near the ribosomes.

VI. CATALYSIS OF PROLYL ISOMERIZATION DURING DE NOVO PROTEIN FOLDING

Early evidence for an involvement of prolyl isomerization and of prolyl isomerases in cellular folding came from studies on the folding of collagen [130]. The formation of the collagen triple helix is limited in rate by successive prolyl isomerizations both in vitro and in vivo. Collagen is resistant to trypsin in the native, but not in the unfolded state, and triple helix formation can thus be followed by protease pulse experiments. Steinmann et al. [130] found that the maturation of collagen in chick embryo fibroblasts is retarded when CsA is present, and they proposed that CsA inhibits a prolyl isomerase of the cyclophilin type that catalyzes collagen folding in the endoplasmic reticulum. A similar effect was found for the folding of luciferase in rabbit reticulocyte lysate [131]. Effects of CsA on proteins other than cyclophilins could not be ruled out in these experiments.

Recently, Rassow et al. [132] and Matouschek et al. [133] presented convincing evidence for a catalysis of prolyl isomerization by a cyclophilin in cellular protein folding. In an "in-organello" strategy, both groups used the mitochondrial protein-import system to investigate the function of cyclophilins as potential catalysts of protein folding. Proteins that are targeted to the mitochondrial matrix must unfold outside the mitochondria, cross the two mitochondrial membranes, and then refold in the matrix. To study this refolding reaction, both groups used a synthetic precursor protein in which the presequence of subunit 9 of the *Neurospora crassa* F_1F_0-ATPase was linked to mouse cytosolic dihydrofolate reductase (Su9-DHFR). This construct is efficiently imported into the matrix of isolated mitochondria and, therefore, it has long been used as a standard substrate to study the mechanism of mitochondrial protein import.

The two groups produced ^{35}S-labeled Su9-DHFR in an in vitro transcription–translation experiment, unfolded it by incubation in 8-M urea, and then diluted it into suspensions of mitochondria from yeast and from *N. crassa.* Su9-DHFR was rapidly imported into the matrix, the presequence was cleaved, and the DHFR moiety refolded inside the mitochondria. DHFR is rapidly degraded by protease K when it is still in nonnative states, but not when it is native-like folded.

When measured by the resistance against proteinase K, the refolding of

DHFR showed half times of about 5 min in yeast mitochondria at 30°C as well as in *N. crassa* mitochondria at 25°C. In both cases, folding was about fivefold decelerated when the mitochondria had been preincubated with 2.5–5 μM CsA to inhibit the prolyl isomerase activity of mitochondrial cyclophilin. The rates of in vitro folding of Su9-DHFR differed to a similar extent in the presence and in the absence of the purified cyclophilin from *N. crassa* mitochondria. Intriguingly, in the yeast mitochondria, folding was also retarded by a nonimmunosuppressive variant of cyclosporine, and stable complexes of cyclophilin with other proteins or refolding precursors could not be detected. This result confirms that cyclophilin acts as an enzyme in mitochondrial folding, rather than as a binding protein (as in immunosuppression).

The refolding of DHFR was also retarded in mitochondria that were derived from yeast and *N. crassa* mutants that lacked a functional mitochondrial cyclophilin, and the kinetics of Su9-DHFR refolding in the mutant mitochondria were almost identical with the refolding kinetics in the presence of CsA in the wild-type mitochondria. As expected, there was no further deceleration of these folding reactions by CsA. This virtually completed the evidence that the mitochondrial cyclophilins acted as prolyl isomerases and, thereby, catalyzed protein folding in organello.

VII. ISOMERIZATION OF PEPTIDE BONDS NOT PRECEDING PROLINE

For normal peptide bonds (not preceding proline) the *trans*-state is favored at least 100-fold over *cis* [18,19], which implies that in unfolded proteins, on average, one incorrect *cis*-peptide bond occurs in every 100 residues. Therefore it would be interesting to know, whether the cis → trans isomerizations of such peptide bonds occur in protein folding and whether they also can be catalyzed by enzymes, such as the prolyl isomerases. Because of the strong preference for the *trans*-state, it is virtually impossible to design peptide substrates with *cis*-peptide bonds to search for such a function. Also, until recently, no protein-folding reactions were known that are limited in rate by the isomerization of normal peptide bonds.

In RNase T1 a *cis*-Tyr–Ala bond could be generated by the replacement of *cis*-Pro-39 by an alanine [112]. This mutation causes a major change in the folding mechanism of this protein, and the very slow *trans* → *cis*-isomerization of the Tyr38–Ala39 peptide bond now determines the refolding rate of the Pro-39–Ala variant [103]. The time constant of this reaction is about 500 s at 25°C. None of the known prolyl isomerases (cyclophilin 18, FKBP12, and parvulin) could catalyze this folding reaction, suggesting that normal peptide bonds are not substrates for these enzymes [C. Scholz and F. X. Schmid, unpublished observations].

Because the *trans*-state of a normal peptide bond is at least 100-fold more highly populated than the *cis*-state, the *cis* → *trans*-reaction is at least 100-fold faster than the trans → *cis*-reaction. This suggests that the *cis* → *trans*-isomerization of the very few peptide bonds that on average, might be *cis* in an unfolded protein, should occur in the time range of a second and, therefore, catalysis is probably not necessary. In contrast, the reverse *trans* → *cis*-isomerization is very slow and should be a rate-limiting step in the folding of the (very few) proteins that contain *cis*-peptide bonds [134] in their native conformations.

VIII. CONCLUDING REMARKS

Proteins differ strongly in size, in physicochemical properties, and also in the rate and the mechanism of folding. Although some small proteins reach the native state extremely fast, in less than a millisecond [135–138], others need hours to complete folding [139–141]. As a consequence, different proteins are expected to require different levels of help during their folding in the cell. When protein folding is investigated in vitro, major problems are often encountered when the temperature is increased. The stability decreases, and folding becomes progressively irreversible, because the aggregation of partially folded intermediates competes with productive folding [142–144]. Undoubtedly, aggregation is also a major, unwanted side reaction in cellular folding [145–147]. It is suppressed by the rapid binding of unfolded or incompletely folded protein chains to chaperones, such as GroEL or DnaK [148,149]. Interestingly, most of the chaperones are heat-shock proteins, as expected if their function is to prevent aggregation of partially folded or partially denatured proteins at elevated temperature.

Another aspect of folding is often overlooked. As every chemical reaction, folding becomes slower with decreasing temperature and, in particular, proline-limited folding reactions are strongly decelerated at low temperature, because prolyl *cis* ⇌ *trans*-isomerization shows a high activation energy of 80–90 kJ mol^{-1} [7,20,36]. The half time of prolyl isomerization is in the range of a few seconds at 37°C, but it increases to several hundred seconds when the temperature is lowered to 10°C. Therefore, a catalysis of proline-limited folding steps should be required at low temperature, rather than under heat-shock conditions, which are normally used to characterize the function of cellular-folding helpers.

It is not yet known whether there is a functional interplay between the prolyl isomerases and the chaperone systems. It is now generally assumed that during Hsp60-mediated folding the proteins go through several cycles of binding and ATP-dependent release. Because they accelerate productive folding, the prolyl isomerases should be able to reduce the number of binding and release cycles and thus make folding more efficient and less energy-consuming [132].

The discovery of the high prolyl isomerase activity of the trigger factors [15] is exciting. These proteins associate with translating ribosomes [16,17], bind well to unfolded proteins, and catalyze proline-limited–folding steps extremely well. In this way, trigger factors seem to combine properties of folding catalysts and of chaperones. Moreover, there is evidence that the trigger factor of *E. coli* functionally interacts with another chaperone GroEL, in a substrate-dependent fashion [129]. This dual location at the ribosome as well as at GroEL might reflect that some polypeptide chains can fold to completion immediately after their synthesis, whereas others require help from the GroE system.

It is now clear that every organism contains a multitude of prolyl isomerases, and undoubtedly, these enzymes have important functions that are not restricted to de novo protein folding. The link with immunosuppression [14], the interaction of a host cyclophilin with the *Gag* polyprotein of HIV-1 [150,151], the stabilization of a calcium-release channel by a FK 506-binding protein [152], and the occurrence of cyclophilin-, FKBP-, and parvulin-like domains in large proteins [153–155], all point to additional functions [see Refs. 12 and 13 for reviews]. Unlike the catalysis of folding, some of these functions seem to involve tight binding to the target molecules. Whether such a binding can change the *cis* → *trans*-isomerism of prolines at the protein surface is not yet known.

In folded proteins proline residues occur most frequently in solvent-exposed turns, they are often preceded by a serine and, in fact, many regulatory protein kinases phosphorylate such serine residues. It is unknown whether, similar to proteases, these kinases show a conformational specificity toward the *cis* or the *trans* isomer of the Ser–Pro bond. It is possible that *cis* → *trans*-prolyl isomerization and its catalysis occur not only during protein folding, but also are used for a tunable conformational switching in the regulation of the activity of native proteins. However, experimental evidence for such proline switches is still lacking.

REFERENCES

1. Stewart, D. E., Sarkar, A., and Wampler, J. E., Occurrence and role of *cis* peptide bonds in protein structures, *J. Mol. Biol. 214*:253–260 (1990).
2. MacArthur, M. W. and Thornton, J. M., Influence of proline residues on protein conformation, *J. Mol. Biol. 218*:397–412 (1991).
3. Stein, R. L., Mechanism of enzymatic and nonenzymatic prolyl *cis–trans* isomerization, *Adv. Protein Chem. 44*:1–24 (1993).
4. Brandts, J. F., Halvorson, H. R., and Brennan, M., Consideration of the possibility that the slow step in protein denaturation reactions is due to *cis–trans* isomerism of proline residues, *Biochemistry 14*:4953–4963 (1975).
5. Kim, P. S. and Baldwin, R. L., Specific intermediates in the folding reactions of small proteins and the mechanism of protein folding, *Annu. Rev. Biochem. 51*:459–489 (1982).
6. Schmid, F. X., Prolyl isomerase—enzymatic catalysis of slow protein-folding reactions, *Annu. Rev. Biophys. Biomol. Struct. 22*:123–143 (1993).

7. Schmid, F. X., Mayr, L. M., Mücke, M., and Schönbrunner, E. R., Prolyl isomerases: Role in protein folding, *Adv. Protein Chem.* *44*:25–66 (1993).
8. Kim, P. S. and Baldwin, R. L., Intermediates in the folding reactions of small proteins, *Annu. Rev. Biochem.* *59*:631–660 (1990).
9. Fischer, G., Bang, H., and Mech, C., Nachweis einer Enzymkatalyse für die *cis–trans*-Isomerisierung der Peptidbindung in prolinhaltigen Peptiden, *Biomed. Biochim. Acta* *43*:1101–1111 (1984).
10. Lang, K., Schmid, F. X., and Fischer, G., Catalysis of protein folding by prolyl isomerase, *Nature* *329*:268–270 (1987).
11. Lin, L.-N., Hasumi, H., and Brandts, J. F., Catalysis of proline isomerization during protein folding reactions. *Biochim. Biophys. Acta* *956*:256–266 (1988).
12. Galat, A. and Metcalfe, S. M., Peptidylproline *cis/trans* isomerases, *Prog. Biophys. Mol. Biol.* *63*:67–118 (1995).
13. Fischer, G., Peptidyl–prolyl *cis/trans* isomerases and their effectors, *Angew. Chem. Int. Ed.* *33*:1415–1436 (1994).
14. Schreiber, S. L., Chemistry and biology of the immunophilins and their immunosuppressive ligands, *Science* *251*:283–287 (1991).
15. Stoller, G., Rücknagel, K. P., Nierhaus, K., Schmid, F. X., Fischer, G., and Rahfeld, J.-U., Identification of the peptidyl–prolyl *cis/trans* isomerase bound to the *Escherichia coli* ribosome as the trigger factor. *EMBO J.* *14*:4939–4948 (1995).
16. Valent, Q. A., Kendall, D. A., High, S., Kusters, R., Oudega, B., and Luirink, J., Early events in preprotein recognition in *E. coli*: Interaction of SRP and trigger factor with nascent polypeptides, *EMBO J.* *14*:5494–5505 (1995).
17. Hesterkamp, T., Hauser, S., Lütcke, H., and Bukau, B., *Escherichia coli* trigger factor is a prolyl isomerase that associates with nascent polypeptide chains, *Proc. Natl. Acad. Sci. USA* *93*:4437–4441 (1996).
18. Ramachandran, G. N. and Mitra, A. K., An explanation for the rare occurrence of *cis* peptide units in proteins and polypeptides. *J. Mol. Biol.* *107*:85–92 (1976).
19. Jorgensen, W. L. and Gao, J., *Cis–trans* energy difference for the peptide bond in the gas phase and in aqueous solution. *J. Am. Chem. Soc.* *110*:4212–4216 (1988).
20. Cheng, H. N. and Bovey, F. A., *Cis–trans* equilibrium and kinetic studies of acteyl-L-proline and glycyl-L-proline. *Biopolymers* *16*:1465–1472 (1977).
21. Grathwohl, C. and Wüthrich, K., NMR studies of the rates of proline *cis–trans* isomerization in oligopeptides. *Biopolymers* *20*:2623–2633 (1981).
22. Evans, P. A., Dobson, C. M., Kautz, R. A., Hatfull, G., and Fox, R. O., Proline isomerism in staphylococcal nuclease characterized by NMR and site-directed mutagenesis, *Nature* *329*:266–268 (1987).
23. Higgins, K. A., Craik, D. J., Hall, J. G., and Andrews, P. R., *Cis–trans* isomerization of the proline residue in insulin studied by ^{13}C NMR spectroscopy, *Drug Design Deliv.* *3*:159–170 (1988).
24. Chazin, W. J., Kördel, J., Drakenberg, T., Thulin, E., Brodin, P., Grundström, T., and Forsén, S., Proline isomerism leads to multiple folded conformations of calbindin D9k: Direct evidence from two-dimensional NMR spectroscopy, *Proc. Natl. Acad. Sci. USA* *86*:2195–2198 (1989).
25. Kördel, J., Forsén, S., Drakenberg, T., and Chazin, W. J., The rate and structural consequences of proline *cis–trans* isomerization in calbindin D9k: NMR studies of

the minor (cisPro43)isoform and the Pro43Gly mutant, *Biochemistry 29*:4400–4409 (1990).

26. Alexandrescu, A. T., Ulrich, E. L., and Markley, J. L., Hydrogen-1 NMR evidence for three interconverting forms of staphylococcal nuclease: Effects of mutations and solution conditions on their distributions, *Biochemistry 28*:204–211 (1989).
27. Alexandrescu, A. T., Hinck, A. P., and Markley, J. L., Coupling between local structure and global stability of a protein: Mutants of stapylococcal nuclease, *Biochemistry 29*:4516–4525 (1990).
28. Eberhardt, E. S., Loh, S. N., Hinck, A. P., and Raines, R. T., Solvent effects on the energetics of prolyl peptide bond isomerization, *J. Am. Chem. Soc. 114*:5437–5439 (1992).
29. Schmid, F. X. and Baldwin, R. L., Acid catalysis of the formation of the slow-folding species of RNase A: Evidence that the reaction is proline isomerization. *Proc. Natl. Acad. Sci. USA 75*:4764–4768 (1978).
30. Fischer, S., Dunbrack, R. L., and Karplus, M., *Cis–trans* imide isomerization of the proline dipeptide, *J. Am. Chem. Soc. 116*:11931–11937 (1994).
31. Van Duyne, G. D., Standaert, R. F., Karplus, P. A., Schreiber, S. L., and Clardy, J., Atomic structures of the human immunophilin FKBP-12 complexes with FK 506 and rapamycin, *J. Mol. Biol. 229*:105–124 (1993).
32. Ke, H. M., Zydowsky, L. D., Liu, J., and Walsh, C. T., Crystal structure of recombinant human T-cell cyclophilin-A at 2.5-A resolution, *Proc. Natl. Acad. Sci USA 88*:9483–9487 (1991).
33. Ke, H. M., Similarities and differences between human cyclophilin-A and other beta-barrel structures—structural refinement at 1.63 angstrom resolution, *J. Mol. Biol. 228*:539–550 (1992).
34. Konno, M., Ito, M., Hayano, T., and Takahashi, N., The substrate-binding site in *Escherichia coli* cyclophilin A preferably recognizes a *cis*-proline isomer or a highly distorted form of the *trans* isomer, *J. Mol. Biol. 256*:897–908 (1996).
35. Garel, J. R. and Baldwin, R. L., Both the fast and slow refolding reactions of ribonuclease A yield native enzyme, *Proc. Natl. Acad. Sci. USA 70*:3347–3351 (1973).
36. Schmid, F. X., Kinetics of unfolding and refolding of single-domain proteins, *Protein Folding* (T. E. Creighton, ed.), W. H. Freeman, New York, 1992, pp. 197–241.
37. Cook, K. H., Schmid, F. X., and Baldwin, R. L., Role of proline isomerization in folding of ribonuclease A at low temperatures, *Proc. Natl. Acad. Sci. USA 76*:6157–6161 (1979).
38. Schmid, F. X. and Blaschek, H., A native-like intermediate on the ribonuclease A folding pathway. 2. Comparison of its properties to native ribonuclease A, *Eur. J. Biochem. 114*:111–117 (1981).
39. Kofron, J. L., Kuzmic, P., Kishore, V., Colón-Bonilla, E., and Rich, D. H., Determination of kinetic constants for peptidyl prolyl *cis–trans* isomerases by an improved spectrophotometric assay, *Biochemistry 30*:6127–6134 (1991).
40. Kofron, J. L., Kuzmic, P., Kishore, V., Gemmecker, G., Fesik, S. W., and Rich, D. H., Lithium chloride perturbation of *cis–trans* peptide bond equilibria—effect on

conformational equilibria in cyclosporin-A and on time-dependent inhibition of cyclophilin, *J. Am. Chem. Soc. 114*:2670–2675 (1992).

41. Garcia-Echeverria, C., Kofron, J. L., Kuzmic, P., Kishore, V., and Rich, D. H., Continuous fluorimetric direct (uncoupled) assay for peptidyl prolyl *cis–trans*-isomerases, *J. Am. Chem. Soc. 114*:2758–2759 (1992).
42. Garcia-Echeverria, C., Kofron, J. L., Kuzmic, P., and Rich, D. H., A continuous spectrophotometric direct assay for peptidyl prolyl *cis–trans*-isomerases, *Biochem. Biophys. Res. Commun. 191*:70–75 (1993).
43. Hübner, D., Drakenberg, T., Forsén, S., and Fischer, G., Peptidyl-prolyl *cis–trans* isomerase activity as studied by dynamic proton NMR spectroscopy, *FEBS Lett. 284*:79–81 (1991).
44. Kern, D., Kern, G., Scherer, G., Fischer, G., and Drakenberg, T., Kinetic analysis of cyclophilin-catalyzed prolyl *cis/trans* isomerization by dynamic NMR spectroscopy, *Biochemistry 34*:13594–13602 (1995).
45. Fischer, G., Wittmann-Liebold, B., Lang, K., Kiefhaber, T., and Schmid, F. X., Cyclophilin and peptidyl–prolyl-*cis/trans*-isomerase are probably identical proteins, *Nature 337*:476–478 (1989).
46. Takahashi, N., Hayano, T., and Suzuki, M., Peptidyl–prolyl *cis–trans* isomerase is the cyclosporin A-binding protein cyclophilin, *Nature 337*:473–475 (1989).
47. Handschumacher, R. E., Harding, M. W., Rice, J., and Drugge, R. J., Cyclophilin: A specific cytosolic binding protein for cyclosporin A, *Science 226*:544–547 (1984).
48. Kallen, J., Spitzfaden, C., Zurini, M. G. M., Wider, G., Widmer, H., Wüthrich, K., and Walkinshaw, M. D., Structure of human cyclophilin and its binding site for cyclosporin-A determined by X-ray crystallography and NMR spectroscopy, *Nature 353*:276–279 (1991).
49. Kallen, J. and Walkinshaw, M. D., The X-ray structure of a tetrapeptide bound to the active site of human cyclophilin-A, *FEBS Lett. 300*:286–290 (1992).
50. Spitzfaden, C., Weber, H. P., Braun, W., Kallen, J., Wider, G., Widmer, H., Walkinshaw, M. D., and Wüthrich, K., Cyclosporin-A–cyclophilin complex formation—a model based on X-ray and NMR data, *FEBS Lett. 300*:291–300 (1992).
51. Mikol, V., Kallen, J., and Walkinshaw, M. D., X-ray structure of a cyclophilin B cyclosporin complex: Comparison with cyclophilin A and delineation of its calcineurin-binding domain, *Proc. Natl. Acad. Sci. USA 91*:5183–5186 (1994).
52. Siekierka, J. J., Hung, S. H. Y., Poe, M., Lin, C. S., and Sigal, N. H., A cytosolic binding protein for the immunosuppressant FK 506 has peptidyl–prolyl isomerase activity but is distinct from cyclophilin, *Nature 341*:755–757 (1989).
53. Harding, M. W., Galat, A., Ueling, D. E., and Schreiber, S. L., A receptor for the immunosuppressant FK 506 is a *cis–trans* peptidyl–prolyl isomerase, *Nature 341*:758–760 (1989).
54. Tropschug, M., Wachter, E., Mayer, S., Schönbrunner, E. R., and Schmid, F. X., Isolation and sequence of an FK 506-binding protein from *N. crassa* which catalyses protein folding, *Nature 346*:674–676 (1990).
55. Michnick, S. W., Rosen, M. K., Wandless, T. J., Karplus, M., and Schreiber, S. L., Solution structure of FKBP, a rotamase enzyme and receptor for FK 506 and rapamycin, *Science 252*:836–842 (1991).
56. Moore, J. M., Peattie, D. A., Fitzgibbon, J., and Thomson, J. A., Solution structure

of the major binding protein for the immunosuppressant FK 506, *Nature 351*:248–250 (1991).
57. Van Duyne, G. D., Standaert, R. F., Schreiber, S. L., and Clardy, J., Atomic structure of the rapamycin human immunophilin FKBP-12 complex, *J. Am. Chem. Soc. 113*:7433–7434 (1991).
58. Harrison, R. K. and Stein, R. L., Substrate specificities of the peptidyl prolyl *cis–trans* isomerase activities of cyclophilin and FK-506 binding protein: Evidence for the existence of a family of distinct enzymes, *Biochemistry 29*:3813–3816 (1990).
59. Liu, J., Farmer, J. D., Lane, W. S., Friedman, J., Weissman, I., and Schreiber, S. L., Calcineurin is a common target of cyclophilin–cyclosporin-A and FKBP–FK 506 complexes, *Cell 66*:807–815 (1991).
60. Stamnes, M. A., Shieh, B. H., Chuman, L., Harris, G. L., and Zuker, C. S., The cyclophilin homolog NINAA is a tissue-specific integral membrane protein required for the proper synthesis of a subset of *Drosophila* rhodopsins, *Cell 65*:219–227 (1991).
61. Colley, N. J., Baker, E. K., Stamnes, M. A., and Zuker, C. S., The cyclophilin homolog NINAA is required in the secretory pathway, *Cell 67*:255–263 (1991).
62. Rutherford, S. L. and Zuker, C. S., Protein folding and the regulation of signaling pathways, *Cell 79*:1129–1132 (1994).
63. Fischer, G., Bang, H., Ludwig, B., Mann, K., and Hacker, J., Mip protein of *Legionella pneumophila* exhibits peptidyl–prolyl-*cis/trans* isomerase (PPlase) activity, *Mol. Microbiol. 6*:1375–1383 (1992).
64. Ludwig, B., Rahfeld, J., Schmidt, B., Mann, K., Wintermeyer, E., Fischer, G., and Hacker, J., Characterization of Mip proteins of *Legionella pneumophila*, *FEMS Microbiol. Lett. 118*:23–30 (1994).
65. Schmidt, B., Rahfeld, J., Schierhorn, A., Ludwig, B., Hacker, J., and Fischer, G., A homodimer represents an active species of the peptidyl–prolyl *cis/trans* isomerase FKBP25mem from *Legionella pneumophila*, *FEBS Lett. 352*:185–190 (1994).
66. Wintermeyer, E., Ludwig, B., Steinert, M., Schmidt, B., Fischer, G., and Hacker, J., Influence of site specifically altered Mip proteins on intracellular survival of *Legionella pneumophila* in eukaryotic cells, *Infect. Immun. 63*:4576–4583 (1995).
67. Rahfeld, J. U., Rücknagel, K. P., Stoller, G., Horne, S. M., Schierhorn, A., Young, K. D., and Fischer, G., Isolation and amino acid sequence of a new 22-kDa FKBP-like peptidyl–prolyl *cis/trans*-isomerase of *Escherichia coli*—similarity to Mip-like proteins of pathogenic bacteria, *J. Biol. Chem. 271*:22130–22138 (1996).
68. Kieffer, L. J., Thalhammer, T., and Handschumacher, R. E., Isolation and characterization of a 40-kDa cyclophilin-related protein, *J. Biol. Chem. 267*:5503–5507 (1992).
69. Kieffer, L. J., Seng, T. W., Li, W., Osterman, D. G., Handschumacher, R. E., and Bayney, R. M., Cyclophilin-40, a protein with homology to the P59 component of the steroid receptor complex. Cloning of the c-DNA and further characterization, *J. Biol. Chem. 268*:12303–12310 (1993).
70. Owens-Grillo, J. K., Hoffmann, K., Hutchison, K. A., Yem, A. W., Deibel, M. R., Handschumacher, R. E., and W. B. Pratt, The cyclosporin A-binding immunophilin Cyp-40 and the FK 506-binding immunophilin hsp56 bind to a common site on

hsp90 and exist in independent cytosolic heterocomplexes with the untransformed glucocorticoid receptor, *J. Biol. Chem. 270*:20479–20484 (1995).
71. Anderson, S. K., Gallinger, S., Roder, J., Frey, J., Young, H. A., and Ortaldo, J. R., A cyclophilin-related protein involved in the function of natural killer cells, *Proc. Natl. Acad. Sci USA 90*:542–546 (1993).
72. Rahfeld, J.-U., Schierhorn, A., Mann, K.-H., and Fischer, G., A novel peptidyl–prolyl *cis/trans* isomerase from *Escherichia coli. FEBS Lett. 343*:65–69 (1994).
73. Rahfeld, J. U., Rücknagel, K. P., Schelbert, B., Ludwig, B., Hacker, J., Mann, K., and Fischer, G., Confirmation of the existence of a third family among peptidyl–prolyl *cis/trans* isomerases—amino acid sequence and recombinant production of parvulin, *FEBS Lett. 352*:180–184 (1994).
74. Kontinen, V. P. and Sarvas, M., The PrsA lipoprotein is essential for protein secretion in *Bacillus subtilis* and sets a limit for high-level secretion, *Mol. Microbiol. 8*:727–737 (1993).
75. Jacobs, M., Andersen, J. B., Kontinen, V., and Sarvas, M., *Bacillus subtilis* PrsA is required in vivo as an extracytoplasmic chaperone for secretion of active enzymes synthesized either with or without pro-sequences, *Mol. Microbiol. 8*:957–966 (1993).
76. Eisenstark, A., Miller, C., Jones, J., and Leven, S., *Escherichia coli* genes involved in cell survival during dormancy: Role of oxidative stress, *Biochem. Biophys. Res. Commun. 188*:1054–1059 (1992).
77. Haandrikman, A. J., Kok, J., Laan, H., Soemitro, S., Ledeboer, A. M., Konings, W. N., and Venema, G., Identification of a gene required for maturation of an extracellular lactococcal serine proteinase, *J. Bacteriol. 171*:2789–2794 (1989).
78. Nissen Meyer, J., Lillehaug, D., and Nes, I. F., The plasmid-encoded lactococcal envelope-associated proteinase is encoded by a chromosomal gene in *Lactococcus lactis* subsp. *cremoris* BC101, *Appl. Environ. Microbiol. 58*:750–753 (1992).
79. Wild, J., Rossmeissl, P., Walter, W. A., and Gross, C. A., Involvement of the DnaK–DnaJ–GrpE chaperone team in protein secretion in *Escherichia coli, J. Bacteriol. 178*:3608–3613 (1996).
80. Missiakas, D., Betton, J. M., and Raina, S., New components of protein folding in extracytoplasmic compartments of *Escherichia coli* SurA, FkpA and Skp/OmpH, *Mol. Microbiol. 21*:871–884 (1996).
81. Crooke, E. and Wickner, W., Trigger factor: A soluble protein that folds pro-OmpA into a membrane-assembly–competent form, *Proc. Natl. Acad. Sci. USA 84*:5216–5220 (1987).
82. Lill, R., Crooke, E., Guthrie, B., and Wickner, W., The "trigger factor cycle" includes ribosomes, presecretory proteins, and the plasma membrane, *Cell 54*:1013–1018 (1988).
83. Crooke, E., Guthrie, B., Lecker, S., Lill, R., and Wickner, W., ProOmpA is stabilized for membrane translocation by either purified *E. coli* trigger factor or canine signal recognition particle, *Cell 54*:1003–1011 (1988).
84. Guthrie, B. and Wickner, W., Trigger factor depletion or overproduction causes defective cell division but does not block protein export, *J. Bacteriol. 172*:5555–5562 (1990).

85. Fraser, C. M., Gocayne, J. D., White, O., et al., The minimal gene complement of *Mycoplasma genitalium* [see comments], *Science 270*:397–403 (1995).
86. Callebaut, I. and Mornon, J. P., Trigger factor, one of the *Escherichia coli* chaperone proteins, is an original member of the FKBP family, *FEBS Lett. 374*:211–215 (1995).
87. Stoller, G., Tradler, T., Rücknagel, J.-U., and Fischer, G., An 11.8 kDa proteolytic fragment of the *E. coli* trigger factor represents the domain carrying the peptidyl–prolyl *cis/trans* isomerase activity, *FEBS Lett. 384*:117–122 (1996).
88. Hesterkamp, T. and Bukau, B., Identification of the prolyl isomerase domain of *Escherichia coli* trigger factor, *FEBS Lett. 385*:67–71 (1996).
89. Kelley, R. F. and Richards, F. M., Replacement of proline-76 with alanine eliminates the slowest kinetic phase in thioredoxin folding, *Biochemistry 26*:6765–6774 (1987).
90. Bächinger, H. P., Bruckner, P., Timpl, R., Prockop, D. J., and Engel, J., Folding mechanism of the triple helix in type-III collagen and type-II pN collagen, *Eur. J. Biochem. 106*:619–632 (1980).
91. Bächinger, H.-P., The influence of peptidyl–prolyl *cys–trans* isomerase on the in vitro folding of type III collagen, *J. Biol. Chem. 262*:17144–17148 (1987).
92. Davis, J. M., Boswell, B. A., and Bächinger, H. P., Thermal stability and folding of type IV procollagen and effect of peptidyl–prolyl *cis–trans* isomerase on the folding of the triple helix, *J. Biol. Chem. 264*:8956–8962 (1989).
93. Matouschek, A., Kellis, J. T., Serrano, L., Bycroft, M., and Fersht, A. R., Transient folding intermediates characterized by protein engineering, *Nature 346*:440–445 (1990).
94. Freskgard, P. O., Bergenhem, N., Jonsson, B. H., Svensson, M., and Carlsson, U., Isomerase and chaperone activity of prolyl isomerase in the folding of carbonic anhydrase, *Science 258*:466–468 (1992).
95. Kern, G., Kern, D., Schmid, F. X., and Fischer, G., Reassessment of the putative chaperone function of prolyl-*cis/trans*-isomerases, *FEBS Lett. 348*:145–148 (1994).
96. Jackson, S. E. and Fersht, A. R., Folding of chymotrypsin inhibitor-2. 2. Influence of proline isomerization on the folding kinetics and thermodynamic characterization of the transition state of folding, *Biochemistry 30*:10436–10443 (1991).
97. Veeraraghavan S. and Nall, B. T., Characterization of folding intermediates using prolyl isomerase, *Biochemistry 33*:687–692 (1994).
98. Veeraraghavan, S., Rodriguez-Gdiharpour, S., MacKinnon, C., Mcgee, W. A., Pierce, M. M., and Nall, B. T., Prolyl isomerase as a probe of stability of slow-folding intermediates, *Biochemistry 34*:12892–12902 (1995).
99. Lilie, H., Rudolph, R., and Buchner, J., Association of antibody chains at different stages of folding: Prolyl isomerization occurs after formation of quaternary structure, *J. Mol. Biol. 248*:190–201 (1995).
100. Mann, C. J., Shao, X., and Matthews, C. R., Characterization of the slow folding reactions of *trp* aporepressor from *Escherichia coli* by mutational analysis of prolines and catalysis by a peptidyl–prolyl isomerase, *Biochemistry 34*:14573–14580 (1995).
101. Pace, C. N., Heinemann, U., Hahn, U., and Saenger, W., Ribonuclease T1: Structure, function, and stability, *Angew. Chem. Int. Ed. Engl. 30*:343–360 (1991).

102. Kiefhaber, T., Quaas, R., Hahn, U., and Schmid, F. X., Folding of ribonuclease T_1. 1. Existence of multiple unfolded states created by proline isomerization, *Biochemistry 29*:3053–3061 (1990).
103. Odefey, C., Mayr, L. M., and Schmid, F. X., Non-prolyl *cis–trans* peptide bond isomerization as a rate-determining step in protein unfolding and refolding, *J. Mol. Biol. 245*:69–78 (1995).
104. Pace, C. N. and Creighton, T. E., The disulphide folding pathway of ribonuclease T1, *J. Mol. Biol. 188*:477–486 (1986).
105. Frech, C. and Schmid, F. X., Influence of protein conformation on disulfide bond formation in the oxidative folding of ribonuclease T1, *J. Mol. Biol. 251*:135–149 (1995).
106. Heinemann, U., and Saenger, W., Specific protein–nucleic acid recognition in ribonuclease T1–2′-guanylic acid complex: An X-ray structure, *Nature 299*:27–31 (1982).
107. Martinez-Oyanedel, J., Choe, H.-W., Heinemann, U., and Saenger, W., Ribonuclease T1 with free recognition and catalytic site: Crystal structure analysis at 1.5 Å resolution, *J. Mol. Biol. 222*:335–352 (1991).
108. Oobatake, M., Takahashi, S., and Ooi, T., Conformational stability of ribonuclease T1. I. Thermal denaturation and effects of salts, *J. Biochem. 86*:55–63 (1979).
109. Pace, C. N., Grimsley, G. R., Thomson, J. A., and Barnett, B. J., Conformational stability and activity of ribonuclease T1 with zero, one, and two intact disulfide bonds, *J. Biol. Chem. 263*:11820–11825 (1988).
110. Kiefhaber, T., Grunert, H.-P., Hahn, U., and Schmid, F. X., Replacement of a *cis* proline simplifies the mechanism of ribonuclease T1 folding, *Biochemistry 29*:6475–6480 (1990).
111. Mayr, L. M., Willbold, D., Landt, O., and Schmid, F. X., Role of the Cys 2–Cys 10 disulfide bond for the structure, stability, and folding kinetics of ribonuclease T1, *Protein Sci. 3*:227–239 (1994).
112. Mayr, L. M., Willbold, D., Rösch, P., and Schmid, F. X., Generation of a non-prolyl *cis* peptide bond in ribonuclease T-1, *J. Mol. Biol. 240*:288–293 (1994).
113. Mayr, L. M., Odefey, C., Schutkowski, M., and Schmid, F. X., Kinetic analysis of the unfolding and refolding of ribonuclease T1 by a stopped-flow double-mixing technique, *Biochemistry 35*:5550–5561 (1996).
114. Kiefhaber, T., Quaas, R., Hahn, U., and Schmid, F. X., Folding of ribonuclease T1. 2. Kinetic models for the folding and unfolding reactions, *Biochemistry 29*:3061–3070 (1990).
115. Kiefhaber, T., Schmid, F. X., Willaert, K., Engelborghs, Y., and Chaffotte, A., Structure of a rapidly formed intermediate in ribonuclease–T1 folding, *Protein Sci. 1*:1162–1172 (1992).
116. Texter, F. L., Spencer, D. B., Rosenstein, R., and Matthews, C. R., Intramolecular Catalysis of a proline isomerization reaction in the folding of dihydrofolate reductase, *Biochemistry 31*:5687–5691 (1992).
117. Mücke, M. and Schmid, F. X., Enzymatic catalysis of prolyl isomerization in an unfolding protein, *Biochemistry 31*:7848–7854 (1992).
118. Mücke, M. and Schmid, F. X., Folding mechanism of ribonuclease T1 in the absence of the disulfide bonds, *Biochemistry 33*:14608–14619 (1994).

119. Mücke, M. and Schmid, F. X., Intact disulfide bonds decelerate the folding of ribonuclease T1, *J. Mol. Biol. 239*:713–725 (1994).
120. Schönbrunner, E. R., Mayer, S., Tropschug, M., Fischer, G., Takahashi, N., and Schmid, F. X., Catalysis of protein folding by cyclophilins from different species, *J. Biol. Chem. 266*:3630–3635 (1991).
121. Scholz, C., Zarnt, T., Kern, G., Lang, K., Burtscher, H., Fischer, G., and Schmid, F. X., Autocatalytic folding of the folding catalyst FKBP12, *J. Biol. Chem. 271*:12703–12707 (1996).
122. Veeraraghavan, S., Holzman, T. F., and Nall, B. T., Autocatalyzed protein folding, *Biochemistry 35*:10601–10607 (1996).
123. Galat, A. and Metcalfe, S. M., Peptidylproline *cis/trans* isomerases, *Prog. Biophys. Mol. Biol. 63*:67–118 (1995).
124. Hani, J., Stumpf, G., and Domdey, H., PTF1 encodes an essential protein in *Saccharomyces cerevisiae*, which shows strong homology with a new putative family of PPIases, *FEBS Lett. 365*:198–202 (1995).
125. Scholz, C., Stoller, G., Zarnt, T., Fischer, G., and Schmid, F. X., Cooperation of enzymatic and chaperone functions of trigger factor in the catalysis of protein folding. *EMBO J. 16*:54–58 (1997).
126. Okazaki, A., Ikura, T., Nikaido, K., and Kuwajima, K., The chaperonin GroEL does not recognize apo-alpha-lactalbumin in the molten globule state, *Nature Struct. Biol. 1*:439–446 (1994).
127. Hayer-Hartl, M. K., Ewbank, J. J., Creighton, T. E., and Hartl, F. U., Conformational specificity of the chaperonin GroEL for the compact folding intermediates of alpha-lactalbumin, *EMBO J. 13*:3192–3202 (1994).
128. Fersht, A., *Enzyme Structure and Mechanism*, W. H. Freeman, New York, 1985.
129. Kandror, O., Sherman, M., Rhode, M., and Goldberg, A. L., Trigger factor is involved in GroEL-dependent protein degradation in *Escherichia coli* and promotes binding of GroEL to unfolded proteins, *EMBO J. 14*:6021–6027 (1995).
130. Steinmann, B., Bruckner, P., and Supertifurga, A., Cyclosporin-A slows collagen triple-helix formation in vivo—indirect evidence for a physiologic role of peptidyl–prolyl *cis–trans*-isomerase, *J. Biol. Chem. 266*:1299–1303 (1991).
131. Kruse, M., Brunke, M., Escher, A., Szalay, A. A., Tropschug, M., and Zimmermann, R., Enzyme assembly after de novo synthesis in rabbit reticulocyte lysate involves molecular chaperones and immunophilins, *J. Biol. Chem. 270*:2588–2594 (1995).
132. Rassow, J., Mohrs, K., Koidl, S., Barthelmess, I. B., Pfanner, N., and Tropschug, M., Cyclophilin 20 is involved in mitochondrial protein folding in cooperation with moelcular chaperones Hsp70 and Hsp60, *Mol. Cell. Biol. 15*:2654–2662 (1995).
133. Matouschek, A., Rospert, S., Schmid, K., Glick, B. S., and Schatz, G., Cyclophilin catalyzes protein folding in yeast mitochondria, *Proc. Natl. Acad. Sci. USA 92*:6319–6323 (1995).
134. Herzberg, O. and Moult, J., Analysis of the steric strain in the polypeptide backbone of protein molecules, *Proteins Struct. Funct. Genet. 11*:223–229 (1991).
135. Sosnick, T. R., Mayne, L., Hiller, R., and Englander, S. W., The barriers in protein folding, *Nature Struct. Biol. 1*:149–156 (1994).
136. Huang, G. S. and Oas, T. G., Submillisecond folding of monomeric lambda repressor, *Proc. Natl. Acad. Sci. USA 92*:6878–6882 (1995).

137. Schindler, T., Herrler, M., Marahiel, M. A., and Schmid, F. X., Extremely rapid folding in the absence of intermediates: The cold-shock protein from *Bacillus subtilis*. *Nature Struct. Biol. 2*:663–673 (1995).
138. Kragelund, B. B., Robinson, C. V., Knudsen, J., Dobson, C. M., and Poulsen, F. M., Folding of a four-helix bundle: Studies of acyl-coenzyme A binding protein, *Biochemistry 34*:7217–7224 (1995).
139. Jaenicke, R., Folding and association of proteins, *Prog. Biophys. Mol. Biol. 49*:117–237 (1987).
140. Jaenicke, R., Protein folding and association: In vitro studies for self-organization and targeting in the cell, *Curr. Topics Cell. Regul. 34*:209–314 (1996).
141. Jaenicke, R., Folding and association versus misfolding and aggregation of proteins, *Philos. Trans. R. Soc. Lond. B. 348*:97–105 (1995).
142. Skrzynia, C., London, J., and Goldberg, M. E., The tryptophanase from *Escherichia coli* K12. 3. Further characterization of hybrids between the apo- and holoenzyme, *J. Biol. Chem. 249*:2325–2326 (1974).
143. Orsini, G. and Goldberg, M. E., The renaturation of reduced chymotrypsinogen A in guanidine HCl. Refolding versus aggregation, *J. Biol. Chem. 253*:3453–3458 (1978).
144. Zettlmeissl, G., Rudolph, R., and Jaenicke, R., Reconstitution of lactic dehydrogenase. Noncovalent aggregation vs. reactivation. 1. Physical properties and kinetics of aggregation. *Biochemistry 18*:5567–5571 (1979).
145. Mitraki, A., Fane, B., Haase-Pettingell, C., Sturtevant, J., and King, J., Global suppression of protein folding defects and inclusion body formation, *Science 253*:54–58 (1991).
146. Speed, M. A., Wang, D. I. C., and King. J., Multimeric intermediates in the pathway to the aggregated inclusion body state for P22 tailspike polypeptide chains, *Protein Sci. 4*:900–908 (1995).
147. King, J., Haase-Pettingell, C., Robinson, A. S., Speed, M., and Mitraki, A., Thermolabile folding intermediates: Inclusion body precursors and chaperonin substrates, *FASEB J. 10*:57–66 (1996).
148. Todd, M. J., Viitanen, P. V., and Lorimer, G. H., Dynamics of the chaperonin ATPase cycle: Implications for facilitated protein folding, *Science 265*:659–666 (1994).
149. Frydman, J. and Hartl, F. U., Principles of chaperone-assisted protein folding: Differences between in vitro and in vivo mechanisms, *Science 272*:1497–1502 (1996).
150. Franke, E. K., Yuan, H. E. H., and Luban, J., Specific incorporation of cyclophilin A into HIV-1 virions, *Nature 372*:359–362 (1994).
151. Thali, M., Bukovsky, A., Kondo, E., Rosenwirth, B., Walsh, C. T., Sodroski, J., and Göttlinger, H. G., Functional association of cyclophilin A with HIV-1 virions, *Nature 372*:363–365 (1994).
152. Timerman, A. P., Wiederrecht, G., Marcy, A., and Fleischer, S., Characterization of an exchange reaction between soluble FKBP-12 and the FKBP ryanodine receptor complex—modulation by FKBP mutants deficient in peptidyl–prolyl isomerase activity, *J. Biol. Chem. 270*:2451–2459 (1995).
153. Yem, A. W., Tomasselli, A. G., Heinrikson, R. L., Zurcherneely, H., Ruff, V. A., Johnson, R. A., and Deibel, M. R., The Hsp56 component of steroid receptor

complexes binds to immobilized FK 506 and shows homology to FKBP-12 and FKBP-13, *J. Biol. Chem. 267*:2868–2871 (1992).

154. Peattie, D. A., Harding, M. W., Fleming, M. A., Decenzo, M. T., Lippke, J. A., Livingston, D. J., and Benasutti, M., Expression and characterization of human FKBP52, an immunophilin that associates with the 90-kDa heat shock protein and is a component of steroid receptor complexes, *Proc. Natl. Acad. Sci. USA 89*:10974–10978 (1992).
155. Galat, A., Lane, W. S., Standaert, R. F., and Schreiber, S. L., A rapamycin-selective 25-kDa immunophilin, *Biochemistry 31*:2427–2434 (1992).

16

Assembly of Oligomers and Multisubunit Structures

Robert Seckler
University of Regensburg, Regensburg, Germany

I. PERSPECTIVES

Although today's best-characterized molecular chaperones of the Hsp60 and Hsp70 classes are now thought to facilitate protein folding, they were originally found as factors required for the formation of multisubunit assemblies, such as bacteriophage capsids [1], chloroplast rubisco [2–4], RNA polymerase [5], or immunoglobulins [6]. Many other proteins that are required for the assembly, but do not become part of the final structure of multisubunit complexes, have been identified. Frequently, their mode of action has remained mysterious, and their description is that of gene products or bands on electrophoresis gels. The present chapter will try to present some exceptions, in addition to illustrating principles of protein assembly in the cell by means of two well-characterized examples. A comprehensive description of all assembly systems studied in vivo is beyond its scope. The reversible, ligand-induced assembly of components of the eukaryotic cytoskeleton is a large field of its own and has been summarized in recent reviews [7–9]. It cannot be detailed here, although chaperones are probably involved in keeping cytoskeleton components competent for assembly. The folding and assembly of collagen is covered in a separate part of this volume [see Chap. 10], as are the functions of the "classic" chaperones of the chaperonin and the Hsp70 families in the assembly of oligomeric proteins, such as rubisco and immunoglobulins [see Chaps. 4–6, 12, 17, and 18]. The important role of the chaperone system, consisting of calnexin, calreticulin, UDP-glucose glycoprotein glucosyltransferase, and the glucosidases, in ensuring the proper assembly of oligomeric membrane and soluble glycoproteins in the endoplasmatic reticulum (ER) will be highlighted in Chapter 17.

In progressing from comparatively small, soluble oligomers to large, multisubunit assemblies and membrane-associated structures, this chapter will focus on systems that have been studied in detail in vivo. Nevertheless, much of what has been learned about their assembly mechanisms results from in vitro reconstitution experiments, and it will be apparent that the principles of folding and assembly as outlined in Chapters 1 and 2, do apply to protein assembly inside living cells.

II. IN VIVO VERSUS IN VITRO ASSEMBLY: THE P22 TAILSPIKE HOMOTRIMER

The P22 tailspike protein, the attachment apparatus of *Salmonella* bacteriophage P22, has been used as a model to study protein folding, assembly, misfolding, and misassembly in the cytoplasm of living cells, and to compare in vivo and in vitro folding and assembly pathways [10]. P22 tailspikes are homotrimers of 72-kDa polypeptides, attached to the viral capsid by their NH_2-terminal domains of 108 residues. The major COOH-terminal part recognizes and also hydrolyzes lipopolysaccharide, the phage receptor at the host cell surface. Crystal structures of both parts of the tailspike protein and of complexes with lipopolysaccharide fragments have been determined at high resolution, and the active site has been identified (Fig. 1) [11–13]. During bacteriophage infection, tailspikes are formed late in the infectious cycle. They assemble in solution and attach to preformed bacteriophage heads [14]. In the absence of head assembly, tailspikes accumulate as soluble proteins in the bacterial cytoplasm. Their assembly pathway, as well as the misassembly reaction leading to the formation of inclusion bodies at high temperature, have been studied in detail, both in vivo and in vitro.

A. Assembly Pathway and Intermediates

Techniques used to elucidate the tailspike assembly pathway in vivo include radioactive pulse labeling, centrifugal separation of soluble and aggregated species, native and sodium dodecyl sulfate (SDS) gel electrophoresis, and subunit hybridization [15,16]. The native tailspike trimer is highly thermostable and resists denaturation by SDS at room temperature, whereas tailspike assembly intermediates and misfolded or misassembled tailspike polypeptides are denatured by the ionic detergent and form complexes with SDS micelles. Thus, SDS–polyacrylamide gel electrophoresis (PAGE) can be used to follow the formation of native tailspike trimers after radiolabeling of newly translated polypeptides. During infection of bacteria with a mixture of wild-type phage and of a mutant phage with a single-amino acid substitution, leading to altered

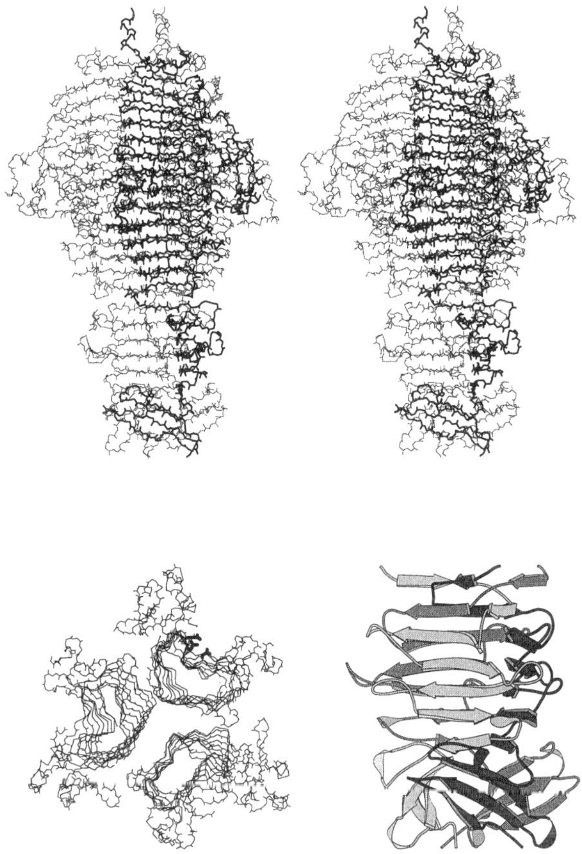

Figure 1 Structure of bacteriophage P22 tailspike protein: The upper part of the figure shows a stereoplot of the tailspike trimer. The protein crystallized lacks the NH_2-terminal, head-binding domains (about 100 residues) that would reside at the top. Only the polypeptide backbone is shown, and one of the three subunits is traced in bold. The major part of a tailspike subunit (residues 143–540) is folded into a right-handed, parallel β-helix, into which several longer loops and a 63-residue domain of largely irregular structure are inserted. A section through the β-helix part of the trimer is shown in the lower left of the figure, illustrating the side-by-side association of the subunits in this part of the protein. Toward the $COOH_2$-termini, however, the subunits are highly interdigitated (bottom of the stereo and ribbon plot in the lower right). In this part of the protein, the native secondary and tertiary structure can be formed only after subunit association. The ribbon plot was prepared using MolScript [95], and the protein database entry code for the tailspike protein is 1TSP. (See Ref. 11.)

electrophoretic mobility of its tailspike, hybrid trimers with intermediate electrophoretic mobility are formed. Comparison of the kinetics of native trimer formation with the time that newly synthesized wild-type and mutant tailspike chains lost the ability to form hybrid trimers defined an assembly intermediate [15]. In this *protrimer* intermediate, the polypeptide chains are firmly associated, but not fully folded, as evident from its sensitivity to proteolysis and to dissociation by SDS. Because the rate-limiting maturation reaction of the trimer is associated with a high activation enthalpy, the protrimer can be trapped in the cold and electrophoretically analyzed. In addition to the protrimer, a monomeric precursor was identified by native gel electrophoresis [15].

Corresponding intermediates were detected when similar techniques were used to analyze the pathway of tailspike reassembly after urea-induced unfolding in vitro, and the rate-limiting trimer-folding reaction occurred with identical kinetics in vitro and in vivo [17–19]. Moreover, tailspike reconstitution yields decreased dramatically with increasing temperature, as do the yields of tailspike folding and assembly in vivo, and single-amino acid substitutions that increased or decreased this temperature sensitivity in vivo exhibited corresponding effects on tailspike reconstitution (Fig. 2) [19,20]. Reconstitution pathways of the

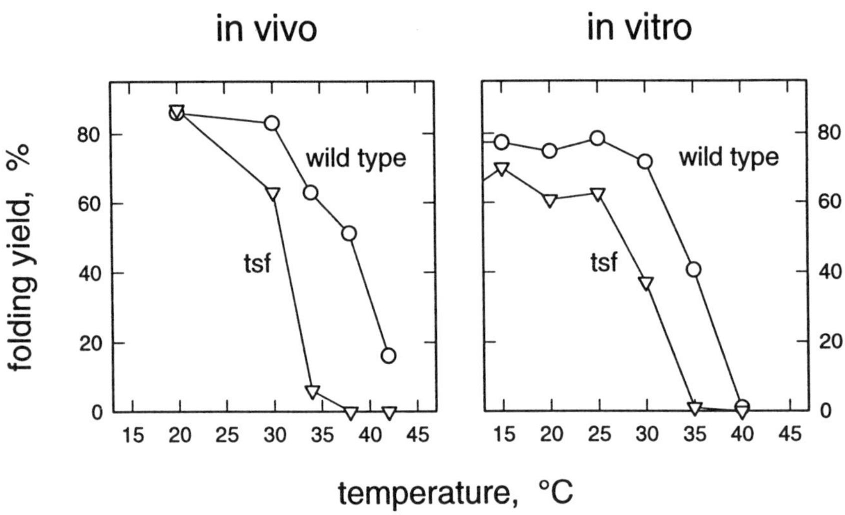

Figure 2 Temperature sensitivity of P22 tailspike assembly: The yield of formation of native tailspike protein trimer as a function of the temperature of expression in vivo (left), and of the temperature of reconstitution from unfolded chains in vitro (right). Several point mutations have been isolated that reduce the folding yield at high temperature, leading to a *temperature-sensitive folding (tsf)* phenotype. (Left: from Ref. 16; right: from Ref. 19.)

complete tailspike protein and of the shortened tailspike lacking the NH_2-terminal head-binding domain were analyzed in detail using spectroscopic and hydrodynamic methods [18,19,21]. They were essentially identical and were affected in a similar fashion by single-amino acid substitutions. Furthermore, the shortened protein efficiently folded and assembled into SDS-resistant trimers when it was expressed from a recombinant plasmid in *Escherichia coli.* These results prove that in vitro reconstitution of tailspikes from unfolded chains is a valid model for the process of folding and assembly of newly synthesized chains inside the bacterial cell. They suggest that cotranslational folding and translational pausing at rare codons downstream from the NH_2-terminal head-binding domain are not important for folding and assembly of the protein.

The folding and assembly pathway of the tailspike protein can be rationalized on the basis of its crystal structure (see Fig. 1). In the main, central part of the tailspike trimer, the subunits form long, right-handed parallel β-helices and are associated side-by-side. Spectroscopic signals, substrate-binding experiments, and the effects of point mutants on refolding kinetics strongly suggest that the β-helices are largely folded before subunit association [18,19,21; U. Baxa and R. Seckler, unpublished observations]. Toward their COOH-termini, however, the chains are strongly interdigitated. Here, the native secondary and tertiary structure cannot form in the isolated subunits, explaining the occurrence of the very slow-folding reaction at the trimer level.

B. Temperature Sensitivity and Folding Mutants

Although temperatures well above 80°C are required to denature the native tailspike trimer in the absence of denaturants, its structural gene is the target of numerous point mutations that prevent phage growth at 40°C [22,23]. These mutations do not affect the synthesis of tailspike polypeptides, and the mutant proteins are perfectly stable and biologically active trimers at 40°C when they are synthesized and assembled at low temperature. After expression at the restrictive temperature, however, the mutant proteins are inactive, sensitive to protease and SDS, and react with antibodies specific for the denatured protein [24]. They have thus been termed *temperature-sensitive for folding (tsf).* A second class of mutations, isolated as *global suppressors (su)* of the *tsf* phenotype have the opposite effect, increasing the yield of correctly assembled tailspikes at high temperature [25–28]. The mode of action of *tsf* and *su* mutations has been elucidated by pulse-labeling studies in vivo [27–29] and by in vitro reconstitution experiments [21,29]. At low temperature, folding and assembly yields and kinetics are identical for the wild-type and all mutant proteins. With increasing temperature of expression or refolding, respectively, the folding yield of the wild-type protein decreases and an increasing fraction of the newly synthesized or refolding chains is incorporated into inactive aggregates (see Fig. 2). The *tsf* mutations decrease, whereas the *su* mutations increase the fraction of tailspike chains reaching the native trimer state at high temperature. The kinetics of folding and assembly at varied temperature and the susceptibility of folding and assembly intermediates to thermal denaturation have been analyzed for wild-type, *tsf* and *su* mutant, and *tsf/su* double mutant proteins in vitro [21,29]. This analysis resulted in a model for the mechanism of action of the mutants that is depicted in the following simplified scheme and outlined in the following:

$$
\begin{array}{lll}
 & \boldsymbol{I} \xrightarrow{k_f}\!\!\rightarrow \boldsymbol{M} \rightarrow\rightarrow\rightarrow \boldsymbol{N} & \\
\boldsymbol{U} \rightarrow & \;\;\downarrow\!\uparrow\; K_{eq} & \\
 & \boldsymbol{I}^* \;\rightarrow\;\rightarrow\; \text{aggregates} &
\end{array}
$$

After release from the ribosome or dilution from a denaturant solution, tailspike polypeptides (***U***) rapidly assume a loosely folded conformation (***I***) containing much secondary structure and partly shielding its aromatic side chains from the solvent. This intermediate is competent to proceed to a state designated a folded monomer (***M***), because its fluorescence spectra and circular dichroism indicate that it contains native-like secondary and tertiary structure. According to the model, the intermediate ***I*** unfolds in a cooperative thermal transition to populate ***I****, a conformation or a population of conformations susceptible to irreversible misfolding or aggregation. The folding mutations would act by destabilizing (*tsf* mutants) or stabilizing (*su* mutants) ***I*** against thermal denaturation. Because the mutations are distributed over a large part of the parallel β-helix and the associated "dorsal fin" domain, the model would require cooperative and rapid folding of these parts of the polypeptide into a native-like conformation. This is in agreement with the effects of the folding mutations on the thermal stability of the native trimer.

Although incompletely folded tailspike polypeptides can be trapped on GroEL in the absence of ATP [30], the chaperone does not appear to play an essential role in tailspike folding and assembly. The presence of a molar excess of the complete GroE system increases tailspike-reconstitution yields only slightly at temperatures higher than 30°C [30], and overexpression of the GroE proteins in *Salmonella* does not suppress *tsf* mutations [31,32]. On recombinant expression in *E. coli,* tailspike mutants impaired in folding or assembly apparently are proteolyzed more rapidly when the GroE proteins are overexpressed, thereby preventing their aggregation in inclusion bodies [S. Miller and R. Seckler, unpublished observations].

C. Misassembly: A Nucleated Linear Polymerization Process

The off-pathway reaction leading to the incorporation of tailspike chains into large aggregates (inclusion bodies) has been studied in vitro by initiating refolding at intermediate denaturant concentrations, at which a large fraction of the refolding chains can be induced to aggregate at temperatures as low as 20°C [20,33]. Under these conditions, the association steps involved are irreversible and slow, allowing the electrophoretic separation of misassembly intermediates and their immunochemical analysis [33–35]. A linear series of dimers, trimers, tetramers, pentamers, and so on, was observed as misassembly intermediates, such that the misassembly reaction can be described as a nucleated linear polymerization process. The reaction is specific such that tailspike polypeptides do not form mixed aggregates with P22 coat protein chains or carbonic anhydrase polypeptides that are induced to aggregate in the same vessel at the same time [34]. This suggests that the aggregates, although reactive with antitailspike antibodies that do not recognize the native trimer, grow by incorporating tailspike chains in a specific, partly folded or misfolded conformation, rather than originating from unfolded chains. The results also explain why inclusion bodies formed on overexpression or heterologous expression of proteins in bacteria largely consist of a single polypeptide species.

III. KINETIC CONTROL OF ASSEMBLY: BACTERIAL LUCIFERASE

Luciferase from *Vibrio harveyi* is a heterodimer of a 40-kDa α- and a 36-kDa β-subunit, each forming a TIM-barrel and likely resulting from gene duplication [36]. Their association in the heterodimer is mediated by two α-helices from each subunit forming a parallel four-helix bundle around the axis of pseudo twofold symmetry relating the two subunits. Bacterial luciferase is a flavin monoxygenase, catalyzing the oxygenation of long-chain aliphatic aldehydes and generating visible light from a hydroperoxyflavin reaction intermediate. Both subunits can be isolated individually in a folded state from recombinant *E. coli* and display low luciferase activity, but the activity of the heterodimer is several orders of magnitude higher [37]. When purified subunits are combined they fail to assemble into active dimers, unless they are denatured and allowed to assemble after refolding [37,38]. Accordingly, formation of active dimers in vivo requires the expression of both structural genes in the same cell and at the same time [38]. In a series of exciting in vivo and in vitro experiments, Baldwin and co-workers have elucidated the folding and assembly pathway of the enzyme, identified the kinetic traps obstructing heterodimer formation, and characterized *tsf* mutations preventing assembly of the native heterodimer at high temperature.

A. Reversible Unfolding, Assembly Pathway, and Cotranslational Folding

Under appropriately selected conditions (neutral buffer, 18°C, about 200 nM protein), the urea-induced unfolding of bacterial luciferase is freely reversible, allowing the assembly pathway to be characterized under equilibrium conditions [38–41]. Luciferase assembly is a multistep process comprising intermediates both preceding and following the association step:

$$\begin{array}{l} \alpha_u \rightarrow \rightarrow \alpha_i \\ \beta_u \rightarrow \rightarrow \beta_i \end{array} \Big\} \longrightarrow \alpha\beta_i \rightarrow \alpha\beta_n$$

Subunit polypeptides (α_u, β_u) fold to assume intermediate conformations (α_i, β_i) sufficiently structured to allow mutual recognition. The primary association product ($\alpha\beta_i$) is enzymatically inactive and incompletely folded, as indicated by its reduced far-UV circular dichroism and increased tryptophan fluorescence relative to the native heterodimer ($\alpha\beta_n$). The incompletely folded heterodimer is in equilibrium with the native protein and with the unfolded subunits and is highly populated at intermediate denaturant concentrations during urea-induced luciferase unfolding. At protein concentrations above 200 nM, the rearrangement of the heterodimer becomes rate-limiting for the formation of active enzyme, whereas the reaction is governed by the association reaction at low protein concentration. Slow subunit folding before association is reflected in a lag period of several minutes, observed in the reactivation kinetics at both high and low protein concentrations.

The contribution of cellular factors and cotranslational folding to luciferase assembly has been assessed in experiments using cell-free translation systems. When luciferase β-subunit was synthesized in *E. coli* lysate in the presence of purified α-subunit, the enzyme acquired activity within 2 min after completion of chain elongation [42]. This was significantly faster than the reactivation of the enzyme occurring after dilution of urea-denatured β-subunit into the same cell-free system in the absence of mRNA. Translation experiments with β-subunit RNA that lacked a stop codon proved that the active heterodimer can form only after completion of chain synthesis and release from the ribosome [42]. This is in agreement with COOH-terminal deletions in the β-subunit impeding heterodimer formation [43]. Nevertheless, the results suggest that cotranslational subunit folding contributes significantly to the overall assembly of bacterial luciferase in vivo.

In experiments targeted at identifying folding and assembly helpers in the eukaryotic cytoplasm, transcripts coding for both subunits of bacterial luciferase were translated in a rabbit reticulocyte lysate system [44,45]. Formation of active luciferase required more than 1 h after completion of polypeptide synthesis and was sensitive to dilution, suggesting that subunit association is rate-limiting under these conditions. The authors found that luciferase assembly was inhibited by ATP depletion and by cyclosporin A and tacrolimus (FK 506) suggesting the involvement of chaperones and peptidyl–prolyl isomerases.

B. Misassembly and Misfolding as Kinetic Traps

The formation of active luciferase heterodimers requires the coexpression of both subunits at the same time, and a mixture of the isolated subunits forms active enzyme only after unfolding and refolding by urea. This suggests that at least one of the subunits is susceptible to a transconformation reaction, irreversible on the time scale of the experiments, rendering it inaccessible to association with the other subunit. In later experiments active heterodimers could be assembled after folding of the β-subunit in the presence of preformed α-subunit, but not vice versa [37,40,46]. On expression in the absence of the α-subunit, the β-subunit is isolated as a β_2-homodimer. Dissociation of β_2 requires high concentrations of denaturant, its half-life extrapolates to more than 1 million years in neutral buffer. Obviously, the formation of β_2 is irreversible on a biologically significant timescale, and the active enzyme only is formed, because heterodimer formation is kinetically preferred ($k_{ass} \approx 3 \times 10^3\ M^{-1}\ s^{-1}$) over the dimerization of β-subunits ($k_{ass} \approx 2 \times 10^2\ M^{-1}\ s^{-1}$). The thermodynamic stabilities and the crystal structures of both dimers are similar [46,47].

At low protein concentration, a second kinetic trap is observed. The yield of heterodimer reconstitution strongly decreases at protein concentrations less than 100 nM [39], and the β-subunit is recovered as an assembly-incompetent monomer (Baldwin, T. O., personal communication). Dimerization-incompetent monomers are also formed on expression of luciferase *tsf* mutants at high temperature. Following expression or refolding at high temperature, such mutants, resulting from modifications or deletions in the COOH-terminal region of the β-subunit, form active heterodimers of wild-type stability. Again, the final conformation of the protein appears to be determined by kinetics and activation enthalpies, rather than by its thermodynamic stability.

IV. MULTISUBUNIT ASSEMBLIES

The problem of the assembly of large, multisubunit structures, such as components of the eukaryotic cytoskeleton, bacterial pili and flagellae, and whole viruses, has long fascinated researchers. Early studies on tobacco mosaic virus

[48], tubulin [49], and other assembly systems proved that these processes can be described as self-assembly reactions driven by burial of the hydrophobic surface area between subunits. Still, accessory proteins that are not incorporated into the final structure facilitate many of these processes, or are even absolutely required for assembly. In microbial systems, the power of genetic techniques has facilitated the discovery of such assembly factors, two of which are the GroE proteins, originally described as host factors required for growth of bacteriophage λ.

A. Assembly of Bacteriophage Capsids

The structure of many animal, plant, and microbial viruses is based on a shell of protein surrounding the viral DNA or RNA genome. The assembly of such capsids has been extensively studied in the bacteriophages [14,50–53], in which in vivo assembly intermediates have been identified [14,50–55], assembly kinetics of purified components have been studied in vitro [14,50,51,53,56–59], and even high-resolution information on the three-dimensional structures of assembly intermediates has recently become available [60–62]. Although, in principle, bacteriophage capsids can form from capsid protein subunits alone [54,59], the assembly of most double-stranded DNA-containing phage is dependent on scaffolding proteins that do not become part of the mature virion. They form multisubunit scaffolds on the inside or on the surface of the primary assembly products, the procapsids, and are later removed concomitant with DNA packaging [14,53,55,58,60–63]. Procapsids containing scaffolding proteins are also observed in the assembly of herpes- and adenoviruses [64,65]. Maturation of the shells often involves proteolysis of scaffolding and capsid proteins, or covalent cross-linking of capsomers, rendering it irreversible. However, phage assembly can occur without such covalent modifications, as exemplified by the assembly pathway of bacteriophage P22 (Fig. 3).

The icosahedral shell of bacteriophage P22 heads consists of 420 copies of a single capsid protein, the 47-kDa coat protein, arranged in 12 pentamers and 60 hexamers [61]. This $T = 7$ symmetry implies that the chemically identical subunits must be flexible enough to adjust to different environments in the icosahedral capsid and still form a sturdy shell [66]. As in other phages, assembly involves the formation of a procapsid that experiences a major conformational change (expansion) when the double-stranded DNA genome is packaged [61,67,68]. The procapsid, which can be isolated from cells infected with phage defective in DNA packaging, has been studied by high-resolution electron microscopy [61]. Its interior contains about 300 copies of the scaffolding protein, which are not packed with icosahedral symmetry. Assembly of a procapsid from coat protein in vivo and in vitro requires nearly 140 copies of the scaffolding protein, but none of the minor protein components of the mature phage head [56]. The reaction proceeds as a nucleated polymerization process, thought to originate at one of the fivefold-

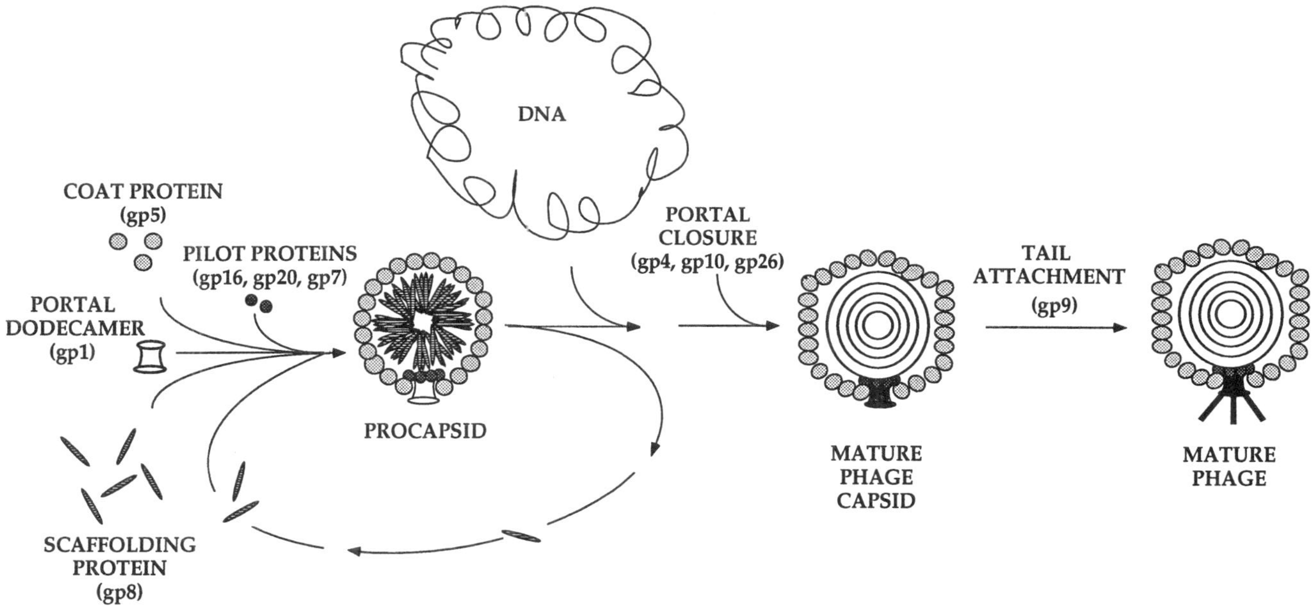

Figure 3 Assembly of bacteriophage P22: The coat protein and the scaffolding protein are sufficient to assemble a procapsid shell in vivo or in vitro, but many other proteins are required to make an infectious phage. The scaffolding protein is removed during DNA packaging and can be recycled. The last step in the assembly is the attachment of the tailspike protein, the viral adhesin (see Fig. 1). Coat protein monomers, competent for assembly in vitro can be obtained after denaturation of empty procapsid shells, but *tsf* mutants will form assembly-incompetent dimers or trimers in this reaction (compare text). (Courtesy of Dr. Carolyn M. Teschke, University of Connecticut.)

vertices, and to progress by addition of coat protein monomers and scaffolding protein dimers or tetramers. In the absence of the scaffolding protein, P22 coat protein assembles into aberrant structures in which pentamers and hexamers are not properly deposited [69]. Apparently, the scaffolding protein does not preform a template onto which the coat is assembled; rather, the geometry of the capsid is determined by local rules originating from the interactions of the two proteins [70]. Concomitant with DNA packaging through the portal vertex, the P22-scaffolding protein escapes from the procapsid, presumably through wide channels present in the centers of coat protein pentamers and hexamers, before these holes close during capsid expansion [67]. Neither scaffolding nor coat protein is proteolyzed during the assembly process, and the scaffolding protein can be reused in subsequent rounds of assembly [71]. In contrast, procapsids of coliphage HK97, which does not use a scaffolding protein, are assembled from preformed coat protein pentamers and hexamers, and head maturation requires proteolytic processing of the coat protein [54].

Because the P22 scaffolding protein acts stoichiometrically, does not become part of the final structure, is not processed, and can function in multiple reaction cycles, it may be considered to be an assembly chaperone, although a highly specialized one. Spectroscopic, hydrodynamic, and thermal unfolding studies have shown that the scaffolding protein is not a classic globular protein [72]. Although highly α-helical, it exhibits a noncooperative thermal-melting transition in the absence of coat protein. Because the protein has a tendency to associate into dimers and higher oligomers, its thermal melting is dependent on the protein concentration. At moderate concentrations (1–2 mg mL^{-1}), the midpoint of the melting transition occurs close to 50°C, and the protein is fully folded only in the vicinity of the freezing point of water. Incorporation into the procapsid significantly stabilizes the scaffolding protein. Its thermal unfolding becomes much more cooperative, and moderate concentrations of guanidinium chloride are required for its reversible extraction from the shells [73], indicating that folding and unfolding of the scaffolding protein are coupled to shell binding and release.

Several amino acid substitutions in the P22 coat protein have been identified that make capsid assembly temperature-sensitive in vivo and lead to the accumulation of coat protein polypeptides in inclusion bodies at high temperature [74]. In contrast with similar mutations in the tailspike protein (see foregoing), coat protein *tsf* mutations are efficiently suppressed by overexpression of the GroE proteins [32]. When the folding of five such *tsf* mutant proteins was analyzed in vitro after guanidine-induced unfolding of shells produced at permissive temperature, all mutants folded into assembly-incompetent dimers or trimers, whereas the wild-type formed monomers that readily assembled [75]. In contrast with the in vivo situation, the dimers or trimers remained soluble, and there was no significant difference between wild-type and mutant proteins in their tendency to form large aggregates at high temperature. How the formation of such kinetically trapped oligomeric species is prevented in vivo and how the *tsf* phenotype is suppressed by the GroE chaperonins remains unknown. The GroE chaperonins were originally identified as host factors required for growth of bacteriophages λ and T4. Their role in the assembly of bacteriophage T4 capsids has been reviewed recently [76].

B. Assembly of Bacteriophage Tail Fibers

Chaperone activities are also required for the assembly of the intriguing structures by which phage T4 and its relatives attach to the host cell surface [77]. The products of T4 genes *38* and *57* are not part of the mature virion and have not been found as parts of assembly intermediates. However, both proteins are required for assembly of the long tail fibers, and short tail fibers cannot be formed without the function of gp57. In the absence of the chaperones, structural proteins of the tail fibers are found in insoluble aggregates [78,79]. The product of gene *63,* a 44-kDa protein which is also not a structural component of the mature phage, appears to function in the attachment of long tail fibers to the baseplate. Mutants in gene *63* attach long tail fibers at low rates and have a cold-sensitive phenotype [80].

The well-known long tail fibers of bacteriophage T4 are made up of the products of genes *34, 35, 36,* and *37,* of which gp34 (140 kDa) forms the proximal part of the fiber, gp36 (23 kDa) and gp37 (109 kDa) form the distal part, and gp35 (30 kDa) is located at the joint (Fig. 4). Mass determinations by scanning transmission electron microscopy have recently identified the fibrous proteins gp34, gp36, and gp37 as trimers [81]. The proteins consist largely of β-sheets and turns, and contain repetitive amino acid sequence stretches. A structural model based on these findings, analogous to the previously proposed structure of the adenovirus fiber [82], envisages the three proteins as largely consisting of triple β-helices (see Fig. 4). The folding problem inherent in the assembly of such a structure from chains of more than 1000 residues is apparent, but it remains completely unclear, how gp57 and gp38 might assist in its formation. The requirement for gp57 in the assembly of T4 tail fibers can be bypassed by mutations in the *E. coli* chromosome, suggesting that host factors can either replace gp57 or are involved in its function [83]. Gp38 can be replaced by the functionally homologous product of bacteriophage λ *tail-fiber assembly* gene *tfa* [84].

Phage T4 gp57 has recently been purified and characterized [85]. Gene *57,* which had previously been misidentified, encodes a 79-amino acid residue polypeptide that oligomerizes in solution and is largely α-helical [85; S. Miller and U. Henning, personal communication]. The protein is highly polar, with 27 charged residues, and is devoid of aromatic amino acids, cysteine, and proline. After recombinant expression in *E. coli,* it forms large, highly ordered fibers that eventually kill the cell. Strikingly, the function of gp57 in T4 tail fiber assembly

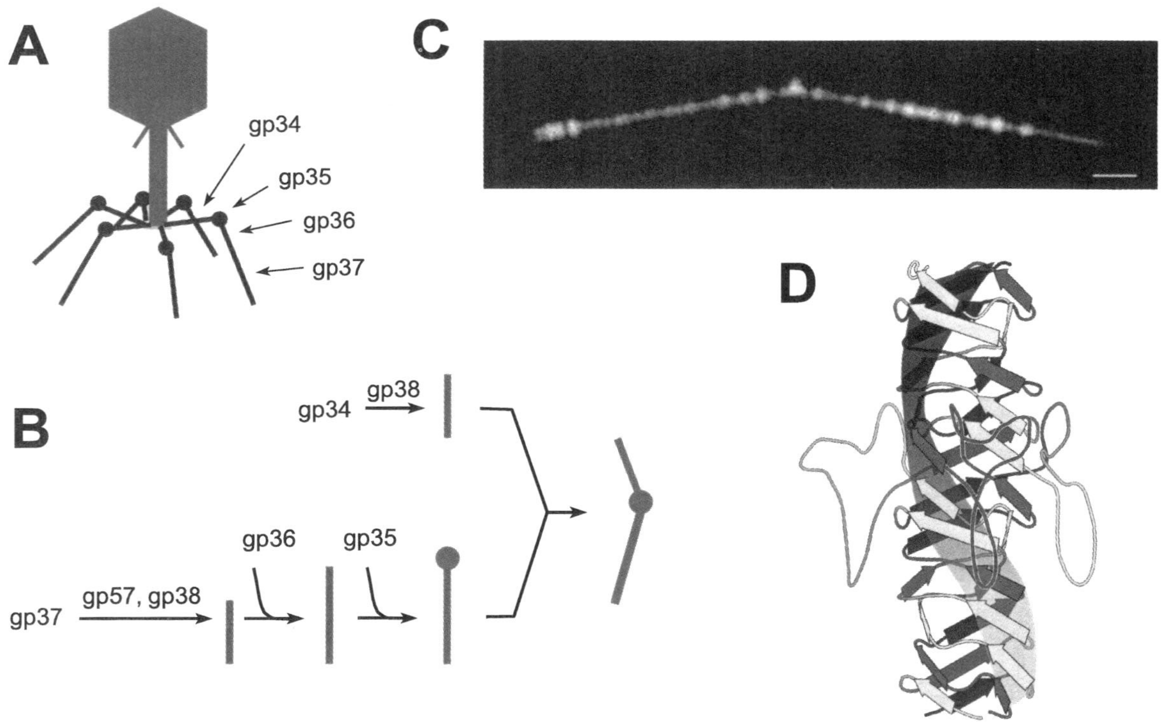

Figure 4 Structure and assembly of phage T4 long tail fibers: (A) Six long tail fibers bound to the base plate of bacteriophage T4 mediate attachment of the phage to the host cell surface receptor. (B) They are assembled from the products of genes *34–37* with the aid of two chaperone proteins, gp38 and gp57. The structure of T4 long tail fibers has been studied by electron microscopy [81]. Results obtained by image processing, similar to that depicted, (C) and of mass determinations by scanning transmission electron microscopy, have led to a structural model (D) for the trimeric proteins making up the fibers. (Assembly pathway drawn after Ref. 77, panels C and D kindly provided by Dr. M. Cerritelli.)

in vivo can be fulfilled—at least in part—by a 26-residue NH_2-terminal fragment of the protein.

C. Assembly of Bacterial Pili and Fimbriae

Attachment fibers called pili or fimbriae are a general feature of pathogenic gram-negative bacteria. Pili and fimbriae are helical structures with a diameter of several nanometers and up to 2 μm in length, anchored in the bacterial outer membrane and frequently containing adhesion proteins at their distal end [86]. Several morphologically and antigenically distinct pili and fimbriae have been identified, at least ten different types in *E. coli* alone. Their assembly uniformly requires the interplay of a specific pair of a periplasmic soluble chaperone and an outer membrane protein termed a *molecular usher,* both of which are part of the gene cluster coding for the structural proteins of the respective fiber. The assembly systems of type 1 and F1C fimbriae encoded in the *fim* and *foc* gene clusters of *E. coli* are reciprocally interchangeable as pairs, but not as single components [87].

The most intensively studied pilus assembly apparatus is the one required to produce P pili in *E. coli* [86,88]. P pili are bipartite structures consisting of a thin tip fibrillum made from four different subunits (PapG, F, E, and K), joined end-to-end with a thick pilus rod of helically arranged PapA which, in turn, is anchored to the cell by the PapH protein. All six subunits interact with PapD, the 24-kDa periplasmic chaperone, on their way to the outer membrane. In the absence of PapC—an 88-kDa outer membrane protein—complexes of PapD with pilus subunits accumulate in the periplasm. A preassembly binary complex of PapD with PapG, the adhesin, is able to bind receptor oligosaccharides, indicating that PapD binds native, folded pilus subunits [89]. Because it suppresses aggregation of pilus subunits in vitro, it is suggested to cap interactive surfaces and, thereby, prevent misassembly. The dissociation of the chaperone complexes on interaction with PapC and the incorporation of pilus subunits into the growing structure occur without the requirement for metabolic energy [90].

The three-dimensional structure of PapD has been determined at 2.5-Å resolution by X-ray crystallography [91]. The two domains of the boomerang-shaped protein are β-barrel structures with a topology resembling an immunoglobulin fold (Fig. 5). The cleft between them is lined on one side by residues conserved among pilus chaperones. Sequence conservation between pilus and fimbria subunits of different bacterial species suggested a COOH-terminal recognition signal, and synthetic peptides corresponding to the COOH-termini of PapG, E, F, and K bound to PapD in an enzyme-linked immunosorbent assay (ELISA) [92]. Moreover, some of the peptides inhibited the formation of a soluble complex between PapD and PapG on dilution from urea. One of the peptides, derived from PapG, was cocrystallized with PapD, and the structure was deter-

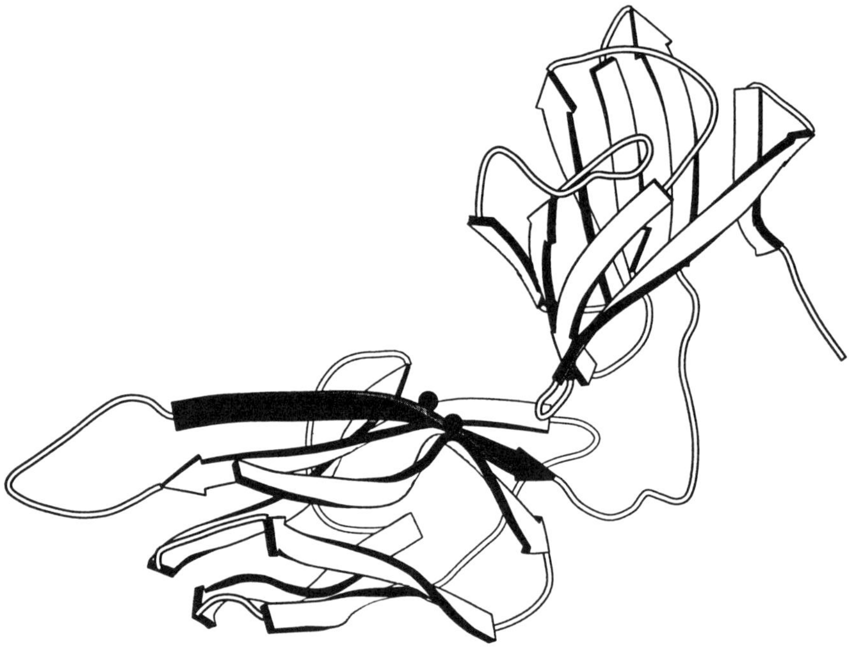

Figure 5 Crystal structure of the PapD chaperone [91]: Both domains of PapD are folded into a β-sandwich structure, similar to an immunoglobulin domain. The deep cleft between the two domains is ready to accept the COOH-terminus of a pilus subunit that will be hydrogen-bonded to β-strand G1 of PapD (black), and anchored by the side chains of two basic residues, Arg-8 and Lys-112, the positions of which are indicated by black dots [92]. The ribbon plot was made with MolScript [95] using coordinates of protein database entry 3DPA.

mined to 3-Å resolution [92]. In the crystal, the COOH-terminal major part of the 19-residue peptide makes a parallel β-strand interaction with Gln-104 to Lys-110 of PapD, extending one of the β-sheets of the NH_2-terminal domain (see Fig. 5), and conserved hydrophobic residues from the peptide and PapD form a hydrophobic zipper. The COOH-terminal residue of the peptide interacts with two conserved, basic residues (Lys-112 and Arg-8) of the protein, mutagenesis of which abolishes the ability of PapD to bind pilus subunits in vivo [93]. Apparently, the COOH-termini of P-pilus subunits play an important role in the binding to PapD. However, the interactions with the periplasmic chaperone extend beyond the COOH-terminus of PapG, as proved by experiments employing fusions of maltose-binding protein (MBP) and PapG [94]. When the COOH-terminal 19 residues (296–314) of PapG were fused to MBP, and the fusion protein was secreted into the periplasm, only weak binding to PapD was observed by affinity chromatography, and formation of pili was largely unaffected. In contrast, strong binding of PapD to an MBP fusion protein, containing residues 175–314, was observed, and secretion of this fusion protein strongly inhibited pilus formation. Because PapD bound to a synthetic peptide corresponding to residues 175–190 in an ELISA, this sequence in PapG may constitute at least part of a secondary chaperone interaction site.

V. CONCLUSIONS

The analyses of the assembly pathways of bacterial luciferase and of the phage P22 tailspike protein have proved that refolding–reassembly after denaturation in vitro remains a valid model reaction for the folding and assembly of even large, oligomeric proteins inside living cells. In both examples, detailed in vitro studies have clarified the mechanisms by which point mutations affect the kinetic distribution of folding polypeptide chains between the productive assembly pathway and competing off-pathway reactions. In vitro assembly reactions for bacteriophage capsids, phage tails, and bacterial pili, although presently at very different stages of development, will help elucidate the mechanism of assembly of such complicated structures in the coming years. Proteins functioning in the assembly reactions, but not becoming part of the final products, are involved in the formation of all three structures. Considering the mode of peptide recognition by PapD, the minimal size of bacteriophage T4 gp57, and the flexibility of phage P22-scaffolding protein, all of them are rather unusual, compared with the classic heat-shock protein chaperones dealt with mostly in this book. Future studies of these unconventional assembly chaperones may provide important insight into mechanisms by which helper proteins may guide folding and assembly in living cells.

REFERENCES

1. Georgopoulos, C. P., Hendrix, R. W., Casjens, S. R., and Kaiser, A. D., Host participation in bacteriophage lambda head assembly, *J. Mol. Biol. 76*:45–60 (1973).
2. Milos, P. and Roy, H., ATP-released large subunits participate in the assembly of RuBP carboxylase, *J. Cell. Biochem. 24*:153–162 (1984).
3. Waldinger, D., Eckerskorn, C., Lottspeich, F., and Clev, H., Amino-acid sequence homology of a polymorphic cellular protein from human lymphocytes and the chaperonins from *Escherichia coli* (groEL) and chloroplasts (rubisco-binding protein), *Biol. Chem. Hoppe Seyler 369*:1185–1189 (1988).
4. Hemmingsen, S. M., Woolford, C., van der Vies, S. M., Tilly, K., Dennis, D. T., Georgopoulos, C. P., Hendrix, R. W., and Ellis, R. J., Homologous plant and bacterial proteins chaperone oligomeric protein assembly, *Nature 333*:330–334 (1988).
5. LeBowitz, J. H., Zylicz, M., Georgopoulos, C., and McMacken, R., Initiation of DNA replication on single stranded DNA templates catalyzed by purified replication

proteins of bacteriophage lambda and *Escherichia coli, Proc. Natl. Acad. Sci. USA 82*:3988–3992 (1985).
6. Haas, I. G. and Wabl, M., Immunoglobulin heavy chain binding protein, *Nature 306*:387–389 (1983).
7. Mandelkow E. and Mandelkow, E. M., Microtubules and microtubule-associated proteins, *Curr. Opin. Cell. Biol. 7*:72–81 (1995).
8. Sun, H. Q., Kwiatkowska, K., and Yin, H. L., Actin monomer binding proteins, *Curr. Opin. Cell. Biol. 7*:102–110 (1995).
9. Carlier, M. F., Valentin-Ranc, C., Combeau, C, Fievez, S., and Pantoloni, D., Actin polymerization: Regulation by divalent metal ion and nucleotide binding, ATP hydrolysis and binding of myosin, *Adv. Exp. Med. Biol. 358*:71–81 (1994).
10. King, J., Haase-Pettingell, C., Robinson, A. S., Speed, M., and Mitraki, A., Thermolabile folding intermediates: Inclusion body precursors and chaperonin substrates, *FASEB J. 10*:57–66 (1996).
11. Steinbacher, S., Miller, S., Baxa, U., Budisa, N., Weintraub, A., Seckler, R., and Huber, R., Phage P22 tailspike protein: Crystal structure of the head-binding domain at 2.3 Å. fully refined structure of the endorhammosidase at 1.56 Å resolution, and the molecular basis of *O*-antigen recognition and cleavage, *J. Mol. Biol. 267*:865–880 (1997).
12. Steinbacher, S., Baxa, U., Miller, S., Weintraub, A., Seckler, R., and Huber, R., Crystal structure of phage P22 tailspike protein complexed with *Salmonella* sp. *O*-antigen receptors, *Proc. Natl. Acad. Sci. USA 93*:10584–10588 (1996).
13. Baxa, U., Steinbacher, S., Miller, S., Weintraub, A., Huber, R., and Seckler, R., Interaction of phage P22 tails with their cellular receptor, *Salmonella O*-antigen polysaccharide, *Biophys. J. 71*:2040–2048 (1996).
14. Prevelige, Jr., P. E. and King, J., Assembly of bacteriophage P22: A model for ds-DNA virus assembly, *Prog. Med. Virol. 40*:206–221 (1993).
15. Goldenberg, D. and King, J., Trimeric intermediate in the in vivo folding and subunit assembly of the tail spike endorhamnosidase of bacteriophage P22, *Proc. Natl. Acad. Sci. USA 79*:3403–3407 (1982).
16. Haase-Pettingell, C. A. and King, J., Formation of aggregates from a thermolabile in vivo folding intermediate in P22 tailspike maturation. A model for inclusion body formation, *J. Biol. Chem. 263*:4977–4983 (1988).
17. Seckler, R., Fuchs-A, A., King, J., and Jaenicke, R., Reconstitution of the thermostable trimeric phage P22 tailspike protein from denatured chains in vitro, *J. Biol. Chem. 264*:11750–11753 (1989).
18. Fuchs, A., Seiderer, C., and Seckler, R., In vitro folding pathway of phage 22 tailspike protein, *Biochemistry 30*:6598–6604 (1991).
19. Danner, M., Fuchs, A., Miller, S., and Seckler, R., Folding and assembly of phage P22 tailspike endorhamnosidase lacking the N-terminal, head-binding domain, *Eur. J. Biochem. 215*:653–661 (1993).
20. Mitraki, A., Danner, M., King, J., and Seckler, R., Temperature-sensitive mutations and second-site suppressor substitutions affect folding of the P22 tailspike protein in vitro, *J. Biol. Chem. 268*:20071–20075 (1993).
21. Danner, M. and Seckler, R., Mechanism of phage P22 tailspike protein folding mutations, *Protein Sci. 2*:1869–1881 (1993).
22. Goldenberg, D. P. and King, J., Temperature-sensitive mutants blocked in the folding

or subunit of the bacteriophage P22 tail spike protein. II. Active mutant proteins matured at 30°C, *J. Mol. Biol. 145*:633–651 (1981).

23. Smith, D. H. and King, J., Temperature-sensitive mutants blocked in the folding or subunit assembly of the bacteriophage P22 tail spike protein. III. Inactive polypeptide chains synthesized at 39°C, *J. Mol. Biol. 145*:653–676 (1981).
24. Goldenberg, D. P., Berget, P. B., and King, J., Maturation of the tail spike endorhamnosidase of *Salmonella* phage P22, *J. Biol. Chem. 257*:7864–7871 (1982).
25. Fane, B. and King, J., Intragenic suppressors of folding defects in the P22 tailspike protein, *Genetics 127*:263–277 (1991).
26. Fane, B., Villafane, R., Mitraki, A., and King, J., Identification of global suppressors for temperature-sensitive folding mutations of the P22 tailspike protein, *J. Biol. Chem. 266*:11640–11648 (1991).
27. Lee, S. C., Koh, H., and Yu, M. -H., Molecular properties of global suppressors of temperature-sensitive folding mutations in P22 tailspike endorhamnosidase, *J. Biol. Chem. 266*:23191–23196 (1991).
28. Mitraki, A., Fane, B., Haase-Pettingell, C., Sturtevant, J., and King, J., Global suppression of protein folding defects and inclusion body formation, *Science 253*:54–58 (1991).
29. Beissinger, M., Lee, S. C., Steinbacher, S., Reinemer, P., Huber, R., Yu, M. -H., and Seckler, R., Mutations that stabilize folding intermediates of phage P22 tailspike protein: Folding in vivo and in vitro, stability, and structural context, *J. Mol. Biol. 249*:185–194 (1995).
30. Brunschier, R., Danner, M., and Seckler, R., Interactions of phage P22 tailspike protein with GroE molecular chaperones during refolding in vitro, *J. Biol. Chem. 268*:2767–2772 (1993).
31. Sather, S. K. and King, J., Intracellular trapping of a cytoplasmic folding intermediate of the phage P22 tailspike using iodoacetamide, *J. Biol. Chem. 269*:25268–25276 (1994).
32. Gordon, C. L., Sather, S. K., Casjens, S., and King, J., Selective in vivo rescue by GroEL/ES of thermolabile folding intermediates to phage P22 structural proteins, *J. Biol. Chem. 269*:27941–27951 (1994).
33. Speed, M. A., Wang, D. I., and King, J., Multimeric intermediates in the pathway to the aggregated inclusion body state for P22 tailspike polypeptide chains, *Protein Sci. 4*:900–908 (1995).
34. Speed, M. A., Wang, D. I. C., and King, J., Specific aggregation of partially folded polypeptide chains: The molecular basis of inclusion body composition, *Nature Biotechnology 14*:1283–1287 (1996).
35. Speed, M. A., Moreshead, T., Wang, D. I. C., and King, J., Conformation of P22 tailspike folding and aggregation intermediates probed by monoclonal antibodies, *Protein Sci. 6*:99–108 (1997).
36. Fisher, A. J., Raushel, F. M., Baldwin, T. O., and Rayment, I., Three-dimensional structure of bacterial luciferase from *Vibrio harveyi* at 2.4 A resolution. *Biochemistry 34*:6581–6586 (1995).
37. Sinclair, J. F., Waddle, J. J., Waddill, E. F., Baldwin, T. O., Purified native subunits of bacterial luciferase are active in the bioluminescence reaction but fail to assemble into the alpha beta structure, *Biochemistry 32*:5036–5044 (1993).

38. Waddle, J. J., Johnston, T. C., and Baldwin, T. O., Polypeptide folding and dimerization in bacterial luciferase occur by a concerted mechanism in vivo, *Biochemistry* *26*:4917–4921 (1987).
39. Ziegler, M. M., Goldberg, M. E., Chaffotte, A. F., and Baldwin, T. O., Refolding of luciferase subunits from urea and assembly of the active heterodimer. Evidence for folding intermediates that precede and follow the dimerization step on the pathway to the active form of the enzyme, *J. Biol. Chem.* *268*:10760–10765 (1993).
40. Baldwin, T. O., Ziegler, M. M., Chaffotte, A. F., and Goldberg, M. E., Contribution of folding steps involving the individual subunits of bacterial luciferase to the assembly of the active heterodimeric enzyme, *J. Biol. Chem.* *268*:10766–10772 (1993).
41. Clark, A. C., Sinclair, J. F., and Baldwin, T. O., Folding of bacterial luciferase involves a non-native heterodimeric intermediate in equilibrium with the native enzyme and the unfolded subunits, *J. Biol. Chem.* *268*:10773–10779 (1993).
42. Fedorov, A. N. and Baldwin, T. O., Contribution of cotranslational folding to the rate of formation of native protein structure, *Proc. Natl. Acad. Sci. USA* *92*:1227–1231 (1995).
43. Sugihara, J. and Baldwin, T. O., Effects of 3′ end deletions from the *Vibrio harveyi* *luxB* gene on luciferase subunit folding and enzyme assembly: Generation of temperature-sensitive polypeptide folding mutants, *Biochemistry* *27*:2872–2880 (1988).
44. Kruse, M., Brunke, M., Escher, A., Szalay, A. A., Tropschug, M., and Zimmermann, R., Enzyme assembly after de novo synthesis in rabbit reticulocyte lysate involves molecular chaperones and immunophilins, *J. Biol. Chem.* *270*:2588–2594 (1995).
45. Brunke, M., Tyedmers, J., and Zimmermann, R., Protein folding within and protein transport into mammalian microsomes are differentially affected by photoaffinity labeling of microsomes with 8-azido-ATP, *Biochem. Biophys. Res. Commun.* *218*:454–460 (1996).
46. Sinclair, J. F., Ziegler, M. M., and Baldwin, T. O., Kinetic partitioning during protein folding yields multiple native states, *Nature Struct. Biol.* *1*:320–326 (1994).
47. Thoden, J. B., Holden, H. M., Fisher, A. J., Sinclair, J. F., Wesenberg, G., Baldwin, T. O., and Rayment, I., Structure of the β_2-homodimer of bacterial luciferase from *Vibrio harveyi*: X-ray analysis of a kinetic protein folding trap, *Protein Sci.* *6*:13–23 (1997).
48. Lauffer, M. A., *Entropy-Driven Processes,* Springer-Verlag, Berlin, 1975, 264 pp.
49. Timasheff, S. N. and Grisham, L. M., In vitro assembly of cytoplasmic microtubules, *Annu. Rev. Biochem.* *49*:565–591 (1980).
50. Casjens, S. and Hendrix, R., Control mechanisms in dsDNA bacteriophage assembly, *The Bacteriophages,* Vol. 1, (Calendar, R., ed.), Plenum Press, New York, 1988, pp. 15–91.
51. Casjens, S. and King, J., Virus assembly, *Annu. Rev. Biochem.* *44*:555–611 (1975).
52. Kellenberger, E. and Wunderli-Allenspach, H., Electron microscopic studies on intracellular phage development—history and perspectives, *Micron* *26*:213–245 (1995).
53. Black, L. W., Showe, M. K., and Steven, A. C., Morphogenesis of the T4 head, *Molecular Biology of bacteriophage T4* (Karam, J. D., ed.), American Society for Microbiology, Washington, DC, 1994, pp. 218–258.
54. Conway, J. F., Duda, R. L., Cheng, N., Hendrix, R. W., and Steven, A. C., Proteolytic

and conformational control of virus capsid maturation: The bacteriophage HK97 system, *J. Mol. Biol. 253*:86–99 (1995).

55. Hayashi, M., Aoyama, A., Richardson, Jr., D. L., and Hayashi, M. N., The biology of bacteriophage ΦX174, *The Bacteriophages,* Vol. 2 (Calendar, R., ed.), Plenum Press, New York, 1988, pp. 1–7.
56. Prevelige, Jr., P. E., Thomas, D., and King, J., Nucleation and growth phages in the polymerization of coat and scaffolding subunits into icosahedral procapsid shells, *Biophys. J. 64*:824–835 (1993).
57. Aoyama, A., Hamatake, R. K., and Hayashi, M., Morphogenesis of phi X174: In vitro synthesis of infectious phage from purified viral components. *Proc. Natl. Acad. Sci. USA 78*:7285–7289 1981).
58. Cerritelli, M. E. and Studier, F. W., Assembly of T7 capsids from independently expressed and purified head protein and scaffolding protein, *J. Mol. Biol. 258*:286–298 (1996).
59. Xie, Z. and Hendrix, R. W., Assembly in vitro of bacteriophage HK97 proheads, *J. Mol. Biol. 253*:74–85 (1995).
60. Ilag, L. L., Olson, N. H., Dokland, T., Music, C. L., Cheng, R. H., Bowen, Z., McKenna, R., Rossmann, M. G., Baker, T. S, and Incardona, N. L., DNA packaging intermediates of bacteriophage phi X174, *Structure 3*:353–363 (1995).
61. Thuman-Commike, P. A., Greene, B., Jakana, J., Prasad, B. V. V., King, J., Prevelige, Jr., P. E., and Chiu, W., Three-dimensional structure of scaffolding-containing phage P22 procapsids by cryomicroscopy, *J. Mol. Biol. 260*:85–98 (1996).
62. Dokland, T., McKenna, R., Ilag, L. L., Bowman, B. R., Incardona, N. L., Fane, B. A., and Rossmann, M. G., Structure of a viral assembly intermediate with molecular scaffolding, submitted.
63. Marvik, O. J., Dokland, T., Nokling, R. H., Jacobsen, E., Larsen, T., and Lindqvist, B. H., The capsid size-determining protein Sid forms an external scaffold on phage P4 procapsids, *J. Mol. Biol. 251*:59–75 (1995).
64. Preston, V. G., al-Kobaisi, M. F., McDougall, I. M., and Rixon, F. J., The herpes simplex virus gene *UL26* proteinase in the presence of the UL26.5 gene product promotes the formation of scaffold-like structures, *J. Gen. Virol. 75*:2355–2366 (1994).
65. Hasson, T. B., Ornelles, D. A., and Shenk, T., Adenovirus L1 52- and 55-kilodalton proteins are present within assembling virions and colocalize with nuclear structures distinct from replication centers, *J. Virol. 66*:6133–6142 (1992).
66. Caspar, D. L. and Klug, A., Physical principles in the construction of regular viruses, *Cold Spring Harbor Symp. Quant. Biol. 27*:1–32 (1962).
67. Prasad, B. V. V., Prevelige, P. E., Marietta, E., Chen, R. O., Thomas, D., King, J., and Chiu, W., Three-dimensional transformation of capsids associated with genome packaging in a bacterial virus, *J. Mol. Biol. 231*:65–74 (1993).
68. Prevelige, Jr., P. E., Thomas, D., Aubrey, K. L., Towse, S. A., and Thomas, Jr., G. J., Subunit conformational changes accompanying bacteriophage P22 capsid maturation, *Biochemistry 32*:537–543 (1993).
69. Earnshaw, W. and King, J., Structure of phage P22 coat protein aggregates formed in the absence of the scaffolding protein, *J. Mol. Biol. 126*:721–747 (1978).

70. Berger, B., Shor, P. W., Tucker-Kellogg, L., and King, J., Local rule-based theory of virus shell assembly, *Proc. Natl. Acad. Sci. USA 91*:7732–7736 (1994).
71. King, J. and Casjens, S., Catalytic head assembling protein in virus morphogenesis, *Nature 251*:112–119 (1974).
72. Tuma, R., Prevelige, Jr., P. E., and Thomas, Jr., G. J., Structural transitions in the scaffolding and coat proteins of P22 virus during assembly and disassembly, *Biochemistry 35*:4619–4627 (1996).
73. Greene B. and King, J., Binding of scaffolding subunits within the P22 procapsid lattice, *Virology 205*:188–197 (1994).
74. Gordon, C. L. and King, J., Genetic properties of temperature-sensitive folding mutants of the coat protein of phage P22, *Genetics 136*:427–438 (1994).
75. Teschke, C. M. and King, J., In vitro folding of phage P22 coat protein with amino acid substitutions that confer in vivo temperature sensitivity, *Biochemistry 34*:6815–6826 (1995).
76. Georgopoulos, C. P. and Linder, C. H., Molecular chaperones in T4 assembly, *Molecular Biology of Bacteriophage T4* (Karam, J. D., ed.), American Society for Microbiology, Washington, DC 1994, pp. 213–217.
77. Wood, W. B., Eiserling, F. A., and Crowther, R. A., Long tail fibers: Genes, proteins, structure, and assembly, *Molecular Biology of Bacteriophage T4* (Karam, J. D., ed.), American Society for Microbiology, Washington, DC, 1994, pp. 282–290.
78. King, J. and Laemmli, U. K., Polypeptides of the tail fibres of bacteriophage T4, *J. Mol. Biol. 62*:465–477 (1971).
79. Ward, S. and Dickson, R. C., Assembly of bacteriophage T4 tail fibers. 3. Genetic control of the major tail fiber polypeptides, *J. Mol. Biol. 62*:479–492 (1971).
80. Wood, W. B. and Conley, M. P., Attachment of tail fibers in bacteriophage T4 assembly: Role of the phage whiskers, *J. Mol. Biol. 127*:15–29 (1979).
81. Cerritelli, M. E., Wall, J. S., Simon, M. N., Conway, J. F., and Steven, A. C., Stoichiometry and domainal organization of the long tail-fiber of bacteriophage T4: a hinged viral adhesin, *J. Mol. Biol. 260*:767–780 (1996).
82. Stouten, P. F., Sander, C., Ruigrok, R. W., and Cusack, S., New triple-helical model for the shaft of the adenovirus fibre, *J. Mol. Biol. 226*:1073–1084 (1992).
83. Revel, H. R., Herrmann, R., and Bishop, R. J., Genetic analysis of T4 tail fiber assembly. II. Bacterial host mutants that allow bypass of T4 gene *57* function, *Virology 72*:255–265 (1976).
84. Montag, D. and Henning, U., An open reading frame in the *Escherichia coli* bacteriophage lambda genome encodes a protein that functions in assembly of the long tail fibers of bacteriophage T4, *J. Bacteriol. 169*:5884–5886 (1987).
85. Hashemolhosseini, S., Stierhof, Y. D., Hindennach, I., and Henning, U., Characterization of the helper proteins for the assembly of tail fibers of coliphage T4 and λ, *J. Bacteriol. 178*:6258–6265 (1996).
86. Hultgren, S. J., Normark, S., and Abraham, S. N., Chaperone-assisted assembly and molecular architecture of adhesive pili, *Annu. Rev. Microbiol. 45*:383–415 (1991).
87. Klemm, P., Jorgensen, B. J., Kreft, B., and Christiansen, G., The export systems of type 1 and F1C fimbriae are interchangeable but work in parental pairs, *J. Bacteriol 177*:621–627 (1995).
88. Hultgren, S. J., Jacob-Dubuisson, F., Jones, C. H., and Branden, C. I., PapD and

superfamily of periplasmic immunoglobulin-like pilus chaperones, *Adv. Protein Chem. 44*:99–123 (1993).
89. Hultgren, S. J., Lindberg, F., Magnusson, G., Kihlberg, J., Tennent, J. M., and Normark, S., The PapG adhesin of uropathogenic *Escherichia coli* contains separate regions for receptor binding and for the incorporation into the pilus, *Proc. Natl. Acad. Sci. USA 86*:4357–4361 (1989).
90. Jacob-Dubuisson, F., Striker, R., and Hultgren, S. J., Chaperone-assisted self-assembly of pili independent of cellular energy, *J. Biol. Chem. 269*:12447–12455 (1994).
91. Holmgren A., and Branden, C. I., Crystal structure of chaperone protein PapD reveals an immunoglobulin fold, *Nature 342*:248–251 (1989).
92. Kuehn, M. J., Ogg, D. J., Kihlberg, J., Slonim, L. N., Flemmer, K., Bergfors, T., and Hultgren, S. J., Structural basis of pilus subunit recognition by the PapD chaperone, *Science 262*:1234–1241 (1993).
93. Slonim, L. N., Pinkner, J. S., Branden, C. I., and Hultgren, S. J., Interactive surface in the PapD chaperone cleft is conserved in pilus chaperone superfamily and essential in subunit recognition and assembly, *EMBO J. 11*:4747–4756 (1992).
94. Xu, Z., Jones, C. H., Haslam, D., Pinkner, J. S., Dodson, K., Kihlberg, J., and Hultgren, S. J., Molecular dissection of PapD interaction with PapG reveals two chaperone-binding sites, *Mol. Microbiol. 16*:1011–1120 (1995).
95. Kraulis, P., MOLSCRIPT: A program to produce both detailed and schematic plots of protein structures, *J. Appl. Crystallogr. 24*:946–950 (1991).

17

Early Events in the Biosynthesis of Secretory Pathway Proteins: Role of Molecular Chaperones

Joseph W. Brewer
and Linda M. Hendershot
St. Jude Children's Research Hospital, Memphis, Tennessee

I. INTRODUCTION

Proteins that are components of the endoplasmic reticulum (ER), Golgi apparatus, nuclear envelope, lysosomes, and endosomes, as well as those proteins that are expressed at the cell surface or secreted from the cell, are recognized as nascent chains, directed to the secretory pathway, and then translocated across the ER membrane. Inside the ER, they begin to fold and often assemble into heterotypic protein complexes. Quality control mechanisms ensure that only "correct" proteins survive and move on. Failure to navigate the control mechanisms results in degradation of the offending protein. It is increasingly clear that molecular chaperones are involved at all steps of these processes and act both to aid in the successful competition of ER protein maturation and to detect and eliminate those proteins that fail to mature properly (Fig. 1).

II. ENTERING THE SECRETORY PATHWAY

Translocation into the ER can occur both co- and posttranslationally. These two pathways are mechanistically distinct, but share many components [1]. In bacteria most protein translocation is posttranslational, in yeast both co- and posttranslational translocation are observed, and in mammals translocation is almost entirely cotranslational.

A. Posttranslational Translocation

Posttranslational protein translocation in yeast is not dependent on either the signal recognition particle (SRP) or the SRP receptor. It does require the activity of cytosolic Hsp70s [2,3], an ER lumenal Hsp70 known as Kar2p/BiP [4,5], and ATP. The addition of urea to in vitro protein translocation systems can substitute for cytosolic Hsp70 proteins, suggesting that they do not push nascent chains through the translocation channel but, instead, act to keep them from folding or aggregating in the cytoplasm [2]. This presumably ensures that nascent proteins enter the translocation channel in an extended form. The energy for translocation may come from Kar2p/BiP pulling the nascent chain into the ER. It appears that Kar2p binds directly to the translocating chain and through interactions with Sec61p and Sec63p, a DnaJ-domain containing protein, undergoes ATP-regulated cycles of binding and release to rachet the protein into the lumen of the ER [5,6]. An ATP transporter was recently identified in the yeast ER that is essential for the translocation of newly synthesized proteins [7].

B. Cotranslational Translocation

In mammalian cells, synthesis of secretory pathway proteins takes place on ribosomes bound to ER membranes [1]. The nascent polypeptide chain emerging from the ribosome enters the lumen of the ER cotranslationally, with the energy for translocation apparently being provided by chain elongation. The targeting of the nascent chain involves recognition by signal recognition particle (SRP) of a hydrophobic signal sequence on the nascent chain emerging from the ribosome. This complex is composed of six protein subunits and one 7S RNA molecule and is responsible for both arresting the elongation of the nascent chain and targeting the ribosome to the ER membrane. Binding of SRP to the SRP receptor on ER membranes releases SRP from the signal sequence and allows translation to resume. The ribosome is then anchored to the membrane by a ribosome receptor(s), after which the nascent chain engages the translocation machinery [8]. Several proteins have been identified by genetic and biochemical means that play a role in translocation. In mammalian cells, the trimeric Sec61 complex (forms a proteinaceous translocation channel and anchors the ribosome to the ER membrane), SRP receptor (targets the nascent chain to the membrane), and translocating-chain-associated membrane protein (TRAM; engages the hydrophobic signal sequence) may be all that is required for translocation of some proteins into the ER [9]. However, studies with reconstituted mammalian ER vesicles have sug-

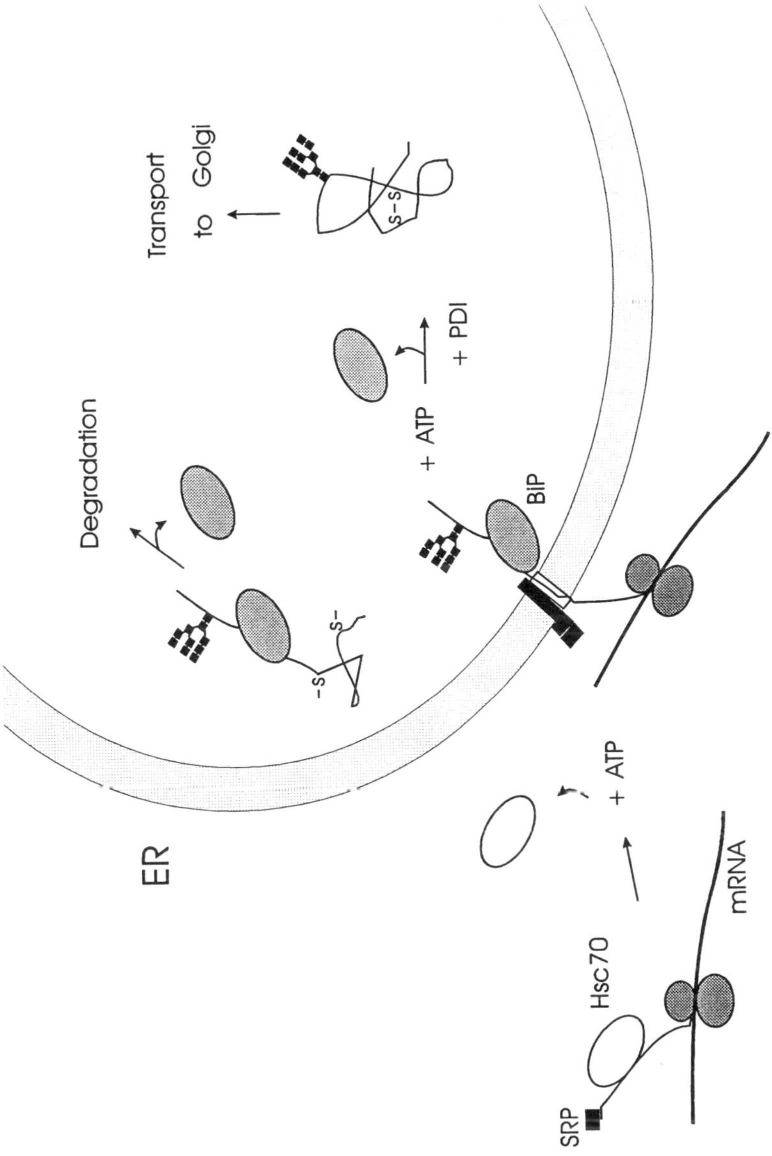

Figure 1 Model for role of Hsp70 proteins in the biosynthesis of ER proteins.

gested that lumenal ER proteins improve the complete translocation of proteins, such as preprolactin and prepro-α factor [10].

Currently, there is no direct evidence that mammalian BiP plays a role in translocation. In fact, neither lowering of cellular ER levels of BiP [11] nor expression of BiP ATPase mutants [12] has any visible effect on the translocation of proteins into the ER of mammalian cells. In addition, no Sec63p homologue or DnaJ domain ER proteins have been identified in mammals, raising the possibility that Hsp70s are much more important in the posttranslational translocation observed in yeast than in the cotranslational translocation found in mammals.

III. FOLDING AND ASSEMBLY OF NEWLY SYNTHESIZED PROTEINS IN THE ENDOPLASMIC RETICULUM

In addition to the problems encountered by all proteins attempting to fold in a cell, the environment of the ER provides additional constraints. The nascent chain translocates into an oxidizing environment containing millimolar concentrations of calcium and a concentrated soup of many different proteins in various stages of folding and assembly. The oxidizing environment causes disulfide bonds to form between cysteine residues that stabilize folding intermediates and attach protein subunits. In many proteins, the correct disulfide linkages do not occur between sequential cysteines but, instead, occur between disparate cysteines that are brought into close proximity after folding commences. Thus, the potential for incorrect disulfide bond formation in a protein that is being extruded one amino acid at a time into the ER seems staggering. In spite of these challenges, protein folding is very efficient in the ER environment. In fact, agents that alter calcium concentrations [13], ATP [14], or the redox potential of the ER [14] deleteriously affect protein folding and secretion.

In the past several years, proteins that interact transiently with incompletely folded and assembled proteins in the ER have been identified and characterized. Some of these are enzymes that catalyze reactions that promote or stabilize protein folding (PDI and PPIase) [15]. Others, termed *molecular chaperones* (e.g., BiP/GRP78, GRP94, calnexin, Hsp47), bind transiently to nascent polypeptides and appear to stabilize protein-folding intermediates by either preventing incorrect interactions between protein domains or by promoting correct ones [15,16]. Two of these chaperones (BiP and GRP94) represent the ER homologues of heat-shock proteins (Hsp70 and Hsp90); thus, information obtained for cytosolic homologues has provided insights into the functions of the ER family members. Others have no homology to heat-shock proteins, but are coordinately regulated by the same stress stimuli that transcriptionally activate BiP and GRP94.

A. The Role of BiP/GRP78 in Protein Folding in the Endoplasmic Reticulum

Immunoglobulin (Ig) heavy chain *binding p*rotein (BiP) was first identified stably and noncovalently associated with unassembled, nontransported Ig heavy chains in Abelson virus-transformed pre-B cell lines [17] and, subsequently, was shown to transiently bind to assembling Ig intermediates, but not to completely assembled Ig molecules [18]. Further studies revealed that BiP associates stably with many other ER proteins that are unable to acquire their transport competent conformation, such as unassembled protein subunits, mutant (misfolded) proteins, and glycosylation-defective proteins [15]. The binding of BiP to wild-type proteins is restricted to incompletely folded intermediates or unassembled subunits, with BiP release occurring just before or at completion of folding and assembly [19,20].

All Hsp70 proteins can be divided into two functional domains: (a) the NH_2-terminal ATP-binding domain and (2) the COOH-terminal protein-binding domain. The binding of ADP in the NH_2-terminal pocket stabilizes protein binding to the COOH-terminal portion, whereas binding of ATP changes the conformation of the protein-binding domain causing the release of bound proteins [21]. That BiP interacts with so many different proteins that do not contain any obvious conserved sequence motifs suggests that BiP does not recognize a unique linear amino acid sequence. Studies using randomly generated peptides demonstrated that peptides containing hydrophobic amino acids are the most effective stimulators of BiP's ATPase activity [22]. Affinity screening of a peptide display library revealed that BiP preferentially bound to peptides with the heptameric motif of HyXHyXHyXHy, where Hy is a bulky aromatic or hydrophobic residue [23]. The alternating pattern of Hy residues in the binding motif is compatible with BiP binding to proteins when they are in an extended conformation, causing the bulky aromatic or hydrophobic side chains to lie on one side of the protein and, presumably, to point into the protein binding site on BiP. These results are consistent with the idea that BiP binds to hydrophobic stretches on nascent or unassembled polypeptides that become inaccessible after completion of folding and assembly. Corroborating data came from experiments in which either wild-type or mutant proinsulin was coexpressed in cells with peroxidase. Peroxidase catalyzed iodination of an exposed tyrosine that could be masked by BiP and that was not exposed in mature, properly folded insulin [24]. In addition, recent structural data obtained for the protein-binding domain of DnaK, the bacterial Hsp70 homologue, revealed that the complexed peptide was in an extended conformation [25]. Thus, Hsp70 proteins can clearly bind to unfolded regions on proteins. However, as BiP also binds protein subunits, it is unclear whether in both cases BiP binds to unfolded regions of the protein, or whether BiP is able to associate with hydrophobic faces of folded protein subunits as well as with linear hydrophobic regions on nascent chains.

Similar to all Hsp70 proteins, BiP binds ATP tightly [26] and has a weak ATPase activity [27] that is stimulated in vitro by small hydrophobic peptides [22] and that induces BiP's release from bound Ig heavy chains [28] and many other proteins [15]. The three-dimensional structure of an ATP-hydrolyzing fragment of Hsc70, which comprises the NH_2-terminal 401 amino acids, has been solved [29]. BiP presumably forms a similar ATP-binding structure, because mutation of BiP residues corresponding to those in Hsc70 that are implicated in nucleotide interactions severely inhibit BiP's ATPase activity [30]. The BiP ATPase mutants bind Ig heavy chains in vivo, but are not released in vitro by ATP [30]. The in vivo expression of mammalian BiP ATPase mutants [12] and yeast BiP (Kar2p) mutants that map to the ATP-binding domain [31] block the in vivo folding of substrate proteins, providing direct evidence for BiP's involvement in protein folding. The stable binding of BiP mutants to wild-type proteins supports the hypothesis that BiP normally undergoes multiple cycles of ATP-mediated binding to and release from unfolded regions on proteins, and that folding occurs while the protein is not bound to BiP. If the protein is unable to fold or has not completed folding, BiP rebinds. This would explain both the transient binding of BiP to wild-type proteins and the stable (or more accurately "continuous") binding to mutant proteins.

B. *N*-Linked Glycosylation and Protein Folding

Most proteins that enter the ER are cotranslationally modified by addition of a complex carbohydrate structure consisting of two *N*-acetylglucosamine, nine mannose, and three glucose residues to the NH_2 group (*N*-linked glycosylation) on the side chain of an asparagine [32]. The glucose molecules are removed one at a time by glucosidase I and II, which reside in the ER. If a protein remains in the ER, a single glucose can be re-added by UDP-glucosyltransferase (UDP-GT) [33], only to be trimmed again by glucosidase II. Alternatively, while the protein is still in the ER, the nonglucosylated sugar complex can be acted on by mannosidase I, which removes four of the nine mannoses. Once more than two of the mannose sugars are removed, the glycan is no longer a substrate for UDP-GT [34]. The reasons for this intricate addition and removal of glucose moieties that are not a component of the mature glycoprotein remained a puzzle until very recently.

Calnexin was simultaneously identified as a transmembrane 90-kDa calcium-binding, ER phosphoprotein [35] and as a molecular chaperone owing to its transient association with various assembling proteins [36–38]. The binding of proteins to calnexin is dependent on both the presence of *N*-linked carbohydrate moieties and on the incomplete folding of the target proteins [39]. Studies using inhibitors of glucosidase I and II, mutant cell lines with defects in the processing of *N*-linked glycans, and direct analyses of the carbohydrates on substrate proteins demonstrated that calnexin recognizes the monoglucosylated intermediate that occurs during ER processing of *N*-linked sugars [40,41]. These data, coupled with the observation that UDP-GT prefers unfolded proteins as substrates [33], led Helenius and colleagues [42] to hypothesize the following scenario: Nascent glycoproteins entering the ER are acted on by glucosidase I and II before translocation and folding are complete. After glucosidase II exposes the monoglucosylated sugar complex, calnexin binds the nascent protein. Removal of the remaining glucose by glucosidase II results in the release of calnexin from the glycoprotien. If the protein is incompletely folded, UDP-GT binds and adds back a single glucose residue, restoring the ability of calnexin to bind the protein. The process continues until folding is complete. In this scheme, the competition between UDP-GT and the ER mannosidase, which irreversibly destroys the UDP-GT recognition substrate, would control the rebinding of calnexin to proteins.

The precise role that calnexin plays in protein biosynthesis is not yet clear. The observations that influenza hemagglutinin (HA), a calnexin substrate, folds normally and is secreted in the presence of glucosidase inhibitors [40] and that the major histocompatibility complex (MHC) class I molecules are expressed and transported normally in both glucosidase II^{-} [43] and calnexin^{-} [44] cells is more compatible with calnexin monitoring protein folding and preventing incompletely folded proteins from leaving the ER, rather than its directly assisting protein folding.

C. Other Resident Endoplasmic Reticulum Proteins May Act as Molecular Chaperones

Calreticulin, a soluble resident ER protein that plays a major role in maintaining ER calcium stores [45], may also act as a molecular chaperone, for it associates transiently with precursors of myeloperoxidase [46], the immature human immunodeficiency virus (HIV) envelope protein gp160 [47], and incompletely folded influenza hemagglutinin (HA) molecules [48]. Calreticulin binding is carbohydrate-dependent and requires the trimming of glucose residues, suggesting that it associates with target proteins by a mechanism similar to that described for calnexin.

GRP94 was originally identified as a protein coordinately regulated by agents that induce BiP synthesis [49,50]. Evidence for its involvement in ER protein folding has been difficult to obtain and is limited to only a few proteins, including Ig heavy and light chains [51], procollagen [52], and the lumenal portion of the *erbB-2* tyrosine kinase [53]. Similar to BiP, GRP94 appears to preferentially associate with unassembled proteins and binds ATP, but unlike BiP, GRP94 is not released from proteins in vitro with ATP [54]. Geldanamycin, a benzoquinoid anisamycin, binds both Hsp90 [55] and GRP94 [53] and inhibits Hsp90 binding to various cytosolic proteins, including v-*src* [55] and the progesterone receptor [56]. In the absence of Hsp90 association, these proteins do not reach their mature form and are inactive. Interestingly, treatment of cells expressing erbB-2 with geldanamycin causes the release of GRP94 from *erbB-2* and results in a rapid down-modulation of *erbB-2* protein [53]. This suggests that one role of GRP94 may be to protect nascent chains from degradation.

Unlike the molecular chaperones discussed thus far, Hsp47 is found only in the ER of collagen-secreting cells. Hsp47 binds transiently to newly synthesized procollagen molecules and more stably to conformationally abnormal ones [57]. The depletion of Hsp47 by antisense oligonucleotides causes a reduction in the amount of full-length procollagen chains and an increase in shorter procollagen peptides that are associated with peptidyl tRNA [58]. This suggests that Hsp47 may be involved in procollagen translocation and is reminiscent of the role of yeast BiP in translocation [4]. Procollagen–Hsp47 complexes can be dissociated in vitro by decreasing the pH [57], suggesting that Hsp47 may dissociate as the complex moves through the increasingly acidic secretory pathway.

D. Endoplasmic Reticulum Folding Enzymes

Protein disulfide isomerase (PDI), a multifunctional resident ER protein that is structurally related to thioredoxin [59], catalyzes the formation and isomerization of disulfide bonds both in vitro [60–62] and in isolated microsomes [63]. Mutational analyses of PDI revealed that the cysteines in the two NH_2-terminal thioredoxin-like regions are required for the catalytic activity of PDI [64]. The protein-binding site resides in a highly acidic region in the COOH-terminus of PDI and not at the active site thioredoxin domains [65]. In vitro refolding experiments revealed that PDI could prevent the aggregation of glyceraldehyde-3-phosphate dehydrogenase, a protein with no disulfide bonds, and promote its proper folding [66], suggesting that PDI may also have chaperonin activity. However, in another study PDI could not promote the folding of denatured Fab molecules that still had intact disulfide bonds, but was able to increase the yield of active Fab molecules that had been both denatured and reduced [61], implying that PDI increases the yield of folded products primarily by catalyzing disulfide bond formation. A yeast homologue of PDI has been identified that catalyzes the formation and isomerization of disulfide bonds and is an essential gene [67]. Two other resident ER proteins, ERp72 and ERp61, have been identified that have homology to PDI and possess disulfide isomerization activity [68,69]. ERp72 can catalyze the reappearance of biologically active Fab fragments and RNase after they have been denatured and reduced [68]. Both mammalian PDI and ERp72 complement yeast PDI mutants, but ERp61 does not [70].

Two families of cytosolic immunophilins have been identified that possess peptidyl prolyl *cis–trans*-isomerase (PPIase) activity [71], an activity that catalyzes a rate-limiting step in protein folding. The ER homologues of both cyclophilin [72] and the FK 506-binding protein [73] have been identified. Although no data has been obtained to implicate either ER immunophilin in protein folding, given their in vitro PPIase activity [72,73] and that cytosolic members stimulate the refolding of Fab molecules [74] and RNase [75], it is anticipated that the ER immunophilins will also have protein-folding activity.

E. Interactions Between Chaperones and Folding Enzymes

It is still unclear whether the various ER chaperones represent a redundant function, with each chaperone recognizing a different structure on unfolded proteins to ensure that all unfolded proteins are recognized, or whether they might act coordinately or sequentially to fold ER proteins. A few recent studies have reported the binding of multiple chaperones to a particular unfolded protein [52,54,76], but it was difficult to assess whether each chaperone bound to a different subset of the unfolded protein, or whether multiple chaperones were bound to a single polypeptide chain. A study on the binding of GRP94 and BiP to Ig light chains concluded that BiP binds to a less-oxidized form of light chain and GRP94 binds to the more-oxidized form [54], providing evidence for sequential action of these chaperones. Similarly, one group reported that BiP binds to a less mature form of vesicular stomatitis virus (VSV) G protein, and calnexin binds later to a more oxidized form [77]. However, another study with reduced thyroglobulin revealed the opposite order for BiP and calnexin association during folding [78]. Although the structure that UDP-GT binds to on unfolded proteins has not been determined, because both BiP and UDP-GT recognize unfolded proteins and probably undergo multiple cycles of binding and release, it is possible that they compete for the same region. Thus, it is not yet at all clear whether chaperones bind independently or in an ordered fashion. In addition, it is not known if proteins differ in the cohort of chaperones with which they associate, or whether chaperones bind in an order determined by the folding pathway of individual proteins.

Some data suggest cooperativity or ordered interactions between an ER chaperone (BiP) and a folding enzyme (PDI). Proteins bound to BiP are often partially oxidized [54,79,80], and alteration of the ER redox potential prevents disulfide bond formation, causing the stable association of proteins with BiP [81]. ATPase mutants of both yeast BiP [31] and mammalian BiP [12] block the in vivo oxidation of their substrate proteins, implying that BiP release must occur for PDI to catalyze disulfide bond formation. In vitro refolding experiments, using reduced and denatured lysozyme, suggest that BiP and PDI can compete for the same site on an unfolded protein and demonstrate that BiP–lysozyme complexes are not good substrates for PDI [60].

IV. IDENTIFICATION AND REMOVAL OF PROTEINS THAT FAIL TO MATURE PROPERLY

A. Recognition of Substrate Proteins

Proteins failing to successfully fold and assemble are detected by a quality control system and then often degraded in the ER [82]. Quality control appears to be partly mediated by prolonged binding of defective proteins to ER chaperones (e.g., BiP and calnexin), suggesting that the processes of assembly and degradation may be interrelated. Some proteins are released from BiP before their degradation [80,83], and both BiP and calnexin can protect certain proteins from proteolysis [24,41]. Thus, these two chaperones can prevent the transport of defective proteins from the ER, but are not directly involved in the degradation process. For some proteins, retention in the ER is dependent on the recognition of free thiols on incompletely assembled subunits [84,85]. Mutation of the appropriate unpaired cysteine residue to alanine allows monomeric IgM and IgA to be secreted [86]. Interestingly, the presence of free thiols has also been implicated as a signal for the degradation of some proteins [84,87]. The cellular machinery for recognizing free thiols has not been identified.

B. Characterization of the Endoplasmic Reticulum Degradation System

Degradation of proteins by an ER system was first described for unassembled T-cell receptor subunits [88]. This degradation is characterized by its resistance to inhibitors of lysosomal proteases and transport inhibitors and its sensitivity to low temperatures [89]. Many of the proteins known to be degraded in the ER are membrane proteins, but several soluble ER proteins also appear to be targets for this pathway. There is substantial diversity in the half-lives of various substrate proteins and in the requirements for their degradation. Some proteins require ATP for degradation; others do not. The requirements for maintenance of normal ER calcium levels and ER redox potential vary, as do the involvement of cysteine, serine, or metalloproteases.

Both ERp61 and ERp72 have cysteine protease activity in vitro [90,91] and are potential candidates for the ER proteolytic machinery for the following reasons: (a) the ER degradation signal for some proteins requires the presence of an unpaired cysteine [87]; (b) both of these proteins may recognize cysteine residues based on their homology to PDI and that both are inhibited by cysteine protease inhibitors [91]; and (c) both proteins are up-regulated in cells with high secretory capacity [69] and by conditions that lead to the accumulation of unfolded proteins in the ER (see later discussion). However, there is as yet no direct evidence that either of these proteins plays a role in protein degradation in vivo.

C. Site of Endoplasmic Reticulum Degradation

Several recent studies have suggested that not all "ER degradation" actually takes place in the ER. In most cases of cystic fibrosis, mutations in the cystic fibrosis transmembrane conductance regulator (CFTR) lead to its failure to mature beyond the ER and its rapid degradation. Degradation of the CFTR can be inhibited by proteasome inhibitors, which induce the accumulation of polyubiquitinated forms of CFTR [92]. CFTR degradation was inhibited by expression of a dominant negative ubiquitin mutant and in cells bearing a temperature-sensitive mutation in the ubiquitin-activating enzyme [92]. In yeast, a ubiquitin-conjugating enzyme (UBC6) has been identified that is associated with the ER membrane. The active site of the enzyme is cytosolic, suggesting that proteasome-mediated degradation of ER proteins may be initiated by ubiquitination of cytosolic lysines. The development of an in vitro ER-associated protein degradation system in yeast demonstrated a requirement of cytosol, ATP, and calnexin for the degradation of unglycosylated prepro-α factor [93]. Cytosol from a yeast strain with a defect in the proteasome is unable to support degradation of pp-αF in this system, demonstrating the involvement of the cytosolic proteasome. Thus, it appears that at least some ER proteins can be degraded by a ubiquitin-dependent proteasome pathway. Whether proteasomal degradation is the fate of most ER transmembrane proteins remains to be determined. It is not known how the remaining lumenal portions of these transmembrane proteins are degraded, or if the proteasome has any role in the degradation of soluble ER proteins that have folded and undergone some covalent subunit assembly, such as Ig heavy chain dimers that are associated with BiP.

V. THE UNFOLDED PROTEIN RESPONSE

Proteins that mature within the secretory pathway enjoy a high success rate for folding and assembly, as reflected by the estimate that more than 95% of newly synthesized proteins achieve a mature functional configuration [15]. However, various physiological conditions can perturb this normally efficient process, causing proteins to remain unfolded and to accumulate within the ER, stably bound to molecular chaperones. Although many unfolded proteins are destined for rapid degradation, they also play a key role in initiating a signaling pathway that ultimately results in an increased synthesis of the ER chaperones, thereby allowing the cell to positively respond to stressful conditions in the ER.

A. Endoplasmic Reticulum Stress and Chaperone Up-Regulation

Increased expression of normal proteins that transit the secretory pathway has only a modest effect on ER chaperone levels, whereas the accumulation of unfolded proteins [94] or protein subunits produced in stoichiometric excess [95] leads to a significant rise in certain ER chaperones, particularly the glucose-regulated proteins (GRPs). Originally identified as proteins that are induced by glucose starvation [50], ER chaperones such as BiP/GRP78, GRP94, ERp72, and PDI are also up-regulated by a variety of agents that negatively affect protein biogenesis in the ER. For example, the reducing agents 2-mercaptoethanol and dithiothreitol, which interfere with the oxidizing environment of the ER; the calcium ionophore calcimycin (A-23187) and the inhibitor of the ER calcium ATPase thapsigargin, which deplete ER calcium stores; and the inhibitor of *N*-linked glycosylation tunicamycin, all result in an accumulation of unfolded proteins and potently induce GRP synthesis [96].

B. BiP as an Endoplasmic Reticulum Sensor

Two lines of evidence indicate that it is the binding of unfolded proteins to BiP, rather than to other ER chaperones, that is critical for up-regulated expression of the GRPs. First, expression of an α_1-antitrypsin mutant that accumulates in the ER bound to calnexin, rather than to BiP, has no effect on BiP synthesis [97]. Second, overexpression of BiP from a heterologous promoter down-regulates both endogenous BiP and the other GRPs and attenuates the cellular response to agents that normally activate increased GRP synthesis [98]. In contrast, ectopic expression of two other chaperones (ERp72 and PDI) has no effect on either the basal level or stress-induced synthesis of ER chaperones [99]. Thus, in addition to its roles in facilitating protein folding, BiP also appears to serve as a sensor for stress conditions within the ER lumen. The mechanism by which BiP may be monitored is not yet clear. Although free BiP apparently exists as dimers and is modified by threonine phosphorylation and ADP ribosylation, BiP that is bound to proteins exists as monomers and lacks these modifications [95]. An intriguing possibility is that posttranslational modification of BiP is part of the mechanism by which ER stress is monitored.

C. Transcriptional Control of the GRPs

Activation of the ER stress response results in a coordinate 5- to 25-fold increase in the transcriptional rates of the GRPs. Current data indicate that these genes are regulated by the same factors through common elements in their promoters [100]. The BiP promoter is the best characterized and consists of a functionally redundant 456-bp segment containing multiple CCAAT- and CCAAT-like elements flanked by GC-rich motifs [101]. Two sequences, the CCAAT element most proximal to the TATA box and the GRP core (a 36-bp motif conserved among GRP genes from yeast to humans), appear to be the most critical for both basal level and stress-induced expression of BiP [96,102]. Nuclear factors bind to these elements, but no stress-regulated binding activities have been described. Indeed, our recent studies indicate that the requirement for new protein synthesis is less stringent for GRP induction by agents that unfold existing ER proteins, suggesting that a stress-inducible factor may not be a requirement for transcriptional up-regulation of the GRPs [J. W. Brewer and L. M. Hendershot, in preparation]. This suggests that the transcriptional machinery responsible for ER stress-induced enhancement of GRP transcription is constitutively present and may either be activated by a posttranslational modification(s), or simply be allowed to function more proficiently after an alteration in the chromatin structure of sequences containing GRP genes.

D. Signaling from the Endoplasmic Reticulum to the Nucleus

An ER transmembrane serine/threonine kinase, the Ire1p kinase, has been identified in yeast that appears to be a proximal component of the Kar2p (yeast BiP) signaling pathway in response to the accumulation of unfolded proteins in the ER [103,104]. The NH_2-terminal half of Ire1p kinase is core-glycosylated and resides in the ER lumen [104]. The kinase domain found in the cytoplasmic COOH-terminal "tail" is essential for its function [104] and, presumably, interacts with or phosphorylates other downstream components of the pathway. Similar to growth factor receptors of higher eukaryotic cells, Ire1p appears to oligomerize and then be phosphorylated in *trans* by other Ire1p molecules when unfolded proteins accumulate in the ER [105]. Whether Ire1p detects unfolded proteins by direct interactions with such ligands, or monitors their presence through an interaction with an ER chaperone like BiP, has not been determined. Although a mammalian homologue for Ire1p has not been identified, experiments in mammalian cells using inhibitors of both serine/threonine kinases [106] and phosphatases [107] have yielded data consistent with a role for a serine/threonine kinase in the mammalian ER stress response.

VI. CONCLUSIONS

The in vivo folding of newly synthesized proteins in the ER requires a delicate balance between allowing folding to occur and preventing incorrect interactions that would ultimately lead to improper folding or aggregation, or both. Although this process is surprisingly efficient, it sometimes fails, and the defective proteins must be identified and eliminated. A cohort of resident ER proteins act to increase the efficiency of protein folding and assembly, detect proteins that fail to mature properly and prevent them traversing the secretory pathway, and if necessary, dispose of damaged proteins. Coordinate regulation of the ER chaperones by conditions that adversely affect protein folding provides both an increased opportunity for proteins to fold under unfavorable conditions and the mechanism(s) to rid the cell of defective proteins.

REFERENCES

1. Walter, P. and Johnson, A. E., Signal sequence recognition and protein targeting to the endoplasmic reticulum membrane, *Annu. Rev. Cell Biol. 10*:87 (1994).
2. Chirico, W. J., Waters, M. G., and Blobel, G., 70K heat shock related proteins stimulate protein translocation into microsomes, *Nature 332*:805 (1988).
3. Deshaies, R. J., Koch, B. D., Werner Washburne, M., Craig, E. A., and Schekman, R., A subfamily of stress proteins facilitates translocation of secretory and mitochondrial precursor polypeptides, *Nature 332*:800 (1988).
4. Vogel, J. P., Misra, L. M., and Rose, M. D., Loss of BiP/GRP78 function blocks translocation of secretory proteins in yeast, *J. Cell Biol. 110*:1885 (1990).
5. Sanders, S. L., Whitfield, K. M., Vogel, J. P., Rose, M. D., and Schekman, R. W., Sec61p and BiP directly facilitate polypeptide translocation into the ER, *Cell 69*:353 (1992).
6. Brodsky J. L. and Schekman, R., A Sec63p–BiP complex from yeast is required for protein translocation in a reconstituted proteoliposome, *J. Cell Biol. 123*:1355 (1993).
7. Mayinger, P., Bankaitis, V. A., and Meyer, D. I., Sac1p mediates the adenosine triphosphate transport into yeast endoplasmic reticulum that is required for protein translocation, *J. Cell Biol. 131*:1377 (1995).
8. Rapoport, T. A., Transport of proteins across the endoplasmic reticulum membrane, *Science 258*:931 (1992).
9. Gorlich, D. and Rapoport, T. A., Protein translocation into proteoliposomes reconstituted from purified components of the endoplasmic reticulum membrane, *Cell 75*:615 (1993).
10. Nicchitta, C. V. and Blobel, G., Lumenal proteins of the mammalian endoplasmic reticulum are required to complete protein translocation, *Cell 73*:989 (1993).
11. Dorner, A. J., Krane, M. G., and Kaufman, R. J., Reduction of endogenous GRP78 levels improves secretion of a heterologous protein in CHO cells, *Mol. Cell. Biol. 8*:4063 (1988).
12. Hendershot, L., Wei, J., Gaut, J., Melnick, J., Aviel, S., and Argon, Y., Inhibition of immunoglobulin folding and secretion by dominant negative BiP ATPase mutants, *Proc. Natl. Acad. Sci. USA 93*:5269 (1996).
13. Suzuki, C. K., Bonifacino, J. S., Lin, A. Y., Davis, M. M., and Klausner, R. D., Regulating the retention of T-cell receptor alpha chain variants within the endoplasmic reticulum: Ca(2+)-dependent association with BiP, *J. Cell Biol. 114*:189 (1991).
14. Braakman, I., Helenius, J., and Helenius, A., Role of ATP and disulphide bonds during protein folding in the endoplasmic reticulum, *Nature 356*:260 (1992).
15. Gething, M. J. and Sambrook, J., Protein folding in the cell, *Nature 355*:33 (1992).
16. Ellis, R. J., Molecular chaperones, *Annu. Rev. Biochem. 60*:321 (1991).
17. Haas, I. G. and Wabl, M., Immunoglobulin heavy chain binding protein, *Nature 306*:387 (1983).

18. Bole, D. G., Hendershot, L. M., and Kearney, J. F., Posttranslational association of immunoglobulin heavy chain binding protein with nascent heavy chains in non-secreting and secreting hybridomas, *J. Cell Biol. 102*:1558 (1986).
19. Hendershot, L. M., Immunoglobulin heavy chain and binding protein complexes are dissociated in vivo by light chain addition, *J. Cell Biol. 111*:829 (1990).
20. de Silva, A. M., Balch, W. E., and Helenius, A., Quality control in the endoplasmic reticulum: Folding and misfolding of vesicular stomatitis virus G protein in cells and in vitro, *J. Cell Biol. 111*:857 (1990).
21. Palleros, D. R., Reid, K. L., Shi, L., Welch, W. J., and Fink, A. L., ATP-induced protein–Hsp70 complex dissociation requires K^+ but not ATP hydrolysis, *Nature 365*:664 (1993).
22. Flynn, G. C., Pohl, J., Flocco, M. T., and Rothman, J. E., Peptide-binding specificity of the molecular chaperone BiP, *Nature 353*:726 (1991).
23. Blond Elguindi, S., Cwirla, S. E., Dower, W. J., Lipshutz, R. J., Sprang, S. R., Sambrook, J. F., and Gething, M. J., Affinity panning of a library of peptides displayed on bacteriophages reveals the binding specificity of BiP, *Cell 75*:717 (1993).
24. Schmitz, A., Maintz, M., Kehle, T., and Herzog, V., In vivo iodination of a misfolded proinsulin reveals co-localized signals for Bip binding and for degradation in the ER, *EMBO J. 14*:1091 (1995).
25. Zhu, X., Zhao, X., Burkholder, W. G., Gragerov, A., Ogata, C. M., Gottesman, M. E., and Hendrickson, W. A., Structural analysis of substrate binding by the molecular chaperone DnaK, *Science 272*:1606 (1996).
26. Wei, J.-Y. and Hendershot, L. M., Characterization of the nucleotide binding properties and ATPase activity of recombinant hamster BiP purified from bacteria, *J. Biol. Chem. 270*:26670 (1995).
27. Kassenbrock, C. K. and Kelly, R. B., Interaction of heavy chain binding protein (BiP/GRP78) with adenine nucleotides, *EMBO J. 8*:1461 (1989).
28. Munro, S. and Pelham, H. R., An Hsp70-like protein in the ER: Identity with the 78 kd glucose-regulated protein and immunoglobulin heavy chain binding protein, *Cell 46*:291 (1986).
29. Flaherty, K. M., DeLuca Flaherty, C., and McKay, D. B., Three-dimensional structure of the ATPase fragment of a 70K heat-shock cognate protein, *Nature 346*:623 (1990).
30. Gaut, J. R. and Hendershot, L. M., Mutations within the nucleotide binding site of immunoglobulin-binding protein inhibit ATPase activity and interfere with release of immunoglobulin heavy chain, *J. Biol. Chem. 268*:7248 (1993).
31. Simons, J. F., Ferro-Novick, S., Rose, M. D., and Helenius, A., BiP/Kar2p serves as a molecular chaperone during carboxypeptidase Y folding in yeast, *J. Cell Biol. 130*:41 (1995).
32. Kornfeld, R. and Kornfeld, S., Assembly of asparagine-linked oligosaccharides, *Annu. Rev. Biochem. 54*:631 (1985).
33. Sousa, M. C., Ferrero Garcia, M. A., and Parodi, A. J., Recognition of the oligosaccharide and protein moieties of glycoproteins by the UDP-Glc:glycoprotein glucosyltransferase, *Biochemistry 31*:97 (1992).
34. Trombetta, S. E. and Parodi, A. J., Purification to apparent homogeneity and partial

characterization of rat liver UDP-glucose:glycoprotein glucosyltransferase, *J. Biol. Chem. 267*:9236 (1992).
35. Wada, I., Rindress, D., Cameron, P. H., Ou, W. J., Doherty, J. J., Louvard, D., Bell, A. W., Dignard, D., Thomas, D. Y., and Bergeron, J. J., SSR alpha and associated calnexin are major calcium binding proteins of the endoplasmic reticulum membrane, *J. Biol. Chem. 266*:19599 (1991).
36. Degen, E. and Williams, D. B., Participation of a novel 88-kD protein in the biogenesis of murine class I histocompatibility molecules, *J. Cell Biol. 112*:1099 (1991).
37. Hochstenbach, F., David, V., Watkins, S., and Brenner, M. B., Endoplasmic reticulum resident protein of 90 kilodaltons associates with the T- and B-cell antigen receptors and major histocompatibility complex antigens during their assembly, *Proc. Natl. Acad. Sci. USA 89*:4734 (1992).
38. David, V., Hochstenbach, F., Rajagopalan, S., and Brenner, M. B., Interaction with newly synthesized and retained proteins in the endoplasmic reticulum suggests a chaperone function for human integral membrane protein IP90 (calnexin), *J. Biol. Chem. 268*:9585 (1993).
39. Ou, W.-J., Cameron, P. H., Thomas, D. Y., and Bergeron, J. J. M., Association of folding intermediates of glycoproteins with calnexin during protein maturation, *Nature 364*:771 (1993).
40. Hammond, C., Braakman, I., and Helenius, A., Role of *N*-linked oligosaccharide recognition, glucose trimming, and calnexin in glycoprotein folding and quality control, *Proc. Natl. Acad. Sci. USA 91*:913 (1994).
41. Kearse, K. P., Williams, D. B., and Singer, A., Persistence of glucose residues on core oligosaccharides prevents association of TCR alpha and TCR beta proteins with calnexin and results specifically in accelerated degradation of nascent TCR alpha proteins within the endoplasmic reticulum, *EMBO J. 13*:3678 (1994).
42. Hebert, D. N., Foellmer, B., and Helenius, A., Glucose trimming and reglucosylation determine glycoprotein association with calnexin in the endoplasmic reticulum, *Cell 81*:425 (1995).
43. Balow, J. P., Weissman, J. D., and Kearse, K. P., Unique expression of major histocompatibility complex class I proteins in the absence of glucose trimming and calnexin association, *J. Biol. Chem. 270*:29025 (1995).
44. Scott, J. E. and Dawson, J. R., MHC class I expression and transport in a calnexin-deficient cell line, *J. Immunol. 155*:143 (1995).
45. Treves, S., De Mattei, M., Landfredi, M., Villa, A., Green, N. M., MacLennan, D. H., Meldolesi, J., and Pozzan, T., Calreticulin is a candidate for a calsequestrin-like function in Ca2(+)-storage compartments (calciosomes) of liver and brain, *Biochem. J. 271*:473 (1990).
46. Nauseef, W. M., McCormick, S. J., and Clark, R. A., Calreticulin functions as a molecular chaperone in the biosynthesis of myeloperoxidase, *J. Biol. Chem. 270*:4741 (1995).
47. Otteken, A. and Moss, B., Calreticulin interacts with newly synthesized human immunodeficiency virus type 1 envelope glycoprotein, suggesting a chaperone function similar to that of calnexin, *J. Biol. Chem. 271*:97 (1996).
48. Peterson, J. R., Ora, A., Van, P. N., and Helenius, A., Transient, lectin-like associa-

tion of calreticulin with folding intermediates of cellular and viral glycoproteins, *Mol. Biol. Cell 6*:1173 (1995).

49. Pouyssegur, J., Shiu, R. P., and Pastan, I., Induction of two transformation-sensitive membrane polypeptides in normal fibroblasts by a block in glycoprotein synthesis or glucose deprivation, *Cell 11*:941 (1977).
50. Lee, A. S., Coordinated regulation of a set of genes by glucose and calcium ionophores in mammalian cells, *Trends Biochem. Sci. 12*:20 (1987).
51. Melnick, J., Aviel, S., and Argon, Y., The endoplasmic reticulum stress protein GRP94, in addition to BiP, associates with unassembled immunoglobulin chains, *J. Biol. Chem. 267*:21303 (1992).
52. Ferreira, L. R., Norris, K., Smith, T., Hebert, C., and Sauk, J. J., Association of Hsp47, Grp78, and Grp94 with procollagen supports the successive or coupled action of molecular chaperones, *J. Cell. Biochem. 56*:518 (1994).
53. Chavany, C., Mimnaugh, E., Miller, P., Bitton, R., Nguyen, P., Trepel, J., Whitesell, L., Schnur, R., Moyer, J., and Neckers, L., p185*erbB2* binds to GRP94 in vivo. Dissociation of the p185*erbB2*/GRP94 heterocomplex by benzoquinone anisamycins precedes depletion of p185erbB2, *J. Biol. Chem. 271*:4974 (1996).
54. Melnick, J., Dul, J. L., and Argon, Y., Sequential interaction of the chaperones BiP and Grp94 with immunoglobulin chains in the endoplasmic reticulum, *Nature 370*:373 (1994).
55. Whitesell, L., Mimnaugh, E. G., De Costa, B., Myers, C. E., and Neckers, L. M., Inhibition of heat shock protein HSP90–pp60 v-*src* heteroprotein complex formation by benzoquinone anisamycins: Essential role for stress proteins in oncogenic transformation, *Proc. Natl. Acad. Sci. USA 91*:8324 (1994).
56. Smith, D. F., Whitesell, L., Nair, S. C., Chen, S., Prapapanich, V., and Rimerman, R. A., Progesterone receptor structure and function altered by geldanamycin, an hsp90 binding agent, *Mol. Cell Biol. 15*:6804 (1995).
57. Nakai, A., Satoh, M., Hirayoshi, K., and Nagata, K., Involvement of the stress protein HSP47 in procollagen processing in the endoplasmic reticulum, *J. Cell Biol. 117*:903 (1992).
58. Sauk, J. J., Smith, T., Norris, K., and Ferreira, L., Hsp47 and the translation–translocation machinery cooperate in the production of alpha 1(I) chains of type I procollagen, *J. Biol. Chem. 269*:3941 (1994).
59. Freedman, R. B., Protein disulfide isomerase: Multiple roles in the modification of nascent proteins, *Cell 57*:1069 (1989).
60. Puig, A. and Gilbert, H. F., Anti-chaperone behavior of BiP during the protein disulfide isomerase-catalyzed refolding of reduced denatured lysozyme, *J. Biol. Chem. 269*:25889 (1994).
61. Lilie, H., McLaughlin, S., Freedman, R., and Buchner, J., Influence of protein disulfide isomerase (PDI) on antibody folding in vitro, *J. Biol. Chem. 269*:14290 (1994).
62. Huth, J. R., Perini, F., Lockridge, O., Bedows, E., and Ruddon, R. W., Protein binding and assembly in vitro parallel intracellular folding and assembly. Catalysis of folding and assembly of the human chorionic gonadotropin alpha beta dimer by protein disulfide isomerase, *J. Biol. Chem. 268*:16472 (1993).
63. Bulleid, N. J. and Freedman, R. B., Defective co-translational formation of disulph-

ide bonds in protein disulphide-isomerase-deficient microsomes, *Nature 335*:649 (1988).
64. Lyles, M. M. and Gilbert, H. F., Mutations in the thioredoxin sites of protein disulfide isomerase reveal functional nonequivalence of the N- and C-terminal domains, *J. Biol. Chem. 269*:30946 (1994).
65. Noiva, R., Freedman, R. B., and Lennarz, W. J., Peptide binding to protein disulfide isomerase occurs at a site distinct form the active sites, *J. Biol. Chem. 268*:19210 (1993).
66. Cai, H., Wang, C. C., and Tsou, C. L., Chaperone-like activity of protein disulfide isomerase in the refolding of a protein with no disulfide bonds, *J. Biol. Chem. 269*:24550 (1994).
67. LaMantia, M. L. and Lennarz, W. J., The essential function of yeast protein disulfide isomerase does not reside in its isomerase activity, *Cell 74*:899 (1993).
68. Rupp, K., Birnbach, U., Lundstrom, J., Van, P. N., and Soling, H. D., Effects of CaBP2, the rat analog of ERp72, and of CaBP1 on the refolding of denatured reduced proteins. Comparison with protein disulfide isomerase, *J. Biol. Chem. 269*:2501 (1994).
69. Kozaki, K., Miyaishi, O., Asai, N., et al., Tissue distribution of ERp61 and association of its increased expression with IgG production in hybridoma cells, *Exp. Cell Res. 213*:348 (1994).
70. Gunther, R., Srinivasan, M., Haugejorden, S., Green, M., Ehbrecht, I. M., and Kuntzel, H., Functional replacement of the *Saccharomyces cerevisiae* Trg1/Pdi1 protein by members of the mammalian protein disulfide isomerase family, *J. Biol. Chem. 268*:7728 (1993).
71. Schreiber, S. L., Chemistry and biology of the immunophilins and their immunosuppressive ligands, *Science 251*:283 (1991).
72. Hasel, K. W., Glass, J. R., Godbout, M., and Sutcliffe, J. G., An endoplasmic reticulum-specific cyclophilin, *Mol. Cell. Biol. 11*:3484 (1991).
73. Jin, Y. J., Albers, M. W., Lane, W. S., Bierer, B. E., Schreiber, S. L., and Burakoff, S. J., Molecular cloning of a membrane-associated human FK 506- and rapamycin-binding protein, *FKBP-13, Proc. Natl. Acad. Sci. USA 88*:6677 (1991).
74. Lilie, H., Lang, K., Rudolph, R., and Buchner, J., Prolyl isomerases catalyze antibody folding in vitro, *Protein. Sci. 2*:1490 (1993).
75. Schonbrunner, E. R. and Schmid, F. X., Peptidyl–prolyl *cis–trans* isomerase improves the efficiency of protein disulfide isomerase as a catalyst of protein folding, *Proc. Natl. Acad. Sci. USA 89*:4510 (1992).
76. Kuznetsov, G., Chen, L. B., and Nigam, S. K., Several endoplasmic reticulum stress proteins, including ERp72, interact with thyroglobulin during its maturation, *J. Biol. Chem. 269*:22990 (1994).
77. Hammond, C. and Helenius, A., Folding of VSV G protein: Sequential interaction with BiP and calnexin, *Science 266*:456 (1994).
78. Kim, P. S. and Arvan, P., Calnexin and BiP act as sequential molecular chaperones during thyroglobulin folding in the endoplasmic reticulum, *J. Cell Biol. 128*:29 (1995).
79. Machamer, C. E., Doms, R. W., Bole, D. G., Helenius, A., and Rose, J. K., Heavy

chain binding protein recognizes incompletely disulfide-bonded forms of vesicular stomatitis virus G protein, *J. Biol. Chem. 265*:6879 (1990).

80. Knittler, M. R., Dirks, S., and Haas, I. G., Molecular chaperones involved in protein degradation in the endoplasmic reticulum: Quantitative interaction of the heat shock cognate protein BiP with partially folded immunoglobulin light chains that are degraded in the endoplasmic reticulum, *Proc. Natl. Acad. Sci. USA 92*:1764 (1995).
81. Kassenbrock, C. K., Garcia, P. D., Walter, P., and Kelly, R. B., Heavy-chain binding protein recognizes aberrrant polypeptides translocated in vitro, *Nature 333*:90 (1988).
82. Hammond, C. and Helenius, A., Quality control in the secretory pathway, *Curr. Opin. Cell Biol. 7*:523 (1995).
83. Knittler, M. R. and Haas, I. G., Interaction of BiP with newly synthesized immunoglobulin light chain molecules: Cycles of sequential binding and release, *EMBO J. 11*:1573 (1992).
84. Fra, A. M., Fagioli, C., Finazzi, D., Sitia, R., and Alberini, C. M., Quality control of ER synthesized proteins: An exposed thiol group as a three-way switch mediating assembly, retention and degradation, *EMBO J. 12*:4755 (1993).
85. Brewer, J. W. and Corley. R. B., Quality control in protein biogenesis: Thio-mediated retention monitors the redox state of proteins in the endoplasmic reticulum, *J. Cell Sci. 109*:2383 (1996).
86. Sitia, R., Neuberger, M., Alberini, C., Bet, P., Fra, A., Valetti, C., Williams, G., and Milstein, C., Developmental regulation of IgM secretion: The role of the carboxy-terminal cysteine, *Cell 60*:781 (1990).
87. Stafford, F. J. and Bonifacino, J. S., A permeabilized cell system identifies the endoplasmic reticulum as a site of protein degradation, *J. Cell Biol. 115*:1225 (1991).
88. Lippincott Schwartz, J., Bonifacino, J. S., Yuan, L. C., and Klausner, R. D., Degradation from the endoplasmic reticulum: Disposing of newly synthesized proteins, *Cell 54*:209 (1988).
89. Bonifacino, J. S. and Lippincott-Schwartz, J., Degradation of proteins within the endoplasmic reticulum, *Curr. Opin. Cell Biol. 3*:592 (1991).
90. Urade, R., Nasu, M., Moriyama, T., Wada, K., and Kito, M., Protein degradation by the phosphoinositide-specific phospholipase C-alpha family from rat liver endoplasmic reticulum, *J. Biol. Chem. 267*:15152 (1992).
91. Urade, R., Takenaka, Y., and Kito, M., Protein degradation by ERp72 from rat and mouse liver endoplasmic reticulum, *J. Biol. Chem. 268*:22004 (1993).
92. Ward, C. L., Omura, S., and Kopito, R. R., Degradation of CFTR by the ubiquitin-proteasome pathway, *Cell 83*:121 (1995).
93. McCracken, A. A. and Brodsky, J. L., Assembly of ER-associated protein degradation in vitro: Dependence on cytosol, calnexin, and ATP, *J. Cell Biol. 132*:291 (1996).
94. Kozutsumi, Y., Segal, M., Normington, K., Gething, M. J., and Sambrook, J., The presence of malfolded proteins in the endoplasmic reticulum signals the induction of glucose-regulated proteins, *Nature 332*:462 (1988).
95. Freiden, P. J., Gaut, J. R., and Hendershot, L. M., Interconversion of three differentially modified and assembled forms of BiP, *EMBO J. 11*:63 (1992).

96. Lee, A. S., Mammalian stress response: Induction of the glucose-regulated protein family, *Curr. Opin. Cell Biol. 4*:267 (1992).
97. Graham, K. S., Le, A., and Sifers, R. N., Accumulation of the insoluble PiZ variant of human alpha 1-antitrypsin within the hepatic endoplasmic reticulum does not elevate the steady-state level of grp78/BiP, *J. Biol. Chem. 265*:20463 (1990).
98. Dorner, A. J., Wasley, L. C., and Kaufman, R. J., Overexpression of GRP78 mitigates stress induction of glucose regulated proteins and blocks secretion of selective proteins in Chinese hamster ovary cells, *EMBO J. 11*:1563 (1992).
99. Dorner, A. J., Wasley, L. C., Raney, P., Haugejorden, S., Green, M., and Kaufman, R. J., The stress response in Chinese hamster ovary cells. Regulation of ERp72 and protein disulfide isomerase expression and secretion, *J. Biol. Chem. 265*:22029 (1990).
100. Li, X. A. and Lee, A. S., Competitive inhibition of a set of endoplasmic reticulum protein genes (*GRP78*, *GRP94*, and *ERp72*) retards cell growth and lowers viability after ionophore treatment, *Mol. Cell. Biol. 11*:3446 (1991).
101. Resendez, E. J., Wooden, S. K., and Lee, A. S., Identification of highly conserved regulatory domains and protein-binding sites in the promoters of the rat and human genes encoding the stress-inducible 78-kilodalton glucose-regulated protein, *Mol. Cell. Biol. 8*:4579 (1988).
102. Mori, K., Sant, A., Kohno, K., Normington, K., Gething, M. J., Sambrook, J. F., A 22 bp *cis*-acting element is necessary and sufficient for the induction of the yeast *KAR2* (BiP) gene by unfolded proteins, *EMBO J. 11*:2583 (1992).
103. Cox, J. S., Shamu, C. E., and Walter, P., Transcriptional induction of genes encoding endoplasmic reticulum resident proteins requires a transmembrane protein kinase, *Cell 73*:1197 (1993).
104. Mori, K., Wenzhen, M., Gething, M. J., and Sambrook, J., A transmembrane protein with a cdc2$^+$/CDC28-related kinase activity is required for signalling from the ER to the nucleus, *Cell 74*:743 (1993).
105. Shamu, C. E. and Walter, P., Oligomerization and phosphorylation of the Ire1p kinase during intracellular signaling from the endoplasmic reticulum to the nucleus, *EMBO J. 15*:3028 (1996).
106. Price, B. D. and Calderwood, S. K., Gadd45 and Gadd153 messenger RNA levels are increased during hypoxia and after exposure of cells to agents which elevate the levels of the glucose-regulated proteins, *Cancer Res. 52*:3814 (1992).
107. Cao, X, Zhou, Y., and Lee, A. S., Requirement of tyrosine- and serine/threonine kinases in the transcriptional activation of the mammalian grp78/BiP promoter by thapsigargin, *J. Biol. Chem. 270*:494 (1995).

18

Roles of Molecular Chaperones in Mitochondrial Protein Import

Toshiya Endo
Nagoya University, Nagoya, Japan

I. INTRODUCTION

Mitochondria are organelles that mediate many cellular processes, including oxidative phosphorylation and synthesis of heme, lipids, amino acids, and nucleotides. They consists of four subcompartments: outer membrane, intermembrane space, inner membrane, and matrix, and each compartment contains a specific set of proteins. According to the endosymbiont hypothesis, mitochondria are derived from bacteria. Eukaryotic cells started out as anaerobic organisms without mitochondria. Then more than a billion years ago, primitive aerobic bacteria were endocytosed by these cells and became protomitochondria. During the evolution of mitochondria, most of the genes carried by the protomitochondria have been either lost or transferred to the host nucleus. Thus, the vast majority of proteins in the present-day mitochondria are encoded by nuclear genes, and only a dozen mitochondrial proteins remain encoded by the mitochondrial genome.

Mitochondrial proteins that are encoded by nuclear genes are synthesized on cytosolic ribosomes as precursor proteins and must be imported into the mitochondria. Mitochondrial precursor proteins usually carry NH_2-terminal presequences that target the proteins to mitochondria and that are cleaved by a matrix-localized processing peptidase. Mitochondrial presequences lack a common primary structure, but instead, display common structural and physical properties [1]. Sequence analyses showed that presequences are rich in positively charged residues and have the potential to form an amphiphilic helical structure. Indeed, chemically synthesized presequences, which exhibit little secondary structure in aqueous solutions, adopted amphiphilic helical structures on binding to lipid bilayers or detergent micelles. The information responsible for the correct targeting of a precursor protein to mitochondria depends on the positive charge and amphiphilicity of the presequence.

Protein import into mitochondria is a multistep process that requires coordinated functions by components in the cytosol and in the mitochondria. Investigation in the past decade has indicated the involvement of several molecular chaperones and their partner proteins in this process. Chaperones in the cytosol facilitate protein targeting to mitochondria, mainly by maintaining transport competence of precursor proteins in the cytosol. Chaperones inside the mitochondria drive vectorial protein translocation across the mitochondrial membranes. In this chapter, I will try to give a current overview of the roles of molecular chaperones in mitochondrial protein import.

II. CYTOSOLIC CHAPERONES

A. Cytosolic Hsp70

The posttranslational targeting of precursor proteins to mitochondria is facilitated by cytosolic Hsp70 (cHsp70). Hsp70 binds to unfolded proteins and releases the bound proteins at the expense of ATP hydrolysis. Cells of the yeast, *Saccharomyces cerevisiae*, contain two subfamilies of Hsp70 in the cytosol. Proteins encoded by the *SSA* gene family (*SSA1*, *SSA2*, *SSA3*, and *SSA4* genes) are essential for cell viability, although each member is interchangeable [2,3]. *SSA1* and *SSA2* are expressed under normal growth conditions, whereas *SSA3* and *SSA4* are expressed only by stress. Cells lacking either *SSA1* or *SSA2* grow as well as wild-type cells, whereas cells lacking *SSA1*, *SSA2*, and *SSA4* are inviable. The latter strain can be rescued by expression of *SSA1* from a galactose-inducible promoter. Slow depletion of the protein encoded by *SSA1* during the growth of these cells results in the accumulation of precursor forms of a mitochondrial protein, the β-subunit of F_1-ATPase (pF_1β) and of a secretory protein, pro-α-factor, in the cytosol. These results suggest that Ssa proteins are involved in the posttranslational transport of proteins to mitochondria and to the ER [4].

How does cHsp70 facilitate posttranslational protein import into mitochondria? Many mitochondrial precursor proteins are prone to aggregate when expressed in *Escherichia coli* cells or when synthesized in a wheat germ cell-free system [5–9; S. Mitsui et al., unpublished results]. Occasionally, however, the addition of purified yeast cHsp70 and an *N*-ethylmaleimide (NEM)-sensitive protein(s) from yeast cytosol could dissociate the aggregated precursor proteins and restore their import competence [6; Mitsui et al., unpublished results]. Because urea could also solubilize the aggregated precursor proteins and overcome the requirement for added chaperones, cytosolic components may function simply by preventing the mitochondrial precursors from aggregating and, thereby, maintaining them in translocation-competent states.

What is the mechanism underlying the aggregation of mitochondrial precursor proteins? Mitochondrial presequences cause aggregation of unfolded proteins [10]. Mitochondrial precursor proteins likely suffer similar consequences of aggregation if the presequence interacts with the mature part of a precursor during the folding process. Therefore, cHsp70 can prevent the aggregation of mitochondrial precursors by two mechanisms. First, because Hsp70 binds to unfolded peptides and proteins, cHsp70 likely binds to unfolded regions of the mature part of the precursor proteins. These regions probably expose hydrophobic patches that are otherwise buried in the molecule and that would promote aggregation of the proteins in the absence of Hsp70. Second, because Ssa1p binds to the amphiphilic presequences, binding of cHsp70 to the presequence of mitochondrial precursor proteins could be an equally effective way to prevent aggregation [11]. The presequence bound to Hsp70 would be prevented from interacting with the newly synthesized, unfolded part of the precursor and would not be able to induce aggregation of these proteins in the cytosol.

To mask the presequence from the mature part of a precursor protein during the folding process, cHsp70 would need to bind to the presequence as the precursor is being synthesized on ribosomes in the cytosol. This requirement appears to be indeed true. Depletion of cHsp70 from reticulocyte lysate by the addition of anti-cHsp70 antibody before translation of the precursor of ornithine transcarbamylase (pOTC) decreased the rate of the in vitro import of this precursor into isolated rat liver mitochondria. Addition of cHsp70 to the Hsp70-depleted lysate after translation and before import assay did not restore the pOTC uptake by mitochondria [12]. Similar treatments of $pF_1\beta$ in yeast lysate also inhibited the import of this precursor into yeast mitochondria [S. Nishikawa et al., unpublished results]. These results indicate that Hsp70 is required for conferring transport competence to mitochondrial precursor proteins during their translation. The extent of association of the precursor of mitochondrial aspartate aminotransferase (pAAT) with Hsp70 along the translation reaction in vitro was directly assessed by immunoprecipitation with anti-Hsp70 antibodies at different times after initiation of translation [13]. Both in reticulocyte lysate and in wheat germ extract, 25% of the translation product of pAAT was associated with Hsp70 at an early stage, possibly cotranslationally. Interestingly, the fraction of translation product in reticulocyte lysate associated with Hsp70 decreased significantly as the translation reaction progressed, whereas in the wheat germ extract, the amount of Hsp70 associated with pAAT remained constant, regardless of the time of translation. Apparently, wheat germ extracts lack a release factor or factors for cHsp70 and pAAT, and this may be the reason for the import incompetence of precursors synthesized in this system.

B. Cytosolic DnaJ-like Protein

It has become evident that Hsp70 does not function alone. The reaction cycle of DnaK, a bacterial member of the Hsp70 family, requires two partner proteins, DnaJ and GrpE [14]. GrpE binds to the nucleotide-free form of DnaK, thereby promoting ADP–ATP exchange of DnaK [15]. DnaJ modulates the affinity of DnaK for unfolded proteins and stimulates the hydrolysis of DnaK-bound ATP [15]. Several DnaJ-like proteins have been identified in various subcellular compartments in eukaryotic cells [16].

Yeast contain at least two cytosolic DnaJ-like proteins, Ydj1p and Sis1p. The *YDJ1* gene (also called *MAS5*) was identified independently by two groups. In one study, the gene was cloned from a yeast expression library using antisera raised against a yeast nuclear subfraction [17]. In the other study, the gene was identified by screening temperature-sensitive yeast mutants for defects in mitochondrial protein import [18]. Cells with a disrupted *YDJ1* grew slowly at 23°C and were inviable at 37°C, indicating that *YDJ1* is essential for growth at the elevated temperature [18]. A temperature-sensitive *ydj1* mutant strain accumulated cytosolic precursor forms of the α-, β-, and γ-subunits of F_1-ATPase and of pro-α-factor at 37°C [19]. These defects were suppressed by overexpression of bacterial DnaJ or yeast Sis1p. The *ydj1* null mutant strain behaved differently; it also showed a defect in the import of $pF_1\beta$ into mitochondria after being shifted to 37°C, but pro-α-factor was normally secreted at this temperature [18]. The reason for the different effects of Ydj1p depletion on mitochondrial proteins and secretory proteins is unclear.

How does Ydj1p facilitate posttranslational protein import into mitochondria? Purified Ydj1p stimulates the ATPase activity of Ssa1p and Ssa2p up to tenfold, but does not stimulate the ATPase activity of Ssb1p and Ssb2p [20]. Also, a purified temperature-sensitive mutant form of Ydj1p exhibits reduced ability for this stimulation [19]. When unfolded rhodanese was diluted out of urea into a buffer containing Ydj1p, Ydj1p formed a complex with the unfolded rhodanese and prevented its agregation [21]. This prevention of aggregation was enhanced in the presence of Ssa proteins and ATP. These observations suggest that, in analogy with bacterial DnaJ and DnaK, Ydj1p couples to the ATP-dependent reaction cycle of cytosolic Hsp70 to prevent aggregation of mitochondrial precursor proteins, and thereby, maintains the import competence of precursor proteins.

The COOH-terminus of Ydj1p is farnesylated, and a fraction of Ydj1p is found in the nuclear membrane as well as in the cytosol [17]. Attachment of the lipid moiety was essential for Ydj1p function at elevated temperature, and a *ydj1* mutant strain, expressing farnesyl-free Ydj1p, was inviable at elevated temperature [22]. This result points to a possibility that Ydj1p may facilitate the release of precursor proteins from cHsp70 at the target membranes. However, because farnesylated Ydj1p was not found on the mitochondrial surface, direct involvement of farnesylation of Ydj1p in protein import into mitochondria remains to be demonstrated.

The foregoing view was recently challenged by the observations obtained with reconstituted cHsp70-dependent protein import into rat liver mitochondria [23]. The purified precursor of adrenodoxin (pAd) formed a complex with cHsp70 from rat liver cytosol after dilution out of urea. Import of cHsp70-bound pAd into rat liver mitochondria and the subsequent release of cHsp70 into the supernatant took place in the absence of ATP outside mitochondria and did not require any additional protein cofactors. This suggests that pAd complexed with cHsp70 does not require DnaJ-like proteins for its in vitro import into mitochondria, but it does not exclude the possibility that the import may be stimulated by a mammalian DnaJ-like protein (Hsp40).

C. Presequence-Binding Proteins

In addition to cytosolic Hsp70, several cytosolic factors that bind to mitochondrial presequences have been identified in mammalian cells. These presequence-binding proteins may well have chaperone functions for mitochondrial precursor proteins, because binding to the presequence will prevent presequence-induced aggregation of precursor proteins.

Presequence-binding proteins have been purified from various cellular lysates. For example, rabbit reticulocyte lysate contains cytosolic factors that have the ability to maintain aggregation-prone precursor proteins in import-competent states [5,24,25]. One of these factors, presequence binding factor (PBF), was purified from reticulocyte lysate as a 50-kDa protein that binds to the precursor form of ornithine carbamoyltransferase (OTC), but not to the mature form of this protein [26]. In vitro import of the precursor of OTC (pOTC) and several precursor proteins into mitochondria was impaired when the precursors were synthesized in reticulocyte lysate that had been depleted of PBF, but addition of purified PBF to the lysates restored the import. The depletion had no effect on the import of 3-oxyacyl-CoA thiolase, a mitochondrial precursor protein that lacks a cleavable presequence [27]. Although the PBF function does not require ATP and is not sensitive to NEM treatments, it acts synergistically with cHsp70 [26].

Another presequence-binding chaperone, mitochondrial import-stimulating factor (MSF), was purified from rat liver cytosol based on the ability of this factor to bind to the presequence of cytochrome oxidase subunit IV (pCox IV) and to stimulate the in vitro import of pAd that was expressed in wheat germ extracts [8]. MSF is a complex between a 32- and a 30-kDa protein that both belong to a family known as 14-3-3 proteins [28]. Members of this family are involved in diverse cellular processes, including signal transduction and cell cycle regulation [29].

Purified MSF bound to pAd that had been denatured with urea and subsequently diluted into a buffer containing MSF [8]. The bound pAd remained import-competent under these conditions. In contrast, urea-denatured pAd that was diluted into buffer lacking MSF rapidly aggregated and lost import competence. The subsequent addition of MSF to this solution solubilized the aggregated pAd and restored its import competence. This dissociation of the complex formed between pAd and MSF required ATP hydrolysis by MSF and, not surprisingly, the presence of aggregation-prone precursor proteins, in turn, stimulated the ATPase activity of MSF. The chaperone function of MSF was not affected by treatment of MSF with NEM. Because MSF bound to synthetic mitochondrial presequences and released them on ATP hydrolysis [30], MSF appears to exert a chaperone function mainly by binding to presequences. Although the affinity of presequences for MSF is weaker than that for cytosolic Hsp70 [11,30], the interactions of presequences with MSF appear to be more specific than those with Hsp70. Ssa1p binds not only to the amphiphilic pCox IV presequence, but also to SynB2, an artificial model presequence that lacks targeting function [11]. In contrast, MSF appears to bind only to functional presequences, but not to mutant presequences that are incapable of targeting proteins to mitochondria [30]. However, MSF may recognize elements other than just the presequence within a mitochondrial protein. For example, the ATPase activity of MSF was stimulated by porin and apocytochrome *c*, but was not stimulated by rhodanese [31]. However, none of these mitochondrial proteins contains a typical, cleavable presequence. A single MSF complex appears to bind to multiple molecules of pAd [31]. The crystal structure of a 14-3-3 protein suggests that the positively charged, amphiphilic presequence of a mitochondrial precursor protein could be accommodated in a wide, negatively charged channel in the 14-3-3-protein dimer [32], but does not readily explain the higher stoichiometry of binding observed with pAd.

Previous studies with chemically pure mitochondrial precursor proteins showed that protein import into isolated mitochondria did not require any cytosolic-targeting factors to recognize the targeting signal in the presequence or to deliver the precursor protein specifically to mitochondrial membranes [7,9,33,34]. Nevertheless, it is still tempting to speculate that, in analogy with the signal recognition particle (SRP) and SRP receptor system for protein targeting to the ER, presequence-binding proteins may also have targeting functions that, with the assistance of their putative receptor protein(s) on the mitochondrial surface, contribute to the specificity of protein targeting to mitochondrial membranes. MSF may have such a function. On the one hand, treatment of MSF with NEM inhibits the ability of MSF to stimulate the in vitro import into isolated rat liver mitochondria of pAd from wheat germ extracts or urea solutions [31]. Conversely, the ability of MSF to maintain aggregation-prone precursor proteins in soluble states is insensitive to treatment with NEM. Therefore, the two functions of MSF, import stimulation and chaperone functions, appear distinct, and the import stimulation activity may well reflect a targeting function of MSF.

III. TRANSFER FROM CHAPERONES TO RECEPTORS

A. Enegy Requirements

Precursor protein recognition at the mitochondrial surface and translocation across the outer membrane are mediated by a multisubunit protein complex referred to as the *translocase of outer mitochondrial membrane* (Tom) [35,36]. Among these proteins, Tom20, Tom70, and Tom71 are anchored to the outer membrane by NH_2-terminal hydrophobic segments and expose large COOH-terminal domains to the cytosol. Tom70 forms a subcomplex with Tom37, a Tom protein with two putative transmembrane segments, and deletion of both Tom70 and Tom37 is lethal in yeast cells, even though either of the two can be deleted without loss of viability. Tom20, the deletion of which leads to a respiration-deficient phenotype, is in a complex with two essential Tom proteins, Tom22 and Tom40. Tom22 has a highly negatively charged NH_2-terminal domain that faces the cytosol. It is anchored to the outer membrane by a hydrophobic segment in the middle of the polypeptide chain, and exposes a hydrophilic COOH-terminal domain to the intermembrane space. Although exact membrane topology of Tom40 is unknown, it is only partly accessible from the cytosol.

Precursor proteins that are kept import-competent by cytosolic chaperones have to be transferred from chaperones to import receptors on the mitochondrial surface. How are the precursor proteins passed from chaperones on to the receptors? It was previously assumed that interactions between a precursor protein and cytosolic chaperones are dynamic, and only the free form of the precursor protein is captured by the mitochondrial import receptors. However, recent studies by Mihara and co-workers showed that a dynamic equilibrium between the precursor protein and cytosolic chaperones is not prerequisite for the effective transfer of the precursor from chaperones to receptors [23]. When purified pAd was diluted out of urea into a buffer containing cHsp70 or MSF, or both, unfolded pAd bound to either cHsp70, MSF, or both. pAd complexed with cHsp70 or MSF was taken up by isolated rat liver mitochondria in the absence of other cytosolic factors. Import of pAd that bound either cHsp70 or MSF, however, differed in its sensitivity to NEM and its dependence on extramitochondrial ATP. When the cHsp70-bound form was incubated with mitochondria in the absence of ATP, cHsp70 was released to the supernatant and pAd was imported into the matrix. Because pAd remained bound to cHsp70 in the absence of ATP, the cHsp70-bound form, but not the free form, of pAd was captured by mitochondrial-import receptors, and docking of the cHsp70–pAd complex onto the receptors led to dissociation of the cHsp70–pAd complex without ATP hydrolysis.

Import of pAd from the cHsp70 complex was insensitive to the presence of NEM. In contrast, when pAd in a complex with MSF was incubated with mitochondria without extramitochondrial ATP, the MSF–pAd complex bound to the mitochondrial surface, but subsequent translocation into the matrix was arrested. Addition of ATP outside the mitochondria dissociated the bound complex and allowed pAd to be imported into the matrix. As with the cHsp70 complex, the MSF-bound form of pAd, but not the free form, was captured by mitochondrial-import receptors. However, in contrast with cHsp70, ATP was needed to dissociate the MSF–pAd complex on the mitochondrial surface. When pAd complexed with both MSF and cHsp70 was incubated with mitochondria without extramitochondrial ATP, only cHsp70 was released from the complex, and MSF together with pAd remained bound to the mitochondrial surface. Addition of ATP outside mitochondria chased pAd into the matrix, and MSF was released from mitochondria. These observations suggest that dissociation of cHsp70–pAd or MSF–pAd complex does not precede binding of the complex to the mitochondrial import receptors and that cHsp70 and MSF are released from mitochondria in an ATP-independent and an ATP-dependent manner, respectively.

B. Docking Sites for Precursors

The docking sites on the mitochondrial surface for pAd (and porin) complexed with MSF or cHsp70 have been identified recently [23,37]. When pAd, in a complex with MSF, was incubated with yeast mitochondrial outer membrane vesicles, the ATPase activity of MSF was quenched. Inhibition of the MSF–ATPase activity was prevented by pretreating the outer membranes with antibodies against Tom70 or Tom37. Similar quenching of the MSF–ATPase activity and inhibition of the quenching with anti-Tom70–Tom37 antibodies were also observed with the solubilized and partially purified Tom70–Tom37 receptor subcomplex. Binding of the pAd–MSF complex and the Tom70–Tom37 complex was further confirmed by immunoprecipitation. These results suggest that the docking site for MSF-bound pAd is Tom70–Tom37 [37]. Similar experiments with rat liver mitochondria showed that the docking site for MSF-bound pAd or porin was a 37-kDa protein in the outer membrane (OM37) [23]. Binding of MSF-bound pAd to deenergized yeast mitochondria depended on Tom70–Tom37, but not on Tom20–Tom22, for anti-Tom70–Tom37 antibodies, but not anti-Tom20–Tom22 antibodies, inhibited the binding. Subsequent reenergizing of the mitochondria could chase the bound pAd to the mitochondrial interior, but this chase was inhibited by anti-Tom70–Tom37 antibodies as well as anti-Tom20–Tom22 antibodies. This suggests that MSF-bound pAd first bound to Tom70–Tom37, then at least a part of pAd, including the presequence, was transferred from Tom70–Tom37 to Tom20–Tom22, and translocation across the mitochondrial membranes followed (Fig. 1, entry site I). The step of precursor transfer from

Entry site	Precursor protein
I	pAd-MSF, porin-MSF, pCyt c_1 AAC, PIC
I or II	pADH III, $pF_1\beta$, pFe/S, denatured pADH III
III	pAd-chsp70, porin-chsp70, pSu9-DHFR, pCyt c_1-DHFR, pCox IV-DHFR, pCpn10, phsp60, denatured pAd

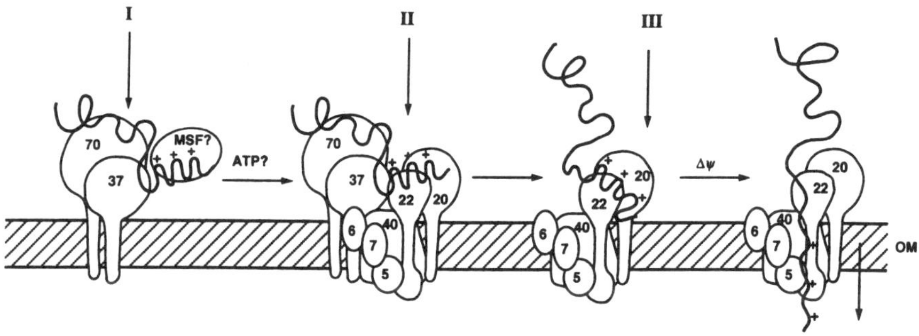

Figure 1 Entry sites for protein import into mitochondria: The numbers x indicate Tom x proteins. pAd-MSF, pAd bound to MSF; pAd-cHsp70, pAd bound to cHsp70; PiC, phosphate carrier; pCpn 10, precursor of chaperonine 10; OM, mitochondrial outer membrane.

Tom70–Tom37 to Tom20–Tom22 appeared to be coupled to the subsequent membrane potential-dependent transmembrane movement of the precursor protein. However, because MSF alone did not bind to mitochondrial outer membrane vesicles, it remains to be determined whether Tom70–Tom37 plays a role of the receptor specific for MSF that is carrying mitochondrial precursor proteins.

In contrast, antibodies against rat Tom20 inhibited the binding as well as the import of cHsp70-bound pAd [23]. The binding of urea-denatured "naked" pAd to the deenergized mitochondria was also inhibited by anti-Tom20–Tom22 antibodies, and neither binding nor chase was affected by anti-Tom70–Tom37 antibodies [37]. These observations suggest that Tom20–Tom22 may be the docking site for cHsp70-bound and naked precursor proteins (see Fig. 1, entry site III).

It has recently been proposed that mitochondrial precursor proteins can be classified into two groups according to their requirement for extramitochondrial ATP [35]. Group 1 precursors include the precursor forms of cytochrome b_2 (pCyt b_2), Hsp60 (pHsp60), and DHFR fusion proteins. These precursor proteins were imported independently of extramitochondrial ATP [38]. On the other hand, group 2 precursors, including the precursor forms of yeast alcohol dehydrogenase III (pADH III), cytochrome c_1 (pCyt c_1), pF$_1$β, and the ADP–ATP carrier (AAC), required extramitochondrial ATP for import [39]. According to the two-pathway model [23,35,40], group 2 precursors are dependent on MSF for import, whereas group 1 precursors require cHsp70 or no cytosolic ATP-depenent chaperones for import. Import of group 1 precursors (pHsp60, DHFR fusion proteins, such as pSu9–DHFR and pCoxIV–DHFR) was inhibited by the inactivation of Tom20–Tom22, but was only slightly influenced by the inactivation of Tom37–Tom70 [42,43]. By contrast, import of group 2 precursors (pCyt c_1 and AAC) was influenced by inactivation of Tom70–Tom37 [43,44]. These results suggest that the docking site for group 2 precursors is Tom70–Tom37 and that for group 1 precursors is Tom20–Tom22.

The two-pathway model is, on the other hand, inconsistent with the observation that the import of urea-denatured pADH III, which should not interact with MSF or any other cytosolic ATP-dependent chaperones and would be expected to enter the import route at Tom20–Tom22, still depended on Tom37–Tom70 significantly [43,44] (see Fig. 1, entry site I or II). In addition, some of the group 2 precursors, pADH III, pF$_1$β, and the precursor of Rieske Fe/S protein (pFe/S), were suggested to bind simultaneously to both Tom37–Tom70 and Tom20–Tom22 in the deenergized yeast mitochondria in the presence of extramitochondrial ATP [42], whereas the MSF-bound pAd (representing an "artificial" group 2 precursor) remained bound to only Tom70–Tom37 in the presence of extramitochondrial ATP until the mitochondria were reenergized [37]. To make the story complicated, the MSF-bound pAd had already been transferred from OM37 to Tom22 in the deenergized rat liver mitochondria when extramitochondrial ATP was present [45]. Clearly, more work is necessary to reveal the roles of the receptor subcomplexes—Tom70–Tom37 and Tom20–Tom22—in the docking of different precursor proteins and those complexed with different cytosolic chaperones.

IV. TRANSLOCATION ACROSS THE MEMBRANES

A. Crossing the Outer Membrane

Precursor proteins bound to Tom20–Tom22 have to insert into or move across the mitochondrial outer membrane [36]. The Tom complex in the outer membrane alone can mediate insertion of outer membrane proteins into the membrane and the import of a few proteins to the intermembrane space. However, import of other mitochondrial proteins also requires the action of the translocase of inner mitochondrial membrane (Tim) complex. For precursor proteins that carry presequences, the Tom complex facilitates the translocation of only the presequence parts of precursor proteins across the outer membrane. Isolated mitochondrial outer membrane vesicles without the inner membrane or soluble intermembrane space components had an ability to translocate the presequence, but not the mature part, of precursor proteins [46]. This study showed that the NH_2-terminal presequence interacted in a sequential and reversible fashion with two specific-binding sites of the outer membrane: *cis*-site on the cytosolic side and *trans*-site on the intermembrane-space side. The *cis*-site is likely provided by the cytosolic domains of Tom20–Tom22, which are rich in negatively charged residues. Transfer of the presequence from *cis*-site to *trans*-site may drive the translocation of the NH_2-terminal part of the precursor protein across the membrane and unfolding of the mature part on the cytosolic side. Because the COOH-terminal intermembrane-space domain of Tom22 has some acidic residues, it may play the role of a *trans*-site. Effects of deletion of the COOH-terminal domain were thus independently examined by three groups [47–49]. On the one hand, Bölliger et al. observed that yeast cells harboring COOH-terminally truncated Tom22 exhibited poor growth and that in vitro import of several precursor proteins into mitochondria with truncated Tom22 was significantly retarded compared with that into wild-type mitochondria [48]. On the other hand, our group and Court et al. observed neither defective growth of yeast or *Neurospora crassa* cells harboring truncated Tom22, nor significantly retarded in vitro import of precursor proteins into mitochondria with truncated Tom22 [47,49]. To find the reason for these divergent results, we and Schatz's group exchanged materials to reexamine the effects of COOH-terminal truncation of Tom22 on yeast cell growth under various conditions. Both groups have now reached essentially the same conclusion; although deletion of the COOH-terminal domain of Tom22 caused some growth defects at elevated temperature, the effects of the deletion on cell growth at physiological temperature depended on host strains [M. Nakai et al., unpublished results; G. Schatz, personal communication]. Although we cannot eliminate a possibility that yeast quickly compensated for the loss of the intermembrane space domain of Tom22 by up-regulating some other protein, these results reconfirm that the intermembrane space domain of Tom22 is not essential for protein import into mitochondria and, therefore, suggest that Tom22 alone cannot provide the *trans*-site for the presequence transfer across the outer membrane.

B. Crossing the Inner Membrane

Protein translocation through the inner membrane into the matrix requires the Tim complex and two essential proteins in the mitochondrial matrix, mitochondrial Hsp70 (Ssc1p in yeast; mHsp70) and Yge1p (Mge1p), a mitochondrial GrpE homologue [50]. The Tim complex consists of at least three essential Tim proteins, Tim17, Tim23, and Tim44. Tim17 and Tim23 are integral membrane proteins and are in proximity to the translocating chain. Although Tim44 displays the alkaline extractability of a peripheral membrane protein, it is believed to span the inner membrane at least once; the COOH-terminus of Tim44 is exposed on the intermembrane side, whereas the bulk of the protein is located on the matrix side. Tim44 appears to play key roles in protein translocation across the Tim channel; Tim44 is cross-linked to a translocating precursor protein in the early stage of import [51], and, as described later, Tim44 interacts with mHsp70 to recruit mHsp70 in the matrix to the site of import [52–54].

In the presence of a translocating precursor chain, the Tim complex interacts with the Tom complex such that efficient transfer of the presequence from the Tom channel to the Tim channel may take place [55,56]. Translocation of the NH_2-terminal presequence through the Tim channel is initiated by a step that depends strictly on the energized inner membrane. The electrical membrane potential $\Delta\psi$ across the inner membrane likely exerts electrophoretic effects on the positively charged presequences, resulting in the sliding of the presequence in the inward direction through the Tim channel [57,58]. Efficient trapping of the presequence in the matrix required both the $\Delta\psi$ and the binding of mHsp70 to the presequence. Neither $\Delta\psi$ nor the mHsp70 binding to the presequence alone was sufficient to hold the precursor stably in the Tim channel [59,60].

Subsequent translocation of the mature part of the precursor chain across the inner membrane requires the hydrolysis of ATP in the matrix (internal ATP), but is independent of $\Delta\psi$. This requirement of internal ATP most likely reflects involvement of mHsp70 in the inward movement as well as in the unfolding of the precursor proteins (see following discussion).

C. Chaperone-Driven Translocation

What drives the inward movement of precursor proteins across the inner membrane? Two key interactions involving mHsp70 need to be considered. First, the interaction of Yge1p (Mge1p), a mitochondrial GrpE homologue, with mHsp70 appears to be essential for protein import into the matrix. More specifically, Yge1p stabilizes the nucleotide-free form of mHsp70 and, thereby, promotes the exchange of the ADP and ATP forms of mHsp70 in vitro [61,62]. Inactivation or genetic depletion of Yge1p rendered mitochondria deficient in protein import into the matrix [61,63,64]. A fraction of the mHsp70 that interacts with a precursor protein in transit across the inner membrane also interacts with Yge1 [65]. Therefore, Yge1p appears to be involved in the ATP-dependent reaction cycle of mHsp70 that is responsible for translocation of precursor proteins. Second, Tim44 recruits a fraction of the mHsp70 at the Tim complex in the inner membrane. The mHsp70 interacts with Tim44 with a 1:1 stoichiometry in an ATP-dependent manner [52–54]. These interactions may be necessary for the appropriate transfer of a translocating precursor protein form the inner membrane to the matrix. On the basis of the foregoing key interactions of mHsp70 with Tim44 and Yge1p, two mechanisms have been proposed to explain the vectorial movement of precursor proteins into the matrix, driven by mHsp70: the "Brownian ratchet" and the "translocation motor" models.

The Brownian ratchet model was originally proposed before the discovery of the interactions between mHsp70 and Tim44 [66,67]. In this model, it is assumed that Brownian motion would cause the translocating polypeptides to fluctuate in both directions through the Tim channel. After a forward movement, mHsp70 would bind to the incoming segment of the precursor chain so that reverse sliding of the precursor would be prevented. The binding of mHsp70 to the precursor chain could thus function as a trap on the *trans*-side of the membrane, and multiple ratchet and binding steps would result in a net displacement of the precursor in the inward direction. Let us call this simple version of the Brownian ratchet model a "soluble trap" model (Fig. 2A). This model was later modified to include the nucleotide-dependent interactions of mHsp70 with Tim44 [52]. In this advanced Brownian ratchet model (see Fig. 2B), the incoming precursor chain is trapped by a fraction of mHsp70 that is anchored to Tim44. Recruitment of mHsp70 at the import site by Tim44 likely increases efficiency of ratcheting by mHsp70. Binding of mHsp70 to the precursor chain prevents its reverse sliding through the Tim channel. The complex between mHsp70 and Tim44 would dissociate while mHsp70 remains bound to the precursor chain. This allows the precursor chain to diffuse further through the Tim channel. The mHsp70 molecule remains bound to the precursor until a second mHsp70 molecule is recruited by Tim44 and binds to a next incoming segment of the precursor chain. Multiple rounds of this reaction will result in unidirectional movement of the precursor chain. If the precursor protein has a folded structure outside mitochondria, breathing conformational fluctuation of the folded domain is a prerequisite for allowing the inward diffusion of the precursor chain through the import channel. The partially unfolded conformation of the precursor protein on the mitochondrial surface is, in turn, stabilized by mHsp70-binding in the matrix. Thus, the Brownian ratchet could drive protein import of folded precursor proteins as well as that of unfolded proteins.

In the translocation motor model [68,69], mHsp70 undergoes a conformational change that generates a directional force on the precursor chain (see Fig. 2C). Because the conformational change of a soluble mHsp70 molecule freely rotating in the matrix cannot generate a directional pulling force, the mHsp70 molecule needs to be fixed in its orientation relative to the Tim channel by a membrane anchor, Tim44. The mHsp70 remains bound to both the incoming precursor chain and Tim44 until mHsp70 undergoes a power stroke, thereby pulling a segment of the chain into the matrix. Then mHsp70 dissociates from the precursor chain and Tim44. The pulling force exerted by mHsp70 will also contribute to the unfolding of precursor proteins outside the mitochondria.

Limitation of the soluble trap model to explain the mechanism of the inward movement of precursor proteins was suggested from experiments with mitochon-

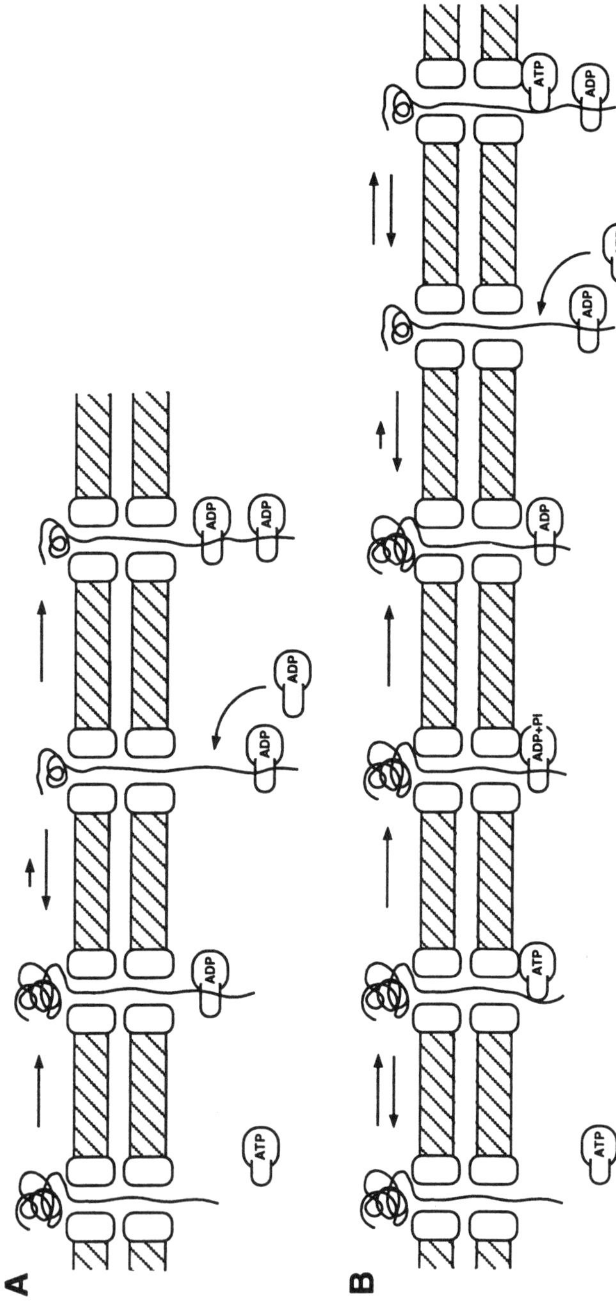
A
B
ATP
ADP
ADP+Pi

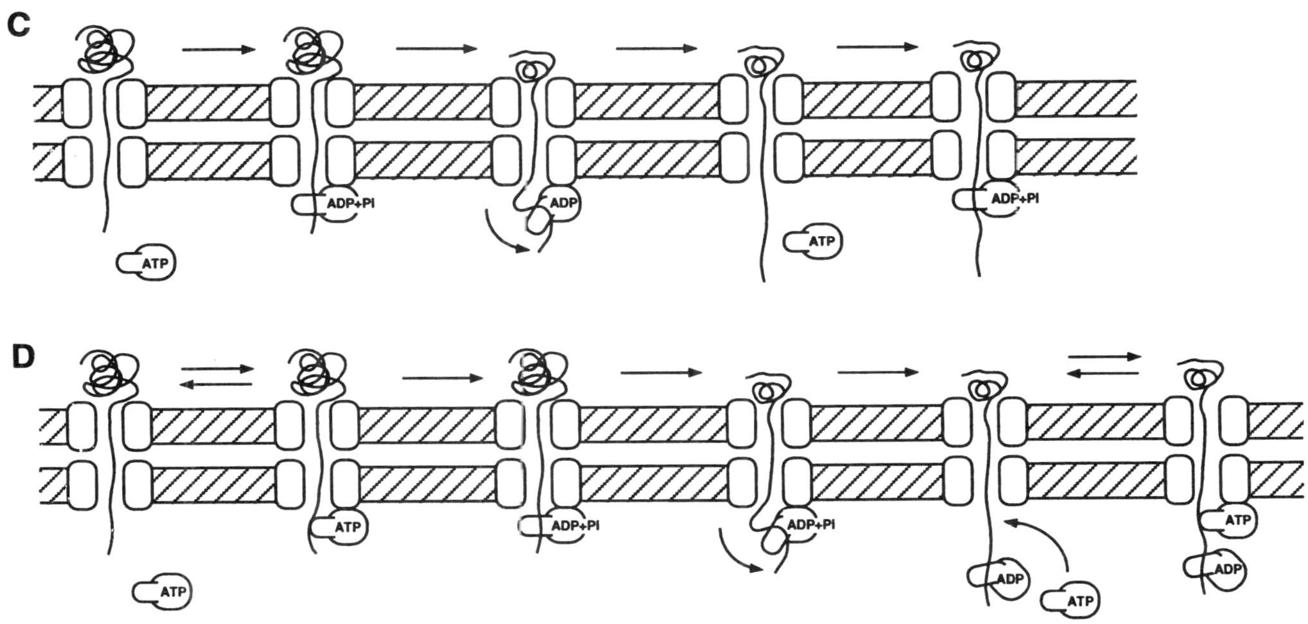

Figure 2 Working models of the roles of mHsp70 in driving protein translocation across the inner membrane: (A) the soluble-trap model (the original Brownian ratchet model); (B) the (advanced) Brownian ratchet model taking into account the Tim44–mHsp70 interactions; (C) the translocation motor model; and (D) the modified translocation motor model that is reconciled with the latest results on the nucleotide-dependent Tim44–mHsp70 interactions.

drial Hsp78, a member of the Clp family. Hsp78 is closely related to bacterial ClpB, and has two consensus ATP-binding motifs. Similar to other Clp-like proteins, Hsp78 binds to unfolded proteins and prevents them from aggregating. Hsp78 can bind to newly imported and misfolded proteins in the mitochondrial matrix, but it does not interact with Tim44 [70]. Interestingly, overexpression of Hsp78 can suppress a defect in protein import into mitochondria when the functions of mHsp70 were partly impaired by the *ssc1–3* mutation in mHsp70. Mitochondria of the *ssc1–3* mutant strain are defective in protein import into mitochondria at both permissive and nonpermissive temperature. At the permissive temperature, Ssc1–3p has reduced affinity for both nucleotides and Yge1p, so that the Ssc1–3p–Tim44 complex is presumably stabilized in the presence of either ATP or ADP. Pretreatment at the nonpermissive temperature caused defects in protein import into the mitochondria isolated from the *ssc1–3* strain, probably owing to the aggregation of Ssc1–3p in the matrix. High-level expression of Hsp78 could overcome the defects in protein import into the *ssc1–3* mutant mitochondria at permissive temperature, suggesting that soluble Hsp78 could substitute for a part of the functions of Ssc1p. However, mitochondria isolated from the *ssc1–3* strain could not import either folded or urea-denatured precursor proteins at nonpermissive temperature, even in the presence of a high level of Hsp78. Therefore, binding of Hsp78 to the translocating polypeptide chain without anchoring to Tim44 alone is not sufficient to drive import of even unfolded proteins. These results indicate that not only the Tim44-dependent binding, but also the Tim44-independent binding of mHsp70 to the incoming polypeptide chain contribute to the unidirectionality in the protein import and that Hsp78 can substitute for only the latter function of mHsp70 when the mHsp70 functions are limited. The simple soluble trap model is thus not sufficient to explain the mechanism of inward movement of mitochondrial precursor proteins.

It has been proposed that the "translocase" function of mHsp70 can be exerted by soluble mHsp70 by the soluble trap mechanism, whereas the "unfoldase" function of mHsp70 requires a dynamic interaction of mHsp70 with Tim44 [71]. This proposal is based on the results of experiments with Ssc1–2p, a mutant yeast Ssc1p, which efficiently bound precursor proteins, but did not show a detectable complex formation with Tim44 at nonpermissive temperature [52,71]. Mitochondria with Ssc1–2p could not import a pCyt b_2 fusion protein with a tightly folded heme-binding domain, but could import another pCyt b_2 fusion protein without the heme-binding domain at nonpermissive temperature [71]. This result suggests that the import of loosely folded precursor proteins may not require the formation of a complex between mHsp70 and Tim44. A likely mechanism for the import of loosely folded precursor proteins is diffusion of the unfolded polypeptide chain in the Tim channel and the subsequent capture of the chain by soluble mHsp70 when it inserts into the matrix. However, this view is not readily compatible with the aforedescribed effects of overexpression of Hsp78 on the *ssc1–3* mutant mitochondria; namely, high-level expression of Hsp78 alone did not overcome the effects of the ssc*1–3* mutation on the import of unfolded precursor proteins [70]. Perhaps even in the *ssc1–2* mutant mitochondria, the mutant mHsp70 could exhibit a weak and transient interaction with Tim44.

One of the essential steps in protein import, the insertion of a presequence into the matrix, can be well explained by the Brownian ratchet model [69]. A DHFR fusion protein containing presequences as short as 12 residues can be imported into mitochondria [72], although at least 50 residues are necessary to span the outer and the inner mitochondrial membranes [73]. The most likely mechanism for the unfolding and initial insertion of these fusion proteins is spontaneous unfolding of the DHFR domain, followed by binding of the presequence by mHsp70 when it enters the matrix. Although a free DHFR molecule is stably folded in solution, DHFR fusion proteins bound on the surface of mitochondria can be at least partially unfolded by acidic lipids [74]; therefore, spontaneous unfolding of the DHFR fusion protein, which allows capture of the presequence by mHsp70, may well work efficiently.

The mechanism for the completion of membrane translocation of mitochondrial precursor proteins is in dispute. The mechanical difference between the Brownian ratchet model and the translocation motor model is reflected in the role of internal ATP in the mHsp70 function. Because the ADP-bound form of Hsp70 has a higher affinity for unfolded proteins than the ATP-bound form, both models assume that ATP hydrolysis results in the formation of a stable complex of mHsp70 with the precursor chain. However, in the Brownian ratchet model, mHsp70–ADP bound to the precursor chain has to dissociate from Tim44 so that the precursor chain can slide back and forth, whereas in the translocation motor model, mHsp70–ADP has to bind to both the precursor chain and Tim44 to translocate the precursor.

The roles of ATP in the formation of a complex between mHsp70 and Tim44 have been investigated extensively, but the results are difficult to compare owing to differences in the experimental conditions used. Variation in the concentrations of sodium and potassium ions, the possible contamination of ATP and other compounds in the commercial preparations of nonhydrolyzable analogues of ATP, and the presence of nucleotide-exchange factor Yge1p can substantially affect the stability of the mHsp70–Tim44 complex [75–77]. The results obtained with chromatographically pure nonhydrolyzable ATP analogues and in the presence of physiological concentrations of potassium ion (about 70 mM) and Yge1p (about 12 μM) are probably the most accurate [77]. In those experiments, Mg–ATP caused the release of mHsp70 from Tim44 independently of the presence of Yge1p. In the absence of Yge1p, the mHsp70–Tim44 complex dissociated with nonhydrolyzable analogues of ATP, AMP–PNP, or ATPγS, but the complex was stable with ADP. In contrast, when Yge1p was present, the complex was stable with AMP–PNP or ATPγS, and it dissociated with ADP.

Inorganic phosphate appeared to stabilize the mHsp70–Tim44 complex in the presence of ADP, and a higher concentration of Yge1p was needed for dissociation of the complex in the presence of Mg–ADP plus P_i than in the presence of Mg–ADP alone. These observations suggest that Yge1p destabilizes the ADP form of the mHsp70–Tim44 complex after release of P_i, and this results in the dissociation of the complex before release of ADP from mHsp70. In other words, mHsp70–ADP remains bound to the precursor chain after dissociation of mHsp70–ADP from Tim44, so that the precursor chain has a good chance to slide further through the Tim channel. This is in accordance with the Brownian ratchet model for protein translocation. However, the translocation motor model can also be reconciled with the regulation of the mHsp70–Tim44 complex formation by ATP in the presence of Yge1p (see Fig. 2D). Because mHsp70 remains bound to Tim44 even after ATP hydrolysis, until release of P_i, mHsp70 has a chance to generate a pulling force on the precursor chain in the meantime.

In Figure 2B and D, the Brownian ratchet model and the translocation motor model reconciled with the currently available experimental data are shown. In both models, the ATP-bound mHsp70 is in an equilibrium between the Tim44-bound form and the soluble form. The Tim44-bound form of mHsp70–ATP can interact with the incoming precursor chain with a low affinity, but with a sufficiently high on-rate. The precursor chain bound to mHsp70 stimulates the ATPase activity of mHsp70, resulting in the tight binding of the precursor chain to the ADP-bound form of mHsp70. In the modified translocation motor model, mHsp70–ADP plus P_i undergoes a conformational change (possibly triggered by tight binding to the precursor chain), thereby pulling a segment of the chain into the matrix. The segment pulled by mHsp70 on a single event of the conformational change has to be long enough to allow the second mHsp70 molecule to bind after dissociation of the mHsp70–Tim44 complex. Next, in both models, the release of P_i from mHsp70 dissociates the mHsp70–Tim44 complex, but mHsp70 remains bound to the precursor chain. In the Brownian ratchet model, the release of mHsp70 from the membrane anchor Tim44, allows the precursor chain to slide further through the TIM channel, whereas the back-sliding of the precursor chain is prevented by the bound mHsp70–ADP. The distance of the inward-sliding has to be large enough to allow the second mHsp70 molecule to bind. Then, in both models, Yge1p promotes nucleotide exchange of mHsp70, and formation of the ATP-bound mHsp70 leads to release of mHsp70 from the precursor chain.

Currently available experimental evidence is thus insufficient to discriminate between the two models for translocation. However, The Brownian ratchet and the translocation motor mechanisms in Figure 2 are not mutually exclusive, but can be compatible with each other. Both mechanisms could potentially operate at different steps of the mHsp70–Tim44 reaction cycle for facilitating the unfolding and inward translocation of precursor proteins across the membranes. Also we cannot exclude that other, equally convincing, model may exist.

D. Insertion into the Inner Membrane

Transport of proteins to various mitochondrial subcompartments and their various topographical arrangements can be achieved by several different intramitochondrial-sorting routes [78,79]. Involvement of mitochondrial chaperones in protein import into the matrix has already been described. Involvement of molecular chaperones in protein transport to the inner membrane and to the intermembrane space has also been recently examined.

Insertion of transmembrane segments into the inner membrane from the cytosolic side can take place in two ways: stop-transfer and anchoring of the transmembrane segment, and insertion of the transmembrane segment in a loop structure. Nuclear-encoded cytochrome oxidase subunit Va (COX Va) spans the inner membrane once in an N_{in}–C_{out} orientation (Fig. 3). Possible interactions of the topogenic signal of COX Va (a hydrophobic segment near the COOH-terminus) with inner membrane components appeared to arrest its complete translocation across the inner membrane. Import of the NH_2-terminal part (up to the topogenic signal) of the precursor of COX Va (pCOX Va) into the matrix required both $\Delta\psi$ across the inner membrane and internal ATP [80], and that into the *ssc1–3* mitochondria, which were defective in the Ssc1p functions, was blocked [81]. This indicates that pCOX Va requires the translocase functions of mHsp70 for the translocation of its NH_2-terminal part into the matrix. Interestingly, a mutant pCOX Va, the topogenic signal of which was placed in the vicinity of the presequence, could now be imported and inserted into the inner membrane correctly, even in the *ssc1–3* mitochondria [81]. The mechanism underlying this observation remains unclear.

The BCS1 protein spans the inner membrane once in an N_{out}–C_{in} orientation with its large COOH-terminal domain exposed to the matrix (see Fig. 3). The BCS1 protein does not have an NH_2-terminal presequence, but instead, a positively charged segment located immediately after the transmembrane domain functions as a targeting signal to the mitochondrial matrix [82]. To function, the positively charged segment was proposed to form a loop structure with the transmembrane segment so that the positively charged segment and the remaining COOH-terminal part of the BCS1 protein cross the inner membrane to achieve the N_{out}–C_{in} orientation. The fusion protein consisting of the NH_2-terminal part of the BCS1 protein and DHFR, which had been arrested on the import pathway in the mitoplast by stabilizing the folded structure of the DHFR moiety with methotrexate, was cross-linked to both the components of the Tim machinery and mHsp70. Therefore, mHsp70 appears to be involved in translocation of at least the COOH-terminal part following the transmembrane segment of the BCS1 protein into the matrix. Whether the transmembrane segment in an N_{out}–C_{in} orientation resists passage across the inner membrane or whether it is also translocated into the matrix and subsequently integrated into the inner membrane

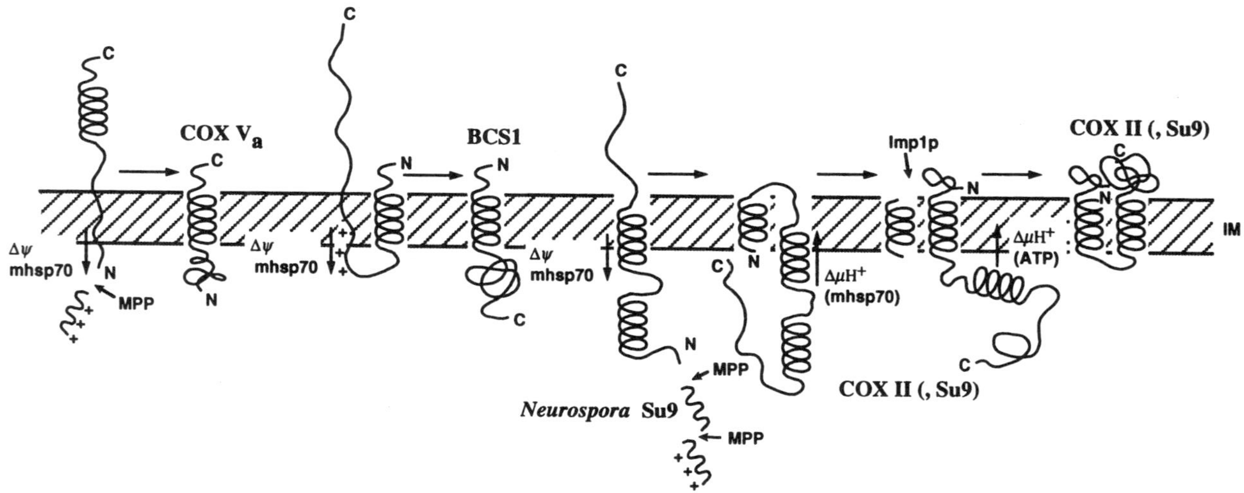

Figure 3 Insertion of COX Va, the BCS1 protein, and COX II into the mitochondrial inner membrane: The insertion process of COX II into the inner membrane includes that of Su9 of *N. crassa*. MPP, matrix processing peptidase; Imp1p, inner membrane protease I; IM, mitochondrial inner membrane.

remains to be clarified. In both cases, translocation of the polypeptide segment following the transmembrane segment across the inner membrane requires $\Delta\psi$ and mHsp70.

Insertion of transmembrane segments into the inner membrane from the matrix side can take place only after a part of the precursor protein, including the transmembrane segment, is imported into the matrix. This import naturally requires $\Delta\psi$ and mHsp70. Insertion from the matrix side into the inner membrane may accompany export of the segment following the transmembrane segment to the intermembrane space. Subunit 9 of the F_0-ATPase (Su9), which is encoded by either the nuclear or the mitochondrial genome, and cytochrome oxidase subunit II (COX II), which is encoded by the mitochondrial genome and synthesized as a precursor containing an NH_2-terminal presequence, have two membrane-spanning segments with both NH_2- and COOH-termini exposed to the intermembrane space (N_{out}–C_{out} orientation; see Fig. 3). A study with a fusion protein consisting of the NH_2-terminal part of nuclear-encoded Su9 (containing the first membrane-spanning segment) and DHFR showed that the fusion protein was imported through both the outer and the inner membranes into the matrix [83]. After processing of the mitochondrial-targeting presequence, the transmembrane domain functioned as a membrane-insertion signal so that the NH_2-terminal hydrophilic segment was exported to the intermembrane space in the N_{out}–C_{in} orientation. This export required a proton motive force ($\Delta\mu H^+$) or ΔpH across the inner membrane and internal ATP, which appeared to be necessary for release of mHsp70 from the fusion protein. When export was inhibited by inactivating a matrix-localized, processing peptidase or by dissipating ΔpH, 75% of the precursor form and 50% of the processed form of the fusion protein were bound to mHsp70, respectively, even after 30 min of incubation. Therefore it is tempting to speculate that mHsp70 binds to hydrophobic segments of Su9, prevents aggregation of the fusion protein, and maintains the protein in a competent conformation necessary for export.

When translated in intact mitochondria, newly synthesized COX II was inserted into the inner membrane in an N_{out}–C_{out} orientation with its NH_2-terminal presequence processed by the Imp1p protease on the external surface of the inner membrane [84]. Export of the NH_2-terminal and the COOH-terminal segments to the intermembrane space required $\Delta\mu H^+$ across the inner membrane. A chimeric protein consisting of a mitochondrial matrix-targeting sequence, the NH_2-terminal part of the precursor form of COX II (including the first membrane-spanning segment), and DHFR was imported into the matrix, similar to the Su9-DHFR fusion protein, and inserted into the inner membrane in an N_{out}–C_{in} topology with its NH_2-terminal segment exported to the intermembrane space. However, in contrast with the Su9–DHFR fusion protein, the COX II–DHFR fusion protein that had been accumulated in the matrix by inhibiting export did not form a stable complex with mHsp70. Therefore, the requirement for mHsp70 in the export and in the proper insertion of the transmembrane segment may depend on the structural properties of the region of the precursor protein that was imported into the matrix.

Insertion of the ADP–ATP carrier (AAC) into the inner membrane is complex. AAC is synthesized without a cleavable presequence, but contains internal-targeting signal(s), and spans the inner membrane six times. Import of AAC required extramitochondrial ATP and $\Delta\psi$ across the inner membrane, but not internal ATP [41]. The extramitochondrial ATP was likely used by cytosolic chaperones. The lack of a requirement for internal ATP suggests that mHsp70 is not involved in the insertion of AAC into the inner membrane, but on the other hand, import of AAC into the *ssc1–2* mitochondria, which was defective in unfoldase function of Ssc1p, was significantly impaired at restrictive temperature [85]. Clearly, additional work is necessary to resolve this apparent discrepancy. A possible mechanism for AAC insertion into the inner membrane without the aid of mHsp70 is that the internal-targeting signals of AAC may respond to $\Delta\psi$ to insert through the Tim channel, but the polypeptide chain is released laterally into the lipid bilayer of the inner membrane instead of entering the matrix. A loop structure similar to that of the BCS1 protein may be formed in the integration process of AAC into the inner membrane.

E. Sorting to the Intermembrane Space

The sorting pathways for directing proteins, including cytochromes b_2 and c_1, to the intermembrane space have been a matter of debate for years [78,79]. The precursors of cytochromes b_2 (pCyt b_2) and c_1 (pCyt c_1) possess cleavable bipartite presequences, the first part of which directs the proteins to mitochondria and the second part of which mediates sorting to the intermembrane space. In the conservative-sorting model, pCyt b_2 and pCyt c_1 are first translocated across both the outer and the inner membranes into the matrix and are then reexported to the intermembrane space (Fig. 4B). In the stop–transfer model, the sorting signal of the presequence arrests translocation through the inner membrane, so that the mature part crosses only the outer membrane to reach the intermembrane space (see Fig. 4A).

In the original form of the conservative-sorting model, mHsp70 and mitochondrial Hsp60 in the matrix were suggested to be involved in the sorting process for pCyt c_1 and pCyt b_2 (see Fig. 4B). Both mHsp70 and mitochondrial Hsp60 may prevent the sorting intermediates of pCyt c_1 and pCyt b_2 from aggregating and keep them in export-competent conformations in the matrix. Involvement of mHsp70 in protein sorting to the intermembrane space was suggested by the observation that the import and sorting of pCyt b_2 into the intermembrane space required internal ATP [38], and that the import and sorting of pCyt c_1 into the intermembrane space of *ssc1–2* mitochondria, which are defective in the Ssc1p functions, was impaired at restrictive temperature [85,86]. However, these results were challenged by the following observations. Import of pCyt c_1 into the intermembrane space was independent of internal ATP [41]. The requirement for internal ATP in the import of pCyt b_2 into the intermembrane space depended on the presence of a tightly folded heme-binding domain of Cyt b_2, which is located in the mature part immediately downstream from the presequence, and when the folded structure of the heme-binding domain was disrupted by mutation or by urea denaturation, correct import and sorting took place in ATP-depleted mitochondria [87,88]. Similarly, the ability of pCyt b_2 to be imported into the *ssc1–2* mitochondria, which lacked unfoldase activity of mHsp70, at restrictive temperature depended on the presence of the folded heme-binding domain of Cyt b_2 [89]. These results are interpreted as follows: In the stop-transfer model, matrix ATP is needed for import of pCyt b_2 because the presequence must be pulled far enough into the mitochondria for the stop-transfer signal to be recognized, but the folded heme-binding domain outside the mitochondria prevents the presequence from such deep insertion without the aid of mHsp70 [87]. In the conservative-sorting model, import of pCyt b_2 lacking the folded structure of the heme-binding domain into the matrix in the absence of ATP or functional mHsp70 was driven by the export system, which would transfer the protein into the intermembrane space in a cotranslocational manner [88] (see Fig. 4C).

When the distance between the presequence and the heme-binding domain was varied, mitochondria could not import pCyt b_2 mutants in which the heme-binding domain was 40 or more residues away from the presequence [90] Complete import of these pCyt b_2 mutants was achieved by unfolding of the mutants before import, or by movement of the heme-binding domain closer to the presequence, indicating that mHsp70 could unfold the heme-binding domains that are 20 residues or fewer away from the presequence. These results suggest that pCyt b_2 disengages the mHsp70-assisted import mechanism in the early stage of import. Possible explanations for the early release of pCyt b_2 from mHsp70 during import are as follows: In the stop-transfer model, because the sorting signal of the presequence arrests translocation through the inner membrane, mHsp70 can no longer exert a pulling force on the heme-binding domain located farther away from the presequence (see Fig. 4A). In the conservative-sorting model, although each part of the precursor protein has to pass through the matrix, recognition of the sorting signal by components in the matrix triggers unloading of pCyt b_2 from mHsp70, and subsequent transfer of the mature part through the matrix takes place independently of mHsp70, but with the aid of the export system in a cotranslocational manner (see Fig. 4C).

Involvement of Hsp60 in the sorting of proteins to the intermembrane space

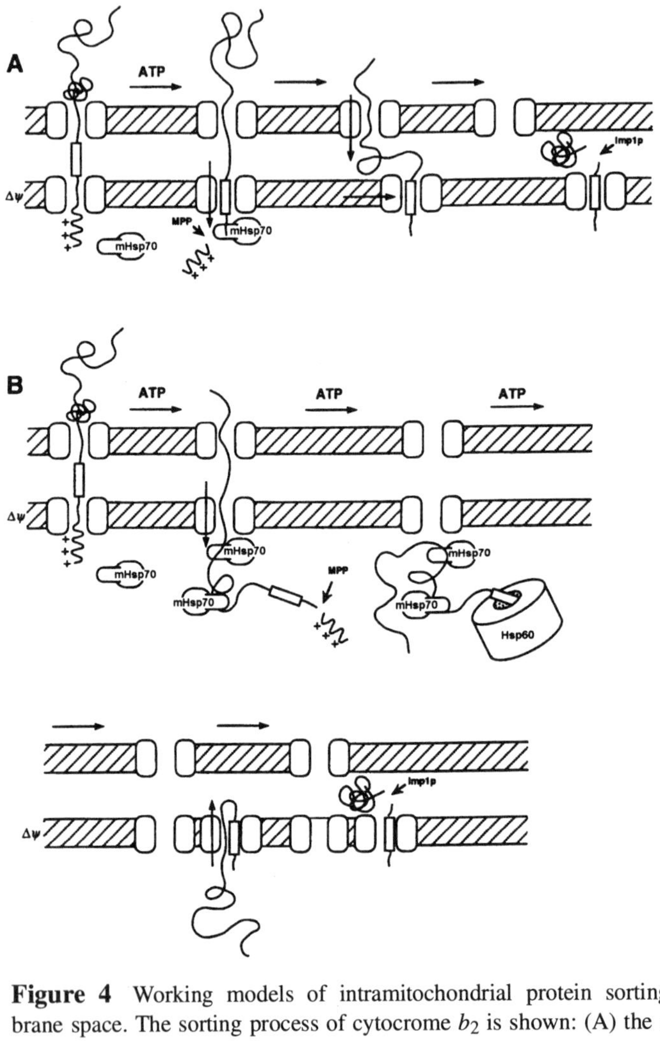

Figure 4 Working models of intramitochondrial protein sorting to the intermembrane space. The sorting process of cytocrome b_2 is shown: (A) the stop–transfer model; (B) the original conservative-sorting model; and (C) the conservative-sorting model in a cotranslocational model. MPP, matrix-processing peptidase; Imp1p, inner membrane protease I.

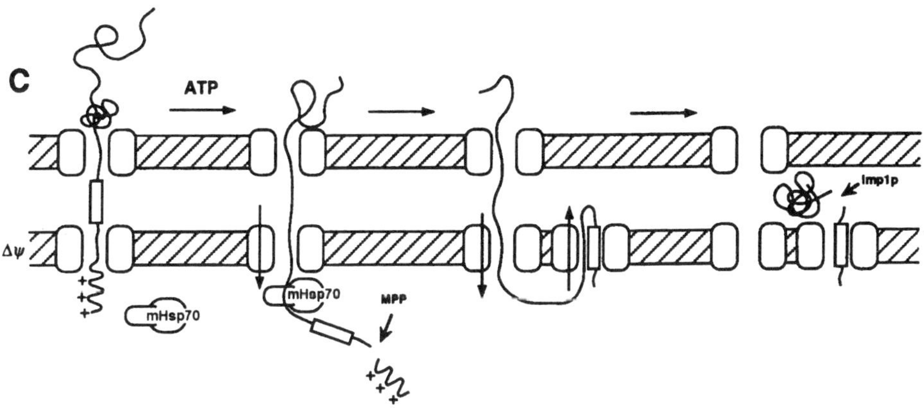

Figure 4 Continued

was originally suggested by the observation that, at nonpermissive temperature, maturation of pCyt b_2 was impaired in the yeast *mif4* strain, which was defective in the functions of mitochondrial Hsp60 [91,92]. In accordance with these results, Cyt b_2 and Cyt b_2–DHFR fusion proteins containing various lengths of the NH_2-terminal segments of pCyt b_2 were associated with mitochondrial Hsp60 during import into isolated mitochondria, as judged by sizing chromatography [92]. However, these observations were challenged by the following results. Different phenotypes of the *mif4* strain relative to Cyt b_2 maturation were reported; although inactivation of mitochondrial Hsp60 in *mif4* cells caused precipitation of newly imported matrix proteins, inhibition of Cyt b_2 maturation was not observed in *mif4* cells at nonpermissive temperature [93], nor did genetic depletion of mitochondrial Hsp60 affect sorting and maturation of Cyt b_2 [94]. Inactivation of yeast mitochondrial Hsp10, a partner protein for Hsp60, resulted in only a moderate effect on maturation of Cyt b_2 in vivo [95]. Interactions between mitochondrial Hsp60 and Cyt b_2–DHFR fusion proteins were not detected by immunoprecipitation using anti-Hsp60 antibodies [96]. Taken together, mitochondrial Hsp60 does not seem to play an essential role in protein sorting to the intermembrane space [79]. Thus, although current experimental evidence is insufficient to distinguish between the two models, both models now accept a similar view on the requirement for mHsp70 in protein sorting to the intermembrane space (see Fig. 4A and C).

V. PERSPECTIVE

I have attempted to summarize the recent progress in understanding the mechanisms of protein transport into mitochondria from the viewpoint of the functions of molecular chaperones. As described here, the last decade has witnessed the identification of many cytosolic and mitochondrial components, including molecular chaperones and their partner proteins, that mediate mitochondrial protein import. Although I did not mention it here, mitochondrial chaperones also mediate protein folding and assembly within mitochondria. Thus, molecular chaperones are involved in every step in the mitochondrial protein import: synthesis of precursor proteins in the cytosol, protein targeting to the mitochondrial membranes, protein translocation across the mitochondrial membranes, and protein folding and assembly inside mitochondria. Molecular chaperones maintain the precursor polypeptides in conformations suitable for reactions in each step and prevent them from undesirable interactions that would lead to their divergence from the correct transport pathways. In this context, the recent hypothesis that the Tom and the Tim complexes may play the roles of "membrane chaperones" [97] is appealing. The Tom and Tim proteins may inhibit undesirable interactions of the translocating polypeptides in extended conformations across the membranes by providing a flexible interface between the lipids, the translocating proteins, and themselves.

In the next few years, investigation on the mitochondrial protein import has to be extended to determine the precise function of each component involved in the complex process of mitochondrial protein import. Such studies will include the characterization of the interplay between the components in well-defined systems reconstituted from purified components, the determination of the three-dimensional structures of the relevant domains of the components, and the application of new biophysical techniques, such as observation and manipulation of single molecules by single-particle tracking, lazer tweezers, atomic force microscopy, and others. We expect that, after completion of such analyses, we will be able to understand the dynamic pictures of mitochondrial protein transport systems operating in cells at the molecular level.

ACKNOWLEDGMENTS

I thank former and current members of the laboratory who contributed to our work. I thank David Roise for critically reading the manuscript and Gottfried Schatz, Martin Horst, Walter Neupert, Michael Brunner, Benedikt Westermann, Katsuyoshi Mihara, and Masataka Mori for communicating results before publication.

REFERENCES

1. Roise, D. and Schatz, G., Mitochondrial presequences, *J. Biol. Chem. 263*:4509 (1988).
2. Craig, E. A. and Jacobsen, K., Mutation of the heat-inducible 70 kilodalton genes of yeast confer temperature-sensitive growth, *Cell 38*:841 (1984).
3. Werner-Washburne, M., Stone, D. E., and Craig, E. A., Complex interactions among members of an essential subfamily of hsp70 genes in *Saccharomyces cerevisiae*, *Mol. Cell Biol. 7*:2568 (1987).
4. Deshaies, R. J., Koch, B. D., Werner-Washburne, M., Craig, E. A., and Schekman, R., A subfamily of stress proteins facilitates translocation of secretory and mitochondrial precursor polypeptides, *Nature 332*:800 (1988).
5. Murakami, K., Amaya, Y., Takeguchi, M., Ebina, Y., and Murakami, M., Reconstitution of mitochondrial protein transport with purified ornithine carbamoyltransferase precursor expressed in *Escherichia coli*, *J. Biol. Chem. 263*:18437 (1988).
6. Murakami, H., Pain, D., and Blobel, G., 70-kD heat shock-related protein is one of at least two distinct cytosolic factors stimulating protein import into mitochondria, *J. Cell Biol. 107*:2051 (1988).
7. Becker, K., Guiard, B., Rassow, J., Söllner, T., and Pfanner, N., Targeting of a chemically pure preprotein to mitochondria does not require the addition of a cytosolic signal recognition factor, *J. Biol. Chem. 267*:5637 (1992).
8. Hachiya, N., Alam, R., Sakasegawa, Y., Sakaguchi, M., Mihara, K., and Omura, T., A mitochondrial import factor purified from rat liver cytosol is an ATP-dependent conformational modulator for precursor proteins, *EMBO J. 12*:1579 (1993).
9. Hajek, P. and Bedwell, D. M., Characterization of the mitochondrial binding and import properties of purified yeast F_1-ATPase β subunit precursor: Import requires external ATP, *J. Biol. Chem. 269*:7192 (1994).
10. Endo, T., Mitsui, S., and Roise, D., Mitochondrial presequences can induce aggregation of unfolded proteins, *FEBS Lett. 359*:93 (1995).
11. Endo, T., Mitsui, S., Nakai, M., and Roise, D., Binding of mitochondrial presequences to yeast cytosolic heat shock protein 70 depends on the amphiphilicity of the presequence, *J. Biol. Chem. 271*:4161 (1996).
12. Terada, K., Ohtsuka, K., Imamoto, N., Yoneda, Y., and Mori, M., Role of heat shock cognate 70 protein in import of ornithine transcarbamylase precursor into mammalian mitochondria, *Mol. Cell Biol. 15*:3708 (1995).
13. Lain, B., Iriarte, A., and Martinez-Carrion, M., Dependence of the folding and import of the precursor to mitochondrial aspartate aminotransferase on the nature of the cell-free translation system, *J. Biol. Chem. 269*:15588 (1994).
14. Frydman, J. and Hartl, F.-U., Molecular chaperone functions of hsp70 and hsp60 in protein folding, *The Biology of Heat Shock Proteins and Molecular Chaperones* (R. I. Morimoto, A. Tissières, and C. Georgopoulos, eds.), Cold Spring Harbor Press, Cold Spring Harbor, NY, 1994, p. 251.
15. Lieberek, K., Marszalek, J., Ang, D., Georgopoulos, C., and Zylicz, M., *Escherichia coli* DnaJ and GrpE heat shock proteins jointly stimulate ATPase activity of DnaK, *Proc. Natl. Acad. Sci. USA 88*:2874 (1991).

16. Cyr, D. M., Langer, T., and Douglas, M. G., DnaJ-like proteins: Molecular chaperones and specific regulators of Hsp70, *Trends Biochem. Sci. 19*:176 (1994).
17. Caplan, A. J. and Douglas, M. G., Characterization of YDJ1: A yeast homologue of the bacterial dnaJ protein, *J. Cell Biol. 114*:609 (1991).
18. Atencio, D. P. and Yaffe, M. P., MAS5, a yeast homolog of DnaJ involved in mitochondrial protein import, *Mol. Cell. Biol. 12*:283 (1992).
19. Caplan, A. J., Cyr, D. M., and Douglas, M. G., YDJ1p facilitates polypeptide translocation across different intracellular membranes by a conserved mechanism, *Cell 71*:1143 (1992).
20. Cyr, D. M., Lu, X., and Douglas, M. G., Regulation of hsp70 function by a eukaryotic DnaJ homolog, *J. Biol. Chem. 267*:20927 (1992).
21. Cyr, D. M., Cooperation of the molecular chaperone Ydj1 with specific Hsp70 homologs to suppress protein aggregation, *FEBS Lett. 359*:129 (1995).
22. Caplan, A. J., Tsai, J., Casey, P. J., and Douglas, M. G., Farnesylation of YDJ1p is required for function at elevated growth temperatures in *Saccharomyces cerevisiae*, *J. Biol. Chem. 267*:18890 (1992).
23. Komiya, T., Sakaguchi, M., and Mihara, K., Cytosolic chaperones determine the targeting pathway of precursor proteins to mitochondria, *EMBO J. 15*:399 (1996).
24. Murakami, K., Tokunaga, F., Iwanaga, S., and Mori, M., Presequence does not prevent folding of a purified mitochondrial precursor protein and is essential for association with reticulocyte cytosolic factor(s), *J. Biochem. 108*:207 (1990).
25. Ono, H. and Tuboi, S., The cytosolic factor required for import of precursors of mitochondrial proteins into mitochondria, *J. Biol. Chem. 263*:3188 (1988).
26. Murakami, K. and Mori, M., Purified presequence binding factor (PBF) forms an import-competent complex with a purified mitochondrial precursor protein, *EMBO J. 9*:3201 (1990).
27. Murakami, K., Tanase, S., Morino, Y., and Mori, M., Presequence binding factor-dependent and -independent import of proteins into mitochondria, *J. Biol. Chem. 267*:13119 (1992).
28. Alam, R., Hachiya, N., Sakaguchi, M., Kawabata, S., Iwanaga, S., Kitajima, M., Mihara, K., and Omura, T., cDNA cloning and characterization of mitochondrial import stimulation factor (MSF) purified from rat liver cytosol, *J. Biochem. 116*:416 (1994).
29. Aitkin, A., 14-3-3 proteins on the MAP, *Trends Biochem. Sci. 20*:95 (1995).
30. Komiya, T., Hachiya, N., Sakaguchi, M., Omura, T., and Mihara, K., Recognition of mitochondria-targeting signals by a cytosolic import stimulation factor, MSF, *J. Biol. Chem. 269*:30893 (1994).
31. Hachiya, N., Komiya, T., Alam, R., Iwahashi, J., Sakaguchi, M., Omura, T., and Mihara, K., MSF, a novel cytoplasmic chaperone which functions in precursor targeting to mitochondria, *EMBO J. 13*:5146 (1994).
32. Xiao, B., Smerdon, S. J., Jones, D. H., Dodson, G. G., Soneji, Y., Aitken, A., and Gamblin, S. J., Structure of a 14-3-3 protein and implications for coordination of multiple signaling pathway, *Nature 376*:188 (1995).
33. Eilers, M. and Schatz, G., Binding of a specific ligand inhibits import of a purified precursor protein into mitochondria, *Nature 332*:228 (1986).

34. Endo, T. and Schatz, G., Latent membrane perturbation activity of a mitochondrial precursor protein is exposed by unfolding, *EMBO J. 7*:1153 (1988).
35. Lithgow, T., Glick, B. S., and Schatz, G., The protein import receptor of mitochondria, *Trends Biochem. Sci. 20*:98 (1995).
36. Lill, R. and Neupert, W., Mechanisms of protein import across the mitochondrial outer membrane, *Trends Cell Biol. 6*:56 (1996).
37. Hachiya, N., Mihara, K., Suda, K., Horst, M., Schatz, G., and Lithgow, T., Reconstitution of the initial steps of mitochondrial protein import, *Nature 376*:705 (1995).
38. Hwang, S. T. and Schatz, G., Translocation of proteins across the mitochondrial inner membrane, but not into the outer membrane requires nucleotide triphosphates in the matrix, *Proc. Natl. Acad. Sci. USA 86*:8432 (1989).
39. Wachter, C., Schatz, G., and Glick, B. S., Protein import into mitochondria: The requirement for external ATP is precursor-specific whereas intramitochondrial ATP is universally needed for translocation into the matrix, *Mol. Biol. Cell 5*:465 (1994).
40. Mihara, K. and Omura, T., Cytoplasmic chaperones in precursor targeting to mitochondria: The role of MSF and hsp70, *Trends Cell Biol. 6*:104 (1996).
41. Wachter, C., Schatz, G., and Glick, B. S., Role of ATP in the intramitochondrial sorting of cytochrome c_1 and the adenine nucleotide translocator, *EMBO J. 11*:4787 (1992).
42. Haucke, V., Lithgow, T., Rospert, S., Hahne, K., and Schatz, G., The yeast mitochondrial protein import receptor Mas20p binds precursor proteins through electrostatic interaction with the positively charged presequence, *J. Biol. Chem. 270*:5565 (1995).
43. Gratzer, S., Lithgow, T., Bauer, R. E., Lamping, E., Patlauf, F., Kohlwein, S. D., Haucke, V., Junne, T., Schatz, G., and Horst, M., Mas37p, a novel receptor subunit for protein import into mitochondria, *J. Cell Biol. 129*:25 (1995).
44. Hines, V. and Schatz, G., Precursor binding to yeast mitochondria: A general role for the outer membrane protein Mas70p, *J. Biol. Chem. 268*:449 (1993).
45. Komiya, T. and Mihara, K., Protein import into mammalian mitochondria: Characterization of the intermediates along the import pathway of the precursor into the matrix, *J. Biol. Chem. 271*:22105–22110 (1996).
46. Mayer, A., Neupert, W., and Lill, R., Mitochondrial protein import: Reversible binding of the presequence at the *trans* side of the outer membrane drives partial translocation and unfolding, *Cell 80*:127 (1995).
47. Nakai, M., Kinoshita, K., and Endo, T., Mitochondrial receptor complex protein: The intermembrane space domain of yeast MAS17 is not essential for its targeting or function, *J. Biol. Chem. 270*:30571 (1995).
48. Bölliger, L., Junne, T., Schatz, G., and Lithgow, T., Acidic receptor domains on both sides of the outer membrane mediate translocation of precursor proteins into yeast mitochondria, *EMBO J. 14*:6318 (1995).
49. Court, D. A., Nargang, F. E., Steiner, H., Hodges, R. S., Neupert, W., and Lill, R., Role of the intermembrane-space domain of the preprotein receptor Tom22 in protein import into mitochondria, *Mol. Cell. Biol. 16*:4035 (1996).
50. Pfanner, N., Craig, E. A., and Meijer, M., The protein import machinery of the mitochondrial inner membrane. *Trends Biochem. Sci. 19*:368 (1994).
51. Blom, J., Kübrich, M., Rassow, J., Voos, W., Dekker, P. J. T., Maarse, A. C., Meijer, M., and Pfanner, N., The essential yeast protein MIM44 (encoded by *MPI1*) is

involved in an early step of preprotein translocation across the mitochondrial inner membrane, *Mol. Cell Biol. 13*:7364 (1993).
52. Schneider, H. C., Berthold, J., Bauer, M. F., Dietmeier, K., Guiard, B., Brunner, N., and Neupert, W., Mitochondrial Hsp70/MIM44 complex facilitates protein import, *Nature 371*:768 (1994).
53. Kronidou, N. G., Oppliger, W., Bolliger, L., Hannavy, K., Glick, B. S., Schatz, G., and Horst, M., Dynamic interaction between ISP45 and mitochondrial hsp70 in the protein import system of yeast mitochondrial inner membrane, *Proc. Natl. Acad. Sci. USA 91*:12818 (1994).
54. Rassow, J., Maarse, A. C., Krainer, E., Kübrich, M., Müller, H., Meijer, M., Craig, E. A., and Pfanner, N., Mitochondrial protein import: Biochemical and genetic evidence for interaction of matrix hsp70 and the inner membrane protein MIM44, *J. Cell Biol. 127*:1547 (1994).
55. Horst, M., Hilfiker-Rothenfluh, S., Oppliger, W., and Schatz, G., Dynamic interaction of the protein translocation systems in the inner and outer membranes of yeast mitochondria, *EMBO J. 14*:2293 (1995).
56. Berthold, J., Bauer, M. F., Schneider, H.-C., Klaus, C., Dietmeier, K., Neupert, W., and Brunner, M., The MIM complex mediates protein translocation across the inner membrane and couples it to the mt-hsp70/ATP driving system, *Cell 81*:1085 (1995).
57. Martin, J., Mahlke, K., and Pfanner, N., Role of an energized inner membrane in mitochondrial protein import: $\Delta\Psi$ drives the movement of presequences, *J. Biol. Chem. 266*:18051 (1991).
58. Maduke, M. and Roise, D., Import of a mitochondrial presequence into protein-free phospholipid vesicles, *Science 260*:364 (1993).
59. Ungermann, C., Neupert, W., and Cyr, D. M., The role of hsp70 in conferring unidirectionality on protein translocation into mitochondria, *Science 266*:1250 (1994).
60. Ungermann, C., Guiard, B., Neupert, W., and Cyr, D. M., The $\Delta\psi$- and Hsp70–MIM44-dependent reaction cycle driving early steps of protein import into mitochondria, *EMBO J. 15*:735 (1996).
61. Nakai, M., Kato, Y., Ikeda, E., Toh-e, A., and Endo, T., Yge1p, a eukaryotic GrpE homolog, is localized in the mitochondrial matrix and interacts with mitochondrial hsp70, *Biochem. Biophys. Res. Commun. 200*:435 (1994).
62. Bolliger, L., Deloche, O., Glick, B. S., Georgopoulos, C., Jenö, P., Kronidou, N., Horst, M., Morishima, N., and Schatz, G., A mitochondrial homolog of bacterial GrpE interacts with mitochondrial hsp70 and is essential for viability, *EMBO J. 13*:1998 (1994).
63. Laloraya, S., Gambill, B. D., and Craig, E. A., A role for a eukaryotic GrpE-related protein, Mge1p, in protein translocation, *Proc. Natl. Acad. Sci. USA 91*:6481 (1994).
64. Westermann, B., Prip-Buus, C., Neupert, W., and Schwarz, E., The role of the GrpE homologue, Mge1p, in mediating protein import and protein folding in mitochondria, *EMBO J. 14*:3452 (1995).
65. Voos, W., Gambill, B. D., Laloraya, S., Ang, D., Craig, E. A., and Pfanner, N., Mitochondrial GrpE is present in a complex with hsp70 and preproteins in transit across membranes, *Mol. Cell. Biol. 14*:6627 (1994).
66. Neupert, W., Hartl, F.-U., Craig, E. A., and Pfanner, N., How do polypeptides cross the mitochondrial membranes? *Cell 63*:447 (1990).

67. Simon, S. M., Peskin, C. S., and Oster, G. F., What drives the translocation of proteins? *Proc. Natl. Acad. Sci. USA 89*:3770 (1992).
68. Glick, B. S., Can hsp70 proteins act as force-generating motors? *Cell 80*:11 (1995).
69. Horst, M., Azem, A., Schatz, G., and Glick, B. S., What is the driving force for protein import into mitochondria? *Biochim. Biophys. Acta 1318*:71 (1997).
70. Schmitt, M., Neupert, W., and Langer, T., Hsp78, a Clp homologue within mitochondria, can substitute for chaperone functions of mt-hsp70, *EMBO J. 14*:3434 (1955).
71. Voos, W., von Ahsen, O., Müller, H., Guiard, B., Rassow, J., and Pfanner, N., Differential requirement for the mitochondrial Hsp70–Tim44 complex in unfolding and translocation of preproteins, *EMBO J. 15*:2668 (1996).
72. Hurt, E. C., Pesold-Hurt, B., Suda, K., Oppliger, W., and Schatz, G., The first twelve amino acids (less than half of the pre-sequence) of an imported mitochondrial protein can direct mouse cytosolic dihydrofolate reductase into the yeast mitochondrial matrix, *EMBO J. 4*:2061 (1985).
73. Rassow, J., Hartl, F.-U., Guiard, B., Pfanner, N., and Neupert, W., Polypeptides traverse the mitochondrial envelope in an extended state, *FEBS Lett. 275*:190 (1990).
74. Endo, T., Eilers, M., and Schatz, G., Binding of a tightly folded artificial mitochondrial precursor protein to the mitochondrial outer membrane involves a lipid-mediated conformational change, *J. Biol. Chem. 264*:2951 (1989).
75. von Ahsen, O., Voos, W., Henninger, H., and Pfanner, N., The mitochondrial protein import machinery: Role of ATP in dissociation of the Hsp70–Mim44 complex, *J. Biol. Chem. 270*:29848 (1995).
76. Horst, M., Oppliger, W., Feifel, B., Schatz, G., and Glick, B., The mitochondrial protein import motor: Dissociation of mitochondrial hsp70 from its membrane anchor requires ATP binding rather than ATP hydrolysis, *Protein Sci. 5*:759 (1996).
77. Schneider, H.-C., Westermann, B., Neupert, W., and Brunner, M., The nucleotide exchange factor MGE exerts a key function in the ATP-dependent cycle of mt-hsp70/Tim44 interaction driving mitochondrial protein import, *EMBO J. 15*:5796 (1996).
78. Glick, B. S., Beasley, E. M., and Schatz, G., Protein sorting in mitochondria, *Trends Biochem. Sci. 17*:453 (1992).
79. Stuart, R. A. and Neupert, W., Topogenesis of inner membrane proteins of mitochondria, *Trends Biochem. Sci. 21*:261 (1996).
80. Cyr, D. M., Stuart, R. A., and Neupert, W., A matrix ATP requirement for presequence translocation across the inner membrane of mitochondria, *J. Biol. Chem. 268*:23751 (1993).
81. Gärtner, F., Voos, W., Querol, A., Miller, B. R., Craig, E. A., Cumsky, M. G., and Pfanner, N., Mitochondrial import of subunit V_a of cytochrome *c* oxidase characterized with yeast mutants: Independence from receptors, but requirement for matrix hsp70 translocase function, *J. Biol. Chem. 270*:3788 (1995).
82. Fölsch, H., Guiard, B., Neupert, W., and Stuart, R. A., Internal targeting signal of the BSC1 protein: A novel mechanism of import into mitochondria, *EMBO J. 15*:479 (1996).
83. Rojo, E. E., Stuart, R. A., and Neupert, W., Conservative sorting of F_0-ATPase subunit 9: Export from matrix requires ΔpH across inner membrane and matrix ATP, *EMBO J. 14*:3445 (1995).

84. Herrmann, J. M., Koll, H., Cook, R. A., Neupert, W., and Stuart, R. A., Topogenesis of cytochrome oxidase subunit II: Mechanisms of protein export from the mitochondrial matrix, *J. Biol. Chem. 270*:27079 (1995).
85. Ostermann, J., Voos, W., Kang, P. J., Craig, E. A., Neupert, W., and Pfanner, N., Precursor proteins in transit through mitochondrial contact sites interact with hsp70 in the matrix, *FEBS Lett. 277*:281 (1990).
86. Kang, P.-J., Ostermann, J., Shilling, J., Neupert, W., Craig, E. A., and Pfanner, N., Requirement for hsp70 in the mitochondrial matrix for translocation and folding of precursor proteins, *Nature 348*:137 (1990).
87. Glick, B. S., Wachter, C., Reid, G. A., and Schatz, G., Import of cytochrome b_2 to the mitochondrial intermembrane space: The tightly folded heme-binding domain makes import dependent upon matrix ATP, *Protein Sci. 2*:1901 (1993).
88. Stuart, R. A., Gruhler, A., van der Klei, I., Guiard, B., Koll, H., and Neupert, W., The requirement of matrix ATP for the import of precursor proteins into the mitochondrial matrix and intermembrane space, *Eur. J. Biochem. 220*:9 (1994).
89. Voos, W., Gambill, B. D., Guiard, B., Pfanner, N., and Craig, E. A., Presequence and mature part of preproteins strongly influence the dependence of mitochondrial protein import on heat shock protein 70 in the matrix, *J. Cell Biol. 123*:119 (1993).
90. Gärter, F., Bömmer, U., Guiard, B., and Pfanner, N., The sorting signal of cytochrome b_2 promotes early divergence from the general mitochondrial import pathway and restricts the unfoldase activity of matrix hsp70, *EMBO J. 14*:6043 (1995).
91. Cheng, M. Y., Hartl, F.-U., Martin, J., Pollock, R. A., Kalousek, F., Neupert, W., Hallberg, E. M., Hallberg, R. L., and Horwich, A., Mitochondrial heat-shock protein hsp60 is essential for assembly of proteins imported into yeast mitochondria, *Nature 337*:620 (1989).
92. Koll, H., Guiard, B., Rassow, J., Ostermann, J., Horwich, A. L., Neupert, W., and Hartl, F.-U., Antifolding activity of hsp60 couples protein import into the mitochondrial matrix with export to the intermembrane space, *Cell 68*:1163 (1992).
93. Glick, B. S., Brandt, A., Cunningham, K., Müller, S., Hallberg, R. L., and Schatz, G., Cytochromes c_1 and b_2 are sorted to the intermembrane space of yeast mitochondria by a stop–transfer mechanism, *Cell 69*:809 (1992).
94. Hallberg, E. M., Shu, Y., and Hallberg, R. L., Loss of mitochondrial hsp60 function: Nonequivalent effects on matrix-targeted and intermembrane-targeted proteins, *Mol. Cell. Biol. 13*:3050 (1993).
95. Höhfeld, J. and Hartl, F.-U., Role of the chaperonin cofactor hsp70 in protein folding and sorting in yeast mitochondria, *J. Biol. Chem. 126*:305 (1994).
96. Rospert, S., Müller, S., Schatz, G., and Glick, B. S., Fusion proteins containing the cytochrome b_2 presequence are sorted to the mitochondrial intermembrane space independently of hsp60, *J. Biol. Chem. 269*:17279 (1994).
97. Roise, D., Recognition and binding of mitochondrial presequences during the import of proteins into mitochondria, *J. Bioenergetics Biomembr. 29*:19 (1997).

19

Protein Folding Mediated by Intramolecular Chaperones

Ujwal Shinde, Jianjun Liu, and Masayori Inouye
Robert Wood Johnson Medical School, Piscataway, New Jersey

I. THE PROTEIN FOLDING PROBLEM

Biologically active proteins need to acquire specifically folded three-dimensional structures. The genetic code specifies only the covalent primary structure of a protein, which is a linear chain of amino acid residues that constitutes the polypeptide backbone. In vitro studies have shown that many purified proteins spontaneously adopt unique, complex tertiary structures after being completely unfolded. This suggested that all information necessary for proteins to adopt these structures resides in their primary sequences. This proposal was made by Anfinsen and coworkers a quarter century ago [1]. Protein folding refers to the study of steps involved in the transition from an inactive linear chain of amino acid residues to a functional protein molecule [2].

However, because the initial proposal two and a half decades ago, physicists, chemists, and biologists have all failed to decode the information that proteins use during their acquisition of complex structures that constitute the basis of biological activity. The protein-folding problem consists of three basic questions [3]: (a) What is the relation between amino acid sequences and protein structures? (b) What are the intermediate states that occur when a linear chain acquires a complex tertiary structure? (c) What are the energy barriers that can exist between various intermediate states and the final state?

II. THE ORIGIN OF ASSISTED PROTEIN FOLDING

Initial studies on protein folding were carried out in vitro because the folding process can be monitored using spectroscopic tools and because of the relative ease by which one can control the environment for protein folding. However, the discovery of specific proteins of "molecular chaperones" that can bind to completely or partially synthesized proteins and, then, help them through the folding process have dramatically altered our views of protein folding [4–6]. Assisting and supervising protein folding, however, does not nullify Anfinsen's hypothesis, but only emphasizes the differences between in vitro and in vivo protein-folding studies [7]. Additional factors, such as prolyl and disulfide isomerases, are also critical for the in vivo-folding process [8,9]. About the time that molecular chaperones were identified, another mechanism of protein folding also emerged [10].

Direct expression of several prokaryotic and eukaryotic proteases did not lead to production of active enzymes [10–13]. These proteins are unable to spontaneously refold from their denatured conditions, even when the forces that induced unfolding were reversed. A careful scrutiny of such proteins has revealed that they are produced as inactive precursors; namely, prepro- or proproteins [14]. The pro-regions are usually attached to the NH_2-terminus of the mature protein and are called propeptides or prosequences and are different from the presequences or signal peptides that are required for secretion across the membrane [15]. The propeptides are critical for correct folding of the mature proteins [16]. Also, they have occasionally been implicated in assisting the signal peptide in correct localization of mature proteins [17]. The presence of such polypeptide extensions has added a new wrinkle to the already complex problem of protein folding. Propeptide-mediated protein folding that was initially demonstrated in subtilisin E, a protein that is homologous with an industrially important protease subtilisin BPN′. Soon, thereafter, α-lytic protease was also shown to employ a similar mechanism for its folding [13].

III. HISTORY OF PROPEPTIDE-MEDIATED PROTEIN FOLDING

Subtilisins are alkaline serine proteases produced in bacilli and because they are industrially important proteins, they have been extensively investigated as targets for enzyme engineering [18]. Several high-resolution crystal structures of subtilisin and subtilisin complexed with polypeptides and inhibitors are available [19–21]. This substantially detailed biochemical and biophysical data about subtilisin and its homologues have immensely benefited propeptide-mediated protein folding research [22].

Subtilisin E is a 275-residue alkaline serine protease that is produced by

Bacillus subtilis 168 [10]. However, when the gene encoding the 275-residue mature protein was cloned and expressed in *Escherichia coli*, it failed to produce an enzymatically active protein [11]. An analysis of the gene sequence coding for this protein showed that the protein is produced as a precursor preprosubtilisin [14]. The pre-region corresponded to the signal peptide that was responsible for correct localization of subtilisin [15]. In addition, there existed a 77-amino acid residue polypeptide (termed the propeptide) that linked the signal peptide to mature subtilisin E [10,23]. Expression of the gene that included the 77-amino acid residue propeptide along with the 275-residue mature protein in *E. coli*, resulted in active subtilisin E and showed that the propeptide was important for enzymatic activity, although it was not a part of the mature protein. This resulted in two fundamentally important questions: (a) How is the propeptide removed from the mature protein? (b) What is the function of the 77-mer propeptide?

A careful dissection of the catalytic properties of the enzyme has shown that enzymatically active subtilisin catalyzes the hydrolysis of peptide and ester bonds [24]. The catalytic triad constitutes Asp-32, His-64 and Ser-221 residues. Catalysis is achieved by specific binding and stabilization of a tetrahedral transition state in which His-64 acts as a base that accepts a proton from Ser-221 as it forms a bond with the substrate carbonyl carbon. The protonated His-64 is stabilized by coloumbic interactions with Asp-32 and the oxyanion by hydrogen bonds from the backbone amide of Ser-221 and amide side chain of Asn-155 [25]. When Asp-32 was mutated to an Asn and the gene encoding subtilisin was expressed, no active subtilisin was produced. Importantly, the protein produced corresponded to the 352-residue precursor pro-Asp-32Asn-subtilisin and showed that the catalytic triad of the mature protein was required for the removal of the propeptide [11,12].

When 352-residue precursor pro-Asp-32Asn-subtilisin was added to denatured mature subtilisin and refolding was initiated, active subtilisin was produced. Because pro-Asp-32Asn-subtilisin does not give active subtilisin, the data suggested that activity resulted from denatured mature subtilisin, and the precursor was probably involved in the facilitating correct folding of denatured subtilisin [15]. Importantly, the 77-residue region that corresponded to the propeptide, when chemically synthesized, could help denatured subtilisin to obtain activity, although it was not covalently linked to the protease domain [26,27]. This clearly indicated that the propeptide facilitated correct folding of the protease domain and was not required for stability of the folded protein.

The 77-residue propeptide of subtilisin also acts as a temporary competitive inhibitor of enzymatic activity and may be required to delay enzymatic activity and prevent autodegradation of the partially folded or unfolded protein molecules [12,28,29]. Studies carried out in trans between the propeptide and mature subtilisin show that the inhibition follows slow-binding kinetics with an inhibition constant in the range of 10^{-9} M [30].

Table 1 Comparison Between IMCs and Molecular Chaperones

Function	IMC	Molecular chaperone
Energy source for mediating folding	None required	ATP hydrolysis
Specificity	Extremely specific	Nonspecific
Association with protein they help fold	Covalently attached to NH_2-terminus	Independent
Catalytic reusability	Not reusable and acts as a single-turnover catalyst	Reusable and functions as a multiturnover catalyst
Interactions with the protein after folding	Temporary competitive inhibitor that is subsequently degraded	None
Mechanism by which they mediate protein folding	a. Helps overcome kinetic barrier along folding pathway b. May impart structural information to mature protein	a. Helps folding by binding to unfolded protein to prevent aggregation b. Does not impart any structural information to the protein "substrate"
Structural organization	Small polypeptide that adopts a single domain	Large polypeptide with multiple domains

IV. THE CONCEPT OF IMCs

Because the propeptides of subtilisin and α-lytic protease facilitate correct folding, they were classified as molecular chaperones [31]. However, significant differences exist between propeptides and molecular chaperones. Therefore, propeptides were termed intramolecular chaperones (IMCs) to differentiate them from molecular chaperones [22,23,32]. Differences between the two are highlighted in Table 1.

V. GENERALIZATION OF THE CONCEPT OF IMCs

After the initial finding that the propeptide of subtilisin E acts as an IMC, similar folding mechanisms were reported in many proteins [22,23,32]. Although most of the well-studied examples include serine proteases, this mechanism is not limited to this class of proteins [19]. Similar mechanisms exist in aspartic [33], cysteine [34], and also metalloproteases [35]. Examples from the nonprotease group include transforming growth factor -β_1 (TFG-β_1) [36], nerve growth factor (NGF) [37], von Willebrand factor [38], activin A [36], amphiregulin [39], bovine pancreatic trypsin inhibitor [40], and others. Although IMC-mediated protein folding is a common phenomenon among proteases, not all proteases require their propeptides for correct folding. Noteworthy exceptions include trypsin, chymotrypsin, elastase and renin [41].

VI. COMMON FEATURES OF IMCs

Although IMC domains of subtilisin, α-lytic protease, carboxypeptidase Y, aqualysin, and proteinase A and B perform similar functions, they display insignificant sequence homology [23]. This suggests that the mechanism of IMC-mediated protein folding may have emerged through convergent evolution [19].

Another important point about bacterial subtilisins is that they belong to a family of proteases that have now been termed as subtilases [42]. More than 100 subtilases are now shown to exist in different species that include archea, bacteria, fungi, yeasts, and higher eukaroytes. Multiple sequence alignments of the catalytic domains of these proteins have allowed a subdivision of four major classes [43]:

A. Bacterial and plant proteases (including subtilisins)
B. Tripeptidyl peptidases
C. Eukaryotic prohormone convertases
D. Proteinase K-like endoproteases found in fungi, yeasts, and gram-negative bacteria

Over 30 proteins are known in classes A, C, and D, whereas fewer than 5 sequences are known in the B class of subtilases. The sequence homology for the catalytic domain is the highest for the C-type and D-type subtilases, whereas the A-type subtilases are extremely heterogeneous in sequence length and homology [see Ref. 43 for details on the sequence alignments].

VII. MECHANISMS OF IMC-MEDIATED PROTEIN FOLDING

The mechanisms of IMC-mediated protein folding have been extensively studied in subtilisin [44,45], α-lytic protease [46], and carboxypeptidase Y [47]. The folding of all these proteins in the absence of their IMC domain does not lead to significant enzymatic activity [48]. Covalent attachment of the IMC domain to the mature protein is not a necessary criterion for facilitating correct folding, and the mature protease domain does not require the IMC domain for its stability [15].

A. Structural Studies on Isolated IMC Domains

Structural studies on the conformations of the IMC domains alone have shown that these domains are unstructured in the absence of the mature protease domains [44,45,48]. However, on interacting with the mature protease domains, they adopt significant secondary structure, as ascertained using differential circular dichroism spectrophotometric techniques [44,45,48]. This has suggested that the interactions that occur between the IMC domain and the mature protein are dynamic [48] and that interactions between the two domains occur only during the late stages of protein folding [45]. In vivo, the NH_2-terminal propeptide is synthesized before the mature protease domain. It is important at this stage to point out that the IMC domain, in addition to its folding function, may also interact with the transport machinery within the cell [19].

B. Intermediates Isolated Along the Protein-Folding Pathway

One mechanism of protein folding was demonstrated by Agard and co-workers [46] and later substantiated by Fersht and co-workers [45]. In this mechanism, the IMC domain was required only to overcome a kinetic barrier along the folding pathway. Therefore, if the protein was folded in the absence of its IMC domain, then it attained a "molten-globule"-like intermediate state [45,46]. Molten globule-like intermediates represent "fluid" states that display significant secondary structure, but little or no tertiary structure [49,50]. These states in subtilisin and α-lytic protease represent trapped intermediates along the folding pathway and depict a secondary structure similar to that of the mature protein. However, no significant tertiary structure could be detected in this intermediate [45,46,51]. Moreover, this state was fairly stable, but catalytically inactive. Addition of the IMC domain to this intermediate state helped it rapidly convert to an active native-like state [45,46] (Fig. 1). These results have strongly suggested that, under the conditions used by these researchers, the IMC domain is required only to circumvent a kinetic block along the folding pathway of these proteins [17].

Refolding studies carried out using precursors of subtilisin have shown that a similar state is formed even in presence of the IMC domain and that the cleavage of the covalent linkage between the IMC domain and the protease domain may occur in this molten–globule-like state [51,52]. Refolding of denatured precursors was initiated by a step-wise dialysis against buffers containing diminishing amounts of the denaturant (here, urea) [53]. Subsequent to the autoprocessing of the IMC domain, the complex (protease domain plus IMC domain) attains a more compact conformation that significantly reduces the solvent accessible hydrophobic surface area [52]. Preliminary results on wild-type prosubtilisin have now shown that the autoprocessing can reach completion under refolding conditions that contain 2.5 M urea (unpublished results).

C. Role of Calcium in Folding of Subtilisin

An alternative method has been employed by Bryan and co-workers to understand the folding mechanisms mediated by IMCs. Mutations were created within the mature domain of subtilisin to disrupt a high-affinity calcium-binding site in

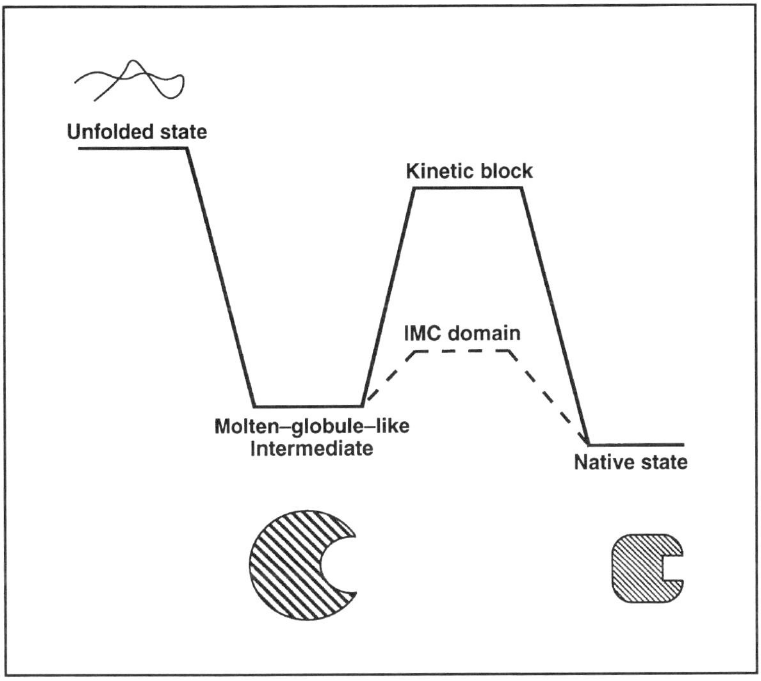

Figure 1 Partial free-energy diagram for IMC-mediated protein folding.

subtilisin [54]. Interestingly, the calcium deletion variant of subtilisin was able to fold into an active form in the absence of the propeptide; also, however, in the presence of high concentrations of salt. This has suggested that the formation of the calcium-binding site in subtilisin may be critical in the folding mechanism and that the presence of the IMC domain somehow delays the formation of this site [55].

D. Molecular Memory During Folding Mediated by an IMC

Recent data now suggest that the IMC of subtilisin facilitates folding by imparting structural information to its protease domain, although it is not part of the domain. Therefore, subtilisin E folded by a mutant IMC (IMC^{PC}) acquires an "alternative" enzymatically active conformation (Sub^{PC}) that differs from wild-type subtilisin E (Sub^{WT}). Both, Sub^{PC} and Sub^{WT} display *memory* of the folding process and, consequently recognize their cognate IMCs (when added in trans) by binding them with a fourfold greater affinity than noncognate IMCs. Significant changes in secondary structures, Thermostability, and substrate specificities of Sub^{WT} and Sub^{PC} are also detected. Although both proteins display identical primary sequences (as ascertained by mass spectroscopy and NH_2-terminal sequencing), the K_{cat}/K_M ratio of Sub^{PC} for *N*-succinyl-AAPF-*p*-nitroanalide is 62% greater than that for Sub^{WT}. Therefore, similar mutations in IMC domains of prohormone convertases may result in their alternative folding that modify substrate specificities and causes hormonal disorders owing to incorrect maturation of hormone precursors (results to be published).

E. Effect of Protein and Salt Concentration on Folding of Subtilisin

Immobilization of native subtilisin on a matrix using chemical cross-linking, followed by unfolding the protein by exposing it to chaotropic denaturants, such as guanidine hydrochloride, and then renaturing it by removing the denaturant, allows quantitative refolding of subtilisin [56]. Immobilization, as just described, reduces the degrees of freedom of the denatured polypeptide to maintain it in a conformation critical for refolding. Also, immobilization reduces intermolecular interactions that may result in aggregation. To this effect, very low concentrations of the denatured subtilisin (1 $\mu g\ mL^{-1}$) can be successfully refolded in the absence of the IMC domain by using 2 M potassium acetate at pH 6.5. Therefore, 2 M potassium acetate can partially compensate for the absence of the IMC domain [57]. In the presence 2 M potassium acetate, the IMC domain further enhances the folding efficiency of denatured subtilisin.

VIII. X-RAY STRUCTURE OF THE IMC DOMAIN COMPLEXED WITH THE PROTEASE DOMAIN OF SUBTILISIN

Recently, the crystal structure of subtilisin (that lacked its calcium-binding site) complexed with the IMC domain was elucidated [58]. In this complex with subtilisin the IMC domain folds into a single compact domain with an antiparallel four-stranded β-sheet and two, three-turn α-helices (Fig. 2). The β-sheet of the IMC domain packs tightly against the two parallel surfaces of the α-helices of subtilisin (residues 104–110 and 133–144). Residues Glu-69 and Asp-71 within the IMC domain form helix caps fo the NH_2-termini of the two parallel α-helices in subtilisin. Another charge dipole interaction exists between the carboxylate of Glu-112 in subtilisin and the peptide nitrogen constituted by residues 42, 43, and 44 within the IMC domain. The COOH-terminal residues 72–77 extend out from the central part of the IMC domain and interact with the substrate-binding loop within the protease domain, in a manner analogous with that of a synthetic substrate. Hence, residues Tyr-77, Ala-76, His-75, and Ala-74 of the IMC domain occupy subsites S1 to S4 of subtilisin. Almost all the interactions of the IMC

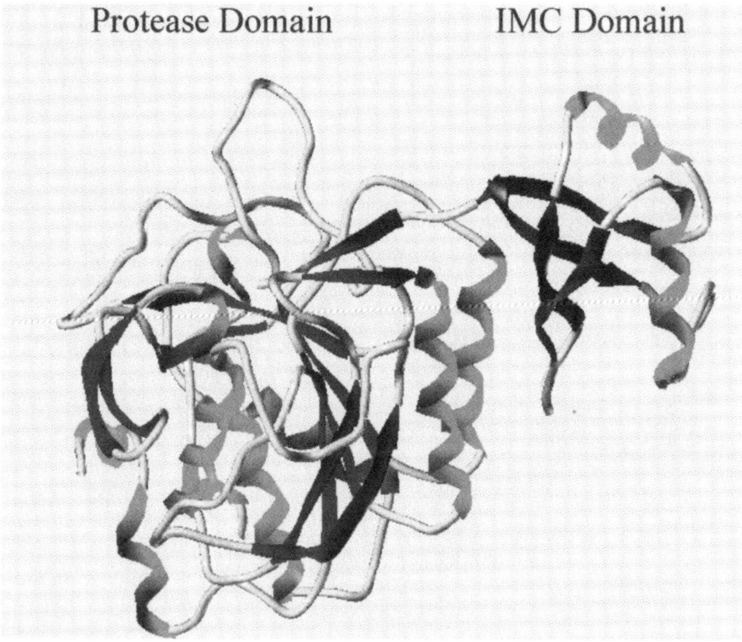

Figure 2 Crystal structure of subtilisin complexed with its IMC domain.

domain with the protease domain are made with residues between 100 and 144 of the protease domain.

IX. SPECULATIONS REGARDING FOLDING MEDIATED BY THE IMC DOMAIN OF SUBTILISIN

The maturation pathway of subtilisin involves four major steps (Fig. 3):

Step 1: Folding of the protease domain mediated by the IMC domain
Step 2: Mechanism of autoprocessing of the IMC domain from the protease domain
Step 3: Structural changes coincident with autoprocessing
Step 4: Degradation of the IMC domain

A. Folding of the Protease Domain Mediated by the IMC Domain

Because folding of mature subtilisin in the absence of its propeptide leads to the formation of a molten globule-like intermediate that binds calcium ions, and

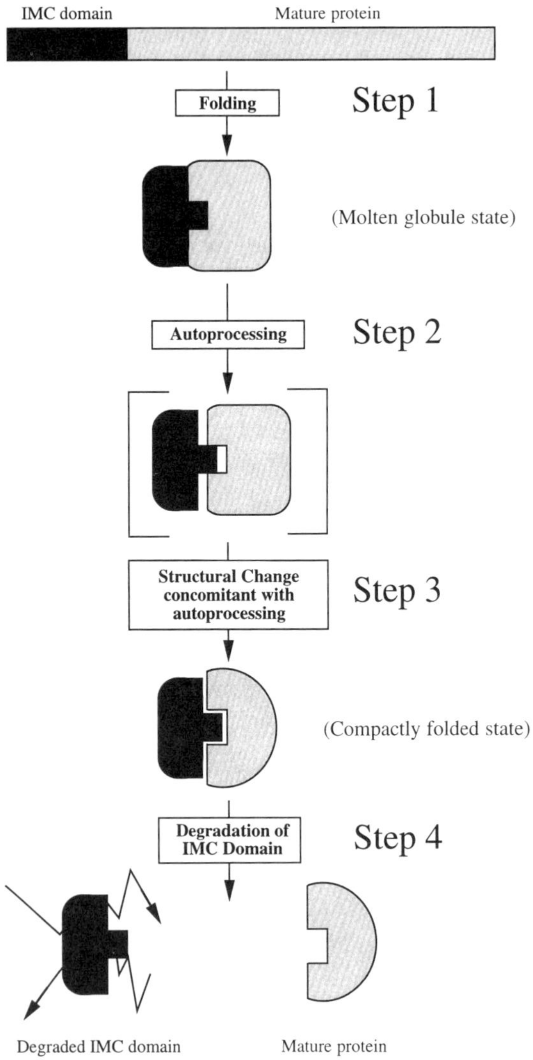

Figure 3 The IMC-mediated maturation pathway of subtilisin.

conversion of this intermediate state to the native state is not observed in the absence of the IMC domain, it was speculated that IMC domain functions only during the late stages of the folding pathway. Deletion of the calcium A site in subtilisin probably destabilizes the intermediate relative to the native state and, therefore, facilitates both catalyzed as well as uncatalyzed folding [59].

Attachment of the IMC domain to subtilisin probably prevents the formation of a stable intermediate by dictating the steps along the folding pathway, such that the calcium-binding site is formed late during the folding pathway. The manner in which the IMC domain binds with the protease domain supports a unimolecular model for the maturation of subtilisin. The COOH-terminal of the IMC binds with the substrate-binding loop of the mature domain. This would involve structural changes around the NH_2-terminal region of the protease domain in an unautoprocessed complex that would enable the cleavage site between the NH_2-terminus of the protease domain and the COOH-terminus of the IMC domain to be oriented into the active site of the protease domain. Modeling studies have shown this to be feasible [51]. In the foregoing conformation, the unautoprocessed protease domain is covalently linked to the COOH-terminus of the IMC domain, and this prevents the ten NH_2-terminal residues of the protease domain from participating in the formation of the calcium-binding site. This is because, in the mature subtilisin, the Gln-2 residue supplies one ligand to calcium and, therefore, is unavailable for complete coordination with the metal ion. However, after autoprocessing the NH_2-terminus swings away from the cleavage site and completes the calcium-binding site. Therefore, in this model, formation of the calcium-binding site occurs only after the autoprocessing of the IMC domain. The initiation of the folding process presumably occurs through the formation of a four-helix bundle that is constituted by two α-helices each from the IMC domain and the protease domain. Four antiparallel β-sheets are sandwiched between this four-helix bundle [59].

B. Mechanism of Autoprocessing of the IMC Domain from the Protease Domain

From the foregoing model, folding is unimolecular; therefore, the autoprocessing of the peptide bond would occur by an intramolecular mechanism. This is the most likely scenario that takes place in vivo. However, in vitro autoprocessing could occur by intramolecular as well as intermolecular mechanisms and is dependent on the protein concentrations. At low protein concentrations, intramolecular autoprocessing dominates, whereas at high concentrations, intermolecular autoprocessing can occur [52,60].

C. Structural Changes Coincident with Autoprocessing

Folding of the prosubtilisin occurs through a molten globular intermediate state. This autoprocessing may occur in this intermediate state. This is followed by a structural reorganization that reduces the solvent-accessible hydrophobic surface area and results in the formation of a more compact IMC–protease domain complex [52].

D. Degradation of the IMC Domain

The propeptide is not essential for the stability of the folded product; therefore, after its autoprocessing, the IMC domain is degraded by the protease domain. This is essential for producing active subtilisin because the propeptide acts as a competitive inhibitor of enzymatic activity [28]. Moreover, in subtilisin, the IMC–protease domain complex is more unstable than the mature protein alone [52]. Release in the inhibition of the protease domain, complexed with the IMC is likely to be the result of a single cut within the IMC domain. This is because various fragments of the IMC were unable to inhibit subtilisin activity. The secondary sites within the IMC domain are still unknown, and identifying these sites will be important in our understanding of the mechanisms by which inhibition is released.

X. OTHER WELL STUDIED EXAMPLES OF IMCs

A. Carboxypeptidase Y

Carboxypeptidase Y (CPY) is an alkaline serine protease produced in *Saccharomyces cerevisiae* and is extensively used as intracellular marker for protein transport and sorting into the vacuole of yeasts [61]. Maturation of CPY is mediated by another vacuolar protease called proteinase A (ProA). Procarboxypeptidase Y (ProCPY) is easily prepared by overexpression of its gene in a strain disrupted in the *PrA* gene [62]. Overexpression leads to production of the zymogen that is secreted into the culture medium owing to lack of the protease required for maturation of proCPY. It is possible to mediate in vitro maturation of proCPY by proteinase K from *Tritirachium album* [63].

Guanidine hydrochloride-denatured proCPY can be rapidly and efficiently refolded by a 60-fold dilution into a suitable buffer. Mature CPY, under identical conditions, fails to refold into an enzymatically active form, suggesting that the propeptide is required for correct folding of the mature protein. Folding of the mature CPY in the absence of its propeptide results in the formation of an inactive molten globule-like intermediate, similar to those observed in α-lytic protease and subtilisin. A comparison of the CPY and proCPY shows that the proregion destabilizes the complex, relative to the mature protein, reinforcing the suggestion that propeptides act by lowering kinetic barriers between folding intermediates and the native protein during the protein-folding pathway [64].

Although the requirement of the propeptide seems essential at low ionic strengths, significant folding can occur under conditions of high ionic strengths. However, unlike subtilisin, folding in the absence of the propeptide does not lead to protein aggregation [65]. Also, other noteworthy differences are that the propeptide binds only weakly with the partially folded intermediate, and it does not cover the active site of the mature protein as seen in subtilisin. Therefore, the active site in proCPY is susceptible to chemical modification. The propeptide of CPY lacks any tertiary structure, although some stable secondary structural elements are observed [66].

B. Proteinase A

Proteinase A (ProA) is a vacuolar aspartic endopeptidase, with a fairly broad substrate specificity. This protein is encoded by the *PEP4* gene in *S. cerevisiae* [67] and is vital for yeast sporulation and for viability during nitrogen starvation. PrA is produced as a precursor containing a 22-amino acid residue signal peptide, 54-residue propeptide, and a 329-amino acid residue protease domain [68]. The signal peptide is removed after translocation into the endoplasmic reticulum, and the propeptide is autoproteolytically removed after it has entered the vacuole. It has not yet been possible to prepare preproPrA because it autoprocesses during purification [69] and, therefore, evidence for the role of the propeptide in folding of PrA has been obtained indirectly. Mutations within the propeptide domain reduce the transport rates of PrA, with a corresponding increase in the half-life of their transit time from the endoplasmic reticulum into the Golgi bodies. One reason for this is attributed to the fact that the mutants do not fold efficiently into a folding-competent state. A mutant PrA precursor, lacking its entire propeptide (prePrA), does not exhibit PrA activity. This protein is translocated into the endoplasmic reticulum, but is subsequently degraded, which may be because the propeptide of PrA acts as a temporary inhibitor of enzymatic activity [70]. However, a mutant preproPrA, in which the active site Asp residue is changed to Glu, rendering PrA inactive, is also degraded at rates similar to the precursor and lacks the propeptide, although this propeptide-lacking variant has rates of folding similar to that of preproPrA. The foregoing studies also show that the degradation activity is not autocatalytic [67].

When the 54-residue propeptide and prePrA (propeptide-deletion mutant) are coexpressed, PrA activity can be recovered in the cell; however, at very low levels (2–3% of wild-type control). This activity is sufficient to facilitate maturation of proCPY. In the absence of the plasmid that encoded the 54-residue propeptide, the proCPY activity was not detected. Overproduction of the propeptide using a multicopy plasmid encoding the propeptide increased the intracellular PrA activity to approximately 8% of that of the wild-type protein [69]. However, most of the prePrA was destined for degradation, even under these conditions. This indicates that the *trans* activity of the propeptide is less effective than the *cis* activity.

Other members of the aspartate family of proteases include well-studied enzymes such as chymosin, pepsin, and cathepsin D [71]. It is interesting that X-ray crystallographic structures have shown that PrA displays secondary and tertiary structures similar to pepsin and cathepsin [70]. Importantly, in spite of the fact that these proteins display significant similarities in their primary sequences of their protease domains, no significant homology is evident within the propeptide domains [72].

C. Aqualysin I

Aqualysin I is a subtilisin-type serine protease that is secreted extracellularly by *Thermus aquaticus* YT-1, an extremely thermophilic, gram-negative bacterium [73]. The enzyme is extremely heat-stable, with an optimum temperature of 80°C for enzymatic activity [74]. Aqualysin I shows substantial sequence homology with proteinase K (43% identity), with subtilisins (37–39%), and thermitase (34%). The enzyme has two disulfide bonds, which seem to be the reason for its heat stability [75].

Aqualysin I is produced as a 51-kDa precursor protein consisting of four distinct domains: an NH_2-terminal signal peptide (14-amino acid residues), an NH_2-terminal prosequence (113 residues), a protease domain (281 residues), and a COOH-terminal prosequence (105 residues) [76,77]. When the gene encoding the entire aqualysin I precursor is expressed in a host–vector system for *Thermus thermophilus*, correctly processed aqualysin I is secreted into the culture medium. On the other hand, in an *E. coli* expression system, a 38-kDa precursor protein (proaqualysin I with the COOH-terminal pro-sequence) is accumulated in the membrane fraction. The aqualysin I precursor can be processed through its own proteolytic activity into a mature and active 28-kDa enzyme on heat-treatment at 65°C [78]. In the active site mutant (S222A), in which Ser-222 (one of the active site residues) is replaced with Ala, a 48-kDa protein was localized in the outer membrane. The 48-kDa protein is a proaqualysin I with both NH_2- and COOH-terminal pro-sequences [77].

The maturation pathway of aqualysin I in *E. coli* can be summarized as follows: First, preproaqualysin I (51-kDa protein) is translocated across the cytoplasmic membrane with the aid of the signal peptide. Second, the signal peptide is cleaved by a signal peptidase of *E. coli*, and a proaqualysin I with NH_2- and COOH-terminal pro-sequences (48-kDa protein) is released into the periplasm and then localized to the outer membrane with the aid of the COOH-terminal pro-sequence. Third, the NH_2-terminal pro-sequence is removed through the proteolytic activity of aqualysin I itself at 37°C, and a proaqualysin I, with the COOH-terminal pro-sequence (38-kDa protein) is produced. Lastly, on heat treatment at 65°C, the COOH-terminal pro-sequence is removed through the proteolytic activity of aqualysin I, and then, the active mature aqualysin I (28-kDa protein) is released from the outer membrane [77,78].

The roles of the NH_2- and COOH-terminal pro-sequences of the aqualysin I precursor in the processing and extracellular secretion of the enzyme have been analyzed. The precursor protein with a deletion in the NH_2-terminal pro-sequence is unstable, whereas the 48-kDa precursor (proaqualysin I with both NH_2- and COOH-terminal pro-sequences) of an active-site mutant (S222A) is more stable, by approximately fourfold. Similar to subtilisin and α-lytic protease, the NH_2-terminal pro-sequence exhibits an inhibitory activity toward aqualysin I. Partial and complete deletions of the COOH-terminal pro-sequence from the COOH-terminus do not affect the production of active aqualysin I in an *E. coli* expression system. Enhancement of the proteolytic activity of the precursors by heat treatment appears to be faster with longer deletions of the COOH-terminal pro-sequence, suggesting that processing and maturation efficiently occur in the deletion mutants. Results suggest that the precursor requires the COOH-terminal pro-sequence for the translocation across the cytoplasmic membrane in *E. coli* cells and functions, along with the signal peptide. The COOH-terminal pro-sequence seems to keep the precursor structure unfolded in the cytoplasm and to facilitate the translocation of the precursor across the cytoplasmic membrane [79,80]. Once the precursor has been translocated across the cytoplasmic membrane, the COOH-terminal pro-sequence probably directly guides the precursor to the outer membrane. In vivo, the COOH-terminal pro-sequence is embedded in the outer membrane and, therefore, loses its unfolding activity toward the precursor. This allows the NH_2-terminal pro-sequence to exert is protein-folding function that results in the expression of enzymatic activity [79].

In *E. coli* cells the noncovalent NH_2-terminal pro-sequence domain functions well as an IMC for folding of the protease domain without the COOH-terminal pro-sequence, to produce active aqualysin I. In contrast, the noncovalent NH_2-terminal pro-sequence does not participate in folding of the protease domain with the COOH-terminal pro-sequence. Even in the presence of the noncovalent NH_2-terminal pro-sequence, the unfolding activity of the COOH-terminal pro-sequence seems to be too strong for the protease domain to fold and take on a stable conformation. These findings seem to support the idea that the COOH-terminal pro-sequence functions as an IMC for unfolding or loosening the conformation of the protease domain [80].

In *T. thermophilus* cells a chaperonin, that is constituted by a complex of chaperonin-60 and chaperonin-10, is found. Therefore, molecular chaperones probably assist in the translocation of aqualysin I across the cytoplasmic membrane in *T. thermophilus* cells. It is possible that the COOH-terminal pro-sequence also functions as an IMC, in this process, together with molecular chaperones, preventing the precursor from taking on an active conformation in the cells. In *E. coli* cells, molecular chaperones are usually found in the cytoplasm. This may also be true in *T. thermophilus* cells. If so, the COOH-terminal pro-sequence would prevent the protease domain from adopting an active conformation in the periplasm. Active aqualysin I in the periplasm as well as in the cytoplasm would be toxic for cell growth. Taking into account these findings and the foregoing discussion, it seems that the COOH-terminal pro-sequence is a specifically evolved molecular chaperone required to facilitate unfolding of a specific protease [77,80].

D. α-Lytic Protease

α-Lytic protease is one of many enzymes secreted by the gram-negative soil bacterium *Lysobacter enzymogenes* to lyse and degrade soil microorganisms. It was first isolated by Whitaker and determined to be a 198-amino acid serine protease, with specificity for peptides with small hydrophobic residues adjacent to the scissile bond [81]. It is clear that α-lytic protease belongs to the same family as the mammalian digestive serine proteases, chymotrypsin, trypsin, and elastase, despite only moderate sequence homology [82]. In fact, high resolution X-ray crystallography has shown that α-lytic protease and chymotrypsin have nearly identical three-dimensional folds [82]. Unlike its mammalian counterparts, α-lytic protease contains only a single histidine, which is located at the active site. This single histidine has made α-lytic protease the model system of choice for spectroscopic studies on the catalytic mechanism of the serine proteases [83].

The crystal structure of α-lytic protease complexed with several high-affinity peptide transition state inhibitors reveals that substrate specificity is governed by direct substrate interactions with side chains of amino acid residues that line the substrate-binding pocket. Also, enzymatic activity is modulated by the ability of these binding pocket residues to accommodate nonoptimal substrates [84,85]. By combining X-ray crystallography with site-directed mutagenesis and functional kinetic analysis, the basis of substrate specificity has been explored [86]. Site-directed mutagenesis has led to the creation of highly active proteases, having specificity profiles dramatically different from the wild-type α-lytic protease [86]. Data from these studies have also been used to develop a new and remarkably accurate theoretical model for predicting the free energy of enzyme–substrate interactions. This, in turn, has been used in the successful design of an α-lytic protease mutant with novel specificity [87].

Molecular cloning of the gene for α-lytic protease revealed that it is synthesized naturally with a large NH_2-terminal precursor domain, referred to as the pro-region or the IMC domain [88]. Several biochemical studies clearly suggest that the IMC domain is not part of the active protease [81,82]. However, *E. coli* expression experiments reveal that the IMC domains are essential for proper activation and secretion of the protease in vivo [22,89].

In vitro experiments have shown that the IMC domain is required for proper folding of the protease and that the pro-region is a potent inhibitor of the active enzyme [90]. Investigations have suggested that the IMC promotes α-lytic protease folding by direct stabilization of the rate-limiting folding transition state [94,95]. Similar results have been seen for the IMC region of subtilisin and carboxypeptidase Y, which are evolutionarily unrelated serine protease [45–48].

The apparent conservation of IMC function across protease families suggests that, similar to the proteases, pro-regions have evolved in multiparallel pathways and may share a similar mechanism of action [91]. Alternatively, it is possible that pro-regions are evolutionarily quite ancient and preceded the formation of the current protease families [92].

XI. IMPLICATIONS OF IMC DOMAINS IN THE MATURATION OF PROTEINS

Why nature has provided IMC domains to facilitate correct folding is an interesting question. The need for an IMC domain is clearly not a general requirement of a serine protease fold. Clues may come from the fact that all secreted bacterial proteases have pro-regions. Additionally, eukaryotic proteases that require IMC domains are usually found in either vacuoles (yeast) or lysozomes, all of which display environments that are extremely harsh.

It is reasonable to believe that secreted proteases have evolved to evade autolysis or inactivation by other proteases during protein folding. Proteases usually cleave at high-mobility sites within their target proteins. Hence, evolving conformationally rigid proteases should increase their resistance to digestion. In these cases, proteolytic stability would emerge in the form of dynamic stability created by tight packing and reduced configurational entropy. A resulting consequence of structural rigidity is the existence of a radically high-energy unfolding transition state, because unfolding would require a global unlocking of a rigid protein.

Unfortunately, this difficulty of simultaneously unlocking the polypeptide during unfolding also exists in the reverse direction, resulting in a high-energy folding transition state. This is evident in stabilities of α-lytic protease, subtilisin, and carboxypeptidase Y towards denaturation [92]. Therefore, the price paid for this dynamic stability is an energy barrier presented during the protein-folding process. In the extreme case, this energy barrier would be so tremendous that folding cannot occur at all. A solution to this problem is to furnish a folding catalyst, in form of an IMC, to enable rapid protein folding and then to destroy the IMC domain, such that it would be unavailable to catalyze the reverse reaction.

In support of this hypothesis, subtilisin and α-lytic protease are extremely stable and resistant to proteolysis. α-Lytic protease can be concentrated to 10 mM with no loss of activity. The crystal structure of native α-lytic protease depicts remarkably little atomic motion, which is evident through very low B-factors (average B = 12) [84]. The free energy of the folding transition state for subtilisin, carboxypeptidase Y, and α-lytic protease, are extremely high in the absence of their IMC domains, and the IMC domains are quite unstable and readily digested by the active proteases. Similar in vitro studies on several different proteases that are made with similar IMC domains have revealed that the refolding rates in the absence of such domains are very slow and sometimes undetectable, indicating the presence of high-energy folding transition states [40,45–48]. The IMC domain stabilizes the folding transition state to rapidly fold the protease and then is proteolytically destroyed so that it cannot catalyze the reverse reaction.

Studies of IMC domains has led to a new understanding of the extreme topology of conformational space and the importance of kinetic pathways in protein folding. The observation of such high-energy kinetic barriers with subtilisin, α-lytic protease, and carboxypeptidase Y raises the possibility that similar barriers may be determining the outcome of other folding reactions. From this it can be inferred that, not only must polypeptides have evolved to specify a stable folded state, but they must also have evolved to specify a pathway to that folded state [92].

The IMC-mediated protein folding adds a new dimension to the folding problem, for it is possible to attribute to a specific region within a polypeptide the function of correct folding. Because this region is not required for the stability of the mature protein, one can create mutations within the IMC domain alone and study their effects on the folding pathways [7].

Moreover, because IMC domains usually function as competitive inhibitors of the proteins they help fold, it may be possible to use sequence information from the IMC domains and create protease inhibitors. This would be especially important for eukaryotic prohormone convertases (PCs), which are proteins similar to bacterial subtilisin. The PCs are required for correct maturation of hormonal precursors; they are also produced as precursors, and results strongly suggest that their propeptides function as IMC domains. Moreover, PCs are also responsible for processing of several viral and bacterial toxins [93].

Recent data suggest that mutations within the IMC domain can affect folding of the protease domain of subtilisin such that it adopts a conformation that is different from that of the wild-type protease. This "alternatively folded protease conformer" displays differences in substrate specificities. Subtilisin is often used as a model for understanding prohormone convertases. The PCs are extremely specific proteases that cut at dibasic or tetrabasic sites. By analogy with subtilisin, it is possible that similar mutations in IMC domains of PCs may result in their alternative folding and cause changes in substrate specificities. This may lead to incorrect hormone maturation and cause hormonal disorders.

REFERENCES

1. Anfinsen, C. R., Principles that govern the folding of protein chains, *Science 181*:223–230 (1973).
2. Richardson, J. S., The anatomy and taxonomy of protein structure, *Adv. Protein Chem. 343*:167–339 (1977).
3. Ptitsyn, O. B., a determinable but unresolved problem, *FASEB J. 10*:3–4 (1996).

4. Ellis, R. J., Proteins as molecular chaperones, *Nature 328*:378–379 (1987).
5. Hendrick, J. P. and Hartl, F.-U., The role of molecular chaperones in protein folding, *FASEB J. 9*:1559–1569 (1995).
6. Hendrick, J. P. and Hartl, F.-U., Molecular chaperone functions of heat shock proteins, *Annu. Rev. Biochem. 62*:349–384 (1993).
7. Shinde, U. P., Li, Y., and Inouye, M., Propeptide mediated protein folding: Intramolecular chaperones, *Intramolecular Chaperones and Protein Folding* (Shinde, U. P. and Inouye, M., eds.), R. G. Landes, p. 1–9 1995.
8. Boston, R. S., Viitanen, P. V., and Vierling, E., Molecular chaperones and protein folding in plants, *Plant Mol. Biol. 32*:191–222 (1996).
9. Luz, J. M. and Lennarz, W. J., Protein disulfide isomerase: A multifunctional protein of the endoplasmic reticulum, *EXS 77*:97–117 (1996).
10. Ikemura, H., Takagi, H., and Inouye, M., Requirement of pro-sequence for the production of active subtilisin E in *Escherichia coli*, *J. Biol. Chem. 262*:7859–7864 (1987).
11. Ikemura, H. and Inouye, M., In vitro processing of pro-subtilisin produced in *Escherichia coli*, *J. Biol. Chem. 263*:12959–12963 (1988).
12. Ohta, Y. and Inouye, M., Pro-subtilisin E: Purification and characterization of its autoprocessing to active subtilisin E in vitro, *Mol. Micro. Biol. 4*:295–304 (1990).
13. Baker, D., Shiau, A. K., and Agard, D. A., The role of pro regions in protein folding, *Curr. Opin. Cell Biol. 6*:966–970.
14. Wong, S. L. and Doi, R. H., Determination of the signal peptidase cleavage site in the preprosubtilisin of *Bacillus subtilis*, *J. Biol. Chem. 261*:10176–10181 (1986).
15. Izard, J. W. and Kendall, D. A., Signal peptides: Exquisitely designed transport promoters, *Mol. Microbiol. 13*:765–773 (1994).
16. Zhu, X., Ohta, Y., Jordan, F., and Inouye, M., Pro-sequence of subtilisin can guide the folding of denatured subtilisin in an intermolecular process, *Nature 339*:483–484 (1989).
17. Eder, J. and Fersht, A. R., Pro-sequence assisted protein folding, *Mol. Microbiol. 16*:609–614 (1995).
18. Kraut, J., Serine proteases: Structure and mechanism of catalysis, *Annu. Rev. Biochem. 46*:331–358 (1977).
19. Bott, R., Ultsch, M., Kossiakoff, A., Graycar, T., Katz, B., and Power, S., The three-dimensional structure of *Bacillus amyloliquefaciens* subtilisin at 1.8 A and an analysis of the structural consequences of peroxide inactivation, *J. Biol. Chem. 263*:7895–7906 (1988).
20. Kelders, H. A., Kalk, K. H., Gros, P., and Hol, W. G. J., Automated protein cyrstallization and a new crystal form of a subtilisin:eglin complex, *Protein Eng. 1*:301–303 (1987).
21. Drenth, J., Hol, W., Jansonius, J. N., and Koekoek, R., Subtilisin novo. The three-dimensional structure and its comparison with subtilisin BPN: *Eur. J. Biochem. 26*:177–181 (1977).
22. Inouye, M., Intramolecular chaperone; The role of the pro-peptide in protein folding, *Enzyme 45*:314–321 (1991).
23. Shinde, U. and Inouye, M., Intramolecular chaperones and protein folding, *Trends Biochem. Sci. 18*:442–446 (1993).

24. Carter, P. and Wells, J. A., Dissecting the catalytic triad of subtilisin, *Nature 237*:394–399 (1987).
25. Wells, J. A. and Estell, D. A., Subtilisin—an enzyme designed to be engineered, *Trends Biochem. Sci. 13*:291–297 (1988).
26. Cash, P. W., Zhu, X., Ohta, Y., Tsao, J., Lackland, H., Nevado, M. D., Inouye, M., Stein, S., Jordan, F., and Tous, G. I., Synthesis of the pro-peptide of subtilisin BPN′, *Pept. Res. 2*:292–296 (1989).
27. Ohta, Y., Hojo, H., Aimoto, S., Kobayashi, T., Zhu, X., Jordan, F., and Inouye, M., Pro-peptide as an Intramolecular Chaperone: Renaturation of Denatured Subtilisin E with a synthetic pro-peptide, *Mol. Microbiol. 5*:1507–1510 (1991).
28. Zhu, X., Ohta, Y., Cash, P. W., Tous, G. I., Stein, S., Inouye, M., and Jordan, F., Refolding of denatured subtilisins with the aid of a synthetic pro-peptide; direct demonstration that the pro-sequence guides the folding of secreted enzymes, *Biochemistry 11*:35–47 (1992).
29. Shinde, U. and Inouye, M., Structural and functional organization of intramolecular chaperones: the N-terminal propeptides that mediate protein folding, *J. Biochem. 115*:629–636 (1994).
30. Li, Y., Hu, Z., Jordan, F., and Inouye, M., Functional analysis of the propeptide of subtilisin E as an intramolecular chaperone for protein folding; purification and characterization of mutant propeptides, *J. Biol. Chem. 270*:25127–25132 (1995).
31. Ellis, R. J. and van der Vies, S. M., Molecular chaperones, *Annu. Rev. Biochem. 60*:321–347 (1991).
32. Shinde, U. P., Li, Y., and Inouye, M., The role of the N-Terminal propeptide in Mediating Folding of Subtilism E., *Intramolecular Chaperones and Protein Folding* (Shinde, U. P. and Inouye M., ed.) R. G. Landes, p. 11–34 (1995).
33. van den Hazel, H. B., Kielland-Brandt, M. C., and Winther, J. R., The propeptide is required for in vivo formation of a stable active yeast proteinase A and can funciton even when it is not covalently attached to the mature protein, *J. Biol. Chem. 268*:18002–18007 (1993).
34. Eakin, A. E., McGrath, M. E., McKerrow, J. H., Fletterick, R. J., and Craik, C. S., Production of crystallizable cruzain, the major cysteine protease from *Trypanosoma cruzi*, *J. Biol. Chem. 268*:6115–6118 (1993).
35. Chang, S. C., Chang, P. C., and Lee, Y. H., The roles of the propeptide in the maturation and secretion of Npr protease from streptomyces, *J. Biol. Chem. 269*:3548–3554 (1994).
36. Gray, A. M. and Mason, A. J., Requirement for activin A and transforming growth factor-β1 pro-regions in homodimer assembly, *Science 247*:1328–1330 (1990).
37. Suter, U., Heymach, J. V., and Shooter, E., Two conserved domains in the NGF propeptide are necessary and sufficient for the biosynthesis of correctly processed and biologically active NGF, *EMBO J. 10*:2395–2400 (1991).
38. Voorberg, J., Fontijn, R., Calafat, J., Janssen, H., van Mourik, J. A., and Pannekoek, H., Biogenesis of von Willebrand factor-containing organelles in heterologous transfected CV-1 cells, *EMBO J. 12*:749–758 (1993).
39. Thorne, B. A. and Plowman, G. D., The heparin-binding domain of amphiregulin necessitates the precursor pro-region for growth factor secretion, *Mol. Cell. Biol. 14*:1635–1646 (1994).

40. Weissman, J. S. and Kim, P. S., The pro region of BPTI facilitates folding, *Cell* *71*:841–851 (1992).
41. Rothwell, V., Kosowski, S., Hadjilambris, O., Baska, R., and Norman, J., Glycosylation of active human renin is necessary for secretion: effect of targeted modifications of Asn-5 and Asn-75, *DNA Cell Biol.* *12*:291–298 (1993).
42. Seizen, R. J., de Vos, W. M., Leunissen, J., and Dijkstra, B. W., Homology modelling and protein engineering strategy of subtilases, the family of subtilisin like resine proteases, *Protein Eng.* *4*:719–737 (1991).
43. Seizen, R. J., Leunissen, J. A. M., and Shinde, U. P., Homology analysis of the propeptides of subtilisin-like serine proteases (subtilases), *Intramolecular Chaperones and Protein Folding* (Shinde, U. P. and Inouye, M., eds.), R. G. Landes, p. 243–257 (1995).
44. Strausberg, S., Alexander, P., Wang, L., Schwarz, F., and Bryan, P., Catalysis of a protein folding reaction: thermodynamic and kinetic analysis of subtilisin BPN′ interactions with its propeptide fragment, *Biochemistry* *32*:8112–8119 (1993).
45. Eder, J., Rheinnecker, M., and Fersht, A., Folding of subtilisin BPN′: characterization of a folding intermediate, *Biochemistry* *32*:18–26 (1993).
46. Baker, D., Sohl, J. L., and Agard, D. A., A protein-folding reaction under kinetic control, *Nature* *356*:263–265 (1992).
47. Winther, J. R. and Sorensen, P., Propeptide of carboxypeptidase Y provides a chaperone-like function as well as inhibition of the enzymatic activity, *Proc. Natl. Acad. Sci. USA* *88*:9330–9334 (1991).
48. Shinde, U., Li, Y., Chatterjee, S., and Inouye, M., Folding pathway mediated by an intramolecular chaperone, *Proc. Natl. Acad. Sci. USA* *90*:6924–6928 (1993).
49. Ptitsyn, O. B., The molten globule state, *Protein Folding* (T. E., Creighton, ed.), W. H. Freeman, New York, 1991, pp. 243–300.
50. Kuwajima, K., The molten globule state of α-lactalbumin, *FASEB J.* *10*:102–109 (1996).
51. Shinde, U. and Inouye, M., Folding mediated by an intramolecular chaperone: Autoprocessing pathway of the precursor resolved via a "substrate assisted catalysis" mechanism, *J. Mol. Biol.* *247*:390–395 (1995).
52. Shinde, U. and Inouye, M., Folding pathway mediated by an intramolecular chaperone: Characterization of the structural changes in pro-subtilisin E coincident with autoprocessing, *J. Mol. Biol.* *252*:25–30 (1995).
53. Li, Y. and Inouye, M., Autoprocessing of pro-thiol-subtilisin E in which the active site serine residue-221 was altered in cysteine, *J. Biol. Chem.* *269*:4169–4174 (1994).
54. Bryan, P., Alexander, P., Strausberg, S., Schwarz, F., Lan, W., Gilliland, G., and Gallagher, D. T., Energetics of folding subtilisin BPN′, *Biochemistry* *31*:4937–4945 (1992).
55. Bryan, P., Subtilisin engineered for facile folding: Analysis of uncatalyzed and prodomain catalyzed folding, *Intramolecular Chaperones and Protein Folding* (Shinde, U. P. and Inouye, M., eds.), R. G. Landes, p. 85–112 (1995).
56. Matsubara, M., Kurimoto, E., Kojima, S., Miura, K., and Sakai, T., Achievement of renaturation of subtilisin BPN′ by a novel procedure using of organic salts and a digestible mutant of *Streptomyces* subtilisin inhibitor, *FEBS Lett.* *342*:193–196 (1994).

57. Hayashi, T., Matsubara, M., Nohara, D., Kojima, S., Miura, K., and Sakai, T., Renaturation of the mature subtilisin BPN′ immobilized on agarose beads, *FEBS Lett.* *350*:109–112 (1994).
58. Bryan, P., Wang, L., Hoskins, I., Ruvinov, S., Strausberg, S., Alexander, P., Gilliland, G., and Gallagher, T., Catalysis of a protein folding reaction: Mechanistic Implications of the 2.0-A structure of the subtilisin-prodomain complex, *Biochemistry* *34*:10310–10318 (1995).
59. Gallagher, T., Gilliland, G., Wang, L., and Bryan, P., The prosegment—subtilisin BPN′ complex: Crystal structure of a specific foldase, *Structure* *3*:907–914 (1995).
60. Li, Y. and Inouye, M., The mechanism of autoprocessing of the propeptide of prosubtilisin E: Intramolecular or intermolecular event? *J. Mol. Biol.* *262*:591–594 (1996).
61. Breddam, K., Serine carboxypeptidases. A review. *Carlsberg Res. Commun.* *51*:83–128 (1986).
62. Woolford, C. A., Daniels, L. B., and Park, F. J., The *PEP4* gene encodes an aspartyl protease implicated in post-translational regulation of *Saccharyomyces cerevisiae* vaculor hydrolases, *Mol. Cell. Biol.* *6*:2500–2510 (1986).
63. Winther, J. R. and Sorensen, P., Propeptide of carboxypeptidase Y provides a chaperone-like function as well as inhibition of the enzymatic activity, *Proc. Natl. Acad. Sci. USA* *88*:9330–9334 (1991).
64. Sorenson, P., Structure–function studies on the pro-region of carboxypeptidase Y, Ph.D. dissertation, University of Copenhagen (1994).
65. Winther, J. R., Sorenson, P., and Kielland-Brandt, M. C., Low affinity binding of carboxypeptidase Y stimulates folding of the enzyme in vitro. *J. Biol. Chem.* *269*:7006–7012 (1994).
66. Sorenson, P., Winther, J. R., and Kaarsholm, N. C., The pro-region required for folding of carboxypeptidase Y is a partially folded domain with little regular structural core, *Biochemistry* *32*:12160–12166 (1993).
67. Dreyer, T., Substrate specificity of proteinase yscA from *Saccharomyces cerevisiae*, *Carlsberg Res. Commun.* *54*:85–97 (1984).
68. Klionsky, D. J., Banta, L. M., and Emr, S. R., Intracellular sorting of yeast vacuolar hydrolase: Proteinase A propeptide contains vacuolar targeting information, *Mol. Cell Biol.* *8*:2105–2116 (1988).
69. van den Hazel, H. B., Kielland-Brandt, M. C., and Winther, J. R., The propeptide is required for in vivo formation of stable active yeast proteinase A and can function even when not covalently linked to the mature region, *J. Biol. Chem.* *268*:18002–18007 (1993).
70. Winther, J. R. and van den Hazel, H. B., The role of propeptides in folding and secretory transport of yeast proteases, *Intramolecular Chaperones and Protein Folding* (Shinde, U. P. and Inouye, M., eds.), R. G. Landes, p. 205–232 (1995).
71. Tang, J. and Wong, R. N., Evolution in the structure and function of aspartic proteases, *J. Cell Biochem.* *33*:53–63 (1987).
72. Lott, T. J., Page, L. S., and Boiron, P., Nucleotide sequence of the *Candida albicans* aspartyl proteinase gene, *Nucleic Acids Res.* *17*:1770–1781 (1989).
73. Matsuzawa, H., Hamaoki, M., and Ohta, T., Production of thermophilic extracellular

proteases (aqualysins I and II) by *Thermus aquaticus* YT-1, an extreme thermophile, *Agric. Biol. Chem. 47*:25–28 (1983).
74. Matsuzawa, H., Tokugawa, K., and Hamaoki, M., Purification and characterization of aqualysin I (a thermophilic alkaline serine protease) produced by *Thermus aquaticus* YT-1, *Eur. J. Biochem. 171*:441–447 (1988).
75. Kwon, S.-T., Terada, I., and Matsuzawa, H., Nucleotide sequence of the gene for aqualysin I (a thermophilic alkaline serine protease) of *Thermus aquaticus* YT-1 and characteristics of the deduced primary structure of the enzyme. *Eur. J. Biochem. 173*:491–497 (1988).
76. Kwon, S.-T., Matsuzawa, H., and Ohta, T., Determination of the positions of the disulfide bonds in aqualysin I (a thermophilic alkaline serine protease) of *Thermus aquaticus* YT-1, *J. Biochem. 104*:557–559 (1988).
77. Terada, I., Kwon, S.-T., and Miyata, Y., Unique precursor structure of an extracellular protease, aqualysin I, with NH_2- and COOH-terminal pro-sequences and its processing in *Escherichia coli*, *J. Biol. Chem. 265*:6576–6581 (1990).
78. Touhara, N., Taguchi, H., and Koyama, Y., Production and extracellular secretion of aqualysin I (a thermophilic subtilisin-type protease) in a host–vector system for *Thermus thermophilus*, *Appl. Environ. Microbiol. 57*:3385–3387 (1991).
79. Lee, Y.-C., Ohta, T., and Matsuzawa, H., A non-covalent NH_2-terminal pro-region aids the production of active aqualysin I (a thermophilic protease) without the COOH-terminal pro-sequence in *Escherichia coli*, *FEMS Microbiol. Lett. 92*:73–78 (1992).
80. Lee, Y.-C., Koike, H., and Taguchi, H., Requirement of a COOH-terminal prosequence for the extracellular secretion of aqualysin I (a thermophilic subtilisin-type protease) in *Thermus thermophilus*, *FEMS Microbiol. Lett. 120*:69–74 (1994).
81. Whitaker, D. R., The alpha-lytic protease of *Myxobacterium. Methods Enzymol. 19*:599–613 (1970).
82. Brayer, G. D., Delbaere, L. T. J., and James, M. N. G., Molecular structure of alpha-lytic protease from *Myxobacter* 495 at 2.8 Å resolution, *J. Mol. Biol. 131*:743–775 (1979).
83. Bachovchin, W. W. and Roberts, J. D., Nitrogen-15 nuclear magnetic resonance-spectroscopy. The state of histidine in the catalytic trial of alpha-lytic protease. Implications for the charge–relay mechanism of peptide-bond cleavage by serine proteases, *J. Am. Chem. Soc. 100*:8041–8097 (1978).
84. Bone, R., Frank, D., Kettner, C. A., and Agard, D. A., Structural analysis of specificity: alpha-Lytic protease complexes with analogues of reaction intermediates. *Biochemistry 28*:7600–7609 (1989).
85. Bone, R., Fujishige, A., Kettner, C. A., and Agard, D. A., Structural basis for broad specificity in alpha-lytic protease mutants, *Biochemistry 30*:10388–10398 (1991).
86. Bone, R., Silen, J. L., and Agard, D. A., Structural plasticity broadens the specificity of an engineered protease, *Nature 339*:191–195 (1989).
87. Wilson, C., Mace, J. E., and Agard, D. A., Computational method for the design of enzymes with altered substrate specificity, *J. Mol. Biol. 220*:495–506 (1991).
88. Fujishige, A., Smith, K. R., Silen, J. L., and Agard, D. A., Correct folding of alpha-lytic protease is required for its extracellular secretion from *Escherichia coli*, *J. Cell Biol. 118*:33–42 (1992).

89. Silen, J. L., McGrath, C. N., Smith, K. R., and Agard, D. A., Molecular analysis of the gene encoding alpha-lytic protease: Evidence for a preproenzyme, *Gene 69*:237–244 (1988).
90. Baker, D., Silen, J. L., and Agard, D. A., Protease pro region required for folding is a potent inhibitor of the mature enzyme, *Proteins 12*:339–344 (1992).
91. Baker, D., Shiau, A. K., and Agard, D. A., The role of pro regions in protein folding, *Curr. Opin. Cell Biol. 5*:966–970 (1993).
92. Sohl, J. L. and Agard, D. A., alpha-Lytic protease: Dynamic stability via kinetic control, *Intramolecular Chaperones and Protein Folding* (Shinde, U. P. and Inouye, M., eds.), R. G. Landes, p. 61–84 (1995).
93. Barr, P. J., Mammalian subtilisins: The long-sought dibasic processing endoproteases, *Cell 66*:1–3 (1991).

20

Molecular Chaperones and Intracellular Protein Degradation with Emphasis on a Selective Lysosomal Pathway of Proteolysis

Ana Maria Cuervo, Sandra A. Hayes, and J. Fred Dice

Tufts University School of Medicine, Boston, Massachusetts

I. INTRODUCTION

Intracellular protein degradation is an important process in all cells from bacterial to mammalian. Cellular proteins are constantly being synthesized and degraded with varying half-lives in response to physiological adaptations and pathological conditions [1]. There are now several known intracellular proteolytic pathways [2–4]. The well-studied ATP–ubiquitin-dependent cytosolic pathway is responsible for the degradation of abnormal proteins as well as many short-lived, normal proteins [3,5]. Other cytosolic pathways for protein degradation appear to be independent of ubiquitin, including a calcium-dependent one. Finally, there are several degradative pathways contained within organelles, such as the endoplasmic reticulum (ER), mitochondrion, and lysosome [4]. The basic requirements of each proteolytic pathway are being worked out in detail, and molecular determinants within proteins that serve to target them for degradation by one or more of these pathways are being investigated [2]. More recently, there is evidence suggesting that molecular chaperones are also involved in targeting proteins for degradation.

Heat shock as well as other physiological and pathological stresses cause proteins to denature and induce the production of heat-shock proteins (Hsps), some of which act as molecular chaperones (Table 1). The role of molecular

Table 1 Prokaryotic and Eukaryotic Molecular Chaperones and Proteases

Factor	Action
Prokaryotes	
DnaK	Hsp70
DnaJ	Stimulates DnaK
GrpE	Cochaperone with DnaJ
GroEL	Hsp60
GroES	Hsp10; cochaperone with GroEL
Trigger factor	Cochaperone with GroEL
La	ATP-dependent protease
Clp	ATP-dependent protease
ClpA	ATPase subunit of Clp; molecular chaperone
ClpP	Proteolytic subunit of Clp
ClpX	ATPase subunit of Clp; molecular chaperone
ClpB	Molecular chaperone
Eukaryotes	
Ssa1, Ssa2	Yeast cytosolic Hsp70s
Mdj1	Yeast mitochondrial DnaJ homologue
Sis	Yeast cytosolic DnaJ homologue
Ydj	Yeast cytosolic DnaJ homologue
Hsc73	Constitutively expressed Hsp70
Pim1	Mitochondrial protease La homologue
Proteasome	Multisubunit ATP-dependent protease

Source: Ref. 9.

chaperones in the stress response is to aid in the refolding of denatured proteins [6–8]. In addition, several molecular chaperones are expressed constitutively, suggesting that they play a role in normal conditions. Some of the other roles of molecular chaperones include (a) catalyzing proper protein folding following synthesis, (b) aiding in the assembly or disassembly of multimeric proteins, and (c) facilitating the translocation of proteins across intracellular membranes [6–8]. It now appears that molecular chaperones are involved in degradation of abnormal and certain normal proteins, either by preventing the formation of massive aggregates, thereby ensuring that the protein remains susceptible to proteolytic attack (Fig. 1), or by specifically targeting substrate proteins to particular proteolytic pathways (Fig. 2). Molecular chaperones may also increase the overall activity of a proteolytic pathway (see Fig. 2). Although the mechanism by which molecular chaperones "refold or degrade" a protein is unclear, there is evidence suggesting that molecular chaperones can be physically associated with proteases, facilitate rate-limiting steps in protein degradation, or facilitate selective transport of proteins into organelles for degradation [9].

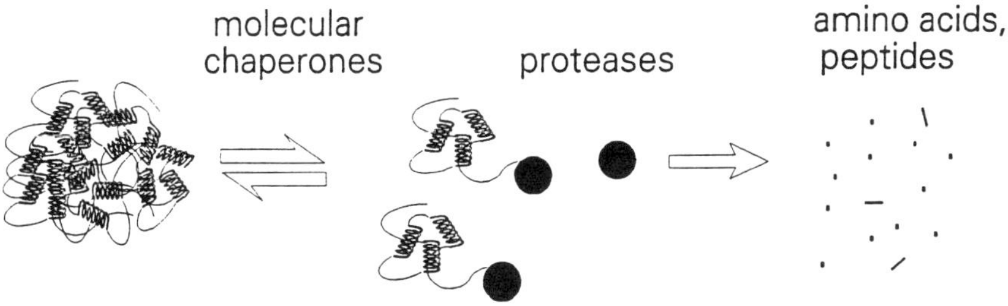

Figure 1 Molecular chaperones prevent aggregation of proteins: Without the actions of molecular chaperones, certain proteins form aggregates that are relatively inaccessible to proteolytic attack.

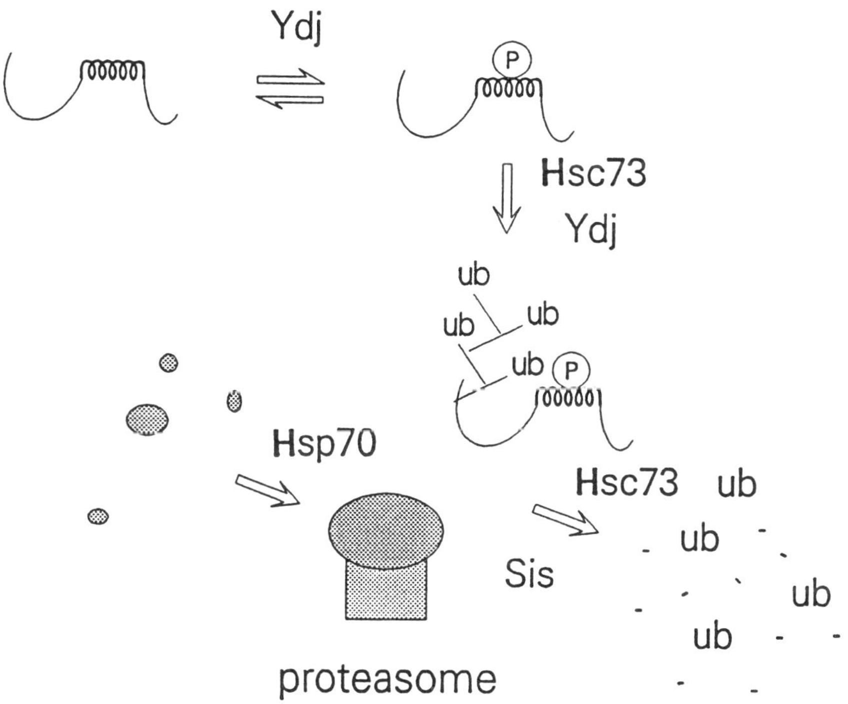

Figure 2 Role of molecular chaperones in the ubiquitin–proteasome proteolytic pathway: Ydj is required for the efficient phosphorylation of a cyclin (the *cln3* gene product) and is also required for the efficient ubiquitination of certain abnormal proteins. Hsc73 is also required for the efficient ubiquitination of certain abnormal proteins, and an Hsp70 family member is required for the assembly of the 26S proteasome. Sis is required for the degradation of certain ubiquitinated protein by the proteasome. Many of the steps shown require ATP. Ubiquitin (ub) and the multiple ubiquitin structure represent a covalent chain of ubiquitins.

II. MOLECULAR CHAPERONES AND PROTEIN DEGRADATION

A. Molecular Chaperones and Bacterial Proteolysis

Degradation of abnormal proteins and certain short-lived regulatory proteins in *Escherichia coli* is carried out by two ATP-dependent proteases, La and Clp [10,11]. Protease La, the product of the *lon* gene, is a serine protease that acts as a tetramer and can degrade proteins in the presence of ATP or a nonhydrolyzable ATP analogue [11]. Protease La is a heat-shock protein and contributes to increased rates of proteolysis during the heat-shock response [10]. The molecular chaperones DnaK, DnaJ, GrpE, GroEL, and GroES (see Table 1) are all required for the protease La-dependent degradation of abnormal proteins [12–14]. A short-lived, nonsecreted form of alkaline phosphatase, PhoA61, is associated with DnaK, GrpE, and protease La [14]. The amount of these chaperones associated with PhoA61 correlates with the rate of degradation of this protein.

Clp protease (or protease Ti) also requires ATP hydrolysis for its activity. However, Clp is made up of two subunits, ClpA or ClpX containing the ATPase activity, and ClpP, possessing the proteolytic activity. The Clp protease acts as a tetradecamer of ClpP assembled with a hexamer of ClpA or ClpX [11,15]. ClpA and ClpX have sequence homology to the 100-kDa class of Hsps and can themselves act as molecular chaperones [16,17]. Clp is also responsible for the degradation of certain abnormal proteins and appears to involve molecular chaperones [16,18]. The degradation of CRAG, a chimeric protein containing 12 amino acids from the *cro* repressor, a truncated form of protein A, and 14 amino acids from β-galactosidase, by Clp requires the presence of GroEL, GroES, and trigger factor [18,19]. Also, the in vitro degradation of the plasmid P1 Rep1 initiator protein by ClpP requires the molecular chaperone activity of ClpA [16]. Similarly, ClpX targets the bacteriophage λ O replication protein for degradation by ClpP [17], and catalyzes the disassembly of Mu virus transposase targeting it for ClpP degradation [20].

B. Molecular Chaperones and Mitochondrial Protein Degradation

Although mitochondria can be taken up and degraded by lysosomes, they also contain their own proteolytic machinery. There are intramitochondrial proteases that are homologous with the prokaryotic proteases and may also require molecular chaperones in the degradation of abnormal proteins [21].

Pim1 protease, a homologue of bacterial protease La, found in yeast and in the mammalian mitochondrial matrix, is heat-inducible and is involved in the degradation of abnormal as well as certain normal mitochondrial proteins [reviewed in 22]. Degradation by Pim1 also requires the molecular chaperones, mitochondrial Hsp70 (mt-Hsp70) and Mdj1, a yeast mitochondrial DnaJ homologue (see Table 1). Without mt-Hsp70 and Mdj1, proteins form large aggregates that can no longer be degraded by Pim1 (see Figure 1) [reviewed in 22]. A homologue of the bacterial Clp protease has also been described in mitochondria. A nuclear-encoded 78-kDa mitochondrial heat-shock protein (Hsp78) is homologous with bacterial ClpB, a heat-shock–inducible member of the Clp family of proteins [22]. Hsp78 contains a mitochondrial leader sequence, nucleotide binding sites, and is found in the mitochondrial matrix [23].

C. Molecular Chaperones in Ubiquitin-Dependent Proteolysis

The ubiquitin-dependent proteolytic system is an ATP-dependent, nonlysosomal pathway that was first identified in reticulocytes [24] and has now been described in other cells [25–27; see Chap. 21]. It is thought to rapidly degrade abnormal and many short-lived normal proteins by conjugation of ubiquitin to the protein [24]. The signal for rapid protein degradation appears to be a polyubiquitin branched-chain added to the amino group of a single-lysine residue [28]. Multiubiquitin-conjugated proteins are then degraded by a cytosolic and nuclear multicatalytic protease called the 26S proteasome [29]. Molecular chaperones are also required for the ubiquitin-dependent degradation of certain abnormal proteins.

The heat-shock cognate protein of 73-kDa (Hsc73) promotes the conjugation of ubiquitin to denatured proteins in reticulocyte extracts. It also appears that Hsc73 is required for the efficient degradation of some multiubiquitinated proteins [A. Ciechanover, personal communication]. In yeast, mutations in the Hsp70 genes, *ssa1* and *ssa2*, and temperature-sensitive mutants for the DnaJ homologues, Ydj and Sis decrease the rates of degradation of abnormal as well as certain short-lived normal proteins [30]. Finally, the degradation of the *cln3* gene product, a yeast cyclin, requires several molecular chaperones (see Fig. 2). The yeast cyclin is degraded by the ubiquitin-dependent system. It appears that the cyclin is phosphorylated before being efficiently ubiquitinated [31], and this phosphorylation is stimulated by Ydj [M. Sherman, personal communication]. Sis is also required for the degradation of certain ubiquitinated proteins [30]. Furthermore, an Hsp70 is required for the proper assembly of the multisubunit proteasome (see Fig. 2) [32].

III. LYSOSOMAL PATHWAYS OF PROTEIN DEGRADATION

A. Introduction

Lysosomes are responsible for the degradation of an important variety of intracellular proteins (Fig. 3). Extracellular proteins, plasma membrane proteins, and certain vesicular proteins are degraded in lysosomes by endocytosis [33]. Secre-

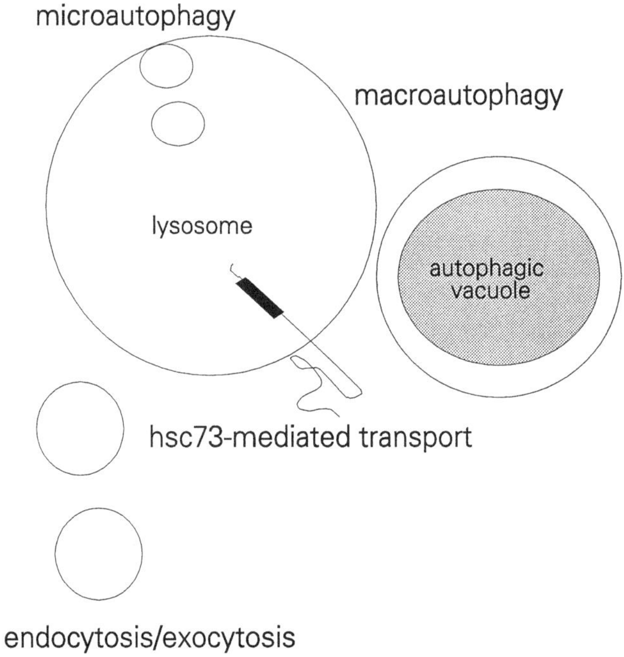

Figure 3 Lysosomal pathways of protein degradation: Different extracellular and intracellular proteins can be delivered to lysosomes for degradation. The best-characterized mechanisms are schematized in this figure. See text for details.

tory proteins, when requirements for their secretion are reduced, can also be degraded in lysosomes by a process called crinophagy [34]. Finally, an important number of cytosolic proteins can be transported into lysosomes by different mechanisms [35,36]. The best-characterized pathways of transport of cytosolic protcins into lysosomcs arc macroautophagy, microautophagy, and a dircct molecule-by-molecule transport pathway. Macroautophagy and microautophagy are usually considered nonselective processes responsible for degradation of complete portions of cytosol, with similar degradation rates for all the proteins in that cytosolic area [35,37]. In contrast, a selective chaperone-mediated transport of cytosolic proteins into lysosomes occurs by mechanisms similar to other organellar protein transport systems [38,39].

Two of the aforementioned lysosomal pathways of degradation of cytosolic proteins, macroatophagy [9,34] and the selective lysosomal proteolytic pathway [9,40], are induced by stress conditions. No specific chaperone has been linked to degradation of proteins by macroautophagy. However, Hsc73 seems to be necessary for degradation of selective cytosolic proteins by lysosomes [41]. A detailed description of the role and mechanisms of action of Hsc73 in that degradative pathway is addressed in the following.

B. A Selective Lysosomal Pathway for Degradation of Cytosolic Proteins

Selective degradation of proteins by lysosomes was considered a possibility some time ago, but the first experimental evidence of such a process has been relatively recent. Several reviews have focused on how this pathway was discovered [36,39,40] and on the KFERQ peptide motif [38] common to protein substrates of this pathway. The reader is referred to these earlier reviews for information on these topics.

Although this pathway of proteolysis was first discovered in cultured human fibroblasts, using microinjected proteins, such as pancreatic ribonuclease A (RNase A) as a substrate, more recent studies indicate a similar selective lysosomal degradation for certain cytosolic proteins in rat liver [42,43]. Specific substrates, such as glyceraldehyde-3-phosphate dehydrogenase (GAPDH) and the proteasome, are degraded selectively in rat liver lysosomes [42–44], and this pathway is activated by prolonged starvation [45], perhaps equivalent to serum deprivation in cultured cells.

1. *Characteristics of the Selective Pathway*

One possible mechanism of degradation of RNase A and GAPDH by lysosomes is a molecule-by-molecule transport of these proteins through the lysosomal membrane, similar to import of precursor proteins for residence in organelles, such as mitochondria. In vitro systems have become very useful for investigating translocation of proteins across organellar membranes [46]; therefore, we tried to reconstitute the selective transport of protein substrates into lysosomes [42,43,47]. Basically, incubation of protein substrates with intact lysosomes from fibroblasts or rat liver in an isotonic medium, allowed us to measure uptake rates and subsequent degradation of proteins transported into the lysosomal matrix.

With the reconstituted system, three main steps of the selective lysosomal pathway have been characterized: binding to the lysosomal membrane, transport into the matrix, and degradation by lysosomal proteases. Briefly, when substrates were incubated with freshly isolated lysosomes, most of these proteins were transported into the lysosomal matrix and completely degraded. However, a small portion of the added protein remained bound to the lysosomal membrane and could be detected by immunoblot analysis [42,43]. When lysosomes were pretreated with cathepsin inhibitors to block degradation, and surface-bound protein was removed by an exogenous protease, the remaining substrate reflected protein that had entered the lysosome lumen [42,43,47]. If labeled proteins and lysosomes without inhibitor treatment were incubated together, degradation of substrates by lysosomes could be followed as the percentage of acid-precipitable radioactivity (protein) transformed into acid-soluble radioactivity (amino acids and small peptides) during the incubation time [42,43,47].

Binding and uptake of substrate proteins, including RNase A and GAPDH, are time- and temperature-dependent, saturable, and stimulated by ATP [42,43,47]. Pretreatment of lysosomes with proteases inhibits the binding and uptake of substrates, suggesting that a protein-containing component of the lysosomal surface, a receptor or polypeptide transporter, is involved in the binding and uptake process [43,47].

Lysosomes from fibroblasts and from rat liver act very similarly in vitro. For example, RNase A, a typical substrate for the lysosomal pathway in fibroblasts, can be selectively transported into rat liver lysosomes, and this uptake can be inhibited by an excess of GAPDH and vice versa (Table 2). Interestingly, uptake of RNase A by rat liver lysosomes appears to involve a kinetic intermediate in which approximately 2 kDa of the COOH-terminal end of the protein remains outside the lysosome [43]. Uptake rates of specific proteins by lysosomes isolated from serum-deprived cells or starved rats were significantly higher than by lysosomes from serum-supplemented cells or normally fed rats [44,47], suggesting an activation of this pathway.

A combination of in vivo and in vitro experiments allowed the identification of substrates for this pathway in addition to those already mentioned. Defined

Table 2 Comparison Between the Mechanisms of Uptake and Degradation of GAPDH and RNase A by Rat Liver Lysosomes

Process	GAPDH	RNaseA
Steps:		
Binding	+	+
Uptake	+	+
Degradation	+	+
Time-dependent (5–10 min lag in uptake compared with binding)	+	+
Temperature-dependent	+	+
Saturable	+	+
Sensitive to protease treatment of lysosomal membrane	+	+
Number of binding sites in the lysosomal membrane (approx.)	7,000	10,000
Kinetic intermediate form	?	+
Effect of		
ATP	↑	↑
Hsc73	↑	↑

substrates include aldolase, phosphoglyceromutase, and phosphoglycerokinase [42; A. M. Cuervo and E. Knecht, unpublished results], the proteasome [44], and the transcription factor c-*fos* [48].

2. A Lysosomal Membrane Receptor for Substrate Proteins

Ongoing studies in our laboratory point to a lysosomal membrane glycoprotein of 96 kDa (referred to as lgp96 or lamp2 in the literature) as a receptor for the selective lysosomal transport and degradation pathway [49]. Substrates of this pathway, RNase A and GAPDH, specifically bind to the cytosolic tail of that integral lysosomal membrane protein (Fig. 4), and binding can be counteracted by specific antibodies to the cytosolic tail, or by a synthetic peptide with the tail amino acid sequence. Importantly, overexpression of lgp96 in CHO cells increases the activity of this selective lysosomal proteolytic pathway two- to three-fold when measured in living cells and in isolated lysosomes incubated in vitro [49]. This result suggests that the receptor is one rate-limiting component of the entire degradation pathway.

3. Role of Molecular Chaperones in the Selective Lysosomal Pathway

Cytosolic Hsc73. The mechanism of transport of proteins into lysosomes by the specific pathway resembles, in many respects, the import of proteins into other organelles, such as the endoplasmic reticulum, mitochondria, chloroplasts, peroxisomes, and the nucleus [46]. In these systems, molecular chaperones play an important role in the transport of proteins through membranes [50,51].

Cytosolic Hsc73 also plays an important role in selective transport of

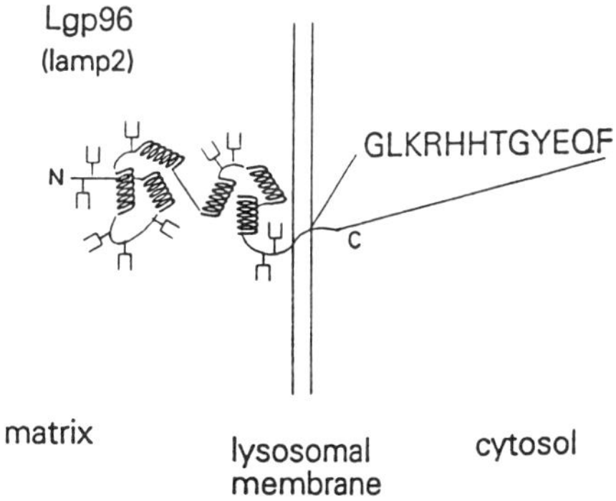

Figure 4 Lgp96, the lysosomal receptor for substrate proteins: Lgp96 is an integral glycoprotein of the lysosomal membrane, with two highly glycosylated lumenal domains, a transmembrane region (20 amino acids), and a cytosolic tail (12 amino acids). The amino acid sequence of the cytosolic tail is shown.

proteins into lysosomes; addition of purified Hsc73 to the in vitro system stimulates binding and import of substrate proteins into fibroblast lysosomes as well as into rat liver lysosomes [43,47] (Fig. 5). Interestingly, when high amounts of Hsc73 were added to the transport system, there was a significant decrease in the amount of transported substrate protein, apparently because Hsc73 is itself a substrate for this degradation pathway. Hsc73 could also bind specifically to the receptor in the lysosomal membrane (A. M. Cuervo and J. F. Dice, unpublished results) and could be transported into lysosomes in the in vitro system [52]. Hsc73 has two KFERQ peptide motifs [53], so it is not surprising that it is also a substrate for this pathway of lysosomal degradation.

Lysosomal Hsc73. Hsp70 family members located within the lumen of organelles are also necessary for the uptake of proteins [54–57]. Lysosomes immunoblotted with different antibodies specific for Hsp70 family members revealed the presence of Hsc73 within this fraction [45,47]. An Hsp70 within lysosomes has also been reported by other authors [58]. Studies using exogenous proteases [59] or immunogold labeling [45] showed that only a small portion of this lysosomal Hsc73 (approximately 17%) was associated with the cytosolic side of the lysosomal membrane, whereas most of the protein (83%) was located in the lysosomal matrix. The presence of Hsc73 in lysosomes was confirmed

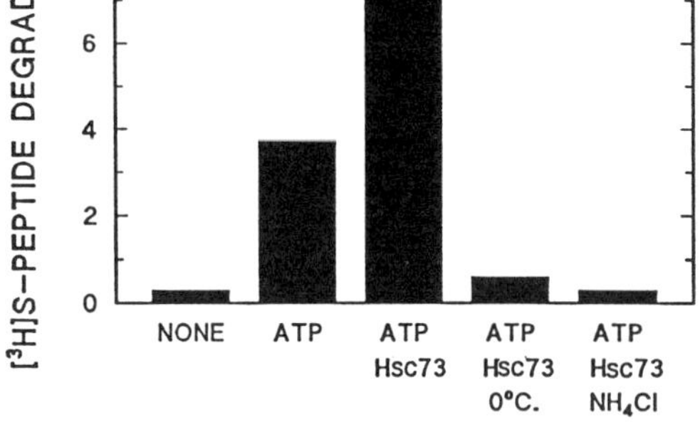

Figure 5 Effect of ATP, cytosolic Hsc73, temperature, and NH_4Cl on in vitro lysosomal degradation of RNase S-peptide (amino acids 1–20 of RNase A): Lysosomes from 2×10^6 fibroblasts were incubated with [^{3}H]RNase S-peptide. Other additions were ATP (10 mM) and an ATP-regenerating system, purified cytosolic Hsc73 (10 μg ml^{-1}) and NH_4Cl (10 mM). The amount of RNase S-peptide degraded into acid-soluble material was quantified.

by colocalization using indirect immunofluorescence of Hsc73 with lysosomal marker enzymes and laser confocal microscopy [59].

Levels of Hsc73 within lysosomes correlate well with the activity of the selective lysosomal proteolysis pathway. The amount of lysosome-associated Hsc73 was higher in serum-deprived cells than in serum-supplemented cells [47], and intralysosomal Hsc73 levels increased in rat liver with prolonged starvation. In addition, electronmicroscopic studies on the distribution of Hsc73 in lysosomes by immunogold labeling showed clear differences in Hsc73 content among lysosomes [45]. When transport of selective substrates into lysosomes was followed using similar techniques, differences in uptake ability among individual lysosomes were also observed [42,45]. In double immunogold labeling with a specific antibody against Hsc73 and a specific antibody for the substrate protein, there was a good correlation between levels of Hsc73 in lysosomes and the amount of RNase A or GAPDH transported into lysosomes [45].

To directly confirm the suggested role of the intralysosomal Hsc73 in the specific transport of proteins into lysosomes, this protein was neutralized in living, cultured fibroblasts by endocytosis of a specific antibody against Hsc73 [59]. This treatment did not affect degradation of proteins in the presence of serum or the lysosomal degradation of an endocytosed protein, but it dramatically decreased the enhanced protein degradation caused by serum deprivation [36,59].

Recently, based on their Hsc73 content, two different lysosomal populations in rat liver have been isolated and characterized. Lysosomes with a high content of Hsc73 are more active in the selective lysosomal degradation pathway than those with lower Hsc73 content [52]. When lysosomes that do not contain Hsc73 are preloaded with cytosolic Hsc73, there is an activation of the selective uptake of cytosolic proteins [52].

Other properties of lysosomal Hsc73 are consistent with it being cytosolic Hsc73. Lysosomal Hsp73 is recognized by two antibodies specific for Hsc73, and it comigrates with one of the cytosolic Hsc73 isoforms in high-resolution two-dimensional gels [59].

Regulation of Lysosomal Hsc73 Content. The explanation for the differences in Hsc73 content between the two different lysosome populations has recently been analyzed. The differences in Hsc73 content are due to differences in their ability to take up Hsc73, combined with differences in its intralysosomal degradation rates [52]. Rates of in vitro uptake of Hsc73 by rat liver lysosomes are lower in the lysosomal population with a low Hsc73 content, consistent with Hsc73 itself being a substrate for this selective proteolytic pathway. In addition, however, Hsc73 degradation in the lysosomal matrix is rapid in lysosomes with low Hsc73 content. Addition of protease inhibitors caused a dramatic increase in this protein in the population with normally low levels [52]. These differences in Hsc73 proteolysis appear to be related to differences in intralysosomal pH between the two populations of lysosomes. Lysosomes with low content of Hsc73 have a higher pH (5.7) than those with high Hsc73 levels (5.3). When intralysosomal pH is artificially increased with ammonium chloride in lysosomes with high Hsc73 content, the level of Hsc73 dramatically decreases owing to an increase in its intralysosomal degradation rate [52].

4. *A Model for the Chaperone-Mediated Lysosomal Pathway of Proteolysis*

Given the results described so far, we propose the following model for the Hsc73-stimulated pathway of lysosomal proteolysis (Fig. 6). In a first step, cytosolic Hsc73, acting as a molecular chaperone, would recognize and interact with specific-targeting sequences (such as the KFERQ-related peptide motif) in the substrate proteins, probably exposed by conformational changes in these proteins. The protein–Hsc73 complex might then be recognized by lgp96, the receptor in the lysosomal membrane (see Fig. 6, step 2). Substrates can interact with lgp96 even in the absence of Hsc73, but this interaction is facilitated by the

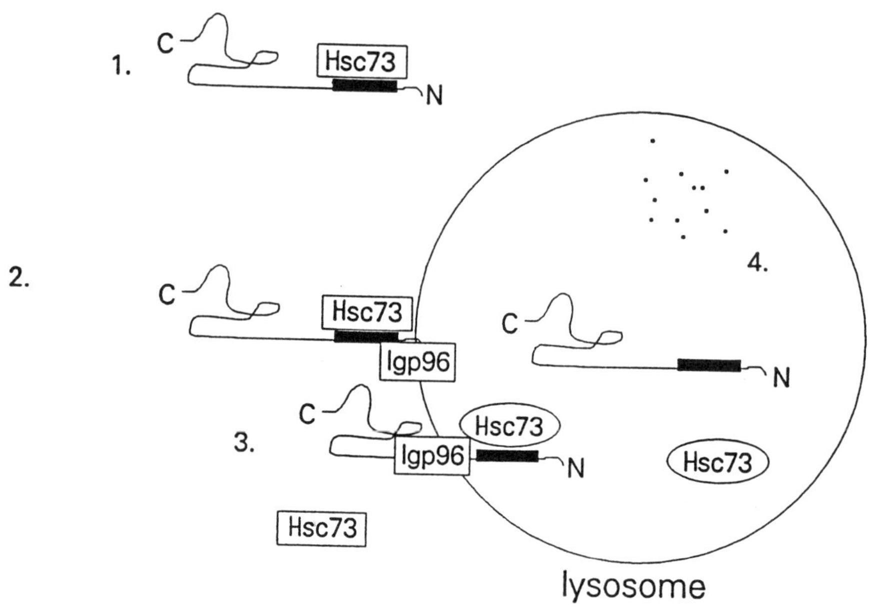

Figure 6 A hypothetical model for the chaperone-mediated selective lysosomal pathway of protein degradation: The Hsc73-dependent lysosomal pathway is able to degrade cytosolic proteins that are specifically recognized by cytosolic Hsc73 (1). After interaction with the receptor, lgp96, in the lysosomal membrane (2), substrate proteins are transported into the matrix helped by the intralysosomal Hsc73 (3), and once there are completely degraded (4).

presence of small amounts of Hsc73 [49]. Binding may increase in the presence of Hsc73 owing to changes in the substrate protein conformation or because Hsc73 participates in the substrate–receptor interaction. Once in the lysosomal membrane, the substrate protein is transported into the matrix (see Fig. 6, step 3) by the action of the intralysosomal Hsc73. This intralysosomal Hsc73, at least in part, reaches the lysosomal matrix by a similar process. Finally, once in the matrix, the substrate protein is degraded to its constituent amino acids (see Fig. 6, step 4).

5. *Physiological Relevance of the Chaperone-Mediated Lysosomal Degradation Pathway*

Prolonged Starvation. The selective lysosomal pathway applies to as much as 30% of cytosolic proteins, and such proteins can be identified by antibodies against the pentapeptide, KFERQ. The loss of rat cytosolic liver proteins containing KFERQ-like peptide sequences occurred primarily during days 2–3 or longer of starvation [60,61]. Early in fasting, the enhanced protein degradation in liver is due to activated macroautophagy [35,37,62].

In our initial studies to characterize the selective lysosomal pathway in rat liver, animals were fasted for 16–24 h. However, we noticed that the activity of rat liver lysosomes in vitro was less than that observed for fibroblast lysosomes [43,47]. Therefore, we examined the selective lysosomal degradation pathway using liver lysosomes from rats that had been fasted for different times (up to 88 h); this pathway is progressively activated at increased times of starvation. Consistent with the activation of the pathway with fasting, there is also an increase in the amount of lysosomal Hsc73 [45].

The physiological importance of this starvation-regulated activation of the selective lysosomal proteolytic pathway could be related to the control of total protein lost with fasting. If the initially activated macroautophagy continued for all the starvation period, approximately 70% of every cytosolic protein would be lost. Instead, activation of the selective lysosomal pathway would spare those proteins not containing KFERQ-like peptide motifs. According to this hypothesis certain proteins contain KFERQ sequences because they are dispensable under conditions of prolonged starvation. This selective degradation of specific proteins during prolonged starvation could also have significant effects in down-regulating transcription (c-*fos* [48]) and the activity of nonlysosomal pathways of proteolysis (proteasome [45]).

Aging. One of the common features of most cells of aged organisms is the accumulation of aberrant proteins [63]. This increase in abnormal proteins is at least partly, a consequence of a decrease in intracellular protein degradation rates with age [64–66]. The chaperone-mediated lysosomal proteolytic pathway is markedly reduced in senescent cells [68], and activity of the selective lysosomal pathway in isolated lysosomes from livers of old rats is lower than in young rats [A. M. Cuervo and J. F. Dice, unpublished results], suggesting that the lysosomal defect in aging could be localized to some step of the chaperone-mediated lysosomal pathway.

The exact mechanisms of the age-related defect in this proteolytic pathway are unclear. Cytosolic levels of Hsc73 in fibroblasts and in rat liver remain constant with age [S. R. Terlecky, A. M. Cuervo, and J. F. Dice, unpublished results], but there is a significant increase in the lysosomal Hsc73 in old rat livers [A. M. Cuervo and J. F. Dice, unpublished results]. Such an increase is expected to lead to more active, rather than to less active, lysosomes. It may be that the function of the lysosomal Hsc73 declines with age, or some other component of the degradative pathway, such as the lysosomal membrane receptor, may be compromised by aging.

Renal Tubule Diseases. Recently, modifications in levels of lysosomal Hsc73 have been reported in renal diseases. Komatsuda et al. [69] found that during gentamicin-induced acute tubular injury, Hsc73 moved from the nucleus to the cytoplasm, and accumulated in granules that were further identified as enlarged lysosomes. Similar lysosomal accumulation of Hsc73 has been described in other acute tubular failures in the kidney [69,70]. These authors propose that Hsc73 accumulation in lysosomes may stimulate the degradation of damaged proteins and, thereby, protect the target cells from nephrotoxicity. This speculation will require experimental testing.

C. Future Directions

Several important aspects of this chaperone-mediated lysosomal degradation pathway remain unclear. The exact role of cytosolic Hsc73 in stimulating the binding of substrate proteins to the lysosomal receptor is unknown. Furthermore, Hsc73 can also enter lysosomes through this same pathway. It is unknown whether such entry is required for substrate proteins to cross the lysosomal membrane. Accordingly, the Hsc73 property of binding to fatty acids [71] may be important.

Once Hsc73 reaches the lysosomal matrix, one remaining question is why it is not degraded by lysosomal proteases. For this, pH-dependent changes in Hsc73 conformation needed to be characterized.

The role of other molecular chaperones in lysosomal degradation needs to be examined. In other intracellular protein-transport systems, more than one chaperone is found in the cytosolic side as well as the lumenal side of the membrane. Very often these proteins work cooperatively, helping complete transport of the substrate protein into the target organelle. It would not be surprising to find DnaJ homologues acting with cytosolic Hsc73 and even perhaps within lysosomes.

Finally, other lysosomal membrane proteins are likely to be required along with the receptor, by analogy with the well-studied mitochondrial and other organellar protein-import machineries. These components may include additional receptors, peptide insertion channels, and other molecular chaperones.

ACKNOWLEDGMENTS

Work in the authors' laboratory was supported by Spain grant Fundacion Ramon Areces and the American Liver Foundation award (AMC), NIH grant DK07542 (SAH), and NIH grant AG06116 (JFD).

REFERENCES

1. Rechsteiner, M., Ubiquitin-mediated pathways for intracellular proteolysis, *Annu. Rev. Cell Biol. 3*:1 (1987).
2. Dice, J. F., Molecular determinants of protein half-lives in eukaryotic cells, *FASEB J. 1*:349 (1987).
3. Hershko, A. and Ciechanover, A., The ubiquitin system for protein degradation, *Annu. Rev. Biochem. 61*:761 (1992).
4. Olson, T. S., Terlecky, S. R., and Dice, J. F., Pathways of intracellular protein degradation in eukaryotic cells, *Stability of Protein Pharmaceuticals, In Vivo Pathways of Degradation and Strategies for Protein Stabilization* (T. J. Ahern and M. C. Manning, eds.), Plenum Publishing, New York, 1992, p. 89.
5. Rechsteiner, M., Natural substrates of the ubiquitin proteolytic pathway, *Cell 66*:615 (1991).
6. Craig, E. A., Gambill, B. D., and Nelson, R. J., Heat shock proteins: Molecular chaperones of protein biogenesis, *Microbiol. Rev. 57*:402 (1993).
7. Parsell, D. A. and Lindquist, S., The function of heat-shock proteins in stress tolerance: Degradation and reactivation of damaged proteins, *Annu. Rev. Genet. 27*:437 (1993).
8. Craig, E. A., Baxter, B. K., Becker, J., Halladay, J., and Ziegelhoffer, T., Cytosolic hsp70s of *Saccharomyces cerevisiae*: Roles in protein synthesis, protein translocation, proteolysis, and regulation, *The Biology of Heat Shock Proteins and Molecular Chaperones* (R. I. Morimoto, A Tissieres, and C. Georgopoulos, eds.), Cold Spring Harbor Laboratory Press, Cold Spring Harbor, NY, 1994, p. 31.
9. Hayes, S. A. and Dice, J. F., Roles of molecular chaperones in protein degradation, *J. Cell Biol. 132*:255 (1996).
10. Goff, S. A., Casson, L. P., and Goldberg, A. L., Heat shock regulatory gene *hptR* influences rates of protein expression of the *lon* gene in *Escherichia coli*, *Proc. Natl. Acad. Sci. USA 81*:6647 (1984).
11. Gottesman, S. and Maurizi, M. R., Regulation by proteolysis: Energy-dependent proteases and their targets, *Microbiol. Rev. 56*:592 (1992).
12. Straus, D. B., Walter, W. A., and Gross, C. A., *Escherichia coli* heat shock gene mutants are defective in proteolysis, *Genes Dev. 2*:1851 (1988).
13. Keller, J. A. and Simon, L. D., Divergent effects of a *dnaK* mutation on abnormal protein degradation in *Escherichia coli*, *Mol. Microbiol. 2*:31 (1988).

14. Sherman, M. Y. and Goldberg, A. L., Involvement of the chaperonin DnaK in the rapid degradation of a mutant protein in *Escherichia coli*, *EMBO J. 11*:71 (1992).
15. Kessel, M., Maurizi, M. R., Kim, B., Kocsis, E., Trus, B. L., Singh, S. K., and Steven, A. C., Homology in structural organization between *E. coli* ClpAP protease and the eukaryotic 26S proteasome, *J. Mol. Biol. 250*:587 (1995).
16. Wickner, S., Gottesman, S., Skowyra, D., Hoskins, J., McKenney, K., and Maurizi, M. R., A molecular chaperone, ClpA, functions like DnaK and DnaJ, *Proc. Natl. Acad. Sci. USA 91*:12218 (1994).
17. Wawrzynow, A., Wojtkowial, D., Marszalek, J., Banecki, B., Jonsen, M., Graves, O., Georgopoulos, C., and Zylicz, M., The ClpX heat-shock protein of *Escherichia coli*, the ATP-dependent substrate specificity component of the ClpPClpX protease is a novel molecular chaperone, *EMBO J. 14*:1867 (1995).
18. Kandror, O., Busconi, L., Sherman, M. Y., and Goldberg, A. L., Rapid degradation of an abnormal protein in *Escherichia coli* involves the chaperones GroEL and GroES, *J. Biol. Chem. 269*:23575 (1995).
19. Kandror, O., Sherman, M., Moershell, R., and Goldberg, A. L., Trigger factor promotes GroEL-dependent proteolysis by enhancing the affinity of GroEL for certain unfolded proteins, *Molecular Chaperones and the Heat Shock Response*, Cold Spring Harbor Press, Cold Spring Harbor, NY, 1996, p. 411.
20. Levchenko, I., Luo, L., and Baker, T., Disassembly of the Mu transposase tetramer by the ClpX chaperone, *Genes Dev. 9*:2399 (1995).
21. Stuart, R. A., Cyr, D. M., and Neupert, W., Hsp70 in mitochondrial biogenesis: From chaperoning nascent polypeptide chains to facilitation of protein degradation, *Experientia 50*:1003 (1994).
22. Tanner, A. J. and Dice, J. F., Batten disease and mitochondrial pathways of proteolysis, *Biochem. Mol. Med. 57*:1 (1996).
23. Leonhardt, S. A., Fearon, K., Danese, P. N., and Mason, T. L., Hsp78 encodes a yeast mitochondrial heat shock protein in the Clp family of ATP-dependent proteases, *Mol. Cell. Biol. 13*:6304 (1993).
24. Hershko, A., Ubiquitin-mediated protein degradation, *J. Biol. Chem. 263*:15237 (1988).
25. Ciechanover, A., Finley, D., and Varshavsky, A., Ubiquitin dependence of selective protein degradation in the mammalian cell cycle mutant ts85, *Cell 37*:57 (1984).
26. Fagan, J. M., Waxman, L., and Goldberg, A. L., Skeletal muscle and liver contain a soluble ATP + ubiquitin-dependent proteolytic system, *Biochem. J. 243*:335 (1987).
27. Haas, A. and Bright, P., The dynamics of ubiquitin pools within cultured human lung fibroblasts, *J. Biol. Chem. 262*:345 (1987).
28. Chau, V., Tobias, J. W., Bachmair, A., Marriott, D., Ecker, D. J., Gonda, D. K., and Varshavsky, A., A multiubiquitin chain is confined to specific lysine in targeted short-lived protein, *Science 243*:1576 (1989).
29. Ciechanover, A., The ubiquitin–proteasome proteolytic pathway, *Cell 79*:13 (1994).
30. Lee, D. H., Sherman, M. Y., and Goldberg, A. L., Involvement of the molecular chaperones (DnaJ-homologs and Hsp70s) in the ubiquitin-dependent degradation of short-lived and abnormal proteins in yeast, *Cold Spring Harbor Symposium on Quantitative Biology*, Vol. 60, Cold Spring Harbor Press, Cold Spring Harbor, 1995, p. 111.

31. Yaglom, J., Linskens, M. H. K., Sadis, S., Rubin, D. M., Futcher, B., and Finley, D., p34Cdc28-mediated control of cln3 cyclin degradation, *Mol. Cell. Biol. 15*:731 (1995).
32. Ohba, M., A 70-kDa heat shock protein suppresses the defects caused by a proteasome mutation in *Saccharomyces cerevisiae*, *FEBS Lett. 351*:263 (1994).
33. Hare, J. F., Mechanisms of membrane protein turnover, *Biochim. Biophys. Acta 1031*:71 (1990).
34. Marzella, L. and Glaumann, H., Autophagy, microautophagy, and crinophagy as mechanisms for protein degradation, *Lysosomes: Their Role in Protein Breakdown* (H. Glaumann and F. J. Ballard, eds.), Academic Press, New York, 1987, p. 319.
35. Dunn, Jr., W. A., Autophagy and related mechanisms of lysosome-mediated protein degradation, *Trends Cell Biol. 4*:139 (1994).
36. Dice, J. F., Agarraberes, F., Kirven-Brooks, M., Terlecky, L. J., and Terlecky, S. R., Heat shock 70-kD proteins and lysosomal proteolysis, *The Biology of Heat Shock Proteins and Molecular Chaperones* (R. I. Morimoto, A. Tissieres, and C. Georgopoulos, eds.), Cold Spring Harbor Laboratory Press, Cold Spring Harbor, NY, 1994, p. 137.
37. Kopitz, J., Kisen, G. O., Gordon, P. B., Bohley, P., and Seglen, P. O., Non-selective autophagy of cytosolic enzymes in isolated rat hepatocytes, *J. Cell Biol. 111*:941 (1990).
38. Dice, J. F., Peptide sequences that target cytosolic proteins for lysosomal proteolysis, *Trends Biochem. Sci. 15*:305 (1990).
39. Dice, J. F., Terlecky, S. R., Chiang, H.-L., Olson, T. S., Isenman, L. D., Short-Russell, S. R., Freundlieb, S., and Terlecky, L. J., A selective pathway for degradation of cytosolic proteins by lysosomes, *Semin. Cell Biol. 1*:449 (1990).
40. Terlecky, S. R., Hsp70s and lysosomal proteolysis, *Experientia 50*:1021 (1994).
41. Chiang, H.-L., Terlecky, S. R., Plant, C. P., and Dice, J. F., A role for a 70-kilodalton heat shock protein in lysosomal degradation of intracellular proteins, *Science 246*:382 (1989).
42. Aniento, F., Roche, E., Cuervo, A. M., Knecht, E., Uptake and degradation of glyceraldehyde-3-phosphate dehydrogenase by rat liver lysosomes, *J. Biol. Chem. 268*:10463 (1993).
43. Cuervo, A. M., Terlecky, S. R., Dice, J. F., and Knecht, E., Selective binding and uptake of ribonuclease A and glyceraldehyde-3-phosphate dehydrogenase by isolated rat liver lysosomes, *J. Biol. Chem. 269*:26374 (1994).
44. Cuervo, A. M., Palmer, A., Rivett, A. J., and Knecht, E., Degradation of proteasomes by lysosomes in rat liver, *Eur. J. Biochem. 227*:792 (1995).
45. Cuervo, A. M., Knecht, E., Terlecky, S. R., and Dice, J. F., Activation of a selective pathway of lysosomal proteolysis in rat liver by prolonged starvation, *Am. J. Physiol. 269*:C12000 (1995).
46. Schatz, G. and Dobberstein, B., Common principles of protein translocation across membranes, *Science 271*:1519 (1996).
47. Terlecky, S. R. and Dice, J. F., Polypeptide import and degradation by isolated lysosomes, *J. Biol. Chem. 268*:23490 (1993).
48. Aniento, F., Papavassiliou, A. G., Knecht, E., and Roche, E., Selective uptake and degradation of c-*fos* and v-*fos* by rat liver lysosomes, *FEBS Lett. 390*:47 (1996).

49. Cuervo, A. M. and Dice, J. F., A receptor for the selective uptake and degradation of proteins by lysosomes, *Science 273*:501 (1996).
50. Chirico, W., Waters, M. G., and Blobel, G., 70K heat shock related proteins stimulate protein translocation into microsomes, *Nature 332*:805 (1988).
51. Deshaies, R. J., Koch, B. D., Werner-Washburne, M., Craig, E. A., and Schekman, R., A subfamily of stress proteins facilitates translocation of secretory and mitochondrial precursor polypeptides, *Nature 332*:800 (1988).
52. Cuervo, A. M., Dice, J. F., and Knecht, E., A population of rat liver lysosomes responsible for the selective uptake and degradation of cytosolic proteins, *J. Biol. Chem. 272*:5606 (1997).
53. Terlecky, S. R., Chiang, H.-L., Olson, T. S., and Dice, J. F., Protein and peptide binding and stimulation of in vitro lysosomal proteolysis by the 73-kDa heat shock cognate protein, *J. Biol. Chem. 267*:9202 (1992).
54. Vogel, J. P., Misra, L. M., and Rose, M. P., Loss of Bip/GRP78 function blocks translocation of secretory proteins in yeast, *J. Cell Biol. 110*:1885 (1990).
55. Nicchitta, C. V. and Blobel, G., Lumenal proteins of the mammalian endoplasmic reticulum are required to complete protein translocation, *Cell 73*:989 (1993).
56. Kang, P. J., Osterman, J., Shilling, J., Neupert, W., Craig, E. A., and Pfanner, N., Requirement for hsp70 in the mitochondrial matrix for translocation and folding of precursor proteins, *Nature 348*:137 (1990).
57. Lill, R. and Neupert, W., Mechanisms of protein import across the mitochondrial outer membrane, *Trends. Cell Biol. 6*:56 (1996).
58. Mayer, R. J., Lowe, J., Landon, M., McDermott, H., Tuckwell, J., Doherty, F., and Lazlo, L., Ubiquitin and the lysosomal system; molecular pathological and experimental findings, *Heat Shock* (B. Maresca and S. Lindquist, eds.), Springer-Verlag, Heidelberg, 1991, p. 299.
59. Agarraberes, F., Terlecky, S. R., and Dice, J. F., Requirement for intralysosomal hsc73 for a selective pathway of lysosomal proteolysis, *J. Cell Biol.*, in press.
60. Chiang, H.-L. and Dice, J. F., Peptide sequences that target proteins for enhanced degradation during serum withdrawal, *J. Biol. Chem. 263*:6797 (1988).
61. Wing, S. S., Chiang, H.-L., Goldberg, A. L., and Dice, J. F., Proteins containing peptide sequences related to KFERQ are selectively depleted in liver and heart, but not skeletal muscle, of fasted rats, *Biochem. J. 275*:165 (1991).
62. Mortimore, G. E., Pösö A. R., and Lardeaux, B. R., Mechanism and regulation of protein degradation in liver, *Diabetes Metab. Rev. 5*:49 (1989).
63. Gershon, H. and Gershon, D., Detection of inactive molecules in aging organisms, *Nature 227*:1214 (1970).
64. Makrides, S., Protein synthesis and degradation during aging and senescence, *Biol. Rev. 58*:393 (1983).
65. Dice, J. F., Altered intracellular protein degradation in aging: A possible cause of proliferative arrest, *Exp. Gerontol. 24*:451 (1989).
66. Dice, J. F., Cellular and molecular mechanisms of aging, *Physiol. Rev. 73*:149 (1993).
67. Hayflick, L., Recent advances in the cell biology of aging, *Mech. Aging Dev. 14*:59 (1980).
68. Dice, J. F. and Goff, S. A., Error catastrophe theory, *Modern Biological Theories of*

Aging (R. Warner, R. N. Butler, R. L. Sprott, and E. L. Schneider, eds.), Raven Press, New York, 1987, p. 155.
69. Komatsuda, A., Wakui, H., Satoh, K., Yasuda, T., Imai, H., Nakamoto, Y., Miura, A. B., Itoh, H., and Tashima, Y., Altered localization of 73-kilodalton heat-shock protein in rat kidneys with gentamicin-induced acute tubular injury, *Lab. Invest. 68*:687 (1993).
70. Komatsuda, A., Wakui, H., Imai, H. et al., Renal localization of the constitutive 73-kDa heat-shock protein in normal and PAN rats, *Kidney Int. 41*:1204 (1992).
71. Guidon, Jr., P. T. and Hightower, L. E., Purification and initial characterization of the 71-kilodalton rat heat-shock protein and its cognate as fatty acid binding proteins, *Biochemistry 25*:3231 (1986).

21

Proteasomes: The Protein Death Machinery

Hiroyuki Kawahara and Keiji Tanaka

The Tokyo Metropolitan Institute of Medical Science, Tokyo, Japan

I. INTRODUCTION

Proteins play fundamental roles in cells; thus, their levels are highly regulated. The synthesis and degradations of intracellular proteins are in dynamic equilibrium, clearly indicating that proteins are degraded continuously in amounts corresponding exactly to those synthesized. Approximately 100,000 species of proteins are present in animal cells, and interestingly, these have individual turnover rates, with half-lives varying from a few minutes to several months. It is not fully known why all proteins synthesized must be degraded, why their half-lives are so different, and what the mechanisms of their degradation are. There are several possible reasons for rapid protein breakdown in living cells. For example, regulatory proteins related to metabolism and cell cycle progression must be degraded in a time- or signal-dependent fashion, as has recently been reported in broad areas of cell biology. In addition, abnormal proteins with aberrant structures, which are generated in cells under various environmental stresses and by genetic mutations, must be degraded immediately. Furthermore, the unassembled protein components of multisubunit complexes should be degraded, even when synthesized normally, if their partner subunits are not synthesized or are broken down for some reason. Thus, cells have acquired a self-surveillance proteolytic system responsible for selective removal of harmful or unnecessary proteins to prevent their accumulation, irrespective of whether these proteins are normal or abnormal. This proteolysis system is presumably required for maintaining cellular homeostasis, and it appears to be an intracellular immune system.

As unregulated protein destruction would lead to cell death, intracellular protein breakdown must be strictly controlled. Various cellular proteins are continuously exposed to nonlysosomal cytoplasmic proteases, hence, it is important to understand how the proteolytic system is regulated in cells and how appropriate target proteins are recognized effectively and selectively. Metabolic energy is required for selective proteolysis in both prokaryotic and eukaryotic cells, but the mechanisms involved are different. In prokaryotic cells, at least five distinct ATP-dependent proteases are present and appear to share biological roles. Studies on these enzymes have contributed to explaining how energy is transferred for intracellular protein degradation. The ATP-dependent proteases use energy from cellular ATP to degrade substrate proteins, whereas other known proteases do not consume energy from ATP or any other source, implying that ATP-dependent proteases have other functions besides simple cleavage of peptide bonds. On the other hand, the eukaryotic ATP-dependent system for proteolysis consists of two distinct processes: first, recognition of target proteins by ubiquitin (Ub) and subsequent degradation of ubiquitinated proteins by the proteasome, with metabolic energy being required for both steps. It is unknown why the proteolytic systems in prokaryotic and eukaryotic cells differ, but these systems seem to have similar biological roles.

In eukaryotic cells, Ub and the proteasome appear to be organized into a major proteolytic system as a multienzymatic cascade system and a supramolecular multisubunit complex, respectively, for rapid and highly regulated degradations of numerous proteins. Therefore, they are assumed to be "a protein death machinery." In this chapter, we summarize current information on the proteasome system, particularly focusing the similarity and cooperation between proteasomes and molecular chaperones in considering why metabolic energy is required for proteolysis.

II. THE UBIQUITIN SYSTEM

Ubiquitin (Ub) is a highly conserved protein, with a molecular mass of 86 kDa. It is encoded by two unique genes: one is the mono-Ub gene, which is fused with two essential ribosomal proteins, and the other is the poly-Ub gene, which encodes multiple Ubs in a single open-reading frame and can produce mature Ub molecules by posttranslational processing [reviewed in Ref. 1]. The expression of the poly-Ub gene, unlike the mono-Ub gene, is greatly enhanced by heat treatment, suggesting that cells must rapidly produce a large amount of Ub protein in response to various stress stimuli. Indeed, the yeast poly-Ub gene is not normally essential, but plays an indispensable role on exposure of the cells to high temperature [2], indicating the necessity for efficient generation of Ub for adaptation to stress. Ub is the smallest heat-shock protein, which contributes to the concept of refolding of unfolded proteins for their removal in the stress response.

Multiple Ub moieties become attached covalently to target proteins by isopeptide linkage through the terminal glycine of Ub and lysine of the proteins, and then repeatedly through lysine of Ub. The resulting poly- or multi-Ub chain acts as a degradation signal for proteolytic attack by the 26S proteasome, as shown schematically in Figure 1 [3]. This polyubiquitination process is performed through a reaction catalyzed by a multienzymatic system, consisting of E1 (Ub-activating), E2 (Ub-conjugating), and E3 (Ub-ligating) enzymes. E1, catalyzing the initial reaction for ubiquitination, is encoded by a single gene. Mutation of this essential gene causes various cellular defects, including abnormal chromosomal condensation in the cell cycle [1]. Mutant cells carrying a temperature-sensitive E1 have been used to examine whether a given protein is actually degraded by the Ub pathway in living mammalian cells.

E2 is a large protein family. Some E2s have the ability to ubiquitinate target proteins directly, as real Ub-conjugating enzymes, but most E2s appear to act as carrier proteins, cooperating with another proteins, termed E3, in ubiquitination [4]. Therefore, the family of E2 proteins can be classified into two subgroups, those with and those without dependence on E3, although they all have a conserved functional domain containing an active cysteine residue to accept Ub from activated E1-Ub to form a thioester bond. It is unknown why some E2s required E3 for protein ubiquitination and others do not. The primary structures of the E2 proteins family cannot explain this, although extensions of NH_2- or COOH-terminal region from the conserved domain are thought to exert specific functions [4]. The yeast E2 gene is named *UBC*, and the twelve UBC proteins identified so far appear for ubiquitinate-distinct target proteins, although some of them show functional redundancy ubiquitinating the same target proteins.

Some Ub-conjugating enzymes are induced in yeast in response to various stresses, such as heat shock or exposure to cadmium [5]. For example, the expressions of the *UBC5* and *UBC7* genes were up-regulated by cadmium treatment, and mutations of these genes result in hypersensitivity to cadmium exposure, indicating that the induced proteins are new types of stress proteins. However, the expressions of these two *UBC* genes differ, depending on the type of stress; expression of *UBC7*, but not *UBC5*, was not induced by heat shock, but responded markedly to cadmium. This suggests that distinct ubiquitin pathways operate in acquirement of cadmium- and heat-resistance, respectively. Thus multiple E2s may contribute to diverse cellular functions, although the exact significance of the molecular diversity of E2s is still unknown.

E3 was originally identified as a protein factor responsible for substrate recognition, and its gene was named *UBR1* (Ub recognition protein) involved in the NH_2-end rule pathway [6]. But later, E3 was found to have ability to bind to Ub, forming a thioester bond, indicating potential ligase functions. Indeed,

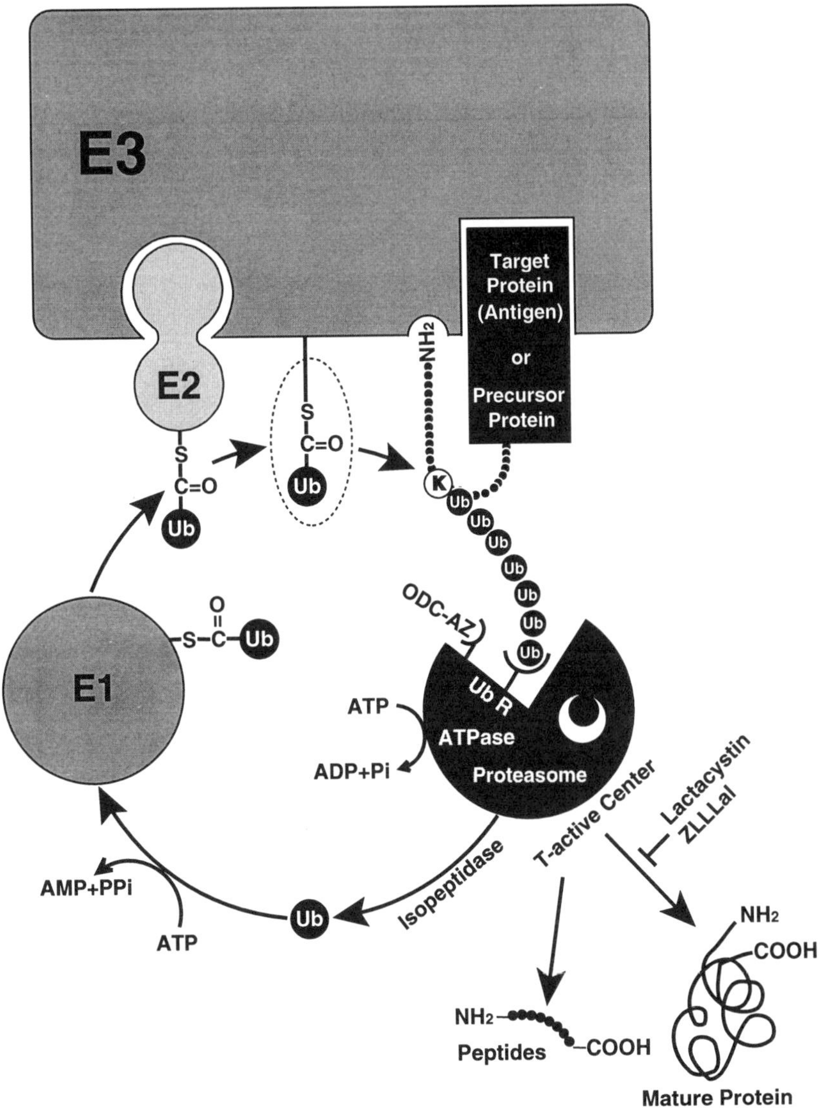

Figure 1 Model of intracellular protein degradation mediated by the ubiquitin–proteasome system: Ub, ubiquitin; E1, Ub-activating enzyme; E2, Ub-conjugating enzyme (also termed carrier protein); E3, Ub-protein ligase (recognition protein); Ub-R, multi-Ub receptor; ODC, ornithine decarboxylase; AZ, antizyme.

multiple functionally distinct E3 species have been found in studies on ubiquitination of different proteins. Of these species, one interesting E3 is E6-AP for ubiquitination of p53 [7], because it consists of a large protein family harboring a conserved Ub-accepting domain named hect (*h*omologous to the *E*6-AP *c*arboxyl *t*erminus), including RSP5 (= Nedd4), Pub1, and Ufd4 [8]. In addition, various new types of E3s for ubiquitinations of c-Fos, NF-κB, actin/troponin–T/MyoD and cyclin B have been discovered very recently. These have large molecular masses of 300–750 kDa, although their exact structures are unknown. The E3 ligase for ubiquitination of cyclin B, named cyclosome or APC (*a*naphase *p*romoting *c*omplex), is also interesting because it is a 20S multicomponent complex consisting of at least eight polypeptides, the functions of which are regulated by phosphorylation and dephosphorylation in a cell cycle-dependent manner and is involved in M-phase traverse, including sister-chromatid separation [9]. Because so many cellular proteins can be degraded by the proteasome pathway, many important E3s may still remain to be found. Identification of these substrate-specific E3s is important for determining the physiological roles of Ub-dependent degradation.

The Ub system has also been implicated in the regulation of various important biological processes, such as the cell cycle, apoptosis, signal transduction, protein processing, immune responses, stress responses, regulations of gene expressions, DNA repair, the biogenesis of ribosomes, peroxisomes, and mitochondria, import of proteins into mitochondria, and the uptake of precursors into neurons [1,8]. Tagging by ubiquitination for selective proteolysis may account for most of these functions, but Ub may also exert other roles, such as in monoubiquitination for posttranslational modification of proteins. Ubiquitination is thought to be a reversible covalent modification of proteins, because there is a large family of Ub isopeptidases responsible for deubiquitinating protein–Ub ligates or poly-Ub protein; 15 genes for these have been isolated from yeast [8]. Studies on these deubiquitinating enzymes should provide an insight into the biological roles of the ubiquitin system.

III. THE PROTEASOME SYSTEM

A. The Cylindrical 20S Proteasome

The 20S proteasome was first identified as an unusually large, inactive form of protease complex, with a molecular mass of about 700 kDa, present in both the nucleus and cytoplasm of various mammalian and yeast cells [10–12]. It appears to be a cylindrical particle, with an apparent diameter of 12 nm and length of 17 nm, and consists of a stack of four rings, with a dimeric structure of two distinct rings. The 20S proteasome is composed of a set of small subunits, with molecular masses of 21–32 kDa. Sequence analysis of the proteasome subunits revealed no similarities with any other known eukaryotic proteins, but considerable inter-subunit homology of proteasomal subunits in the same species, and high evolutional conservation in various eukaryotes form yeast to humans [13,14]. The archaebacterium *Thermoplasma acidophilum* 20S proteasome is a cylindrical particle consisting of four rings, each consisting of seven identical α- or β-subunits [15], and the 20S proteasome of the eubacterium *Rodococcus* consists of four subunits, two α-type subunits, α_1 and α_2, and two β-type subunits, β_1 and β_2 [16]. The quaternary structures of *Thermoplasma* and *Rodococcus* proteasomes are indistinguishable from those of eukaryotic proteasomes, suggesting high evolutional conservation of their gross structures. Eukaryotic proteasome genes can also be divided into two distinct subgroups, α and β, with high similarities to archaebacterial α- and β-subunits, respectively [13,17]. Phylogenetic analyses of the various eukaryotic proteasomal genes show that the α-type and β-type subunits can each be classified into seven subgroups [12,17]. Immunoelectron microscopic examination demonstrated that the α-subunits constitute rings at each end of the cylinder, whereas the β-subunits form the two inner rings. Thus, the 20S proteasome subunits seem to be assembled as an α7β7β7α7 complex [15,18], as shown in Figure 2 (upper panel).

Consistent with the assumed subunit organization of the α-ring, seven human cDNAs for α-type subunits have been cloned, each showing high homology with its counterpart in the budding yeast. Some α-subunits contain a cluster of basic amino acid residues that presumably acts as a nuclear localization signal (NLS), and a complementary NLS (cNLS) sequence, consisting of an acidic amino acid cluster, which may antagonize the NLS-dependent nuclear import of proteasomes [19]. Recently, the NLSs of human and *Thermoplasma* proteasomal α-subunits were found to be functional in vivo [20], suggesting that the α-subunits contribute to determination of the intracellular location of proteasomes. Eukaryotic α-type subunits have a highly conserved NH_2-terminal α-helical extension of approximately 35 residues, and this helix is situated near the entrance gate to the channel. Archaebacterial proteasome α-subunits assemble spontaneously to form seven-membered rings, but deletion of their first 35 amino acids abolishes this assembly, indicating a critical role of the NH_2-terminal extension in α-subunit ring assembly [21].

Ten human cDNAs of β-subunits have now been found, although seven genes for β-subunits have been cloned from yeast, consistent with the molecular structure of the β-ring [12,13]. The existence of ten β-subunits is curious, considering that one β-type ring is formed from seven subunits in vertebrates. Given this discrepancy with the number in the β-ring, we proposed the replacements of three pairs of highly homologous β-type subunits in response to extracellular signals, such as that of interferon gamma (IFN-γ) [for details see Ref. 22]. We have proposed that proteasomes containing these IFN-γ-inducible subunits should be named "immunoproteasomes" to emphasize their specialized

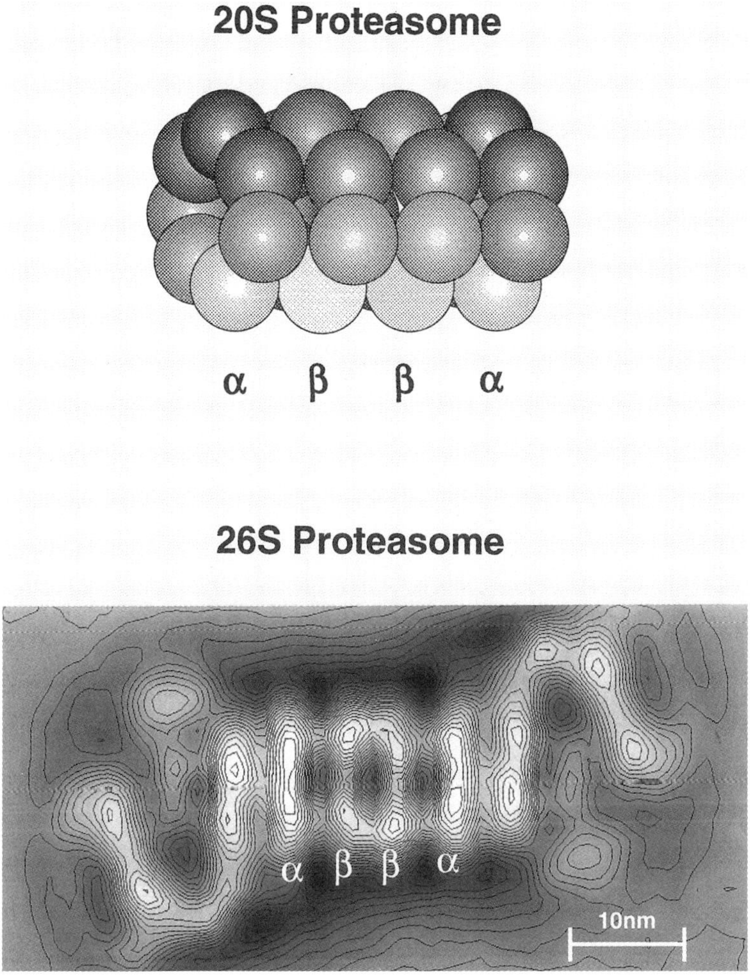

Figure 2 Structural features of 20S cylindrical and 26S dumbbell-shaped proteasomes: (upper panel) hypothetical model of the 20S proteasome; (lower panel) the 26S proteasome complex. (Photograph courtesy of W. Baumesiter.)

functions in antigen presentation [22]. Immunoproteasomes may be better suited for immunological processing of endogenous antigens. The altered molecular organization of the proteasome induced by IFN-γ may be responsible for acquisition of its functional changes in a variety of heterogeneous complexes. The total number of β-type subunits may have increased in adaptation to various environmental circumstances during evolution.

In the highly ordered structure of the *Thermoplasma* proteasome, the 20S proteasome has a central cavity. Multiple catalytic sites are exposed on the inner surface of the cavity, composed of β-type subunits, indicating that substrate proteins have to go through the narrow gates of the α-subunit rings [21]. The outer gates of α-rings are only 1.3 nm in diameter. Because this is only slightly larger than the diameter of an α-helix, protein substrates in their native conformation should be unable to reach the active site. This explains the low proteinase activity of the intact 20S proteasome on nondenatured protein substrates [17]. Therefore, a protein must be unfolded to be degraded by this complex [15]. Peroxidized proteins were degraded efficiently by the 20S proteasome, probably owing to partial unfolding of their tertiary structure, whereas most native proteins were poor substrates. Access to the inner catalytic site is now thought to be controlled by a peptide channel operating by size exclusion. Baumeister and colleagues proved this directly by mixing nanogold-labeled insulin and 20S proteasomes [22]. Small peptides, such as insulin, are usually degraded rapidly by 20S proteasomes, but they demonstrated that increase in size of insulin by nanogold labeling prevented it from entering proteasomes and being degraded. Furthermore, they showed that the nanogold accumulated in the center of the α-rings, plugging the channel. These results indicate the necessity for unfolding of substrate proteins before their entry into the inner cavity for their destruction.

Interestingly, there is apparent homology in structural organization between proteasomes and chaperonin-family proteins, such as bacterial GroEL and the eukaryotic cytoplasmic chaperonin termed CCT (chaperonin containing TCP-1; *t*-complex polypeptide 1) [24,25]. GroEL, an 850-kDa complex consisting of 14 subunits arranged in two heptameric rings stacked back to back, which participates in ATP-dependent polypeptide chain folding. Folding is achieved through cycles of protein release and rebinding that are dependent on ATP hydrolysis and are regulated by GroES, a single heptameric ring of about 10-kDa subunits, suggesting that the native state is reached by repeated rounds of folding and structural rearrangement. The GroEL chaperonin has a cylinder-like structure with sevenfold symmetry, and a central channel. It is believed that protein folding mediated by the GroEL is performed inside the cavity of the GroEL complex, and also that this cavity is of prime importance for its chaperonin activity [26]. In contrast, the eukaryotic cytoplasmic chaperonin termed CCT is hetero-oligomeric, consisting of at least eight homologous subunits sharing approximately 30% identity, which are highly conserved in eukaryotes from yeast to mammals [25].

Despite the lack of sequence similarity between proteasomes and chaperonins, these multimeric complexes show considerable similarity in overall dimensions and have a bipolar, multiple-ring, symmetric structure as their central element, implying that both play critical roles in the life cycles of all cells, and suggesting functional similarity in the underlying biochemical mechanisms by which these proteins bind, unfold, and deliver targeted protein substrates.

The sevenfold-symmetry structures of the 20S proteasome and GroEL may provide some advantages in the catalytic efficiency or processing of enzymes [24]. However, the functions of these two particles are opposite: chaperonins function in protein folding and promoting protein assembly, whereas proteasomes constitute a multisubunit machinery contributing to destruction of target proteins with aberrant structures.

B. The Dumbbell-Shaped 26S Proteasome

The 26S proteasome appears to be a dumbbell-shaped particle, with two large V-like or U-like terminal subunit complexes in opposite orientations attached to a smaller, four-layered central 20S proteasome (see Fig. 2, lower panel). It is a large protease complex, with a molecular mass of 2000 kDa, consisting of the 20S proteasome as a catalytic core, and terminal subsets of heterogeneous proteins of 25–112 kDa, which possibly have regulatory functions [17,18]. Little is known about the structures of the regulatory subunits, but molecular clonings of their cDNAs are now in progress [13,27]. So far, cDNAs encoding about 20 subunits of the human 26S proteasomal regulatory subunits have been isolated, and have been separated into two functionally different subgroups: a group of six ATPases of relatively homogeneous sizes of 42–55 kDa and a group of heterogeneous non-ATPase subunits. Identification of the multiple ATPase subunits accounts for the intrinsic ATPase activity of in the 26S proteasome, and provides a molecular basis for the ATP-requirement for proteolysis. These findings strongly suggest that the 26S proteasome is a eukaryotic ATP-dependent protease.

Presumably, one role of ATPase is to supply energy continuously for the selective degradations of ubiquitinated proteins by the active 26S proteasome complex. However, the mechanism by which the energy is consumed during proteolysis is unknown. One possible function of the ATPase subunit is to induce the unfolding of target proteins to allow their accesses into the catalytic cavity of the complex, as shown diagrammatically in Figure 3. The multiple ATPases in proteasomes of eukaryotes, ranging from yeast to mammals, have common features, but it is unclear why so many ATPases are associated with the 26S proteasome complex. A single ATPase subunit should be sufficient to supply energy for proteolysis; therefore, these ATPases may have other functions besides supplying energy for proteolysis. Alternatively, each ATPase may be associated with a distinct 26S complex, each with a different function. Namely, 26S proteasomes may be present in cells as a heterogeneous mixture of 26S proteasomes, each with a different ATPase, but may be copurified from the cells because of their very high structural similarities.

Why so many non-ATPase subunits are associated with the 26S proteasome is also unclear. One subunit, named S5a or MBP1 (*m*ulti-Ub *b*inding *p*rotein), that specifically binds to Ub-protein conjugates functions as a Ub-receptor to trap

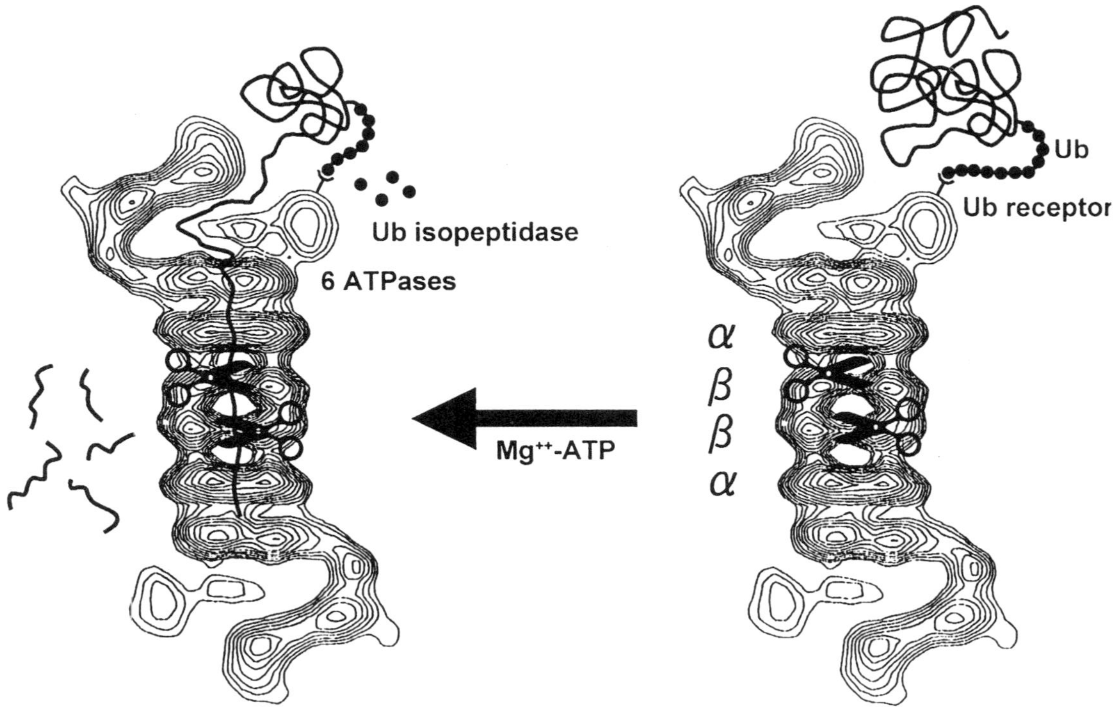

Figure 3 Schematic view of degradation of ubiquitinated proteins by the 26S proteasome: α and β, α- and β-subunit rings, respectively, of the 20S proteasome; Ub, ubiquitin; Ub receptor, multi-Ub-binding protein. For details, see text.

target proteins [28,29]. Because the 26S proteasome is responsible for the degradations of a variety of functional proteins, with a multi-Ub chain as a degradation signal [3,8], the discovery of the Ub receptor provides a molecular basis for selection of target proteins (see Fig. 3). Moreover, the 26S proteasome also degrades nonubiquitinated ornithine decarboxylase (ODC) when associated with its specific inhibitor protein, named antizyme, which may expose the degradation signal of the ODC moiety [30]. Therefore, the regulatory subunits may be able to recognize various potential degradation signals of diversed target proteins for proteolytic attack (see Fig. 1). Moreover, a deubiquitinating enzyme, generating the Ub moiety from the multi-Ub chain for reutilization, was reported to be associated with 26S proteasomes from various sources [8]. Recently, the cDNA of Ub isopeptidase associated with the 26S proteasome has been cloned. Interestingly, the genes of several non-ATPase subunits are not essential for yeast cell proliferation, whereas almost all 20S proteasome and six ATPase subunit genes are essential. In addition, some non-ATPase subunits identified in humans are not present in the yeast genome. Therefore, non-ATPase subunits may have been altered considerably during evolution, suggesting functional diversity of the 26S proteasome in different organisms. Further study is required to clarify the relation of ATPase and non-ATPase subunits in the 26s proteasome complex.

The Ub-proteasome system is responsible for not only catalysis of exhaustive degradation of target proteins, but also for generation of antigenic peptides for presentation of the major histocompatibility complex (MHC) class I-restricted antigen, implying a distinct role of the proteasome in processing as well as protein degradation. But the latter function may be obligatory for use of products by the immune system [22]. Moreover, the proteasome has a new function as a real processing enzyme, namely for activation of transcriptional factor NF-κB through conversion of the p105 precursor to the mature p50 protein [31]. Transcriptional factor NF-κB1 is synthesized as a p105 precursor, which is processed by degradation of its COOH-terminal domain containing ankylin repeats, and the resulting p50 NH_2-terminal domain is associated with a partner p65 protein to function as the active transcriptional factor NF-κB. It is still unknown how the proteasome specifically cleaves the COOH-terminal half [32]. Notch is also an integral membrane protein receptor, which on association with another membrane-spanning ligand protein, delta, activates various genes responsible for determination of a cell's fate during development. The cytoplasmic portion of the Notch receptor containing ankylin repeats becomes detached from plasma membranes in a signal-dependent fashion, and the mediator is generated to be transported into the nucleus to activate transcription of target genes, suggesting that proteolytic cleavage plays a critical role in Notch signaling [32]. Generation of the cytoplasmic domain of Notch was strongly inhibited by MG132 (Z-LLL-CHO), suggesting that the proteasome is involved in the Notch receptor-signaling pathway through limited proteolysis [33]. It is interesting to consider how the 26S proteasome can catalyze the limited cleavage reaction. Multiubiquitination occurs before the digestion of NF-κB, although this has not yet been demonstrated for the Notch receptor, so the degradation is mediated by trapping with the 26S proteasome. Therefore, the regulatory subunit may have another function to promote the folding of proteins escaping form degradation. Alternatively, the NH_2-terminal part of the NF-κB may have a structure that is unable to unfold, preventing its transfer into the 26S proteasome complex for breakdown. However, the latter possibility cannot explain the limited digestion of the Notch receptor without degradation of the cytosolic tail domain, when the protein is actually processed by the proteasome.

C. The Football-Shaped Proteasome

The 20S proteasome has latent peptidase activity, but its assembly into the 26S proteasome complex results in full peptidase activity [34]. A 20S proteasome activator protein has been isolated and named PA28 (also named the 11S regulator) [35,36]. ATP energy is required for formation of the 26S proteasome, whereas activation by PA28 is ATP-independent. Thus PA28 can reversibly associate with the 20S proteasome in the absence of ATP and dramatically stimulate its peptide-hydrolyzing activities, but it does not support the breakdown of large proteins, even when the proteins have already been denatured or ubiquitinated. PA28 association with the proteasome resulted in both an increase in the V_{max} and a decrease in the K_m of the peptidase activities of the 20S proteasome, suggesting that it functions as a positive allosteric effector.

PA28 is composed of subunits of 28–32 kDa, but the native molecule has a molecular weight of about 170–180 kDa, suggesting that it is an oligomeric complex consisting of at least six copies of the PA28 subunit [37]. Purified PA28 contains two 28- to 32-kDa polypeptide isoforms, α and β. PA28α and PA28β are approximately 47% identical and are highly conserved between the rat and human [38]. Surprisingly, PA28α and PA28β are homologous with a previously described protein, the Ki antigen, which was identified as a nuclear protein detected with autoantibodies in sera of patients with systemic lupus erythematosus [39]. The primary structures of these three proteins, of similar sizes and charges, are approximately 40% identical (Fig. 4, lower panel), suggesting that they constitute a new protein family generated by duplication of a common ancestral gene. Although it remains to be determined whether Ki antigen is actually a component of PA28, an antibody specific for Ki antigen reacted appreciably with purified PA28 preparations [N. Tanahashi et al., to be published], suggesting that it is a component of PA28; therefore, Ki antigen was renamed PA28γ. Thus, the PA28 activator is now thought to be present in cells as an heteropolymer, composed of two or three nonidentical, but homologous, subunits. The PA28α subunit expressed in *Escherichia coli* markedly stimulated latent 20S

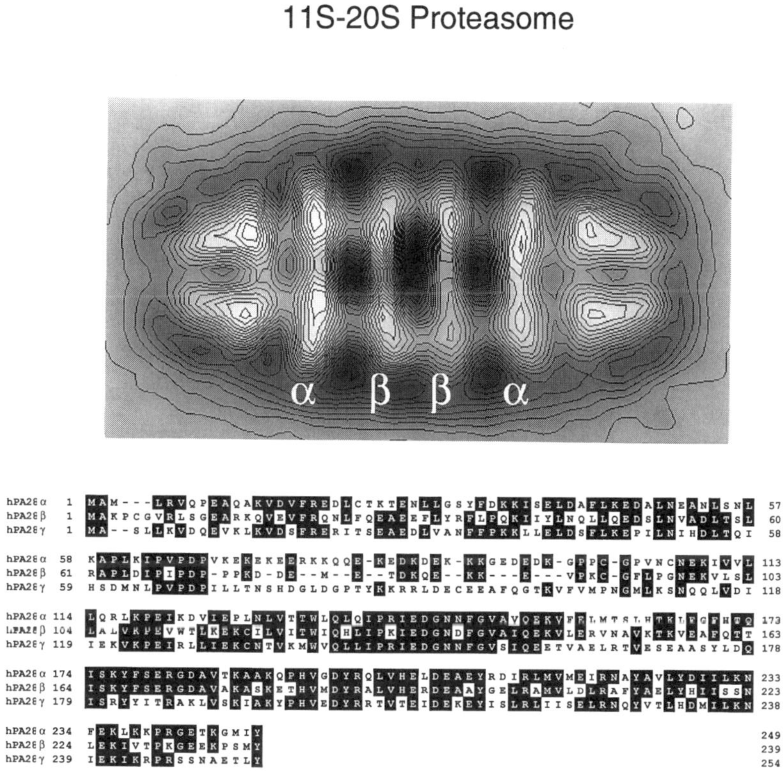

Figure 4 Structural features of the football-shaped proteasome: (upper panel) the 11S proteasome activator–20S proteasome complex. (Photograph courtesy of W. Baumesiter.) (Lower panel) sequence alignment of the human PA28 activator family proteins. Dark blocks indicate identical amino acid residues. The sequences of human PA28α, PA28β, and PA28γ are taken from Refs. 40, 38, and 39, respectively.

proteasome activity, suggesting that the homopolymeric complex of PA28α is also active. Conversely, PA28β alone had no appreciable effect, but appreciably reduced the K_m, but not the V_{max}, of PA28α-activated proteasome activity when added simultaneously with PA28α, suggesting an important functional interaction between these two PA28 subunits. The enzymatic function of PA28γ is currently unknown, because it had no effect on the proteasomal activity with or without PA28α or PA28β.

An average projection image of PA28 associated with the 20S proteasome is shown in Figure 4 (upper panel). PA28 forms a ring-shaped particle, about 10 nm wide at the base and 7 nm long from base to tip [37], and caps that are associated with the 20S proteasome at both or either end to give a football-like appearance. The overall structural organization of the football-like proteasome somewhat resembles that of the 26S dumbbell-shaped proteasome (see Fig. 3), indicating that PA28 occupies the same site on the 20S core particle as the 26S proteasomal regulator does on the 26S proteasome complex.

Interferon gamma markedly induces expressions of the mRNAs for PA28α and PA28β, leading to large increases in the amounts of their proteins [38,40]. This suggests that PA28 may be important in generation by the 20S proteasome of antigenic peptides capable of binding with MHC class I molecules. No PA28 activity was found in extracts of higher plants and the budding yeast. No gene homologous with a PA28 gene was found in the yeast genome, suggesting that PA28 genes may have been generated during evolution, in accordance with the appearance of immunoproteasomal genes and perhaps MHC/TAP genes. Thus, these newly identified IFN-γ-regulated activator genes, in combination with the three pairs of IFN-γ-regulated immunoproteasome genes, perhaps act synergically in antigen presentation [22]. Indeed transfections of cDNA for PA28α and for MHC-encoded proteasomal subunits increased the spectrum of peptides generated in antigen-presenting cells [41]. On the other hand, PA28γ, unlike PA28α and PA28β, may be responsible for functions unrelated to immunity because its protein level was very high in certain, but not all, tumor cells, implying its involvement in cell proliferation [N. Tanahashi et al., to be published]. However, the exact function of PA28γ is still unknown. The PA28 activator does not affect the rate of hydrolysis of native proteins, and thus, it does not function in regulation of initial cleavages of proteins, suggesting its cooperation in primary digestion by the 26S dumbbell-shaped ATP-dependent proteasome.

D. Similarity to Bacterial ATP-Dependent Proteases

Heat treatment of prokaryotic cells or their exposure to an amino acid analogue increases their productions of several proteases and induces massive proteolysis. This is presumed to be essential for the rapid removal of damaged proteins to maintain homeostasis of the living cells. Two well-studied stress proteases have been documented in *E. coli*: *lon* protease and Clp protease [42,43]. The *lon* protease (also named protease La) is an ATP-dependent protease and is a typical heat-shock protein with a seryl-protease domain and ATPase domain within its single polypeptide in a tetrameric complex. Clp protease (also called protease Ti) is a two-component ATP-dependent protease complex, consisting of ClpP, a stress-inducible serine catalytic subunit, and ClpA or ClpX, both of which are ATPase subunits. They form two ATP-dependent proteases, ClpAP and ClpXP [43], that share the proteolytic subunit ClpP, but the two enzymes degrade different proteins in vivo and in vitro, indicating that ATPase subunits confer ability for substrate recognition. ClpA contains two highly conserved ATP-binding motifs, one responsible for oligomerization and the other for ATP hydrolysis, whereas ClpX contains only one ATP-binding motif. Interestingly, both ClpA and ClpX alone act as molecular chaperones to promote protein folding [44,45], implying that Clp ATP-dependent protease is a noncovalent complex of seryl-protease and molecular chaperone. These components may have bifunctional roles, catalyzing refolding and destruction of target proteins with aberrant structures. ClpP is not only stress-inducible [43], but also catalyzes the bacterial NH_2-end rule pathway [6], implying its involvement in various cellular processes.

ClpP, with an average diameter of 11 nm and a height of 10 nm, is composed of two rings of seven subunits superimposed in bipolar fashion along the axis of rotational symmetry [46]. This structure is similar to that formed by the β-subunits of proteasomes and is reminiscent of the bipolar sevenfold symmetric structure of GroEL. ClpAP complexes contain a tetradecamer of ClpP, flanked at both ends by a hexamer of ClpA [46]. In the complex, the ClpP and ClpA rings are aligned axially to form a barrel-like structure, strikingly similar to that of the 26S dumbbell proteasome or the football proteasome complex (see Figs. 2 and 4). In ClpAP protease, ATPase complexes are attached to one or both ends of the protease complex, and their complex formation is dependent on ATP and $MgCl_2$, as in the assembly of the 26S proteasome complex [18,34,42].

FtsH is a bacterial ATP-dependent protease important in stress-induced degradation of transcription factor ρ^{32} [47]. It is an essential ATP-dependent protease in *E. coli*. This single polypeptide contains a membrane-anchoring domain, a metal-binding site in the protease domain, and an ATPase domain, indicating that it is a membrane-bound ATP-dependent metaloprotease. An interesting feature of FtsH is that its ATPase domain is closely related to some of the regulatory ATPase subunits of the eukaryotic 26S proteasome, both ATPase domains belonging to the AAA (*A*TPase *a*ssociated with a variety of cellular *a*ctivities) family [48]. The molecular chaperones DnaK, DnaJ, and GrpE are also involved in ρ^{32} degradation, and act cooperatively in presenting a substrate to FtsH protease. Furthermore, FtsH is involved in protein assembly into and through membranes, suggesting that it has chaperone activity similar to that of ClpA and ClpX [49].

Very recently, the *HslU–HslV* complex was found to be an ATP-dependent protease in *E. coli*. These two genes, encoded within the same operon, were originally found as heat-shock–responsive genes [50,51]. HslV is a 19-kDa protein similar to proteasome β-subunits, and HslU, a 50-kDa protein, shares significant homology with the ATPase subunits of eukaryotic 26S proteasomes and bacterial ClpX. Both HslV and HslU appear to form homopolymers, which then form a complex. Electron micrographs of active fractions containing only HslV or HslU revealed a homogeneous population of ring-shaped particles, with an approximate diameter of 11 nm [50]. The particles have six- to sevenfold rotational symmetry and viewed from the top, are very similar to the 20S proteasome and ClpAP proteases.

Thus, partial sequence similarity has recently been found between eukaryotic proteasomes and several bacterial ATP-dependent proteases, despite the absence of overall sequence homology, and recent analysis has further indicated striking similarities in their gross structures, such as in their sizes and shapes, suggesting their close functional relations.

IV. PROTEOLYSIS AND MOLECULAR CHAPERONES

The main unsolved problem in the proteolysis is how individual proteins in cells are degraded at different rates. As discussed here, the fate of each protein in cells seems closely related to the functions of molecular chaperones, from governing the folding of newly synthesized proteins to their destruction, because most bacterial ATP-dependent proteases belong to a family of so-called stress proteins, or have cooperative roles with other molecular chaperones [52,53]. Chaperone proteins are believed to prevent nonspecific aggregation of damaged or misfolded proteins by their binding to hydrophobic regions exposed in partially undenatured proteins. This may occur by buffering and reducing the concentration of unfolded protein. In addition, chaperones may be involved in the process of proper folding of newly synthesized protein. This process contains successive repeats of foldings and refolding of the target protein. Thus, the chaperone has the rather paradoxical function of "unfolding" to facilitate proper "folding" of the target protein. From the viewpoint of protein destruction, the unfolding ability of the chaperone molecule is of particular interest. For example, it was proposed that if the GroEL–GroES chaperonin fails to catalyze the proper folding of a protein, the molecular chaperones may, in turn, accelerate the rate of its degradation by maintaining the unfolded state, or even by actively unfolding polypeptides, thereby facilitating access to the ATP-dependent proteolytic system in *E. coli* [53].

In addition, certain chaperones are required for both efficient ubiquitination and subsequent degradation in eukaryotic cells. The yeast DnaJ homologue Ydj facilitates phosphorylation-dependent ubiquitination of G1 cyclin Cln3, and Sis—another yeast DnaJ homologue—and Hsc73 promotes the destruction of ubiquitinated Cln3 by the 26S proteasome [54; for review see Ref. 55]. Moreover, immunodepression of Hsc73 from a reticulocyte lysate diminished both the efficiency of ubiquitination and the rate of degradation of certain ubiquitinated substrates [55]. These findings indicate that molecular chaperones may be involved in multiple steps in the destruction of proteins in a highly selective proteolysis pathway mediated by Ub and the proteasome. However, it is unknown whether these chaperone proteins are involved in the degradations of other proteins by the Ub pathway, or of proteins with as yet uncharacterized, degradation signals other than the multi-Ub chain.

On the other hand, the catalytic site of the 26S proteasome exists deep inside the hole of the 20S proteasome, so unfolding of target proteins must be necessary for polypeptide import to the active site. Thus, the 26S proteasomal regulatory complex may have chaperone-like activity, inducing protein unfolding, and resulting in polypeptide transport to the catalytic site. The energy required for ATP-dependent proteolysis is thought to be used for unfolding and translocating the appropriate targets, as well as for formation of the protease complex itself.

V. PERSPECTIVES

The life-spans of all proteins in the cell differ, suggesting that individual proteins have distinct susceptibility for degradation. If so, what is the biochemical basis of the heterogeneity in their degradation rates? The primary structures of proteins code not only genetic information to replicate polypeptides exactly, but also subsequent information for constructing the three-dimensional structure of functional proteins in cells. Clarification of the latter mechanism is currently defined as the experimental project to resolve "the second genetic code." The heterogeneities in the half-lives of intracellular proteins also may imply that this amino acid sequences encode additional information for determining their stabilities, although it is unlikely that each protein has a distinct degradation signal or that the proteases alone determines the various rates of degradation of target proteins. Studies on the structural basis of instability imposed in the primary structure may be thought as the experimental project for deciphering the "third genetic code."

Determination of the life spans of given proteins and methods for their alterations would be useful in many fields of life science. This may be possible, because, during evolution, such alterations of proteins have occasionally occurred during adaptation to environmental conditions. For instances, c-Fos and c-Jun are both immediate-early gene products required for G_1 to S transition in the cell cycle, and both have very short half-lives of 10–30 min. The instabilities of these short-lived proteins are essential for normal cell cycle progression. c-Fos has the PEST sequence in its COOH-terminal region [56], and c-Jun has the δ-domain in its NH_2-terminal region [57]. Both of these sites lead to rapid degradation mediated by the Ub-proteasome pathway, although the PEST sequence is thought to be the common degradation signal in many short-lived proteins [58]. Interestingly, the primary structures of the corresponding oncoprotein forms, v-Fos and v-Jun, resemble those of c-Fos and c-Jun, respectively, except for the sequences for instability, such as PEST and the δ-domain. Thus, their oncoproteins alter their life span by frame-shirt mutations to maintain their longevity in cells, suggesting a novel oncogenic pathway. Therefore, if the principles for instability imposed in the genetic code were known, changes in the life span of desired proteins in cells would be possible.

Unfortunately, little is known about how cellular proteins are degraded in cells. It is not obvious whether degradation of proteins is due to aging corresponding to their usage, other than their impairment induced by stresses, because it has not yet been demonstrated that old proteins are degraded faster than newly synthesized proteins with the same sequences in cells. Conversely, most proteins are reported to be degraded by a first-order reaction, suggesting that their breakdown may depend on their concentrations in cells, irrespective of whether they are old or newly synthesized. Because these points are so important in considering the life spans of proteins in cells, studies are required on molecular changes of proteins with age, and changes in stability of proteins dependent on the times of their syntheses. However, as discussed here, we believe that molecular chaperones help or support not only the formation of the tertial structure of proteins, but also the selective destruction of intracellular proteins. The cooperation of molecular chaperones with a supramolecular proteolysis system, such as the Ub-proteasome pathway, in the protein death machinery is thought to contribute to deciding the fate of proteins in cells, although it is still unknown how the chaperone-mediated proteolysis system is responsible for determining the life span of proteins.

REFERENCES

1. Finley, D. and Chau, V., Ubiquitination, *Annu. Rev. Cell Biol. 7*:25–69 (1991).
2. Finley, D., Ciechanover, A., and Varshavsky, A., Thermolability of ubiquitin-activating enzyme from the mammalian cell cycle mutant ts85, *Cell 37*:43–55 (1984).
3. Hershko, A. and Ciechanover, A., The ubiquitin system for protein degradation, *Annu. Rev. Biochem. 61*:761–807 (1992).
4. Jentsch, S., The ubiquitin-conjugation system, *Annu. Rev. Genet. 26*:179–207 (1992).
5. Jungmann, J., Reins, H.-A., Schobert, C., and Jentsch, S., Resistance to cadmium mediated by ubiquitin-dependent proteolysis, *Nature 361*:369–371 (1993).
6. Varshavsky, A., The N-end rule, *Cell 69*:725–735 (1992).
7. Huibregtse, J. M., Scheffner, M., Beaudenon, S., and Howley, P. M., A family of proteins structurally and functionally related to the E6-AP ubiquitin–protein ligase, *Proc. Natl. Acad. Sci. USA 92*:2563–2567 (1995).
8. Hochstrasser, M., Ubiquitin, proteasomes, and the regulation of intracellular protein degradation, *Curr. Opin. Cell Biol. 7*:215–223 (1995).
9. Murray, A. W., Cyclin ubiquitination: The destructive end of mitosis, *Cell 81*:149–152 (1995).
10. Orlowski, M., The multicatalytic proteinase complex, a major extralysosomal proteolytic system, *Biochemistry 29*:10289–10297 (1990).
11. Rivett, A. J., Proteasomes: Multicatalytic proteinase complexes, *Biochem. J. 291*:1–10 (1993).
12. Hilt, W. and Wolf, D. H., Proteasomes: Destruction as a programme, *Trends Biochem. Sci. 21*:96–102 (1995).
13. Tanaka, K., Molecular biology of proteasomes, *Mol. Biol. Rep. 21*:21–26 (1995).

14. Tanaka, K. and Tanahashi, N., The 20S proteasome: Subunits and functions, *Adv. Mol. Cell. Biol.* (in press, 1997).
15. Lupas, A., Koster, A. J., and Baumeister, W., Structural features of 26S and 20S proteasomes, *Enzyme Protein 47*:252–273 (1993).
16. Tamura, T., Nagy, I., Lupas, A., Lottspeich, F., Cejka, Z., Schoofs, G., Tanaka, K., Mot, R. D., and Baumeister, W., The first characterization of a eubacterial proteasome: The 20S complex of *Rhodococcus* sp. strain NI86/21, *Curr. Biol. 5*:766–774 (1995).
17. Coux, O., Tanaka, K., and Goldberg, A. L., Structure and functions of the 20S and 26S proteasomes, *Annu. Rev. Biochem.* 65:801–847 (1996).
18. Peters, J.-M., Proteasomes: Protein degradation machines of the cells, *Trends Biochem. Sci. 19*:377–382 (1994).
19. Tanaka, K., Yoshimura, T., Tamura, T., Fujiwara, T., Kumatori, A., and Ichihara, A., Possible mechanism of nuclear translocation of proteasomes, *FEBS Lett. 271*:41–46 (1990).
20. Nederlof, P. M., Wang, H.-R., and Baumeister, W., Nuclear localization signals of human and *Thermoplasma* proteasomal α subunits are functional in vitro, *Proc. Natl. Acad. Sci. USA 92*:12060–12064 (1995).
21. Löwe, J., Stock, D., Jap, B., Zwickl, P., Baumeister, W., and Huber, R., Crystal structure of the 20S proteasome from the archaeon *T. acidophilum* at 3.4 Å resolution, *Science 268*:533–539 (1995).
22. Tanaka, K., Tanahashi, N., Tsurumi, C., Yokota, K., and Shimbara, N., Proteasomes and antigen processing, *Adv. Immunol.* 64:1–38 (1996).
23. Wenzel, T. and Baumeister, W., Conformational constraints in protein degradation by the 20S proteasome, *Nature Struct. Biol. 2*:199–204 (1995).
24. Weissman, J. S., Sigler, P. B., and Horwich, A. L., From the cradle to the grave: Ring complexes in the life of a protein, *Science 268*:523–524 (1995).
25. Kubota, H., Hynes, G., and Willison, K., The chaperonin containing *t*-complex polypeptide (TCP-1). Multisubunit machinery assisting in protein folding an assembly in the eukaryotic cytosol, *Eur. J. Biochem. 230*:3–16 (1995).
26. Hartl, F. U., Molecular chaperones in cellular protein folding, *Nature 381*:571–580 (1996).
27. Dubiel, W., Ferrell, K., and Rechsteiner, M., Subunits of the regulatory complex of the 26S proteasome, *Mol. Biol. Rep. 21*:27–34 (1995).
28. van Nocker, S., Deveraux, Q., Rechsteiner, M., and Vierstra, R., *Arabidopsis* MBP1 gene encodes a conserved ubiquitin recognition component of the 26S proteasome, *Proc. Natl. Acad. Sci. USA 93*:856–860 (1996).
29. Ferrell, K., Deveraux, Q., Nocker, S. V., and Rechsteiner, M., Molecular cloning and expression of a mutiubiquitin chain binding subunit of the human 26S protease, *FEBS Lett. 381*:143–148 (1996).
30. Murakami, Y., Matsufuji, S., Kameji, T., Hayashi, S., Igarashi, K., Tamura, T., Tanaka, K., and Ichihara, A., Ornithine decarboxylase is degraded by the 26S proteasome without ubiquitination, *Nature 360*:597–599 (1992).
31. Palombella, V. J., Rando, O. J., Goldberg, A. J., and Maniatis, T., The ubiquitin–proteasome pathway is required for processing the NF-kappa B1 precursor protein and the activation of NF-kappa B, *Cell 78*:773–785 (1994).

32. Pahl, H. L. and Baeuerle, P. A., Control of gene expression by proteolysis, *Curr. Opin. Cell Biol. 8*:340–347 (1996).
33. Kopan, R., Schroeter, E. H., Weintraub, H., and Nye, J. S., Signal transduction by activated mNotch: Importance of proteolytic processing and its regulation by the extracellular domain, *Proc. Natl. Acad. Sci. USA 93*:1683–1688 (1996).
34. Orino, E., Tanaka, K., Tamura, T., Sone, S., Ogura, T., and Ichihara, A., ATP-dependent reversible association of proteasomes with multiple protein components to form 26S complexes that degrade ubiquitinated proteins in human HL-60 cells, *FEBS Lett. 284*:206–210 (1991).
35. Ma, C.-P., Vu, J. H., Slaughter, C. A., and DeMartino, G. N., Identification, purification, and characterization of a protein activator (PA28) of the 20S proteasome (macropain), *J. Biol. Chem. 267*:10515–10523 (1992).
36. Dubiel, W., Pratt, G., Ferrell, K., and Rechsteiner, M., Purification of an 11S regulator of the multicatalytic protease, *J. Biol. Chem. 267*:22369–22377 (1992).
37. Gray, C. W., Slaughter, C. A., and DeMartino, G. N., PA28 activator protein forms regulatory caps on protease stacked rings, *J. Mol. Biol. 236*:7–15 (1994).
38. Ahn, J. Y., Tanahashi, N., Akiyama, K., Hisamatsu, H., Noda, C., Tanaka, K., Chung, C. H., Shimbara, N., Willy, P. J., Mott, J. D., Slaughter, C. A., and DeMartino, G. N., Primary structure of two homologous subunits of PA28, a γ-interferon-inducible protein activator of the 20S proteasome, *FEBS Lett. 366*:37–42 (1995).
39. Nikaido, T., Shimada, K., Shibata, M., Hata, M., Sakamoto, M., Takasaki, Y., Sato, C., Takahashi, T., and Nishida, Y., Cloning and nucleotide sequence of cDNA for Ki antigen, a highly conserved nuclear protein detected with sera from patients with systemic lupus erythematosus, *Clin. Exp. Immunol. 79*:209–214 (1990).
40. Realini, C., Dubiel, W., Pratt, G., Ferrell, K., and Rechsteiner, M., Molecular cloning and expression of a γ-interferon-inducible activator of the multicatalytic protease, *J. Biol. Chem. 269*:20727–20732 (1994).
41. Groettrup, M., Soza, A., Eggers, M., Kuehn, L., Dick, T. P., Schild, H., Rammensee, H.-G., Koszinowski, U. H., and Kloetzel, P.-M., A role for the proteasome regulator PA28α in antigen presentation, *Nature 381*:166–168 (1996).
42. Goldberg, A. L., The mechanism and functions of ATP-dependent proteases in bacterial and animal cells, *Eur. J. Biochem. 203*:9–23 (1992).
43. Gottesman, S. and Maurizi, M. R., Regulation by proteolysis: Energy-dependent proteases and their targets, *Microbiol. Rev. 56*:592–621 (1992).
44. Wawrzynow, A., Wojtkowiak, D., Marszalek, J., Banecki, B., Jonson, M., Graves, B., Georgopoulos, C., and Zylicz, M., The ClpX heat-shock protein of *Escherichia coli*, the ATP-dependent substrate specificity component of the ClpP–ClpX protease, is a novel molecular chaperone, *EMBO J. 14*:1867–1877 (1995).
45. Wickner, S., Gottesman, S., Skowyra, D., Hoskins, J., McKenny, K., Maurizi, M. R., A molecular chaperone, ClpA, functions like DnaK and DnaJ, *Proc. Natl. Acad. Sci. USA 91*:12218–12222 (1994).
46. Kessel, M., Maurizi, M. R., Kim, B., Kocsais, E., Trus, B. L., Singh, S. K., and Steven, A. C., Homology in structural organization between *E. coli* ClpAP protease and the eukaryotic 26S proteasome, *J. Mol. Biol. 250*:587–594 (1995).
47. Tomoyasu, T., Gamer, J., Bukau, B., Kanemori, M., Mori, H., Rutman, A. J., Oppenheim, A. B., Yura, T., Yamanaka, K., Niki, H., Hiraga, S., and Ogura, T.,

Escherichia coli FtsH is a membrane-bound, ATP-dependent protease which degrades the heat shock transcription factor ρ^{32}, *EMBO J.* *14*:2551–2560 (1995).
48. Confalonieri, F. and Duguet, M., A 200-amino acid ATPase module in search of a basic function, *BioEssay* *17*:639–650 (1995).
49. Akiyama, Y., Shirai, Y., and Ito, K., Involvement of FtsH in protein assembly into and through the membrane, *J. Biol. Chem.* *269*:5225–5229 (1994).
50. Rohrwild, M., Coux, O., Huang, H.-C., Moerschell, R. P., Yoo, S. J., Seol, J. H., Chung, C. H., and Goldberg, A. L., HslV–HslU: A novel ATP-dependent protease complex in *Escherichia coli* related to the eukaryotic proteasome, *Proc. Natl. Acad. Sci. USA* *93*:5808–5813 (1996).
51. Yoo, S. J., Soel, J. H., Shin, D. H., Rohrwild, M., Kang, M.-S., Tanaka, K., Goldberg, A. L., and Chung, C. H., Purification and characterization of the heat shock proteins HslV and HslU that form a new ATP-dependent protease in *Escherichia coli*, *J. Biol. Chem.* *271*:14035–14040 (1996).
52. Sherman, M. Y. and Goldberg, A. L., Involvement of the chaperonin DnaK in the rapid degradation of a mutant protein in *Escherichia coli*, *EMBO J.* *11*:71–77 (1992).
53. Kandror, O., Busconi, L., Sherman, M., and Goldberg, A. L., Rapid degradation of an abnormal protien in *Escherichia coli* involves the chaperones GroEL and GroES, *J. Biol. Chem.* *269*:23575–23582 (1994).
54. Yaglom, J. A., Goldberg, A. L., Finley, D., and Sherman, M. Y., The molecular chaperone Ydj1 is required for the p34cdc28-dependent phosphorylation of the cyclin Cln3 that signals its degradation, *Mol. Cell. Biol.* *16*:3679–3684 (1996).
55. Hayes, S. A. and Dice, J. F., Roles of molecular chaperones in protein degradation, *J. Cell Biol.* *132*:255–258 (1996).
56. Tsurumi, C., Ishida, N., Tamura, T., Kakizuka, A., Nishida, E., Okumura, E., Kishimoto, T., Inagaki, M., Okazaki K., Sagata, N., Tanaka, K., and Ichihara, A., Degradation of c-Fos by the 26S proteasome is accelerated by c-Jun and multiple protien kinases, *Mol. Cell. Biol.* *15*:5682–5687 (1995).
57. Baichwal, V. R. and Tjian, R., Control of c-Jun activity by interaction of a cell-specific inhibitor with regulatory domain delta: Differences between v- and c-Jun, *Cell* *63*:815–825 (1990).
58. Rechsteiner, M. and Rogers, S. W., PEST sequences and regulation by proteolysis, *Trends Biochem. Sci.* *21*:267–271 (1996).

22

Structure and Function of Small Heat-Shock Proteins

Monika Ehrnsperger
and Johannes Buchner
Universität Regensburg, Regensburg, Germany

Matthias Gaestel
Max-Delbrück-Centrum für Molekulare Medizin, Berlin, Germany

I. INTRODUCTION

Small heat-shock proteins (sHsps) represent an abundant and ubiquitous family of stress proteins. They have been detected in virtually all types of organisms; however, the number of members in different species is variable. Although the overall homology between different sHsps is rather low, they are grouped together based on (a) conserved regions in the COOH-terminal half of the protein [1–3], (b) their increased expression after heat shock [4–9], and (c) their monomeric size (15–40 kDa). In this review we will also include αA-crystallin. Because of its structural and functional similarity, αA-crystallin is considered a member of the sHsp family, although no increased expression with heat shock has yet been detected. Despite their low monomeric molecular weights, sHsps exist within the cell as large oligomeric complexes of up to 50 subunits. Molecular masses of 150–800 kDa have been reported under physiological conditions [10–13]. Following heat shock or several other stress conditions, some sHsps become phosphorylated and, in parallel, the oligomeric size undergoes significant changes, leading to either smaller or larger complexes [14–19].

The abundance of sHsps under physiological conditions varies significantly, depending on cell type and organism studied [20,21]. Furthermore, the expression of sHsps can vary, depending on development, growth conditions, differentiation, and the oncogenic status of the cell [8,22–27]. Under heat-shock conditions, the level of sHsps increases drastically 10- to 20-fold [28,29], which suggests functional importance in supporting survival at higher temperatures [30,31]. Interestingly, overexpression of virtually all mammalian sHsps conveys a significant degree of thermoresistance to cells [6,32–34], indicating a general thermoprotective function of these proteins.

The sHsps function in several different, seemingly unrelated processes, ranging from RNA stabilization [16] to elastase inhibition [35,36], interaction with the cytoskeleton [37–41], and involvement in the regulation of the platelet factor XIII [42]. In addition, sHsps are involved in several pathological conditions [26], including multiple sclerosis [43] and apoptosis [44,45].

In vitro sHsps, including α-crystallin, which was previously thought to play only a structural role in the eye lens, exhibit molecular chaperone properties, similar to the model chaperone GroEL [3,35,46,47]. However, sHsps chaperone protein folding in an ATP-independent manner [12,35,46,47]. Little is known about the molecular mechanisms of this chaperone function and their integration into the chaperone system of the cell for which only the coordinated function of several chaperone systems seems to guarantee survival under heat-shock conditions.

II. CHARACTERISTICS OF SMALL HEAT SHOCK PROTEINS

A. The sHsp Family

The sHsps constitute one of the four most dominant groups of Hsps [31,48,49]. Whereas in plants, many different classes of sHsps coexist in the cytosol as well as in organelles [50], yeasts (*Saccharomyces cerevisiae*) possess three putative sHsps (Hsp12, -26, -42) [51], and mammalian cells have only one sHsp in addition to the homologous αA and αB-crystallin [28]. In rat and humans another homologous protein (p20) has been identified [52]. In *Escherichia coli*, two related proteins (IbpA and IbpB) were described in association with overexpressed recombinant protein [53,54]. In plants, several classes of sHsps are found in different organelles (chloroplast, mitochondria, and endoplasmic reticulum, ER), as well as the cytosol [29,50–56], whereas in other eukaryotes, mostly cytosolic representatives are characterized. However, *Drosophila* Hsp22 and the Hsp30 of *Neurospora crassa* are localized in mitochondria [28,57], *Drosophila* Hsp27 is predominantly localized in the nuclear region [58], and recently, a mitochondrial sHsp has been identified in *Drosophila* [R. Tanguay, personal communication]. Furthermore, periplasmic homologues have been found in bacteria [59].

In contrast with other Hsp families, such as Hsp70, in which 50% identity can be observed between prokaryotic and eukaryotic proteins [48], sHsps show little sequence homology, even in closely related species [3,56,60]. Avian αB-crystallin has been reported to exhibit sequence homology not only with sHsps, but also with the Hsp70 family [61]. Interestingly, the numerous sHsps in plants have undergone significant diversification, so that substantial differences can be observed within members of one species [50,56].

For a compilation of representative members of the sHsp family see Table 1.

B. Expression and Regulation of sHsps

Except in the eye lens, the highest sHsp levels of about 0.1% of the total protein are reached in heart muscle cells under physiological conditions [20,80].

Studies on the expression of sHsps in *Drosophila* as well as mammalian cells have shown that this group of Hsps is among the most strongly induced heat-shock proteins after elevation of temperature and in response to other stress factors [7,8,9,28,81]. Final yields can be up to 1% of total protein, depending on the cells and tissue analyzed [28,29]. The regulation of sHsps induction is not yet fully understood. Apart from αA-crystallin, which is mainly expressed in the eye lens and is not heat-inducible, it is believed that transcriptional activation is important. As in all species yet tested, the regulatory heat-shock element (HSE) [82] was upstream from the sHsp genes [72,83–86]. The HSE motif is typical for heat-shock proteins, it consists of at least three inverted repeats of the sequence (nnGAAn), and activation of transcription is induced after binding of the trimeric heat-shock transcription factor (HSF) to the HSE. However, differences have been observed between sHsps and other Hsps relative to stress-induced transcriptional activation, suggesting that additional mechanism may be involved. Induction of sHsps is stimulated at slightly lower temperatures than Hsp70 [87], but in NIH 3T3 cells the stress-mediated accumulation of sHsp mRNA proceeds with slower kinetics than the Hsp70 mRNA [8], and the strong induction of Hsp27 in chicken embryo cells is, in contrast with other induced Hsps, not due to transcriptional activation [88]. A similar situation was described for the induction of Hsp25 in Ehrlich ascites tumor cells by the anticancer drug cisplatin. Here, Hsp25 synthesis on the level of mRNA translation is enhanced without changes in gene transcription [89].

Interestingly, the transcriptional regulation of Hsp27 is not only triggered by HSF, but also correlates with hormone levels. In this context, human Hsp27 was described as an estrogen-regulated 24- to 29-kDa protein [90] that is heat-inducible [91] and localizes in granules [92]. The identity of the estrogen 24- to 29-kDa protein has subsequently been established by cDNA analysis [91] and immunochemical methods [93,94]. Interestingly, the mouse Hsp25- and the human Hsp27 promoter contain an estrogen-responsive half-palindromic motif [72] that could be functionally active by interacting with the TATA box [95]. The regulation of Hsp25 by estrogen is probably responsible for its cell type-specific

Table 1 Representative Members of the sHsp Family

	Protein	Organism	Comments	Ref.
Prokaryotes	IbpA/IbpB	*E. coli*	Induced by heat shock, found associated with inclusion bodies	53,54
	Hsp16.3	*M. tuberculosis*	Immunodominant surface antigen, trimer of trimers; chaperone activity	13,62
Eukaryotes Yeast	Hsp26	*S. cerevisiae*	Inhibits actin polymerization, deletion leads to no phenotype	63,64
	Hsp42	*S. cerevisiae*	Expressed in unstressed cells, putative Rap1p-interacting protein	51
	Hsp9	S. pombe	Complements a temperature-sensitive mutant	65,66
Invertebrates	Hsp27	*D. melanogaster*	Constitutively expressed, hormone- and stress-induced, differentially expressed during development	28,67–69
Vertebrates	αA-/αB-crystallin	All vertebrates	Major lens protein, also present in nonlenticular tissue, only αB–stress-inducible, involved in thermotolerance and several diseases, (auto-) phosphorylated, chaperone function	20,34,46,47,70
	Hsp25/27	Mammals	Involved in thermotolerance, phosphorylated, induced in several diseases, chaperone function	26,32,35,47,71–73
	p20	Mammals	Homologues to Hsp27, forms oligomers, redistribution after heat shock	52
Plants	Hsp22	*Glycine max*	Endomembrane localized, ER retention signal	74,75
	Hsp18.1	*P. sativum*	Cytosolic, class I, chaperone activity	12,29,56,76
	Hsp22	*P. sativum*	Mitochondrial, induced only under extreme growth conditions	55,77,78
	Hsp21	*A. thaliana*	Chloroplast, putative "methionine bristle" structure motif	79

expression in the rat uterus during pregnancy [96]. Differential expression of sHsps in *Drosophila melanogaster* (Dm) during development has been reported to correlate with the concentration of the hormone ecdysterone [97]. Mutation studies on the gene structure of sHsps revealed that different promoter regions were responsible for heat- and hormone-induced up-regulation of Dm-Hsp26 [98,99]. On the other hand, the enhancement of stress-induced Hsp27 and αB-crystallin synthesis by modulators of the arachidonic acid cascade, is regulated by the HSF–HSE system [100]. Furthermore, the 5′ flanking regions of sHsp genes contain sequences similar to those found in an osmoregulated *E. coli* gene [85], and a consensus target sequence for the transcription factor NF αB, which is thought to be involved in tumor necrosis factor-alpha (TNF-α) response [101]. Not only is the control of sHsp expression diverse, but the regulation of αB-crystallin and Hsp27 expression can be either independent or coordinated, according to the stress factors applied [101]. The sHsps thus seem to be regulated by several independent systems, indicating a wide set of functions, which need further characterization in the future.

C. Structural Features of sHsps

1. *The α-Crystallin Domain*

Conservation of the Carboxy-Terminal Part of sHsps. The homology between different sHsps is mostly restricted to the COOH-terminal half of the protein, including a conserved, flexible COOH-terminal extension [1,2,102–104]. This putative domain, consisting of about 100 residues, is sometimes termed the "α-crystallin domain" (Fig. 1). Recently, however, short stretches with sequence homology have been identified in the most NH_2-terminal part of several sHsps [76] (Fig. 2).

By comparing the sequences of several sHsps, Wistow [102] concluded that sHsps are probably derived from an ancestral gene by gene duplication and, based on the exon–intron structure, the protein is most likely composed of two domains, each comprising two structurally related motifs. According to this hypothesis, domain I of sHsps corresponds to exon I of α-crytallin (residues 1–65) or to residues 1–87 of human Hsp27. The NH_2-terminal part of the sHsps was

```
(1)  ELxVxVxDxxVLxIxGxR-x(12-24)-FxRRFxLP-x(7-9)-VxAxLxxDGVLTVxxPK
(2) EEVxVxVxxxxxLQIxGxx-x(12-24)-FxRRFxLP-x(7-9)-VxxxxxxxGVLTVxxPK
     DL I I     VEV              Y KKY I          I        M SI  R
          L     I                                 L           L
```

Figure 1 Conserved residues within the COOH-terminal domain of sHsp family members: The motif is shown as formulated by Plesofsky-Vig et al. (1) or by de Jong et al. (2). Alternate common residues at several positions as indicated in the latter reference are also shown.

```
E.coli IbPA 16.0                   1-MR...NFDLSPLYRSAIGFDRLFNHLENNQSQSNG
E.coli IbPB 16.0                   1-MR...NFDLSPLMRQWIGFDKLANALQNAGESQSF
Arabidopsis 17.6         1-MSLIPSIFGGRR.TNVFDPFSL....DVFDPFEGFLTPSGLANAP
Citrus maxima 20.0       1-MSLIPSIFGGRR.TNVFDPFSL....DVWDPFDGFLSSALTNAPS
Chenopodium rubrum 1-MSLIPNNWFNTGRR.SNIFDPFSL...DEIWDPFFGLPSTLSTVPRS
Carrot 17.7              1-MSIIPSFFGGRR.SNVFDPFSL....DVWDPFKDFPLVTSSASEF
Soybean 18.5             1-MSLIPNFFGGRR.NNVFDPFSL....DVWDPFKDFPFPNTLSSAS
Soybean 22.3            25-ALMPYTRSTLWDMM.LQSEDPFGI....LEQNPFNNIPNIRGGAETL
Sunflower 17.6           1-MSIIPSFFTSKR.SNIFDPFSL....DTWDPFQGIISTEPARETA
Barley 17.0                     1-MSIVRR.SNVLDPFAD....LWADPLDTFRSIFPAISGG
Tomato                   1-MSLIPRIFGDRRSSSMFDPFSI....DVFDPFRELGFPSTNSGES
Alphaalpa 18.2           1-MSLIPSFFGGRR.SNVFDPFSL....DVWDPFKDFPFNNSALSAS
Maize 17.2                      1-MSLVRR.SNVFDPFSM....DLWDPFDTMFRSIVPSATS
Tobacco 18.0             1-MAMIPSFFGGRR.SNIFDPFSL....DIFDPFEGFPFSGTVANVP
P.sativum 17.9            2-IIPRVFGTGRR.TNAFDPFSL....DLWDPFQNFQLARSATGTT
P.glaucum 16.9                  1-MSLVRR.GNVFDPFSM....DLWDPFDNMFRSIVPSSSS
Douglas fir               1-MSIIPSFFGGRRSSSAFDPFSL....DVWDPFRAFTDLSGGGPSG
Opium                    1-MSIIPSFFSNQR.SNVFDPFSL....DIWDPFQGFPFSTGALTAN
Rice 16.9                       1-MSLVRR.SNVFDPFSL....DLWDPFDSVFRSVVPATSD
Wheat 16.9                      1-MSIVRR.SNVFDPFAD....LWADPFDTFRSIVPAISGG
N.crassa 30                 1-MALFPRGFYG.SYGSDPSFT.NLFRLLDDFDTYTREVQGSAPE
Gallus 25                          1-MAERR.....VPFTF.LTSPSWEPFRDWYHGSRLFDQS
Mouse 25                           1-MTERR.....VPFSL.LRSPSWEPFRDWYPAHSRLFDQ
Human 27                           1-MTERR.....VPFSL.LRGPSWDPFRDWYP.HSRLFDQ
Mouse αB-cry             1-MDIAIHHPWIRR...PFFPFHS..PSRLFDQFFGEHLLESDLFST
Human αB-cry             1-MDIAIHHPWIRR...PFFPFHS..PSRLFDQFFGEHLLESDLFPT

                                RR....FDPFSL......WDPF
                                      F           FE
```

Figure 2 Amino acid sequence alignment of the NH_2-terminal region of selected sHsps: A phenylalanine–proline-rich homology region could be identified. The consensus motif proposed is indicated below the sequence. (From M. Gaestel, unpublished data).

obviously more sensitive to evolutionary changes. However, some phenylalanine- and proline-rich sequences are conserved in the most NH_2-terminal part of a wide variety of sHsps (see Fig. 2.).

Function and Stability of the α-Crystallin Domain. The function of the conserved COOH-terminal domain of the sHsps is not yet clear. It has been suggested to be involved in the chaperone activity of the proteins. However, studies with isolated fragments, as well as truncation and mutation experiments, have led to the consensus that the α-crystallin domain might indeed be involved, but is not sufficient for this protective function. [36,105–107; M. Ehrnsperger, M. Gaestel, and J. Buchner, unpublished results] (see Sec. III.C). The COOH-terminal part of sHsps was suggested to be also important for complex formation [102]. Indeed, isolated fragments form defined oligomeric complexes [107,108], depending on a conserved COOH-terminal sequence [51]. Taken together, these findings indicate that the α-crystallin domain may be important, but is not sufficient, for organizing quaternary structure and, thereby, the functionality of sHsps.

An interesting feature of the α-crystallin domain is a COOH-terminal flexible extension of variable length [102–104]. The function of this extension is still unclear; however, truncation of the COOH-terminal 18 residues from mouse Hsp25 did not lead to significant changes in its oligomerization behavior or in vitro chaperone activity [M. Ehrnsperger, A. Kotlyarov, M. Gaestel, and J. Buchner, unpublished results]. Nuclear magnetic resonance (NMR) spectroscopic measurements with α-crystallin have shown that the flexible tail of the lens protein is not involved in intra- or intermolecular contacts of the native oligomeric structures [103,109]. The flexible tails were also proposed to be involved in membrane binding [110,111]. However, more recent observations challenged this hypothesis, for truncation of the COOH-terminal extension in α-crystallin did not lead to reduced membrane binding [112]. Le Breton and Carver [109] suggested that the polar extension might be a means of enhancing the solubility of the protein.

An explanation for the selective conservation of the α-crystallin domain was proposed to be the extreme thermal stability of the protein [102] because α-crystallin did not denature even at 100°C [113,114]; however, recent data indicate otherwise. From infrared spectroscopy, circular dichroism (CD), and calorimetry measurements, as well as size exclusion chromatography, Surewitz and Olesen [115] and Gesierich and Pfeil [116] showed that α-crystallin loses its native β-sheet structure in the temperature range between 50° and 70°C and, thus, seems no more stable than other crystallins [117,118]. This notion is consistent with the fact that the protein does not show exceptional stability when chemically denatured. It unfolds in 2.5–3 M guanidine hydrochloride and 4 M urca [119,120].

In summary, the function of the conserved COOH-terminal domain of sHsps is still unknown, but there are indications that it is involved in functional as well as structural aspects of the sHsps.

2. *Secondary and Tertiary Structure*

Although the primary sequences of many sHsps are known, their structural organization on a higher level is still poorly understood. The CD measurements have shown that the secondary structure of α-crystallin and mammalian sHsps are comparable, in that both show predominantly β-sheet structure, with less than 5% α-helix [35,115,121,122]. In addition, Fourier-transformed infrared analysis indicated 40–50% β-sheet and 10–20% α-helical structure [123].

Alignment of the hydrophobic residues of α- and β/γ-crystallins has led to a model for the tertiary structure of α-crystallin with six folding units arranged in a two- or three-domain structure [124]. Sequence comparison between different crystallins, together with secondary structure prediction and CD measurements, led to a four-motif structure for α-crystallin [121,125]. After analyzing gene structure and internal similarity Wistow [102] proposed that the structure of α-crystallin comprised two domains: a globular NH_2-terminal domain, consisting of two symmetry-related motifs, and a larger COOH-terminal domain also with two motifs and a flexible COOH-terminal extension. The NH_2-terminal domain is encoded by exon 1, whereas the COOH-terminal part corresponds to exons 2 and 3. Which of these models resembles sHsp secondary structure most closely remains to be determined, but unfolding studies with α-crystallin support Wistow's two-domain hypothesis [119].

Recently, a structural prediction for the COOH-terminal domain [60] was proposed in which the first half of the domain is composed of a short α-helix followed by three stretches of β-strands. This αβββ-motif is then repeated in the second half of the domain. A periplasmic molecular chaperone, Caf1M, of *Yersinia pestis* has significant sequence homology with the bacterial sHsps IbpA and IbpB, especially in its NH_2-terminal domain [59]. Caf1M belongs to a superfamily of periplasmic bacterial proteins, for one of which (PapD of *E. coli*) the three-dimensional structure has been solved by X-ray analysis [126,127]. An alignment of the known primary structure of the Ibps to this structure of PapD led to the suggestion that sHsps might form an immunoglobulin-like fold [59]. Furthermore, in the NH_2- and COOH-terminal regions of some sHsps, amphiphilic α-helices with a high hydrophobic moment were observed [57]. These might be involved in the interaction with membranes and other proteins.

3. Quaternary Structure

Oligomerization. A striking feature of sHsps is their tendency to form large oligomeric structures. No conclusive data have yet been obtained on the quaternary structure of the protein family. α-Crystallin has been suggested to exist as a heterogeneous and highly oligomeric species of 40–50 subunits under physiological conditions [128,129]. The water-soluble oligomers are mostly of an average molecular weight of 800 kDa, although complexes ranging from 280 kDa to 2 MDa have also been isolated [130,131]. The conserved COOH-terminal part of sHsps was suggested to be important for this complex formation [102]. Although the COOH-terminal domain of sHsps was proposed to be located on the surface of the native oligomers [108], Merck et al. [107] found that isolated domains of Hsp25, αA- and αB-crystallin form defined oligomers. These tetramers might represent a basic building block of sHsp oligomers. Furthermore, for a related protein, Hsp42 of *Saccharomyces cerevisiae*, Wotton et al. [51] showed that a conserved COOH-terminal sequence was necessary for multimerization. However, truncation of the COOH-terminal 18 amino acids from mouse Hsp25 did not alter the oligomerization behavior of the protein [M. Ehrnsperger, A. Kotlyarov, and J. Buchner, unpublished observations]. These findings indicate that indeed the α-crystallin domain might be important, but is not sufficient for organizing quaternary structure.

Proposed Models for Quaternary Structure. Because no X-ray structural data are yet available, proposals for the quaternary structure of α-crystallin and other sHsps are based solely on indirect methods [70]. Given that sulfydryl groups in the α-crystallin complex exist in three different environments, a three-layer model was proposed [132,133]. Subunits in this model appear surface-exposed, partially exposed, or completely buried.

This model was supported by spectroscopic and ultracentrifugation measurements leading to a structure prediction with three layers, comprising 43 subunits when some positions in the outer layer are still unoccupied, and 51 subunits when tightly packed [132–135]. From X-ray and quasi-elastic light-scattering measurements this model was modified to a tetrahedral symmetry, in which the core layer consists of 12 subunits and the second and third layers of 24 subunits each [128].

This hypothesis was challenged by Thomson and Augusteyn [136,137] because the different subunits within the complex seem to be in equivalent positions [138] and are interchangeable [139]. Heterogeneity of complexes and subunits were explained with conformational changes and modification after aging. Indeed, there is only one class of sulfhydryl groups in fetal α-crystallin [140]. Given these data and the findings that a low molecular weight complex (≈ 300 kDa) [136,137] is predominant under physiological temperatures, a dodecamer was proposed as the quaternary structure of α-crystallin in which the subunits are arranged in a tetrahedral symmetry [139].

This structure prediction, however, is not sufficient to explain the heterogeneity of α-crystallin complexes under different physicochemical conditions [134,136]. To account for this Augusteyn and Koretz [141] proposed a flexible dynamic micelle structure, with equivalent distribution of the subunits. The possible existence of micellar α-crystallin was subsequently confirmed [103,142]. The high degree of flexibility of the oligomer was consistent with the finding that subunits were exchanged between complexes under physiological conditions [143]. Walsh [114] found a compromise between the three-layer and the micelle model by proposing a three-layer structure for the protein in which the inner layer formed a micelle of 12 subunits. In this model, the apolar NH_2-terminal regions are directed inward, forming a hydrophobic core, the second layer consists of 6 subunits with their hydrophobic faces directed toward the hydrophobic core, and a further layer comprises 6 subunits. From the assumption that among a variety of dynamic structures at least one stable structure is a roughly spherical oligomer of 800–900 kDa and from studies by Merck et al. [144] that showed that isolated COOH-terminal regions of α-crystallin were able to form dimers and tetramers, whereas the NH_2-terminal domain forms less-defined multimers, Wistow [108] suggested that tetramers form the basic building block for larger α-crystallin complexes with fourfold symmetries. A stable complex form was proposed to be a rhombic dodecahedron with 12 faces and 48 subunits. Here, all subunits are in equivalent surroundings with the COOH-terminal halves being exposed, whereas the more hydrophobic NH_2-terminal domains constitute a stabilizing core. With changes of physicochemical conditions, Wistow's model makes allowance for changes in conformation and size, so that another possible structure comprises 6 tetrameric units forming a cubic particle.

In summary, the available data on the quaternary structure of α-crystallin do not allow a conclusive view, partly owing to the proposed dynamic nature of the aggregates [108,143] and the different data sets obtained under various in vitro conditions [134,145].

Because there are fewer data available, the structure of other sHsps is even less understood than that of α-crystallin. However, its ability to assemble into high molecular complexes has been reported in virtually all organisms investigated. The observed oligomeric sizes of mammalian sHsp vary between organisms, and according to physicochemical conditions, from 125 kDa [146] to more than 2 MDa after heat treatment [14]. Cytosolic plant sHsps 17.7 and 18.1 seem to comprise 12 subunits [12] and chloroplast Hsp21 forms oligomers of comparable molecular masses [147]. For the related surface antigen of *Mycobacterium tuberculosis* nine subunits have been proposed, comprising a trimer of trimers. Dudich et al. [148] have shown by difference adiabatic-scanning calorimetry that the minimum cooperatively melting structure of Hsp25/27 is a dimer. Comparison of α-crystallin and mammalian sHsp by electron microscopy showed similar globular or spherical structures, with a typical diameter of 10–18 nm [14,16,149,150]. For recombinant murine Hsp25, these were further analyzed by Behlke et al. [11]. By analytical ultracentrifugation, sedimentation coefficients of 20–22S were obtained, and the complexes were calculated to consist of 32 subunits. Given these data and use of electron microscopy, the quaternary structure of Hsp25 was postulated to be sphere-like, with subunits arranged in hexagonal packing. Under the conditions chosen (Tris $MgCl_2$, pH 7.5) the complex was stable and did not show dynamic properties. In contrast with these findings, in HEPES pH 7.5, Hsp25 assembles into at least two species (tetramer and 16–20 mer) which are in a concentration-dependent equilibrium [M. Ehrnsperger and J. Buchner, unpublished data]. These data are reminiscent of the finding that Hsp27 exists as oligomers of 100, 400, and 800 kDa in arsenite-treated cells [146].

D. Dynamics and Distribution of sHsp Complexes

Interestingly, hetero-oligomeric complexes between αB-crystallin and mammalian Hsp27, with a molecular weight of about 700 kDa, can be found in vivo [151,152] and reconstituted in vitro [35], indicating that even though the sequence homology is restricted, the mechanisms of complex formation must be similar. These hetero-oligomeric complexes dissociate with heat treatment [151]. In contrast with this observation, α-crystallin and homo-oligomeric sHsp complexes were repeatedly reported to increase significantly in size after heat shock and to redistribute from the cytosol to the perinuclear space or the nucleus itself [14–16,38,58,106].

This phenomenon seems to be modulated by the metabolic state [153] and the degree of thermoresistance of the cell [14]; for example, during heat shock, Dm-Hsp27 forms insoluble superoligomers (larger than 1 MDa) that are redistributed to the nucleus [14], whereas pretreated, thermotolerant cells do not show these effects [14,154]. In contrast, chemical stress, such as arsenite treatment, or stress caused by serum starvation, lead to a decreased level of Hsp27 oligomerization, shifting the molecular weight to about 200 kDa [18,146]. Exception to these findings seems to be yeast Hsp26, which has been found only in a high molecular complex (about 550 kDa), regardless of the growth conditions [153,155]. Stress-inducible granules formed in chicken embryo fibroblasts are found solely in the cytosol [15]. In heat-shocked tomato cell cultures, sHsps were large heat-shock granules in the nuclear periphery associated with a set of untranslated mRNAs [16]. Investigations in other cell types, however, did not show similar RNA-binding [15]. The chloroplast-localized sHsp Hsp21, stays soluble during an initial heat shock and insolubilizes during additional or prolonged heat treatment [156].

In all examples described so far, changes in Hsp27 oligomerization after stress were reversible, with the complexes relocalizing to their physiological compartment during recovery.

E. Phosphorylation

At the posttranslational level, several modifications of sHsps have been detected, including phosphorylation [5,157], deamidation, acylation, as well as mixed intermolecular disulfide formation, oxidation, and glycation [70]. One of the most striking modifications of sHsps is their phosphorylation, which will be reviewed here in detail.

1. *Phosphorylated Amino Acid Residues*

Covalent modification of sHsps by phosphorylation has been detected for bovine αA-crystallin [157] and αB-crystallin [158], for rat Hsp25 [159], human Hsp27 [160], and mouse Hsp25 [161]. In all these proteins, serines are the exclusively phosphorylated amino acid residues. αA-Crystallin is phosphorylated at S122 [157] and αB-crystallin at S45, S59, [158] and also at S19, S45, after in vivo labeling [162]. Mouse Hsp25 is phosphorylated at two sites, S15 and S86 [71]. Hence, Hsp25 can exist in four different isoforms [S15,86;S15P;S86P;S15,86P]. Human Hsp27 possesses three phosphorylation sites, S15, S78; and S82 [163], leading to eight possible, differentially phosphorylated isoforms. The phosphorylation sites of mouse and human sHsp are located in a protein kinase (PK) recognition motif RXXS(P).

A heat-shock and ecdysterone-dependent appearance of two phosphorylated isoforms of *Drosophila* Hsp27 has been described [164]. The phosphorylated amino acids and the phosphorylation site in Dm-Hsp27 have not yet been identified; however, Dm-Hsp27 can be phosphorylated by the purified hamster Hsp27-kinase [165], indicating similar phosphorylation sites.

2. *Stimuli*

Phosphorylation of α-crystallin could be observed in mouse heart [C. E. M. Voorter and W. W. de Jong, unpublished data], in Alexander disease brain [166], and in a differentiation-dependent manner in lens fiber cells [167]. While relatively stable in the mature lens fiber, α-crystallin phosphorylation could be reversed by dephosphorylation in the lens epithelium [168].

In contrast with the α-crystallins in which no significant changes in the degree of phosphorylation in response to extracellular stimuli could be observed, an intriguing feature of several, in particular mammalian sHsps, is the phosphorylation in response to a wide variety of extracellular stimuli. Rat Hsp25/28 is rapidly phosphorylated following heat shock or addition of either phorbol ester (PMA), the calcium ionophore calcimycin (A23187) or serum [5]. Further investigations revealed that sHsps are also phosphorylated in response to mitogenic signals, such as platelet-derived growth factor (PDGF), acid fibroblast growth factor (FGF) [169], and basic FGF [170]. More strikingly, human Hsp27 or mouse Hsp25 were rapidly phosphorylated after treatment of cells with inflammatory cytokines, such as TNF-α and interleukin-1 (IL-1) [171,172], or agents inducing chemical or oxidative stress, such as arsenite and H_2O_2 [173,174]. Because pretreatment of mammalian cells with antioxidants could completely inhibit TNF-α-, H_2O_2-, or arsenite-induced sHsp-phosphorylation, it was suggested that reactive oxygen species are involved in the process [174]. Differentiation-inducing agents, such as PMA and leukemia inhibitory factor–D-factor, lead to an increased sHsp phosphorylation that in some cases even correlates with the differentiation process [175–178]. The sHsp-phosphorylation has also been described as a result of a variety of other stimuli, including bradykinin [169], thrombin [93], histamine [179], ifosfamide [180], ocadaic acid [181], IL-3, granulocyte–macrophage colony-stimulating factor (GM-CSF) [182] and IL-6 [183].

3. *Enzymes and Pathways*

The serine residues S121 in αA- and S19,45,59 in αB-crystallin can be phosphorylated in vitro by PKA [184], an enzyme that is also present in the lens. For αA-crystallin, weak autophosphorylation, which can be enhanced by dissociation of oligomers, has also been reported at a site between amino acids 131 and 145 [185,186]. Hsp25 and Hsp27 were only poorly phosphorylated by PKA and PKC [71] and show no autophosphorylation [K. Engel and M. Gaestel, unpublished observations].

A first direct indication of the kinase involved in sHsp phosphorylation came from the observation that rabbit mitogen-activate protein kinase–activated protein (MAPKAP) kinase 2 efficiently phosphorylates Hsp25 and Hsp27 in vitro and that this kinase copurifies with the Hsp25 kinase activity during the whole purification process from rabbit muscle [187]. This kinase shared all properties now described for the sHsp-kinase [188,189]. It is activated by heat shock [190] and possesses a phosphorylation site recognition motif LXRXXS(P) [191] that is in agreement with the sites on Hsp25 and Hsp27. Recently, a kinase very similar to MAPKAP kinase 2 has been identified, which is designated 3pK [192] or MAPKAP kinase 3 [193]. This protein kinase very efficiently phosphorylates Hsp25 and Hsp27 at the same sites as MAPKAP kinase 2 [194].

The main activator of MAPKAP kinase 2 and 3pK is the p38 MAPK [195,196], which phosphorylates MAPKAP kinase 2 at several regulatory sites [190,197]. p38 MAPK is a central component of a stress and inflammatory cytokine-activated signaling cascade that parallels the classic MAPK cascade and that is activated by IL-1, TNF-α, heat shock, osmotic stress, UVC, anisomycin, and arsenite [198]. In addition, 3pK is also activated by the classic, mitogen- and growth factor-stimulated MAP kinase cascade [194]. Hence, MAPKAP kinase 2 and 3pK together can explain sHsp-phosphorylation as a result of most of the stimuli and stress conditions mentioned earlier. Only PMA-induced sHsp phosphorylation could not be explained by these cascades per se. However, a cross-talk between PMA-induced isoforms of PKC and upstream components of the p38 cascade seems likely [199; H. Schultz and M. Gaestel, unpublished data].

Because p38 MAPK plays a central role in the signaling processes leading to sHsp phosphorylation, a specific inhibitor for p38 MAPK, the pyridyl imidazole SB203580 [200], could be successfully used to suppress sHsp phosphorylation in vivo [199,201]. This inhibitor might be able to contribute to our further understanding of the biological role of sHsp phosphorylation in the future.

It is probable that the degree of sHsp phosphorylation is regulated not only by phosphorylating enzymes, but also by specific protein phosphatases. For mouse Hsp25 dephosphorylation by the calcium-dependent protein phosphatase 2B (calcineurin) has been described [202]. Heat shock increases the intracellular calcium level; therefore, this enzyme could be of physiological relevance for the regulation of sHsp phosphorylation.

III. FUNCTIONAL ASPECTS OF sHsps

A. Involvement of sHsps in Cellular Protection

1. Thermo- and Chemoresistance

Experiments performed in *Drosophila* tissue cultures as well as mammalian cell lines and plants have shown that sHsps are among the most strongly induced heat-shock proteins under a variety of stress conditions. This overexpression of sHsps has conveyed thermoresistance in various organisms and cell types. This phenomenon has been shown in mammalian cells for human Hsp27 [6], murine Hsp25 [32], Dm-Hsp27 [106,203], αB-crystallin [33], and finally also for αA-crystallin [34], indicating a general thermoprotective function of these proteins. Plant sHsps are strongly induced during heat shock and according to the developmental status of the cells [204,205], but thermal protection has not yet been proved. Surprisingly, the yeast homologue does not seem to have a demonstrable function in thermoprotection or during development [63,206]. Overexpression of Hsp27 has been correlated with not only thermotolerance, but it also confers resistance to oxidative stress [69,174,207] as well as several cytotoxic agents, including cytochalasin D [39,146], tumor necrosis factor [69], and some anticancer drugs [208].

2. Mechanism of Protection and Functional Role of Phosphorylation

No clear picture on the mechanism of sHsps in cellular protection has yet emerged. On the one hand, sHsps have repeatedly been reported to act as ATP-independent molecular chaperones [209] (see later). On the other hand, there is growing evidence that sHsps can specifically interact with certain components of the cytoskeleton such as actin [37,38,40,210] or intermediate filaments [41]. Both, the general chaperone properties, leading to stabilization of thermolabile intracellular proteins under stress conditions, and the specific interaction with the cytoskelcton, possibly preserving cell shape and integrity, could contribute to cellular protection against stress. Furthermore, it should be taken into account that the increased resolubilization of nuclear protein, connected with overexpression of Hsp27 [211,212], the correlation between intracellular sHsp level and glutathion concentration [213], the stabilization of RNA [16], the block of the apoptotic pathway [45], and the inhibition of elastase [36] by sHsps, all could contribute to cellular stress resistance. Interestingly, the protective capacity against oxidative stress seems to be based on the conserved ability of sHsps to raise the intracellular concentration of glutathion, thereby abolishing the burst of reactive oxygen species [213].

Another interesting, but still poorly understood, aspect of sHsps is the functional role of phosphorylation. A key role of phosphorylation in cellular protection seems unlikely, for α-crystallins that are not phosphorylated in response to heat treatment show heat protection in a manner comparable with Hsp25 or Hsp27 [33,34]. In addition, transient overexpression of a nonphosphorylateable mutant of Hsp25 led to virtually the same thermal protection as the wild-type protein [214]. In attachment-dependent hamster fibroblast CCL39 cells, overexpression of a Hsp27 triple phosphorylation mutant, has given only weak protection against heat shock [146], whereas in the tumorigenic, nonattachment-dependent variant of CCL39 (cell line 023), Hsp27 phosphorylation site mutants and wild-type protein increase heat resistance to a comparable degree [E. Hickey, D. Latour, D. Egeneder, and L. A. Weber, personal communication].

How can these differences be explained? The in vitro chaperone properties of Hsp25 seem to be phosphorylation-independent [214] and could be the basis for the phosphorylation-independent thermoprotective properties of sHsps in the attachment independent cells. On the other hand, evidence is growing that sHsps' interaction with the cytoskeleton is phosphorylation-dependent: (a) Hsp27 enhances growth factor-induced F-actin accumulation, whereas the phosphorylation mutant Hsp27-S15,78,82G showed a dominant negative effect on F-actin formation [38]; (b) nonphosphorylated low molecular weight species of Hsp25, but not phosphorylated Hsp25, inhibit actin polymerization in vitro [40]; (c) only phosphorylated Hsp27 associates with the activation-dependent cytoskeleton in human platelets [210]; (d) ATP-depletion within the cell leads to a dissociation of Hsp27 from the actin filaments [215]; and (e) only wild-type Hsp27, but not the phosphorylated mutant, could protect against an inhibitor of actin polymerization (cytochalasin D) [146] and actin fragmentation by oxidative stress [207]. Taking into account that in attachment-dependent cells the actin-modulating function of sHsps may be of greater importance for stress resistance than in cells that grow attachment-independent and are able to form colonies in soft agar, the phosphorylation-dependence of the protection of CCL39 cells by Hsp27 is plausible. Hence, the protective activity of sHsps may be the sum of phosphorylation-independent chaperone properties and phosphorylation-mediated stabilization of the cytoskeleton and may depent on the attachment behavior of the cell type analyzed.

In addition, the phosphorylation-dependent actin-modulating function of Hsp27 suggests a role of at least some mammalian sHsps in signal transduction in response to extracellular stimuli by MAP kinase cascades to the cytoskeleton. In this context, interestingly, the enzymes responsible for sHsp phosphorylation, MAPKAP kinase 2 and 3pK (see foregoing), are downstream components of the p38 MAP kinase cadcade, which could be activated through the p21-activated kinase 1 (Pak1) by the Rho family GTPase Rac and Cdc42 [216]. These small GTPases are also responsible for actin filament concentration at the plasma membrane, forming membrane ruffles [217] and the formation of actin-based filopodia [218], suggesting a possible involvement of sHsp phosphorylation in this process.

Influence of sHsp phosphorylation has been described to be associated with changes in the oligomeric structure of sHsp complexes: Depending on the cell type, extracellular stimulus, and the time after stimulation, phosphorylation of Hsp27 correlates with either the dissociation or further increase in size of the oligomers. In U251 MG human glioma cells, IL-1, TNF-α, okadaic acid, as well as chemical stresses, such as arsenite, induce phosphorylation and dissociation of Hsp27. The oligomeric size of the Hsp27 complexes changed from 300–400 kDa to structures smaller than 70 kDa [17]. In hamster CCL39 cells, a reduction of the multimeric size of transfected human Hsp27 in response to mitogenic stimuli and arsenite is observed only for the wild-type protein, but not for the phosphorylation site mutant [146], indicating that phosphorylation is necessary for this phenomenon. In contrast, in starved HeLa cells, Hsp27 is dephosphorylated and quantitatively recovered as small structures (less than 200 kDa), whereas the protein is phosphorylated after serum stimulation, leading to an increased molecular mass (up to 700 kDa) [18]. The situation is even more complicated in TNF-α: Following TNF-α treatment of HeLa cells, Hsp27 is rapidly phosphorylated and detected in larger oligomers (over 700 kDa), but after 4 h of treatment, the large complexes dissociate into smaller species (less than 200 kDa). During the whole process, the phosphorylated isoforms of Hsp27 remained in the smaller oligomeric form of the protein (under 300 kDa) [19]. Interestingly, in stably transfected L929 cells a nonphosphorylatable mutant of Hsp25 is also shifted to large oligomeric complexes after TNF-α treatment, but in contrast with the wild-type protein, it stays stably hyperaggregated over the whole time of treatment [X. Preville, U. Knauf, M. Gaestel, and P. Arrigo, unpublished data].

From the foregoing data, one may conclude that phosphorylation is necessary, but not sufficient, for dissociation of the sHsp oligomers. Other factors may contribute to the regulation of the oligomeric size of sHsps and thus modulate function by structural changes. Whereas when chaperone activity of sHsps seems to be phosphorylation-independent, signal transduction and the interaction with the cytoskeleton might be important phosphorylation-regulated processes.

B. Involvement of sHsps in Disease

Although the main focus of research up to now was directed at the expression and function of sHsps under heat and stress conditions, α-crystallin and other mammalian sHsps are also expressed under physiological conditions [8,20,21,27, 80,219–221]. Changes in sHsp expression and phosphorylation have repeatedly been reported in the context of several diseases and pathological conditions.

In agreement with the notion that sHsps are synthesized during embryogenesis (see foregoing) [28], several studies have associated increased sHsp expression with the presence of teratogens and teratogenic conditions (e.g., heat-shock [222–225]). In rodents, Hsp25 is induced by retinoic acid, a substance inducing fetal malformations, which led to the proposal that Hsp25 might be actively involved in the process of malformation [225,226]. Changes of sHsp expression have been reported in several cancers. In the pluripotent embryonal carcinoma cell line P19 and embryonic stem cell line BLC6, retinoic acid-induced differentiation leads to a dramatic increase of the intracellular Hsp25 level [227]. Similarly, in Ehrlich ascites tumor (EAT) cells, Hsp25 is expressed predominantly in the stationary phase of the tumor [23], and EAT growth can be inhibited by Hsp25 overexpression [32]. These data indicate an inverse correlation between the intracellular sHsp level and cell growth. This idea was also supported by the finding that Hsp27 expression is inversely correlated with the oncogenicity of adenovirus-transformed cells [228]. However, in various tumors, including breast cancer and leukemias, the opposite has also been reported; that is, an increased expression of sHsps [26,229,230]. The Hsp27 present in tumor cells and observed in normal cells are identical, apart from their phosphorylation pattern. Diminished phosphorylation of Hs27 has been found in acute lymphoblastic leukemia cells [231]. In some cancers, sHsp expression seems to be controlled by hormones, whereas other cancer cell lines show induction of Hsp27 even when no steroid receptors are present in the cells [91].

The sHsps have been discussed as being of prognostic value for different tumors [232,233]. However, immunohistochemical studies on tissues of node-positive breast cancer suggested that Hsp27 expression is not predictive of the outcome of the disease [234]. Furthermore, whereas patients with malignant fibrous histiocytoma who show increased expression of Hsp27 have a better chance of recovery and no drug resistance is observed [233], high levels of Hsp27 in breast cancer are correlated with a poor survival rate [232,235], and in vitro studies showed increased resistance to cytotoxic drugs [236]. This increased anticancer drug resistance as well as resistance to hyperthermia or starvation in the context of Hsp27 accumulation might be correlated with resistance to apoptosis [44,45,237]. One factor that induces necrotic or apoptotic cell death is TNF-α. Overexpression of several sHsps may covey resistance to TNF-α, possibly by altering the glutathion levels of the affected cells [19,69,213].

The sHsps have also been suggested to play a role in the inflammatory response, for Hsp27 and αB-crystallin were induced by modulators of the arachidonic acid cascade [100], and their phosphorylation is induced by pro-inflammatory cytokines (see foregoing). Arachidonic acid is a precursor for eicosanoids, members of which play an important role in the inflammatory response [238].

An increasing number of diseases have been described in which mutations and modifications exert their pathological effect by altering protein folding or assembly. Increased expression of sHsps and, especially, α-crystallin has been observed in several neurodegenerative disorders, such as Alzheimer's disease [239,240], Alexander disease [241,242], Creutzfeld–Jakob disease [243,244], supranuclear palsy [245], and others. The phosphorylation patterns of Hsp25 and αB-crystallin in brains of patients with Alexander disease have been reported to be distinct from control brains, indicating that posttranslational modification might play a role in the progress of the disease [242]. A common feature of all these neurodegenerative disorders is the deposition of improperly folded protein in fibers, inclusion bodies, or plaques in the nevous system [246]. sHsps are typically found in association with the insoluble protein aggregates and ubiquitin [247,248]. Increased expression of both α-crystallin and Hsp27 has been observed in the context of these diseases. However, the total amount and the distribution of the proteins vary in different cell types and compartments [239,249]. Interestingly, in scrapie-infected mouse neuroblastoma cells, induction of Hsp27 and Hsp72 by metabolic stress is blocked, but Hsp73 and Hsp90 expression is enhanced [250]. This observation could indicate that Hsp27, together with Hsp72, is specifically involved in the pathogenesis of prion disease. Furthermore, sHsps have been associated with functions in the immune system. Prominent antigens of *Mycobacterium tuberculosis* and *M. leprae* show sequence homology with the sHsp family [251]. These antigens are probably located on the surface of the bacteria [252] and are, among others, the cause for the immune response in host cells [13]. sHsps not only exhibit antigenic function in host organisms, but also αB-crystallin causes a dominant immune response in multiple sclerosis patients and might constitute one of the major autoantigenes [43].

In summary, the function of altered sHsp expression in the course of several seemingly unrelated diseases is still enigmatic. A common feature of many of the described disorders is the accumulation of misfolded protein [246]. The unregulated growth of tumor cells might lead to problems in the protein-folding processes, and deposition of misfolded protein in the neuronal system is the main symptom in neuronal disorders [253]. Thus, the chaperone function of sHsps might be the dominant feature explaining their involvement in diseases.

C. sHsps as Molecular Chaperones

1. *Chaperone Function in General*

There is a wide diversity of factors that lead to enhanced expression of sHsps and Hsps in general, such as elevated temperatures, heavy metals, various ionophores, amino acid analogues, oxidants, metabolic poisons interfering with ATP production, and changes in developmental stages of cells and organisms [254]. Many of these agents or factors affect the proper conformation and, therefore, the function of proteins in general, leading to the accumulation of nonnative protein in cells [255,256]. It was initially speculated that Hsps are part of the protein machinery supervising folding [257]. Not surprisingly, most classes of heat-shock proteins have been identified as molecular chaperones in recent years [31]. Molecular chaperones support the correct folding of proteins during translation and translocation as well as during stress periods when cellular proteins are in danger of irreversible aggregation. To achieve this aim, chaperones must be able to differentiate between native and nonnative forms of target proteins. The binding of unfolded polypeptide chains to the chaperone reduces the probability of irreversible interactions of exposed hydrophobic sequence stretches and, thereby, allows folding or maintenance of protein structure under nonpermissive conditions. Molecular chaperones are divided into two main groups according to their dependency on energy. Both, ATP-dependent and ATP-independent chaperones (including sHsps) seem to act in concert to guarantee survival of the cell under stress situations.

The molecular chaperone concept is no contradiction to the principle of spontaneous folding [258], as molecular chaperones do not convey any specific information for folding, but rather, support correct structure by preventing improper or unproductive reactions that could lead to protein misfolding or aggregation. Without becoming part of the native structure, molecular chaperones, therefore, ensure high fidelity in the protein folding and assembly process.

Interestingly, sHsps, which are sometimes termed junior chaperones, compose the most effective chaperone family concerning the efficiency of substrate binding per oligomeric complex [209] (see the following sections).

2. *sHsps Interact with Nonnative Protein*

One of the first indications of sHsps being implicated in chaperoning protein folding was the finding that an ammonium sulfate fraction, enriched in plant sHsps, could prevent soluble proteins from thermal denaturation in vitro [259]. Mammalian sHsps and α-crystallin have subsequently been shown to be able to suppress aggregation of thermally denaturing proteins, such as β- and γ-crystallin, in the case of α-crystallin, and model enzyme substrates [35,46,47]. Generally, sHsps seem to be rather promiscuous in their binding behavior, for complex formation with a variety of heat-denatured substrates has been reported. Interaction with heat-inactivated citrate synthase (CS) was shown for Hsp27, Hsp25, α-crystallin, cytosolic sHsps from pea, and Hsp16.3 from *Mycobacterium tuberculosis* [12,13,47,73]. The recombinant mouse Hsp25 can form complexes with several nonnative substrates, ranging from insulin B-chain (3.5 kDa) to dimeric CS (100 kDa) [M. Ehrnsperger, M. Gaestel, and J. Buchner, unpublished data]. Promiscuity in binding was also shown for the cytosolic plant sHsp 18.1 [76] and for α-crystallin, which forms stable complexes with various nonnative proteins, such as carbonic anhydrase [260], rhodanese [261], and γ-, β_L-, β_H-crystallin [261–263]. Interestingly, α-crystallin was more efficient in preventing aggregation of other lens crystallins than of model enzymes [46].

The chaperone function of sHsps is not restricted to the prevention of thermal protein aggregation because renaturation of chemically denatured protein was also reported. Renaturation of γ-crystallin was supported by α-crystallin [46], and the yields of reactivation of different model enzymes were significantly increased in the presence of Hsp25, Hsp27, or α-crystallin [47]. In this context, the mode of action was almost indistinguishable among the different sHsp species [35,47].

The described interaction of sHsps with nonnative proteins has never been associated with ATP-binding or hydrolysis. Similar to Hsp90, sHsps are thus believed to function independently of energy (see the following).

3. Complex Formation and Substrate Range

The underlying mechanism of sHsps chaperoning protein folding remains enigmatic. Until recently, only very stable complexes between sHsps and nonnative substrate proteins were detected [12,261,264]. For example, the interaction between α-crystallin and carbonic anhydrase is stable at 4°C as well as at room temperature for at least 18 h [264]. The stoichiometry of the complexes between sHsps and nonnative protein seems to be dependent on the substrate investigated [76]. However, several reports suggest a maximum-binding capacity of one denatured protein molecule per subunit of the sHsp oligomers [76,263,265; M. Ehrnsperger and J. Buchner, unpublished data]. The structural requirements for proteins to be recognized and bound by sHsps are not yet entirely clear. For rhodanese bound to α-crystallin, tryptophan fluorescence suggests that the protein is more native than complexed with GroEL [261]. Similar binding patterns were observed when α-crystallin was complexed with alcohol dehydrogenase and γ-crystallin [266]. In contrast with other chaperones, such as GroEL, DnaK, and SecB, α-crystallin does not interact with stably unfolded, hydrophobic proteins, such as carboxymethylated α-lactalbumin and α-casein [266]. From these data it was suggested that sHsps preferentially recognize early labile intermediates, which are most sensitive to aggregation [261,266]. In contrast with other chaperones, such as Hsp70 [267] or Hsp90 [268], sHsps do not seem to alter the inactivation kinetics of proteins [46,73,264,269], indicating that the different chaperone classes accomplish distinct tasks during a heat-shock response. In the context of the substrate range of sHsps, under physiological temperatures α-crystallin interacts with the reduced and aggregation-prone B chains of insulin and melittin, a peptide from bee venom that adopts an unordered conformation [265]. Furthermore, sHsps seem to be able to protect glucose-6-phosphate dehydrogenase and 6-phosphogluconate dehydrogenase from modification-induced inactivation [270,271]. In comparison with GroEL, sHsps are more effective in this protective function [76,270]. Major physiologically relevant factors to induce protein aggregation of lens proteins include UV radiation and oxidative stress. Although these damages are not likely to cause extensive unfolding of protein, nevertheless, they may perturb tertiary protein structure and eventually lead to protein aggregation [272]. Interestingly, α-crystallin protects lenticular proteins against the harmful effects of these agents [263,272,273]. Although α-crystallin is present in several other tissues, it is most abundant in the eye lens, where it can constitute 30–40% of the soluble protein. Because no turnover occurs in lenses, the conformational preservation of the set of proteins present is a serious problem. This task of protein maintenance is, at least partly, performed by α-crystallin: α-crystallin protects lens proteins from irreversible denaturation, whereas removal of the chaperone leads to the aggregation of most of the lens proteins with heat stress [269]. The ability of lens α-crystallin to prevent heat-induced aggregation of β-crystallin in vitro decreases with age [46], which might be one of the reasons for cataract formation in elderly lenses. Because α-crystallin is post-translationally modified by oxidation, deamidation, racemization, acetylation, glycation, phoshorylation, and age-dependent truncation [70], these modifications might be responsible for decreased chaperone activity in the aging lens and resulting cataractogenesis [274,275].

4. *The Substrate-Binding Site*

Where the binding sites of nonnative proteins is located in the large oligomeric structures of sHsps is not yet fully understood. Boyle et al. [276] suggested binding to a central cavity, as in GroEL. However, the ability of binding stoichiometric amounts of substrate per sHsp monomer argues against this idea (see foregoing). Furthermore, recent studies, using electron microscopy, strongly suggest that the substrate protein is bound on the surface of the oligomeric sHsp structure [73,76]. Theoretical studies with an open micellar structure and surface binding of protein as a working model for α crystallin [277] are in good agreement with experimental data [263,265,278].

The conserved α-crystallin domain of sHsps was suggested to be involved in the chaperone activity of the proteins. Removal of parts of the COOH-terminal domains of α-crystallin and Hsp27 from *Drosophila* have led to a marked reduction of the ability of the protein to suppress thermal aggregation and to protect against thermal stress in vivo [105,106], and isolated COOH-terminal domains of mammalian sHsps were sufficient to inhibit elastase activity [36]. In agreement with these findings, Lee et al. [76] could show by cross-linking that bound substrates were in contact with a conserved region in the COOH-terminal part of the protein. This consensus sequence is found in all members of the sHsp family [1,2,56] (see Fig. 1). However, studies with the isolated α-crystallin domain of Hsp25 did not provide any evidence for in vitro chaperone activity [144; M. Ehrnsperger, M. Gaestel, and J. Buchner, unpublished data]. The flexible COOH-terminal tail, which is found in virtually all sHsps, with variable sequence and length [102–104], does not seem to be involved in chaperone activity, for truncation of the COOH-terminal 18 residues of Hsp25 has no effect on protective function of the protein [M. Ehrnsperger, M. Gaestel, A. Katlyarov, and J. Buchner, unpublished data]. These findings indicate that the conserved α-crystallin domain might be important, but is not sufficient, for the chaperone activity of sHsps. Apparently, other or additional parts of the proteins are required.

Nonnative protein not only could be cross-linked to COOH-terminal consensus sequences, but also to a hydrophobic NH_2-terminal region, indicating that, in combination with the COOH-terminal domain, conserved NH_2-terminal sequence stretches (see Fig. 2) might be possible candidates for substrate binding [76].

5. *Mechanism of Substrate Binding*

Not only is the binding site on sHsp oligomers still unknown, but also the mechanism of interaction with denatured substrate molecules remains enigmatic. It is widely believed that hydrophobic contacts contribute to binding. This is mostly based on data indicating that the hydrophobicity profile of the chaperones changes under stress conditions. α-Crystallin binds increasing amounts of the hydrophobic dyes 8-anilino-1-naphthalene sulfonate (ANS) [266,279] and bis-ANS [280] at temperatures higher than 30°C, suggesting that α-crystallin undergoes a temperature-dependent structural change that increases surface hydrophobicity.

Raman and Rao [262,272,279] even suggested that a structurally perturbed conformation of α-crystallin is necessary for optimal chaperone activity. However, data alluding to possible binding sites of plant sHsps seem to contradict the hypothesis that only hydrophobic interactions are involved in substrate binding [76]. The putative-binding site in this study contains several highly conserved aliphatic residues with intervening charged amino acids. This alternating polar–nonpolar sequence suggests that efficient binding of denatured protein is mediated by the hydrophobic residues interacting with exposed nonpolar amino acids of the substrate, whereas charged amino acids might serve to optimally space the residues and aid in hydration of the region [76]. Previous studies on the binding of the spin-labeled insulin B chain and melittin to α-crystallin support the notion that polar residues are involved in substrate binding to sHsps [265]. Electron spin resonance (ESR) analysis of the labeled peptides show that the nitroxide side chains are immobilized in a polar environment on α-crystallin and that they are separated by at least 25 Å, indicating that they are not clustered on the sHsp. The bound insulin chains are not in a fully extended conformation, and melittin does not appear to bind to hydrophobic surfaces on α-crystallin as an amphiphatic helix, as it does to membranes and some other proteins [265]. Smulders et al. [282] could show that the D69S mutant of αA-crystallin shows decreased chaperone activity, indicating that the charged Asp69 might also be involved in protein binding. The mechanism of substrate binding to sHsps is not yet fully understood. However, recent data challenging the previous hypothesis that interaction of sHsps with nonnative protein is due to only hydrophobic interactions, might help to approach this problem in the near future.

6. sHsps in the Context of the Cellular Chaperone Machinery

Taken together, sHsps seem to be potent and efficient chaperones in that they are able to selectively bind nonnative proteins in large quantities per oligomeric sHsp complex. However, although sHsps have been reported to bind to several denatured substrates no release of bound substrate has been observed until recently. For α-crystallin, it was suggested that irreversible binding of other lens crystallins, which might occur in the aging, lens, led to decreased protective activity of the chaperone [282]. Along these lines, Lee et al. [12] suggested that in the plant sHsps, bound substrate was held in a soluble form for subsequent degradation. However, in the living cell it seems detrimental to entertain a chaperone system that traps nonnative proteins irreversibly without allowing reactivation of denatured proteins once a stress situation is over. A first hint that sHsp–substrate complexes were not dead end, but were productive intermediates, was the discovery that CS could be dissociated and reactivated from Hsp25–CS complexes by oxaloacetate (OAA), a stabilizing ligand of CS [73]. Thus, Hsp25 is able to bind to thermally inactivated protein and traps it in a soluble and folding competent state until permissive folding conditions are restored. Although the aforementioned experiments used the ligand OAA, in the cell a physiological release factor would be required. Because Hsp70 similar to the sHsps, is strongly induced under heat-shock conditions and acts as an ATP-dependent chaperone in the cytosol, it is a likely partner protein for Hsp25. Indeed, in concert with the energy-dependent chaperone, Hsp70 reactivation of the sHsp-bound substrate is possible in the presence of ATP [73]. In agreement with these findings, Kampinga et al. [211,284] and Stege et al. [212] could demonstrate that overexpression of Hsp27 in several cell lines did not decrease the inactivation of nuclear proteins, but the marker proteins were insolubilized together with Hsp27. This is reminiscent of the finding that prokaryotic sHsps are found specifically associated with inclusion bodies, intracellular aggregates of overexpressed proteins [53]. For Hsp27, the in vivo experiments showed solubilization of bound substrate proteins and a strongly accelerated recovery after return to physiological conditions.

From these data, a model that allows integration of sHsps in the cellular chaperone machinery was proposed (Fig. 3). It seems likely that the abundant energy-independent sHsps trap nonnative proteins during stress conditions, thereby protecting them from irreversible aggregation and providing a reservoir of refoldable protein, until more active, ATP-dependent chaperones, such as Hsp70, can complete the refolding of cellular proteins once physiological conditions are reestablished.

In summary, research on the chaperone activity of sHsps has revealed that they are able to bind to a variety of nonnative proteins more efficiently than, for example, the model substrate GroEL. The binding is productive, because, in cooperation with Hsp70 and ATP, bound protein can be released and reactivated.

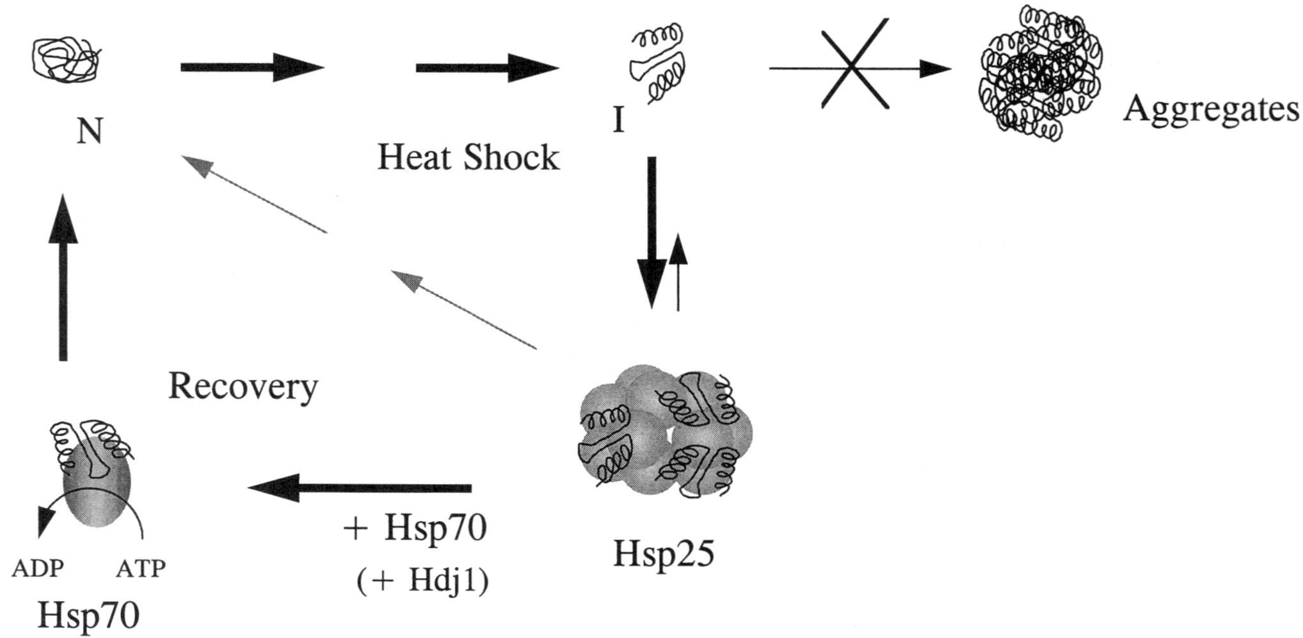

Figure 3 Model for the chaperone function of sHsps under heat-shock conditions: Unfolding of native protein (N) leads to the formation of nonnative, aggregation-prone intermediates (I). In the presence of sHsps, these unfolding intermediates can be bound in high quantity, in a soluble and refoldable state. Even at permissive temperatures the nonnative protein stays bound. However, in the presence of Hsp70, ATP, and possibly cofactors such as Hdj-1, sHsp-bound protein can be released and refold into its native state.

Whether other factors, such as Hsp70 cochaperones, or other Hsp families are involved in this process remains to be seen. Another open question is whether the model proposed for mammalian sHsp action (see Fig. 3) applies to other organisms such as plants or yeast.

D. Conclusions

With good reason sHsps can be called "the forgotten chaperones," because although they had been implicated to be involved in several fundamental cellular processes, in addition to the heat-shock response, their function remained enigmatic. Recent experiments, using in vitro protein-unfolding assays, allowed us to define functional properties of sHsps under stress conditions. These results clearly showed that sHsps belong to the class of ATP-independent chaperones. The current evidence suggests that a specific function of sHsps is to effectively trap aggregation-prone folding intermediates and maintain them in a refoldable conformation. Structural information for sHsps is still rather poor. Thus, the underlying conformational features responsible for the stable binding of non-native proteins remain elusive. In the eye lens, stable binding of α-crystallin to other crystallins seems to be sufficient to protect the lens efficiently from irreversible aggregation processes that could cause turbidity. However in the cytosol and potentially in orgenelles, cooperation with other ATP-dependent chaperone systems seems to be required to achieve productive folding, even under permissive conditions.

Furthermore, it remains to be seen how the chaperone properties of sHsps are linked to pathological conditions and diseases.

REFERENCES

1. Plesofsky-Vig, N., Vig, J., and Brambl, R., Phylogeny of the α-crystallin-related heat-shock proteins, *J. Mol. Evol. 35*:537 (1992).
2. de Jong, W. W., Leunissen, J. A., and Voorter, C. E., Evolution of the α-crystallin/small heat-shock protein family, *Mol. Biol. Evol. 10*:103 (1993).
3. Jakob, U. and Buchner, J., Assisting spontaneity: The role of Hsp90 and small Hsps as molecular chaperones, *Trends Biochem. Sci. 19*:205 (1994).
4. Lindquist, S., DiDomenico, B. J., Bugaisky, G. E., Kurtz, S., and Petko, L., Regulation of the heat shock response in *Drosophila* and yeast, *Heat Shock from Bacteria to Man* (M. J. Schlesinger, M. Asburner, and A. Tissières, eds.), Cold Spring Harbor Press, Cold Spring Harbor, NY, 1982, p. 167.
5. Welch, W. J., Phorbol ester, calcium ionophore, or serum added to quiescent rat embryo fibroblast cells all result in the elevated phosphorylation of two 28,000-dalton mammalian stress proteins, *J. Biol. Chem. 260*:3058 (1985).
6. Landry, J., Chrétien, P., Lambert, H., Hickey, E., and Weber, L. A., Heat shock resistance conferred by expression of the human HSP27 gene in rodent cells, *J. Cell Biol. 109*:7 (1989).

7. Klemenz, R., Fröhli, E., Steiger, R. H., Schafer, R., and Aoyama, A., αB-crystallin is a small heat shock protein, *Proc. Natl. Acad. Sci. USA 88*:3652 (1991).
8. Klemenz, R., Andres, A. C., Frohli, E., Schafer, R., and Aoyama, A., Expression of the murine small heat shock protein hsp25 and αB crystallin in the absence of stress, *J. Cell Biol. 120*:639 (1993).
9. Inaguma, Y., Shinohara, H., Goto, S., and Kato, K., Translocation and induction of αB crystallin by heat shock in rat glioma (GA-1) cells, *Biochem. Biophys. Res. Commun. 182*:844 (1992).
10. Chiesa, R., McDermott, M. J., Mann, E., and Spector, A., The apparent molecular size of native α-crystallin B in non-lenticular tissues, *FEBS Lett. 268*:222 (1990).
11. Behlke, J., Lutsch, G., Gaestel, M., and Bielka, H., Supramolecular structure of the recombinant murine small heat shock protein hsp25, *FEBS Lett. 288*:119 (1991).
12. Lee, G. J., Pokala, N., and Vierling, E., Structure and in vitro molecular chaperone activity of cytosolic small heat shock proteins from pea, *J. Biol. Chem. 270*:10432 (1995).
13. Chang, Z., Primm, T. P., Jakana, J., Lee, I. H., Serysheva, I., Chiu, W., Gilbert, H. F., and Quiocho, F. A., *Mycobacterium tuberculosis* 16-kDa antigen (Hsp16.3) functions as an oligomeric structure in vitro to suppress thermal aggregation, *J. Biol. Chem. 271*:7218 (1996).
14. Arrigo, A.-P., Suhan, J. P., and Welch, W. J., Dynamic changes in the structure and intracellular locale of the mammalian low-molecular-weight heat shock protein, *Mol. Cell. Biol. 8*:5059 (1988).
15. Collier, N. C., Heuser, J., Levy, M. A., and Schlesinger, M. J., Ultrastructural and biochemical analysis of the stress granules in chicken embryo fibroblasts, *J. Cell Biol. 106*:1131 (1988).
16. Nover, L., Scharf, K. D., and Neumann, D., Cytoplasmic heat shock granules are formed from precursor particles and are associated with a specific set of mRNAs, *Mol. Cell. Biol. 9*:1298 (1989).
17. Kato, K., Hasegawa, K., Goto, S., and Y. Inaguma, Dissociation as a result of phosphorylation of an aggregated form of the small stress protein, hsp27, *J. Biol. Chem. 269*:11274 (1994).
18. Mehlen, P. and Arrigo, A.-P., The serum-induced phosphorylation of mammalian hsp27 correlates with changes in its intracellular localization and levels of oligomerization, *Eur. J. Biochem. 221*:327 (1994).
19. Mehlen, P., Mehlen, A., Guillet, D., Preville, X., and Arrigo, A.-P., Tumor necrosis factor-α induces changes in the phosphorylation, cellular localization and oligomerization of human hsp27, a stress protein that confers cellular resistance to this cytokine, *J. Cell. Biochem. 58*:248 (1995).
20. Bhat, S. P. and Nagineni, C. N., αB Subunit of lens-specific protein α-crystallin is present in other ocular and non-ocular tissues, *Biochem. Biophys. Res. Commun. 158*:319 (1989).
21. Kato, K., Shinohara, H., Kurobe, N., Inaguma, Y., Shimizu, K., and Ohshima, K., Tissue distribution and developmental profiles of immunoreactive αB crystallin in the rat determined with a sensitive immunoassay system, *Biochim. Biophys. Acta 1074*:201 (1991).

22. Bond, U. and Schlesinger, M. J., Heat-shock proteins and development, *Adv. Genet. 24*:1–29 (1987).
23. Gaestel, M., Gross, B., Benndorf, R. et al., Molecular cloning, sequencing and expression in *Escherichia coli* of the 25-kDa growth-related protein of Ehrlich ascites tumor and its homology to mammalian stress proteins, *Eur. J. Biochem. 179*:209 (1989).
24. Crête, P. and Landry, J., Induction of HSP27 phosphorylation and thermoresistance in Chinese hamster cells by arsenite, cycloheximide, A23187, and EGTA, *Radiat. Res. 121*:320 (1990).
25. Pauli, D., Tonka, C. H., Tissières, A., and Arrigo, A.-P., Tissue-specific expression of the heat shock protein HSP27 during *Drosophila melanogaster* development, *J. Cell Biol. 111*:817 (1990).
26. Ciocca, D. R., Oesterreich, S., Chamness, G. C., McGuire, W. L., and Fuqua, S. A. W., Biological and clinical implications of heat shock protein 27,000 (Hsp27): A review, *J. Natl. Cancer Inst. 85*:1558 (1993).
27. Gernold, M., Knauf, U., Gaestel, M., Stahl, J., and Kloetzel, P. M., Development and tissue-specific distribution of mouse small heat shock protein hsp25, *Dev. Genet. 14*:103 (1993).
28. Arrigo, A.-P. and Landry, J., Expression and function of the low-molecular-weight heat shock proteins, *The Biology of Heat Shock Proteins and Molecular Chaperones* (R. I. Morimoto, ed.), Cold Spring Harbor Laboratory Press, Plainview, NY, 1994, p. 335.
29. Waters, E. R., Lee, J. L., and Vierling, E., Evolution, structure and function of the small heat shock proteins in plants, *J. Exp. Bot. 47*:325 (1996).
30. Tissières, A., Mitchell, H. K., and Tracy, U. M., Protein synthesis in salivary glands of *Drosophila melanogaster*: Relation to chromosome puffs, *J. Mol. Biol. 84*:389 (1974).
31. Buchner, J., Supervising the fold: Functional principles of molecular chaperones, *FASEB J. 10*:10 (1996).
32. Knauf, U., Bielka, H., and Gaestel, M., Over-expression of the small heat shock protein hsp25 inhibits growth of Ehrlich ascites tumor cells, *FEBS Lett. 309*:297 (1992).
33. Aoyama, A., Fröhli, E., Schäfer, R., and Klemenz, R., αB-crystallin expression in mouse NIH 3T3 fibroblasts: Glucocorticoid responsiveness and involvement in thermal protection, *Mol. Cell. Biol. 13*:1824 (1993).
34. van den IJssel, P. R., Overkamp, P., Knauf, U., Gaestel, M., and de Jong, W. W., αA-Crystallin confers cellular thermoresistance, *FEBS Lett. 355*:54 (1994).
35. Merck, K. B., Groenen, P. J., Voorter, C. E., de Haard-Hoekman, W. A., Horwitz, J., Bloemendal, H., and de Jong, W. W., Structural and functional similarities of bovine α-crystallin and mouse small heat-shock protein. A family of chaperones, *J. Biol. Chem. 268*:1046 (1993).
36. Voorter, C. E., de Haard-Hoekman, W., Merck, K. B., Bloemendal, H., and de Jong, W. W., Elastase inhibition by the C-terminal domains of α-crystallin and small heat-shock protein, *Biochim. Biophys. Acta 1204*:43 (1994).
37. Miron, T., Vancompernolle, K., Vandekerckhove, J., Wilchek, M., and Geiger, B., A

25-kD inhibitor of actin polymerization is a low molecular mass heat shock protein, *J. Cell Biol. 114*:255 (1991).
38. Lavoie, J. N., Hickey, E., Weber, L. A., and Landry, J., Modulation of actin microfilament dynamics and fluid phase pinocytosis by phosphorylation of heat shock protein 27, *J. Biol. Chem. 268*:24210 (1993).
39. Lavoie, J. N., Gingras-Breton, G., Tanguay, R. M., and Landry, J., Induction of Chinese hamster HSP27 gene expression in mouse cells confers resistance to heat shock. HSP27 stabilization of the microfilament organization, *J. Biol. Chem. 268*:3420 (1993).
40. Benndorf, R., Hayess, K., Ryazantsev, S., Wieske, M., Behlke, J., and Lutsch, G., Phosphorylation and supramolecular organization of murine small heat shock protein HSP25 abolish its actin polymerization-inhibiting activity, *J. Biol. Chem. 269*:20780 (1994).
41. Nicholl, I. D. and Quinlan, R. A., Chaperone activity of α-crystallins modulates intermediate filament assembly, *EMBO J. 13*:945 (1994).
42. Zhu, Y., Tassi, L., Lane, W., and Mendelsohn, M. E., Specific binding of the transglutaminase, platelet factor XIII, to Hsp27, *J. Biol. Chem. 269*:22379 (1994).
43. van Noort, J. M., van Sechel, A. C., Bajramovich, J. J., el Quagmiri, M., Polman, C. H., Lassmann, H., and Ravid, R., The small heat shock protein αB-crystallin as candidate autoantigen in multiple sclerosis, *Nature 375*:798 (1995).
44. Gabai, V. L., Zamulaeva, I. V., Mosin, A. F., Makarova, Y. M., Mosina, V. A., Budagova, K. R., Malutina, Y. V., and Kabakov, A. E., Resistance of Ehrlich tumor cells to apoptosis can be due to accumulation of heat shock proteins, *FEBS Lett. 375*:21 (1995).
45. Mehlen, P., Schulze-Osthoff, K., and Arrigo, A.-P., Small stress proteins as novel regulators of apoptosis. Heat shock protein 27 blocks Fas/APO-1- and staurosporine-induced cell death, *J. Biol. Chem. 271*:16510 (1996).
46. Horwitz, J., α-Crystallin can function as a molecular chaperone, *Proc. Natl. Acad. Sci. USA 89*:10449 (1992).
47. Jakob, U., Gaestel, M., Engel, K., and Buchner, J., Small heat shock proteins are molecular chaperones, *J. Biol. Chem. 268*:1517 (1993).
48. Lindquist, S. and Craig, E. A., The heat-shock proteins, *Annu. Rev. Genet. 22*:631 (1988).
49. Morimoto, R. I., Tissières, A., and Gcorgopolous, C., Progress and perspectives on the biology of heat shock proteins and molecular chaperones, *The Biology of Heat Shock Proteins and Molecular Chaperones* (R. I. Morimoto, A. Tissiéres, and C. Georgopolous, eds.), Cold Spring Harbor Laboratory Press, Cold Spring Harbor, NY, 1994, p. 1.
50. Vierling, E., The role of heat shock proteins in plants, *Annu. Rev. Plant Physiol. Plant Biol. 42*:579 (1991).
51. Wotton, D., Freeman, K., and Shore, D., Multimerization of Hsp42p, a novel heat shock protein of *Saccharomyces cerevisiae*, is dependent on a conserved carboxyl-terminal sequence, *J. Biol. Chem. 271*:2717 (1996).
52. Kato, K., Goto, S., Inaguma, Y., Hasegawa, K., Morishita, R., and Asano, T., Purification and characterization of a 20-kDa protein that is highly homologous to αB crystallin, *J. Biol. Chem. 269*:15302 (1994).

53. Allen, S. P., Polazzi, J. O., Gierse, J. K., and Easton, A. M., Two novel heat shock genes encoding proteins produced in response to heterologous protein expression in *Escherichia coli*, *J. Bacteriol. 174*:6938 (1992).
54. Chuang, S. E., Burland, V., Plunkett, G., Daniels, D. L., and Blattner, F. R., Sequence analysis of four new heat-shock genes constituting the *hslTS/ibpAB* and *hslVU* operons in *Escherichia coli*, *Gene134*:1 (1993).
55. Lenne, C. and Douce, R., A low molecular weight heat-shock protein is localized to higher plant mitochondria, *Plant Physiol. 105*:1255 (1994).
56. Waters, E. R., The molecular evolution of the small heat-shock proteins in plants, *Genetics 141*:785 (1995).
57. Plesofsky-Vig, N. and Brambl, R., Gene sequence and analysis of hsp30, a small heat shock protein of *Neurospora crassa* which associates with mitochondria, *J. Biol. Chem. 265*:15432 (1990).
58. Beaulieu, J.-F., Arrigo, A.-P., and Tanguay, R. M., Interaction of *Drosophila* 27 000 M_r heat-shock protein with the nucleus of heat-shocked and ecdysone-stimulated culture cells, *J. Cell Sci. 92*:29 (1989).
59. Zav'yalov, V. P., Zav'yalova, G. A., Denesyuk, A. I., Gaestel, M., and Korpela, T., Structural and functional homology between periplasmic bacterial molecular chaperones and small heat shock proteins, *FEMS Immunol. Med. Microbiol. 11*:265 (1995).
60. Caspers, G. J., Leunissen, J. A., and de-Jong, W. W., The expanding small heat-shock protein family, and structure predictions of the conserved "α-crystallin domain," *J. Mol. Evol. 40*:238 (1995).
61. Lee, D. C., Kim, R. Y., and Wistow, G. J., An avian αB-crystallin. Non-lens expression and sequence similarities with both small (HSP27) and large (HSP70) heat shock proteins, *J. Mol. Biol. 232*:1221 (1993).
62. Verbon, A., Hartskeerl, R. A., Schuitema, A., Kolk, A. H. J., Young, D. B., and Lathigra, R., The 14,000-molecular-weight antigen of *Mycobacterium tuberculosis* is related to the α-crystallin family of low-molecular-weight heat shock proteins, *J. Bacteriol. 174*:1352 (1992).
63. Petko, L. and Lindquist, S., Hsp26 is not required for growth at high temperatures, nor for thermotolerance, spore development, or germination, *Cell 45*:885 (1986).
64. Rahman, D. R., Bentley, N. J., and Tuite, M. F., The *Saccharomyces cerevisiae* small heat shock protein Hsp26 inhibits actin polymerisation, *Biochem. Soc. Trans. 23*:77 (1995).
65. Orlandi, I., Cavadini, P., Popolo, L., and Vai, M., Cloning, sequencing and regulation of a cDNA encoding a small heat-shock protein from *Schizosaccharomyces pombe*, *Biochim. Biophys. Acta 1307*:129 (1996).
66. Jang, Y. J., Park, S. K., and Yoo, H. S., Isolation of an HSP12-homologous gene of *Schizosaccharomyces pombe* suppressing a temperature-sensitive mutant allele of cdc4, *Gene 172*:125 (1996).
67. Wheeler, J. C., Bieschke, E. T., and Tower, J., Muscle-specific expression of *Drosophila* hsp70 in response to aging and oxidative stress, *Proc. Natl. Acad. Sci. USA 92*:10408 (1995).
68. Sandaltzopoulos, R., Mitchelmore, C., Bonte, E., Wall, G., and Becker, P. B., Dual

regulation of the *Drosophila* hsp26 promoter in vitro, *Nucleic Acids Res. 23*:2479 (1995).

69. Mehlen, P., Preville, X., Chareyron, P., Briolay, J., Klemenz, R., and Arrigo, A.-P., Constitutive expression of human hsp27, *Drosophila* hsp27, or human αB-crystallin confers resistance to TNF- and oxidative stress-induced cytotoxicity in stably transfected murine L929 fibroblasts, *J. Immunol. 154*:363 (1995).
70. Groenen, P. J. T. A., Merck, K. B., de Jong, W. W., and Bloemendal, H., Structure and modifications of the junior chaperone α-crystallin: From lens transparency to molecular pathology, *Eur. J. Biochem. 225*:1 (1994).
71. Gaestel, M., Schroder, W., Benndorf, R., Lippmann, C., Buchner, K., Hucho, F., Erdmann, V. A., and Bielka, H., Identification of the phosphorylation sites of the murine small heat shock protein hsp25, *J. Biol. Chem. 266*:14721 (1991).
72. Gaestel, M., Gotthardt, R., and Müller, T., Structure and organisation of a murine gene encoding small heat-shock protein hsp25, *Gene 128*:279 (1993).
73. Ehrnsperger, M., Gräber, S., Gaestel, M., and Buchner, J., Binding of nonnative protein to Hsp25 during heat shock creates a reservoir of folding intermediates for reactivation, *EMBO J.* (in press; 1997).
74. Helm, K. W., LaFayette, P. R., Nagao, R. T., Key, J. L., and Vierling, E., Localization of small heat shock proteins to the higher plant endomembrane system, *Mol. Cell. Biol. 13*:238 (1993).
75. Helm, K. W., Schmeits, J., and Vierling, E., An endomembrane-localized small heat-shock protein from *Arabidopsis thaliana*, *Plant Physiol. 107*:287 (1995).
76. Lee, G. J., Roseman, A. M., Saibil, H. R., and Vierling, E., A small heat shock protein stably binds heat-denatured model substrates and can maintain a substrate in a folding competent state, *EMBO J.* (in press, 1997).
77. Lenne, C., Block, M. A., Garin, J., and Douce, R., Sequence and expression of the mRNA encoding HSP22, the mitochondrial small heat-shock protein in pea leaves, *Biochem. J. 311*:805 (1995).
78. LaFayette, P. R., Nagao, R. T., O'Grady, K., Vierling, E., and Key, J. L., Molecular characterization of cDNAs encoding low-molecular-weight heat shock proteins of soybean, *Plant. Mol. Biol. 30*:159 (1996).
79. Chen, Q. and Vierling, E., Analysis of conserved domains identifies a unique structural feature of a chloroplast heat shock protein, *Mol. Gen. Genet. 226*:425 (1991).
80. Kato, K., Shinohara, H., Kurobe, N., Goto, S., Inaguma, Y., and Ohshima, K., Immunoreactive αA crystallin in rat non-lenticular tissues detected with a sensitive immunoassay method, *Biochim. Biophys. Acta 1080*:173 (1991).
81. Landry, J., Chrétien, P., Laszlo, A., and Lambert, H., Phosphorylation of HSP27 during development and decay of thermotolerance in Chinese hamster cells, *J. Cell Physiol. 147*:93 (1991).
82. Morimoto, R. I., Tissiéres, A., and Georgopolous, C., The stress response, function of the proteins, and perspectives, *Stress Proteins in Biology and Medicine* (R. I. Morimoto, Tissiéres, A., and Georgopolous, C., eds.), Cold Spring Harbor Laboratory Press, Cold Spring Harbor, NY, 1990, p. 1.
83. Hickey, E., Brandon, S. E., Potter, R., Stein, G., Stein, J., and Weber, L. A., Sequence

and organization of genes encoding the human 27 kDa heat shock protein, *Nucleic Acids Res. 14*:4127 (1986).
84. Klemenz, R., Fröhli, E., Aoyama, A., Hoffmann, S., Simpson, R. J., Moritz, R. L., and Schafer, R., αB crystallin accumulation is a specific response to Ha-*ras* and v-*mos* oncogene expression in mouse NIH 3T3 fibroblasts, *Mol. Cell. Biol. 11*:803 (1991).
85. DasGupta, S., Hohman, T. C., and Carper, D., Hypertonic stress induces αB-crystallin, *Exp. Eye Res. 54*:461 (1992).
86. Fröhli, E., Aoyama, A., and Klemenz, R., Cloning of the mouse hsp25 gene and an extremely conserved hsp25 pseudogene, *Gene 128*:273 (1993).
87. Yost, H. J., Petersen, R. B., and Lindquist, S., Posttranslational regulation of heat shock protein synthesis in *Drosophila, Stress Proteins in Biology and Medicine*, (R. I. Morimoto, A. Tissiéres, and Georgopolous, C., eds.), Cold Spring Harbor Laboratory Press, Cold Spring Harbor, NY, 1990, p. 379.
88. Edington, B. V. and Hightower, L. E., Induction of a chicken small heat shock (stress) protein: Evidence of multilevel posttranscriptional regulation, *Mol. Cell. Biol. 10*:4886 (1990).
89. Gotthardt, R., Neininger, A., and Gaestel, M., The anti-cancer drug cisplatin induces H25 in Ehrlich ascites tumor cells by a mechanism different from transcriptional stimulation influencing predominantly H25 translation, *Int. J. Cancer 66*:790 (1996).
90. Edwards, D. P., Adams, D. J., Savage, N., and McGuire, W. L., Estrogen induced synthesis of specific proteins in human breast cancer cells, *Biochem. Biophys. Res. Commun. 93*:804 (1980).
91. Fuqua, S. A., Blum-Salingaros, M., and McGuire, W. L., Induction of the estrogen-regulated "24K" protein by heat shock, *Cancer Res. 49*:4126 (1989).
92. Ciocca, D. R., Adams, D. J., Edwards, D. P., Bjercke, R. J., and McGuire, W. L., Estrogen-induced 24K protein in MCF-7 breast cancer cells is localized in granules, *Breast Cancer Res. Treat. 4*:261 (1984).
93. Mendelsohn, M. E., Zhu, Y., and O'Neill, S., The 29-kDa proteins phosphorylated in thrombin-activated human platelets are forms of the estrogen receptor-related 27-kDa heat shock protein, *Proc. Natl. Acad. Sci. USA 88*:11212 (1991).
94. Ciocca, D. R. and Luque, E. H., Immunological evidence for the identity between the hsp27 estrogen-regulated heat shock protein and the p29 estrogen receptor-associated protein in breast and endometrial cancer, *Breast Cancer Res. Treat. 20*:33 (1991).
95. Tora, L., Gaub, M. P., Mader, S., Dierich, A., Bellard, M., and Chambon, P., Cell-specific activity of a GGTCA half-palindromic oestrogen-responsive element in the chicken ovalbumin gene promoter, *EMBO J. 7*:3771 (1988).
96. Ciocca, D. R., Stati, A. O., Fanelli, M. A., and Gaestel, M., Expression of heat shock protein 25,000 in rat uterus during pregnancy and pseudopregnancy, *Biol. Reprod. 54*:1326 (1996).
97. Thomas, S. R. and Lengyel, J. A., Ecdysteroid-regulated heat-shock gene expression during *Drosophila melanogaster* development, *Dev. Biol. 115*:434 (1986).
98. Cohen, R. S. and Meselson, M., Separate regulatory elements for the heat-inducible and ovarian expression of the *Drosophila* hsp26 gene, *Cell 43*:737 (1985).

99. Hoffman, E. and Corces, V., Sequences involved in temperature and ecdysterone-induced transcription are located in separate regions of a *Drosophila melanogaster* heat shock gene, *Mol. Cell. Biol. 6*:663 (1986).
100. Ito, H., Hasegawa, K., Inaguma, Y., Kozawa, O., and Kato, K., Enhancement of stress-induced synthesis of hsp27 and αB crystallin by modulators of the arachidonic acid cascade, *J. Cell Physiol. 166*:332 (1996).
101. Head, M. W., Corbin, E., and Goldman, J. E., Coordinate and independent regulation of αB-crystallin and HSP27 expression in response to physiological stress, *J. Cell. Physiol. 159*:41 (1994).
102. Wistow, G. J., Domain structure and evolution in α-crystallins and small heat-shock proteins, *FEBS Lett. 181*:1 (1985).
103. Carver, J. A., Aquilina, J. A., Truscott, R. J. W., and Ralston, G. B., Identification by ^{1}H NMR spectroscopy of flexible C-terminal extensions in bovine lens α-crystallin, *FEBS Lett. 311*:143 (1992).
104. Carver, J. A., Esposito, G., Schwedersky, G., and Gaestel, M., ^{1}H NMR spectroscopy reveals that mouse Hsp25 has a flexible C-terminal extension of 18 amino acids, *FEBS Lett. 369*:305 (1995).
105. Takemoto, L., Emmons, T., and Horwitz, J., The C-terminal region of α-crystallin: Involvement in protection against heat induced denaturation, *Biochem. J. 294*:435 (1993).
106. Mehlen, P., Briolay, J., Smith, L., Diaz-Latoud, C., Fabre, N., Pauli, D., and Arrigo, A.-P., Analysis of the resistance to heat and hydrogen peroxide stresses in COS cells transiently expressing wild type or deletion mutants of the *Drosophila* 27-kDa heat-shock protein, *Eur. J. Biochem. 215*:277 (1993).
107. Merck, K. B., Horwitz, J., Kersten, M., Overkamp, P., Gaestel, M., Bloemendal, H., and de Jong, W. W., Comparison of the homologous carboxy-terminal domain and tail of α-crystallin and small heat shock proteins, *Mol. Biol. Rep. 18*:209 (1993).
108. Wistow, G., Possible tetramer-based quaternary structure for α-crystallins and small heat shock proteins, *Exp. Eye Res. 56*:729 (1993).
109. Le Breton, E. R. and Carver, J. A., Solution conformation of bovine lens α- and βB2-crystallin terminal extensions, *Int. J. Pept. Protein Res. 47*:9 (1996).
110. Ifeanyi, F. and Takemoto, L., Differential binding of α-crystallins to bovine lens membrane, *Exp. Eye Res. 49*:143 (1989).
111. Ifeanyi, F. and Takemoto, L., Involvement of the N-terminal region in α crystallin lens membrane recognition, *Exp. Eye Res. 53*:305 (1991).
112. Cenedella, R. J. and Chandrasekher, G., High capacity binding of α-crystallins to various bovine lens membrane preparations, *Curr. Eye Res. 12*:1025 (1993).
113. Maiti, M., Kono, M., and Chakrabarti, B., Heat-induced changes in the conformation of α- and β-crystallins: Unique thermal stability of α-crystallin, *FEBS Lett. 236*:109 (1988).
114. Walsh, M. T., Sen, A. C., and Chakrabarti, B., Micellar subunit assembly in a three-layer model of oligomeric α-crystallin, *J. Biol. Chem. 266*:20079 (1991).
115. Surewicz, W. K. and Olesen, P. R., On the thermal stability of α-crystallin: A new insight from infrared spectroscopy, *Biochemistry 34*:9655 (1995).
116. Gesierich, U. and Pfeil, W., The conformational stability of α-crystallin is rather low: Calorimetric results, *FEBS Lett. 393*:151 (1996).

117. Kono, M., Sen, A. C., and Chakrabarti, B., Thermodynamics of thermal and a thermal denaturation of γ-crystallins: Changes in conformational stability upon glutathion reduction, *Biochemistry 29*:464 (1990).
118. Steadman, L., Trautman, P. A., Lawson, E. Q., Raymond, M. J., Mood, D. A., Thomson, J. A., and Middaugh, C. R., A differential scanning calorimetric study of the bovine lens crystallins, *Biochemistry 28*:9653 (1989).
119. van den Oetelaar, P. J. M. and Hoenders, H. J., Folding–unfolding and aggregation-dissociation of bovine α-crystallin subunits; evidence for unfolding intermediates of the αA subunits, *Biochim. Biophys. Acta 995*:91 (1989).
120. Santini, S. A., Mordente, A., Meucci, E., Miggiano, G. A. D., and Martorana, G. E., Conformational stability of bovine α-crystallin. Evidence for a destabilizing effect of ascorbate, *Biochem. J. 287*:107 (1992).
121. Siezen, R. J. and Argos, P., Structural homology of lens crystallins. III. Secondary structure estimation from circular dichroism and prediction from amino acid sequences, *Biochim. Biophys. Acta 748*:56 (1983).
122. Surewicz, W. K., Mantsch, H. H., and Chapman, D., Determination of protein secondary structure by Fourier transform infrared spectroscopy: A critical assessment, *Biochemistry 32*:389 (1993).
123. Lamba, O. P., Borchman, D., Sinha, S. K., Shah, J., Renugopalakrishnan, V., and Yappert, M. C., Estimation of the secondary structure and conformation of bovine lens crystallins by infrared spectroscopy: Quantitative analysis and resolution by Fourier self-deconvolution and curve fit, *Biochim. Biophys. Acta 1163*:113 (1993).
124. Siezen, R. J., Reflections on the internal primary, secondary and tertiary structure homology of the eye lens proteins α-, β-, and γ-crystallin, *FEBS Lett 133*:1 (1981).
125. Argos, P. and Siezen, R. J., Structural homology of lens crystallins. A method to detect protein structural similarity from primary sequences, *Eur. J. Biochem. 131*:143 (1983).
126. Holmgren, A. and Brändén, C.-I., Crystal structure of chaperone protein PapD reveals an immunoglobulin fold, *Nature 342*:248 (1989).
127. Holmgren, A., Kuehn, M. J., Brändén, C.-I., and Hultgren, S. J., Conserved immunoglobulin-like features in a family of periplasmatic pilus chaperones in bacteria, *EMBO J. 11*:1617 (1992).
128. Tardieu, A., Laporte, D., Licinio, P., Kropp, B., and Delaye, M., Calf lens α-crystallin quaternary structure: A three-layer tetrahedral model, *J. Mol. Biol. 192*:711 (1986).
129. Andreasi-Bassi, F., Arcovito, G., De Spirito, M., and Mordente, A., Self-similarity properties of α-crystallin supramolecular aggregates, *Biophys. J. 69*:2720 (1995).
130. Spector, A., Li, L.-K., Augusteyn, R. C., Schneider, A., and Freund, T., α-Crystallin. The isolation and characterization of distinct macromolecular fractions, *Biochem. J. 124*:337 (1971).
131. Clauwaert, J., Ellerton, H. D., Koretz, J. F., Thomson, K., and Augusteyn, R. C., The effect of temperature on the renaturation of α-crystallin, *Curr. Eye Res. 8*:397 (1989).
132. Bindels, J. G., Siezen, R. J., and Hoenders, H. J., A model for the architecture of α-crystallin, *Opthalmic Res. 11*:441 (1979).
133. Siezen, R. J., Coenders, F. G., and Hoenders, H. J., Three classes of sulphydryl

group in bovine α-crystallin according to reactivity to various reagents, *Biochim. Biophys. Acta 537*:456 (1978).

134. Siezen, R. J., Bindels, J. G., and Hoenders, H. J., The quaternary structure of bovine α-crystallin. Effects of variation in alkaline pH, ionic strength, temperature and calcium ion concentration, *Eur. J. Biochem. 111*:435 (1980).
135. Siezen, R. J. and Bindels, J. G., Stepwide dissociation/denaturation and reassociation/renaturation of bovine α-crystallin in urea and guanidine hydrochloride: Sedimentation, fluorescence, near-ultraviolet and far-ultraviolet circular dichroism studies, *Exp. Eye Res. 34*:969 (1982).
136. Thomson, J. A. and Augusteyn, R. C., a_m-Crystallin: The native form of the protein? *Exp. Eye Res. 37*:367 (1983).
137. Thomson, J. A. and Augusteyn, R. C., On the structure of a_m-crystallin. The reversibility of urea dissociation, *J. Biol. Chem. 259*:4339 (1984).
138. Hendriks, W., Weetink, H., Voorter, C. E. M., Sanders, J., Bloemendal, H., and de Jong, W. W., The alternative splicing product α Ains-crystallin is structurally equivalent to αA and αB subunits in the rat α-crystallin aggregate, *Biochim. Biophys. Acta 1037*:58 (1990).
139. Thomson, J. A., *A model for multisubunit protein assemblies*, PhD Dissertation, University of Melbourne, 1985.
140. Augusteyn, R. C., Hum, T. P., Putilin, T. P., and Thomson, J. A., The location of sulphydryl groups in α-crystallin, *Biochim. Biophys. Acta 915*:132 (1987).
141. Augusteyn, R. C. and Koretz, J. F., A possible structure for α-crystallin, *FEBS Lett. 222*:1 (1987).
142. Augusteyn, R. C., Ghiggino, K. P., and Putilina, T., Studies on the location of aromatic amino acids in α-crystallin, *Biochim. Biophys. Acta 1162*:61 (1993).
143. van den Oetelaar, P. J. M., van Someren, P. F., Thomson, J. A., Siezen, R. J., and Hoenders, H. J., A dynamic quaternary structure of bovine α-crystallin as indicated from intermolecular exchange of subunits, *Biochemistry 29*:3488 (1990).
144. Merck, K. B., de Haard-Hoekman, W. A., Oude-Essink, B. B., Bloemendal, H., and de Jong, W. W., Expression and aggregation of recombinant αA-crystallin and its two domains, *Biochim. Biophys. Acta 1130*:267 (1992).
145. van den Oetelaar, P. J. M., Clauwaert, J., van Laethem, M., and Hoenders, H. J., The influence of isolation conditions on the molecular mass of bovine α-crystallin, *J. Biol. Chem. 260*:14030 (1985).
146. Lavoie, J. N., Lambert, H., Hickey, E., Weber, L. A., and Landry, J., Modulation of cellular thermoresistance and actin filament stability accompanies phosporylation-induced changes in the oligomeric structure of heat shock protein 27, *Mol. Cell. Biol. 15*:505 (1995).
147. Chen, Q., Osteryoung, K., and Vierling, E., A 21-kDa chloroplast heat shock protein assembles into high molecular weight complexes in vivo and in organelle, *J. Biol. Chem. 269*:13216 (1994).
148. Dudich, I. V., Zav'yalov, V. P., Pfeil, W., Gaestel, M., Zav'yalova, G. A., Denesyuk, A. I., and Korpela, T., Dimer structure as a minimum cooperative subunit of small heat-shock proteins, *Biochim. Biophys. Acta 1253*:163 (1995).
149. Arrigo, A.-P. and Pauli, D., Characterization of HSP27 and three immunologically

related polypeptides during *Drosophila* development, *Exp. Cell. Res. 175*:169 (1988).

150. Longoni, S., Lattonen, S., Bullock, G., and Chiesi, M., Cardiac α-crystallin. II. Intracellular localization, *Mol. Cell. Biol. 97*:121 (1990).

151. Zantema, A., Verlaan-De Vries, M., Maasdam, D., Bol, S., and van der Eb, A., Heat shock protein 27 and αB-crystallin can form a complex, which dissociates by heat shock, *J. Biol. Chem. 267*:12936 (1992).

152. Kato, K., Shinohara, H., Goto, S., Inaguma, Y., Morishita, R., and Asano, T., Copurification of small heat shock protein with αB crystallin from human skeletal muscle, *J. Biol. Chem. 267*:7718 (1992).

153. Rossi, J. M. and Lindquist, S., The intracellular location of yeast heat shock protein 26 varies with metabolism, *J. Cell. Biol. 108*:425 (1989).

154. Arrigo, A.-P., Cellular localization of hsp23 during *Drosophila* development and following subsequent heat shock, *Dev. Biol. 122*:39 (1987).

155. Bentley, N. J., Fitch, I. T., and Tuite, M. F., The small heat-shock protein Hsp26 of *Saccharomyces cerevisiae* assembles into a high molecular weight aggregate, *Yeast 8*:95 (1992).

156. Osteryoung, K. W. and Vierling, E., Dynamics of small heat shock protein distribution within the chloroplasts of higher plants, *J. Biol. Chem. 269*:28676 (1994).

157. Voorter, C. E., Mulders, J. W., Bloemendal, H., and de Jong, W. W., Some aspects of the phosphorylation of α-crystallin A, *Eur. J. Biochem. 160*:203 (1986).

158. Chiesa, R., Gawinowicz-Kolks, M. A., Kleiman, N. J., and Spector, A., The phosphorylation sites of the B2 chain of bovine α-crystallin, *Biochem. Biophys. Res. Commun. 144*:1340 (1987).

159. Kim, Y. J., Shuman J., Sette, M., and Przybyla, A., Nuclear localization and phosphorylation of three 25-kilodalton rat stress proteins, *Mol. Cell. Biol. 4*:468 (1984).

160. Arrigo, A.-P. and Welch, W. J., Dynamic changes in the structure and intracellular locale of the mammalian low-molecular-weight heat shock protein. Characterization and purification of the small 28,000-dalton mammalian heat shock protien, *J. Biol. Chem. 262*:15359 (1987).

161. Benndorf, R., Kraft, R., Otto, A., Stahl, J., Bohm, H., and Bielka, H., Purification of the growth-related protein p25 of the Ehrlich ascites tumor and analysis of its isoforms, *Biochem. Int. 17*:225 (1988).

162. Voorter, C. E., de Haard-Hoekman, W. A., Roersma, E. S., Meyer, H. E., Bloemendal, H., and de Jong, W. W., The in vivo phosphorylation sites of bovine αB-crystallin, *FEBS Lett. 259*:50 (1989).

163. Landry, J., Lambert, H., Zhou, M., Lavoie, J. N., Hickey, E., Weber, L. A., and Anderson, C. W., Human HSP27 is phosphorylated at serines 78 and 82 by heat shock and mitogen-activated kinases that recognize the same amino acid motif as S6 kinase II, *J. Biol. Chem. 267*:794 (1992).

164. Rollet, E. and Best-Belpomme, M., HSP 26 and 27 are phosphorylated in response to heat shock and ecdysterone in *Drosophila melanogaster* cells, *Biochem. Biophys. Res. Commun. 141*:426 (1986).

165. Marin, R., Landry, J., and Tanguay, R. M., Tissue-specific posttranslational modifi-

cation of the small heat shock protein HSP27 in *Drosophila*, *Exp. Cell Res.* *223*:1 (1996).

166. Mann, E., McDermott, M. J., Goldman, J., Chiesa, R., and Spector, A., Phosphorylation of α-crystallin B in Alexander's disease brain, *FEBS Lett.* *294*:133 (1991).
167. Chiesa, R., McDermott, M. J., and Spector, A., Differential synthesis and phosphorylation of the α-crystallin A- and B-chains during bovine lens fiber cell differentiation, *Curr. Eye Res.* *8*:151 (1989).
168. Chiesa, R. and Spector, A., The dephosphorylation of lens α-crystallin A chain, *Biochem. Biophys. Res. Commun.* *162*:1494 (1989).
169. Saklatvala, J., Kaur, P., and Guesdon, F., Phosphorylation of the small heat-shock protein is regulated by interleukin-1, tumor necrosis factor, growth factors, bradykinin and ATP, *Biochem. J.* *277*:635 (1991).
170. Zhou, M., Lambert, H., and Landry, J., Transient activation of a distinct serine protein kinase is responsible for 27-kDa heat shock protein phosphorylation in mitogen-stimulated and heat-shocked cells, *J. Biol. Chem.* *268*:35 (1993).
171. Kaur, P., Welch, W. J., and Saklatvala, J., Interleukin 1 and tumour necrosis factor increase phosphorylation of the small heat shock protein. Effects in fibroblasts, Hep G2 and U937 cells, *FEBS Lett.* *258*:269 (1989).
172. Arrigo, A.-P., Tumor necrosis factor α induces the rapid phosphorylation of the mammalian low molecular weight heat shock protien hsp28, *Mol. Cell. Biol.* *10*:1276 (1990).
173. Oesterreich, S., Benndorf, R., and Bielka, H., The expression of the growth-related 25kDa protein (p25) of Ehrlich ascites tumor cells is increased by hyperthermic treatment (heat shock), *Biomed. Biochim. Acta* *49*:219 (1990).
174. Huot, J., Lambert, H., Lavoie, J. N., Guimond, A., Houle, F., and Landry, J., Characterization of 45-kDa/54-kDa HSP27 kinase, a stress-sensitive kinase which may activate the phosphorylation-dependent protective function of mammalian 27-kDa heat-shock protein HSP27, *Eur. J. Biochem.* *227*:416 (1995).
175. Regazzi, R., Eppenberger, U., and Fabbro, D., The 27,000 dalton stress proteins are phosphorylated by protein kinases C during the tumor promoter-mediated growth inhibition of human mammary carcinoma cells, *Biochem. Biophys. Res. Commun.* *152*:62 (1988).
176. Michishita, M., Satoh, M., Yamaguchi, M., Hirayoshi, K., Okuma, M., and Nagata, K., Phosphorylation of the stress protein hsp27 is an early event in murine myelomonocytic leukemic cell differentiation induced by leukemia inhibitory factor/D-factor, *Biochem. Biophys. Res. Commun.* *176*:979 (1991).
177. Spector, N. L., Ryan, C., Samson, W., Levine, H., Nadler, L. M., and Arrigo, A.-P., Heat shock protien is a unique marker of growth arrest during macrophage differentiation of HL-60 cells, *J. Cell. Physiol.* *156*:619 (1993).
178. Minowada, G. and Welch, W., Variation in the expression and/or phosphorylation of the human low molecular weight stress protein during in vitro cell differentiation, *J. Biol. Chem.* *270*:7047 (1995).
179. Santell, L., Bartfeld, N. S., and Levin, E. G., Identification of a protein transiently phosphorylated by activators of endothelial cell function as the heat-shock protein HSP27. A possible role for protein kinase C, *Biochem. J.* *284*:705 (1992).

180. Issels, R. D., Meier, T. H., Muller, E., Multhoff, G., and Wilmanns, W., Ifosfamide induced stress response in human lymphocytes, *Mol. Aspects Med. 14*:281 (1993).
181. Kasahara, K., Ikuta, T., Chida, K., Asakura, R., and Kuroki, T., Rapid phosphorylation of 28-kDa heat-shock protein by treatment with okadaic acid and phorbol ester of BALB/MK-2 mouse keratinocytes, *Eur. J. Biochem. 213*:1101 (1993).
182. Ahlers, A., Engel, K., Sott, C., Gaestel, M., Herrmann, F., and Brach, M. A., Interleukin-3 and granulocyte–macrophage colony-stimulating factor induce activation of the MAPKAP kinase 2 resulting in in vitro serine phosphorylation of the small heat shock protein (Hsp 27), *Blood 83*:1791 (1994).
183. Belka, C., Ahlers, A., Sott, C., Gaestel, M., Herrmann, F., and Brach, M. A., Interleukin (IL)-6 signaling leads to phosphorylation of the small heat shock protien (Hsp)27 through activation of the MAP kinase and MAPAKAP kinase 2 pathway in monocytes and monocytic leukemia cells, *Leukemia 9*:288 (1995).
184. Spector, A., Chiesa, R., Sredy, J., and Garner, W., cAMP-dependent phosphorylation of bovine lens α-crystallin, *Proc. Natl. Acad. Sci. USA 82*:4712 (1985).
185. Kantorow, M. and Piatigorsky, J., α-Crystallin/small heat shock protein has autokinase activity, *Proc. Natl. Acad. Sci. USA 91*:3112 (1994).
186. Kantorow, M., Horwitz, J., van Boekel, M. A., de Jong, W. W., and Piatigorsky, J., Conversion from oligomers to tetramers enhances autophosphorylation by lens αA-crystallin. Specificity between αA- and αB-crystallin subunits, *J. Biol. Chem. 270*:17215 (1995).
187. Stokoe, D., Engel, K., Campbell, D. G., Cohen, P., and Gaestel, M., Identification of MAPKAP kinase 2 as a major enzyme responsible for the phosphorylation of the small mammalian heat shock proteins, *FEBS Lett. 313*:307 (1992).
188. Stokoe, D., Campbell, D. G., Nakielny, S., Hidaka, H., Leevers, S. J., Marshall, C., and Cohen, P., MAPKAP kinase-2; a novel protein kinase activated by mitogen-activated protein kinase, *EMBO J. 11*:3985 (1992).
189. Cano, E., Doza, Y. N., Ben-Levy, R., Cohen, P., and Mahadevan, L. C., Identification of anisomycin-activated kinases p45 and p55 in murine cells as MAPKAP kinase-2, *Oncogene 12*:805 (1996).
190. Engel, K., Schultz, H., Martin, F., Kotlyarov, A., Plath, K., Hahn, M., Heinemann, U., and Gaestel, M., Constitutive activation of mitogen-activated protein kinase-activated protein kinase 2 by mutation of phosphorylation sites and an A-helix motif, *J. Biol. Chem. 270*:27213 (1995).
191. Stokoe, D., Caudwell, B., Cohen, P. T., and Cohen, P., The substrate specificity and structure of mitogen-activated protein (MAP) kinase-activated protein kinase-2, *Biochem. J. 296*:843 (1993).
192. Sithanandam, G., Latif, F., Duh, F. M. et al., 3pK, a new mitogen-activated protein kinase-activated protein kinase located in the small cell lung cancer tumor suppresser gene region, *Mol. Cell. Biol. 16*:868 (1996).
193. McLaughlin, M. M., Kumar, S., McDonnell, P. C., Van-Horn, S., Lee, J. C., Livi, G. P., and Young, P. R., Identification of mitogen-activated protein (MAP) kinase-activated protein kinase-3, a novel substrate of CSBP p38 MAP kinase, *J. Biol. Chem. 271*:8488 (1996).
194. Ludwig, S., Engel, K., Hoffmeyer, A., Sithanandam, G., Palm, D., Gaestel, M., and

Rapp, U., 3pK, a novel MAP kinase activated protein kinase is targeted by three MAP kinase pathways, *Mol. Cell. Biol.* (in press, 1996).

195. Rouse, J., Cohen, P., Trigon, S., Morange, M., Alonso-Llamazares, A., Zamanillo, D., Hunt, T., and Nebreda, A. R., A novel kinase cascade triggered by stress and heat shock that stimulates MAPKAP kinase-2 and phosphorylation of the small heat shock proteins, *Cell 78*:1027 (1994).
196. Freshney, N. W., Rawlinson, L., Guesdon, F., Jones, E., Cowley, S., Hsuan, J., and Saklatvala, J., Interleukin-1 activates a novel protein kinase cascade that results in the phosphorylation of Hsp27, *Cell 78*:1039 (1994).
197. Ben-Levy, R., Leighton, I. A., Doza, Y. N., Attwood, P., Morrice, N., Marshall, C. J., and Cohen, P., Identification of novel phosphorylation sites required for activation of MAPKAP kinase 2, *EMBO J. 14*:5920 (1995).
198. Kyriakis, J. M. and Avruch, J., Protein kinase cascades activated by stress and inflammatory cytokines, *BioEssays 18*:567 (1996).
199. Schultz, H., Engel, K., Rogella, T., Lee, J. C., and Gaestel, M., The protein kinase inhibitor SB 203580 uncouples PMA-induced differentiation of HL-60 cells from phosphorylation of Hsp27. *Cell Stress Chaperones* (in press, 1996).
200. Lee, J. C., Laydon, J. T., McDonnell, P. C. et al., A protein kinase involved in the regulation of inflammatory cytokine biosynthesis, *Nature 372*:739 (1994).
201. Cuenda, A., Rouse, J., Doza, Y. N., Meier, R., Cohen, P., Gallagher, T. F., Young, P. R., and Lee, J. C., SB 203580 is a specific inhibitor of a MAP kinase homologue which is stimulated by cellular stresses and interleukin-1, *FEBS Lett. 364*:229 (1995).
202. Gaestel, M., Benndorf, R., Hayess, K., Priemer, E., and Engel, K., Dephosphorylation of the small heat shock protein hsp25 by calcium/calmodulin-dependent (type 2B) protein phosphatase, *J. Biol. Chem. 267*:21607 (1992).
203. Rollet, E., Lavoie, J. N., Landry, J., and Tanguay, R. M., Expression of *Drosophila*'s 27 kDa heat shock protein into rodent cells confers thermal resistance, *Biochem. Biophys. Res. Commun. 185*:116 (1992).
204. DeRocher, A. E., Helm, K. W., Lauzon, L. M., and Vierling, E., Expression of a conserved family of cytoplasmic low molecular weight heat shock proteins during heat stress and recovery, *Plant Physiol. 96*:1038 (1991).
205. DeRocher, A. E. and Vierling, E., Developmental control of small heat shock protein expression during pea seed maturation, *Plant J. 5*:93 (1994).
206. Susek, R. E. and Lindquist, S. L., hsp26 of *Saccharomyces cerevisiae* is related to the superfamily of small heat shock proteins but is without a demonstrable function, *Mol. Cell. biol. 9*:5265 (1989).
207. Hout, J., Houle, F., Spitz, D. R., and Landry, J., HSP27 phosphorylation-mediated resistance against actin fragmentation and cell death induced by oxidative stress, *Cancer Res. 56*:273 (1996).
208. Oesterreich, S., Weng, C. N., Qiu, M., Hilsenbeck, S. G., Osborne, C. K., and Fuqua, S. A., The small heat shock protein hsp27 is correlated with growth and drug resistance in human breast cancer cell lines, *Cancer Res. 53*:4443 (1993).
209. Jaenicke, R. and Creighton, T. E., Junior chaperones, *Curr. Biol. 3*: 234 (1993).
210. Zhu, Y., O'Neill, S., Saklatvala, J., Tassi, L., and Mendelsohn, M. E., Phosphory-

lated HSP27 associates with the activation-dependent cytoskeleton in human platelets, *Blood 84*:3715 (1994).
211. Kampinga, H. H., Brunsting, J. F., Stege, G. J. J., Konings, A. W. T., and Landry, J., Cells overexpressing Hsp27 show accelerated recovery from heat-induced nuclear protein aggregation, *Biochem. Biophys. Res. Commun. 204*:1170 (1994).
212. Stege, G. J. J., Brunsting, J. F., Kampinga, H. H., and Konings, A. W. T., Thermotolerance and nuclear protein aggregation: Protection against initial damage or better recovery? *J. Cell. Physiol. 164*:579 (1995).
213. Mehlen, P., Kretz-Remy, C., Préville, X., and Arrigo, A.-P., Human hsp27, *Drosophila* hsp27 and human αB-crystallin expression-mediated increase in glutathion is essential for the protective activity of these proteins against TNFα-induced cell death, *EMBO J. 15*:2695 (1996).
214. Knauf, U., Jakob, U., Engel, K., Buchner, J., and Gaestel, M., Stress and mitogen-induced phosphorylation of the small heat shock protein Hsp25 by MAPKAP kinase-2 is not essential for chaperone properties and thermoresistance, *EMBO J. 13*:54 (1994).
215. Loktionova, S. A., Ilyinskaya, O. P., Gabai, V. L., and Kabakov, A. E., Distinct effects of heat shock and ATP depletion on distribution and isoform patterns of human Hsp27 in endothelial cells, *FEBS Lett. 392*:100 (1996).
216. Zhang, S., Han, J., Sells, M. A., Chernoff, J., Knaus, U. G., Ulevitch, R. J., and Bokoch, G. M., Rho family GTPases regulate p38 mitogen-activated protein kinase through the downstream mediator Pak1, *J. Biol. Chem. 270*:23934 (1995).
217. Ridley, A. J., Paterson, H. F., Johnston, C. J., Diekmann, D., and Hall, A., The small GTP-binding protein rac regulates growth factor-induced membrane ruffling, *Cell 70*:401 (1992).
218. Nobes, C. D. and Hall, A., Rho, rac, and cdc42 GTPases regulate the assembly of multimolecular focal complexes associated with actin stress fibers, lamellipodia, and filopodia, *Cell 81*:53 (1995).
219. Dubin, R. A., Wawrousek, E. F., and Piatigorsky, J., Expression of the murine αB-crystallin gene is not restricted in the lens, *Mol. Cell. Biol. 9*:1083 (1989).
220. Iwaki, T., Kume-Iwaki, A., and Goldman, J. E., Cellular distribution of α-B crystallin in non lenticular tissues, *J. Histochem. Cytochem. 38*:31 (1990).
221. Hanash, S. M., Strahler, J. R., Chan, Y. et al., Data base analysis of protein expression pattern during T-cell ontogeny and activation, *Proc. Natl. Acad. Sci. USA 90*:3314 (1993).
222. Buzin, C. H. and Bournias-Vardiabasis, N., Teratogens induce a subset of small heat shock proteins in *Drosophila* primary embryonic cell cultures, *Proc. Natl. Acad. Sci. USA 81*:4075 (1984).
223. Mirkes, P. E., Hyperthermia-induced heat shock response and thermotolerance in postimplantation rat embryos, *Dev. Biol. 119*:115 (1987).
224. Hansen, D. K., Anson, J. F., Hinson, W. G., and Pipkin, Jr., J. L., Phenytoin-induced stress protein synthesis in mouse embryonic tissue, *Proc. Soc. Exp. Biol. Med. 189*:136 (1988).
225. Anson, J. F., Laborde, J. B., Pipkin, J. L., Hinson, W. G., Hansen, D. K., Sheehan, D. M., and Young, J. F., Target tissue specificity of retinoic acid-induced stress proteins and malformations in mice, *Teratology 44*:19 (1991).

226. LaBorde, J. B., Pipkin, Jr., J. L., Hinson, W. G., Anson, J. F., Sheehan, D. M., Young, J. F., and Hansen, D. K., Retinoid acid-induced stress protein synthesis in the mouse, *Life Sci. 56*:1767 (1995).
227. Stahl, J., Wobus, A. M., Ihrig, S., Lutsch, G., and Bielka, H., The small heat shock protein hsp25 is accumulated in P19 embryonal carcinoma cells and embryonic stem cells of line BLC6 during differentiation, *Differentiation 51*:33 (1992).
228. Zantema, A., de Jong, E., Lardenoije, R., and van der Eb, A. J., The expression of heat shock protein hsp27 and a complexed 22-kilodalton protein is inversely correlated with oncogenicity of adenovirus-transformed cells, *J. Virol. 63*:3368 (1989).
229. Fuller, K. J., Issels, R. D., Slosman, D. O., Guillet, J.-G., Soussi, T., and Polia, B. S., Cancer and the heat shock response, *Eur. J. Cancer 30*:1884 (1994).
230. Love, S. and King, R.-J., A 27 kDa heat shock protein that has anomalous prognostic powers in early and advanced breast cancer, *Br. J. Cancer 69*:743 (1994).
231. Strahler, J. R., Kuick, R., and Hanash, S. M., Diminished phosphorylation of a heat shock protein (HSP 27) in infant acute lymphoblastic leukemia, *Biochem. Biophys. Res. Commun. 175*:134 (1991).
232. Thor, A., Benz, C., Moore, D., Goldman, E., Edgerton, S., Landry, J., Schwartz, L., Mayall, B., Hickey, E., and Weber, L. A., Stress response protein (srp-27) determination in primary human breast carcinomas: Clinical, histologic, and prognostic correlations, *J. Natl. Cancer. Inst. 83*:170 (1991).
233. Tetu, B., Lacasse, B., Bouchard, H. L., Lagace, R., Huot, J., and Landry, J., Prognostic influence of HSP-27 expression in malignant fibrous histiocytoma: A clinicopathological and immunohistochemical study, *Cancer Res. 52*:2325 (1992).
234. Tetu, T., Brisson, J., Landry, J., and Huot, J., Prognostic significance of heat-shock protein-27 in node-positive breast carcinoma: An immunohistochemical study, *Breast Cancer Res. Treat. 36*:93 (1995).
235. Fuqua, S. A., Oesterreich, S., Hilsenbeck, S. G., Von Hoff, D. D., Eckardt, J., and Osborne, C. K., Heat shock proteins and drug resistance, *Breast Cancer Res. Treat. 32*:67 (1994).
236. Ciocca, D. R., Fuqua, S. A. W., Lock-Lim, S., Toft, D. O., Welch, W. J., and McGuire, W. L., Response of human breast cancer cells to heat shock and chemotherapeutic drugs, *Cancer Res. 52*:3648 (1992).
237. Kabakov, A. E., Molotkov, A. O., Budagova, K. R., Makarova, Y. M., Mosin, A. F., and Gabai, V. L., Adaptation of Ehrlich ascites carcinoma cells to energy deprivation in vivo can be associated with heat shock protein accumulation, *J. Cell. Physiol. 165*:1 (1995).
238. Shimizu, T. and Wolfe, L. S., Arachidonic cascade and signal transduction, *J. Neurochem. 55*:1 (1990).
239. Shinohara, H., Inaguma, Y., Goto, S., Inagaki, T., and Kato, K., αB-crystallin and HSP28 are enhanced in the cerebral cortex of patients with Alzheimer's disease, *J. Neurol. Sci. 119*:203 (1993).
240. Renkawek, K., Voorter, C. E., Bosman, G. J., van Workum, F. P., and de Jong, W. W., Expression of αB-crystallin in Alzheimer's disease, *Acta Neuropathol. Berl. 87*:155 (1994).
241. Iwaki, T., Iwaki, A., Tateishi, J., Sakaki, Y., and Goldman, J. E., αB-crystallin and

27-kd heat shock protein are regulated by stress conditions in the central nervous system and accumulate in Rosenthal fibers, *Am. J. Pathol. 143*:487 (1993).

242. Head, M. W., Corbin, E., and Goldman, J. E., Overexpression and abnormal modifications of the stress proteins αB-crystallin and HSP27 in Alexander disease, *Am. J. Pathol. 143*:1743 (1993).
243. Renkawek, K., de Jong, W. W., Merck, K. B., Frenken, C. W., van Workum, F. P., and Bosman, G. J., αB-crystallin is present in reactive glia in Creutzfeldt–Jakob disease, *Acta Neuropathol. Berl. 83*:324 (1992).
244. Kato, S., Hirano, A., Umahara, T., Llena, J. F., Herz, F., and Ohama, E., Ultrastructural and immunohistochemical studies on ballooned cortical neurons in Creutzfeldt–Jakob disease: Expression of αB-crystallin, ubiquitin and stress-response protein 27, *Acta Neuropathol. Berl. 84*:443 (1992).
245. Higuchi, Y., Iwaki, T., and Tateishi, J., Neurodegeneration in the limbic and paralimbic system in progressive supranuclear palsy, *Neuropathol. Appl. Neurobiol. 21*:246 (1995).
246. Thomas, P. J., Qu, B.-H., and Pedersen, P. L., Defective protein folding as a basis of human disease, *Trends Biochem. Sci. 20*:456 (1995).
247. Lowe, J., Mayer, R. J., and Landon, M., Ubiquitin in neurodegenerative diseases, *Brain Pathol. 3*:55 (1993).
248. Cooper, P. N., Jackson, M., Lennox, G., Lowe, J., and Mann, D. M., Tau, ubiquitin and αB-crystallin immunohistochemistry define the principal causes of degenerative frontotemporal dementia, *Arch. Neurol. 52*:1011 (1995).
249. Kato, S., Hirano, A., Umahara, T., Kato, M., Herz, F., and Ohama, E., Comparative immunohistochemical study on the expression of αB crystallin, ubiquitin and stress-response protein 27 in ballooned neurons in various disorders, *Neuropathol. Appl. Neurobiol. 18*:335 (1992).
250. Tatzelt, J., Zuo, J., Voellmy, R., Scott, M., Hartl, U., Prusiner, S. B., and Welch, W. J., Scrapie prions selectively modify the stress response in neuroblastoma cells, *Proc. Natl. Acad. Sci. USA 92*:2944 (1995).
251. Nerland, A. H., Mustafa, A. S., Sweetser, D., Godal, T., and Young, R. A., A protein antigen of *Mycobacterium leprae* is related to a family of small heat shock proteins, *J. Bacteriol. 170*:5919 (1988).
252. Schoningh, R., Verstijnen, C. P., Kuijper, S., and Kolk, A. H., Enzyme immunoassay for identification of heat-killed mycobacteria belonging to the *Mycobacterium tuberculosis* and *Mycobacterium avium* complexes and derived from early cultures, *J. Clin. Microbiol. 28*:708 (1990).
253. Cheetham, M. E., Cell stress genes and chronic neurodegenerative disorders, *Neuropathol. Appl. Neurobiol. 21*:486 (1995).
254. Hightower, L. E., Heat shock, stress proteins, chaperones, and proteotoxicity, *Cell 66*:191 (1991).
255. Hightower, L. E., Cultured animal cells exposed to amino acid analogues or purumycin rapidly synthesize several polypeptides, *J. Cell. Physiol. 102*:407 (1980).
256. Ananthan, J., Goldberg, A. L., and Voellmy, R., Abnormal proteins serve as eukaryotic stress signals and trigger the activation of heat shock genes, *Science 232*:522 (1986).

257. Pelham, H. R. B., Speculations on the functions of the major heat shock and glucose-regulated proteins, *Cell 46*:959 (1986).
258. Anfinsen, C. B., Principles that govern the folding of protein chains, *Science 181*:223 (1973).
259. Jinn, T. L., Yeh, Y. C., Chen, Y. M., and Lin, C. Y., Stabilization of soluble proteins in vitro by heat shock proteins-enriched ammonium sulfate fraction from soybean seedlings, *Plant Cell. Physiol. 30*:463 (1989).
260. Rao, P. V., Horwitz, J., and Zigler, Jr., J. S., α-Crystallin, a molecular chaperone, forms a stable complex with carbonic anhydrase upon heat denaturation, *Biochem. Biophys. Res. Commun. 190*:786 (1993).
261. Das, K. P., Petrash, J. M., and Surewicz, W. K., Conformational properties of substrate proteins bound to a molecular chaperone α-crystallin, *J. Biol. Chem. 271*:10449 (1996).
262. Raman, B., Ramakrishna, T., and Rao, C. M., Rapid refolding studies on the chaperone-like α-crystallin. Effect of α-crystallin on refolding of β- and γ-crystallins, *J. Biol. Chem. 270*:19888 (1995).
263. Wang, K. and Spector, A., α-Crystallin can act as a chaperone under conditions of oxidative stress, *Invest. Ophthalmol. Vis. Sci. 36*:311 (1995).
264. Rao, P. V., Horwitz, J., and Zigler, Jr., J. S., Chaperone-like activity of α-crystallin. The effect of NADPH on its interaction with zeta-crystallin, *J. Biol. Chem. 269*:13266 (1994).
265. Farahbakhsh, Z. T., Huang, Q.-L., Ding, L.-L., Altenbach, C., Steinhoff, H.-J., Horwitz, J., and Hubbell, W. L., Interaction of α-crystallin with spin-labeled peptides, *Biochemistry 34*:509 (1995).
266. Carver, J. A., Guerreiro, N., Nicholls, K. A., and Truscott, R. J. W., On the interaction of α-crystallin with unfolded protein, *Biochim. Biophys. Acta 1252*:251 (1995).
267. Stege, G. J. J., Li, G. C., Li, L., Kampinga, H. H., and Konings, A. W. T., On the role of hsp72 in heat-induced intranucleas protein aggregation, *Int. J. Hyperthermia 10*:659 (1994).
268. Jakob, U., Lilie, H., Meyer, I., and Buchner, J., Transient interaction of Hsp90 with early unfolding intermediates of citrate synthase. Implications for heat shock in vivo, *J. Biol. Chem. 270*:7288 (1995).
269. Rao, P. V., Huang, Q. L., Horwitz, J., and Zigler, Jr., J. S., Evidence that α-crystallin prevents non-specific protein aggregation in the intact eye lens, *Biochim. Biophys. Acta 1245*:439 (1995).
270. Ganea, E. and Harding, J. J., Molecular chaperones protect against glycation-induced inactivation of glucose-6-phosphate dehydrogenase, *Eur. J. Biochem. 231*:181 (1995).
271. Ganea, E. and Harding, J. J., Inhibition of 6-phosphogluconate dehydrogenase by carbamylation and protection by α-crystallin, a chaperone-like protein, *Biochem. Biophys. Res. Commun. 222*:626 (1996).
272. Raman, B. and Rao, C. M., Chaperone-like activity and quaternary structure of α-crystallin, *J. Biol. Chem. 269*:27264 (1994).
273. Hook, D. W. and Harding, J. J., α-Crystallin acting as a molecular chaperone protects catalase against steroid-induced inactivation, *FEBS Lett. 382*:281 (1996).

274. Cherian, M. and Abraham, E. C., Decreased molecular chaperone property of α-crystallins due to posttranslational modifications, *Biochem. Biophys. Res. Commun. 208*:675 (1995).
275. Takemoto, L. J., Oxidation of cysteine residues from α-A crystallin during cataractogesis of the human lens, *Biochem. Biophys. Res. Commun. 223*:216 (1996).
276. Boyle, D., Gopalakrishnan, S., Takemoto, L., Localization of the chaperone binding site, *Biochem. Biophys. Res. Commun. 192*:1147 (1993).
277. Singh, K., Groth-Vasselli, B., Kumosinski, T. F., and Farnsworth, P. N., α-Crystallin quaternary structure: Molecular basis for its chaperone activity, *FEBS Lett. 372*:283 (1995).
278. Wang, K. and Spector, A., The chaperone activity of bovine α-crystallin. Interaction with other lens crystallins in native and denatured states, *J. Biol. Chem. 269*:13601 (1994).
279. Raman, B., Ramakrishna, T., and Rao, C. M., Temperature dependent chaperone activity of α-crystallin, *FEBS Lett. 365*:133 (1995).
280. Das, K. P. and Surewicz, W. K., Temperature-induced exposure of hydrophobic surfaces and its effect on the chaperone activity of α-crystallin, *FEBS Lett. 369*:321 (1995).
281. Das, K. P. and Surewicz, W. K., On the substrate specificity of α-crystallin as a molecular chaperone, *Biochem. J. 311*:367 (1995).
282. Smulders, R. H. P. H., Merck, K. B., Aendekerk, J., Horwitz, J., Takemoto, L., Slingsby, C., Bloemendal, H., and de Jong, W. W., The mutation Asp69 → Ser affects the chaperone-like activity of αA-crystallin, *Eur. J. Biochem. 232*:834 (1995).
283. Takemoto, L. and Boyle, D., Binding of denatured protein decreases the chaperone properties of α crystallin, *Arch. Biochem. Biophys. 315*:133 (1994).
284. Kampinga, H. H., Brunsting, J. F., Stege, G. J. J., Burgman, P. W. J. J., and Konings, A. W. T., Thermal protein denaturation and protein aggregation in cells made thermotolerant by various chemicals: Role of heat shock proteins, *Exp. Cell. Res. 219*:536 (1995).

23

Aggregation and Accessory Proteins in the Pathology of Protein-Folding Diseases

***Ronald Wetzel**
Temple University, Philadelphia, Pennsylvania

I. INTRODUCTION

A major function of molecular chaperones is to retard, and sometimes reverse, folding-related aggregation that can occur in cells under stress as well as under normal conditions [1]. With the emergence of a group of human diseases that are related to protein misfolding and misassembly [2–5], it is reasonable to consider whether there is a natural or therapeutic role in these diseases for accessory proteins that retard or enhance the rate or extent of aberrant folding. In this chapter I will review the evidence for the involvement of protein misfolding and, in particular, protein aggregation, in different classes of disease, as well as the evidence for the role of chaperones or other accessory proteins in modulating the aggregation or disease process.

Most chaperone activities described to date involve the aggregation of proteins in the process of folding. In contrast, some protein-folding diseases seem to involve aggregation as part of a degradative process after folding is completed [2]. For those folding diseases in which intracellular misassembly is implicated, one might expect that members of the classic chaperone families could become involved in modulating disease. For those diseases involving extracellular mis-

*Current affiliation: University of Tennessee Medical Center at Knoxville, Knoxville, Tennesse

assembly, the involvement of accessory proteins would be more surprising and, if demonstrated, might be expected to represent previously uncharacterized proteins or protein functions. However, this attempt to posit two broad classes of misfolding diseases may prove simplistic. For example, the most extensively studied chaperone in the modulation of the yeast prion state [*PSI*$^+$] is Hsp104, a protein capable of the *disaggregation* of already aggregated proteins [6]. The prion protein, PrP^C, is apparently folded and secreted, then reabsorbed into a nonacidic compartment of the cell, before formation of its infectious, toxic state, PrP^{Sc} [7].

Although the aggregation of a specific polypeptide molecule into a highly insoluble precipitate is a feature of many mammalian diseases (Table 1), it was only relatively recently that the analogy was drawn between these diseases and previously characterized folding and assembly-related aggregation and precipitation reactions encountered in microbial cells and in folding experiments in vitro [2,8–11]. In some mammalian protein deposition phenomena it remains unclear whether the observed aggregation is important to the etiology of the disease process. However, the importance of folding stability, folding intermediates, and folding transformations in several of these disease-related protein deposition states is now well established [4,5,12]. The recent discovery that two previously characterized nonmendelian traits in yeast genetics have attributes suggesting a relation with mammalian prion diseases [13–15] may be a major breakthrough in the understanding of protein deposition diseases. Our current understanding of these yeast genetic traits implicates protein folding and aggregation in both their "disease" phenotypes and in the unusual inheritance patterns of these markers. As discussed in this chapter, it now seems conceivable that both transmissibility and neurotoxicity in the spongiform encephalopathies may ultimately be traced to very similar protein aggregation processes.

Table 1 Protein Deposition Conditions

Disease, condition, or pathological feature	Ref.
Amyloidosis	16
Prion disease	101
Z variant α_1-antitrypsin deficiency	139
Inclusion body myositis; distal myopathy	140–142
Immunoglobulin deposition diseases	143
Lewy body dementia	144
Paired helical filaments	145
Amyotrophic lateral sclerosis	134,146
Mallory bodies	147
Intranuclear inclusion body disease	148
Rosenthal fibers; Alexander disease	133,135

II. AMYLOID DISEASES

Amyloid formation constitutes the class of folding and aggregation diseases that probably generally involve catabolism or unfolding of a mature protein [2]. Although this is not the class of conditions most likely to involve classic chaperones, significant work has been done to characterize both the aggregation process and the involvement of novel accessory proteins such as apolipoprotein E. Because the issues of aggregation selectivity and promiscuity encountered in amyloid biochemistry may be featured in other folding abnormalities, significant space is devoted here to a discussion of amyloid and its own special class of accessory proteins.

A. Common Features

There are over 15 polypeptides associated with amyloid formation in humans [16]. In addition, other polypeptides are known that are capable of forming amyloid fibrils in vitro, but they are not known to be involved—at least as independent peptides—in human disease [17–19]. These proteins and peptides, disease-related or not, do not exhibit broad primary sequence homology, although at least one subclass exists exhibiting limited sequence homology [18], and most amyloidogenic sequences lack proline residues [20]. Many—but not all—are rich in β-structure in the native state.

What unifies these proteins is their common ability to form amyloid fibrils with remarkably similar features. In any of these disease states, fibrils are preponderantly composed of one particular protein, although substantial amounts of other proteins, such as serum amyloid P component [21] and protcoglycans [22], and lesser amounts of antichymotrypsin [23] and apolipoprotein E [24,25], may also be associated with fibrils. Occasionally, fibrils indistinguishable from those isolated from human tissue can be grown in vitro using only the component protein or peptide [17,26,27]. Notwithstanding that there exist such sequences that are intrinsically "amyloidogenic," it is clear from some features of amyloid diseases—such as late and sporadic onset—that there must also be additional factors that determine if and when fibril formation develops in a particular individual.

Fibrils exhibit birefringence [16], suggesting that these protein aggregates have a regular repeat structure and significant order. Electron microscopy shows that amyloid fibrils are of indeterminate length and are 70–130 Å in cross section. Fibrils have an associated twist, which apparently derives from their substructure of a lateral association of smaller units called β-crystallites or protofilaments [28,29]. It is possible that different amyloid sequences form fibrils that contain different numbers of bundled protofilaments [28,29]. All fibrils are rich in β-sheets that contain a fundamental structural organization known as "cross-β," consisting of stacks of antiparallel β-sheets. A model for a fundamental β-sheet helix molecular structure in the protofilaments of transthyretin amyloid has been proposed based on X-ray diffraction data [30]. Some fibrils also exhibit measurable content of non–β-sheet secondary structure [16], which perhaps exists in those domains of the constituent proteins not intimately involved in the fibril structure.

Amyloid fibrils are generally found in the extracellular space. They can be found in the proximity of lysosomal compartments [26,31,32], and moderately low pH values, such as those found in lysosomes, can enhance fibril formation from some proteins in vitro [11]; consequently, it has been suggested that fibril formation might occur, or at least be initiated, in a low-pH compartment of the cell [11]. Details of the pH-dependent aggregation of Aβ argue against this mechanism in Alzheimer's disease, however [33]. Certainly, amyloid formation is not likely to occur at an early stage of the cellular folding process: protein isolated from natural fibrils consistently lack the signal peptide required for secretion into the endoplasmic reticulum (ER), and also contain posttranslational modifications, such as glycosylation and disulfide bonds.

B. Unfolding Intermediates and Amyloid Formation

Most of the polypeptides associated with amyloid formation are globular proteins or domains of globular proteins. In several of these disease families, significant evidence has been obtained for the involvement of folding and assembly intermediates in the amyloid formation process (Fig. 1). In vitro studies of the unfolding and aggregation of transthyretin—a homotetrameric protein that carries thyroxine in the bloodstream—suggest that a nonnative state is a key precursor of fibril formation by this molecule [11]. In addition, transthyretin molecules harboring mutations associated with familial amyloidosis lower the folding stability of the protein, thereby allowing fibril formation to occur under less-denaturing conditions than required by the wild-type protein [34]. The working model for the intermediate involved in transthyretin fibril formation is that it is a monomer retaining the basic-folding pattern of the native subunit, but with perhaps as little as a single β-hairpin unfolded [35,36].

Identification of disease-related sequence changes in the immunoglobulin light-chain form of amyloidosis is more difficult, owing to the sequence diversity intrinsic to this family of molecules. Nonetheless, rare amino acid replacements have been identified in several amyloid-associated light-chain sequences, and when these replacements are introduced into the benign V_L background sequence REI, they render the protein less thermodynamically stable and increase its susceptibility to aggregate formation in vitro [37,38]. Because amino acid substitutions in the loops of the V_L domain can be highly destabilizing [39], it is possible that replacements in the hypervariable loops of this immunoglobulin domain—which are even more difficult to evaluate than replacements in the framework regions of the variable domains—might be responsible for some cases of light-chain amyloidosis. Interestingly, the mutations that destabilize REI and make it more susceptible to formation of ordered aggregates in vitro also increase the level of inclusion body formation when these proteins are produced in *Escherichia coli* [40].

A point mutation in cystatin C associated with amyloid disease is also highly destabilizing to the folding of the protein [41].

The partially folded state responsible for amyloid formation may contain a significant degree of native structural elements. Several lines of evidence support a model for the amyloidogenic conformation of transthyretin that consists of the native fold lacking one specific β-hairpin [35,36]. Nuclear magnetic resonance (NMR) characterization of a nonnative dimer of cystatin C, which may be on the path of amyloid formation, shows that it retains much, but not all, of the spectral features of the native state [41]. In the immunoglobulin V_L domain REI, mutations associated with amyloid formation confer an intermediate level of destabilization onto the native state, whereas a mutation associated with formation of less-ordered, punctate deposits in light-chain deposition disease (LCDD) confers more extensive destabilization [38]. One interpretation of this data is that the highly destabilized LCDD mutant is incapable of accessing the amyloidogenic intermediate, which, by inference, must therefore be more highly ordered and possibly native-like. Occasionally, more significant rearrangements may be required to support amyloid formation. In particular, two proteins capable of forming amyloid fibrils—serum amyloid A and apolipoprotein A-I—that are expected to be highly α-helical in the native state [42,43] might have to undergo more extensive structural rearrangements to support the cross-β structure that is the basis of amyloid fibril formation.

The model that emerges from these studies is that nonnative folding states, populated in vivo owing to a combination of destabilizing changes in molecular structure and a destabilizing environment, are the building blocks of amyloid formation. The process is thus similar to the aggregation that takes place when globular proteins are exposed in vitro to denaturing conditions, which also often involve unfolding intermediates [9].

C. Nucleation of Fibril Formation in Amyloidogenic Peptides

Several disease-related amyloid depositions involve relatively small peptides, such as amylin and the Alzheimer's peptide Aβ. Although these peptides may have a preferred conformation in solution, the assumption is that the energy differences between favored conformations and the energy barriers between them are small. It is perhaps because there is no major impediment in these peptides to accessing the amyloidogenic conformation that nucleation of amyloid fibril formation has been observed in peptide systems [18,44–47], but not yet in

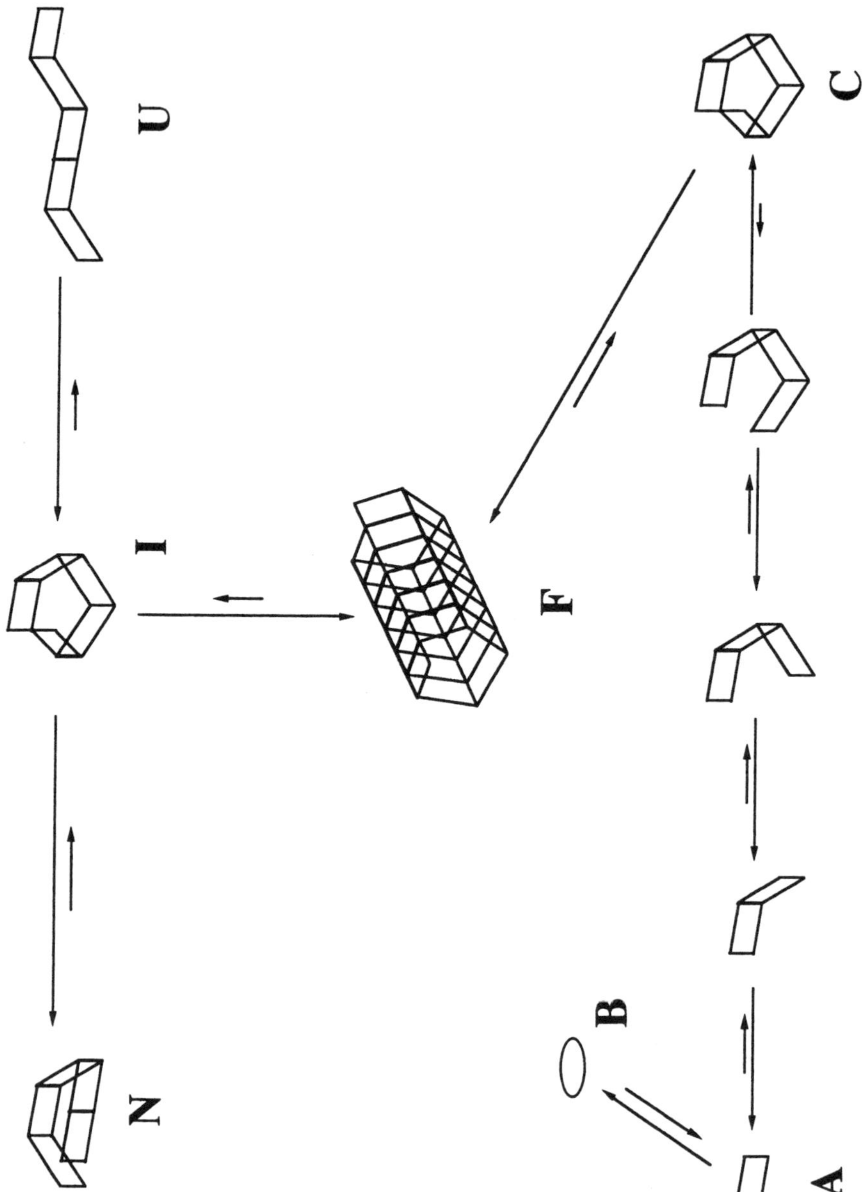
U
I
N
F
C
B
A

globular protein amyloid systems. In a nucleated growth mechanism, oligomerization is energetically unfavorable until the growing aggregate reaches a critical size that has some internal stability, rather like the internal stability of a folded globular protein. When this critical size is reached, extension of the aggregate occurs readily (see Fig. 1). The time lag to formation of the nucleus, or nidus, is highly concentration-dependent and somewhat variable. According to this model (see Fig. 1), globular proteins may thus face two barriers to efficient fibril formation: formation of a partially folded intermediate, and a nucleation event dependent on the concentration of that intermediate. Alternatively, because the partially folded state of a globular protein may itself have sufficient mass to support stabilizing internal-packing interactions in the misfolded state, and hence, a significant lifetime for the misfolded state in solution, the nucleus for fibril formation in such systems may require only one or a few molecules of rearranged protein.

To date, there have been no reports of facilitated fibril formation by seeding in amyloid formation involving globular proteins. Likewise, there have been no reports of a significant energy barrier between an amyloidogenic conformation and populated nonamyloidogenic conformation(s) of a peptide. At the same time, it seems formally possible that there may be some examples of amyloid formation requiring both misfolding and multimeric nucleus formation.

A requirement for nucleation as part of the aggregation mechanism may be taken as an indication of the specificity of the process. Thus, Aβ(1–40) rapidly aggregates at pH 5.8, near its isoelectric point, but the aggregate formed appears

---

Figure 1 Models for amyloid fibril formation by proteins and peptides. The top part of the figure shows the unfolding of a globular protein from the native state **N**, through a folding intermediate **I**, to the unfolded state **U**. The unfolding intermediate **I** possesses an amyloidogenic conformation, allowing it to pack into the stable, ordered fibril **F**. High concentrations of protein or destabilizing mutations increase the concentration of **I** under given conditions and promote increased formation of **F**. The bottom part of the figure illustrates one possible model for a nucleated growth mechanism involving an amyloidogenic peptide. Peptide monomers **A** weakly self-associate until the assembly reaches a critical size to constitute the fibril formation nucleus **C**. This assembly has a relatively long lifetime in solution and is likely to add more **A** monomers rather than disassemble. Once **C** is formed, fibril formation proceeds rapidly. In principle, globular proteins might also require formation of a multimeric nucleus, but such a requirement for nucleation by amyloidogenic proteins has not been demonstrated; the misfolded monomer **I** may have sufficient solution stability to constitute the nucleus for fibril formation. In principle, peptides may also exhibit amyloidogenic **A** and nonamyloidogenic **B** conformations in solution, but the energy barriers between conformation are generally not expected to be large. Details of the fibril-formation mechanism will determine aspects of reaction kinetics, including the presence or absence of a time lag for nucleation, and the concentration dependence of the lag and of the rate of fibril formation. (Adapted from Ref. 5.)

to have a decidedly nonfibrillar structure by electron microscopy [33]. In contrast, the same peptide at pH 7.4 forms a fibrillar aggregate, but with a long time lag that is clearly associated with a nucleation phenomenon [48]. Specificity is suggested not only by the time lag, but also by cross-seeding experiments. Despite being composed of the chemically identical peptide, the pH 5.8 aggregate not only does not nucleate fibril formation at pH 7.4, but it also inhibits the ability of bona fide fibrils to seed at this pH [33]. Similarly, the ability of fibrils derived from one peptide to serve as seeds for the fibril formation of another [47] must have important things to say about the molecular structure of the fibril.

D. Accessory Proteins in Amyloid Formation

1. *Apolipoprotein E*

A possible role for apolipoprotein E (ApoE) in amyloid deposition was first suggested by the immunochemical detection of ApoE as a component of amyloid plaque in the brains of persons with Alzheimer's disease (AD) [24,25]. A few years later, it was reported that ApoE is a risk factor in Alzheimer's disease: there are three common alleles in the human population (*E2*, *E3*, and *E4*), and individuals with one or two copies of the relatively rare *E4* allele are more likely to develop various forms of the disease [49–63]. Although it is not a certainty that this genetic risk is mediated by a direct ApoE–Aβ interaction, it is intriguing and supportive of this hypothesis that the *E4* allele may also be a risk factor for the deposition of Aβ amyloid fibrils [64–68], but not other aggregates [69], in other pathological settings. The simplest explanation for the genetics of ApoE is thus that Aβ fibril formation is an important causative factor in the development of AD, and that different isoforms of ApoE influence amyloid formation to different extents in vivo.

Many in vitro studies have been conducted to explore the possibility of ApoE effects on Aβ behavior in vitro. These studies differ substantially in many respects, including the structures, sources, and qualities of the ApoE and Aβ used, and in the techniques used to monitor Aβ–ApoE interactions. Perhaps as a consequence of these differences, the reported results of these studies also differ substantially. In assessing these studies, it is important to remember that there are really two separate, but related, issues: (a) Is there a fundamental interaction between Aβ and any isoform of ApoE under conditions sufficiently close to physiological to be considered biologically relevant, and can the result of that interaction be rationalized according to a model for the development of AD? (b) Is there a significant difference in how the three major isoforms of ApoE undergo this interaction, thereby offering an explanation for the observed genetics?

Binding of Aβ to Apolipoprotein E. Preliminary studies from several laboratories gave an early indication of a strong-binding interaction in vitro between Aβ and ApoE [70–74]. Thus, Aβ-peptides covalently immobilized onto membranes are capable of binding ApoE from cerebrospinal fluid (CSF); trace amounts of ApoE were detected immunochemically after washing the membrane with sodium dodecyl sulfate (SDS) or guanidine hydrochloride to release other CSF proteins that were bound less tightly [70,71]. Alternatively, SDS gels, on coincubation of mixtures of Aβ and ApoE in aqueous SDS, also demonstrate a low stoichiometry SDS-stable complex [72–74]. ApoE noncovalently bound to plastic is also capable of binding Aβ peptides [70].

Several studies have also been performed to monitor the interaction of Aβ and ApoE in the solution phase in native buffer [75–77]. The results confirm the preliminary reports and provide important kinetic and binding stoichiometry data for the interaction under physiological conditions. In one study, incubation of micromolar amounts of Aβ and ApoE generated a complex, that consisted of approximately four molecules of Aβ per ApoE tetramer, within minutes at room temperature (Fig. 2A) [78]. On further incubation (1–20 h), a soluble, high molecular weight (> 500 kDa) complex, consisting of approximately 100–150 molecules of Aβ per ApoE tetramer, is generated. In another study, incubation of Aβ with ApoE generated fibrillar material by electron microscopy (EM), but the aggregates appear to be monofibrillar (i.e., protofibrils or β-crystallites; see Sec. II.A) [75]. Similar conditions were used to generate a complex that was partially purified by gel filtration chromatography in 70% formic acid [76]. In this instance, the fibrils obtained did not appear to be monofibrillar, but of normal amyloid dimensions, indistinguishable from synthetic Aβ fibrils and from fibrils isolated from AD brain amyloid plaques [76]. In both EM studies, the fibrils obtained were shown by immunogold staining to contain ApoE dispersed along the fibril length [75,76], confirming their coaggregate nature and indicating a correspondence to the coaggregates isolated by native gel filtration chromatography [77]. It is unclear why the two EM studies obtained different structures for the aggregates [75,76]; perhaps ApoE blocks the bundling of protofilaments into amyloid fibrils, whereas formic acid treatment facilitates the annealing of these ApoE-bound protofilaments into fibrils.

These results are consistent with the following scheme: There is a fairly rapid-binding interaction between ApoE and Aβ in native buffer, with a 1:1 stoichiometry (this ratio could be accomplished by various binding scenarios [78]). On longer incubation, a very stable, higher molecular weight fibrillar complex is generated that contains many molecules of Aβ per molecule of ApoE. This complex formation could occur either by the displacement of the previously bound low molecular weight Aβ by an Aβ aggregate, or by the addition of new Aβ to the first bound Aβ. Because the time course of formation of SDS-stable complex is also hours [72,74], rather than minutes, it is probably the high molecular weight aggregate—not the low molecular weight complex—that is being detected by SDS–wash methods [75].

The ability of immobilized Aβ to selectively bind ApoE from CSF [70], and

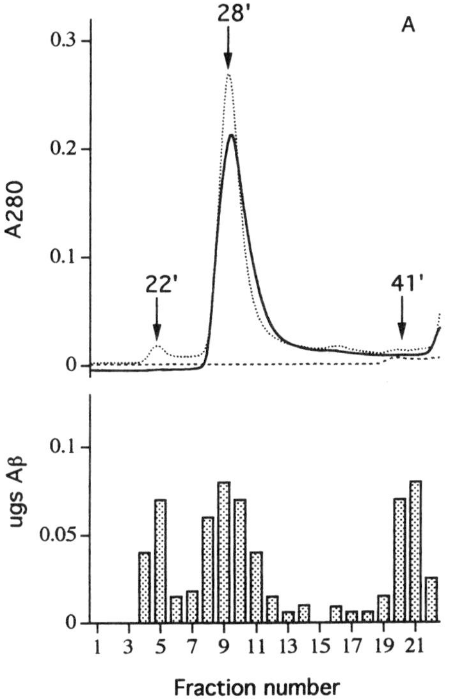
28'
A
0.3
0.2
0.1
0
A280
22'
41'
ugs Aβ
0.1
0.05
0
1 3 5 7 9 11 13 15 17 19 21
Fraction number

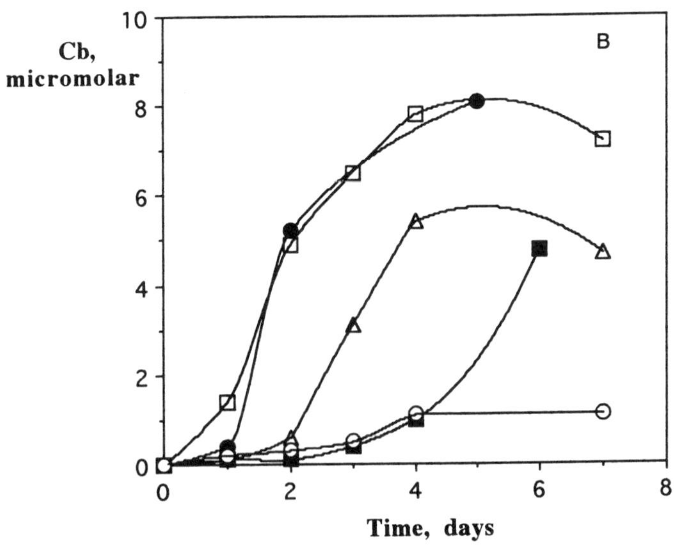
10
B
Cb, micromolar
8
6
4
2
0
0 2 4 6 8
Time, days

the resemblance of Aβ–ApoE complexes generated in vitro to fibrillar co-aggregates isolated and purified from brain plaques [76], suggests that the phenomena measured in vitro have some relevance to what is happening in the brain. More clues to the nature of this interaction are found in studies on the effect of ApoE on the kinetics of fibril formation.

Effect of Apolipoprotein E on Aβ Aggregation Kinetics. Several studies have been reported in which ApoE seems to accelerate Aβ aggregation [79–81], consistent with the term "pathological chaperone" [25]. However, in the context of the aggregation stimulation data, this term would seem to be a misnomer. First, molecular chaperones are normally thought of as inhibitors of protein aggregation, not accelerators of aggregation. Second, the ability to observe ApoE promotion of aggregation is now known to be highly dependent on experimental conditions, with other conditions leading to dramatic inhibition of aggregation by ApoE [46,48,78]. Thus, although ApoE probably functions differently from most molecular chaperones, under some circumstances it displays the accepted general function of this class of molecules: an inhibition of aggregation. However, if this is true, and if Aβ aggregation indeed has toxic effects, including promotion of AD, then the inhibition of Aβ fibril formation would interfere with AD pathology, rather than promote it.

A necessary prerequisite for fibril growth from Aβ and other small peptides is the development of a nucleus or the presence of fibril seeds (see Fig. 1). In a fibril-formation reaction from aggregate-depleted Aβ, the addition of ApoE has the effect of extending the time lag associated with nucleus formation [46,48] (see

Figure 2 Interactions of Aβ with recombinant ApoE subtype 3 in vitro: (A) The 1–40 version of Aβ was incubated with ApoE for 2 h, then submitted to gel-filtration chromatography. (Top panel) chromatographic runs of ApoE3 alone (———), Aβ alone (- - -), and an incubation mixture of ApoE3 and Aβ (········), showing that some new A_{280} = absorbing material elutes in the void volume when the comixture is run. (Bottom panel) estimates of Aβ content of each chromatography fraction based on a dot-blot analysis of fractions using an anti-Aβ antibody, showing that Aβ comigrates with the main ApoE peak as well as with the A_{280}-absorbing high molecular weight peak. (B) ApoE3 was added to fibril formation reactions with disaggregated Aβ (1–40): Fibril formation was quantified by Congo red binding. Some reactions were seeded with a constant dilution of fibrils of Aβ that had been made previously. Shown are a seeded reaction (●) and an unseeded reaction (■), as well as seeded reactions containing low (□), intermediate (△), and high (○) levels of ApoE. The data shows that low concentrations of ApoE have no effect on the fibril reaction lag time, but intermediate levels of ApoE3 delay onset of aggregation by a day, whereas higher concentrations of ApoE3 (still much lower than the molar concentration of Aβ in solution) shut down fibril formation beyond the time seen for an unseeded reaction. (A: adapted from Ref. 77; B: S. J. Wood, W. Chan, and R. Wetzel, unpublished; and Ref. 48).

Fig. 2B). If Aβ fibrils are added as seeds to a fibril formation reaction, addition of sufficient ApoE blocks the ability of the added seeds to reduce or eliminate the time lag [46,48]. These phenomena exhibit dose–response effects. Higher amounts of added seed require higher amounts of ApoE to provide protection. The molar amount of ApoE required to provide protection is substantially lower than the molar concentration of total Aβ in the reaction, suggestion that ApoE is acting at the level of seed, rather than complexing with all available Aβ [82].

At first blush, this observed inhibition of fibril growth seems at odds both with observations that ApoE supports Aβ aggregation [79–81], and with the observation of ApoE–Aβ coaggregate formation in native buffer in solution [75–77]. However, all the results are fairly consistent with the following hypothesis, summarized in Figure 3: (a) ApoE–Aβ complex formation occurs by the binding of ApoE, in a defined stoichiometry [78], to the aggregated Aβ seed as the latter forms. (b) The addition of ApoE to a growing Aβ fibril inhibits the ability of the fibril to support further Aβ additions, retarding and, under some conditions, essentially inhibiting seeding and fibril extension [48,78]. (c) At relatively low concentrations of ApoE, formation of ApoE–Aβ complexes may be difficult to detect, even though the functional effect of ApoE—inhibition of Aβ fibril growth—is clearly observed. At higher concentrations of ApoE, ApoE–Aβ complex formation can be observed by isolation of the complex or by methods normally used to monitor Aβ fibril formation, such as dye binding.

This hypothesis is supported by the following observations: Isolated ApoE–Aβ complex is a poor seed for fibril formation [78]. Isolated ApoE–Aβ complex gives as strong a thioflavin T signal as does an equivalent amount of Aβ aggregated into monocomponent fibrils [78]. Fibrils containing both Aβ and ApoE are formed when the two molecules are incubated at relatively high concentrations [75–77], but, when ApoE is limiting and the Aβ solution is initially depleted of seeds, inhibition of seeding [48] and nucleation [46,48] by ApoE is observed.

The fibrils generated at high concentrations of Aβ and ApoE are clearly coaggregates of ApoE and Aβ, and as such may have different properties, including altered neurotoxicity, compared with monocomponent Aβ fibrils. The retarded growth rate of fibril forms of Aβ in the presence of ApoE might also be associated with a retarded generation of neurotoxicity. In cell culture, generation of Aβ cell toxicity is coincident with the generation of Aβ aggregates (so-called aging). In the presence of ApoE, the development of Aβ neurotoxicity in cell culture is suppressed [83].

Most studies of the ApoE–Aβ interaction were probably initiated in hopes of not only identifying a strong interaction, but also of defining isoform-dependent differences in that interaction, consistent with the ApoE genetics of AD risk. However, this literature is as diverse as the literature on the fundamental nature of the interaction. Groups who find that ApoE stimulates Aβ aggregation also find that the high-risk isoform E4 is a better stimulant than E3 [79,80]. When complex formation is measured at high concentrations in native buffer, and ApoE inhibition of aggregation is observed at lower concentrations, no appreciable difference in the activities of the major ApoE isoforms is observed [48,77]. It is possible that the isoforms have equivalent specific activities in interacting with Aβ, but exhibit different steady-state levels in the brain, or have different abilities to form covalent ApoE aggregates of altered specific activities in vivo [46,48]. Finally, in one set of experiments based on measuring SDS-stable ApoE–Aβ aggregation, ApoE isoforms were equipotent at binding Aβ unless the ApoE was bound up in lipoprotein particles, as it exists in the circulation [74].

The manner by which ApoE binds to the growing fibril to inhibit its extension is of considerable interest. The growing face of the fibril is probably quite hydrophobic, so that ApoE might bind to the face by its exposed amphiphilic helices, in analogy with models for apolipoprotein binding to lipoprotein particles [84]. Alternatively, a portion of the ApoE sequence may rearrange into an extended chain configuration that might add to the growing fibril by extending the β-sheet network, at the same time not providing the correct packing interface to allow subsequent additions of Aβ—essentially, an amyloid cap. Several observations are consistent with this model. Several apolipoproteins are themselves capable of fibril formation [16], even though they are highly α-helical in the native state, suggesting the ability of native α-helix elements to support β-sheet formation. In fact, most of the 20 standard amino acids are regularly found in both α-helices and β-sheets [85]. Finally, of three proteins tested as controls for the ApoE effect, the only one that exhibits any appreciable ability to block seeding is the immunoglobulin V_L domain REI [48], a β-sheet–rich protein, which in mutated form is capable of forming fibril-like aggregates [37].

According to the foregoing model, ApoE is capable of neither disaggregating Aβ fibrils, nor of facilitating the folding of Aβ into a nonaggregating form. As such, ApoE in the context of Aβ fibril formation does not behave as a classic molecular chaperone molecule, which cycles as it saves proteins that are in the process of folding from entering an irreversible aggregation pathway. Rather, the molecule has the effect of slowing the rate of the aggregation pathway by becoming incorporated into the growing fibril. Thus, although ApoE acts at concentrations significantly lower than that of the Aβ in a fibril formation reaction, free ApoE molecules are lost during its function, so that it is not a true catalyst.

The implications for the role of ApoE in Alzheimer's disease are unclear. The amyloid hypothesis posits that amyloid plaques play a causative role in neurodegeneration [86,87], and this is supported by studies showing that fibrillar Aβ is toxic to cells in culture [88]; whereas ApoE—a risk factor for AD—inhibits the toxicity of Aβ on cultured cells [83]. The recognition that brain-derived amyloid is a coaggregate of Aβ and ApoE [76] adds one complication to the amyloid–ApoE hypothesis. It is possible that monocomponent Aβ fibrils and

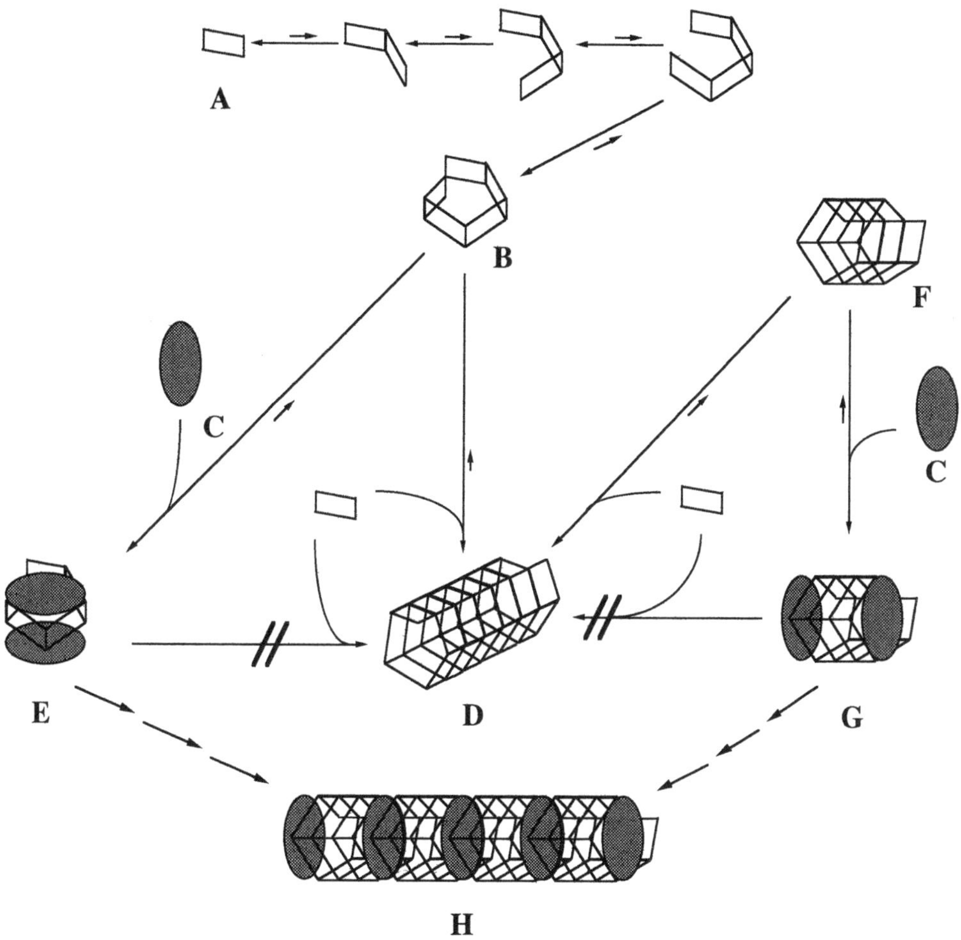
A
B
C
C
D
E
F
G
H

Aβ–ApoE coaggregates are equipotent in their role in AD etiology, but that Aβ–ApoE fibrils grow more slowly and, in this way, disease progress is modulated. Alternatively, the pathotoxicity of fibrils might decrease as their ApoE content increases. Studies on the physical and biological properties of these coaggregates will be required to better assess the connection between amyloid plaque and AD, and how ApoE impinges on this system.

2. *Other Possible Accessory Proteins for Amyloid Fibril Formation*

Similar to apolipoprotein E, other proteins can be detected in association with amyloid fibrils, and it is worth considering whether any of these, as does ApoE, provide some functional modulation of fibril growth and structure characteristics.

Serum amyloid P component (SAP) is a member of the pentraxin family of proteins that undergoes an ubiquitous association with amyloid fibrils in vivo [89]. Recently, Janciauskiene et al. [90] reported that SAP, at low concentrations, is capable of inhibiting fibril formation by Aβ. The reaction has many similarities to the properties of ApoE described in the foregoing. The inhibition takes place in a dose-dependent manner, and the product of an inhibited reaction is a coaggregate exhibiting morphological characteristics different from those of monocomponent Aβ fibrils. These results are interpreted as indicating the binding of SAP to the growing fibril to modulate its growth and structure, in a mechanism similar to that described for ApoE. SAP is rich in β-sheet [91]; thus, the mechanism by which it binds to the growing fibril may be by incorporating into the fibril's β-sheet network, in analogy to the model presented here for ApoE and REI. Another activity of SAP that might be relevant to the development of amyloid abnormality is its ability, when bound to amyloid fibrils, to protect them from proteolysis [92].

Antichymotrypsin (ACT) is another protein that can be identified in brain plaque amyloid fibrils [23]. Similar to ApoE, there are diametrically opposed observations in the literature on the effect of ACT on Aβ fibril formation in vitro.

Figure 3 A model for the interactions of apolipoprotein E with the Alzheimer's peptide Aβ: Fibril formation kinetics by Aβ and other peptides support a nucleated growth mechanism, one version of which is shown here and in Figure 1. Formation of fibrils **D** takes place rapidly in the presence of either an endogenously formed nucleus **B** or the addition of exogenous fibril seed **F**. Nucleation and seeding of the extension of monocomponent Aβ fibrils **D** can be inhibited by addition of sufficient ApoE to bind to the growth faces of the nucleus or seed. However, coaggregates **H** of ApoE and Aβ grow instead, albeit with different rates and concentration requirements compared with the growth of **D**. The coaggregate **H** gives a thioflavin T signal identical with that of the same amount of monocomponent Aβ fibrils **D**. Isolated **G** is an ineffective nucleus for extension of monocomponent Aβ fibrils **D** [78].

According to one report, ACT stimulates fibril formation [80]. Two independent laboratories report the interesting finding that ACT not only inhibits fibril formation, but it also promotes the disaggregation of preformed fibrils [93,94]. Antichymotrypsin is also rich in β-sheet and thus might be capable of incorporating into a growing Aβ fibril. In fact, peptide fragments of ACT are capable of forming amyloid fibrils in vitro [95].

Transthyretin, another amyloidogenic, β-sheet–rich peptide, has also been reported to inhibit Aβ fibril formation [96], but in this case, the form of Aβ used was the 1–28 version, which is much less amyloidogenic than the physiologically relevant 1–40 and 1–42 forms.

At the other end of the spectrum is the protein known as NAC (non-Aβ component), a peptide found in brain amyloid plaque [97], which is itself capable of forming amyloid fibrils in vitro [98]. The NAC fibrils seed fibril formation by Aβ, and vice versa [47].

Thus, it would seem that there are many proteins that might be considered accessory proteins for the formation of amyloid. At least some of these proteins probably function by adding to the growth point of the Aβ fibril, extending the β-sheet network. Depending on the packing characteristics of the element of the accessory protein left exposed when it adds to the fibril, the rate of addition of more Aβ molecules can be enhanced, unaffected, or restricted, and the structure of the resulting coaggregate may be altered from that of a monocomponent Aβ fibril.

The implications of these results for the biological consequences of Aβ amyloid formation are unclear. Some of these proteins might influence Alzheimer's progression by altering the rate of fibril growth or the structural and functional properties of fibrils. On the other hand, these "fellow travelers," which we know about because they are relatively abundantly represented in fibrils, may be "innocent" of any biological implications. The really important fibril components may ultimately prove to be relatively rare, difficult-to-detect proteins that have important functions in brain biology and for which depletion, by becoming entrapped in the growing fibril in analogy with examples discussed here, contributes to neurodegeneration.

Coaggregate formation may thus be an important general mechanism in protein deposition diseases, influencing not only the growth rate, but also the biological consequences, of aggregate formation. As discussed in the following sections, this is also suggested by our current knowledge of the mechanism of aggregate toxicity and transmissibility of the microbial prions.

III. PRION DISEASES

The term *prion* was invented [99] in the context of studies on the mammalian neurodegenerative diseases known as spongiform encephalopathies, to indicate a class of transmissible disease in which the agent of transmission is not a nucleic acid [100], but a protein. Recently, as this concept began to gain wider acceptance, possibly related phenomena have been noted in two unusual yeast genetic markers [13–15]. The major conceptual challenge of the prion hypothesis—how a protein isoform can be self-replicating—exists equally in these microbial settings. Another difficult concept—how a prion might survive passage through the gut and be transported into the bloodstream and brain—is not an issue in these yeast systems, in which the phenomenon is not transmission from one animal to another, but rather, transmission to spores or daughter cells. These yeast systems will be considered here in perhaps as much detail as the mammalian diseases because, although genetically complicated, they may be biochemically more simple than the spongiform encephalopathies and also may provide opportunities to observe interactions with classic chaperones. Consideration of commonalities between the two known yeast prion systems allow formulation of a general prion hypothesis, elements of which may even extend to the mammalian prions (Fig. 4).

A. The Spongiform Encephalopathies and Accessory Proteins

1. The Spongiform Encephalopathies

The spongiform encephalopathies are a family of mammalian neurodegenerative diseases characterized by varying degrees of dementia and ataxia, but with three common pathological features: (a) degeneration of neurons by vacuolization of nerve cell processes, (b) profound astrocytic proliferation, and (c) amyloid plaque formation [101]. There are sporadic, genetic, and iatrogenic forms of the disease. It is the latter mode of induction that makes these diseases particularly interesting. In this form, there exists a transmissible agent that allows the disease to be transferred from one animal to another by oral ingestion (ritualistic cannibalism—kuru), or by injection (iatrogenic Creutzfeldt–Jakob disease in growth hormone-treated patients; transfer of scrapie to mice in laboratory experiments).

In studies of the sheep disease scrapie, infectious particles were purified and characterized. Despite a large battery of tests over the course of three decades, no evidence for nucleic acid has been found in the active, purified infectious fractions. However, a major protein species, now known as PrP^{Sc}, is present in an infectious preparation. These experiments led to the proposal that the infectious agent was composed of protein only, without the nucleic acid expected in a virus-like activity, and that the agent, therefore, must possess a novel mechanism for self-replication. This new class of infectious agent was termed prions [99].

The partial amino acid sequence of PrP^{Sc} made possible the cloning of the gene for the precursor protein to PrP^{Sc}. In turn, this leads to the startling

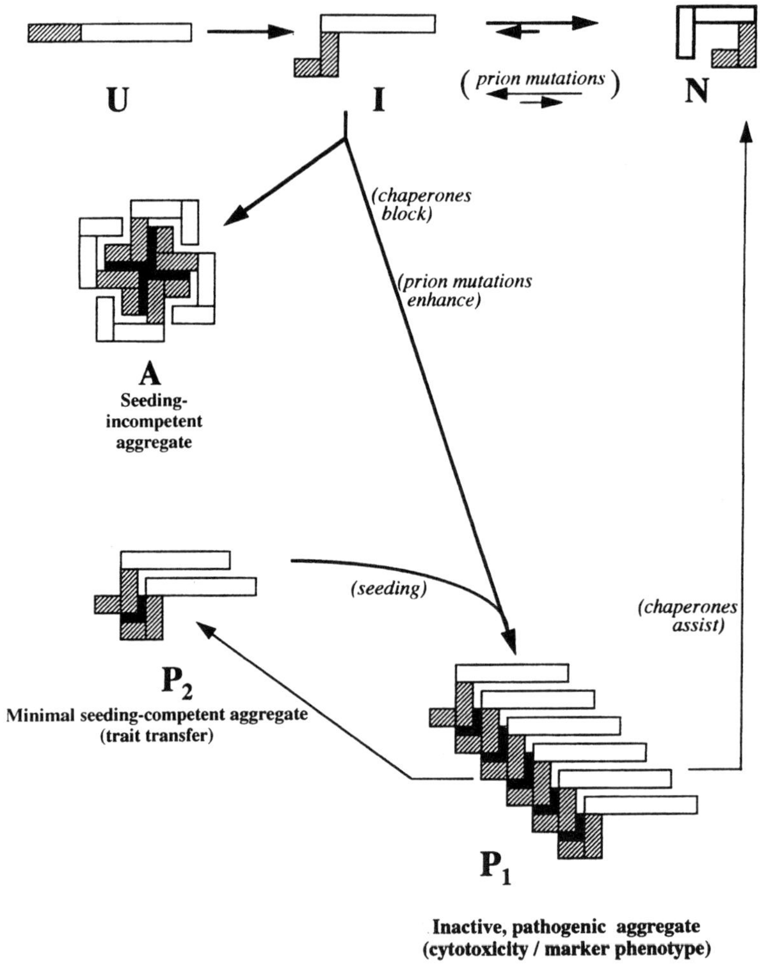

Figure 4 A unified mechanistic model of prions consistent with most of the data from both mammalian and microbial prions, and with recent data on mechanisms of aggregate formation by misfolded proteins. The prion precursor protein emerges from the ribosome in a disordered state (**U**) and proceeds to fold *via* one or more folding intermediates. A critical intermediate (**I**) is susceptible to formation of one or more types of aggregate by virtue of the surface exposure and conformation of an *aggregation module* (shaded). Once **N** is folded, the aggregation module cannot support aggregation. High concentrations of the prion precursor protein enhance aggregation by producing high concentrations of **I** (sporadic formation of **P**). Mutations within the aggregation module enhance formation of **P** by enhancing aggregate packing. Mutations outside the aggregation module enhance formation of **P** by either slowing the rate of formation of **N** form **I** (mutational effect on

finding that this precursor, PrP^C, is encoded in the genomes, and transcribed in the cells, of normal individuals [102,103]. This implies that if the PrP^{Sc} component of infectious prion preparations is responsible for infectivity, there must be at least two states of the protein, a benign state and a toxic, infectious state [7,101,104]. Individuals producing the protein in the benign state have a very low chance of experiencing conversion to the toxic form; exposed to the infectious form of the protein, chances of developing the disease increase dramatically, with affected individuals producing, from their own PrP^C, high titers of PrP^{Sc}. This model is consistent with transgenic mouse studies [105,106] and other experiments, but the molecular nature of the proliferative conversion of PrP^C to Prp^{Sc} is unknown.

The resemblance of this model to the nucleated growth mechanism for protein polymerization, as well as the possible involvement of a PrP^C aggregation process, has been noted [45,107]. In fact, PrP^{Sc} has several features consistent with involvement of a highly aggregated state of a protein. It is recognized primarily by a resistance to proteolysis, which leads to the formation of a highly protease-resistant core, instead of the complete digestion experienced by PrP^C; aggregated proteins are often resistant to proteolysis. PrP^{Sc} is poorly soluble and appears as amorphous particles in electron micrographs. Infectious preparations exhibit high β-sheet content, even though the native protein is rich in α-helix [108]; protein aggregation is often accompanied by an increase in β-sheet content. In fact, PrP^C produced in *E. coli* is highly α-helical when first isolated, but is capable of converting to β-sheet during purification [109].

If the aggregation state of PrP is the key to its pathological properties, then it should be possible to effect the conversion of benign to infectious PrP in the test tube, using conventional techniques for manipulating protein conformation and solubility. Such an experiment, in fact, is the Holy Grail of prion research. However, it has not yet been possible to convert isolated PrP^C to infectious PrP^{Sc} in vitro [110]. It has been possible to demonstrate the assisted formation of PrP^{Sc}

---

folding kinetics), or by destabilizing **N** with respect to **I** (mutational effect on thermostability). Chaperones suppress the build-up of **P** either by suppressing aggregation of **I** or by facilitating disaggregation of **P** to **N**. Trait transfer is modeled as a seeding phenomenon. Some aggregated forms of **I** ($\mathbf{P_1}$ and $\mathbf{P_2}$) can seed further, prion-related aggregation, while other aggregated forms of **I** (**A**) cannot seed prion-related aggregation. In microbial prions, one type of aggregate ($\mathbf{P_1}$) may be responsible for both phenotype and genetic transfer. In mammalian prions, it may be necessary to invoke a seeding-competent, derivative form of $\mathbf{P_1}$ (designated $\mathbf{P_2}$ in the figure) with properties that allow it to be transported intact into the brain from the gut and/or circulation. The phenotype or toxicity produced in the prion-infected organism is modeled as a loss of normal prion precursor protein function due to loss of active **N** to the aggregation pathway.

from PrP^C, by exogenously added PrP^{Sc}, but only by the criterion of protease resistance [111,112], a criterion that may prove to be a flawed heuristic for the infectious species [113]. In these experiments, no new infectivity could be demonstrated, perhaps because the high levels of "catalytic" PrP^{Sc}, which had to be used to effect the conversion, masked the relatively small expected additional infectivity.

Why has it not been possible to create infectious prions in vitro by PrP^C misfolding reactions? One possible explanation is that infectious particles are not only misfolded, but also contain a chemical alteration that stabilizes the misfolded state. Although no chemical alterations have been found in PrP^{Sc} preparations [7], it is possible that the preparations are a mixture of benign and infectious PrP^{Sc}, and that the signals for chemical alterations in the infectious component are simply below detection. It is also possible that PrP^{Sc} is misfolded in a way that is not simple to reproduce in vitro from PrP^C. Either of these scenarios suggests the possibility of a required cofactor, such as another protein, that is required to complete the conversion of PrP^C to PrP^{Sc} in infected individuals. If this facilitated conversion is purely of a protein-folding nature, then the analogy with molecular chaperones may be apt.

2. *Evidence for Accessory Proteins in PrP Diseases*

The obtaining of strong evidence for the involvement of accessory proteins in scrapie and related diseases is particularly challenging, given the limitations of the available experimental systems. Confirmation of the role of a particular protein or secondary factor with the aid of a mouse model will likely prove extremely difficult. If infectious particles can be generated from benign protein in a defined in vitro refolding reaction or in cell culture (in the process validating the prion hypothesis), this would make available a convenient experimental system with which to look for accessory proteins. In fact, however, achieving this system in the first place may ultimately depend on the discovery of just such a cofactor [114]. This said, the involvement of an additional protein, besides PrP, in the development of infectious prions was recently suggested, based on the results of transgenic mouse experiments [114]. These results and arguments are briefly summarized here.

A fundamental observation from transgenic mouse experiments is that mice expressing the normal genetic complement of mouse PrP (MoPrP) exhibit an intermediate sensitivity to cerebrally injected mouse prions (PrP^{Sc}). Mice expressing high amounts of $MoPrP^c$ from a transgene are more susceptible, and PrP gene-ablated mice expressing no MoPrP are resistant.

In the new experiments, mouse lines were created that express relatively high levels of the human homologue of PrP (HuPrP), or a human–mouse chimeric PrP (Mo–HuPrP) [114]. Mice expressing the normal chromosomal complement of MoPrP, in addition to one of the transgenes, exhibit a modulated sensitivity, compared with the homologous transgene experiment summarized in the foregoing. When the transgene is HuPrP, the mice are resistant to injected Hu prions; when the transgene is the Mo–Hu hybrid, the mice are sensitive to injected Hu prions. In contrast, mice lacking the normal chromosomal MoPrP are sensitive to Hu prions whether the expressed transgene is Hu or Hu–Mo. This suggests that the identity (Mo versus Hu) of a portion of the PrP sequence must be homologous between PrP^{c} and the inducing PrP^{Sc} for the disease to proliferate, and that related, but nonidentical, sequence forms can be inhibitory.

Given these results, a model was proposed invoking a second protein "X," which is critical for the conversion of PrP^{c} to PrP^{Sc} and for which species specificity is related to a sequence-dependent binding complementarity between the protein X and PrP^{c} [114]. The implication of this model is that the role of protein X is to accelerate the conversion of PrP^{c} to PrP^{Sc}; that is, to accelerate the formation of a misfolded, nonnative form of PrP. If the key sequence exhibits insufficient homology in the animal, however, then endogenous levels of protein X can block or slow down prion propagation or toxicity.

This scenario depends on a limited promiscuity in protein aggregation not unlike some phenomena in the Alzheimer's Aβ literature. Thus, amyloid fibrils of Aβ itself, or of the sequence-related peptide NAC, are capable of accelerating fibril formation by a seeding effect [47]. At the same time, other sequence-related peptides inhibit fibril growth [115]. In fact, because the PrP sequences in the transgenic experiments described earlier are themselves different in sequence, it seems possible that the PrP sequences themselves might control the observed effects, and that invoking a separate "protein X" is unnecessary. Prusiner and co-workers consider this possibility (i.e., that protein X is PrP itself), but deem it unlikely based on their general model for prion infectivity and propagation, which does not invoke formation of large aggregates such as amyloid fibrils [114]. It is possible, however, that the aggregated, rod form of prions serve as the major mode by which the prion state proliferates in an infected animal, even though it may not constitute the infectious form of the protein. A model explaining the hybrid transgenic mice experiments that invokes PrP aggregation specificity is relatively simple, and should perhaps not be excluded from further experimental tests.

In the following sections, the recently "rediscovered" microbial prions will be discussed. For both of these genetic markers and, in particular, for [*PSI*$^{+}$], there is some evidence for the involvement of classic molecular chaperones, operating in the cytoplasm, either during folding or after aggregation has occurred. Although PrP^{Sc} formation apparently takes place within the cell, it does so only after PrP^{C} is first folded and secreted, then taken up by a nonacidic cellular compartment [7]. Thus, one might expect that if chaperones are involved in modulating mammalian prion disease, they may be a novel class of accessory molecule.

B. Yeast Prions and Chaperones

1. Yeast Prions

Recently, there have been reports leading to a new appreciation of two previously poorly understood phenomena in yeast genetics, the nonmendelian transfer of traits in the genetic markers [*PSI*$^+$] and [*URE3*]. The central feature of this newly understood pathway is the recognition of prion-like behavior in these yeast systems [13–15]. These systems are of significant interest for several reasons and will be discussed here at some length. First, the role of protein aggregation in mediating [*PSI*$^+$] dominance as well as phenotype is now quite clear. Second, recent results in the PSI system offer an intriguing and surprising look at the role of molecular chaperones in mediating a protein-folding pathology. [*PSI*$^+$] and [*URE3*] offer wonderful model systems to study a perhaps more primitive sort of prion than those implicated in mammalian dementias. It is possible that the protein biophysics of the yeast systems may actually mirror that of the spongiform encephalopathies quite closely, and that studying the yeast systems may lead us on a fast track through the complexities of mammalian genetics and brain biochemistry to an improved understanding of human disease.

Broadly considered, prions have two features. One feature is the actual phenotype, or abnormality, associated with the condition. In the yeast prion effects, these phenotypes can be explained by the ability of protein aggregation to remove a required protein from solution, rendering it inactive. In the spongiform encephalopathies, the molecular mechanisms of toxicity are not at all understood, and in this sense, the yeast systems are of questionable value in understanding the human diseases. At the same time, some very recent reports on the biological consequences of PrP gene ablation [116,117] might be used to argue that the underlying mechanism of mammalian prion pathology—namely, loss of protein function by aggregation, is quite similar to that of the microbial forms (see Sec. V).

Even so, it is their unique ability to confer a genotype without the aid of nucleic acid effects that makes prions of special interest. With the spongiform encephalopathies, diseases that also exhibit sporadic and inheritable modes, it is the ability of prions to transmit the disease between hosts that has provided the major mechanistic challenge to biologists. In the yeast prion phenotypic conditions (which can also exhibit sporadic onset and mendelian genetic transfer), the parallel phenomenon to scrapie infectivity is the nonmendelian transfer of traits in mating experiments. Yeast prion systems may thus also provide clues to the basis for non–DNA-mediated transfer of phenotype in the mammalian prion diseases.

2. [*PSI*$^+$]

Normal [*psi*$^-$] cells containing a gene that harbors a deleterious nonsense mutation either exhibit loss of function of that gene product, or require a suppressor tRNA to allow maintenance of that function. The [*PSI*$^+$] phenotype is the ability of yeast cells possessing this trait to suppress nonsense mutations, presumably without the aid of a suppressor tRNA [118]. The manner in which [*PSI*$^+$] achieves its phenotype is understood; it does so by an alteration in the activity level of the protein Sup35. Sup35, along with Sup45, make up a translation-release factor that normally recognizes and aborts translation complexes stalled at a nonsense mutation. Loss of functional Sup35–Sup45 complex from cells allows such complexes to linger until the nonsense mutation is suppressed, presumably by codon–anticodon misrecognition by a normal cellular aminoacyl-tRNA. Thus, the [*PSI*$^+$], omnipotent nonsense suppression, *phenotype* is synonymous with the loss of active Sup35. In fact, it can be achieved by inactivating mutations in Sup35, and such [*Sup35*$^-$] mutations transfer according to standard mendelian genetics. However, [*PSI*$^+$], is essentially uniformly transferred to daughter cells, a decidedly nonmendelian genetic effect.

The manner in which nonmendelian transfer of [*PSI*$^+$] is achieved is also beginning to be understood [118]. In the [*PSI*$^+$] state, Sup35 activity is lost, even though Sup35 protein is not. The inactivation of Sup35 in this case is apparently due to its aggregation. Aggregation also accounts for the proliferation of the state. Thus, in analogy with the *specificity* of aggregation discussed in an earlier section, aggregated Sup35 is apparently capable of selectively incorporating newly synthesized Sup35 into a growing aggregate. As long as these conditions persist, active Sup35 will not accumulate and normal function of Sup35 will not be observed. Thus, Lindquist and co-workers observe a good correlation between the aggregation state of Sup35 in the cell and the cell's [*PSI*$^+$]/[*psi*$^-$] status [118]. In analogy with aggregation in in vitro-folding experiments, in some mammalian amyloid diseases, in spongiform encephalopathy animal models, and in bacterial inclusion body formation, high-concentration–over-production of Sup35 can lead to high levels of aggregated Sup35 in the cell and to spontaneous generation of the [*PSI*$^+$] marker [118].

The specific aggregation property of Sup35 is located in its NH_2-terminal region (the "NPD," or NH_2-terminal prion-determining domain), whereas its ribosomal-release activity is located in the COOH-terminal region [118]. Thus, a point mutation in the NH_2-terminal region leads to the PNM, or "psi-no-more," phenotype, the total abrogation of [*PSI*$^+$] [119]. (This mutation, the conversion of glycine to asparagine in the tandem repeat unit PQGGYQQYN in the NH_2-terminal region, is of added interest for its connection to other protein aggregation phenomena. This sequence is homologous with a repeat unit, PHGGGWGQ, found in the prion protein [15]. The ability of a Gly → Asp mutation to inactivate this sequence is particularly striking, given the frequent occurrence of Gly–Gly sequences in aggregation-prone polypeptides, including the Alzheimer's peptide Aβ [18]). Overexpression of the NPD by itself, or of an NPD-containing fusion protein, can also induce the [*PSI*$^+$] marker, by aggregate formation and subsequent depletion of newly synthesized Sup35 [118] through coaggregation (see Sec. II.D.1).

Although protein aggregation clearly plays a crucial role in defining [*PSI*$^+$], self-replication is apparently not a property found in all aggregates of Sup35. When [*PSI*$^+$] is cured from a cell containing a mutated Hsp104 (as discussed later), a significant amount of aggregated Sup35 remains in the cell, but for some reason it is not capable of facilitating aggregation-mediated depletion of newly synthesized Sup35 (i.e., is not [*PSI*$^+$]). Such diversity of behavior is not unprecedented in protein aggregation. Monocomponent fibrils of the Alzheimer's peptide Aβ are capable of seeding solutions of Aβ to induce fibril formation, but two other types of Aβ aggregates do not share this property: (a) coaggregates of Aβ and ApoE [78], and (b) pH 5.8 monocomponent aggregates of Aβ [33].

*Molecular Chaperones and [*PSI$^+$]. Because chaperones have the ability to suppress or reverse protein aggregation, it is reasonable to expect that some chaperones might be able to modulate a condition of intracellular protein aggregation such as [*PSI*$^+$]. Growth of [*PSI*$^+$] cells in the protein denaturant–disaggregant guanidine hydrochloride (Gdn-HCl) cures the cells of the [*PSI*$^+$] marker, and this result has been interpreted as evidence for the involvement of protein aggregation in the generation of [*PSI*$^+$]. However, as pointed out by Chernoff et al. [14], the concentration of Gdn-HCl that is effective at curing [*PSI*$^+$], 5 mM, is far below the concentrations of Gdn-HCl typically required for protein unfolding or protein disaggregation effects. However, growth of yeast cells in 5–10 mM Gdn-HCl also elevates the levels of the chaperone Hsp104 by three- to fivefold [14], suggesting that the Gdn-HCl activity may be mediated by a chaperone effect. There is now a very good case for the interaction of molecular chaperones with the [*PSI*$^+$] system, although some of the specific data are somewhat surprising and difficult to understand.

Most of the work on the effect of chaperones on [*PSI*$^+$] is with Hsp104, a member of the Hsp100 (Clp) chaperone family [6]. Despite its homology with the Clp protease, the stress-relieving activity of yeast Hsp104 is apparently due to its ability to disaggregate, rather than hydrolyze, previously aggregated proteins [6]. Modulation of Hsp104 activity has parallel effects on the [*PSI*$^+$] marker and on the Sup35 aggregation state in yeast [118].

Thus, when high levels of Hsp104 are induced in [*PSI*$^+$] cells, the cells are converted to [*psi*$^-$], and the amount of aggregated Sup35 in the cells is decreased [118]. Likewise, elimination of Hsp104 by a deletion mutation also converted [*PSI*$^+$] to [*psi*$^-$] and leads to decreased aggregated Sup35. Intermediate levels of Hsp104 activity, generated by point mutations in Hsp104, cause intermediate [*PSI*$^+$] activities (loss of omnipotent suppression without [*PSI*$^+$] curing) and intermediate solubilization of Sup35 [118].

Although the correlation between Hsp104 activity and Sup35 aggregation and [*PSI*$^+$] stringency is, in one sense, pleasing, it also introduces a conundrum:

Why is the least favorable state of the yeast cell, in terms of Sup35 aggregation, an *intermediate* level of Hsp104? In contrast with this unusual effect, aggregation of NPD, the NH_2-terminal region of Sup35, decreases in a monophasic dose–response manner, relative to Hsp104 levels, that is more typical for chaperone activities [118]. When the COOH-terminus is included in the protein, however, to make full-length Sup35, aggregation is most evident at intermediate Hsp104 levels. Lindquist and co-workers discuss this confusing result in terms of the possibility of the COOH-terminal domain altering the way Hsp104 interacts with the NPD, and the way that Sup35 packs into an aggregate [118].

The observation that significant levels of aggregated Sup35 can be present in cells "cured" of [*PSI*$^+$] by growth with mutated Hsp104 [118], and the observation that cells with intermediate levels of chaperone activity are the least able to eradicate [*PSI*$^+$], suggests that disease will not always be monotonically related to the degree of aggregation. That is, some polypeptide sequences might be capable of producing more than one kind of insoluble aggregate [5,33,38,77], and only a subpopulation of total-aggregated protein may be responsible for a biological effect. Just as some aggregated states of Sup35 are incapable of promoting the further aggregation of freshly synthesized Sup35 to sustain [*PSI*$^+$] [118], one form of aggregated Aβ is incapable of seeding the growth of Aβ fibrils [33].

3. [URE3]

The second known example of the prion phenomenon in microbes is the [*URE3*] marker in yeast. Wild-type yeast strains can take up ureidosuccinate from the growth medium to supplement or replace the ureidosuccinate normally produced by aspartate transcarbamylase, but this uptake is repressed by ammonium ions. The URE strains are capable of taking up ureidosuccinate even in the presence of ammonium. Some URE markers are chromosomally encoded and follow mendelian genetics but, as shown by Wickner, the [*URE3*] marker has several of the prion attributes described earlier for [*PSI*$^+$] [13,120]. First, in mating experiments, [*URE3*] is passed on to progeny at a much higher frequency than expected for a chromosomal factor [13]. Second, the URE phenotype can be replicated by (mendelian) mutations in a gene *URE2*, the ablation of which also renders cells incapable of acquiring the [*URE3*] factor [13]. Third, overexpression of the *URE2* gene to generate high cellular levels of the product enzyme Ure2p, renders cells much more sensitive to acquiring [*URE3*] [13,120]. Fourth, overexpression of a 65-amino acid NH_2-terminal "prion-inducing domain" also greatly increases the frequency of [*URE3*] induction [120]. Fifth, the Ure2p protein isolated from [*URE3*] cells is in the particulate fraction and does not appear to be a chemically modified form of the enzyme [13]. Sixth, growth of [*URE3*] cells in 5-mM levels of Gdn-HCl cures the [*URE3*] factor, suggesting the involvement of chaperones [13] (see Sec. III.B.1). Wickner also showed that [*URE3*] can be passed to cells by cytoplasmic mixing experiments [13], another result suggesting the lack of involvement of DNA and a further analogy to mammalian prions, which can transmit disease iatrogenically.

Thus, in analogy with [*PSI*$^+$], the results suggest that Ure2p, an enzyme involved in regulation of nitrogen catabolism enzyme synthesis, can be inactivated by aggregation, and that this aggregation can be specific and self-perpetuating, explaining both the phenotype and the nonmendelian genetics of [*URE3*].

C. General Features of the Prion Systems

Studies on protein deposition in vivo have revealed some important common features and principles. These concepts may well extend beyond the known cases of amyloid diseases and bacterial inclusion body formation, to other states involving protein misassembly, whether the aggregates involved are formally defined as amyloid, or are some other kind of large, insoluble aggregate, or a smaller, but thermodynamically stable, oligomer. The common features include (a) the importance of thermodynamic stability in determining the susceptibility of a globular protein to aggregation, (b) the ability of protein aggregation to be accelerated by seeding with an aggregate of the same or a similar sequence, and (c) the ability of aggregation to inactivate proteins, but also to stabilize them against proteolysis. If these features are also shared by the prion proteins, it may provide a relatively simple unifying hypothesis for the transmissibility and toxicity of prion conditions.

Masison and Wickner showed that deletions in the COOH-terminus of Ure2p make the protein much more capable of generating prions than the full-length wild-type protein [120]. From these results, they draw an analogy to the dispersion of PrP mutations responsible for inheritable spongiform encephalopathies [104], and suggest that the absence of structural similarity and mutational patterns between Sup35, Ure2p, and PrP^C speak against a common mechanism for prion generation [120].

However, there may be a simple common link among these three systems: the relation between protein stability, unfolding, and aggregation [9]. The established relation between stability and amyloid formation was discussed earlier in this chapter [4,5,12]. In turn, over a decade of protein-engineering experiments have shown that proteins can be destabilized [121,122], or have their folding compromised [123], by mutations at many points along the linear sequence. Prion proteins may generally possess a highly aggregation-prone sequence that is sequestered in the normally folded state, out of harm's way; folding or destabilizing mutations elsewhere in the sequence would then be expected to allow increased exposure of the aggregation element, in turn, increasing the frequency of induction of the prion state.

PrP loss by aggregation may explain at least some prion pathology, in analogy with the two yeast systems. Ataxia has been observed both in Prp gene-ablated mice [116] as well as in one form of prion disease [124]. Fatal familial insomnia is another form of prion disease [125], and some PrP gene-ablated mice also exhibit sleep disorders [117].

Consideration of these apparent common features allows formulation of a working generic model for prion diseases (see Fig. 4).

IV. OTHER FOLDING-RELATED DISEASES

The focus of this chapter is on conditions of aberrant protein folding that feature a buildup of protein aggregates. In at least some of the prion conditions, the phenotype is probably created by the effect of the aggregation process on removing active protein from the environment; in other examples of protein deposition, aggregates themselves may display some novel toxicity.

Other protein-folding conditions exist that may not involve stable aggregate formation, but, nonetheless, feature an associated abnormality that is sufficiently explained by an inability to produce active protein [3]. This inability may stem from glitches in the folding pathway, leading to proteolysis or improper localization of the gene product. For example, mutations in the cystic fibrosis transconductance regulator (CFTR) lead to the inability of a specific domain [126] and, hence, the whole protein [127], to fold efficiently. Reduced folding efficiency, in turn, leads to detrimental changes in the complex maturation and trafficking pathway of this membrane-bound, multidomain protein. Similarly, mutations in an epidermal growth factor-like domain of fibrillin render the domain incapable of folding [128], and, hence, may be sufficient to explain the connective tissue disorder Marfan's syndrome.

There are several hints of chaperone effects in this other class of folding diseases. For example, anti-Hsp70 antibodies coimmunoprecipitate both wild-type and cystic fibrosis mutants of CFTR from producing cells, suggesting an involvement of Hsp70 in maturation [129]. In two diseases, the effect of mutation may, in fact, be the alteration of the nature of the interaction between the maturing protein and a molecular chaperone. In both hereditary medium-chain acyl-CoA dehydrogenase deficiency (MCAD) [130] and p53 oncogenesis [131], mutations introduce a diminished ability of the partially folded protein to be released by Hsp60 or Hsp70, respectively, In spite of these suggestions of a negative effect of chaperones, chaperone overproduction improves proper folding and suppresses toxicity in both these examples. Thus, overproduction of Hsp60 improves the proper folding of the MCAD dehydrogenase [3]. In p53, overproduction of Hsp70 reduces the oncogenicity of the mutated protein [132]. These examples provide some intriguing parallels to the yeast [*PSI*$^+$] data described earlier, in which intermediate levels of Hsp104 are detrimental, compared with either overproduction or nonproduction of chaperone.

Similar immunochemical data exist implicating chaperone involvement in other diseases involving detectible protein aggregation. For example, αB-crystallin and Hsp27 have been identified as components of Lewy bodies, Rosenthal fibers, and Mallory bodies [133,134]. It is possible that, in some conditions, such as Alexander disease, deposition of chaperones may arise from their overproduction in response to some other stress, rather than as the result of a rescue effort mounted against other deposited proteins [135,136].

V. CONCLUSIONS

Mammalian diseases and microbial phenotypes involving misfolding and selective aggregation and deposition of proteins is an important new area of biology [149], and various types of accessory proteins play major roles influencing the molecular course of events in many of these conditions.

Classic chaperones are involved in the microbial prion diseases. Whether chaperones or other accessory proteins are also involved in mammalian prion conditions is not yet known. Amyloid formation can be modulated by accessory proteins such as ApoE. However, the mechanism of action of ApoE is not the same as a classic chaperone. Rather, it seems to bind to amyloid protofilaments and fibrils to retard their growth and change their seeding properties. Although ApoE has also been observed to associate with PrP plaques [24], at best, reports are mixed on whether there is a genetic relation between ApoE and the risk of acquiring a spongiform encephalopathy [69,137,138].

Further work on discovering and elucidating the molecular mechanisms of accessory proteins involved in protein deposition conditions is important for many reasons. Some kind of accessory protein may be required to generate a viable in vitro system for generating or replicating prions. Helper proteins can serve as tooks for studying the aggregation process and how aggregates effect biology. Proteins that have a natural role in modulating protein deposition phenotypes may also serve as models of small-molecule therapeutics.

REFERENCES

1. Hartl, F. U., Molecular chaperones in cellular protein folding, *Science 381*:571 (1996).
2. Wetzel, R., Protein aggregation in vivo: Bacterial inclusion bodies and mammalian amyloid, *Stability of Protein Pharmaceuticals: In Vivo Pathways of Degradation and Strategies for Protein Stabilization* (T. J. Ahern and M. C. Manning, eds.), Plenum Press, New York, 1992, p. 43.
3. Thomas, P. J., Qu, B.-H., and Pedersen, P. L., Defective protein folding as a basis of human disease, *Trends Biochem. Sci. 20*:456 (1995).
4. Kelly, J. W., Alternate conformations of amyloidogenic proteins govern their behavior, *Curr. Opin. Struct. Biol. 6*:11 (1996).
5. Wetzel, R., For protein misassembly, it's the "I" decade, *Cell 86*:699 (1996).

6. Parsell, D. A., Kowal, A. S., Singer, M. A., and Lindquist, S., Protein disaggregation mediated by heat-shock protein Hsp104, *Nature 372*:475 (1994).
7. Prusiner, S. B. and DeArmond, S. J., Prion protein amyloid and neurodegeneration, *Amyloid 2*:39 (1995).
8. Mitraki, A. and King, J., Protein folding intermediates and inclusion body formation, *Biotechnology 7*:690 (1989).
9. Wetzel, R., Perry, L. J., Mulkerrin, M. G., and Randall, M., Unfolding and inactivation: Genetic and chemical approaches to the stabilization of T4 lysozyme and interferon-gamma against irreversible thermal denaturation, *Protein Design and the Development of New Therapeutics and Vaccines; Proceedings of the Sixth Annual Smith, Kline and French Research Symposium* (G. Poste and J. B. Hook, eds.), Plenum, New York, 1990, p. 79.
10. Colon, W. and Kelly, J. W., Transthyretin acid induced denaturation is required for amyloid fibril formation in vitro, *Applications of Enzyme Biotechnology* (J. W. Kelly and T. O. Baldwin, eds.), Plenum, New York, 1991, p. 99.
11. Colon, W. and Kelly, J. W., Partial denaturation of transthyretin is sufficient for amyloid fibril formation in vitro, *Biochemistry 31*:8654 (1992).
12. Wetzel, R., Mutations and off-pathway aggregation, *Trends Biotechnol. 12*:193 (1994).
13. Wickner, R. B., [URE3] as an altered URE2 protein: Evidence for a prion analog in *Saccharomyces cerevisiae* [see comments], *Science 264*:566 (1994).
14. Chernoff, Y. O., Lindquist, S. L., Ono, B., Inge, V. S., and Liebman, S. W., Role of the chaperone protein Hsp104 in propagation of the yeast prion-like factor [Psi^+], *Science 268*:880 (1995).
15. Tuite, M. F., Psi no more for yeast prions, *Nature 370*:327 (1994).
16. Sipe, J. D., Amyloidosis, *Annu. Rev. Biochem. 61*:947 (1992).
17. Gustavsson, A., Engstrom, U., and Westermark, P., Normal transthyretin and synthetic transthyretin fragments form amyloid-like fibrils in vitro, *Biochem. Biophys. Res. Commun. 175*:1159 (1991).
18. Jarrett, J. T. and Lansbury, P. J., Amyloid fibril formation requires a chemically discriminating nucleation event: Studies of an amyloidogenic sequence from the bacterial protein OsmB, *Biochemistry 31*:12345 (1992).
19. Gasset, M., Baldwin, M. A., Lloyd, D. H., Gabriel, J.-M., Holtzman, D. M., Cohen, F., Fletterick, R., and Prusiner, S. B., Predicted α-helical regions of the prion protein when synthesized as peptides form amyloid, *Proc. Natl. Acad. Sci. USA 89*:10940 (1992).
20. Wood, S. J., Wetzel, R., Martin, J. D., and Hurle, M. R., Prolines and amyloidogenicity in fragments of the Alzheimer's peptide β/A4, *Biochemistry 34*:724 (1995).
21. Pepys, M. B., Amyloid P component and the diagnosis of amyloidosis, *J. Intern. Med. 232*:519 (1992).
22. Kisilevsky, R., Proteoglycans, glycoaminoglycans, amyloid-enhancing factor, and amyloid deposition, *J. Intern. Med. 232*:515 (1992).
23. Abraham, C. R., Selkoe, D. J., and Potter, H., Immunochemical identification of the serine protease inhibitor alpha 1-antichymotrypsin in the brain amyloid deposits of Alzheimer's disease, *Cell 52*:487 (1988).
24. Namba, Y., Tomonaga, M., Kawasaki, H., Otomo, E., and Ikeda, K., Apolipoprotein

E immunoreactivity in cerebral amyloid deposits and neurofibrillary tangles in Alzheimer's disease and kuru plaque amyloid in Creutzfeldt-Jakob disease, *Brain Res. 541*:163 (1991).
25. Wisniewski, T. and Frangione, B., Apolipoprotein E: A pathological chaperone protein in patients with cerebral and systemic amyloid, *Neurosci. Lett. 135*:235 (1992).
26. Glenner, G. G., Ein, D., Eanes, E. D., Bladen, H. A., Terry, W., and Page, D. L., Creation of "amyloid" fibrils from Bence Jones proteins in vitro, *Science 174*:712 (1971).
27. Kirschner, D. A., Inouye, H., Duffy, L. K., Sinclair, A., Lind, M., and Selkoe, D. J., Synthetic peptide homologous to β protein from Alzheimer disease forms amyloid-like fibrils in vitro, *Proc. Natl. Acad. Sci. USA 84*:6953 (1987).
28. Inouye, H., Fraser, P. E., and Kirschner, D. A., Structure of beta-crystallite assemblies formed by Alzheimer beta-amyloid protein analogues: Analysis by X-ray diffraction, *Biophys. J. 64*:502 (1993).
29. Serpell, L. C., Sunde, M., Fraser, P. E., Luther, P. K., Morris, E. P., Sangren, O., Lundgren, E., and Blake, C. C., Examination of the structure of the transthyretin amyloid fibril by image reconstruction from electron micrographs, *J. Mol. Biol. 254*:113 (1995).
30. Blake, C. and Serpell, L., Synchrotron X-ray studies suggest that the core of the transthyretin amyloid is a continuous beta-sheet helix, *Structure 4*:989 (1996).
31. Shirahama, T. and Cohen, A. S., Intralysosomal formation of amyloid fibrils, *Am. J. Pathol. 81*:101 (1975).
32. Westermark, P., Amyloid and polypeptide hormones: What is their relationship? *Amyloid 1*:47 (1994).
33. Wood, S. J., Maleeff, B., Hart, T., and Wetzel, R., Physical, morphological and functional differences between pH 5.8 and 7.4 aggregates of the Alzheimer's peptide Aβ, *J. Mol. Biol. 256*:870 (1996).
34. McCutchen, S. L., Lai, Z., Miroy, G. J., Kelly, J. W., and Colon, W., Comparison of lethal and nonlethal transthyretin variants and their relationship to amyloid disease, *Biochemistry 34*:13527 (1995).
35. Terry, C. J., Serpell, L., and Blake, C. C. F., Understanding the structure of the amyloid fibril in transthyretin related amyloidoses, *Amyloid and Amyloidosis 1993* (R. Kisilevsky, et al., eds.), Parthenon, New York, 1994, p. 510.
36. Kelly, J. W. and Lansbury, Jr., P. T., A chemical approach to elucidate the mechanism of transthyretin and β-protein amyloid fibril formation, *Amyloid 1*:186 (1994).
37. Hurle, M. R., Helms, L. R., Li, L., Chan, W., and Wetzel, R., A role for destabilizing amino acid replacements in light chain amyloidosis, *Proc. Natl. Acad. Sci. USA 91*:5446 (1994).
38. Helms, L. R. and Wetzel, R., Specificity of abnormal assembly in immunoglobulin light chain deposition disease and amyloidosis, *J. Mol. Biol. 257*:77 (1996).
39. Helms, L. R. and Wetzel, R., Destabilizing loop swaps in the CDRs of an immunoglobulin V_L domain, *Protein Sci. 4*:2073 (1995).
40. Chan, W., Helms, L. R., Brooks, I., Lee, G., Ngola, S., McNulty, D., Maleeff, B., Hensley, P., and Wetzel, R., Mutational effects on inclusion body formation in the

periplasmic expression of the immunoglobulin V_L domain REI, *Folding Design 1*:77 (1996).
41. Ekiel, I. and Abrahamson, M., Folding-related dimerization of human cystatin C, *J. Biol. Chem. 271*:1314 (1996).
42. McCubbin, W. D., Kay, C. M., Narindrasorasak, S., and Kisilevsky, R., Circular-dichroism studies on two murine serum amyloid A proteins, *Biochem. J. 256*:775 (1988).
43. Nolte, R. T. and Atkinson, D., Conformational analysis of apolipoprotein A-1 and E-3 based on primary sequence and circular dichroism, *Biophys. J. 63*:1221 (1992).
44. Jarrett, J. T., Berger, E. P., and Lansbury, Jr., P. T., The carboxy terminus of the β amyloid protein is critical for the seeding of amyloid formation: Implications for the pathogenesis of Alzheimer's disease, *Biochemistry 32*:4693 (1993).
45. Jarrett, J. T. and Lansbury, P. J., Seeding "one-dimensional crystallization" of amyloid: A pathogenic mechanism in Alzheimer's disease and scrapie? [Review], *Cell 73*:1055 (1993).
46. Evans, K. C., Berger, E. P., Cho, C.-G., Weisgraber, K. H., and Lansbury, Jr., P. T., Apolipoprotein E is a kinetic but not a thermodynamic inhibitor of amyloid formation: Implications for the pathogenesis and treatment of Alzheimer disease, *Proc. Natl. Acad. Sci. USA 92*:763 (1995).
47. Han, H., Weinreb, P. H., and Lansbury, P. T., The core Alzheimer's peptide NAC forms amyloid fibrils which seed and are seeded by β-amyloid: Is NAC a common trigger or target in neurodegenerative disease? *Chem. Biol. 2*:163 (1995).
48. Wood, S. J., Chan, W., and Wetzel, R., Seeding of Aβ fibril formation is inhibited by all three isotypes of apolipoprotein E, *Biochemistry 35*:12623 (1996).
49. Corder, E. H., Saunders, A. M., Strittmatter, W. J., Schmechel, D. E., Gaskell, P. C., Small, G. W., Roses, A. D., Haines, J. L., and Pericak-Vance, M. A., Gene dose of apolipoprotein E type 4 allele and the risk of Alzheimer's disease in late onset families, *Science 261*:921 (1993).
50. Mayeux, R., Stern, Y., Ottman, R., Tatemichi, T. K., Tang, M. X., Maestre, G., Ngai, C., Tycko, B., and Ginsberg, H., The apolipoprotein epsilon 4 allele in patients with Alzheimer's disease, *Ann. Neurol. 34*:752 (1993).
51. Saunders, A. M., Strittmatter, W. J., Schmechel, D. et al., Association of apolipoprotein E allele epsilon 4 with late-onset familial and sporadic Alzheimer's disease, *Neurology 43*:1467 (1993).
52. Chartier, H. M., Parfitt, M., Legrain, S. et al., Apolipoprotein E, epsilon 4 allele as a major risk factor for sporadic early and late-onset forms of Alzheimer's disease: Analysis of the 19q13.2 chromosomal region, *Hum. Mol. Genet. 3*:569 (1994).
53. Corder, E. H., Saunders, A. M., Risch, N. J. et al., Protective effect of apolipoprotein E type 2 allele for late onset Alzheimer disease, *Nature Genet. 7*:180 (1994).
54. Lucotte, G., Visvikis, S., Leininger, M. B., David, F., Berriche, S., Reveilleau, S., Couderc, R., Babron, M. C., Aguillon, D., and Siest, G., Association of apolipoprotein E allele epsilon 4 with late-onset sporadic Alzheimer's disease, *Am. J. Med. Genet. 54*:286 (1994).
55. Sorbi, S., Nacmias, B., Forleo, P., Latorraca, S., Gobbini, I., Bracco, L., Piacentini, S., and Amaducci, L., ApoE allele frequencies in Italian sporadic and familial Alzheimer's disease, *Neurosci. Lett. 177*:100 (1994).

56. Clair, D. St., Norrman, J., Perry, R., Yates, C., Wilcock, G., and Brookes, A., Apolipoprotein E epsilon 4 allele frequency in patients with Lewy body dementia, Alzheimer's disease and age-matched controls, *Neurosci. Lett. 176*:45 (1994).
57. Tsai, M. S., Tangalos, E. G., Petersen, R. C., Smith, G. E., Schaid, D. J., Kokmen, E., Ivnik, R. J., and Thibodeau, S. N., Apolipoprotein E: Risk factor for Alzheimer disease, *Am. J. Hum. Genet. 54*:643 (1994).
58. van Duijn, C. M., de, K. P., Cruts, M., Wehnert, A., Havekes, L. M., Hofman, A., and Van, B. C., Apolipoprotein E4 allele in a population-based study of early-onset Alzheimer's disease [see comments], *Nature Genet. 7*:74 (1994).
59. West, H. L., Rebeck, G. W., and Hyman, B. T., Frequency of the apolipoprotein E epsilon 2 allele is diminished in sporadic Alzheimer disease, *Neurosci. Lett. 175*:46 (1994).
60. Zubenko, G. S., Stiffler, S., Stabler, S., Kopp, U., Hughes, H. B., Cohen, B. M., and Moossy, J., Association of the apolipoprotein E epsilon 4 allele with clinical subtypes of autopsy-confirmed Alzheimer's disease, *Am. J. Med. Genet. 54*:199 (1994).
61. Lehtimaki, T., Pirttila, T., Mehta, P. D., Wisniewski, H. M., Frey, H., and Nikkari, T., Apolipoprotein E (ApoE) polymorphism and its influence on ApoE concentrations in the cerebrospinal fluid in Finnish patients with Alzheimer's disease, *Hum. Genet. 95*:39 (1995).
62. Locke, P. A., Conneally, P. M., Tanzi, R. E., Gusella, J. F., and Haines, J. L., Apolipoprotein E4 allele and Alzheimer disease: Examination of allelic association and effect on age at onset in both early- and late-onset cases, *Genet. Epidemiol. 12*:83 (1995).
63. St. Clair, D., Rennie, M., Slorach, E., Norrman, J., Yates, C., and Carothers, A., Apolipoprotein E epsilon 4 allele is a risk factor for familial and sporadic presenile Alzheimer's disease in both homozygote and heterozygote carriers, *J. Med. Genet. 32*:642 (1995).
64. Frisoni, G. B., Calabresi, L., Geroldi, C., Bianchetti, A., D'Acquarica, A. L., Govoni, S., Sirtori, C. R., Trabucchi, M., and Franceschini, G., Apolipoprotein E epsilon 4 allele in Alzheimer's disease and vascular dementia, *Dementia 5*:240 (1994).
65. Garlepp, M. J., Tabarias, H., Van, B. F., Zilko, P. J., Laing, B., and Mastaglia, F. L., Apolipoprotein E epsilon 4 in inclusion body myositis, *Ann. Neurol. 38*:957 (1995).
66. Greenberg, S. M., Rebeck, G. W., Vonsattel, J. P., Gomez, I. T., and Hyman, B. T., Apolipoprotein E epsilon 4 and cerebral hemorrhage associated with amyloid angiopathy, *Ann. Neurol. 38*:254 (1995).
67. Premkumar, D. R., Cohen, D. L., Hedera, P., Friedland, R. P., and Kalaria, R. N., Apolipoprotein E-epsilon 4 alleles in cerebral amyloid angiopathy and cerebrovascular pathology associated with Alzheimer's disease, *Am. J. Pathol. 148*:2083 (1996).
68. Nicoll, J. A., Roberts, G. W., and Graham, D. I., Amyloid beta-protein, APOE genotype and head injury, *Ann. N. Y. Acad. Sci. 777*:271 (1996).
69. Saunders, A. M., Schmader, K., Breitner, J. C. et al., Apolipoprotein E epsilon 4 allele distributions in late-onset Alzheimer's disease and in other amyloid-forming diseases [see comments], *Lancet 342*:710 (1993).
70. Wisniewski, T., Golabek, A., Matsubara, E., Ghiso, J., and Frangione, B., Apolipo-

protein E: Binding to soluble Alzheimer's beta amyloid, *Biochem. Biophys. Res. Commun. 192*:359 (1993).

71. Strittmatter, W. J., Saunders, A. M., Schmechel, D., Pericak-Vance, M., Enghild, J., Salvesen, G. S., and Roses, A. D., Apolipoprotein E: High-avidity binding to β-amyloid and increased frequency of type 4 allele in late-onset familial Alzheimer's disease, *Proc. Natl. Acad. Sci. USA 90*:1977 (1993).
72. Strittmatter, W. J., Weisgraber, K. H., Huang, D. Y., Dong, L.-M., Salvesen, G. S., Pericak-Vance, M., Schmechel, D., Saunders, A. M., Goldgraber, D., and Roses, A. D., Binding of human apolipoprotein E to synthetic amyloid β peptide: Isoform-specific effects and implications for late-onset Alzheimer's disease, *Proc. Natl. Acad. Sci. USA 90*:8098 (1993).
73. LaDu, M. J., Falduto, M. T., Manelli, A. M., Reardon, C. A., Getz, G. S., and Frail, D. E., Isoform-specific binding of apolipoprotein E to β-amyloid, *J. Biol. Chem. 269*:23403 (1994).
74. LaDu, M. J., Pederson, T. M., Frail, D. E., Reardon, C. A., Getz, G. S., and Falduto, M. T., Purification of apolipoprotein E attenuates isoform-specific binding to β-amyloid, *J. Biol. Chem. 270*:9039 (1995).
75. Sanan, D. A., Weisgraber, K. H., Russell, S. J., Mahley, R. W., Huang, D., Saunders, A., Schmechel, D., Wisniewski, T., Frangione, B., Roses, A. D., and Strittmatter, W. J., Apolipoprotein E associates with β amyloid peptide of Alzheimer's disease to form novel monofibrils, *J. Clin. Invest. 94*:860 (1994).
76. Naslund, J., Thyberg, J., Tjernberg, L. O., Wernstedt, C., Karlstrom, A. R., Bogdanovic, N., Gandy, S. E., Lannfelt, L., Terenius, L., and Nordstedt, C., Characterization of stable complexes involving apolipoprotein E and the amyloid β peptide in Alzheimer's disease brain, *Neuron 15*:219 (1995).
77. Chan, W., Fornwald, J., Brawner, M., and Wetzel, R., Native complex formation between ApoE isoforms and the Alzheimer's disease peptide Aβ, *Biochemistry 35*:7123 (1996).
78. Wood, S. J., Chan, W., and Wetzel, R., An inhibition complex between apolipoprotein E and β-amyloid fibril nucleation seeds, *Chem. Biol. 3*:949 (1996).
79. Wisniewski, T., Castano, E. M., Golabek, A., Vogel, T., and Frangione, B., Acceleration of Alzheimer's fibril formation by apolipoprotein E in vitro, *Am. J. Pathol. 145*:1030 (1994).
80. Ma, J., Yee, A., Brewer, H. J., Das, S., and Potter, H., Amyloid-associated proteins alpha 1-antichymotrypsin and apolipoprotein E promote assembly of Alzheimer beta-protein into filaments [see comments], *Nature 372*:92 (1994).
81. Castano, E. M., Prelli, F., Wisniewski, T., Golabek, A., Kumar, R. A., Soto, C., Frangione, B., Fibrillogenesis in Alzheimer's disease of amyloid β peptides and apolipoprotein E, *Biochem. J. 306*:599 (1995).
82. Wood, S. J., MacKenzie, L., Maleeff, B., Hurle, M. R., and Wetzel, R., Selective inhibition of βA4 fibril formation, *J. Biol. Chem. 271*:4086 (1996).
83. Whitson, J. S., Mims, M. P., Strittmatter, W. J., Yamaki, T., Morrisett, J. D., and Appel, S. H., Attenuation of the neurotoxic effect of Aβ amyloid peptide by apolipoprotein E, *Biochem. Biophys. Res. Commun. 199*:163 (1994).
84. Weisgraber, K. H., Apolipoprotein E: Structure-function relationships, *Advances in*

Protein Chemistry (V. N. Schumaker, ed.), Academic Press, San Diego, CA, 1994, p. 249.
85. Creighton, T. E., *Proteins: Structures and Molecular Properties*, W. H. Freeman, New York, 1984.
86. Selkoe, D. J., The molecular pathology of Alzheimer's disease, *Neuron 6*:487 (1991).
87. Price, D. L., Walker, L. C., Martin, L. J., and Sisodia, S. S., Amyloidosis in aging and Alzheimer's disease, *Am. J. Pathol. 141*:767 (1992).
88. Pike, C. J., Burdick, D., Walencewicz, A. J., Glabe, C. G., and Cotman, C. W., Neurodegeneration induced by β-amyloid peptides in vivo: The role of peptide assembly state, *J. Neurosci. 13*:1676 (1993).
89. Baltz, M. L., Caspi, D., Evans, D. J., Rowe, I. F., Hind, C. R., and Pepys, M. B., Circulating serum amyloid P component is the precursor of amyloid P component in tissue amyloid deposits, *Clin. Exp. Immunol. 66*:691 (1986).
90. Janciauskiene, S., de Frutos, P. G., Carlemalm, E., Dahlback, B., and Eriksson, S., Inhibition of Alzheimer beta-peptide fibril formation by serum amyloid P component, *J. Biol. Chem. 270*:26041 (1995).
91. Emsley, J., White, H. E., O'Hara, B. P., Oliva, G., Srinivasan, N., Tickle, I. J., Blundell, T. L., Pepys, M. B., and Wood, S., *Nature 367*:338 (1994).
92. Tennent, G. A., Lovat, L. B., and Pepys, M. B., Serum amyloid P component prevents proteolysis of the amyloid fibrils of Alzheimer disease and systemic amyloidosis, *Proc. Natl. Acad. Sci. USA 92*:4299 (1995).
93. Fraser, P. E., Nguyen, J. T., McLachlan, D. R., Abraham, C. R., and Kirschner, D. A., alpha 1-Antichymotrypsin binding to Alzheimer A beta peptides is sequence specific and induces fibril disaggregation in vitro, *J. Neurochem. 61*:298 (1993).
94. Eriksson, S., Janciauskiene, S., and Lannfelt, L., alpha 1-Antichymotrypsin regulates Alzheimer beta-amyloid peptide fibril formation, *Proc. Natl. Acad. Sci. USA 92*:2313 (1995).
95. Janciauskiene, S., Carlemalm, E., and Eriksson, S., In vitro fibril formation from alpha 1-antitrypsin-derived C-terminal peptides, *Biol. Chem. Hoppe Seyler 376*:415 (1995).
96. Schwarzman, A. L., Gregori, L., Vitek, M. P., Lyubski, S., Strittmatter, W. J., Enghilde, J. J., Bhasin, R., Silverman, J., Weisgraber, K. H., Coyle, P. K., Zagorski, M. G., Talafous, J., Eisenberg, M., Saunders, A. M., Roses, A. D., and Goldgaber, D., Transthyretin sequesters amyloid β protein and prevents amyloid formation, *Proc. Natl. Acad. Sci. USA 91*:8368 (1994).
97. Ueda, K., Fukushima, H., Masliah, E., Xia, Y., Iwai, A., Yoshimoto, M., Otero, D. A., Kondo, J., Ihara, Y., and Saitoh, T., Molecular cloning of cDNA encoding an unrecognized component of amyloid in Alzheimer disease, *Proc. Natl. Acad. Sci. USA 90*:11282 (1993).
98. Iwai, A., Yoshimoto, M., Masliah, E., and Saitoh, T., Non-A beta component of Alzheimer's disease amyloid (NAC) is amyloidogenic, *Biochemistry 34*:10139 (1995).
99. Prusiner, S. B., Novel proteinaceous infectious particles cause scrapie, *Science 216*:136 (1982).

100. Alper, T., Cramp, W. A., Haig, D. A., and Clarke, M. C., Does the agent of scrapie replicate without nucleic acid? *Nature 214*:764 (1967).
101. Prusiner, S. B., Genetic and infectious prion diseases, *Arch. Neurol. 50*:1129 (1993).
102. Chesbro, B., Race, R., Wehrly, K., Nishio, J., Bloom, M., Lechner, D., Bergstrom, S., Robbins, K., Mayer, L., Keith, J. M., Garon, C., and Haase, A., Identification of scrapie prion protein-specific mRNA in scrapie-infected and uninfected brain, *Nature 315*:331 (1985).
103. Oesch, B., Westaway, D., Walchli, M., McKinley, M. P., Kent, S. B. H., Aebersold, R., Barry, R. A., Tempst, P., Teplow, D. B., Hood, L. E., Prusiner, S. B., and Weissman, C., A cellular gene encodes scrapie PrP 27–30 protein, *Cell 40*:735 (1985).
104. Prusiner, S. B., Chemistry and biology of prions, *Biochemistry 31*:12277 (1992).
105. Prusiner, S. B., Groth, D., Serban, A., Koehler, R., Foster, D., Torchia, M., Burton, D., Yang, S.-L., and DeArmond, S. J., Ablation of the prion protein (PrP) gene in mice prevents scrapie and facilitates production of anti-PrP antibodies, *Proc. Natl. Acad. Sci. USA 90*:10608 (1993).
106. Bueler, H., Aguzzi, A., Sailer, A., Greiner, R.-A., Autenried, P., Aguet, M., and Weissman, C., Mice devoid of PrP are resistant to scrapie, *Cell 73*:1339 (1993).
107. Griffith, J. S., Self-replication and scrapie, *Nature 215*:1043 (1967).
108. Riek, R., Hornemann, S., Wider, G., Billeter, M., Glockshuber, R., and Wuthrich, K., NMR structure of the mouse prion protein domain PrP (121–231), *Nature 382*:180 (1996).
109. Mehlhorn, I., Groth, D., Stockel, J., Moffat, B., Reilly, D., Yansura, D., Willett, W. S., Baldwin, M., Fletterick, R., Cohen, F. E., Vandlen, R., Henner, D., and Prusiner, S. B., High-level expression and characterization of a purified 142-residue polypeptide of the prion protein, *Biochemistry 35*:5528 (1996).
110. Mestel, R., Putting prions to the test, *Science 273*:184 (1996).
111. Kocisko, D. A., Come, J. H., Priola, S. A., Chesebro, B., Raymond, G. J., Lansbury, P. T., and Caughey, B., Cell-free formation of protease-resistant prion protein, *Nature 370*:471 (1994).
112. Kocisko, D. A., Priola, S. A., Raymond, G. J., Chesebro, B., Lansbury, P. J., and Caughey, B., Species specificity in the cell-free conversion of prion protein to protease-resistant forms: A model for the scrapie species barrier, *Proc. Natl. Acad. Sci. USA 92*:3923 (1995).
113. Lasmezas, C. I., Deslys, J.-P., Robain, O., Jaegly, A., Beringue, V., Peyrin, J.-M., Fournier, J.-G., Hauw, J.-J., Rossier, J., and Dormont, D., Transmission of the BSE agent to mice in the absence of detectable abnormal prion protein, *Science 275*:402 (1997).
114. Telling, G. C., Scott, M., Mastrianni, J., Gabizon, R., Torchia, M., Cohen, F. E., DeArmond, S. J., and Prusiner, S. B., Prion propagation in mice expressing human and chimeric PrP transgenes implicates the interaction of cellular PrP with another protein, *Cell 83*:79 (1995).
115. Tjernberg, L. O., Naslund, J., Lindqvist, F., Johansson, J., Karlstrom, A. R., Thyberg, J., Terenius, L., and Nordstedt, C., Arrest of β-amyloid fibril formation by a pentapeptide ligand, *J. Biol. Chem. 271*:8545 (1996).
116. Sakaguchi, S., Katamine, S., Nishida, N., Moriuchi, R., Shigematsu, K., Sugimoto,

T., Nakatani, A., Kataoka, Y., Houtani, T., Shirabe, S., Okada, H., Hasegawa, S., Miyamoto, T., and Noda, T., Loss of cerebellar Purkinje cells in aged mice homozygous for a disrupted PrP gene, *Nature 380*:528 (1996).

117. Tobler, I., Gaus, S. E., Deboer, T., Achermann, P., Fischer, M., Rulicke, T., Moser, M., Oesch, B., McBride, P. A., and Manson, J. C., Altered circadian activity rhythms and sleep in mice devoid of prion protein, *Nature 380*:639 (1996).
118. Patino, M. M., Liu, J.-J., Glover, J. R., and Lindquist, S., Support for the prion hypothesis for inheritance of a phenotypic trait in yeast, *Science 273*:622 (1996).
119. Doel, S. M., McCready, S. J., Nierras, C., and Cox, B. S., The dominant $PNM2^-$ mutation which eliminates the ψ factor of *Saccharomyces cerevesiae* is the result of a missense mutation in the SUP35 gene, *Genetics 137*:659 (1994).
120. Masison, D. C. and Wickner, R. B., Prion-inducing domain of yeast Ure2p and protease resistance of Ure2p in prion-containing cells, *Science 270*:93 (1995).
121. Shortle, D., Mutational studies of protein structures and their stabilities, *Q. Rev. Biophys. 25*:205 (1992).
122. Matthews, B. W., Structural and genetic analysis of protein stability, *Annu. Rev. Biochem. 62*:139 (1993).
123. Jennings, P. A., Saalau-Bethell, S. M., Finn, B. E., Chen, X., and Matthews, C. R., Mutational analysis of protein folding mechanisms, *Molecular Design and Modeling: Concepts and Applications. Part A, Proteins, Peptides and Enzymes.* (J. J. Langone, ed.), Academic Press, San Diego, CA, 1991, p. 113.
124. Mastrianni, J. A., Curtis, M. T., Oberholtzer, J. C., Da, C. M., DeArmond, S., Prusiner, S. B., and Garbern, J. Y., Prion disease (PrP-A117V) presenting with ataxia instead of dementia, *Neurology 45*:2042 (1995).
125. Gambetti, P., Parchi, P., Petersen, R. B., Chen, S. G., and Lugaresi, E., Fatal familial insomnia and familial Creutzfeldt–Jakob disease: Clinical, pathological and molecular features, *Brain Pathol. 5*:43 (1995).
126. Qu, B. H. and Thomas, P. J., Alteration of the cystic fibrosis transmembrane conductance regulator folding pathway, *J. Biol. Chem. 271*:7261 (1996).
127. Thomas, P. J., Shenbagamurthi, P., Sondek, J., Hullihen, J. M., and Pedersen, P. L., The cystic fibrosis transmembrane conductance regulator. Effects of the most common cystic fibrosis-causing mutation on the secondary structure and stability of a synthetic peptide, *J. Biol. Chem. 267*:5727 (1992).
128. Wu, Y.-S., Bevilacqua, V. L. H., and Berg, J. M., Fibrillin domain folding and calcium binding: Significance to Marfan syndrome, *Chem. Biol. 2*:91 (1995).
129. Yang, Y., Janich, S., Cohn, J. A., and Wilson, J. M., The common variant of cystic fibrosis transmembrane conductance regulator is recognized by hsp70 and degraded in a pre-Golgi nonlysosomal compartment, *Proc. Natl. Acad. Sci. USA 90*:9480 (1993).
130. Saijo, T., Welch, W. J., and Tanaka, K., Intramitochondrial folding and assembly of medium-chain acyl-CoA dehydrogenase (MCAD). Demonstration of impaired transfer of K304E-variant MCAD from its complex with hsp60 to the native tetramer, *J. Biol. Chem. 269*:4401 (1994).
131. Hainaut, P. and Milner, J., Interaction of heat-shock protein 70 with p53 translated in vitro: Evidence for interaction with dimeric p53 and for a role in the regulation of p53 conformation, *EMBO J. 11*:3513 (1992).

132. Yehiely, F. and Oren, M., The gene for the rat heat-shock cognate, hsc70, can suppress oncogene-mediated transformation, *Cell Growth Differ. 3*:803 (1992).
133. Groenen, P. J., Merck, K. B., de Jong, W. W., and Bloemendal, H., Structure and modifications of the junior chaperone alpha-crystallin. From lens transparency to molecular pathology [Review], *Eur. J. Biochem. 225*:1 (1994).
134. Lowe, J., New pathological findings in amyotrophic lateral sclerosis, *J. Neurol. Sci. 124*(Suppl):38 (1994).
135. Head, M. W., Corbin, E., and Goldman, J. E., Overexpression and abnormal modification of the stress proteins alpha B-crystallin and HSP27 in Alexander disease, *Am. J. Pathol. 143*:1743 (1993).
136. Iwaki, T., Iwaki, A., Tateishi, J., Sakaki, Y., and Goldman, J. E., alpha B-crystallin and 27-kd heat shock protein are regulated by stress conditions in the central nervous system and accumulate in Rosenthal fibers, *Am. J. Pathol. 143*:487 (1993).
137. Amouyel, P., Vidal O., and Launay, J. M., The apolipoprotein E alleles as major susceptibility factors for Creutzfeldt-Jakob disease, *Lancet 344*:1315 (1994).
138. Dranovsky, A., Goldfarb, L., Gajdusek, D. C., and Goldgaber, D., Apolipoprotein E ε4 allele does not correlate with amyloid plaques in transmissible spongiform encephalopathies, *Amyloid 2*:36 (1995).
139. Lomas, D. A., Evans, D. L., Finch, J. T., and Carrell, R. W., The mechanism of Z alpha 1-antitrypsin accumulation in the liver [see comments], *Nature 357*:605 (1992).
140. Chou, S. M., Inclusion body myositis, *Baillieres Clin. Neurol. 2*:557 (1993).
141. Askanas, V. and Engel, W. K., New advances in inclusion-body myositis, *Curr. Opin. Rheum. 5*:732 (1993).
142. Murakami, N., Ihara, Y., and Nonaka, I., Muscle fiber degeneration in distal myopathy with rimmed vacuole formation, *Acta Neuropathol. 89*:29 (1995).
143. Buxbaum, J. N., Chuba, J. V., Hellman, G. C., Solomon, A., and Gallo, G. R., Monoclonal immunoglobulin deposition disease: Light chain and light and heavy chain deposition diseases and their relation to light chain amyloidosis, *Ann. Intern. Med. 112*:455 (1990).
144. Pollanen, M. S., Bergeron, C., and Weyer, L., Detergent-insoluble cortical Lewy body fibrils share epitopes with neurofilament and tau, *J. Neurochem. 58*:1953 (1992).
145. Goedert, M., tau Protein and the neurofibrillary pathology of Alzheimer's disease, *Trends Neurosci. 16*:460 (1993).
146. Mather, K., Watts, F. Z., Carroll, M., Whitehead, P., Swash, M., Cairn, N., and Burke, J., Antibody to an abnormal protein in amyotrophic lateral sclerosis identifies Lewy body-like inclusions in ALS and Lewy bodies in Parkinson's disease, *Neurosci. Lett. 160*:13 (1993).
147. Zatlaukal, K., Fesus, L., Denk, H., Tarcsa, E., Spurej, G., and Bock, G., High amount of epsilon-(gamma-glutamyl)lysine cross-links in Mallory bodies, *Lab. Invest. 66*:774 (1992).
148. Weidenheim, K. M. and Dickson, D. W., Intranuclear inclusion bodies in an elderly woman: A form of intranuclear inclusion body disease, *Clin. Neuropathol. 14*:93 (1995).
149. *Protein Misassembly, Advances in Protein Chemistry* (Wetzel, R., ed.), Vol. 50, Academic Press, San Diego, CA, 1997, in press.

Index